HR 14319
22.08.79

ST. ANDREWS UNIVERSITY LIBRARY

This book was issued on

20 OCT 1986

15 DEC 1986

04 MAY 1987

UNIVERSITY OF ST. ANDREWS
DAVID RUSSELL HALL

Human Physiology

HUMAN PHYSIOLOGY

Robert S. Shepard, Ph.D.

*Professor, Department of Physiology and Pharmacology,
Wayne State University School of Medicine,
Detroit*

J. B. Lippincott Company

Philadelphia · Toronto

UNIVERSITY OF ST. ANDREWS

DAVID RUSSELL HALL

Copyright © 1971 by J. B. Lippincott Company

This book is fully protected by copyright and, with the exception of brief excerpts for review, no part of it may be reproduced in any form, by print, photoprint, microfilm, or any other means, without the written permission of the publishers.

ISBN 0-397-52052-2

Library of Congress Catalog Card Number 72-109945

Printed in the United States of America

3 5 4 2

Dedicated to
Connie, Karen, Mark, Steve, and Garry

76-DRH-051

PREFACE

I have written this book with the intent of presenting a well-integrated, comprehensive, current picture of the physiology of man in health and disease. Realizing the importance of good communication in education, I have tried to keep it within the scope of the average medical student—while not slighting important material. It is not designed as an encyclopedia, though it should prove a useful reference. The early chapters cover material readily understood by readers with a limited background and provide the foundations of physiology, anatomy, chemistry, physics, and mathematics used in later chapters. I have tried to develop the subject logically, so that sequential chapters are clearly related. Chapters 1 through 4, for example, dealing with excitability and control, form a basis for the discussion of skeletal, cardiac, and smooth muscle in Chapters 5 through 17, while the concepts of fluid dynamics developed in Chapters 10 through 16 provide a basis for the discussion of the respiratory and renal systems in Chapters 17 through 20. Also included are numerous clinical references for the benefit of students intending to practice medicine.

I have purposely limited the scope of this volume so that it can be read and comprehended in less than one school year. All too frequently, I suspect, the nature of the learning experience in physiology renders the subject a formidable obstacle to medical practice, rather than a means toward understanding the individual in health and disease.

Again, this text is primarily an introduction to physiology, not a review. Few areas are dealt with in summary fashion, though I have not, on the other hand, chosen to emphasize how man's excretory system differs from that of a dalmatian, or how the control of his heart differs from that of the cat. Such areas are important, perhaps, in experimental studies, but merely distract the average medical student.

This is not to say that I have totally ignored experimental animal work; in areas where inadequate human data are available we must, at least for the present, depend on information through animal research. Finally, I have placed little emphasis on nutrition and the anatomy of the brain, as these subjects are generally well covered in chemistry and anatomy textbooks.

Hopefully the numerous illustrations and tables will both clarify important written material and aid the student in reviewing.

Robert S. Shepard

ACKNOWLEDGMENTS

The writing and publication of a book such as this one is a formidable undertaking and does not occur in a vacuum. It is an outgrowth of the perspectives and enthusiasm of my former teachers, an attitude of investigative zeal and cooperation from my co-workers and students, a spirit of understanding from my family, and the patience and helpfulness of the people at the J. B. Lippincott Company. I would particularly like to thank Dr. Edward J. Masoro and Miss Bonnie L. Briggs, both of whom read the entire manuscript and made many helpful suggestions, and Mrs. Judy Henry Haverty, who acted as typist and critic. My wife, Connie, is responsible for most of the original finished illustrations. Mr. Stephen W. Kmetz and I developed the rough sketches from which she worked. I would also like to thank Miss Hilda Griscom (Chaps. 1 through 27), Dr. Rupert L. Green (Chaps. 28 through 40), Dr. John Burton (Chaps. 2 and 3), Dr. Eberhard Mammen (Chaps. 5, 6, 9, 14, and 15), Dr. Ewa Marciniak (Chap. 14), Dr. Walter Seegers (Chap. 14), Dr. Genesio Murano (Chap. 14), Dr. Demetrios Triantaphyllopoulos (Chaps. 14 and 16 through 19), Dr. Jeanne Riddle (Chap. 15), and Dr. Stanley L. Rosenthal (Chaps. 17 through 19) for their many helpful suggestions for revising the original manuscript.

CONTENTS

Part One: Introduction 1
1. **Physiologic Principles** 3
   Language 3
   The Cell........................... 6
   The Cell Membrane 8
   The Endoplasmic Reticulum
   and Ribosomes..................... 12
   The Golgi Complex 14
   Mitochondria....................... 14
   Lysosomes......................... 14
   Filaments 14

*Part Two: Control — The Peripheral
Nervous System* 17
2. **Organization of the Nervous System** .. 19
   The Neurilemma.................... 20
   Sensory Neurons 22
   The Synapse 25
   Motor Neurons 27
   The Neuromuscular Junction 28
   The Reflex Arc 29
3. **The Transmembrane Potential**........ 36
   The Equilibrium Potential........... 37
   The Action Potential 38
   Conduction 42
   Receptor Mechanisms 51
   Electrical Stimulation............... 53
4. **The Autonomic Nervous System** 55
   Anatomy........................... 55
   Function 56
   The Sympathetic Nervous System 58
   The Parasympathetic Nervous System . 65
   Autonomic Tone 66

Blocking Agents 67
Decentralization 68

Part Three: Contraction 71
5. **Skeletal Muscle** 73
   Structure.......................... 73
   Muscle Twitch 77
   Muscle Tetanus 81
   Muscle Performance (Contractility).... 86
   Chemistry 86
6. **The Heart — Electrical Properties** 92
   General Principles 92
   The Electrocardiogram 96
7. **The Heart — Contraction** 110
   Review 110
   Cardiac Performance 110
   Cardiac Nutrition.................. 114
   Semilunar Valves................... 117
   Atrioventricular (A-V) Valves......... 118
   Cardiac Cycle 119
   Comparison of the Left and
   Right Heart 121
   Cardiac Catheterization.............. 123
   Heart Sounds 123
   Murmurs........................... 126
8. **The Heart — Cardiac Output** 129
   Determination of Cardiac Output...... 130
   Stroke Volume 131
   Heart Rate 142
9. **Smooth Muscle**..................... 148
   Electrical Properties 148
   Activation-Contraction Coupling 149
   Multi-Unit Smooth Muscle 151

Contents

9. Smooth Muscle—*Continued*
   Single-Unit Smooth Muscle 151
   Control 153

Part Four: Circulation 163

10. Arteries, Arteriovenous Anastomoses and Veins 165
   Fluid Dynamics..................... 165
   Hemodynamics...................... 168

11. Circulatory Control 186
   Cardiac Output..................... 186
   Distribution of Cardiac Output 190
   Blood Volume 202
   Central Nervous System 206

12. Capillaries and Sinusoids 212
   Distribution 212
   Flow 212
   Structure.......................... 212
   Permeability 216
   Filtration 219
   The Lymphatic System 222
   Arteriovenous Differences........... 224

13. Extravascular Fluid Systems........ 226
   Cerebrospinal Fluid 226
   Intraocular Fluid 232
   Endolymph and Perilymph—The Labyrinthine Fluids 235
   Amnionic Fluid 237

14. Hemostasis........................ 242
   Clot formation 242
   Formation of a Platelet Plug.......... 249
   Response of the Blood Vessels 251
   Tests for Hemostatic Function........ 251

15. Blood 254
   Sedimentation Rate................. 254
   Hematocrit......................... 254
   Plasma............................ 254
   Transfusion 259
   Resistance to Infection 266
   Formed Elements 270

16. The Erythrocyte 275
   Structure.......................... 275
   Fragility 277
   Transport of Gases 278
   Oxygen Transport 279
   Carbon Dioxide Transport........... 282
   Buffering Action 283

Part Five: Nutrients and Waste Products... 289

17. Respiration........................ 291
   Cellular Respiration 291
   Inspiration......................... 292
   Expiration......................... 297
   Ventilation........................ 298
   Diffusion.......................... 314

18. Respiratory Control 318
   Medullary Centers 318
   Pontine Centers 322
   Other Centers 323
   Neurological Lesions in Man 325
   Chemoreceptors 326
   Stretch Receptors.................. 331
   Receptors in the Upper Respiratory Tract 336
   Response of Receptors to Injected Agents........................... 338
   Other Receptors 339

19. Respiratory Adjustments 340
   Exercise 341
   Breath-Holding.................... 343
   Diving............................ 344
   Drowning 346
   Hyperventilation 347
   Barometric Pressure 348
   Oxygen Toxicity 353

20. Urine Formation 355
   Anatomy........................... 355
   Filtration 357
   Reabsorption...................... 361
   Secretion 371
   Clinical Evaluation of Kidney Function. 377

21. Renal Regulation of Water and Electrolyte Balance................... 379
   Renal Blood Flow 379
   Electrolyte Balance................. 385
   Water Balance 390

22. Elimination of Acids, Bases, Waste Products, and Foreign Substances 397
   Acid-Base Balance 397
   Foreign Substances................. 404
   Uremic Syndrome 405
   Micturition........................ 405

23. The Digestive System (Motility)..... 410
   Digestion 410
   Anatomy.......................... 410
   Chewing (Mastication) 412
   Swallowing (Deglutition) 412
   The Stomach...................... 415
   The Small Intestine 420
   The Colon 423
   Defecation 424
   Colectomy 425
   Megacolon 426

24. Digestion and Absorption 427
Carbohydrates 428
Proteins 429
Fat 431
Water and Electrolytes 435
Vitamins 440
Protection of the Digestive Tract 440
Feces 441

25. Secretion and Its Control 443
Control of the Salivary Glands 443
Control of Gastric Secretion 447
Control of Pancreatic Secretion 451
Control of Bile Secretion 452
The Gallbladder 455
The Intestine 456

26. Energy Balance 457
Catabolism 457
Anabolism 463
Calorimetry 465
Caloric Intake 469

27. Regulation of Body Temperature 472
Heat Loss 472
The Core Temperature 476
The Control of Body Temperature 477

Part VI: Control—The Endocrine System . . . 487

28. The Pituitary Gland 489
Control 489
The Posterior Pituitary 492
The Anterior Pituitary 493

29. The Thyroid Gland 503
Formation of Triiodothyronine and
  Thyroxine 503
Secretion 504
Plasma Transport 504
Functions of the Thyroid Hormones . . . 506
Clinical Correlates 508

30. Endocrine Functions of the Pancreas . 511
Formation of Insulin 511
Secretion of Insulin 511
Actions of Insulin 514
Destruction of Insulin 515
Diabetes Mellitus 515
Glucagon 519

31. The Adrenal Glands 521
Development of the Adrenal Cortex . . . 521
Development of the Adrenal Medulla . 521
Accessory Adrenocortical and
  Adrenal Medullary Tissue 522
The Adrenocorticoid Steroids 522
The Adrenal Medulla 534

32. Calcium Metabolism, The Parathyroid Glands, and Other Endocrine Glands 536
Calcium Metabolism 536
The Parathyroid Gland 538
Thyrocalcitonin 539
Other Endocrine Glands 540

33. Growth, Development, and Reproduction 542
Bone 545
Development of the Reproductive
  System 546
The Adult Male 551

34. The Adult Female 560
The Ovary 561
The Uterus 563
The Cervix of the Uterus 567
The Vagina 568
The Oviduct 568
The Climacteric and Menopause 568
Sexual Response 569

Part VII: Sensation and Integration 577

35. Vision 579
Light 579
The Refractive Elements of the Eye . . . 581
The Retina 586
Binocular Vision 590

36. Neural Mechanisms in Vision 592
Visual Fields and the Optic Pathways . 592
The Striate Area of the Occipital Cortex . 594
Other Visual Centers 595
Visual Reflexes 595
Light Deprivation 597

37. Hearing 599
Sound 599
The Ear 601
Neural Mechanisms 605

38. The Vestibular Apparatus 609
The Semicircular Canals 610
The Otolithic Organs 615

39. Taste and Smell 617
Taste 617
Olfaction 619

40. The Forebrain 622
Touch, Pressure, and Kinesthetic Tracts . 622
Pain Tracts 623
Temperature Tracts 624
The Thalamus 624
The Cerebral Cortex 625

Index 639

Human Physiology

Part I

INTRODUCTION

Physiology applies the concepts and tools of physics and chemistry to the study of the living organism. Its approach, for the most part, is a *mechanistic* one. The physiologist might look at the increased respiratory rate of an individual after he has been exposed to an environment of low oxygen concentration and proclaim that the hypoxia has stimulated chemoreceptors, which have in turn stimulated sensory neurons to the inspiratory center, which have in some way modified the output of this center so as to increase the respiratory rate. Note how different this is from the *teleological* approach, i.e., "During hypoxia we breathe more rapidly because we need more oxygen."

Similarly, the physician's knowledge of the role of chemoreceptors in hypoxia enables him to understand the consequences of their depression or loss. He may even go so far as to replace a defective natural mechanism with a man-made one. It is becoming increasingly more common, for example, to replace a defective natural pacemaker with a pair of electrodes implanted in the ventricle of the heart. These electrodes are in turn connected to an implanted stimulator and its associated batteries.

Part 1

INTRODUCTION

Chapter 1

LANGUAGE
THE CELL
THE CELL MEMBRANE
ENDOPLASMIC RETICULUM
THE GOLGI COMPLEX
MITOCHONDRIA
LYSOSOMES
FILAMENTS

PHYSIOLOGIC PRINCIPLES

LANGUAGE

The CGS System

The language of physiology is made up, in part, of terms of measurement and quantification. In this text we will rely heavily on the cgs (centimeter, gram, second) system; in it, all variables can be expressed in terms of the three fundamental parameters—*length* (meters or centimeters), *mass* (grams), and *time* (seconds). By using a prefix in combination with each of these parameters we can readily indicate the power to which each parameter is raised. Thus 1 kilogram (kg.) represents 10^3 grams (gm.) or 1,000 gm.; 1 milligram (mg.), 10^{-3} gm. (1/1,000 gm.); 1 microgram (μg.), 10^{-6} gm.; 1 nanogram (ng.), 10^{-9} gm.; 1 picogram (pg.), 10^{-12} gm.; 1 femtogram, 10^{-15} gm.; and 1 attogram, 10^{-18} gm. Two peculiarities in usage, however, are (1) the generally accepted tendency to refer to what some call the micrometer (μm.) as a micron (μ), and (2) the tendency to call 0.1 nanometers an angstrom (A).

Even with the large number of units indicated

Parameter (Symbol and/or Formula)	Units (cgs)	Electrical Equivalent
1. Length (L)	1 meter = 10^2 cm. = 10^3 mm. = 10^6 μ = 10^{10} A	
Area: $(L_1 \cdot L_2) = A$	1 meter^2 = 10^4 cm.^{2} = 10^6 mm.^{2} = 10^{12} μm.^{2}	
Volume: $(A \cdot L_3) = Q$	1 cm.^{3} = 1 ml. = 10^{-3} liters	Coulombs
2. Mass (m)	10^{-3} kg. = 1 gm. = 10^3 mg. = 10^6 μg.	
3. Duration (t)	1 sec. = 10^3 msec.	
4. Derived Units		
Velocity: $(L/t) = v$	cm./sec.	
Acceleration: $(v/t) = a$	cm./sec./sec.	
Density: $(m/Q) = \rho$	gm./ml.	
Force: $(m \cdot a) = F$	gm.-cm./sec./sec. = dynes	
Tension: $(F/L) = T$	dynes/cm.	
Pressure: $(F/A) = P = V$	dynes/cm.^{2} = (1/1330) mm. Hg = 9.87×10^{-7} atmospheres	Volts
Flow: $(L^3/t) = I$	ml./sec.	Amperes
Resistance = $(P/I) = \Omega$	(mm. Hg)/(ml./sec.) = 1 P.R.U. = 1330 dyne-sec./cm.^{5}	Ohms
Compliance = $(Q/P) = c$	ml./mm. Hg	Capacitance
Viscosity: $(F/A)/(v/L) = \eta$	(dyne-sec.)/cm.^{2} = 1 poise	
Work: $(F \cdot L) = W$	dyne-cm. = 1 erg = 10^{-7} joules = 2.39×10^{-8} calories	
Kinetic energy: $(1/2\ mv^2) = KE$	(gm.-cm.^{2})/sec.^{2} = 1 erg	
Potential energy: $(P \cdot Q) = PE$	(gm.-cm.^{2})/sec.^{2} = 1 erg	
Potential energy: $(m \cdot L \cdot g) = PE$ (g = gravitational constant)	(gm.-cm.^{2})/sec.^{2} = 1 erg	
Power: $(W/t) = P$	ergs/sec. = 10^{-7} watts	Watts

Table 1-1. Some parameters and units used in physiology.

above, however, the language of physiology would be cumbersome if we did not derive some additional ones. For example, if we caused a mass of 1 gm. to accelerate 1 cm. per second per second, we would be exerting a force of 1 gm.-cm. per second per second on the mass. We generally find it more convenient, however, to refer to this force as 1 dyne. Thus, for the sake of brevity and convenience, a number of units have been derived (but remain part of the cgs system). These are listed in Table 1-1. Also listed in Table 1-1 are some of the symbols we shall be using for our derived and basic parameters, as well as some of the electrical equivalents with which we will be concerned. You will perhaps remember from your previous studies of electricity that one can calculate the resistance (Ω) in a wire by applying a certain voltage (V) to the wire and measuring the current (I) that results:

$$\Omega = \frac{V}{I}$$

Similarly, one can calculate the resistance to air flow or blood flow (Ω) between two points in a tube (bronchus or artery) from the pressure difference between those two points (P) and the flow of blood or air (I) that results:

$$\Omega = \frac{P}{I}$$

Finally, let me point out that in physiology we are frequently satisfied with 2- or 3-place accuracy. This limited accuracy sometimes permits us to assume, for example, that 1 cm.³ of water has a mass of 1 gm. rather than the more accurate 0.99336 gm. at 37°C.

Concentration

The external and internal environments of the cell play an essential role in its function. We are concerned in physiology not only with how many grams of a particular substance are in solution in and around the cell, but also with how many particles the substance forms in that solution. Important too is the charge carried by these particles and their effect on the hydrogen ion (H^+) concentration (Table 1-2).

Per cent. Two of the most common ways of expressing concentration are in terms of grams per cent (grams per 100 gm.) and milligrams per cent (milligrams per 100 gm.). A 0.9 per cent saline solution would, for example, be equivalent to a 900 mg. per cent solution of NaCl or one of 0.9 gm. of NaCl per 100 gm. of solution. Concentration is also sometimes expressed as milliliters of solute per 100 ml. of solution.

Mols. It is also useful to characterize a solution by the number of molecules it contains. The basic unit in this system is the mol. One mol represents 6.02×10^{23} molecules (Avogadro's number). In the case of NaCl, 1 mol would weigh 58.5 gm. (the gram-molecular weight of NaCl being 23 + 35.5) and in the case of $CaCl_2$, 111.0 gm. (40 + 35.5 + 35.5). Or, the number of mols of a particular substance added to a solution can be calculated by dividing the weight of that substance in grams by its gram-molecular weight. For example, 9 gm. of NaCl would represent 0.154 mols or 154 millimols (mM.):

$$\text{Mols} = \frac{\text{weight in grams}}{\text{gram-molecular weight}} = \frac{9 \text{ gm.}}{58.5 \text{ gm.}} = 0.154$$

If the 9 gm. of NaCl were now added to a liter of water we would have a 0.154 molar solution (0.9 per cent saline). To find the molarity of a solution, in other words, all we must do is divide the number of mols of solute dissolved in the solution by the solution's volume in liters:

$$\text{Molarity} = \frac{\text{mols}}{\text{liters}}$$

Parameter	Definition
Grams per cent	Grams of solute/100 gm. of solution
Milligrams per cent	Milligrams of solute/100 gm. of solution
Mol	Grams of substance/gram-molecular weight of substance
Molarity	Mols of solute/liter of solution
Equivalents	Gram-molecular weight of a particle/valence of particle
Normality	Chemical equivalents of anion or cation/liter of solution
Osmol	Mols of solute/dissociation constant for solute
Osmolarity	Osmols of solute/liter of solution
Osmolality	Osmols of solute/kilogram of solution
pH	$-$ Log of the H^+ concentration

Table 1-2. Some parameters used in characterizing a concentration. The mol is used to indicate the concentration of a molecule; the equivalent, the concentration of a submolecular species; and the osmol, the concentration of a group of different particles.

Equivalents. Usually more important to the physiologist than the molarity of a solution is the number and character of the reactive units in the solution. A reactive unit may be a molecule, an ion, or an atom. For example, in the chemical reaction

$$HCl + NaOH \rightarrow NaCl + HOH,$$

H, Cl, Na, and OH are the reactive units. In this reaction each gram of H is capable of replacing 23 gm. of Na. Each gram of H is also capable of combining with 35.5 gm. of Cl or 17 gm. of OH. Or, 1 gm. of H can be thought of as equivalent to 23 gm. of Na, 35.5 gm. of Cl, or 17 gm. of OH. These values are generally referred to as the gram-equivalent weights (loosely, the equivalent weight, the chemical equivalent, or the gram-equivalent) for their respective reactive units. Any quantity in grams of a particular reactive unit that combines with, displaces, or otherwise plays the part of 1 gm. of H or H^+ is called a gram-equivalent weight.

In the case of the monovalent ion, the gram-equivalent weight is equal to the gram-ionic weight, but in the case of the divalent ion (Ca^{++}, Mg^{++}, SO_4^{--} for example) the gram-equivalent weight equals the gram-ionic weight divided by 2. For example, in the chemical reaction

$$Ca(OH)_2 + 2\ HCl \rightarrow CaCl_2 + 2\ HOH,$$

40 gm. of Ca is capable of displacing 2 gm. of H. Therefore the gram-equivalent weight of Ca is 40/2 or 20 gm.

In the above discussion I have concentrated on the *chemical equivalent*. Here I have emphasized that 1 equivalent of H can displace 1 equivalent of Ca and can combine with 1 equivalent of OH. We are also concerned in physiology with the number of equivalents a particle releases when dissolved in water. We find, for example, that 9 gm. of NaCl in 1 liter of water produces a 0.154-molar solution (9 gm./(23 gm. + 35.5 gm.) = 0.154). This solution contains 0.154 equivalents (154 milliequivalents) of Na^+ and 0.154 equivalents of Cl^-. These are sometimes called *electrical equivalents*. On the other hand a 1-molar solution of $CaCl_2$ contains 2 equivalents of Ca^{++} plus 2 equivalents of Cl^- per liter of solution. In both cases the salt has completely dissociated. Some molecules, however, either do not dissociate completely or dissociate in several different ways. Phosphate in water at a pH of 7.4 forms both $H_2PO_4^-$ (20 per cent) and $HPO_4^=$ (80 per cent). This results in the phosphate's having a virtual valence of 1.8 ((0.2 × 1) + (0.8 × 2)) and the phosphate-P's having an equivalent weight of 17 (31/1.8).

Osmols. The osmol is a unit related to the number of particles dissolved in a solvent. It is equal to the number of mols of solute divided by the number of particles formed by a single molecule of solute:

$$\text{Osmol} = \frac{\text{mols of solute}}{\text{dissociation constant for solute}}$$

In calculating the number of osmols in solution, you will note that we are concerned with neither the valence nor the charge of the dissolved particles. The dissociation constant for NaCl is approximately 2; for glucose, approximately 1; and for Na_2SO_4, 3. Therefore 1 mol of NaCl would produce 2 osmols, and 1 mol of glucose 1 osmol.

The *osmolarity* of a solution is the concentration of particles in osmols per liter, and can be measured by the degree to which it depresses the freezing point of water. For example, a solution of 1 osmol per liter has a freezing point 1.86°C. lower than that of pure water. The formula for osmols per liter in a solution, then, is:

$$\text{Osmols/liter} = \frac{\text{(freezing-point depression)}}{1.86}$$

A 0.9 per cent saline solution has an osmolarity of 0.310 osmols per liter (310 milliosmols per liter). Urine, on the other hand, may vary from a concentration of 40 to 1,500 milliosmols per liter.

Since osmolar solutions involve osmols per liter of solution, their characteristics are dependent upon temperature. Increases in temperature increase the volume of the solution and therefore decrease the osmolar concentration. If, however, we express the concentration in terms of osmols per kilogram of solvent, we have a parameter, *osmolality*, which does not change with changes in temperature.

pH. The concentration of H^+ in an aqueous solution is usually expressed in terms of the pH of the solution. This is defined as minus the log of the H^+ concentration:

$$\text{pH} = -\log(H^+)$$

In an aqueous solution at 25°C., the product of the equivalents per liter of H^+ and OH^- is a constant, $1/10^{14}$ (or 10^{-14}):

$$(H^+ \text{ concentration}) \cdot (OH^- \text{ concentration}) = 10^{-14}$$

In a neutral solution, the concentrations of H^+ and OH^- both equal $10^{-7.0}$ and the solution is said to have a pH of 7.0. In an acid solution the concentration of H^+ exceeds that of OH^- and the pH is less than 7.0. In an alkaline solution the pH is greater than 7.0. The pH of the plasma is normally about 7.4 (H^+ concentration = $10^{-7.4}$). The pH of the urine, on the other hand, may range from 4.4 (acid) to 8.0 (alkaline).

THE CELL

Each of us begins life as a single fertilized egg. This egg, like its parent, has a cell membrane, cytoplasm, and a nucleus. The cell membrane acts as an important barrier between the intracellular and intercellular environments of the cell; the cytoplasm contains the atoms, monomers, macromolecules, and organelles

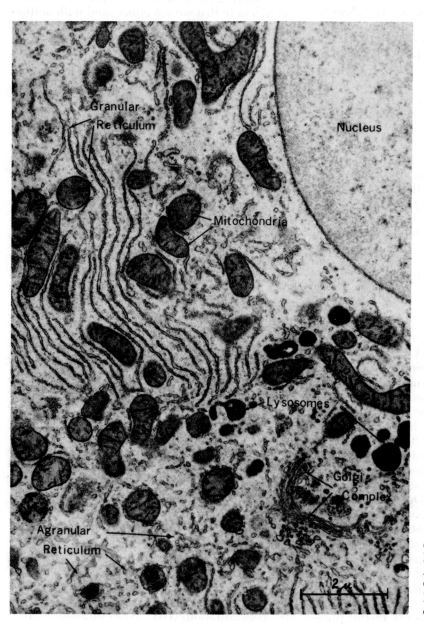

Fig. 1-1. Ultrastructure of part of a liver cell from a rat. (Porter, K. R.: The endoplasmic reticulum. *In* Goodwin, T. W., and Lindberg, O. (eds.): Biological Structure and Function. P. 130. New York, Academic Press, 1961)

(endoplasmic reticulum, Golgi complex, mitochondria, filaments, and lysosomes) essential for the cell's function; and the nucleus plays a major role in cell division and protein anabolism.

As cells divide, some become progressively more distant from their nutritive supply and begin to differentiate. This results in the different types of tissues, organs, and systems found in the body. Specialization of cells occurs throughout intrauterine life, childhood, and senescence. Some cells (skeletal-muscle) elongate, become multinuclear, develop an extensive sarcoplasmic reticulum, and form large quantities of the contractile proteins actin and myosin. Others (neurons) elongate, contain one nucleus, and have little or no actomyosin. Cells are formed (red blood cells) which contain no endoplasmic reticulum, lose their nucleus, and accumulate large quantities of hemoglobin. Cells called megakaryocytes develop which later explode into hundreds of anuclear cell fragments. Some cells (osteoblasts, pancreatic acinar cells, and liver parenchymal cells) specialize for the production of large intercellular molecules. Others (histiocytes) retain the capacity to differentiate.

All living cells, however, have one thing in common: they are constantly in a state of change. In some organs this change is associated with (1) hyperplasia (cell division) and cell destruction, (2) hypertrophy (increase in cell size) and atrophy (decrease in cell size), or (3) a combination of these. In young healthy persons, hyperplasia and hypertrophy are the predominant phenomena; in elderly, inactive ones, cell destruction and atrophy may predominate.

Many cells lose their ability to divide. In Figure 1-2, for example, we note that this occurs in the fetus during the second and third trimesters of pregnancy in the case of nerve, skeletal-muscle, and heart-muscle cells. At this time additional seminiferous tubules also stop forming. Other structures (blood cells, hepatic cords, thyroid follicles, and so on) continue being formed throughout life and in many cases serve to replace destroyed ones. There is, on the other hand, no known mechanism capable of replacing destroyed neurons, skeletal-muscle fibers, or nephrons in the adult. The function of destroyed neurons is either compensated by the branching or growth of remaining neurons, or lost. Compensation for lost muscle fibers occurs through the laying down of more contractile proteins in the remaining muscle fibers. Both phenomena are forms of hypertrophy.

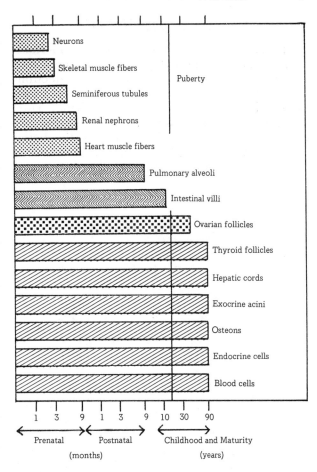

Fig. 1-2. Duration of hyperplasia (cell division) in various structures in man. Organs whose substructures cease to multiply before maturity have determinate numbers of structural units and restricted capacities for growth beyond the adult size. Those retaining their hyperplastic abilities can grow without limit, since they have indeterminate numbers of structural units. (*Modified from* Gross, R. J.: Hypertrophy versus hyperplasia. Science, *153*:1616; copyright September, 1966 by the American Association for the Advancement of Science)

Regardless of the degree of hypertrophy or hyperplasia occurring in the body, there is constant utilization of oxygen (O_2) and production of carbon dioxide (CO_2). Macromolecules are produced and broken down; ions and molecules diffuse and are actively transported into and out of the cell. Also associated with active transport and other energy-requiring reactions is heat production by the cell.

THE CELL MEMBRANE

The cell membrane constitutes an important barrier between the cytoplasm of the cell and its external environment. Under the electron microscope (Fig. 1-3) it appears as a light band bounded on either side by a dark band. This trilaminar structure is referred to as a *unit membrane* and surrounds all the cells of the body (as well as those of plants, animals, and microorganisms, with the exception of the viruses).

The cell membrane is generally regarded as containing phospholipids and proteins and possibly some polysaccharides, but its exact structure is a matter of speculation (Fig. 16-2). Since other characteristics of the cell membrane vary from one cell type to the next, however, it is quite likely there are important differences in the structure of different cell membranes.

One suggestion is that the unit membrane is a phospholipid sandwich—an inner layer of phospholipid between two layers of protein. The lipid part of the phospholipid does not ionize and hence does not dissolve in water. It is said to be *hydrophobic* (water-hating) and to be more attracted to other lipids than to water. The phosphate portion of the phospholipid, on the other hand, tends to ionize and orient itself toward water and the proteins, forming the surfaces of the sandwich. It represents the *hydrophilic* (water-loving) portion of the phospholipid.

Permeability

It was suspected, before the cell had been analyzed chemically, that its membrane consisted of lipid or lipoprotein. This hypothesis was based on the low permeability of the membrane to ions as compared with lipid-soluble substances. The unit membrane is approximately 10,000 to 100,000 times as impermeable to water as a similar thickness (50 to 100 A) of water would be; it is even less permeable to NaCl, urea, and glucose (Table 12-2). On the other hand, substances which are both water- and fat-soluble (O_2 and CO_2) penetrate the membrane readily.

An important consequence of this phenomenon is that the intracellular fluid composition can be and is, in the living state, markedly different from that of the intercellular fluid (Fig. 1-4). Within the cell, for example, there is a higher concentration of potassium cations (K^+), magnesium cations (Mg^{++}), phosphate anions ($H_2PO_4^-$, HOP_4^{--}, PO_4^{---}), and protein anions, while the intracellular fluid contains a lower concentration of sodium cations (Na^+), calcium cations (Ca^{++}), chloride anions (Cl^-), and bicarbonate anions (HCO_3^-) than the interstitial fluid.

This variation in the concentrations of charged particles inside and outside the cell results in a small but important electric potential between the intercellular and intracellular fluids. In the case of some skeletal-muscle fibers this potential

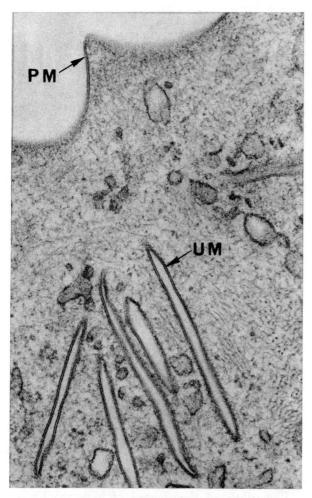

Fig. 1-3. Electron micrograph of part of an epithelial cell from the urinary bladder of a mouse. (× 69,000) Note that a trilaminar membrane about 100 A in thickness forms the outer surface of the cell (PM, plasma membrane), as well as the outer surface of some vesicles in the cytoplasm (UM, unit membrane). (Porter, K. R., and Bonneville, M. A.: An Introduction to the Fine Structure of Cells and Tissues. 3rd ed., Plate 27. Philadelphia, Lea & Febiger, 1968)

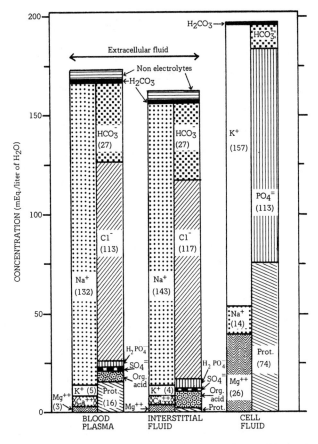

Fig. 1-4. Composition (in mEq. per liter of water) of body fluids. Some of the body fluids whose composition differs from the above are discussed in Chapter 13 and include cerebrospinal fluid, aqueous humor (Table 13-1), endolymph (Table 13-3), and amnionic fluid. (*Redrawn from* Leaf, A., and Newburgh, L. H.: Significance of Body Fluids in Clinical Medicine. 2nd ed., p. 8. Springfield, Charles C Thomas, 1955)

The Cell Membrane

A good commercial insulator such as rubber generally has a *dielectric strength* of less than 10^6, or 1,000,000 volts per meter. By comparison, we see that the cell membrane constitutes an effective barrier for maintaining the specific internal environment and transmembrane potential of the cell. This internal environment is essential for the functioning of the cell; if too much of the external environment moves into it, the cell will die. If too much of the internal environment escapes into the blood, the heart will stop pumping blood and the organism will die. The transmembrane potential, on the other hand, is essential for the excitability of the cell. Should a nerve- or muscle cell lose its transmembrane potential, it would become inexcitable (refractory).

Diffusion

Were the plasma membrane completely impermeable to non-lipid-soluble substances it could not accumulate additional water, K^+, Mg^{++}, etc., and therefore could not hypertrophy. The point of the preceding discussion is not that the cell membrane is completely impermeable to these substances, however, but that they do not penetrate the membrane as readily as O_2 and CO_2; hence the cell maintains certain concentration differences in the case of the small particles of the body (Na^+) as well as the larger ones (protein).

One of the mechanisms by which particles enter and leave the cytoplasm is diffusion—a random movement of particles in liquids and gases which tends to produce an equal concentration of dissolved particles throughout a solution. If, for example, we place a water-soluble blue dye in an aqueous solution, we note that the dye is initially concentrated in one part of the solution. After 10 or 20 minutes, however, it is spread evenly throughout the solution. This is because the Brownian-like movement of the molecules in the solution has caused a random redistribution of the dye and water molecules, eventually producing a homogeneous solution.

In the case of the cell there is also a constant movement or passive diffusion of particles from one point to another. The number of particles that diffuse into the cell (influx) or out of the cell (efflux) (L. *ex*, from; plus *fluxus*, flow) depends on (1) the character of the particle (its size, charge, and solubility in lipid), (2) the character of the plasma membrane, and (3) the electrochemical difference or gradient between the intracellular and intercellular fluids. The ease

is of the order of 80 millivolts (mv.); in the case of other cells it may be less. The so-called resting cell (unstimulated cell), because of the slightly higher concentration of anions in intracellular fluid, is negatively charged with respect to intercellular fluid.

In the case of a cell with a *transmembrane potential* of 80 mv. (0.08 volts) and a plasma membrane thickness of 80 A (80×10^{-10} meters), the electric field across the cell membrane would be 10,000,000 volts per meter:

$$\frac{0.08 \text{ volts}}{80 \times 10^{-10} \text{ meters}} = 10^7 \text{ volts/meter}$$

with which a given solute passes through the cell membrane can be expressed in terms of a single number, the *permeability coefficient* (k), calculated by determining the net movement in mols per second (Q) of the particle; the surface area of the membrane in cm.^{2} (A); and the concentration difference for the particle between the inside and outside of the cell in mols per cm.^{3} ($C_1 - C_2$):

$$k = \frac{Q}{(A) \cdot (C_1 - C_2)} = \frac{\text{mols/sec.}}{(\text{cm.}^2) \cdot (\text{mols/cm.}^3)} = \text{cm./sec.}$$

Using the above formula we find the approximate permeability coefficient for water to be 10^{-2}; for Cl^-, 10^{-4}; for K^+, 10^{-8}; and for Na^+, 10^{-10}. Looking solely at the concentration gradient across the cell membrane for Cl^- and K^+, one would expect these ions to be more evenly distributed than they are; but you will note that the electrical gradient across the membrane is such that it interferes with the efflux of K^+ and the influx of Cl^-. Or, the presence in the cell of negatively charged protein particles which cannot penetrate the intact cell membrane tends to keep (attract) K^+ in the cell and to keep (repel) Cl^- out of the cell.

Active Transport

The electrochemical gradients discussed above serve well to explain the intracellular and extracellular concentrations of Cl^- and in part those of K^+, but fail to explain the concentration differences for Na^+. It should be emphasized in this regard that neither Na^+ nor K^+ exist in solution as isolated ions, but rather in combination with water as hydrated ions. The size of the hydrated Na^+ is about 1.5 times that of the hydrated K^+; even this does not explain the markedly higher concentration of Na^+ outside the cell however. The likeliest explanation for the Na^+ concentrations in the body is that there is a system in the cell that extrudes Na^+ from the cell (Fig. 1-5). This mechanism is variously referred to as a sodium pump or a sodium-potassium pump (Fig. 20-12), since it also moves K^+ into the cell.

The pump mechanisms of the body are responsible not only for the maintenance of a low concentration of Na^+ in the cytoplasm of the cell, but also for the production of such special body fluids as cerebrospinal fluid, endolymph, and urine. Pump mechanisms will be discussed more thoroughly in Chapter 20 on the kidney (see "Active Reabsorption") and elsewhere, but we

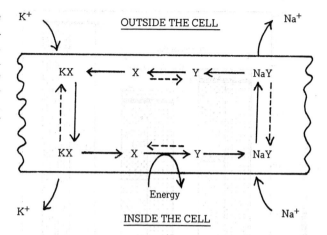

Fig. 1-5. Hypothetical scheme for the sodium-potassium pump. It is suggested here that a molecule (Y) is formed in the central part of the cell membrane which has an affinity for Na^+ and which diffuses to the outer surface of the membrane; there it is changed to a molecule (X) having a low affinity for Na^+ and a high affinity for K^+. The change of X back to Y is illustrated as an energy-requiring reaction. (*Reprinted with permission from* Glynn, I. M.: The ionic permeability of the red cell membrane. Progr. Biophys., 8:292, 1957. New York, Pergamon Publishing Co.)

shall take a somewhat superficial look at the sodium pump here. It should be noted that the sodium pump, like many others, is a mechanism for moving a particle against an electrochemical concentration gradient (uphill transport), and that it requires energy, acts on a specific particle or on a limited number of particles, and can be saturated or inhibited.

One suggestion is that a molecule which has an affinity for Na^+ (an Na^+ carrier) is formed in the cytoplasm or cell membrane. The resulting Na^+ complex diffuses away from the area of formation (the area in which it is most concentrated), and complexes coming into contact with the enzymes at the surface of the cell membrane are broken down; as a result, the Na^+ is released into the interstitial fluid. The molecule resulting from this metabolism has no affinity for Na^+ but may combine with K^+, diffuse into the cytoplasm, and be changed back into a molecule with an affinity for Na^+ (but not K^+).

Thus the sodium pump is regarded as a mechanism whereby certain carrier molecules are formed in the cytoplasm and broken down at the cell surface. Needless to say, it is an active mechanism: energy is required to either produce or catabolize the carrier molecule. It is also apparent that the amount of carrier formed determines the number of particles transported. Thus there is a limit to the rate at which Na$^+$ can be carried out or K$^+$ carried into the cell.

Pinocytosis and Phagocytosis

Another means of getting a particle into a cell is pinocytosis (Gk. *pineo*, to drink). In this process, invagination of the cell membrane occurs and a small vesicle containing part of the interstitial fluid surrounded by a unit membrane is formed (Fig. 1-6). It has been shown in certain cells that large molecules like ferritin can get into the cell by this method, but it is doubtful that pinocytosis is an important transport mechanism for smaller particles.

Probably more important then pinocytosis is the process of phagocytosis (Gk. *phagein*, to eat). In this mechanism the cell surrounds another cell or a large particle, thus incorporating the substance into its cytoplasm (where it can be acted upon by the cell's enzymes). This is one means the body uses to destroy invading microorganisms and to remove worn-out cells and debris from the blood and interstitial fluid. Important phagocytes include the microphages (neutrophils) and the macrophages (histiocytes).

Osmosis

If a cell is placed in distilled water, more water diffuses into the cell than out and the cell enlarges. If a mammalian cell is placed in 0.9

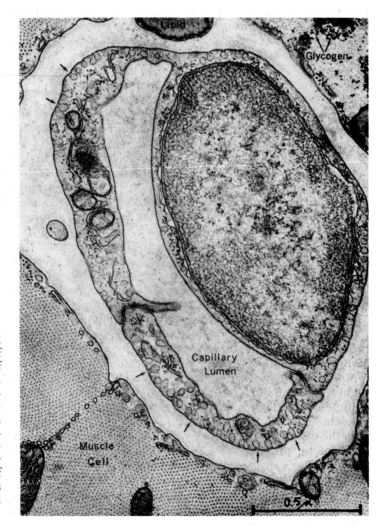

Fig. 1-6. Pinocytosis in an endothelial cell from a capillary of the myocardium of a cat. Apparently small invaginations of both the luminal and perivascular plasma membranes occur, and result in the pinching off of vesicles within the cytoplasm. These vesicles are found in abundance in endothelial cells, smooth and striated muscle fibers, and adipose cells. The importance of this form of imbibition is poorly understood. (Fawcett, D. W.: An Atlas of Fine Structure. The Cell, Its Organelles and Inclusions. P. 403. Philadelphia, W. B. Saunders, 1966)

per cent NaCl (310 milliosmols per liter), the influx of water will equal the efflux. If, on the other hand, the cell is placed in 1.8 per cent NaCl (620 milliosmols per liter), there is a net efflux of water from the cell and the cell decreases in size. In other words we find that water (the solvent) moves through a semipermeable membrane from an area of low concentration of solute particles to an area of high concentration. This phenomenon is called osmosis. As far as we know no cell has an active transport system for water; rather, osmosis results from (1) the diffusion of water through the plasma membrane and (2) the greater permeability of this membrane to water than to most other substances. If the concentration of water outside the cell is greater than inside, more water will diffuse in than out.

Since a 0.9 per cent saline solution results in no net influx or efflux of water in mammals, it is referred to as an *isotonic* (Gk. *isos*, equal; plus *tonos*, tension) solution. A concentration less than 0.9 per cent is *hypotonic* (Gk. *hypo*, below) and one greater than 0.9 per cent, *hypertonic* (Gk. *hyper*, above). A hypertonic solution, in other words, has a concentration of dissolved particles greater than that of the cell cytoplasm. A 5 per cent dextrose solution is also isotonic to the mammalian cytoplasm, since the concentration of dissolved particles (expressed in osmols per liter) in 0.9 per cent saline is the same as the concentration of dissolved particles in 5 per cent dextrose.

Osmosis can also be studied under conditions involving no change in volume. If, for example, we have 2 rigid boxes separated by a membrane permeable to water but not to the particles dissolved in the water, and if we place isotonic dextrose in one box and water in the other, we find that at 37°C. the pressure that develops in the box containing isotonic dextrose is about 7.6 atmospheres (5,800 mm. Hg) greater than that in the box containing just water. In other words, the tendency of more water to pass through the membrane into the dextrose solution can either (1) increase the pressure in the solution with the higher concentration of solute, (2) increase the volume of the dextrose solution, or (3) do some of each.

In the example given above we have measured the osmotic pressure of a 5 per cent dextrose solution. One can also calculate osmotic pressure (π) using the concentration of particles in a solution (C in mols per liter), the temperature of the solution (T in degrees Kelvin), and the universal gas constant (R in liter-atmospheres per mol × deg. K.). For example, a 0.9 per cent saline solution has a molarity of 0.15:

Molarity (mols/liter) =

$$\frac{C \text{ (gm./liter)}}{\text{molecular weight (gm./mol)}} =$$

$$\frac{9 \text{ gm./liter}}{58.5 \text{ gm./mol}} = 0.15 \text{ mols/liter}$$

We find, however, that each NaCl particle in solution forms approximately 2 particles (NaCl $\rightleftharpoons$ Na$^+$ + Cl$^-$). Hence there will be 0.30 mols of osmotically active particles per liter:

$$(0.15 \text{ mols of NaCl/liter}) \cdot 2 =$$
$$0.30 \text{ mols of particles/liter} = n$$

On the basis of the above data we find that the osmotic pressure of 0.9 per cent saline at body temperature (37°C. or 310° K.) is equal to about 7.6 atmospheres:

$$\pi = nRT = (0.30) \cdot (8.2 \times 10^{-2}) \cdot (3.1 \times 10^2) = 7.6$$

THE ENDOPLASMIC RETICULUM AND RIBOSOMES

In 1945 Porter, Claude, and Fullam, using the electron microscope, noted what they called the endoplasmic reticulum—a series of strands and vesicles in the central portion of the cytoplasm (endoplasm) of a cell grown in tissue culture. This reticulum was later found to extend, in some cases, from the cell membrane to the perinuclear space, and to form an extensive intracellular labyrinth (Fig. 1-7). The membranes forming the reticulum have a trilaminar structure similar to that of the plasma membrane.

In the case of secretory cells much of the reticulum is bounded on the cytoplasmic side by *ribosomes*—granules rich in ribonucleic acid. The *deoxyribonucleic acid* (DNA) on the chromosomes in the nucleus of the cell acts as a template for the production of *ribonucleic acid* (RNA), which moves to the cytoplasm and there regulates the synthesis of protein. These proteins in turn control the metabolism of the cell. Some of the RNA molecules attach to the ribosomes of the endoplasmic reticulum and, in certain glands, act as templates for the production of proteins that accumulate in the reticulum and eventually pass out of the cell. Such proteins

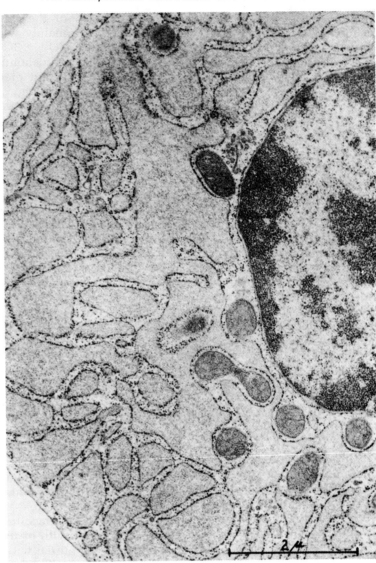

Fig. 1-7. Endoplasmic reticulum (granular type) of a plasma cell in the bone marrow of a guinea pig. In this specimen the cisternae of the reticulum have become greatly distended by the accumulation of products from the cytoplasm. Note the labyrinthine character of the reticulum and the accumulation of ribosomes (small granules) adjacent to it. Note also the numerous mitochondria surrounded by reticulum, and the nucleus at the right of the picture. (Fawcett, D. W.: An Atlas of Fine Structure. The Cell, Its Organelles and Inclusions. P. 155. Philadelphia, W. B. Saunders 1966)

include the enzymes released into the digestive tract, some of the hormones, and the plasma proteins produced by the liver.

In starvation, the ribosomes of the endoplasmic reticulum slowly disappear, apparently due to their metabolism by the cell. There are, however, numerous cells in the body whose reticulum never contains ribosomes, despite good nutrition. This type, an *agranular reticulum*, is found in striated muscle, where it seems to serve primarily as an intracellular conduction pathway. Characteristic of muscles containing an extensive endoplasmic or sarcoplasmic reticulum (skeletal muscle) is rapid activation of the contractile units of the muscle cell, while muscles without an extensive reticulum (smooth muscle) show slow movement of the wave of activation through the cell. This results in less synchronous activation of the contractile units, and therefore in a weaker and more prolonged contraction.

Other functions attributed to the agranular reticulum are the transport of lipid (intestine), the synthesis of steroids (testis, adrenal cortex, and ovary), detoxification (liver), and the metabolism of lipid and fat (liver). In short, the trilaminar membranes of the granular and agranular endoplasmic reticulum, like the plasma membranes, appear to constitute an effective barrier between the cytoplasm and a second medium. Their cisternae apparently serve in some cases as transport channels; in other cases as areas for metabolic activity; and in still other cases as areas of storage.

THE GOLGI COMPLEX

In 1898 Camillo Golgi identified a structure in the cytoplasm which he called the internal reticular apparatus (Fig. 1-1). This is now called the Golgi complex or -apparatus and is thought to store and act on some of the products of the endoplasmic reticulum. In some cases the protein stored in the endoplasmic reticulum moves into the Golgi complex and is there combined with carbohydrate produced in the complex itself. This combination results in the formation of semisolid zymogen granules, which become concentrated in the complex. During secretion these granules are released to the surface of the cell. Although the complex is particularly concentrated in secretory cells, it is also found in cells without a secretory function. In some of the absorptive cells of the intestine, for example, lipid is frequently seen concentrated in Golgi cisternae.

MITOCHONDRIA

Mitochondria are delicate rods, filaments, or granules found in the cytoplasm of all animal cells. Their contents are segregated from the rest of the cytoplasm by 2 trilaminar limiting membranes, the inner one producing folds or cristae (Fig. 1-8). Attached to the outside surface of the mitochondrion are enzymes concerned mainly with oxidative reactions (pyruvic acid + $3O_2 \rightarrow 3CO_2 + 3H_2O$). On the cristae inside the mitochondrion are packets of enzymes (granules called electron transport particles) which catalyze the production of the high-energy phosphate compound adenosine triphosphate (ATP). Apparently the reactions at the mitochondrion's surface supply the raw materials for the formation of ATP inside the mitochondrion. The breakdown of ATP is the immediate source of energy for the cell's various processes (active transport, contraction, secretion, conduction); thus the mitochondria seem to be its ultimate source of fuel (ATP). We might compare the mitochondrion to a factory producing and selling gasoline (ATP) for our automobile (cell). The cost of the fuel in milliliters of O_2 and grams of carbohydrate, lipid, and protein depends on the efficiency of the factory. Characteristically, cells which are highly active metabolically have the greatest concentration of mitochondria.

LYSOSOMES

Lysosomes too have a trilaminar membrane segregating their contents from the surrounding cytoplasm. They are digestive vesicles which contain enzymes capable of lysing (breaking down) RNA, DNA, phosphate esters, proteins, glycosides, polysaccharides, and sulfate esters. Lysosomes differ from zymogen granules in that they contain enzymes for intracellular digestion, whereas zymogen granules contain enzymes for extracellular digestion. Apparently they can fuse with or incorporate into themselves certain material with which they come in contact.

The lysosomal membrane has been known to break down during a marked decrease in the O_2 concentration in the cell, causing the destruction of the cell. This has led some to call these digestive vesicles "suicide bags."

FILAMENTS

Fine threads of protein material 40 to 100 A thick are also found in the cytoplasm of most cells. In many cells they appear to serve as a cytoskeleton. In skeletal and cardiac muscle, on the other hand, they consist of a thick filament (100 A) — myosin, and adjacent thin filaments (60 A) — actin, which interact chemically to produce muscle tension. And in smooth muscle they

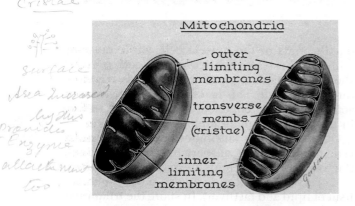

Fig. 1-8. Diagram of two mitochondria. (Ham, A. W.: Histology. 6th ed., p. 155. Philadelphia, J. B. Lippincott, 1969)

consist primarily of actin, probably the contractile protein of smooth muscle.

REFERENCES

Bennett, H. S.: The concepts of membrane flow and membrane vesiculation as mechanisms for active transport and ion pumping. J. Biophys. Biochem. Cytol., 2:99-103, 1956.

Fawcett, D. W.: An Atlas of Fine Structure. The Cell, Its Organelles and Inclusions. Philadelphia, W. B. Saunders, 1966.

Glynn, I. M.: The ionic permeability of the red cell membrane. Progr. Biophys., 8:241-307, 1957.

Goss, R. J.: Hypertrophy versus hyperplasia. Science, 153:1615-1620, 1966.

Ham, A. W.: Histology. 6th ed. Philadelphia, J. B. Lippincott, 1969.

Korn, E. D.: Structure of biological membranes. Science, 153:1491-1498, 1966.

Leaf, A., and Newburgh, L. H.: Body Fluids in Clinical Medicine. 2nd ed. Springfield, Charles C Thomas, 1955.

Loewenstein, W. R. (ed.): Biological Membranes: Recent Progress. Vol. 137, pp. 403-1048. New York, New York Academy of Sciences, 1966.

Porter, K. R.: The endoplasmic reticulum. *In* Goodwin, T. W., and Lindberg, O. (eds.): Biological Structure and Function. Vol. I, pp. 127-156. New York, Academic Press, 1961.

Porter, K. R., and Bonneville, M. A.: An Introduction to the Fine Structure of Cells and Tissues. 3rd ed. Philadelphia, Lea & Febiger, 1968.

Robertson, J. D.: The membrane of the living cell. Sci. Am., 206:64-72, 1962.

Part II

CONTROL – THE PERIPHERAL NERVOUS SYSTEM

Characteristically, the cytoplasm of the body's living cells is electrically negative in comparison with the interstitial fluid. This is due to (1) its high concentration of negatively charged protein particles, (2) the impermeability of the cell membrane to these proteins, (3) the high intracellular concentration of K^+ and low concentration of Na^+, and (4) the higher permeability of the plasma membrane to the intracellular cation (K^+) than to the extracellular cation (Na^+). Thus, although the active transport system pumps approximately equal quantities of K^+ and Na^+ into and out of the cell, respectively, the higher permeability of the plasma membrane to K^+ permits a greater net diffusion of cations out than in. Or, K^+ diffuse out of the cell along a concentration gradient until an electrical gradient is established that balances the concentration gradient. This electrical gradient is variously referred to as the transmembrane potential, the resting potential, and the steady potential (E_S).

If we apply a sufficiently strong electric current to an excitable cell membrane, the membrane's permeability to Na^+ increases (increased Na^+ conductance) and Na^+ will move into the cytoplasm at a rate sufficient to make the cytoplasm positive with respect to the interstitial fluid. This electrical change is called an *action potential*, and is conducted throughout the plasma membrane. In some structures it may even spread from one cell to the next (intercellular conduction). In peripheral neurons, however, when an action potential reaches the cell's terminal portion a chemical (neurotransmitter) is released which affects neighboring cells. In certain nerve cells and in smooth-muscle, cardiac-muscle, and gland cells the neurotransmitter may hyperpolarize, depolarize, or otherwise modify the activity of the effector cell. In the case of a skeletal-muscle cell, the neurotransmitter initiates an end-plate potential, which produces an action potential. This is followed by an intracellular release of Ca^{++}, which catalyzes a chemical reaction that produces contraction (production of tension). The neurotransmitter may also initiate a series of reactions involving glandular secretion and the contraction of certain smooth muscles. An action potential in neurons and skeletal-muscle cells is characteristically of short duration (0.4 to 10 msec.), and is quickly followed by a return of the transmembrane potential to its earlier state (repolarization).

Some cells have an unstable transmembrane potential: in the basal condition Na^+ leaks into the cytoplasm at a rate sufficient to produce a gradual depolarization of the membrane. Eventually the potential reaches the *threshold value*, at which point there is a sudden marked increase in the permeability of the cell membrane to Na^+, an influx of Na^+, and a resultant action potential. These cells are sometimes called *pacemaker* cells, since they are independent of extrinsic factors for their depolarization. Other cells in the body have a stable transmembrane potential and depolarize only in response to extrinsic stimuli. These are *follower* cells.

The role of the peripheral nervous system is to (1) initiate action potentials in some of the follower cells, and (2) modify transmembrane potentials in some of the follower and pacemaker cells. In skeletal muscle, the peripheral motor neurons seem to act merely as initiators of events producing action potentials, while in other structures they play a major role in modifying *excitability*. Some neurons decrease the excitability of the cells they innervate. They may also decrease the frequency of action potentials from a pacemaker cell. Other neurons increase the excitability of follower cells, as well as the frequency of action potentials from a pacemaker cell.

In this section we will be concerned primarily with the role of the nervous system in body control. We will emphasize that *somatic efferent neurons* innervate skeletal muscle and are responsible for initiating their contraction. The role of *sympathetic* and *parasympathetic neurons* in the control of the heart, smooth muscle, and glands will also be noted. Here we shall see that these neurons in some cases initiate a series of events producing action potentials and in other cases merely modify the transmembrane potential. Nervous control of the heart is one example of the second type. Here, sympathetic neurons (1) facilitate a more rapid firing and hence a more rapid heart rate (positive chronotropic action) by their action on the cardiac pacemaker cells, (2) facilitate conduction of the action potential (positive dromotropic action), by their action on the follower cells, and (3) produce a more forceful contraction (positive inotropic action).

This is not to say that the nervous system is the body's only control system. Various glands in the body (endocrine glands) release chemical compounds (hormones) which also initiate events producing action potentials (depolarization) or changes in excitability. Some of the hormones control the metabolism of the cell as well as its internal and external environment. The hormone testosterone, for example, facilitates the laying down of the contractile proteins actin and myosin in the muscle cell.

Some cells respond to changes in the concentration of electrolytes, H^+, O_2, etc. in their environment. Increases in the extracellular concentrations of Ca^{++}, for example, decrease the excitability of the cell and, within limits, facilitate the activation of the contractile mechanism, thereby increasing the force of contraction of a muscle fiber in reponse to stimulation.

Cells also respond to physical changes in their environment, such as changes in their temperature and length. Most muscle cells contract more forcefully if stretched prior to stimulation; some produce action potentials in response to stretch.

In summary, then, there are many levels of control in the body. The pacemaker cell is self-stimulating and functions without the help of nerves, hormones, or marked changes in the physical or chemical environment. Other cells act as pacemakers only if an extrinsic factor (such as a hormone) decreases their transmembrane potential (e.g., from 80 to 60 mv.). Still other cells are completely dependent upon extrinsic factors for action potentials. The role of the nervous system is to integrate these diverse cells to meet the needs of the organism, specifically, the maintenance of a fairly constant internal environment (homeostasis) and of an almost continuous blood supply to the heart and brain. The nervous system performs these functions by (1) the initiation, (2) the facilitation, and (3) the inhibition of cellular activity.

Chapter 2

THE NEURILEMMA
SENSORY NEURONS
THE SYNAPSE
MOTOR NEURONS
THE NEUROMUSCULAR JUNCTION
THE REFLEX ARC

ORGANIZATION OF THE NERVOUS SYSTEM

The nervous system consists of about 10 billion nerve cells which conduct impulses (1) to the central nervous system (afferent or sensory neurons), (2) within the central nervous system (interneurons and the cortical neurons of the cerebral cortex), or (3) to effector organs (central effector neurons and peripheral effector neurons, Fig. 2-1). Each neuron is made up of a cell body, which contains the nucleus, and cytoplasmic extensions of the cell body. All sensory neurons are bipolar, which is to say that they contain a single long cylindrical process carrying impulses to the cell

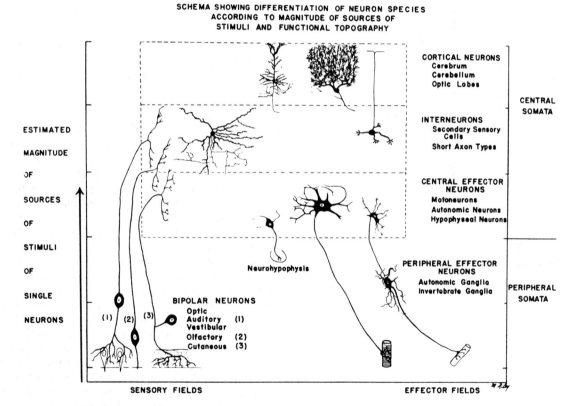

Fig. 2-1. Organization of the nervous system. Impulses enter the central nervous system via bipolar neurons (*left*) and there synapse with interneurons or effector neurons. The interneurons synapse with either other interneurons, cortical neurons, or central effector neurons which connect with peripheral effector neurons (postganglionic neurons) or effector organs. (Bodian, D.: Introductory survey of neurons. Cold Spring Harbor Symposium on Quantitative Biology, *17*:3, 1952; Cold Spring Harbor: Long Island Biological Association, Inc.)

body, and another carrying impulses away from it. These processes are called axons. All other neurons are multipolar (Fig. 2-2). They contain numerous short processes called dendrites (Gk. *dendritēs*, treelike) which carry impulses to the cell body, and an axon which carries impulses from the cell body. In the case of the afferent (L. *ad*, toward; plus *fero*, carry) neuron, the axon leading to the cell body may extend as far as from the toe to the vertebral column. The axons of neurons forming the corticospinal tract, on the other hand, may extend from the cerebral cortex to the sacral spinal cord, while those of somatic efferent (L. *ex*, from; plus *fero*, carry) or motor neurons (central effector neurons) may extend from the lumbar cord to the toe.

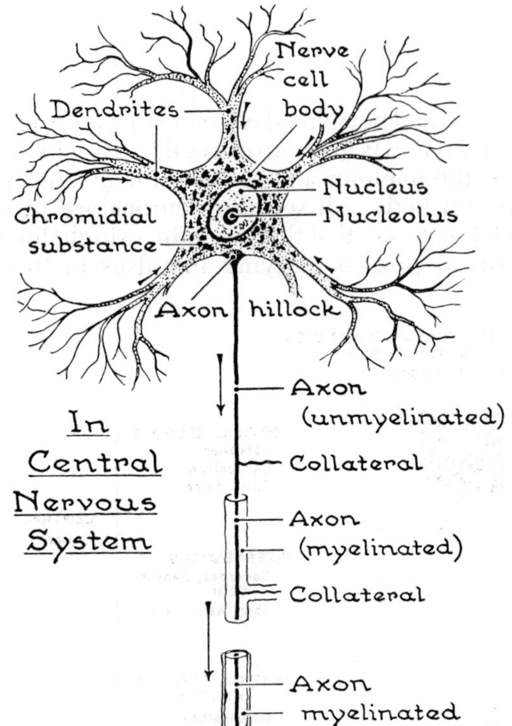

THE NEURILEMMA

All axons outside the central nervous system are covered with a thin protoplasmic sheath called the *neurilemma* or *sheath of Schwann*. In a crushed or severed axon, that part of the fiber separated from its cell body degenerates and disappears, while the attached portion, if the cell body is intact, regenerates. The sheath of Schwann serves as a conduit, leading the regenerating axon back to the area it previously innervated.

The neurilemma contains a lipid called *myelin*. In some axons (myelinated neurons) the myelin is quite thick; in others there is only a trace of it. (Observing no myelin at all in the latter variety, light-microscopists called them unmyelinated.) In myelinated axons, the myelin periodically disappears and exposes the fiber. These points of exposure are called nodes of Ranvier, and the area between two adjacent nodes, an internode.

The myelin serves to insulate the neuron from its environment. As a result, an action potential set up at one node of Ranvier produces a current in the extracellular fluid, which, in turn, initiates an action potential in an adjacent node. Hence in myelinated neurons the action potential seems to jump from one node to the next. This type of conduction is called *saltatory* (L. *saltatio*, to dance or leap) *conduction*. Since the current in the intercellular fluid travels much more rapidly than the action potential in the neuron (Table 2-2, p. 29), myelinated neurons generally conduct nerve impulses more rapidly (about 6 meters per second per μ of diameter) than unmyelinated neurons (about 1.7 meters per second per μ).

In the central nervous system the myelinated neurons too have nodes of Ranvier. Here, however, there are satellite cells instead of Schwann cells. The brain and spinal cord can be divided into two main parts. One part, the white matter (Fig. 2-3), consists of myelinated and unmyelinated fiber tracts from one area of the central

Fig. 2-2. Diagram of a myelinated multipolar neuron. (Ham, A. W.: Histology. 5th ed., p. 526. Philadelphia, J. B. Lippincott, 1965)

GRAY AND WHITE MATTER OF SPINAL CORD

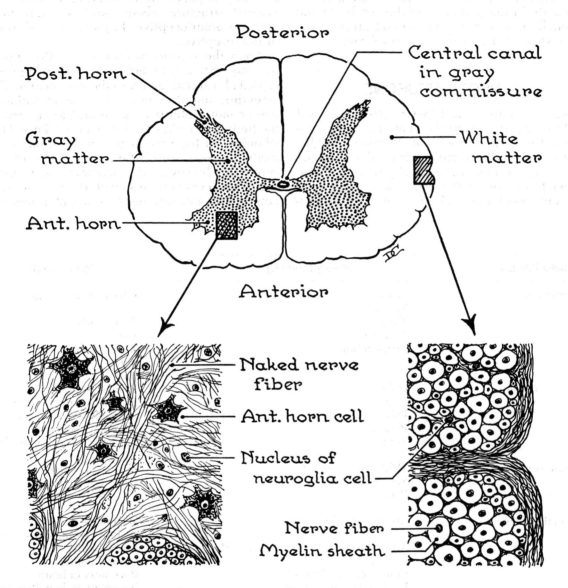

Fig. 2-3. Diagram of a cross-section of the spinal cord. The grey matter is characterized as an area of synapse or integration and the white matter as an area of conduction. (Ham, A. W.: Histology. 6th ed., p. 488. Philadelphia, J. B. Lippincott, 1969)

nervous system to another. The spinothalamic tract, for example, conducts impulses from the cord to the thalamus. The second part, the gray matter, is made up primarily of cell bodies and dendrites and represents an area in which one cell influences other cells. In other words it is an area of synapse (Gk. *synaphe*, contact) and integration. It is here that a cell receives facilitatory and inhibitory impulses from other cells. Here too the sensory neurons synapse with motor neurons and interneurons, and interneurons synapse with other interneurons and motor neurons. In the spinal cord, the gray matter is concentrated in a central "H." In the brain it lies primarily in the periphery of the cerebrum and cerebellum and in various central areas of the brain stem, diencephalon, and cerebrum.

SENSORY NEURONS

Sensory neurons are bipolar cells whose cell bodies lie just outside the central nervous system in the spinal (dorsal root) or cranial ganglia. Those conducting impulses from the skin (somatic afferent neurons) generally receive stimulation from outside the body (pain, heat, cold, pressure, touch, etc.) and are said to be exteroceptive (L. *exterus*, outside; plus *capere*, take). Those carrying impulses from muscle, tendon, and visceral structures (body position, pain, heat, etc.) are proprioceptive (L. *proprius*, one's own) or interoceptive.

When the sensory neuron enters the central nervous system it branches and sends impulses cephalad, caudad, and into the gray matter. The ascending impulses may, after the activation of one or more interneurons, stimulate an area in the brain that makes the subject conscious of the stimulus. In the gray matter of the spinal cord, medulla, pons, cerebellum, and parts of the cerebrum, however, the activation of neurons occurs at an unconscious level. Here the sensory neuron activates interneurons or motor neurons.

Classification	Sensory Modality	Sense Organ
Special senses	Smell	Olfactory membrane
	Vision	Eye
	Taste	Taste buds
	Audition	Ear
	Acceleration	Semicircular canals
		Utricle
Cutaneous senses	Cold	Krause end-bulbs (?)
	Heat	Ruffini end-organs (??)
	Pain	Free nerve endings (?)
	Pressure	Pacinian corpuscles
	Touch	Meissner's corpuscles (?)
Muscle senses	Mild muscle stretch	Muscle spindles
	Severe tendon stretch	Golgi tendon organs
	Pain	Free nerve endings
Visceral senses	Arterial blood pressure	Wall of aortic arch
		Wall of carotid sinus
	Arterial plasma P_{O_2}, P_{CO_2}, and pH	Aortic bodies
		Carotid bodies
	Cerebrospinal fluid pH	Receptors in brain
	Plasma osmotic pressure	Receptors in hypothalamus
	Plasma temperature	Cells in hypothalamus
	Urinary bladder distention	Stretch receptors
	Lung distention	Smooth-muscle endings (?)
		Unencapsulated endings (?)
	Cold	?
	Heat	?
	Pain	Free nerve endings

Table 2-1. A partial list of the sensory modalities and sense organs hypothesized for man. Most and possibly all the organs listed for the cutaneous senses are probably of historical interest only. In many cases their existence, function, or importance is in dispute. Some of the modalities listed act at both conscious and unconscious levels (vision, muscle stretch, bladder distention, etc.). Others act only at an unconscious level (arterial P_{O_2}, etc.).

Receptors

Man distinguishes both consciously and unconsciously the stimuli to which he is exposed. One stimulus he characterizes as red, another as hot; one stimulus elicits the contraction of a muscle group, another its relaxation. Unfortunately, our understanding of the mechanisms by which we distinguish one stimulus from another is fragmentary. We know, for example, that there is an area in the cerebral cortex which, when stimulated by an electrical impulse, elicits the sensation of light. We also know that light shined into the retina of the eye stimulates specialized end-organs which stimulate neurons leading to this area.

Because of these and similar observations, many investigators have speculated that all sensation is based on the stimulation of specialized receptors having a lower threshold to one particular modality. It has been further speculated that these receptors in turn, stimulate neurons having a private line to some area in the cortex and to certain internuncial and motor neurons. Hypothetically, in other words, the significant factor in sensation and motor response is not the character of the stimulus, but rather which neurons are stimulated. This theory was formulated by Johannes Müller (1801-1858) and is known as the *doctrine of specific nerve energies*.

In Table 2-1 we have a partial list of the different sensory modalities and of the end-organs which some have suggested are important in our response to these modalities. No doubt many of the end-organs listed for the cutaneous senses are of chiefly historical interest. Sinclair states that there is no specialized receptor for heat and that many of the Krause end-bulbs reported to produce the sensation of cold are artifacts. He summarizes our knowledge of cutaneous sensation as follows: "Few people outside the field of sensory investigation are aware of the extent of our ignorance about cutaneous sensation, as may be seen by the confident—and often groundless—statements made about it in current textbooks." He further submits that the likeliest theory of cutaneous sensation is the modified pattern concept proposed by Melzack and Wall in 1962. According to this concept a single stimulus to the skin produces impulses which travel to the cortex in different fibers at different rates. Thus the message arriving at the cortex is a pattern of impulses dispersed in time and space. The modified pattern concept also states that a single cutaneous neuron can participate in relaying patterns that elicit either pressure- or heat sensations. The important factor, then, is not which neuron is stimulated (specific nerve energies), but rather what combination of neurons is stimulated.

Nerve Impulses

Sensory axons from skeletal muscle vary in *diameter* from 1 μ to 22 μ; those from the skin, from 1 μ to 14 μ (Fig. 2-4). The larger axons (11 to 22 μ) are called Group I fibers; for the most part, they are neurons from stretch receptors in muscle. Group II fibers are about 7 μ in di-

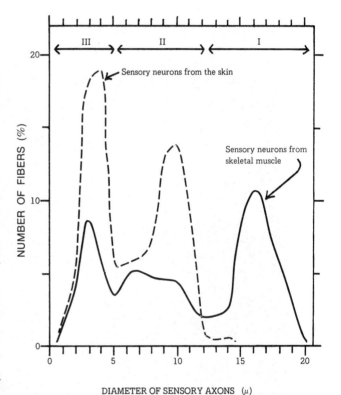

Fig. 2-4. Distribution of axons of different diameters in the afferent fibers of a "demotored" muscle nerve (*solid line*) and in a cutaneous nerve (*interrupted line*). Note that fibers centered at a diameter of 17 μ (group I) are characteristic of the skeletal muscle nerve. Fibers centered at 8 μ (group II) and 3 μ (group III) are found in both the cutaneous and muscle nerves. The diameters given are those for axons close to the spinal cord, where the diameter is maximal. (*Modified from* Lloyd, D. P. C.: *In* Fulton, J. F. (ed.): Textbook of Physiology. 17th ed., p. 54. Philadelphia, W. B. Saunders, 1955)

ameter and in the case of the skin carry impulses associated with the senses of touch and pressure. Group III fibers, about 3 μ in diameter, carry impulses from the free nerve endings.

If one stimulates the central end of a sectioned sensory nerve with a single super-threshold stimulus and records the action potentials a few centimeters central to the point of stimulation, one finds that the action potentials reach the recording electrode at different times. The most rapidly conducting fibers (conduction velocity equals 5 to 120 meters per second) are called A fibers. The least rapidly conducting ones (0.6 to 2 meters per second) are called C fibers. The *A fibers* can be further broken down to alpha (α), beta (β), gamma (γ), and delta (δ) subdivisions (Fig. 2-5). All A fibers are myelinated, have diameters between 1 and 22 μ, and carry sensory impulses from skeletal muscle and the skin, as well as motor impulses to skeletal muscles. In other words they include most of the somatic afferent and efferent neurons. All *B fibers* are also myelinated, but are part of the visceral efferent system (preganglionic neurons). *C fibers* are unmyelinated and carry the sensation of slow pain to the central nervous system, as well as motor impulses to smooth- and cardiac muscle and glands (postganglionic neurons).

The diameter and degree of myelination of each axon or dendrite in part determines the

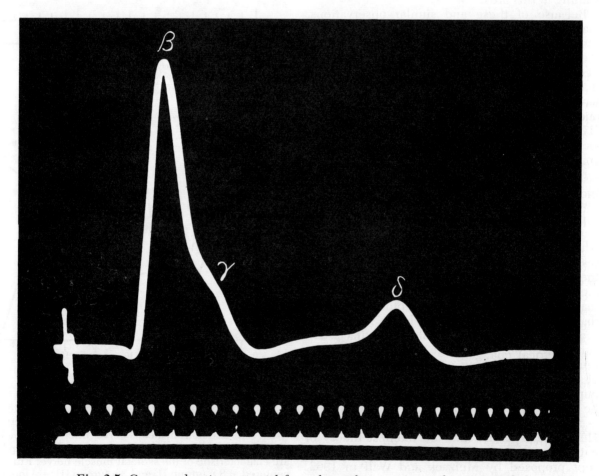

Fig. 2-5. Compound action potential from the saphenous nerve of a cat recorded 54 mm. from the locus of stimulation. Since the saphenous nerve lacks α fibers from stretch receptors no α deflection is noted. The elevation due to the stimulation of C fibers is not shown. The time line contains 5,000 deflections per second. (Rose, J. E., and Mountcastle, V. B.: Touch and kinesthesis. *In* Field, J. (ed.): Handbook of Physiology. Sec. 1 (Neurophysiology), vol. I, p. 393. Washington, American Physiological Society, 1959)

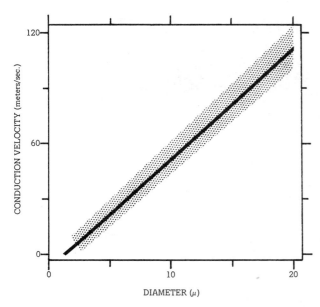

Fig. 2-6. Relation between conduction rate in meters per second and the diameter of mammalian type A nerve fibers. (*Redrawn from* Hursh, J. B.: Conduction velocity and diameter of nerve fibers. Am. J. Physiol., *127*:136, 1939)

electrical properties of that structure. In the case of the A fibers (myelinated somatic afferent and efferent neurons), for example, increases in diameter are associated with (1) increases in the velocity of conduction and (2) increases in excitability (decreases in threshold). As we see in Figure 2-6, the speed of conduction increases about 6 meters per second for each 1-μ increase in overall diameter. The value is 8.7 meters per second per 1-μ increase in the diameter of the axon (diameter of the fiber minus its myelin sheath).

Afterpotentials. Figure 2-7 is a diagram of the electrical changes occurring in an axon in response to stimulation. The predominant response, of course, is the spike potential. After the spike, however, there is a wave (lasting about 10 msec. in this case) representing a period of delayed repolarization. It is called the negative afterpotential and is associated with a period of *supernormal excitability* or hyperexcitability. In A fibers it lasts an average of 50 msec., and in autonomic C fibers an average of 60 msec. It is found in neither dorsal root C- nor B fibers.

Following the negative afterpotential, or in those fibers without a negative afterpotential following the action potential, there is a period of subnormal excitability associated with a positive afterpotential. In A fibers the positive afterpotential lasts 40 to 60 msec., in B fibers 100 to 300 msec., and in autonomic C fibers 300 to 1,000 msec. As can be seen from Figure 2-8, the character of this potential varies with the frequency of stimulation.

THE SYNAPSE

The synapse is an area of functional contact between two neurons. It is the point at which one neuron inhibits or facilitates the formation of an action potential in a second neuron. By stretching a muscle we can, through the stretch receptors, produce action potentials in sensory neurons. A typical sensory neuron conducts impulses in a mixed nerve to the dorsal (posterior) root nerve and into the central nervous system, where it branches (*divergence*), sending impulses to many different areas (Fig. 2-9). Some of the branches of an afferent neuron from a stretch receptor synapse with somatic efferent neurons (anterior-horn cells); others synapse with interneurons. Before the advent of the electron microscope, anatomists were unable to de-

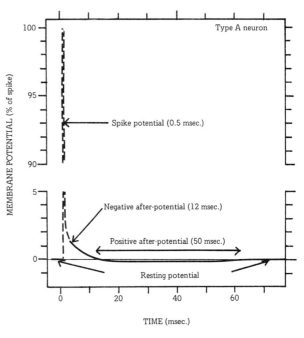

Fig. 2-7. Diagram of the action potential of a large myelinated nerve fiber in a cat. (*After* Gasser, H. S.: The classification of nerve fibers. Ohio J. Sci., *41*: 149, 1941)

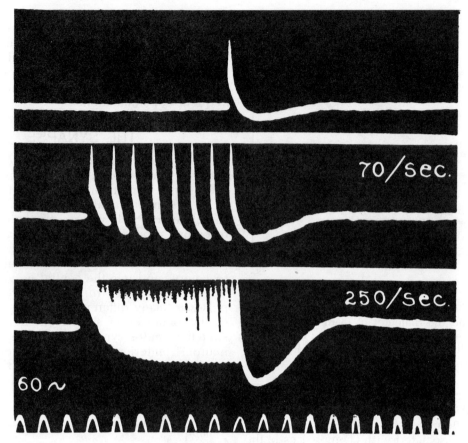

Fig. 2-8. Afterpotentials in the phrenic nerve stimulated at different frequencies. The peak of each spike potential is not shown due to the high amplification of the signal. Note that with increasing frequencies there is an increase in the positive afterpotential (shift of baseline downward) which results in a decrease in excitability (hyperpolarization). Time scale is 60 cycles/sec. or 16.7 msec./cycle. (Gasser, H. S., 1938. *In* Ruch, T. C., and Patton, H. D.: Medical Physiology and Biophysics. 19th ed., p. 79. Philadelphia, W. B. Saunders, 1965)

termine if the synapse was an area of protoplasmic continuity (part of a syncytium) between the afferent and efferent neurons. Using the electron microscope, however, they have demonstrated a space (synaptic gap) of about 200 A between the plasma membranes of two synapsing neurons.

If we stimulate sensory neurons in the dorsal root, we initiate a series of events in these neurons and in the central nervous system that produces action potentials in the motor neurons of the ventral root. If, on the other hand, we stimulate fibers going to the cord in the ventral root (motor neurons), no action potentials are produced in the dorsal (sensory) root. In short, there is a *unidirectional movement* of information across the synapse. Though sensory neurons can stimulate motor neurons, the reverse is not true.

A second important observation concerning the synapse is that the presynaptic volley of impulses is not faithfully reproduced by the postsynaptic neuron. A single afferent action potential may (1) elicit no propagated postsynaptic response or (2) produce a volley of action potentials. Presynaptic stimulation, that is, may result in either facilitation or inhibition of postsynaptic activity. The synapse is also an area of delay (0.5 to 0.9 msec.). We find, for example, that the transmission time increases as the number of synapses in series with each other in-

Motor Neurons

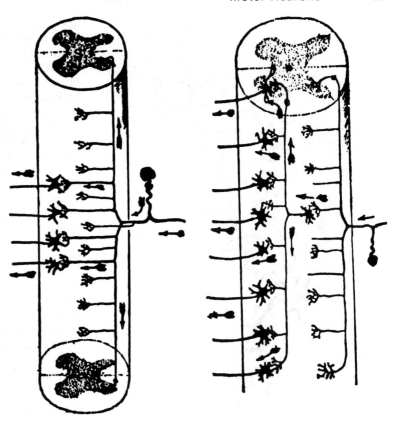

Fig. 2-9. Diagram of the monosynaptic (*left*) and disynaptic (*right*) reflex arc. In the former the afferent neuron synapses directly with the efferent neuron. In the latter the afferent neuron synapses with an interneuron which synapses with the efferent neuron. All reflexes except the stretch reflex involve more than one synapse in the central nervous system. (Ramón y Cajal, S.: Histologie du système nerveux de l'homme et des vertébrés. Vol. 1, pp. 531, 532. Madrid, Instituto Ramón y Cajal, 1952)

creases, even though the lengths of the neuron chains are equal.

Integrative Function of the Synapse

Sensory neurons entering the cord or brain branch and form many synapses. In contrast to this divergence by sensory neurons, we have in the case of the motor neurons a marked convergence (Fig. 2-10) of axons from both afferent and internuncial neurons on (1) its dendrites, (2) its cell body (soma), and (3) the initial segment of its unmyelinated axon (Fig. 2-2). These synapses on the motor neuron can either facilitate or inhibit the production of action potentials. The integration of these facilitatory and inhibitory influences by the motor neuron determines whether the axon conducts an impulse to its effector organ or not.

The synaptic organization discussed above is an *axodendrosomatic system* and is characteristic of the innervation of the motor neurons, the cortical neurons, and the neurons of Clarke's column. In the cerebellar cortex, however, we find "basket" endings which encase the soma of the cell (*axosomatic*), and in the olfactory glomeruli we note an axodendritic system of synapses.

MOTOR NEURONS

The cell bodies of skeletal-muscle motor neurons lie in the central nervous system, where they are influenced by many impinging interneurons and sensory neurons. Their axons leave the brain in the cranial nerves, and the spinal cord in the ventral (anterior) roots. At their point of exit each axon has a diameter between 1 and 20 μ.

Alpha Fibers

Axons having a diameter between 12 and 20 μ are generally classified as alpha (α) *fibers* of the A group (Table 2-2) and innervate muscle fibers outside the muscle spindle (a fusiform stretch

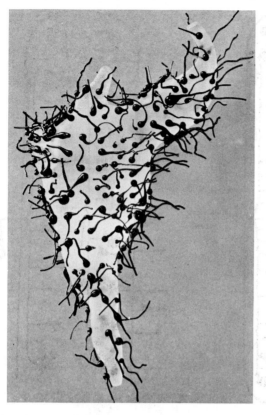

Fig. 2-10. Model showing the convergence of many axon end-bulbs (end-feet) on a neuron in the dorsal horn of the cat's spinal cord. The model is based upon the analysis of many serial sections of a spinal cord. (Haggar, R. A., and Barr, M. L.: Quantitative data on the size of synaptic end bulbs in the cat's spinal cord. J. Comp. Neurol. 93:35, 1950)

receptor in skeletal muscle). These muscle fibers are called *extrafusal fibers* and are responsible for the tension a muscle exerts. In the case of the extrinsic muscles of the eye, 1 neuron innervates as few as 3 muscle fibers, but in the soleus muscle, 1 neuron may innervate as many as 150 muscle fibers. The motor neuron plus the muscle fibers it innervates is called a *motor unit*. The total number of motor units in a muscle varies from 200 to 600.

Gamma Fibers

Each skeletal muscle in the body contains groups of 2 to 10 muscle fibers enclosed in a capsule which attaches at either end to the connective tissue of the endomysium, perimysium, aponeurosis, or tendon. This structure is called a *muscle spindle* (Fig. 2-11) and the muscle fibers it contains are termed intrafusal fibers (L. *fusus*, spindle). Each fiber in the muscle spindle contains a center with a sparse concentration of contractile molecules (myofilaments). It also contains the cell's nuclei, as well as its annulospiral and sometimes its flower-spray endings, which are innervated by large sensory neurons (Group I) and intermediate-size sensory neurons (Group II), respectively. Each end of the intrafusal fiber contains numerous myofilaments and is innervated by a gamma (γ) efferent neuron.

The Group I sensory fibers from the *annulospiral endings* fire in response to distending forces of 2 gm. or more, and the Group II fibers from the *flower-spray endings* (myotube or secondary endings) fire in response to forces of 20 gm. or more. The gamma efferent neuron, on the other hand, initiates the contraction of the peripheral ends of the intrafusal fibers. The contraction contributes little to the tension exerted by the muscle, but is an important mechanism for stretching the annulospiral and flower-spray endings (Fig. 2-12).

B and C Fibers

Also leaving the ventral root in the area of the thoracic and lumbar cord are myelinated fibers (preganglionic sympathetic neurons) with a diameter of about 3 μ (B fibers). These neurons synapse with unmyelinated C fibers (postganglionic sympathetic neurons) in the lateral chain of sympathetic ganglia on either side of the vertebral column, or in collateral ganglia. Some also innervate the adrenal medulla. The B fibers are called preganglionic sympathetic neurons. The fibers they innervate (postganglionic sympathetic neurons) carry impulses to the heart, smooth muscle, and glands of the body.

THE NEUROMUSCULAR JUNCTION

The large A fibers which innervate skeletal muscle branch after entering the muscle; each of these axon branches terminates in a trough in a small part of a single muscle fiber as an unmyelinated axon of 1 go 2 μ in diameter. As in the case of the synapse, there is at the neuromuscular junction of skeletal muscle a space between the axon and the membrane it innervates. Apparently there is also a release of neuro-

Fibers	A	B	d.r.C	a.C
Somatic efferent neurons	x	None	None	None
Most somatic afferent neurons	x	None	Some	None
Preganglionic autonomic neurons	None	x	None	None
Postganglionic autonomic neurons	None	None	None	x
Slow pain	None	None	x	None
Myelinated	x	x		
Unmyelinated	None	None	x	x
Fiber diameter (microns)	1-22	< 3	0.4-1.2	0.3-1.3
Conduction (meters per second)	5-120	3-15	0.6-2.0	0.7-2.3
Conduction/diameter ratio (meters per second per microns)	6	?	1.73	1.7
Spike duration (milliseconds)	0.4-0.5	1.2	2.0	2.0
Absolute refractory period (milliseconds)	0.4-1.0	1.2	2.0	2.0
Negative afterpotential	x	None	None	x
Amplitude (per cent of spike)	3-5			3-5
Duration (milliseconds)	10-20			50-80
Positive afterpotential				
Amplitude (per cent of spike)	0.2	1.5-4.0	10-30	1.5
Duration (milliseconds)	40-60	100-300		300-1000
Order of susceptibility to asphyxia	2	1	3	3

Table 2-2. Characteristics of A, B, dorsal root C (d.r.C), and autonomic C (a.C) nerve fibers.

transmitter substance, which in skeletal muscle causes a series of events resulting in the formation of an action potential that is transmitted throughout the muscle.

The C fibers innervating smooth and cardiac muscle and glands, unlike the A fibers innervating skeletal muscle, do not form a well-defined muscle end-plate (neuromuscular junction). In the muscle or gland they branch to form fibers ranging in diameter from 0.1 to 1.3 μ. Two to 8 of these fibers may be enclosed by one Schwann cell, but the Schwann cell is frequently interrupted at areas of the neuron loaded with vesicles. Presumably a neurotransmitter is released at these points. The transmitter then diffuses to many of the surrounding cells, where it may depolarize or hyperpolarize cell membranes or modify some other function of the cells. At the neuromuscular junction of skeletal-muscle fibers, on the other hand, the neurotransmitter is for the most part restricted to the muscle end-plate, where it has but one known function, depolarization. Another important difference between somatic and autonomic innervation is that in somatic innervation a single action potential in a neuron is followed by only one action potential in each of the muscle fibers it innervates. In the case of the autonomic neuron, a single action potential may be followed by numerous action potentials in the structures it innervates.

THE REFLEX ARC

Most stimuli originating outside the body and many originating inside produce contraction or secretion by means of the reflex arc. The reflex arc consists of (1) a transducer to convert one form of energy (heat, for example) to another (a nerve impulse), (2) a sensory neuron, (3) a central nervous system, (4) a motor neuron or neurons, and (5) an effector organ (muscle or gland). Some of the better known sensory transducers are listed in Table 2-1, though free nerve endings too, as we have noted, may act as transducers.

We have indicated that the sensory neurons are specialized to transmit impulses to the central nervous system, where they branch to many different areas and synapse with either motor neurons or interneurons. We have noted that they can be divided into Group I neurons (fibers of 11 to 22 μ in diameter, coming from stretch receptors in muscle), Group II neurons (fibers of about 7 μ in diameter, coming from sensory areas throughout the body), and Group III neurons (fibers of about 3 μ in diameter). Another classification divides them into A (large-diameter

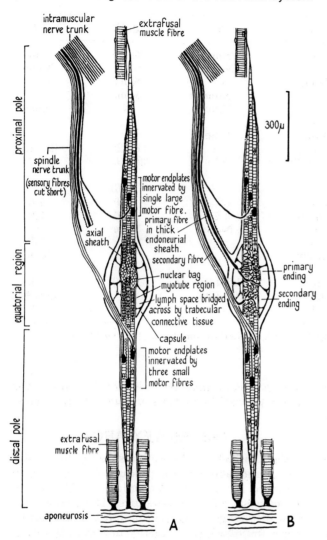

Fig. 2-11. The muscle spindle. In A we note the innervation of the intrafusal fibers by gamma efferent neurons. In B we see, in addition, the sensory fibers from the annulospiral endings and flower-spray endings in the equatorial region. (Barker, D.: The innervation of the muscle spindle. Quart. J. Micr. Sci., 89:156, 1948)

myelinated), B (small-diameter myelinated), and C (small-diameter unmyelinated) fibers.

The motor neurons receive impulses from many different sensory and internuncial neurons. A single sensory neuron, for example, might send impulses directly to a motor neuron in the spinal cord (stretch reflex), while at the same time one of its branches is conducting an impulse to the contralateral cord, medulla, pons, and thalamus. The ascending branches of the sensory neuron might then activate a series of interneurons, thereby stimulating descending neurons which impinge on the same motor neuron. This system of overlapping arcs of control permits (1) rapid response to a stimulus (short reflex arc) and (2) the integration of this response with other body activities (long reflex arc). Some of the ascending impulses to the thalamus, for example, are relayed to areas of the cerebral cortex which produce consciousness of the stimulus, enabling integration of the reflex act with the desires of the subject. Some of the ascending impulses to the pons, on the other hand, are relayed to the cerebellum, where they are unconsciously integrated with patterns controlling equilibrium. These ascending impulses, then, diverge to many different reflex centers, which send impulses back to the motor neurons.

The efferent neurons innervate skeletal muscle, the adrenal medulla, and neurons which have their cell bodies in ganglia (collections of cell bodies) outside the central nervous system.

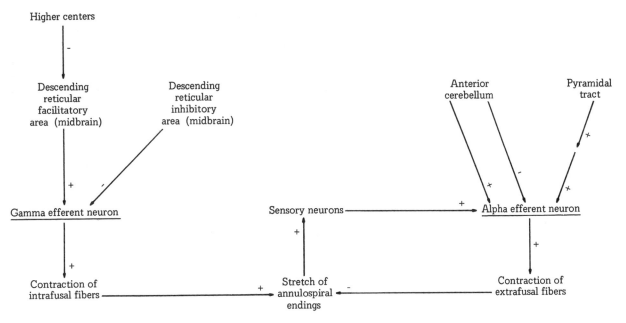

Fig. 2-12. Control of alpha and gamma efferent neurons.

Efferent fibers synapsing with neurons in *ganglia* outside the central nervous system are called preganglionic neurons, and the fibers they innervate, postganglionic neurons. The postganglionic fibers innervate the heart, smooth muscle, and glands. Motor neurons range in diameter from 1 to 20 μ and are subdivided into 3 groups, A (α, β, γ, and δ), B, and C fibers.

Monosynaptic Reflex Arc

The simplest reflex arc in the body is the monosynaptic reflex arc (sensory neuron, plus synapse, plus motor neuron). The stimulus for this reflex is stretch, and the receptor is the *annulospiral* ending in the *muscle spindle*. Group I neurons carry impulses to the central nervous system during the stretch of the intrafusal fiber (located in the spindle). Here these neurons excite somatic efferent neurons to the same muscle from which the stimulus originated (homonymous neurons) (Gk. *homonymous*, of the same name). The efferent neurons are of the A (α) group and cause a contraction of some of the extrafusal fibers—until the stretch of the intrafusal fibers is eliminated. This reflex might continue for a few milliseconds (twitch) or for hours, depending on whether the resultant contraction eliminates the stretch of the intrafusal fiber.

Under most conditions stimulation of the annulospiral endings causes only one obvious response—contraction of the stretched muscle. As can be shown, however, there is also an inhibition of ipsilateral antagonistic motor neurons and facilitation of ipsilateral synergistic motor neurons. The extent and degree of facilitation and inhibition depend upon the amount of stretch and its duration.

Gamma efferent neurons. The stretch reflex may be initiated either by (1) an external, or (2) an internal distending force. The internal distending force is produced by a gamma efferent neuron which causes the contraction of the two ends of the intrafusal fiber containing the annulospiral endings. The significance of the gamma efferent mechanism is not fully understood, though there is evidence that it is responsible for much of the muscle tone seen in conscious persons. Apparently the following sequence of events occurs.

1. Gamma efferent neurons are stimulated.
2. This causes the contraction of intrafusal muscle fibers.
3. The annulospiral endings in series with the intrafusal fibers are stretched (by the contraction of the intrafusal fibers).
4. Sensory neurons are activated.
5. Alpha efferent neurons are stimulated.

6. Extrafusal muscle fibers contract to produce muscle tone.

Evidence for the importance of the above reactions in the maintenance of tone is that (1) selective blocking of stretch receptors, or (2) section of the dorsal roots containing the afferent neurons from the annulospiral endings, produces a decrease in muscle tone. Either procedure also causes loss of the ability of an affected appendage to support the individual's weight.

Stimulation of the gamma efferent neuron. Much of the facilitation or inhibition of gamma efferent neuron activity apparently comes from areas in the brain (Fig. 2-12). Facilitation is most pronounced in the antigravity muscles. We find, for example, on section of the midbrain between the superior and inferior colliculi (*decerebration*), that a marked increase in extensor muscle tone (hypertonicity) and an exaggeration of the stretch reflexes (hyperreflexia) of extensor muscles result. These symptoms can be eliminated by either a high transection of the cervical cord or a transection of the dorsal roots. Apparently parts of the brain above the inferior colliculi inhibit a center (*descending reticular facilitatory area*) which facilitates gamma efferent neuron activity. (This accounts for the effectiveness of decerebration in producing hypertonicity.) The center apparently lies between the superior colliculi and the spinal cord. There is also evidence for the existence of a descending reticular inhibitory area in this region which inhibits the gamma efferent neurons of extensor muscles.

This is not to say, however, that all hypertonicity or all facilitation of motor activity is due to the facilitation of neurons to the intrafusal fibers (gamma efferent neurons). We know that

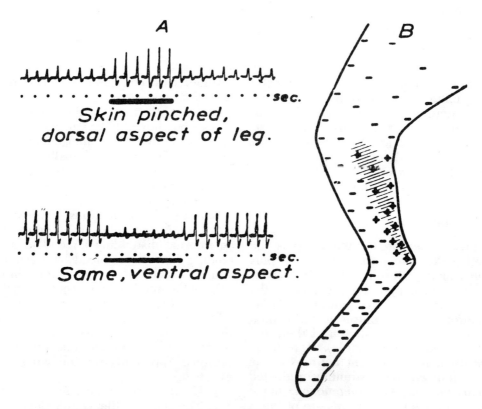

Fig. 2-13. Facilitation and inhibition of a monosynaptic reflex. In these experiments the gastrocnemius muscle was stretched and at the same time different areas of the skin were pinched. In the upper record in part A we see a facilitation of the reflex due to pinching the skin that lies over the muscle. In the lower record we see an inhibition of the reflex by pinching the skin elsewhere on the appendage. (Hagbarth, K. E.: Excitatory and inhibitory areas for flexor and extensor motoneurones. Acta Physiol. Scand., 26:23, 24, 1952)

neurons in the pyramidal tract (corticospinal tract) activate interneurons which directly facilitate alpha motor neuron activity in either the presence or the absence of the dorsal roots. Apparently the anterior cerebellum also contains fibers that activate interneurons, which in turn directly facilitate or inhibit alpha efferent neuron discharge. We find too that anterior cerebellar destruction produces a hypertonicity not relievable by dorsal root section.

Cutaneous stimuli such as light touch, pressure, or a pinprick also play a role in the stretch reflex (Fig. 2-13). Such stimuli can facilitate gamma efferent fibers to ipsilateral flexor muscles and inhibit those to ipsilateral extensor muscles. Possibly this is an important means of maintaining tension on the intrafusal fibers while the extrafusal fibers are shortening during, for example, a withdrawal reflex. It is important to remember, in this regard, that shortening the extrafusal fibers decreases the stretch on the intrafusal fibers unless accompanied by the contraction of intrafusal fibers. Thus a simultaneous contraction of intrafusal and extrafusal fibers maintains the facilitation of homonymous motor neurons by afferent fibers from the annulospiral endings during muscle shortening.

Knee jerk. One of the methods the diagnostician uses to study the stretch reflex is that of stretching a muscle suddenly by means of a sharp blow to a tendon. This results in a short-lived synchronous contraction of the stretched muscle. There are probably three reasons for the short duration of this response. First, in the knee jerk (a monosynaptic reflex) extensive multiple chain and closed chain pathways do not exist (Fig. 2-14). Second, the stretch-induced contraction of extrafusal fibers relieves the stretch on the adjacent intrafusal fibers. Third, the alpha efferent neuron, before emerging in the ventral root, gives off recurrent collaterals which stimulate an interneuron (Renshaw cell), which in turn inhibits the alpha efferent neuron. The tapping of the patellar tendon, the resultant knee jerk, and the period of electrical silence in the dorsal and ventral root as the muscle contracts is an example of this.

Multisynaptic Reflexes

The multisynaptic reflexes are more diffuse than the monosynaptic reflexes, in both origin and effect, and are considerably more difficult to study. They are initiated by impulses from (1)

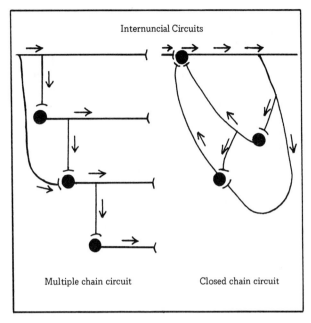

Fig. 2-14. Diagram of 2 types of internuncial circuits, multiple chain and closed chain circuits. (*After* Lorente de Nó, R. J.: Analysis of the activity of the chains of internuncial neurons. J. Neurophysiol., 1:210, 1938)

structures in muscle, (2) structures in the tendon, joint, or periosteum, and (3) the skin or viscera. Excitation in multisynaptic reflexes, unlike that in monosynaptic ones, generally elicits a response far outlasting the stimulus. The mechanisms responsible for this after-discharge are indicated in Figure 2-14. In the multiple chain circuit we note part of a reflex arc which involves parallel chains of interneurons. Since each path involves a different number of synaptic delays (about 0.5 msec. per synapse), each path delivers its impulse at a different time. In the closed chain circuit (reverberating circuit) we note a second means of prolonging the response. Here the efferent impulse stimulates neurons that feed back to the original neuron.

The usual response to *electrical stimulation* of sensory neurons other than those in Group I is (1) facilitation of the motor neurons to the ipsilateral flexor muscles and (2) inhibition of the motor neurons of the ipsilateral extensor muscles. On the contralateral side of the body there is (3) inhibition of the motor neurons to the flexor muscles and (4) facilitation of the motor neurons to the extensor muscles. The degree of facilitation and inhibition of the various motor neurons depends on which sensory neurons are stimulated (local sign) and the strength and character of the stimulus.

The nociceptive and extensor thrust reflex. A number of the multisynaptic reflexes seen in humans do not conform to the pattern of ipsilateral contraction of flexors and inhibition of extensors noted above in the experiments on electrical stimulation of sensory nerves. One that is consistent with this pattern is the nociceptive (L. *noceo*, to injure) or withdrawal reflex; one that is inconsistent with the pattern is the extensor thrust reflex, demonstrable in animals whose cord has been transected at the cervical level. In such preparations, light pressure or stretch applied between the toe pads produces extension of the limb. This is called the extensor thrust reflex and is believed to be an important mechanism in walking and in maintaining a standing posture. The sensory limb for this reflex is contained in the plantar nerves.

Autogenic inhibition. In our discussion of the monosynaptic reflex it was emphasized that distending forces as small as 2 gm. can stimulate annulospiral endings, thereby eliciting reflex contraction of extrafusal muscle fibers. This, in turn, can eliminate the original stretch stimulus. We find, however, that if a distending force of 200 gm. or more is applied to a muscle a second type of stretch receptor is stimulated. This receptor (the *Golgi tendon organ*), unlike the muscle spindle, lies in series with the extrafusal

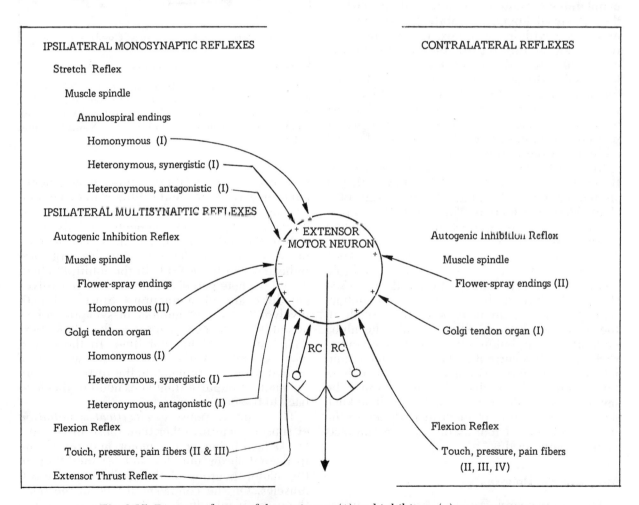

Fig. 2-15. Diagram of some of the excitatory (+) and inhibitory (−) neurons converging on a single somatic efferent neuron to an extensor muscle. Numerals in parentheses indicate groups to which sensory neurons belong. RC: Renshaw cell. Homonymous: contained in the same muscle. Heteronymous: contained in a different muscle.

fibers and therefore increasingly stimulates afferent neurons as contraction progresses. Both the muscle spindle (the low-threshold stretch receptor) and the Golgi tendon organ (the high-threshold stretch receptor) are innervated by Group I sensory neurons. The former are referred to as Group I-A, and the latter are referred to as Group I-B.

The stimulation of the I-B fibers results in multisynaptic inhibition of the homonymous motor neurons and stimulation of antagonistic neurons, apparently preventing damage to the muscle when its contraction meets excessive resistance. It might be characterized as an inverse stretch reflex. If sufficient stretch occurs in any Golgi tendon organ, as in Indian wrestling, this reflex can prevent the homonymous neurons from firing at all. This, then, is one of many reflexes impinging on the motor neuron in the central nervous system. A few of these influences are summarized in Figure 2-15.

Other reflexes. There are of course numerous other reflexes. Some of these involve control of the diaphragm, of the urinary bladder, the heart, the smooth muscle of blood vessels, the salivary glands, the sweat glands, the ciliary muscle of the eye, the iris, etc. Some of the reflexes elicit secretion and contraction. Others modify the character of the secretion and contraction. Apparently in smooth-muscle reflexes as well as skeletal-muscle multisynaptic reflexes, synergists are stimulated at the same time that antagonists are inhibited. In the case of the gastrointestinal tract, for example, reflexes producing waves of contraction that propel the intestinal contents also elicit relaxation of the sphincters, which prevent propulsion. In the urinary bladder, reflexes that elicit contraction of the detrusor muscle of the bladder also elicit relaxation of the sphincters of the urethra. Reflexes that elicit contraction of the ciliary muscle of the eye also facilitate motor neurons to the circular muscle of the iris and inhibit motor neurons to the radial fibers of the iris. Reflexes eliciting an increase in cardiac output (heart minute output) also cause constriction of the arterioles. The two actions work together to produce an increase in arterial pressure.

Chapter 3

THE EQUILIBRIUM POTENTIAL
THE ACTION POTENTIAL
CONDUCTION
RECEPTOR MECHANISMS
ELECTRICAL STIMULATION

THE TRANSMEMBRANE POTENTIAL

It is now possible to construct ultramicroelectrodes with tips less than 1 μ in diameter. These tips can be filled with a KCl solution and inserted into the cytoplasm of a cell without damaging the cell. Using such a procedure, investigators have found that the cytoplasm of most cells is electrically negative to the extracellular fluid. The electromotive force that exists across the unstimulated cell membrane varies from −20 to −100 mv. and is variously referred to as the resting transmembrane potential, the steady potential, or E_S. By convention, E_S is written as a minus value, since the cytoplasm is negative to the extracellular fluid (reference fluid). A change of E_S from a normal resting value to a more negative value is called hyperpolarization, and that to a more positive value, depolarization. Unfortunately, confusion occasionally results from the terminology; a change in the transmembrane potential from −80 to −40 or to +40 mv., for example, is characterized as a *decrease in potential* or *depolarization*, while a change from 0 to −80 mv. is called an *increase in potential* or *repolarization*.

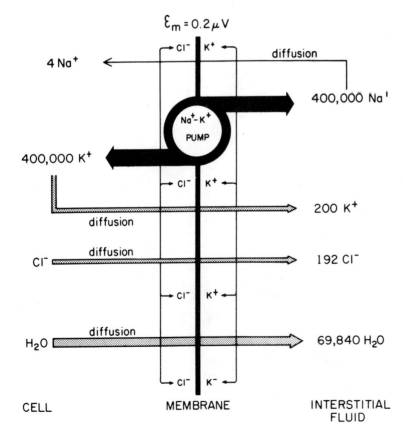

Fig. 3-1. Scheme of ion and water movements that would initially occur in a hypothetical cell which had just had its sodium-potassium pump activated. The pump is characterized as actively moving Na+ out of the cell and K+ into the cell. The concentration gradient that results causes a small amount of Na+ to diffuse back into the cell (4 Na+) and K+ (200 K+) to diffuse out of the cell. The larger efflux of K+ than influx of Na+ is due to the greater permeability of the cell membrane to K+ (represented by a thicker arrow for K+ diffusion than for Na+ diffusion). The greater net movement of cations out of the cell results in a negatively charged cytoplasm. This, in turn, causes an efflux of those anions to which the cell membrane is permeable, namely Cl−. There is also an efflux of water in response to the osmotic gradient which results from the net efflux of ions. The changes noted above result in a transmembrane potential (E_m) of 0.2 μV. (Ruch, T. C., and Patton, H. D.: Medical Physiology and Biophysics. 19th ed., p. 21. Philadelphia, W. B. Saunders, 1965)

The transmembrane potential results from the difference in concentration of positively and negatively charged ions in the intracellular and extracellular fluids. This difference, in turn, results from the extrusion of Na$^+$ from, and the infusion of K$^+$ into the cell by the sodium-potassium pump, and from the variable *permeability* of the plasma membrane. Under resting conditions the plasma membrane constitutes a complete barrier to the large, negatively charged organic ions formed in the cell and a partial barrier to Na$^+$, K$^+$, H$^+$, Cl$^-$, and HCO$_3^-$. Also under resting conditions, the pump maintains a high intracellular concentration of K$^+$ and a low intracellular concentration of Na$^+$ by moving K$^+$ into the cell and Na$^+$ out. Because of its high intracellular concentration, the K$^+$ tends to diffuse back into the interstitial fluid (Fig. 3-1), while Na$^+$, as a result of its concentration gradient, tends to diffuse into the cytoplasm. The resting cell membrane, however, is over fifty times more permeable to K$^+$ than to Na$^+$. Hence K$^+$ diffusion through the membrane occurs much more readily than Na$^+$ diffusion, even though the sodium-potassium pump transports approximately equal quantities of Na$^+$ and K$^+$. Or, more K$^+$ diffuse out of the cell than Na$^+$ into it when the sodium-potassium pump is active. As the net loss of positively charged ions from the cell continues, the cytoplasm becomes increasingly negative, until the electrical gradient balances the K$^+$ concentration gradient (i.e., until the negatively charged cytoplasm impedes the further diffusion of K$^+$ out of the cell). The negativity of the cytoplasm also results in a higher concentration of anions to which the membrane is permeable in the extracellular fluid than in the cytoplasm. We find, for example, that the concentration of Cl$^-$ is 30 times, and HCO$_3^-$ 3.4 times higher outside the cell than inside. Many of these facts are summarized in Table 3-1.

THE EQUILIBRIUM POTENTIAL

We have noted above that the diffusion of K$^+$ from an area of high K$^+$ concentration to one of low K$^+$ concentration contributes to the development of a negatively charged cytoplasm. The transmembrane potential resulting from this process (E_K) is due to the concentration gradient for K$^+$ ($K_o/K_i = K_{outside}/K_{inside}$) and the charge carried by each potassium ion (Z_K). Its magnitude can be calculated by the *Nernst equation* as follows:

$$E_K = \frac{RT}{FZ_K} \log_e (K_o/K_i)$$

In this formula, R is the gas constant, T the absolute temperature, and F the Faraday number (number of coulombs of charge per mol of particle). At 37°C. the formula is:

$$E_K = 61.5 \log_{10} (K_o/K_i)$$

Using the data in Table 3-1, we obtain a value for E_K of 97 mv. This transmembrane potential, in other words, would result from the transmembrane concentration gradient for K$^+$ once K$^+$ effusion equalled K$^+$ infusion. You will note in the table that this value (the equilibrium potential

(ions)	Interstitial Fluid (μmols/cm.^{3})	Intracellular Fluid (μmols/cm.^{3})	(C_o/C_i)	E_{ion} (mv.)
Cations	154	167		
Na$^+$	145	12	12.1	+ 66
K$^+$	4	155	1/39	− 97
H$^+$	3.8 × 10^{-5}	13 × 10^{-5}	1/3.4	− 32
Others	5			
Anions	154	167		
Cl$^-$	120	4	30	− 90
HCO$_3^-$	27	8	3.4	− 32
Others	7	155		
pH	7.43	6.9		
Potential	0	− 90 mv.		− 90 mv.

Table 3-1. Steady state ion concentrations in a mammalian muscle and the resulting equilibrium potential. (C_o/C_i), concentration gradient between the outside and inside of a cell; E_{ion}, equilibrium potential for the listed ions.

for K^+) is close to the measured transmembrane potential (-90 mv.).

The equilibrium potential for Na^+ (E_{Na}) is calculated using the same formula. The value obtained by this calculation is $+66$ mv. If the membrane's permeability to K^+ were 0 and its permeability to Na^+ considerably higher, Na^+ would infuse into the cell and produce a transmembrane potential of $+66$ mv. The resting transmembrane potential is negative rather than positive because of the membrane's greater permeability to the cation that tends to diffuse out of the cell (K^+) than to the one that tends to diffuse into it (Na^+).

You will also note in Table 3-1 that the equilibrium potential for Cl^- is the same as the transmembrane potential. The Nernst equation for a negatively charged particle (in this case Cl^-) at 37°C. is:

$$E_{Cl} = 61.5 \log_{10} (Cl_i/Cl_o)$$

Apparently the plasma membrane is sufficiently permeable with respect to Cl^- to let Cl^- reach an equilibrium; apparently too, there is no active transport process or intracellular production of Cl^- sufficiently important to distort this equilibrium.

THE ACTION POTENTIAL

As noted above, the resting transmembrane potential in some cells is dependent upon the low permeability of the cell membrane to Na^+. Increasing the permeability of the cell membrane to Na^+ produces an influx of positively charged ions (Na^+ current), thereby decreasing the negativity of the cytoplasm; that is, increased permeability to Na^+ depolarizes the cell.

There are a number of *stimuli* capable of producing depolarization in particular cells. These include (1) mechanical stimuli (pinching, stretching, heating, etc.); (2) chemical stimuli (acetylcholine, norepinephrine, etc.); and (3) electrical stimuli. In most of the studies reported in this chapter we shall be dealing with the response of excitable cells to electrical stimuli—this type being preferred in physiological studies because its strength and the duration of its application are readily controlled. In many cases, however, mechanical and chemical stimuli produce results similar to those of electrical stimuli. In other cases, the same stimulus (acetylcholine, for example) that produces depolarization in one type of cell produces hyperpolarization in another.

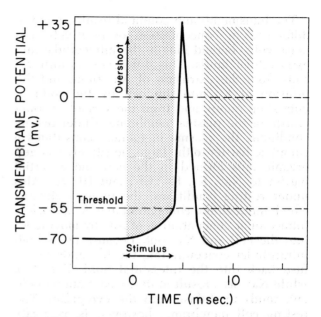

Fig. 3-2. Response of a nerve fiber to a gradual increase in stimulus strength. Note the resting transmembrane potential (*clear area*), the gradual depolarization as the stimulus strength is increased (*shaded*), the action potential once threshold is obtained (*clear*), and the positive after-potential (*shaded*).

Response to an Electrical Stimulus

In Figure 3-2 we note the change in the transmembrane potential of a nerve fiber as increasing voltage is applied to the surface of its plasma membrane. This change is recorded near the cathode of the stimulating electrode and is similar to that which we find in the muscle fiber. We see that initially there is a gradual decrease in the negativity of the cytoplasm as the strength of the stimulus (voltage) is increased. Eventually, however, a point is reached at which rapid depolarization occurs, called the spike or action potential. As may be seen, the peak of the spike does not stop at a transmembrane potential of zero, but reaches a higher value. A stimulus just strong enough to initiate the spike potential is called a *threshold stimulus*; one that is too weak to initiate a spike potential is called a *subthreshold* stimulus, and one stronger than threshold is called a *superthreshold* stimulus. You will note that although a subthreshold stimulus does not initiate an action potential, it can produce changes in the transmembrane potential. Any decrease in the negativity of the cytoplasm is characterized as a *depolarization*, whether it results in an action potential or not.

In other words, the response of an excitable cell to an adequate stimulus (a threshold or super-threshold stimulus) results in a decrease in the transmembrane potential to a critical level (firing or threshold level) and in the formation of an action potential associated with the change of the cytoplasm from electronegativity to -positivity (the overshoot). This does not occur simultaneously throughout the cell, but rather at one locus on it. Normally, however, the action potential is propagated from this locus to all parts of the cell membrane. The act of propagation will be discussed more fully later.

Sodium Conductance

We have described the plasma membrane of the resting cell as having a permeability to K^+ greater than that to Na^+, and have indicated that this, plus the sodium-potassium pump, is in part responsible for the resting transmembrane potential. We have also noted that one can decrease the transmembrane potential by electrical stimulation and that if the depolarization reaches a certain critical value (the threshold) an action potential occurs. Bernstein, in 1902, suggested that stimulation produced the action potential by a general increase in the permeability of the cell membrane. He believed that the membrane was selectively permeable to K^+ in the resting condition and became permeable to all ions upon stimulation. In 1940, however, intracellular electrodes were used to measure transmembrane potentials and it was noted that depolarization is associated not merely with a loss of charge, but with a reversal of charge—an *overshoot*. In 1949, Hodgkin and Katz suggested that this overshoot was due to a brief and specific increase in the permeability of the membrane to Na^+. This increase in permeability to Na^+, it was hypothesized, permitted the concentration and voltage gradients (Table 3-1) to drive sufficient amounts of Na^+ into the cytoplasm to move the membrane charge away from the equilibrium potential for K^+ (−97 mv.) and toward that for Na^+ (+66 mv.), or, that stimulation caused a selective increase in Na^+ conductance (Fig. 3-4).

Much of the evidence for the above hypothesis comes from a technique developed by Hodgkin, Huxley, and Katz. In this technique, electrodes are passed down the length of the cytoplasm, the entire cell is stimulated, and the movement of current through the membrane is recorded. Upon depolarization past the threshold level, there is an early inward flow of current due to the influx of Na^+. This is followed by an outflow of current attributed to the efflux of K^+. The inward current (Na^+ current) does not occur if depolarization (1) does not reach threshold or (2) exceeds the equilibrium potential for Na^+. The inward current can be reduced if the equilibrium potential for Na^+ is reduced by decreasing the concentration of Na^+ in the external environment (see Nernst equation). The inward current can be eliminated if the Na^+ in the external environment is replaced by choline. In Figure 3-3 we see the decrease in the action potential that results from lowering the concentration of external Na^+.

In summary, then, all available evidence is consistent with the view that a threshold stimulus initiates a 500-fold increase in Na^+ conductance, and that this change is responsible for (1) the inward current recorded once a threshold depolarization has been reached and (2) the overshoot of the spike potential. The concentration of Na^+ in the external environment and the fact that Na^+ is the only biologically occurring

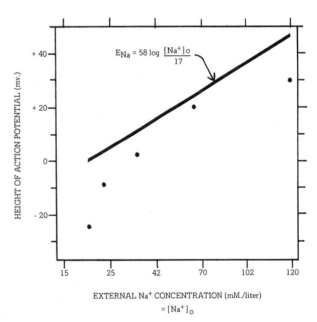

Fig. 3-3. Changes in the height of the action potential of frog sartorius muscle fibers in response to changes in the external Na^+ concentration. The dots represent experimental results and the line is a plot of the equilibrium potential for Na^+. The change in Na^+ concentration was produced by replacing the NaCl in solution with choline chloride. (*After* Nastuk, W. L., and Hodgkin, A. L.: The electrical activity of single muscle fibers. J. Cell. Comp. Physiol., 35:65, 1950)

ion which can penetrate the membrane and which has a positive equilibrium potential also lends credibility to this view. So far, however, we know very little of the molecular mechanism responsible for the two separate conductance systems, i.e., for K+ and Na+. The Na+ conductance system might be visualized as one containing carrier molecules (Fig. 1-5) which are released as depolarization occurs and sequestered at the end of depolarization. These carriers, when released, would combine with Na+ at the membrane surface, and the resulting complex would then diffuse inward and release the Na+ into the cytoplasm.

The all-or-none nature of the action potential. As one increases the strength of a stimulus to a threshold value, he progressively depolarizes the cell membrane at the cathode. Once the threshold has been reached, the action potential develops explosively; one obtains the same type of action potential (in terms of amplitude and duration) whether he applies a threshold or superthreshold stimulus. A single subthreshold stimulus, on the other hand, yields no action potential. This relationship between the stimulus strength and the action potential is variously referred to as an *all-or-none* or trigger relationship. There are numerous examples of such a relationship both in the body and in the world in general. When one fires a gun, for example, the energy imparted to the bullet is independent of the pressure applied to the trigger, provided

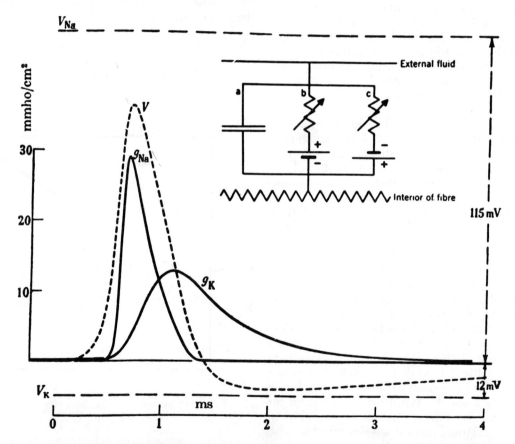

Fig. 3-4. Theoretical action potential (V) and change in membrane conductance for Na+ (g_{Na}) and K+ (g_K) derived by Hodgkin and Huxley for the giant axon at 18.5°C. Also shown is a characterization of the cell membrane as a (a) capacitor with associated variable resistance channels for (b) K+ and (c) Na+. (Hodgkin, A. L., and Huxley, A. F.: A quantitative description of membrane current and its application to conduction and excitation in nerve. J. Physiol., *117*:530, 1952; and Huxley, A. F.: Electrical processes in nerve conduction. *In* Clarke, H. T. (ed.): Ion Transport Across Membranes. P. 32. New York, Academic Press, 1954)

the pressure is threshold or greater. Other all-or-none relationships include (1) pressure and flushing a water closet, (2) friction and lighting a match, (3) pressure and springing a mouse trap, and (4) pressure and turning on a light. The fact that there is a trigger relationship between the stimulus strength and the action potential does not mean, however, that the character of the action potential is fixed. In Figure 3-3 we have shown that it can be modified by changes in the environment of the cell, for example.

Apparently the action potential is an intrinsic property of many cells. Once the critical potential has been reached, there is an increase in the permeability of the membrane to Na^+, causing the entry of Na^+ into the cytoplasm; this, in turn, further depolarizes the membrane, thereby increasing Na^+ conductance:

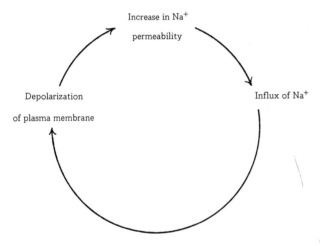

In short, an adequate stimulus initiates a vicious cycle which culminates in the reversal of the transmembrane potential. As this process continues, the permeability of the membrane to K^+ also starts to increase (Fig. 3-4), and eventually Na^+ conductance returns to its low resting value. It is the decrease in Na^+ conductance and the increase in K^+ conductance that results in the rapid disappearance of the action potential and the return of the transmembrane potential to the equilibrium potential for K^+. The net effect at the end of the action potential is that a small amount of Na^+ (about 3 parts per million) has entered the cytoplasm and an equivalent amount of K^+ has left. The sodium-potassium pump eventually restores the cytoplasm to its prestimulation concentration however.

Potassium Conductance

You will note in Figure 3-4 that potassium conductance (g_K) begins to increase at about the same time as the sodium conductance (g_{Na}), but does not increase as rapidly or as much. Its increase, however, far outlasts that for g_{Na}. This is probably why the transmembrane potential comes closest to the equilibrium potential for K (V_K) during elevated K^+ conductance; that is, the hyperpolarization (positive after-potential) following the action potential probably results from increased K^+ conductance.

Just as the Na^+ concentration in the external environment tends to control the height of the action potential (Fig. 3-3), so apparently does the K^+ concentration in the intercellular fluid control the resting transmembrane potential. In Figure 3-5, for example, we note that as we increase the extracellular K^+ (K_o), we decrease the concentration gradient for K^+ and in this way facilitate an inward movement of K^+ and a decrease in the negativity of the cytoplasm. In the

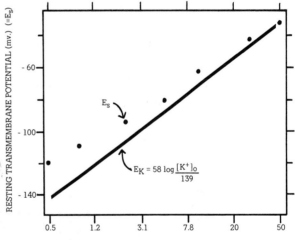

Fig. 3-5. Changes in the transmembrane potential (E_S) in response to changes in the extracellular concentration of K^+ ((K^+)$_O$). Note that as the extracellular K^+ concentration goes up, the equilibrium potential for K^+ (E_K) and the transmembrane potential (E_S) become less negative. Apparently the resultant influx of K^+ reduces the negativity of the cytoplasm. These experiments were done on excised frog muscle at room temperature. (*Redrawn from* Adrian, R. H.: The effect of internal and external potassium concentration on the membrane potential of frog muscle. J. Physiol., *133*:641, 1956)

terms of the Nernst equation, if we increase the ratio of K_o to K_i, we increase E_K (58 $\log_{10}$ (K^+_o/K^+_i)) at room temperature. You will note in Figure 3-5 that as the potassium equilibrium potential (E_K) increases, so also does the transmembrane potential (E_S).

CONDUCTION

In most parts of the unmyelinated neuron and muscle fiber, an action potential at one locus on the plasma membrane initiates action potentials in successive portions of the membrane. In an axon, for example, the action potential might represent a change in transmembrane potential of 110 mv., whereas the critical depolarization necessary for adjacent areas is only about 12 per cent of that value. Hence we would expect to see a wave of depolarization pass from the point of stimulation at the cathode throughout the fiber (Fig. 3-6).

In the myelinated neuron, the myelin sheath effectively insulates the neuron from depolarization at all points except the nodes of Ranvier, where there is little or no myelin. In these neurons the action potential jumps from one node to the next. Normally an action potential at one node sets up a current in the interstitial fluid which is capable of producing depolarization at an adjacent node 4 to 7 times greater than the threshold value for that node. In fact, these currents can travel past as many as 3 narcotized

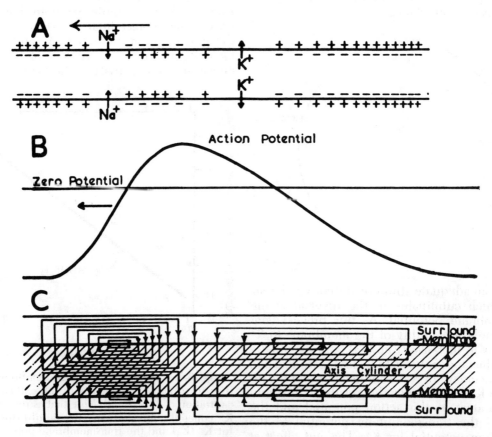

Fig. 3-6. Conduction of an action potential. In this diagram the action potential is characterized as moving from right to left and as being associated with a surface negativity due to an inward Na+ current. This inward current is followed by a K+ efflux. Parts A, B, and C are drawn so that the action potential is at the same position in each. (Eccles, J. C.: Neuron physiology—Introduction. *In* Field, J. (ed.): Handbook of Physiology. Sec. 1 (Neurophysiology), vol. I, p. 64. Washington, American Physiological Society, 1959)

nodes to produce stimulation in a given neuron. On the other hand, currents initiated by one neuron, although they may change the threshold of adjacent fibers, seldom stimulate these fibers.

This leaping of the action potential from one node to the next is called *saltatory conduction.* The activation of each node requires a short period for the critical potential to be reached, but is a more rapid system of activation than is found in unmyelinated neurons. Unmyelinated neurons conduct the action potential at a velocity of from 0.6 to 2 meters per second (1.7 meters/sec./μ of diameter), whereas myelinated fibers conduct at 3 to 120 meters per second (6 meters/sec./μ).

Intercellular Conduction

At present we do not completely understand how the action potentials of many of the body's cells initiate action potentials in other cells. In the case of the heart, however, we know that at certain intercalated discs, the plasma membrane of one cell seems to fuse with that of another. This apparently constitutes an intercellular conduction pathway. It should be emphasized, though, that this pathway has properties different from those of the plasma membrane, since it is apparently more susceptible to blockage by drugs such as quinidine than is the plasma membrane itself.

Another type of conduction pathway is the neuromuscular junction, characteristic of skeletal muscle, and the synapse, characteristic of the autonomic nervous system (visceral efferent neurons). In each of these a definite space (junction or synapse) exists between two cells. It is into this space that a neurotransmitter (in these cases acetylcholine) is liberated, which is important in causing an action potential in the receptor cell.

Neuromuscular Transmission

The importance of "animal electricity" in initiating the contraction of muscle was emphasized by Aloysio Galvani (1737-1798) and has since been a prime factor in the thinking of physiologists. The importance of chemical transmitters, on the other hand, was less readily acknowledged. One of the first to effectively experiment on neuromuscular transmission was Claude Bernard. In the middle of the nineteenth century, he used the South American Indians' arrow poison, *curare,* on experimental animals, and showed that it blocked transmission of the nerve impulse to skeletal muscle by acting at or near the muscle. It apparently had no effect on transmission in the spinal cord or peripheral nerves.

In 1921, Otto Loewi reported that stimulation of the vagus nerve in a frog produced cardiac slowing not only in that frog but also in a second frog which received blood by means of cross-circulation from the first animal. He concluded that the vagus releases a chemical, which he called "vagus stuff," responsible for the cardiac slowing. In 1936, Sir Henry Dale stimulated nerves to skeletal muscle and collected a fluid which, when applied to the acetylcholine-sensitive muscles of a leech, caused the muscles to contract. It was soon concluded that both the "vagus stuff" of Loewi and the secretion collected by Dale was *acetylcholine.*

It was later noted that certain drugs, *physostigmine* and *neostigmine*, prolonged and intensified the action of acetylcholine applied on a muscle or released by a neuron (Fig. 3-7). Apparently they acted by inhibiting enzymes

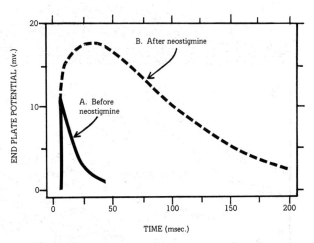

Fig. 3-7. Effect of an antiacetylcholine esterase drug on the response of the motor end-plate to the stimulation of a motor nerve. The development of action potentials in the muscle fiber was prevented by placing the fiber in a Na^+-deficient bath. Record A was taken before and B after the injection of an antiacetylcholine esterase (neostigmine). Note the increased amplitude and duration of the end-plate potential after the antiacetylcholine esterase. (Redrawn from Fatt, P., and Katz, B.: An analysis of the end-plate potential recorded with an intra-cellular electrode. J. Physiol., 115:338, 1951)

(*acetylcholine esterases*) normally found in the tissues which destroy acetylcholine. These drugs are classified as antiacetylcholine esterases.

More recently, Fatt and Katz have studied the relationship between a single axon and the muscle fiber it innervates, and have reported that a single nerve impulse causes the release of about 10^{-8} mols of acetylcholine at each neuromuscular junction. It, on the other hand, requires about 5×10^{-16} mols to produce contraction if the acetylcholine is applied at the neuromuscular junction by a micropipette. In summary, then all the above data, plus the observation that the electrical currents set up by the nerve endings in skeletal muscle are not adequate to stimulate the muscle fiber, lead to the conclusion that the action potential in the somatic efferent neuron causes the release of a neurotransmitter substance, acetylcholine, which initiates a series of contraction-producing events in the muscle fiber. It is further suggested that the acetylcholine normally (i.e., in the absence of physostigmine) remains at an effective concentration at the postjunctional surfaces for only about 1 msec. because of its rapid destruction by acetylcholine esterase. This short period of acetylcholine exposure results in only a single action potential in the skeletal-muscle fiber.

The release of acetylcholine. Acetylcholine is apparently stored in bound form in vesicles in the terminals of the somatic efferent neurons (Fig. 3-8). It is speculated that when the action potential reaches that part of the neuron with

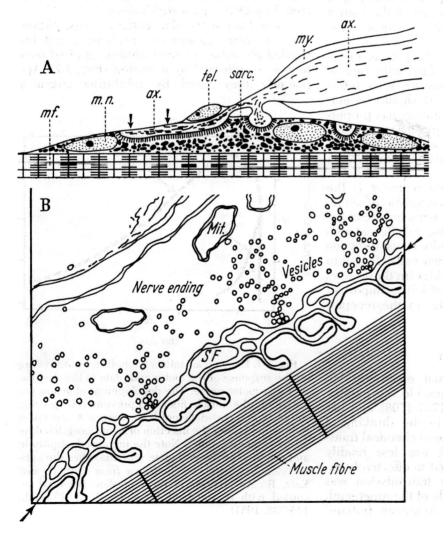

Fig. 3-8. Schematic drawing of a motor end-plate. At A we note that as the axon (ax, axoplasm and mitochondria) approaches the muscle fiber it loses its myelin sheath (my) and terminates in "synaptic troughs" adjacent to the sarcoplasm (sarc.). Note also the teloglia (tel.), muscle nuclei (m.n.), and myofibrils (mf.). At B we see a tracing of the frog neuromuscular junction ($\times$ 26,000). The arrows show the line of the synaptic cleft. Note above the cleft the vesicles and mitochondria of the axoplasm and below it the muscle fiber. (A: Couteaux, R.: Morphological and cytochemical observations on the postsynaptic membrane at motor end-plates and ganglionic synapses. Exp. Cell. Res., Suppl. 5:296, 1958; B: Birks, R., Huxley, H. E., and Katz, B.: The fine structure of the neuromuscular junction of the frog. J. Physiol., *150*:137, 1960)

the acetylcholine-containing vesicles, the intracellular concentration of Ca^{++} increases, causing the release of the neurotransmitter.

Part of the evidence for this is that the amount of acetylcholine released in response to an action potential is proportional to the concentration of Ca^{++} in the external environment. It has also been speculated that the action potential in muscle likewise causes an increase in intracellular Ca^{++} and that it is this Ca^{++} which activates the contractile mechanism. In the case of skeletal and cardiac muscle, the increase in Ca^{++} in the cytoplasm is thought to be due primarily to the release of Ca^{++} sequestered in the cell. In the case of smooth muscle, most of the Ca^{++} diffuses into the cell during the action potential. In muscle, therefore, the increase in intracellular Ca^{++} is characterized as the *activation-contraction coupling mechanism*, and in nerve, as the *activation-secretion coupling mechanism*.

The end-plate potential. The acetylcholine (ACh) released at the neuromuscular junction can combine with either a receptor on the muscle end-plate or an acetylcholine esterase (AChE). In the case of the former, the combination results in depolarization of the postjunctional tissue. This depolarization is called the end-plate potential and is the result of an increase in the permeability of the end-plate to Na$^+$ and K$^+$. It is a localized, nonpropagated phenomenon, which, within limits, is proportional to the amount of ACh released and inversely proportional to the speed at which the ACh is hydrolyzed by the enzyme AChE. Normally, sufficient ACh combines with the neuromuscular junction that the end-plate potential initiates a propagated depolarization in the rest of the cell membrane. This propagated potential, unlike the end-plate potential, is an all-or-none phenomenon.

The relationship between the motor neuron and the end plate can be modified by a number of agents. For example, the injection of the anticholinesterase, physostigmine, decreases the rate of destruction of the ACh released at the neuromuscular junction. *Botulinum toxin*, on the other hand, interferes with the release of ACh by the neuron, and *d*-tubocurarine (curare) interferes with the activation of the muscle end-plate by ACh.

Curare is a competitive inhibitor of ACh at the neuromuscular junction; apparently both can combine with the receptor protein of the end-plate. Curare, however, does not produce an increase in the permeability of the membrane to Na$^+$ and K$^+$, but rather prevents ACh from doing so. The effectiveness of curare in blocking the action of ACh is directly proportional to the ratio of the concentration of curare to ACh at the

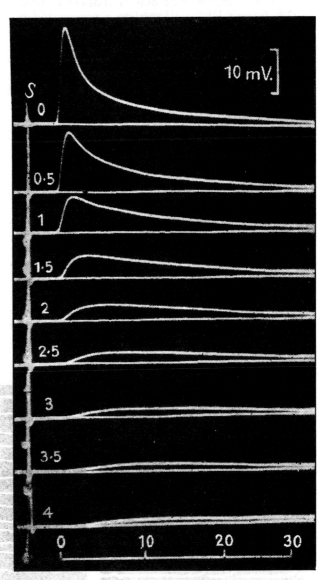

Fig. 3-9. Change in the intracellular potential of a curarized muscle fiber in response to the stimulation of its motor nerve (S, stimulus artifact). The dose of curare used prevented the EPP from reaching a value sufficient to produce an action potential. Note that the end-plate potential is greatest at the end-plate (0 mm.) and almost nonexistent at a point 4 mm. away. Abscissa time is in msec. (Fatt, P., and Katz, B.: An analysis of the end-plate potential recorded with an intracellular electrode. J. Physiol., *115*:326, 1951)

postjunctional surface (muscle end-plate). Thus, at relatively low concentrations of curare the end-plate potential (EPP) following the stimulation of a motor nerve may be reduced, for example, to 70 mv., whereas at higher concentrations of curare the EPP may be reduced to 10 mv. The former EPP will initiate a propagated potential in the rest of the cell membrane, while the latter will not. By experiments such as these it has been found that reductions in the EPP by curare to values of from 30 to 50 mv. result in failure of the end-plate to initiate a propagated potential and therefore in failure to initiate a muscle contraction.

In summary, then, ACh is apparently released by the motor neuron within about 1 μ of the motor end-plate. The ACh causes an increase in the permeability of the end-plate to both Na^+ and K^+, causing a depolarization of the end-plate (EPP). The EPP then produces an action potential, which travels throughout the muscle fiber and initiates an intracellular release of Ca^{++} that brings about muscle contraction.

Miniature end-plate potentials. In 1952 Fatt and Katz reported what they called miniature end-plate potentials (mepp's). These (Fig. 3-10) were noted while the experimenters were recording at the muscle end-plate, and consisted of small, randomly occurring depolarizations with a frequency of about 1 per second and an amplitude of 0.5 mv. They were reduced in amplitude by curare and increased in amplitude and duration by neostigmine (an antiacetylcholine esterase). They were also reduced by hyperpolarization of the motor neuron. On the basis of these and other experiments they suggested that the mepp's were due to the release of one quantum of ACh (several thousand ACh ions) and its subsequent action on the muscle end-plate. They developed the theory that a normal EPP (one that produces an action potential) is due to the release of from 100 to 200 packets or quanta of ACh.

Synaptic Transmission

In the previous section we characterized the synapse as an area of functional contact between two neurons, and indicated that synapses lie in the central nervous system and in autonomic ganglia in the periphery. The synapse, unlike the neuromuscular junction, is an area of integration, i.e., an area in which numerous facilitatory and inhibitory impulses impinge upon a single neuron. A single facilitatory input from a single neuron does not usually elicit an action potential in the postsynaptic neuron as it does in the post-

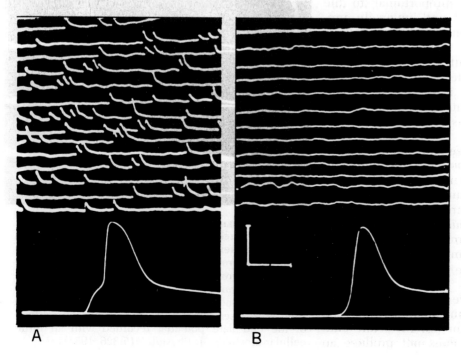

Fig. 3-10. Intracellular recordings from a single frog muscle fiber. Upper recordings (miniature end-plate potentials) were obtained prior to and lower recordings (action potential) after the stimulation of the muscle's motor nerve. The recordings at A were obtained at the motor end-plate and those at B some distance away from the end-plate. (Fatt, P., and Katz, B.: Spontaneous subthreshold activity at motor nerve endings. J. Physiol., *117*:110, 1952)

junctional cell (muscle fiber) of the neuromuscular junction.

There are certain similarities, however, between the neuromuscular junction and the synapse. It is believed that in both cases the prejunctional neuron releases a neurotransmitter. In the case of the neuromuscular junction this chemical is acetylcholine, its role being to produce an end-plate potential which initiates an action potential that is conducted throughout the muscle fiber. In the case of the synapse, the presynaptic cells release numerous different *transmitters*. There is evidence that some presynaptic cells release the excitatory transmitter acetylcholine and that it, like the acetylcholine released at the neuromuscular junction, increases the permeability of the postjunctional cell to both K^+ and Na^+. The resulting depolarization is called an excitatory postsynaptic potential (EPSP) and can, if sufficiently strong, trigger an action potential. It seems unlikely, however, that acetylcholine is the excitatory transmitter at all synapses. For example, analysis shows that there is very little acetylcholine or choline-acetylase (an enzyme that helps to acetylate choline) in the dorsal root fibers which synapse on motor neurons.

In the synapses of the central nervous system

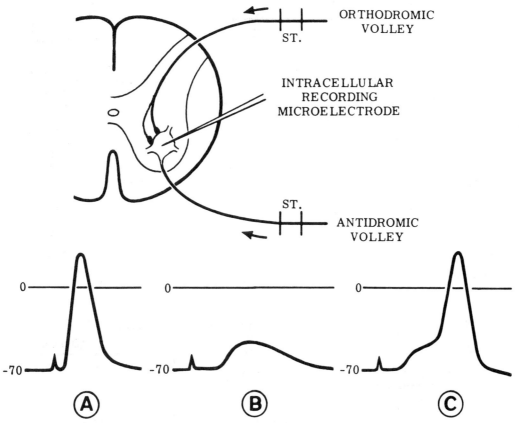

Fig. 3-11. Changes in the intracellular potential of the cell body of a motor neuron to antidromic and orthodromic stimuli. At A stimulation of axon elicits an action potential (*large deflection*) in the cell body of the same neuron (antidromic stimulation). At B stimulation of sensory neurons with a weak stimulus produces an excitatory postsynaptic potential (EPSP) in the cell body of a motor neuron (orthodromic stimulation). At C strong sensory stimulation elicits an EPSP which produces an action potential in the cell body of the motor neuron. The initial small peak seen in the 3 traces is the stimulus artifact. Note the greater interval between the beginning of the stimulus artifact and the action potential at C than at A. The difference between these 2 intervals is called the synaptic delay. (Selkurt, E. E. (ed.): Physiology. 2nd ed., p. 137. Boston, Little, Brown & Co., 1966)

we also note inhibition. One of the mechanisms involved is apparently the release of a neurotransmitter producing increased permeability of the postsynaptic neuron to K+. This, in turn, causes the transmembrane potential to move closer to the equilibrium potential for K+ (i.e., to hyperpolarize). Gamma-aminobutyric acid (GABA), it has been suggested, is one of numerous inhibitory transmitters.

The excitatory postsynaptic potential (EPSP). By placing a microelectrode in the cell body of a motor neuron (Fig. 3-11), we can record the electrical changes in the cytoplasm of the cell body in response to stimulation of its own axon (antidromic volley) or of a neuron which synapses with it (orthodromic volley). Under these experimental conditions, we find that any action potential produced in the axon is conducted throughout the axon and cell body, while action potentials produced in presynaptic cells elicit a depolarization of the postsynaptic surface (EPSP) that is limited to the cell body and that may or may not be followed by an action potential. Generally the depolarization of the postsynaptic cell,

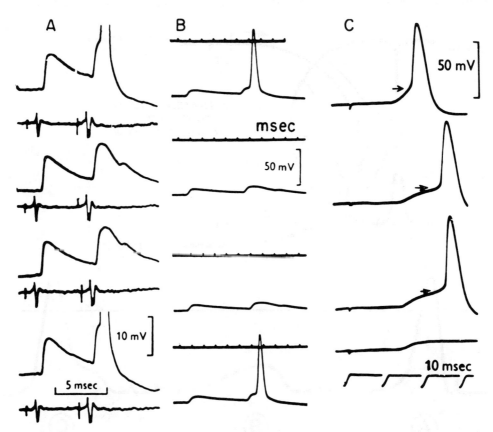

Fig. 3-12. The excitatory postsynaptic potential (EPSP). At A is a series of intracellular potentials recorded from a cat motor neuron (high-amplitude tracings). The EPSP's seen above were initiated by the stimulation (low-amplitude tracings) of impinging neurons. Note that in some cases 2 different EPSP's fuse (temporal summation) to produce a threshold depolarization and a resultant action potential. At B the same experiments were recorded at one-fifth the sensitivity. At C progressively stronger stimuli were applied to the preganglionic nerve of a rabbit and recordings were obtained from a postganglionic sympathetic neuron. Note that as the strength of stimulation progressively increased the critical firing level was reached sooner. (A and B: Brock, L. G.: The recording of potentials from motoneurones with an intracellular electrode. J. Physiol., *117*:447, 1952; C: Eccles, R. M.: Intracellular potentials recorded from a mammalian sympathetic ganglion. J. Physiol., *130*:578, 1955)

if it is greater than 10 mv., elicits an action potential. Whether only an EPSP or both an EPSP and an action potential are formed depends upon the strength of the orthodromic (Gk. *orthos* plus *dromos*, straight running) volley.

In summary, then, the response of the cell body to an antidromic volley is different from that to an orthodromic volley. In the case of the former there is protoplasmic continuity between the cell body and its axon, and simple conduction; in the case of the latter, there is neither protoplasmic continuity nor simple conduction between the presynaptic and postsynaptic surfaces. Here the postsynaptic surface apparently responds to a neurotransmitter rather than to an action potential. Whether this neurotransmitter produces a sufficient EPSP to initiate an action potential depends upon (1) the amount of neurotransmitter released, (2) the state of polarization of the cell membrane, and (3) the environment of the cell membrane.

The EPSP in a motor neuron involved in a monosynaptic reflex generally begins about 0.5 msec. after the sensory impulse enters the spinal cord. It then rises to a peak within 1.0 to 1.5 msec. and starts to decline exponentially. This can be seen in the intracellular recording pictured in Figure 3-12. Note also in this figure that when two stimuli are applied to the sensory root within 6 msec. of each other, the second EPSP produced sums with what is left of the first EPSP. It is possible, in other words, for two EPSP's which by themselves are incapable of eliciting an action potential to sum, thereby producing a critical depolarization and a resulting action potential. Summation may result from (1) two or more different subliminal (below threshold) stimuli arriving at different loci on the cell body simultaneously (spatial summation), (2) two or more different subliminal stimuli arriving at the same loci at different times (temporal summation), or (3) various combinations of temporal and spatial summation.

The above concepts help to explain a number of experimental observations. We find, for example, that stimulation of a certain sensory nerve activates 100 motor neurons, and that stimulation of another sensory nerve activates 80 motor neurons. When, on the other hand, we stimulate both sensory nerves together, we might find that we activate 250 motor fibers. The simultaneous stimulation of the two sensory nerves, in other words, elicits a response greater than the sum of the individual responses produced by the sensory nerves alone. Apparently the stimulation of each of these nerves elicits action potentials in some motor fibers and only an EPSP in others. The motor fibers responding with only an EPSP are said to constitute a *subliminal fringe*. Apparently in the case cited above the two sensory nerves studied had overlapping subliminal fringes and their simultaneous activation resulted in the summation of subthreshold stimuli in the motor neurons in these areas. Some observations on the expansion of the subliminal fringe and discharge zones (areas producing action potentials) with increased sensory stimulation are presented in Figure 3-13.

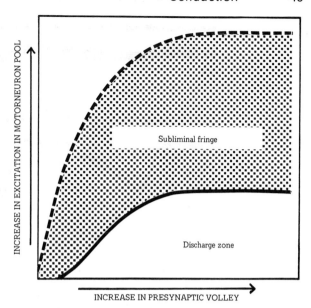

Fig. 3-13. Relative sizes of the discharge zone and the subliminal fringe as presynaptic stimulation increases. (*Redrawn from* Lloyd, D. P. C.: Reflex action in relation to pattern and peripheral source of afferent stimulation. J. Neurophysiol., 6:116, 1943)

Repeating the above experiment on different sensory nerves, however, we do not always obtain data consistent with the mechanism of a summation of subliminal fringes. Sometimes we find that simultaneous stimulation of two sensory nerves elicits a response equaling less than the sum of the individual responses. This phenomenon is called *occlusion* and is apparently the result of an overlap of discharge zones rather than subliminal fringes.

The inhibitory postsynaptic potential (IPSP). We noted in the previous chapter that if we stimulate Type I heteronymous antagonistic sensory neurons to a motor neuron we produce in-

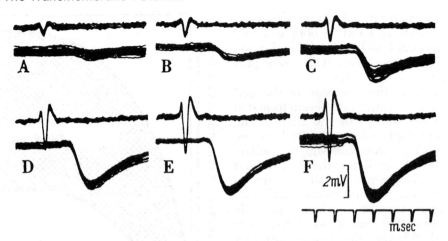

Fig. 3-14. Inhibitory postsynaptic potentials (IPSP). Each upper tracing (*thinner line*) is from a surface electrode on a dorsal root containing sensory neurons from the quadriceps muscles and each lower tracing (*thicker line*) is from an intracellular electrode in a biceps-semitendinosus motor neuron. From A through F progressively stronger stimuli were delivered to the sensory nerves, resulting in progressively greater hyperpolarizations (*lower tracing*) in the motor neuron (downward deflections represent increased negativity of the cytoplasm). In other words by stimulating certain sensory neurons (heteronymous antagonistic neurons) we can produce an IPSP in certain motor neurons. Each record represents about 4 superimposed tracings. (Eccles, J. C.: The behaviour of nerve cells. *In* Wolstenholme, G. E. W., and O'Connor, C. M. (eds.): Ciba Foundation Symposium on the Neurological Basis of Behaviour. P. 38. Boston, Little, Brown & Co., 1958)

hibition in that motor neuron (Figs. 2-15 and 3-14). Apparently the sensory neuron releases a transmitter (possibly gamma-aminobutyric acid or glycine) which produces a selective increase in the permeability of the postsynaptic surface to K^+ and a movement of the cytoplasmic potential toward the K^+ equilibrium potential. This usually has two effects, the stabilization of the transmembrane potential and the hyperpolarization of the cell. The former response makes it more difficult to change the transmembrane potential, and the latter means that a greater depolarization is required to reach the critical level for the production of an action potential. The amplitude of the IPSP, unlike that of the EPSP, rarely exceeds 5 mv. It also takes somewhat longer for the IPSP to develop, but it, like the EPSP, shows both temporal and spatial summation (Fig. 3-15).

Presynaptic inhibition. Frank and Fuortes have reported that the stimulation of inhibitory sensory neurons, though it decreases the firing of motor neurons, does not always do this by producing an IPSP. They have noted, for example, that the stimulation of inhibitory afferent neurons which do not elicit an IPSP can reduce by as much as 50 per cent the EPSP produced by the stimulation of facilitatory neurons. Apparently the mechanism for this lies in the inhibition of the presynaptic fibers. Supporting this concept are some electron microscope studies which show densely packed presynaptic fibers in close contact with one another, and other studies showing that the stimulation of presynaptic inhibitory fibers can elicit prolonged depolarization of intramedullary presynaptic facilitatory fibers.

The amount of transmitter released by a presynaptic neuron depends upon (1) the environment of that neuron, and (2) the height of the action potential. Both factors may be changed by increasing or decreasing the resting transmembrane potential by a few millivolts. In presynaptic inhibition, for example, a slight depolarization results in a decrease in the height of the action potential and a resulting decrease in the amount of transmitter released. Small decreases in the concentration of Ca^{++} in the intercellular fluid, or increases in that of Mg^{++}, also decrease the amount of transmitter released.

RECEPTOR MECHANISMS

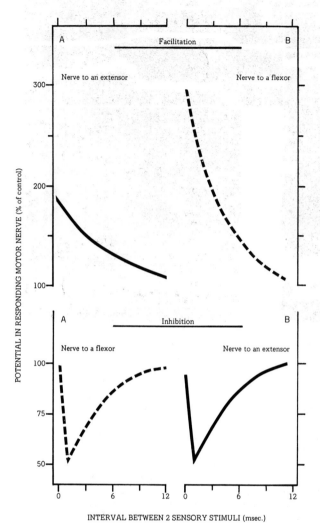

Fig. 3-15. Time course of facilitation and inhibition. Upper graphs show facilitation in a nerve to (A) an extensor and (B) a flexor muscle. Lower graphs show inhibition in a nerve to (A) a flexor and (B) an extensor muscle. The data were obtained by stimulating a sensory nerve and recording the potential produced in an associated motor nerve (control). The sensory stimulus was then applied again after preconditioning of the motor neuron by the stimulation of another facilitatory or inhibitory sensory nerve. Note that the simultaneous stimulation of 2 different facilitatory sensory nerves (0 interval) gives a maximum motor response. As the separation of facilitatory stimuli increases, the motor response decreases. Note that the optimal inhibition occurs when the inhibitory stimulus precedes the facilitatory one by about 0.5 msec. (*Modified from* Lloyd, D. P. C.: Facilitation and inhibition of spinal motoneurons. J. Neurophysiol., 9:426, 427, 430, 431, 1946)

Until now we have been concerned with how an action potential or action potentials cause another action potential to develop. In this section we shall be concerned with the more difficult problem of how one form of energy (heat, light, stretch, chemicals) elicits another form of energy (the action potential) in the axon. The structural element responsible for this transition is called a transducer (L. *trans*, plus *ducere*, lead). We have already listed, in Chapter 2, Table 2-1, many of the stimuli eliciting action potentials, along with their associated transducers (olefactory membrane, muscle spindles, etc.). We have stated that, at least in some cases, the receptors act as filters, allowing certain stimuli to elicit an action potential in a specific neuron while preventing other stimuli from activating that neuron. We have also noted that some areas in the body are sensitive to certain types of stimuli for which no receptor can be found.

The results of studies thus far conducted are consistent with the view that stimuli act on the body's transducers in one of three ways. They may act (1) directly on the neuron to produce a *generator potential*, (2) on a receptor to produce a *receptor potential*, or (3) on a receptor to produce a *chemical or physical change* which in turn results in a receptor or generator potential. In the eye, for example, light first breaks down a visual pigment, and in the ear, sound waves first set up vibrations which eventually lead to action potentials in sensory neurons.

The generator potential and the receptor potential are probably similar in character to the end-plate potential and the excitatory postsynaptic potential (EPSP). That is to say they are all nonpropagated graded depolarizations caused by a stimulus-induced increase in the permeability of the plasma membrane to Na^+, K^+, and certain other ions. In Figure 3-16A and -B we see some receptor potentials recorded from a muscle spindle. In these experiments the spike-generating mechanism was blocked by the drug *procaine*, an agent which at low concentrations has no effect on either the generator or the receptor potential. Note in A that although the stretch was applied for well over 20 msec. (stretch is upper deflection, and potential, lower deflection) the depolarization became maximal in the first 2 msec. and had returned to resting polarity about 12 msec. later. This decline in a receptor potential during a sustained stimulus is called

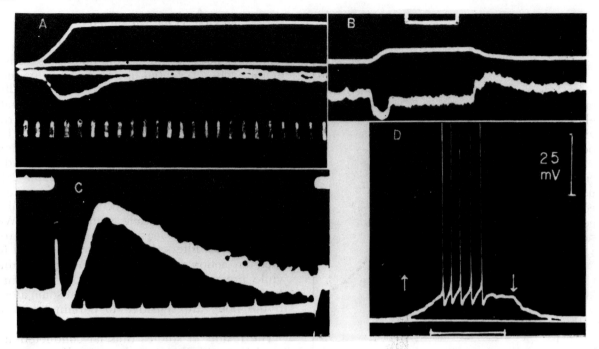

Fig. 3-16. Receptor potentials from the muscle spindle and Pacinian corpuscle. In A and B we note the potentials (*lower tracing*) during stretch (*upper tracing*) from a procainized frog muscle spindle. In A the lower vertical lines are 2 msec. apart and in B the upper square wave represents 0.1 sec. Record C is from a procainized cat Pacinian corpuscle. Upper trace shows the duration of the stimulus (pressure) and the lower trace gives a signal every 1 msec. Record D is from a crayfish stretch receptor (not procainized). The arrows signal the beginning and end of stretch. A number of invertebrates such as the crayfish have "giant axons" which make the study of transmembrane potentials considerably easier than in vertebrates. (A and B: Katz, B.: Depolarization of sensory terminals and the initiation of impulses in the muscle spindle. J. Physiol., *111*:267 and 275, 1950; C: Gray, J. A. B., and Sato, M.: Properties of the receptor potential in Pacinian corpuscles. J. Physiol., *122*:620, 1953; D: Eyzaguirre, C., and Kuffler, S. W.: Processes of excitation in the dendrites and in the soma of single isolated sensory nerve cells of the lobster and crayfish. J. Gen. Physiol., *39*:98, 1955)

adaptation. Some of the receptors in the body adapt rapidly and are known as phasic receptors. Others adapt slowly and are called tonic receptors.

It is also interesting to note in Figure 3-16B that the application of stretch produced a depolarization followed by adaptation and that the release of stretch produced a temporary hyperpolarization. In Figure 3-16D we see the effect of stretch upon a stretch receptor that has not been exposed to procaine. Note that this stretch elicited a receptor potential which brought and kept the membrane at a critical firing level for about 0.8 sec. Note, too, that the interval between action potentials consistently increases as the stretch is maintained (i.e., the frequency of action potentials decreases as the stimulus is maintained). This too is referred to as adaptation. An example of this type of adaptation is seen in Figure 3-17.

In general we find that the character of the response of a receptor is, within limits, dependent upon (1) the strength of the stimulus, (2) the rate at which the stimulus is applied (velocity of stretch in mm./sec. in the case of a stretch receptor), and (3) the duration of stimulation. In tonic receptors, that is, one can increase the receptor potential (Figure 3-18) and, in so doing, increase the frequency of action potentials up to 100 to 200 firings per seconds by increasing the stimulus strength and the velocity of application of the stimulus. In the case of the Pacinian corpuscle (a phasic receptor), on the other hand, a single stimulus, regardless of how long or how

Fig. 3-17. Response of a cat muscle spindle to different degrees of prolonged stretch. (*Redrawn from* Matthews, B. H. C.: Nerve endings in mammalian muscle. J. Physiol., 78:10, 1933)

rapidly it is applied or how strong it is, seldom elicits more than one action potential, whereas vibratory stimuli can elicit up to 1000 action potentials per second.

It should also be borne in mind that as we increase the stimulus strength we may increase not only the frequency of firing in a single receptor and neuron, but also the number of receptors and neurons stimulated. This latter phenomenon is sometimes called *recruitment* (Fig. 11-10).

ELECTRICAL STIMULATION

When voltage is applied via two electrodes—a cathode (negative electrode) and an anode (positive electrode)—to an excitable cell, a current passes through the intercellular and intracellular fluids from the region of the anode to that

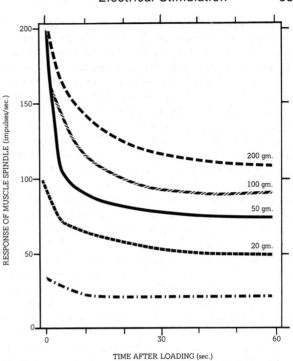

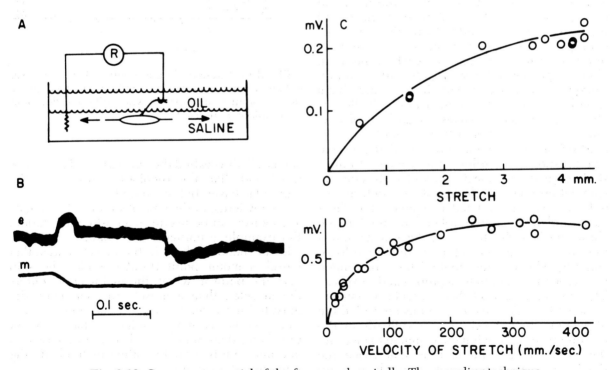

Fig. 3-18. Generator potential of the frog muscle spindle. The recording technique is shown at A, one of the records obtained at B (e, electrical changes; m, mechanical changes), and 2 graphs of factors modifying the amplitude (in millivolts) of the generator potential at C and D. (*After* Katz, B.: J. Physiol., *111*:261-282, 1950. *In* Ruch, T. C., and Patton, H. D.: Medical Physiology and Biophysics. 19th ed., p. 100. Philadelphia, W. B. Saunders, 1965)

of the cathode. This results in depolarization of the cell membrane at the cathode and hyperpolarization at the anode. The amount of depolarization depends, in part, on the resistance offered by the cell membrane and on the cross-sectional area of the intracellular conducting medium, the cytoplasm. Because of these factors, higher voltages are necessary to trigger an action potential in skeletal muscle (higher membrane resistance) than in nerve, and higher voltages are needed to initiate an action potential in small-diameter axons (higher cytoplasmic resistance) than in large-diameter axons.

The main reason the electrical stimulus is preferred in most clinical and experimental studies is the ease with which one can control the three important characteristics of any electrical stimulus, (1) its strength, (2) the rate of change of its strength, and (3) its duration. As for (2), many good electronic stimulators have a stimulus which reaches its full strength almost instantaneously—called a *square* or *rectangular wave*.

Using a square-wave stimulus, then, one can study the importance of the stimulus-strength and the duration of stimulation in the initiation of a propagated action potential. Figure 3-19 represents such a study. Here the stimulus-strength is represented on the ordinate, and the minimum time necessary for each stimulus to elicit a propagated action potential is represented on the abscissa. Note in this study that a stimulus-strength of less than one volt did not elicit a propagated response regardless of how long it was applied, and that a stimulus of less than 0.2 msec. duration did not elicit a propagated response regardless of its voltage. In most studies on the responsiveness of a structure to stimulation, however, the investigator determines only one strength-related parameter, the rheobase, and one duration-related parameter, the chronaxie. The rheobase (Gk. *rheos*, a stream) is the minimum strength of a stimulus in volts or amperes which, when applied for a prolonged time (100 msec.), initiates a propagated response. The minimum period that a rheobasic current must act to initiate such a response is called the utilization time. Since this is difficult to determine accurately, it is more common to measure the minimum amount of time that twice the rheobasic current must act to produce a re-

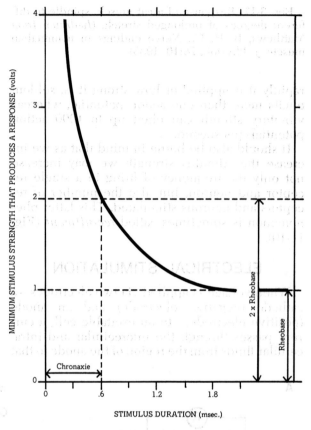

Fig. 3-19. Strength-duration curve. The curve was obtained from a study of the minimum voltage necessary to produce a response when different durations of stimulation are used.

sponse. This is called the *chronaxie* (Gk. *chronos*, time) and, like the rheobase, varies with the particular tissue being studied.

A knowledge of the type of stimulus that elicits a response can be useful diagnostic information. For example, a normally innervated muscle has a lower rheobase, both in the muscle and at that muscle's motor point (located under the skin near the point at which the motor nerve enters the muscle), than a muscle that has been denervated for several days. This, of course, is because nerve is more excitable (has a lower threshold) than muscle, so that after denervation one must apply a greater stimulus to obtain the same muscle response.

ANATOMY
FUNCTION
THE SYMPATHETIC NERVOUS SYSTEM
THE PARASYMPATHETIC NERVOUS SYSTEM
AUTONOMIC TONE
BLOCKING AGENTS
DECENTRALIZATION

Chapter 4

THE AUTONOMIC NERVOUS SYSTEM

In the preceding chapters we have concentrated on (1) the facilitation and inhibition of action potentials in the somatic efferent neuron (anterior-horn cell) and on (2) how the somatic efferent neuron initiates a series of changes that lead to action potentials in skeletal muscle. We have emphasized the importance of sensory and internuncial impulses in both facilitating and inhibiting action at the synapses of the spinal cord and brain and have indicated that the effect one neuron has on another depends, at least in part, on the type and quantity of neurotransmitter released by the presynaptic fiber. We have also noted that, although facilitation and inhibition are characteristic of central nervous system activity, the role of the somatic efferent neuron is neither to facilitate nor to inhibit the contraction of skeletal muscle, but rather to initiate its activation by the release of acetylcholine at the neuromuscular junction.

ANATOMY

In the autonomic nervous system we are dealing with efferent neurons which innervate the glands, smooth muscle, and cardiac muscle of the body. For the most part the sensory and internuncial neurons which synapse with the autonomic fibers are structurally and functionally similar to those which synapse with the somatic efferent neurons (Fig. 4-1); in both motor systems there is considerable facilitation and inhibition of activity in the central nervous system. The major centers integrating total skeletal-muscle function and those integrating total autonomic function, however, are quite different. In the former, the (1) *basal ganglia* of the forebrain and (2) *cerebellum* are of prime importance, whereas in the case of autonomic integration, certain *hypothalamic areas* take over this role (Fig. 4-2).

The main difference between the autonomic and somatic efferent systems lies in the structure and function of the motor neurons. Unlike the somatic efferent neurons, all autonomic neurons except those going to the adrenal medulla synapse outside the central nervous system with other neurons. A typical autonomic pathway, in other words, consists of (1) a myelinated B fiber, with its cell body in the central nervous system; (2) a peripheral synapse; and (3) an unmyelinated C fiber, with its cell body outside the central nervous system. The cell bodies of the more peripheral neurons are generally found in clusters called ganglia. For this reason the more central autonomic neurons are appropriately called *preganglionic neurons,* and the more peripheral ones (less appropriately), postganglionic neurons (ganglionic neurons would be more accurate).

The cell bodies of all preganglionic neurons are in either (1) the brain stem, (2) the lateral gray horn of the thoracolumbar spinal cord (T-1 through L-3), or (3) the lateral gray horn of the sacral cord (S-2 through S-4). The pre- and post-

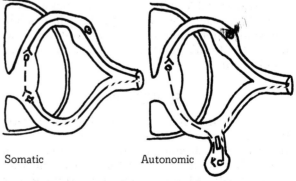

Fig. 4-1. Diagram of a somatic and an autonomic reflex arc. (*Redrawn from* Ruch, T. C., and Patton, H. D.: Medical Physiology and Biophysics. 19th ed., p. 227. Philadelphia, W. B. Saunders, 1965)

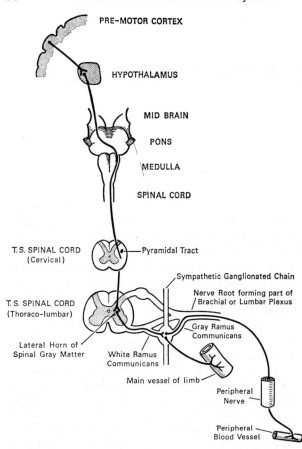

Fig. 4-2. Diagram of sympathetic vasomotor pathways to the limbs. T. S., transection. (Richards, R. L.: The Peripheral Circulation in Health and Disease. P. 18. Edinburgh, E. & S. Livingstone, 1946)

ganglionic neurons most closely associated with the thoracolumbar cord are called sympathetic or orthosympathetic fibers, and those most closely associated with the brain stem and sacral cord, craniosacral or parasympathetic fibers. The parasympathetic and sympathetic systems differ both structurally and functionally from each other (Table 4-1).

FUNCTION

From a functional point of view, the sympathetic and parasympathetic systems differ from the somatic efferent system in the manner in which they control the structures they innervate. Each somatic efferent neuron represents the exclusive source of innervation for a series of muscle fibers. Each somatic efferent neuron and the skeletal-muscle fibers it innervates constitute a functional unit called the motor unit. The relationship between motor neuron and skeletal-muscle fiber is an all-or-none relationship. Under normal conditions (in the absence of pharmacological blocking agents such as curare, and of pathological states such as myasthenia gravis) a single action potential in a single efferent neuron produces a single short-lived contraction (twitch) in each of the skeletal-muscle fibers it innervates (Fig. 4-6).

The role of the autonomic nervous system in controlling the structures it innervates is not so simple and straightforward as the role of the somatic efferent neurons. Many, but not all, the structures innervated by the autonomics contract and secrete and function fairly normally in the absence of the nervous system. The completely decentralized heart continues to contract and to pump blood either in the body or outside (see heart-lung preparation in later chapters). The small intestine completely removed from the body continues to show peristalsis and to respond to stretch and other stimuli even though entirely disconnected from the central nervous system (see Fig. 23-8). The diaphragm (a skeletal muscle), on the other hand, stops contracting if its phrenic nerves are cut or if parts of the brain or cervical cord are markedly depressed. Many of the structures innervated by the autonomic nervous system, in other words, exhibit what is called *autoregulation*. Unlike skeletal muscle, they are not dependent upon the central nervous system for their activity.

Apparently in many structures the role of the autonomic nervous system is not to initiate contraction or secretion but to modify the activity already performed by the organ innervated. Thus in the case of the heart, sympathetic fibers increase (1) the rate, (2) the force of contraction, (3) the velocity at which the action potential travels over the fibers, and (4) the irritability of the myocardial cells, but do not stimulate the muscle to contract. Parasympathetics to the heart, on the other hand, slow the heart rate. These antagonistic effects on heart rate are seen in Figure 4-4. Note in Part C that the parasympathetics hyperpolarize the cardiac pacemaker cell, and in Part D that the sympathetics bring about a more rapid depolarization; at least some of the peripheral autonomic facilitation and inhibition is probably brought about by an action on the cell membrane, possibly a modification of Na^+ conductance, K^+ conductance, or both (Fig. 4-5). The radial fibers of the iris (smooth muscle), on the other hand, are activated only by postganglionic sympathetic neurons, and the circular

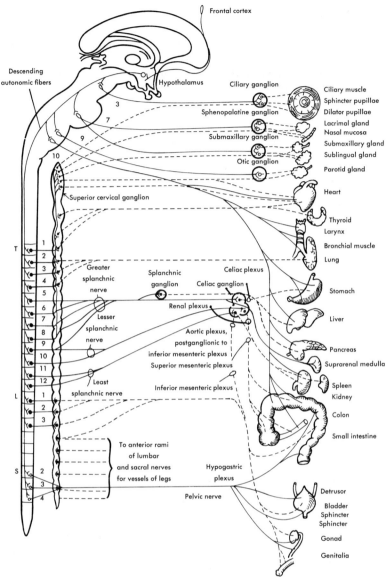

Fig. 4-3. Diagram showing the innervation of glands, smooth muscle, and the heart. These structures are innervated by postganglionic neurons (*dotted lines*) which, in turn, are innervated by neurons from the thoracolumbar cord (preganglionic sympathetic fibers) or brain stem and sacral cord (parasympathetic fibers). Note that the sympathetic ganglia lie nearer the central nervous system than the parasympathetic ganglia. Note also that the adrenal medulla is innervated by a preganglionic sympathetic fiber. (Mettler, F. A.: Neuroanatomy. *In* Anthony, C. P.: Textbook of Anatomy and Physiology. 7th ed., p. 233. St. Louis, C. V. Mosby, 1967)

fibers of the iris (smooth-muscle fibers antagonistic to the radial fibers) are activated only by postganglionic parasympathetic neurons.

Another characteristic of the autonomic nervous system is its lack of well-defined neuromuscular neuroglandular endings. Apparently in most and possibly all cases the neurotransmitter released by the postganglionic ending is not restricted to a single cell, but diffuses and influences many cells. Because of (1) this lack of well-defined, discrete innervation, and because (2) any single area might contain numerous postganglionic fibers each releasing a different neurotransmitter substance, we find that autonomic control is not only associated with central inhibition and facilitation but also with peripheral facilitation and inhibition. In the case of the heart, for example, the postganglionic sympathetic fibers release a neurotransmitter, *norepinephrine*, which increases heart rate, and the

	Sympathetic Fibers	Parasympathetic Fibers	Somatic Efferent Fibers
Structures Innervated	Glands, smooth muscle, cardiac muscle	Glands, smooth muscle, cardiac muscle	Skeletal muscle
Overlap of Innervation	Characteristic	Characteristic	Absent
Peripheral Facilitation and Inhibition	Characteristic	Characteristic	Absent
Cholinergic Motor Fibers	Present	Characteristic	Characteristic
Preganglionic Neurons Release	Short, myelinated ACh	Long, myelinated ACh	Absent
Postganglionic neurons Release	Long, unmyelinated Norepinephrine or ACh	Short, unmyelinated ACh	
Neuromuscular junction	Absent	Absent	Present
Functions			
Heart rate	Increase	Decrease	
Pupil size	Increase (radial m. contraction)	Decrease (circ. m. contraction)	
Salivary glands	Produce viscous secretion	Produce watery secretion	
Arterioles of skeletal muscle	Produce constriction and dilation	No innervation	
Pancreas	No innervation to glands	Produce secretion	

Table 4-1. Comparison of sympathetic, parasympathetic, and somatic efferent neurons.

postganglionic parasympathetic fibers release a substance, *acetylcholine*, which decreases heart rate. In the case of the arterioles of skeletal muscle, one type of postganglionic sympathetic fiber (*adrenergic sympathetic*) releases norepinephrine and facilitates contraction of the smooth muscles of the arterioles, while another type of postganglionic sympathetic fiber (cholinergic sympathetic) releases acetylcholine and inhibits arteriolar contraction.

It should be reemphasized that the peripheral control of glands, smooth muscle, and cardiac muscle is not restricted to the synergistic and antagonistic influences of the autonomic nervous system. We have already alluded to the fact that these structures respond to changes in their *environment*. The adrenal cortex, for example, secretes aldosterone in response to a decrease in the concentration of Na^+ in its environment. Many arterioles contract in response to stretch and dilate in response to hypoxia or increased carbon dioxide concentration. And finally, we should remember that glands either increase or decrease their secretions, and smooth and cardiac muscle either increase or decrease their contractions in response to certain chemicals called *hormones*. These substances are produced and secreted by glands and carried by the blood and lymph to all parts of the body where they, like the neurotransmitters, influence function.

In summary, then, the autonomic nervous system is a group of neurons which in many cases initiates contraction or secretion, but in many other cases makes up but one of numerous systems influencing the function of a single cell, organ, or system.

THE SYMPATHETIC NERVOUS SYSTEM

The preganglionic sympathetic neurons leave the thoracolumbar cord to (1) synapse in the paravertebral sympathetic chain of ganglia on either side of the vertebral column, (2) pass through the paravertebral chain and synapse in the collateral ganglia which extend down the abdominal aorta and its major branches to the iliac arteries, or (3) pass through the lateral chain to the adrenal medulla. The adrenal medulla is

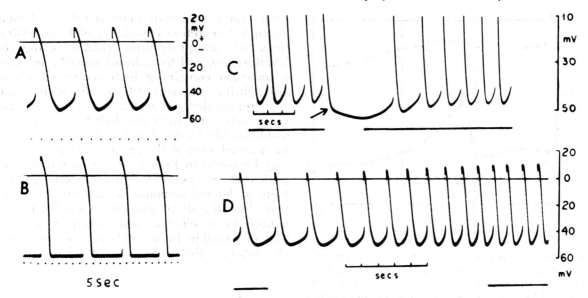

Fig. 4-4. The transmembrane potentials from the cells of the frog's atrium. In A we see the transmembrane potential from a pacemaker cell (a cell that depolarizes in the absence of an extrinsic stimulus). In B is the transmembrane potential of a follower cell. Note its stable (horizontal) potential until stimulated. In C we note the response (hyperpolarization) of the pacemaker cell to vagal stimulation (*break in line*). In D we see its response (more rapid depolarization) to sympathetic stimulation (*break in line*). (Hutter, O. F., and Trautwein, W.: Vagal and sympathetic effects on pacemaker fibers in sinus venosus of the heart. J. Gen. Physiol., 39:718, 720, 728, 1956)

the only effector organ directly innervated by preganglionic neurons.

The paravertebral ganglia contain one ganglion for each segmental nerve from the lumbar and thoracic cord. In the sacral region there are four pairs of ganglia, and in the cervical region there are a superior cervical ganglion, a middle cervical ganglion, and an inferior cervical ganglion, the latter two ganglia usually fusing to form the stellate ganglion. As can be seen in Figure 4-3, the preganglionic fibers entering the various ganglia have a diffuse distribution, one preganglionic neuron sending branches to as many as nine different ganglia. It is at the synapses in the ganglia that the preganglionic neurons release their neurotransmitter, acetylcholine. In response to this the postganglionic neurons develop action potentials which travel (1) from the superior cervical ganglion to the heart, the smooth muscle, and the glands of the head and neck, (2) from the celiac ganglion to the stomach, intestine, liver, etc., (3) from the inferior mesenteric ganglion to the colon, bladder, sex organs, etc., or (4) from the sacral ganglia to the blood vessels and sweat glands of the leg and thigh.

The Postganglionic Sympathetic Neuron

The postganglionic sympathetic neurons release any one of a number of secretions at their endings. The best understood and probably the most common secretion is norepinephrine (adrenergic fibers). Postganglionic fibers to the sweat glands, and a few of the postganglionic fibers to the blood vessels of skeletal muscle, however, release acetylcholine (cholinergic sympathetic fibers). A number of authors have also presented evidence for the existence of other cholinergic sympathetic fibers, as well as fibers that release histamine and numerous other substances. We will in this text, however, limit our discussion of postganglionic sympathetic neurons to the cholinergic and adrenergic ones.

The postganglionic neurons innervate the atria and ventricles of the heart and most of the

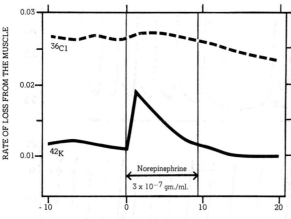

Fig. 4-5. Changes in the rate of loss of ^{42}K and ^{36}Cl from the taenia muscle of the guinea pig colon in response to noradrenalin. (*Redrawn from* Jenkinson, D. H., and Morton, I. K. M.: The effect of noradrenalin on the permeability of depolarized intestinal smooth muscle to inorganic ions. J. Physiol., *188*:379, 1967)

glands and smooth muscle of the body. Adrenergic sympathetic fibers are important in maintaining muscle tone in most of the arterioles of the body and are important in increasing muscle tone, the heart rate, and the force of contraction of the heart in response to low arterial pressure. During exercise, on the other hand, cholinergic postganglionic sympathetic fibers to some of the arterioles of skeletal muscle are activated and tend to inhibit smooth-muscle tone in these arterioles.

Stimulation of adrenergic sympathetics can also (1) facilitate the breakdown of glycogen in the liver to blood sugar (glycogenolysis), (2) facilitate the contraction of the sphincters of the gastrointestinal tract and inhibit gastrointestinal contractions and secretions elsewhere (Fig. 4-5), (3) decrease the pressure in the urinary bladder, (4) inhibit contraction of the nonpregnant uterus, and (5) dilate the pupil. Stimulation of cholinergic sympathetic fibers to the sweat glands can cause the secretion of sweat.

Epinephrine and Norepinephrine

We have emphasized that the preganglionic sympathetic neurons, through the release of acetylcholine, activate (1) postganglionic sympathetic neurons and (2) the adrenal medulla. In response to this activation, most of the postganglionic sympathetic neurons release at their endings a secretion containing high concentrations of norepinephrine (noradrenaline or levarterenol), and the adrenal medulla releases a secretion containing high concentrations of epinephrine (adrenaline). Generally epinephrine or norepinephrine injected into the blood remains active for from one half to two minutes (considerably longer than the duration of action for acetylcholine at the neuromyal junction). As may be noted in Figure 4-6, the stimulation of postganglionic neurons also frequently gives a more prolonged response in their effector organ than does the stimulation of a somatic efferent neuron in its effector organ (skeletal muscle). This may be due, in part, to lower concentrations of enzymes that destroy acetylcholine, nor-

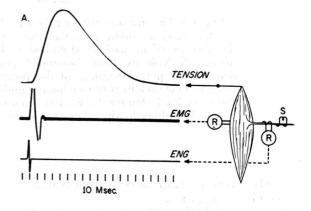

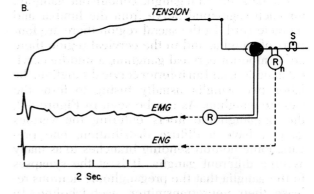

Fig. 4-6. Comparison of (A) the response of a skeletal muscle to the stimulation of its motor nerve with (B) the response of the nictitating membrane to the stimulation of its sympathetic nerve. EMG, electromyogram; ENG, electroneurogram; R, recorder; S, stimulator. Note the different time scales in A and B. (*Partly after* Eccles, J. C., and Magladery, J. W. *In* Ruch, T. C., and Patton, H. D.: Medical Physiology and Biophysics. 19th ed., p. 232. Philadelphia, W. B. Saunders, 1965)

epinephrine, and epinephrine in some glands, in smooth muscle, and in the heart than in skeletal muscle.

Formation and destruction. Epinephrine and norepinephrine are formed from L-phenylalanine or L-tyrosine in the presence of suitable catalysts. Their production and destruction occur in the following manner:

Phenylalanine
↓
Tyrosine
↓
Dihydroxyphenylalanine (= dopa)
↓
Dopamine
↓
Norepinephrine
| N-methyl transferase
↓
Epinephrine

Catechol-O-methyl transferase → Normetanephrine
Monoamine oxidase → Dihydroxymandelic acid
Metanephrine
↓ (Catechol-O-methyl transferase / Monoamine oxidase)
3-Methoxy-4-hydroxymandelic acid
(Commonly called vanillylmandelic acid or VMA)
(2–8 mg. of VMA appear in urine per day)

The norepinephrine produced by the adrenergic neuron is stored in vesicles in the cytoplasm of the neuron. These vesicles also actively pick up and store any norepinephrine in their environment, so that the norepinephrine in the vesicle may either have been produced in the adrenergic neuron or picked up from its environment (blood and intercellular fluid). The norepinephrine released by the neuron (1) combines with the receptor of an effector cell, (2) diffuses into the surrounding intercellular space and into the circulation, or (3) moves into a storage vesicle in the same or a different adrenergic neuron. Most of the norepinephrine destroyed in the intercellular space and blood is destroyed by the enzyme catechol-O-methyl transferase. Most of that destroyed in the adrenergic neuron itself is destroyed in the mitochondria by the enzyme monoamine oxidase.

Since o-dihydroxybenzene is also called catechol, all the molecules listed above which include the dihydroxybenzene ring and an amine are called *catecholamines* (dopa, dopamine, norepinephrine, and epinephrine). You will note that the catecholamine norepinephrine, in the adrenal medulla in the presence of N-methyl transferase, is converted to epinephrine; or, norepinephrine differs from epinephrine by one methyl group. The prefix "nor-" derives from the German, *Nitrogen ohne Radikal*. Literally, then, norepinephrine is the catecholamine whose nitrogen contains no methyl radical. Apparently the reason the adrenal medulla releases considerably more epinephrine than norepinephrine is its large quantities of N-methyl transferase, while the reason the postganglionic neurons produce and release so little of it is the small amounts of this catalyst present in and around the postganglionic sympathetic neurons.

Sympathomimetic actions. Epinephrine and norepinephrine do many of the same things (Table 4-2). Both produce arteriolar vasoconstriction in the skin and viscera, increase cardiac force, increase the blood sugar levels, and have certain effects on the smooth muscles and glands of the respiratory, digestive, and reproductive systems which we will not enumerate at this time. The similarities between the responses of the body to stimulation of the sympathetic nervous system and to the injection of epinephrine or norepinephrine have led pharmacologists to classify the two substances as sympathomimetic agents (that is, their injection produces changes which in many cases mimic the actions of the sympathetic nervous system). Under most cir-

	Epinephrine	Norepinephrine
Heart		
Cardiac rate	+	−*
Stroke volume	++	++
Cardiac output	+++	0*
Arrhythmias	++++	++++
Coronary blood flow	++	+++
Circulation		
Systolic arterial pressure	+++	+++
Mean arterial pressure	+ or −	++
Diastolic pressure	+ or −	++
Mean pulmonary pressure	++	++
Total peripheral resistance	−	++
Muscle blood flow	++	0
Cutaneous blood flow	−−	0
Renal blood flow	−	−
Blood sugar	+++	+

* Direct action of norepinephrine on the heart masked by reflex response to hypertension.

Table 4-2. Response of man to the intravenous infusion of 0.1 to 0.4 μg. of catecholamine per kilogram of body weight per minute. Norepinephrine can be characterized as a general vasoconstrictor which produces an increase in arterial pressure and a reflex decrease in heart rate. Epinephrine, on the other hand, tends to produce vasoconstriction in certain areas (kidneys and skin) and vasodilation in other areas (skeletal muscle). Both epinerphrine and norepinephrine tend to increase the force and irritability of the heart. (Data modified from Goldenberg, M., Aranow, H., Jr., Smith, G. G., and Faber, M.: Pheochromocytoma and essential hypertensive vascular disease. Arch. Int. Med., 86:823-836, 1950)

cumstances the amount of epinephrine released by the adrenal medulla is small and seems to have a minor role in body control. During mild exercise, for example, changes in heart rate, heart force, and vasoconstriction are elicited more by changes in the activity of postganglionic sympathetic neurons than by a release of epinephrine from the adrenal medulla. During extreme exercise, on the other hand, increased sympathetic activity is associated with the release of substantial quantities of epinephrine from the adrenal medulla. In many ways the release of epinephrine seems to serve as a back-up mechanism to further facilitate certain sympathetic functions.

Alpha and beta adrenergic receptors. In 1906 Sir Henry Dale reported a drug, ergot, which produced "epinephrine reversal." He noted that prior to the administration of ergot, epinephrine caused an increase in arterial pressure, whereas after its administration, it caused a decrease in arterial pressure (Fig. 4-7). From this and later work it became increasingly clear that the response of an organ to an agent such as epinephrine depends not only on the character of the agent, but also on the character of the receptor mechanisms available to that agent; before the ergot was administered, epinephrine was producing in some structures changes that tended to increase the arterial pressure (arteriolar vasoconstriction), and in other structures, changes that tended to decrease the pressure (vasodilation), though the net effect of its actions was an increase in arterial pressure. After the ergot, most of the vasoconstriction, occurring in response to epinephrine was prevented; as a result, epinephrine produced a decrease in arterial pressure. The observation that part of the body's response to an adrenergic agent (vasoconstriction) could be blocked by ergot—without blocking the other part of the response (vasodilation)—led to the hypothesis that there are different types of adrenergic receptors. The receptor blocked by ergot was called the alpha receptor, and that which was left intact after ergot but was still capable of being stimulated by epinephrine was called the beta receptor.

In time it was learned that the alpha receptor could be blocked by phenoxybenzamine, phentolamine, and other agents as well. Each of these

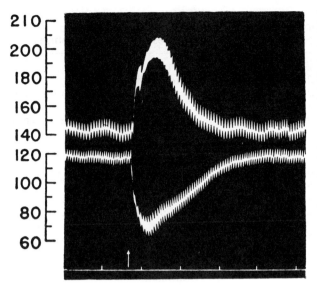

Fig. 4-7. Changes in arterial pressure in response to epinephrine (2.5 μg./kg.) before (*upper pattern*) and after (*lower pattern*) alpha blockade with dibenamine (15 mg./kg.). The drugs were injected intravenously into an anesthetized cat. The ordinate is the arterial pressure in mm. Hg and the abscissa is time in minutes. (Nickerson, M.: Drugs inhibiting adrenergic nerves and structures innervated by them. *In* Goodman, M. A., and Gilman, A. (eds.): Pharmacologic Basis of Therapeutics. 4th ed., p. 553. New York, Macmillan, 1970)

agents, like ergot, left the beta receptor intact. It was discovered too that the drug phenylephrine stimulated alpha receptors while having little or no direct effect on the beta receptors. Norepinephrine, on the other hand, lay somewhere between phenylephrine and epinephrine in its effects. It stimulated all alpha receptors but only a fraction of the beta receptors.

The beta receptor was quickly defined as a mechanism that could be (1) blocked by dichloroisoproterenol (DCI) and propranolol, and (2) stimulated by isoproterenol as well as epinephrine, with DCI, propranolol, and isoproterenol having little or no direct action on the alpha receptor. With alpha and beta blocking agents and alpha and beta stimulators available, the investigator began characterizing various structures by the alpha and beta receptors they apparently contained (Table 4-3). The increases in heart rate, force, conduction velocity, and irritability that were produced by epinephrine and norepinephrine were found to be caused by the beta-stimulator isoproterenol and prevented by the beta-blocker, DCI. The heart, in other words, was characterized as having beta receptors, while the blood vessels of the skin seemed to have primarily alpha receptors (adrenergic receptors blocked by phenoxybenzamine), and the vessels of skeletal muscle, both alpha and beta receptors. The blood vessels of skeletal muscle respond to strong alpha stimulators such as phenylephrine and norepinephrine with vasoconstriction, but to epinephrine and the beta stimulator isoproterenol with net vasodilation. After the administration of DCI, however, epinephrine produces vasoconstriction in skeletal muscle. The blood vessels of skeletal muscle, that is, contain important quantities of both alpha and beta receptors and therefore have the capacity to either dilate or constrict in response to an adrenergic agent. The net response depends on the character of the adrenergic agent.

It has recently become apparent that there are at least two types of beta receptors. Both can be blocked by DCI and stimulated by isoproterenol. The first type, however, is different from the second in that it is also blocked by 4-(2-hydroxy-3-isopropyl aminopropoxy) acetanilide. The second type is blocked by dimethyl isopropylmethoxamine, but not by 4-(2-hydroxy-3-isopropyl aminopropoxy) acetanilide. The first type has been called the $beta_1$ receptor and the excitatory beta receptor ($beta_E$). The second type is called the $beta_2$ receptor and the inhibitory receptor ($beta_I$).

$Beta_E$ receptors are found in the heart, where their stimulation has been shown to increase heart rate, force, conduction velocity and irritability, and in the liver, where they facilitate glycogenolysis. $Beta_I$ receptors are apparently the beta receptors found elsewhere in the body. This particular concept is a useful one to help us understand the action of norepinephrine, since we can now characterize this catecholamine as (1) a strong stimulator of alpha receptors, (2) a strong stimulator of $beta_E$ receptors, and (3) a very weak stimulator of $beta_I$ receptors.

In summary, then, we have noted that a single agent, epinephrine, has the capacity to facilitate smooth-muscle contraction in some structures and smooth-muscle relaxation in others. We have noted that we can block either or both of these actions (Table 4-4). On this basis we have postulated the existence of three types of receptor mechanisms for adrenergic agents (epinephrine, norepinephrine, etc.): alpha, $beta_I$, and $beta_E$.

We have further suggested that a single cell

Effector Organ	Type of Receptor	Response to Epinephrine
Heart		
Sino-atrial node (pacemaker)	β	Increased heart rate
Atria	β	Increased contractility and conduction velocity
Ventricles	β	Increased contractility, conduction velocity, and automaticity
Blood Vessels		
Pulmonary	α	Constriction
Salivary glands	α	Constriction
Skin and mucosa	α	Constriction
Abdominal viscera	α, β	Constriction and dilation
Skeletal muscle	α, β	Dilation and constriction
Eye		
Radial muscle of iris	α	Contraction (mydriasis)
Ciliary muscle	β	Relaxation (slight effect)
Skin		
Pilomotor muscle	α	Contraction
Sweat glands	α	Sweating in palms of hands
Spleen Capsule	α	Contraction
Submaxillary and Sublingual Glands	α	Viscous secretion
Lungs		
Bronchial muscle	β	Relaxation
Stomach		
Sphincters	α	Contraction
Motility and tone	β	Decreased
Intestine		
Sphincters	α	Contraction
Motility and tone	α, β	Decreased
Urinary Bladder	α, β	Changes in tone
Uterus	α, β	Varies with menstrual cycle
Adipose Tissue	β	Lipolysis

Table 4-3. Response of various organs to epinephrine. An alpha receptor is defined as one that is strongly stimulated by phenylephrine and norepinephrine and blocked by phenoxybenzamine. A beta receptor is one that is strongly stimulated by isoproterenol and blocked by dichloroisoproterenol. Epinephrine stimulates both alpha and beta receptors and therefore produces the effects of both norepinephrine and isoproterenol.

may have none, one, or two of these receptors. If, for example, a cell has both the alpha and $beta_I$ receptors, it has the capacity to respond to both alpha and $beta_I$ stimulators. Its response to an agent that stimulates both alpha and $beta_I$ receptors will depend on (1) the character of the agent, (2) the concentration of the agent which reaches the receptor site, (3) the sensitivity of the receptor site, (4) the relative concentration of the alpha and beta receptors present, and (5) the way the receptors interact. We find, for example, that low concentrations of epinephrine produce mainly vasodilation in the body, whereas higher concentrations produce vasoconstriction as well. In the case of the intestine and urinary bladder there is some evidence that norepineph-

rine and postganglionic sympathetic neurons act directly on some postganglionic parasympathetic neurons to inhibit their secretion of acetylcholine. In other areas, however, catecholamines probably act solely on the muscle or gland cells rather than on their innervation.

It should also be remembered that the alpha, $beta_I$, and $beta_E$ concept is used as an explanation only for the action of adrenergic agents. Agents such as acetylcholine, Ca^{++}, CO_2, H^+, serotonin, histamine, and so on may also affect the functioning of muscles and glands. Their actions, however, apparently occur through other receptor mechanisms — since they are not blocked by DCI, phenoxybenzamine, or any of the other adrenergic blocking agents discussed above.

	Adrenergic Receptors		
	Alpha	Beta$_I$	Beta$_E$
Produce sweating and salivation	xxxx		
Produce vasoconstriction except in the coronary and cerebral vessels	xxxx		
Produce contraction of splenic muscle and of the dilator muscle of the pupil	xxxx		
Inhibit contraction of bronchi, uterus, and some blood vessels		xxxx	
Increase heart rate, force and irritability			xxxx
Produce liver glycogenolysis			xxxx
Blocked by phenoxybenzamine, phentolamine, ergot	xxxx		
Blocked by dichloroisoproterenol and propranolol		xxxx	xxxx
Blocked by 4-(2-hydroxy-3-isopropyl aminopropoxy) acetanilide			xxxx
Blocked by dimethyl isopropylmethoxamine		xxxx	
Stimulated by phenylephrine	xxxx		
Stimulated by norepinephrine	xxxx		xxxx
Stimulated by isoproterenol		xxxxxx	xxxxxx
Stimulated by epinephrine	xxxx	xxxx	xxxx

Table 4-4. Characteristics of the alpha and beta receptors.

THE PARASYMPATHETIC NERVOUS SYSTEM

The preganglionic parasympathetic neurons originate from the second, third, fourth, and fifth sacral spinal segments and the brain stem nuclei (clusters of cell bodies in the central nervous system) of cranial nerves III, VII, IX, and X. Cranial nerves III, VII, and IX innervate structures in the head and neck, while cranial nerve X contains preganglionic parasympathetic fibers to the heart, stomach, urinary bladder, and many other abdominal organs. Its extensive distribution has earned it the name vagus nerve, from the Latin word meaning wandering. The sacral parasympathetics send fibers to various abdominal structures, as well as to the sex organs and external genitalia (Fig. 4-3).

In many cases the parasympathetic nerves innervate the same structures as the sympathetics. In fact the overlap in control can be so pronounced that a single cell is acted upon by vagal parasympathetics, sacral parasympathetics, and various sympathetic nerves. For example, the sino-atrial node of the heart receives numerous parasympathetics from both the right and the left vagus nerves and from both right and left sympathetic nerves. The overlap is so effective that unilateral vagotomy and unilateral sympathectomy have no apparent effect on cardiac function. Bilateral vagotomy, however, has a profound effect on cardiac function.

Some of the areas of the body where sympathetic and parasympathetic fibers, through multiple innervation, have antagonistic actions are (1) the right and left atria of the heart, (2) the blood vessels of the salivary glands, (3) the bronchial muscles, (4) the muscles of the intestine, (5) the gallbladder and its ducts, and (6) the internal sphincter of the urinary bladder. On the other hand there are structures in which sympathetic fibers are either the dominant or the only nervous control mechanism. These areas are (1) the ventricles of the heart (dominant mechanism), (2) the radial muscle of the iris, (3) most of the blood vessels of the body, (4) the organs associated with ejaculation, (5) the pilomotor muscles of the skin, (6) the smooth muscle of the spleen, and (7) the structures in the liver responsible for the breakdown of glycogen (glycogenolysis).

Areas in which parasympathetic fibers are the dominant or sole nervous control mechanism include (1) sphincter muscles of the iris, (2) the ciliary muscle of the eye, (3) the glands of the stomach, (4) the blood vessels associated with erection, (5) the detrusor muscle of the urinary bladder, (6) the secretory cells of the pancreas,

(7) the lacrimal glands, (8) the nasopharyngeal glands, (9) the parotid glands, and (10) the muscles of the colon and rectum. There is dual innervation of the submaxillary and sublingual glands too, but the sympathetic and parasympathetic fibers are not antagonistic; they both stimulate the production of saliva, the sympathetic stimuli eliciting a viscous saliva rich in mucin, and the parasympathetic stimuli eliciting a watery saliva rich in the enzyme salivary amylase.

Parasympathetic fibers, in other words, act in a manner antagonistic to sympathetic fibers in some structures, and in others are the main control mechanism. For example, in the eye it is the parasympathetics which elicit the increased convexity of the lens, allowing us to see close objects distinctly (accommodation). Parasympathetics are also primarily responsible for the acts of urination and defecation.

AUTONOMIC TONE

The autonomic nervous system, like the somatic efferent neurons, exerts two possible types of influence upon the structures it innervates, a tonic influence and an acute influence. The adrenergic sympathetic neurons to the arterioles of skeletal muscle, for example, are tonically active in the resting individual, maintaining a degree of vasoconstriction in skeletal muscle, while parasympathetic fibers to the ciliary muscle of a normal eye are activated only when we try to focus on an object less than 20 feet away. One way of studying the importance of autonomic tone is to administer a blocking

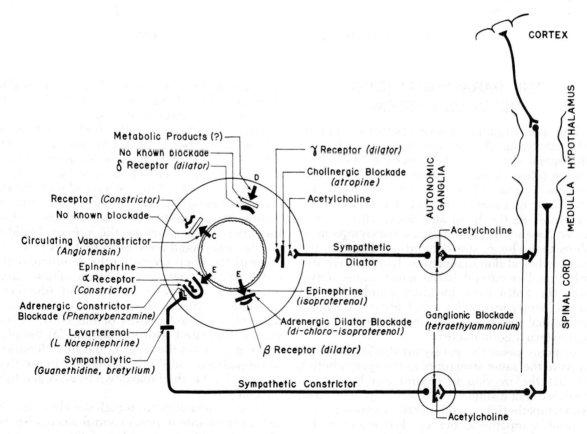

Fig. 4-8. Diagram of hypothetical receptor sites in an arteriole of skeletal muscle. (Green, H. D., Rapela, C. E., and Conrad, M. C.: Resistance (conductance) and capacitance phenomena in terminal vascular beds. *In* Hamilton, W. F. (ed.): Handbook of Physiology. Sec. 2 (Circulation), vol. II, p. 949. Washington, American Physiological Society, 1963)

agent which prevents the activation of postganglionic neurons by preganglionic neurons. Such substances (hexamethonium chloride and tetraethylammonium chloride) are called *ganglionic blocking agents*.

In man, ganglionic blocking agents eliminate autonomic tone and in so doing (1) raise the heart rate (increase arterial pressure), (2) dilate arteries (lower arterial pressure), (3) dilate veins, (4) dilate the pupils (mydriasis), and (5) decrease gastrointestinal motility (produce constipation). Or, the parasympathetic tone in man results in (1) a slower heart rate, (2) a smaller pupil, and (3) greater GI motility. Sympathetic tone, on the other hand, is responsible for maintained contractions in the arteries and veins. Because of autonomic tone we have two nervous mechanisms for increasing heart rate: (1) decreased parasympathetic tone to the heart, and (2) increased sympathetic tone to the heart. We also have two nervous mechanisms for producing arteriolar vasodilation: (1) decreased adrenergic sympathetic tone to the arterioles, and (2) stimulation of cholinergic sympathetics to the arterioles. These mechanisms are illustrated in Figure 11-1.

The second effect that ganglionic blocking agents have on man is interference with numerous phasic (nontonic) reflexes. The absence of a functioning autonomic nervous system may have the following effects: (1) a constant blood flow to the brain (orthostatic hypotension) is not maintained when one changes from a reclining to a standing position, causing loss of consciousness; (2) focusing on close objects is impossible (cycloplegia); (3) micturation in response to moderate bladder distention does not occur; (4) defecation does not occur in response to moderate distention of the rectum; (5) erection and ejaculation do not occur in response to sexual stimuli; (6) salivation is impaired (xerostomia); and (7) the capacity to sweat is decreased (anhidrosis).

BLOCKING AGENTS

We have in the preceding portions of this book been concerned with the phenomena of (1) inhibition, (2) facilitation, and (3) activation. In the control of skeletal muscle we have noted that facilitation and inhibition are central phenomena and that the activation of the contractile mechanism is probably due to an intracellular release of Ca^{++} in the muscle fiber. In the case of control of the glands, smooth muscle, and heart, we have found that inhibition and facilitation occur both centrally and peripherally. We have also suggested that here too secretion and contraction are the results of the intracellular release of Ca^{++}.

We have further indicated that many of the responses of receptor structures to the autonomic nervous system are due to their interaction with the neurotransmitters (acetylcholine and norepinephrine) released at postganglionic endings, and to the hormone released by the adrenal medulla (epinephrine) in response to the stimulation of certain preganglionic sympathetic fibers. These agents probably perform many of their functions by modifying Na^+ conductance, K^+ conductance, or both in the plasma membrane, and by modifying the intracellular release of Ca^{++} in response to an action potential.

Finally, we have reported that the response of a muscle cell or gland cell to a certain chemical varies from one cell to the next, and have suggested that there are a multitude of different receptor sites or receptor molecules. In Figure 4-8, for example, it is suggested that there may be (1) receptor sites on certain arterioles which, when stimulated by epinephrine or norepinephrine, elicit contractions (alpha receptors), (2) other receptor sites which, when stimulated by epinephrine, elicit dilation (beta receptors), (3) still others which, when stimulated by acetylcholine, produce dilation (gamma receptors), and finally (4) sites which, when stimulated by CO_2, hypoxia, and so on, produce dilation (delta receptors). You will note too in the diagram the suggestion that there are receptor sites sensitive to acetylcholine at the cell body of the postganglionic neuron.

Much of the information given above is based on data obtained by the use of drugs that interfere with one or more phases of the conduction-activation processes. Some of these blocking agents are listed in Table 4-5. You will note, for example, that such agents as tetrodotoxin prevent the development of an action potential without preventing the development of an inhibitory postsynaptic potential (IPSP) or an excitatory postsynaptic potential (EPSP), while strychnine and picrotoxin block inhibition in the central nervous system. There are drugs such as hexamethonium chloride which block the action of acetylcholine on the postganglionic neuron but nowhere else; atropine blocks the action of ACh primarily at glands, smooth muscle, and the heart, whereas curare blocks ACh primarily at skeletal muscle; other drugs block the alpha

Blocking Agents	Example	Site of Action
Adrenergic		
Alpha	Phenoxybenzamine	Alpha receptors of glands and smooth muscle
Beta	Dichloroisoproterenol	Beta receptors of glands, smooth muscle, and the heart
Postganglionic, adrenergic neuron	Guanethidine and bretylium	Terminal part of adrenergic neurons (?)
Cholinergic		
Ganglionic	Hexamethonium and tetraethylammonium	Postsynaptic membrane in autonomic ganglia
Postganglionic	Atropine	Gamma receptors of glands, smooth muscle, and the heart
Neuromuscular junction	Curare	Postjunctional membrane of skeletal muscle
Conduction	Procaine	Na^+ and K^+ conductance
	Tetrodotoxin	Na^+ conductance
Central	Strychnine	Postsynaptic inhibitory receptor site (does not block stimulation of Renshaw cell)
	Picrotoxin and tetanus toxin	Presynaptic inhibitory receptor site (?)

Table 4-5. Some common blocking agents. For additional information on alpha and beta blocking agents, see Table 4-4.

effect of catecholamines, or merely the beta effect, and still others prevent the release of norepinephrine by postganglionic adrenergic fibers.

DECENTRALIZATION

One of the advantages of the use of blocking agents is that they produce only a temporary disruption of nervous function, while severing or crushing a nerve may cause permanent separation of the innervated organs from the central nervous system. The various axons, in response to severe damage, degenerate peripheral to the injury and for two or three internodes central to the injury. If the central degeneration reaches the cell body, the entire neuron will be destroyed, but if it does not, what remains of the axon will begin to grow. In the case of an untreated severed nerve, attempted regeneration is usually unsuccessful, since the regenerating axon generally can not penetrate the scar tissue resulting from the injury. In the case of a nerve crush, on the other hand, little or no scar tissue may form and the injured neurons grow into the peripheral neurilemmal sheaths and back to the organ they innervated.

During the weeks required for regeneration, decentralized skeletal muscle shows marked *atrophy*. In some cases it may take so long for reinnervation that by the time it occurs there are no contractile elements left, but merely connective tissue and blood vessels. In these cases muscle function is lost forever. This type of atrophy (denervation atrophy) can be slowed by electrical stimulation of the muscle and by other forms of physical therapy (heat and massage treatments, etc.), but it cannot be totally prevented.

Another response of cells to decentralization is a *hypersensitivity* to such agents as acetylcholine and catecholamines. This is true of skeletal muscle as well as smooth muscle, the heart, and glands. Some have suggested that the hypersensitivity is due to the disappearance of acetylcholine esterase, monoamine oxidase, and other enzymes important in the destruction of these agents, but apparently this is not the sole mechanism. In the case of skeletal muscle, the hypersensitivity is reflected in periodic asynchronous contractions of the muscle fibers (fibrillation), apparently in response to small concentrations of acetylcholine in its environment. Sometimes one also sees contractions of denervated skeletal muscle in response to the stimulation of postganglionic cholinergic neurons (autonomic neurons).

In the case of the organs innervated by the

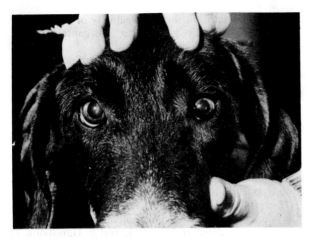

Fig. 4-9. Unilateral Horner's syndrome in the dog. One day prior to this photograph the sympathetic fibers to the head and face were severed in the left side of the neck. These neurons send adrenergic sympathetics to the blood vessels, salivary glands, and radial muscle of the eye, and cholinergic sympathetics to the nictitating membrane. As a result of the unilateral loss of this sympathetic tone the left ear and face became warmer (vasodilation), the left pupil constricted (miosis), and the nictitating membrane became elevated. (Shepard, R. S., and Whitty, A. J.: Bilateral cervical vagotomy: A long term study on the unanesthetized dog. Am. J. Physiol., *206*:267, 1964)

autonomic nervous system, we find that severing preganglionic neurons produces less marked hypersensitivity than the section of postganglionic neurons. This is sometimes referred to as (Walter) *"Cannon's law of denervation."* We find, for example, that if we sever the preganglionic fibers to the right superior cervical ganglion, as well as the postganglionic fibers from the left superior cervical ganglion, there is an initial bilateral constriction of the pupil. After a week, however, the right pupil (preganglionic decentralization) is smaller than the left (postganglionic decentralization). These differences become even more marked when the animal is excited or when one injects into him small amounts of epinephrine. Postganglionic decentralization, that is, produces a more pronounced hypersensitivity to epinephrine than does preganglionic decentralization.

These observations become important in various types of autonomic surgery. In Raynaud's disease, for example, spasms of the blood vessels (usually in the fingers or toes) may occur, resulting in intense pain and even gangrene. Where these spasms are localized in the hand one can sometimes provide temporary relief by a procaine block of the stellate ganglion (superior and middle cervical ganglion). When procaine block relieves the symptoms the next step is usually partial sympathectomy. If the stellate ganglion is removed, however, one not only produces some unfortunate changes in the head and face (Horner's syndrome), but in time the symptoms of Raynaud's disease return (due to an increased sensitivity to catecholamines), particularly when the subject becomes excited. The preferred operation, therefore, is section of the sympathetic chain between the third and fourth ganglia and section at the gray rami in the second and third thoracic segments. This provides a preganglionic sympathectomy in which hypersensitivity is less pronounced and the sympathetic fibers to the head are left intact.

The Postganglionic Neuron

The postganglionic neuron, the heart, the smooth muscles, and the glands, unlike the skeletal-muscle fiber, characteristically do not atrophy in response to decentralization. The heart continues to pump blood, some smooth muscle continues to contract in response to certain hormones, to stretch, and to other changes in its environment, and at least some postganglionic neurons continue to play an integrative role. We find, for example, that a decentralized intestine continues to show waves of contraction (peristalsis) — due to its intrinsic autonomic fibers.

REFERENCES

Adrian, R. H.: The effect of internal and external potassium concentration on the membrane potential of frog muscle. J. Physiol., *133*:631-658, 1956.

Ahlquist, R. P.: Agents which block adrenergic β-receptors. Ann. Rev. Pharmacol., 8:259-272, 1968.

Bain, W. A.: A method of demonstrating the humoral transmission of the effects of cardiac vagus stimulation in the frog. Quart. J. Exp. Physiol., 22:269-274, 1932.

Barker, D.: The innervation of the muscle spindle. Quart. J. Micr. Sci., 89:143-186, 1948

Bodian, D.: Introductory survey of neurons. Cold Spring Harbor Symposium on Quantitative Biology, *17*:1-13, 1952.

Bolme, P., Ngai, S. H., and Rosell, S.: Influence of vasoconstrictor nerve activity on the cholinergic vasodilator response in skeletal muscle in the dog. Acta Physiol. Scand., *71*:323-333, 1967.

Boullin, D. J.: The action of extracellular cations on the release of the sympathetic transmitter from peripheral nerves. J. Physiol., *189*:85-99, 1967.

Decandia, M., Provini, L., and Tabarikova, H.: Presynaptic inhibition of the monosynaptic reflex following the stimulation of nerves to extensor muscles of the ankle. Resp. Physiol., *2*:2-42, 1967.

Dunlop, D., and Shanks, R. G.: Selective blockage of adrenoceptive beta receptors in the heart. Brit. J. Pharmacol., *32*:201-218, 1968.

Eccles, J. C.: Neuron physiology—Introduction. In Field, J. (ed.): Handbook of Physiology. Sec. 1 (Neurophysiology), vol. I, pp. 59-74. Washington, American Physiological Society, 1959.

Eccles, J. C.: The physiology of synapses. Berlin, Springer-Verlag, 1964.

Eccles, J. C., and Magladery, J. W.: The excitation and response of smooth muscle. J. Physiol., *90*:31-67, 1937.

Eccles, J. C., and Magladery, J. W.: Rhythmic responses of smooth muscle. J. Physiol., *90*:68-99, 1937.

Fatt, P., and Katz, B.: An analysis of the end-plate potential recorded with an intra-cellular electrode. J. Physiol., *115*:320-370, 1951.

Feigl, E. O.: Sympathetic control of coronary circulation. Circ. Res., *20*:262-271, 1967.

Gasser, H. S.: The classification of nerve fibers. Ohio J. Sci., *41*:145-159, 1941.

Gershon, M. D.: Inhibition of gastrointestinal movement by sympathetic nerve stimulation: The site of action. J. Physiol., *189*:317-327, 1967.

Goldenberg, M., Aranow, H., Jr., Smith, G. G., and Faber, M.: Pheochromocytoma and essential hypertensive vascular disease. Arch. Int. Med., *86*:823-836, 1950.

Goodman, L. S., and Gilman, A.: The Pharmacologic Basis of Therapeutics. 4th ed. New York, Macmillan, 1970.

Gray, J. A. B.: Initiation of impulses at receptors. In Field, J. (ed.): Handbook of Physiology. Sec. 1 (Neurophysiology), vol. I, pp. 123-145. Washington, American Physiological Society, 1959.

Green, H. D., Rapela, C. E., and Conrad, M. C.: Resistance and capacitance phenomena in terminal vascular beds. In Hamilton, W. F. (ed.): Handbook of Physiology. Sec. 2 (Circulation), vol. II, pp. 935-960. Washington, American Physiological Society, 1963.

Grundfest, H.: Synaptic and ephaptic transmission. In Field, J. (ed.): Handbook of Physiology. Sec. 1 (Neurophysiology), vol. I, pp. 147-197. Washington, American Physiological Society, 1959.

Hutter, O. F., and Trautwein, W.: Vagal and sympathetic effects on the pacemaker fibers in the sinus venosus of the heart. J. Gen. Physiol., *39*:715-733, 1956.

Jenkinson, D. H., and Morton, I. K. M.: The effect of noradrenalin on the permeability of depolarized intestinal smooth muscle to inorganic ions. J. Physiol., *188*:373-386, 1967.

Katz, B.: Depolarization of sensory terminals and the initiation of impulses in muscle spindle. J. Physiol., *111*:261-282, 1950.

Lands, A. M., Luduena, F. P., and Buzzo, H. J.: Differentiation of receptors responsive to isoproterenol. Life Sci., *6*:2241-2249, 1967.

Lehman, R. A., and Hayes, G. J.: Degeneration and regeneration in peripheral nerve. Brain, *90*:285-301, 1967.

Lloyd, D. P. C.: Reflex action in relation to pattern and peripheral source of afferent stimulation. J. Neurophysiol., *6*:111-120, 1943.

Lloyd, D. P. C.: Spinal mechanisms involved in somatic activities. In Field, J. (ed.): Handbook of Physiology. Sec. 1 (Neurophysiology), vol. II, pp. 929-949. Washington, American Physiological Society, 1960.

Mantner, H. G.: The molecular basis of drug action. Pharmacol. Rev., *19*:107-144, 1967.

Nastuk, W. L.: Fundamental aspects of neuromuscular transmission. Invest. Ophthal., *6*:235-252, 1967.

Nastuk, W. L., and Hodgkin, A. L.: The electrical activity of single muscle fibers. J. Cell. Comp. Physiol., *35*:39-74, 1950.

Obats, K., Ito, M., Ochi, R., and Sato, N.: Pharmacological properties of the postsynaptic inhibition by Purkinje cell axons and the action of γ-aminobutyric acid on Deiters' neurons. Exp. Brain Res., *4*:43-57, 1967.

Root, W. S., and Hofmann, F. G.: Physiologic Pharmacology. Vol. IV. New York, Academic Press, 1967.

Rose, J. E., and Mountcastle, V. B.: Touch and kinesthesis. In Field, J. (ed.): Handbook of Physiology. Sec. 1 (Neurophysiology), vol. I, pp. 387-429. Washington, American Physiological Society, 1959.

Ruch, T. C., and Patton, H. D.: Medical Physiology and Biophysics. 19th ed. Philadelphia, W. B. Saunders, 1965.

Selkurt, E. E. (ed.): Physiology. 2nd ed. Boston, Little, Brown, 1966.

Shepard, R. S., and Whitty, A. J.: Bilateral cervical vagotomy: A long term study on the unanesthetized dog. Am. J. Physiol., *206*:265-269, 1964.

Sinclair, D.: Cutaneous Sensation. London, Oxford University Press, 1967.

Tasaki, I.: Conduction of the nerve impulse. In Field, J. (ed.): Handbook of Physiology. Sec. 1 (Neurophysiology), vol. I, pp. 75-121. Washington, American Physiological Society, 1959.

Uvnäs, B.: Central cardiovascular control. In Field, J. (ed.): Handbook of Physiology. Sec. 1 (Neurophysiology), vol. II, pp. 1131-1162. Washington, American Physiological Society, 1960.

Youmans, W. B.: Fundamentals of Human Physiology. P. 567. Chicago, Year Book Publishers, 1957.

Part III

CONTRACTION

Contraction is one of the phenomena by which skeletal muscle, cardiac muscle, and smooth muscle develop *tension* (expressed in units of dynes per centimeter of muscle length). When, for example, the brachialis muscle contracts, it exerts a force (expressed in dynes) on the humerus and ulnar bones. When this force plus the force exerted by synergistic muscles is greater than all the opposing forces (forces due to the antagonistic muscles, for example), the brachialis muscle shortens and the forearm is flexed. When contraction of a muscle is associated with shortening of that muscle, the contraction is called *isotonic*.

All contraction is not, however, associated with shortening. When we use the brachialis muscle to support an object, the muscle is exerting a force (in dynes) equal to the weight (in dynes) of the object (*isometric contraction*) — without changing its own length. A muscle can also exert tension while *lengthening*. This may occur when both the brachialis muscle and its antagonistic muscles contract at the same time, but the antagonistic muscles exert a greater force.

In all cases contraction is associated with the muscle exerting a force on a resistance. If, for example, the resistance which is to be lifted, supported, or lowered has a *mass of 1 gm.*, it will weigh *980.7 dynes*. In order to support the resistance, the muscle must exert an equal but opposite force (980.7 dynes). Although the gram is a unit of mass, it is also frequently used to characterize a force (gram) or tension (grams per centimeter). The student should not, therefore, be surprised to hear an author talk about a force of 1 gm. or 980.7 dynes. When he says that the muscle exerts a force of 1 gm. he means that the muscle is exerting a force capable of supporting a mass of 1 gm. in the earth's gravitational field.

The tension produced by muscle is used by the body in many ways. It moves solids (organs, feces, feti), liquids (blood, lymph, bile, urine, and semen), and gases (air and flatus), and it resists gravity, inertia, and forces which can produce hernia, rupture, or dislocation. Contractions of some skeletal muscles (diaphragm, intercostal muscles, transversus abdominus muscle, etc.) are used to change the pressure (dynes per square centimeter) within the thoracic or abdominal cavities. The contraction of the heart is used to increase the pressure within the heart. The contraction of the smooth muscle in the arterioles is used to increase the arterial pressure.

Contraction is also associated with the release of heat. Under some conditions heat is eliminated from the body as a waste product. Under other conditions (*shivering* in a cold environment) heat is retained to maintain a constant body temperature. Thus, muscle serves as a physiological transducer, converting nutritive energy (carbohydrates, lipids, and proteins) into (1) *work* (measured in dyne-centimeters, or ergs), (2) *tension* (grams per centimeter or dynes per centimeter), (3) *pressure* (dynes per square centimeter or millimeters of mercury), and (4) *heat* (calories).

The method the body uses to stimulate its muscle is varied. *Skeletal muscle* is stimulated by *nervous impulses* coming from the central nervous system. The neurons which carry these impulses release *acetylcholine* at various neuromuscular junctions, resulting in a *depolarization* of that part of each muscle-cell membrane at each junction. This phase in the initiation of contraction in skeletal muscle is called *neuromuscular transmission*. Next we have *activation-contraction coupling*. Here the depolarization is conducted throughout the cell membrane and results in the intracellular release of a substance (probably Ca^{++}) which initiates the contractile reaction (an interaction between actin and myosin).

Cardiac muscle, unlike skeletal muscle, can contract in the absence of the central nervous system or in the absence of nerve fibers. It contains a cell (a *pacemaker* cell) which depolarizes in the absence of extrinsic stimuli. This depolarization is conducted throughout the cell membrane and, unlike skeletal muscle, is also conducted from one cell to the next (*intercellular conduction*). This intercellular conduction normally results in the stimulation of every muscle cell in the heart each time the pacemaker cell fires. This too is different from skeletal muscle. In skeletal muscle we find that the firing of one neuron usually results in the stimulation of less than 1 per cent of the muscle cells in the muscle. The function of the efferent nerves to the heart, on the other hand, is not to *initiate* contraction, but to *modify* the activity of the pacemaker and the response of the various muscle fibers to that pacemaker. Thus the stimulation of cardiac parasympathetic fibers can decrease the rate of firing of the pacemaker, and stimulation of cardiac sympathetic fibers can increase the rate of firing of the pacemaker and increase the force of contraction.

In the case of skeletal muscle and cardiac muscle, we have good evidence that contraction results from an interaction between *actin* and *myosin* molecules, which results in a *sliding* of actin toward the center of the myosin. In the case of smooth muscle, we may be dealing with a different contractile mechanism. Smooth muscle does, however, resemble skeletal and cardiac muscle in some respects. In many cases its contractions are, for example, initiated by the depolarization of its cell membrane. In some smooth muscle (ciliary muscle) this depolarization is initiated by a neuron (postganglionic parasympathetic neuron, in the case of the ciliary muscle). In other smooth muscle (that of the intestine) it may be initiated by a pacemaker cell. In still other types of smooth muscle (uterus), pacemaker cells are usually absent but may develop in response to a chemical in the blood (estrogen in the case of the uterus). Many smooth muscles seem to be more strongly affected by *hormones* such as estrogen than is either skeletal or cardiac muscle. Hormones may, for example, change the *irritability* of a cell by modifying the resting transmembrane potential. They may make a smooth muscle either more or less responsive to stimulation by nerves.

Chapter 5

STRUCTURE
MUSCLE TWITCH
MUSCLE TETANUS
CONTROL
MUSCLE PERFORMANCE
(CONTRACTILITY)
CHEMISTRY

SKELETAL MUSCLE

STRUCTURE

Skeletal muscle consists of a series of continuous connective tissue sheaths (epimyseum, perimyseum, endomyseum, and tendon), associated blood vessels, lymph vessels, nerves, and numerous elongate muscle cells. A single muscle cell or fiber varies in length from 5 mm. in the case of the multifidus muscle to 400 mm. in the sartorius muscle. In those muscles in which the fibers are parallel, contraction may result in a 60 per cent decrease in muscle length. Where the fibers are pinnate (featherlike in structure), shortening is less. Each fiber is between 50 and 100 micrometers (μ) thick and is surrounded by a semipermeable cell membrane or *sarcolemma*, which forms an effective barrier between the intra- and extracellular environments. Within the cell are numerous peripheral nuclei, muscle cytoplasm (*sarcoplasm*), and membrane-lined channels. In Figures 5-1 and 5-2 we see some of the elements contained in a muscle fiber. The sarcoplasm consists of hundreds of elongate myofibrils, certain organelles called mitochondria or sarcosomes, glycogen granules, lipid droplets, and pinocytotic vesicles.

The *mitochondria* carry out many reactions essential to the cell. One of these is *oxidative phosphorylation*. In this reaction the energy released by the oxidation of carbohydrates, lipids, and amino acids is used to produce adenosine triphosphate (ATP) from adenosine diphosphate (ADP) and inorganic phosphate (see Chap. 26). It has been suggested that the closely associated oxygen-containing pigment myoglobin is important in this process. Myoglobin, which contains iron, is a porphyrin-protein similar to another red pigment, hemoglobin. It, however, becomes saturated at lower O_2 concentrations (that is at a PO_2 of 40 mm. Hg) and becomes desaturated at a PO_2 of about 20 mm. Hg. It is found in smooth, cardiac, and skeletal muscle, its average concentration in human skeletal muscle being about 700 mg. per 100 gm. of wet muscle. This concentration of myoglobin, however, would contain at best less than a minute's supply of O_2 for the muscle. Besides this storage function, myoglobin may also serve as a vehicle for the intracellular transport of O_2. The latter suggestion is appealing but is, up to now, based on relatively little evidence.

It has been hypothesized that the intracellular membrane-lined channels constitute a communications system from the sarcolemma to the interior of the cell. We know, for example, that large molecules such as ferritin get into the transverse tubules (T system). This, along with other evidence, is consistent with the view that the T system is an invagination of the plasma membrane which forms a patent pathway to the interior of each sarcomere. It is apparently not continuous with the horizontal component (*sarcoplasmic reticulum*), but lies near the terminal sacs (cisternae) of the reticulum and probably activates the reticulum to release Ca^{++}. The reticulum itself branches and anastomoses (joins) around the myofibrils. It may not only initiate contraction by the release of Ca^{++}, but also terminate it by removing Ca^{++} from the immediate environment of the myofibrils.

The intracellular *myofibril* is composed of myofilaments, which are made up of *myosin* (molecular weight approximately 500,000) and *actin* (molecular weight approximately 50,000). A third protein, tropomyosin, is also found in the fiber, but its function and location are poorly understood. In Figure 5-3 we see a diagram of the relationship between myosin and actin. Also shown in this diagram are the cross-bridges, which extend from the thick myosin filament to the thin actin filament. With the exception of the

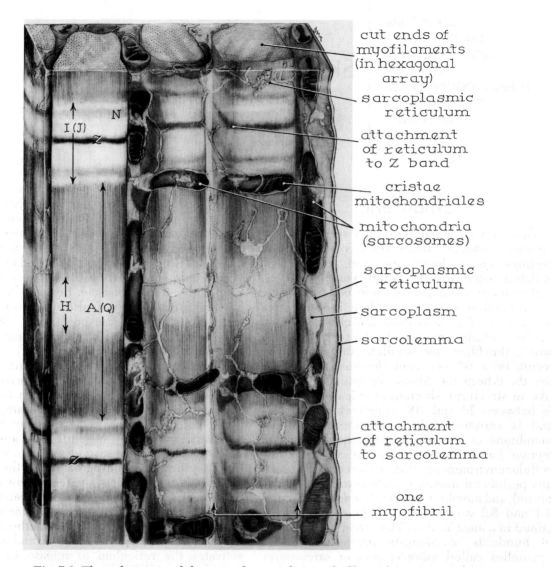

Fig. 5-1. Three-dimensional drawing of a part of a muscle fiber. That segment of the myofibril between 2 Z-bands is called a sarcomere and is approximately 2.5 μ long. A more detailed drawing is seen in Figure 5-3. (Ham, A. W., and Leeson, T. S.: Histology. 4th ed., p. 425. Philadelphia, J. B. Lippincott, 1961)

central 0.15 μ, these bridges are found every 0.04 μ (400 angstroms or 40 nanometers) along the myosin.

The I and A bands of the fibers were first described by light-microscopists, but have been more clearly defined by the use of the electron microscope. The A, or anisotropic band, is the darker or more highly refractory part. It sometimes consists of a central light area, the H-zone (Ger. hell, light), composed primarily of myosin, and a peripheral dark area containing both myosin and actin. The I bands are composed primarily of actin.

In muscle contraction the actin is pulled closer to the center of the myosin molecule by the crossbridges. It is theorized that the myosin bridge attaches to an active site on the actin moiety, pulls the actin a short distance, and then attaches

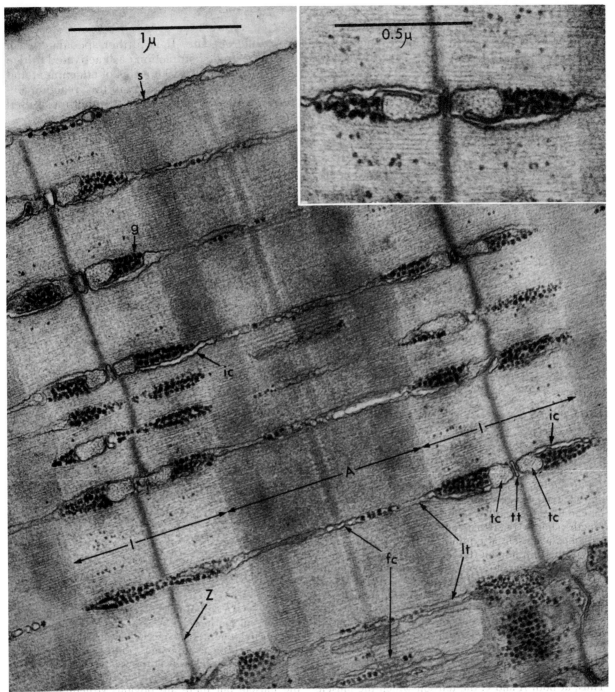

Fig. 5-2. Electron micrograph of a longitudinal section of a portion of a frog sartorius muscle. A part of the sarcolemma (s) is seen in the upper left of the picture. Enclosed by the sarcolemma are the myofibrils which consist of dense A bands and light I bands. The light bands are bisected by the Z line. In the lower right of the picture we can see a part of a transverse tubule (tt). This lies close to a horizontal component, the sarcoplasmic reticulum. A magnification of these elements is inserted at the upper right of the micrograph. The sarcoplasmic reticulum lies between the myofibrils and consists of the terminal cisternae (tc) seen next to the transverse tubule, the intermediate cisternae (ic), the longitudinal tubules (lt), and the fenestrated collars (fc). The transverse tubule plus the terminal cisternae on either side constitute a triad (tc, tt, tc). The dark staining granules are glycogen (g). (Peachey, L. D.: Transverse tubules in excitation-contraction coupling. Fed. Proc., 24:1126, 1965)

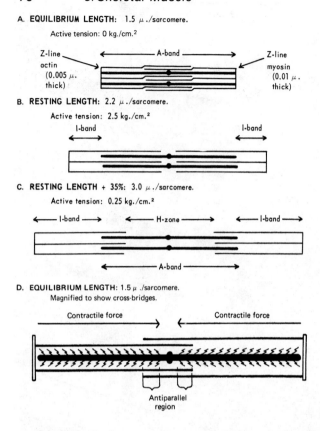

Fig. 5-3. Distension and shortening of skeletal muscle. Skeletal muscle after tenotomy or after being removed from the body exists at a length (A, equilibrium length) shorter than that found in the body. As one stretches a muscle at equilibrium length, the central end of the thin filament (the end of actin furthest from the Z-line) ceases to overlap adjacent thin filaments (B), and eventually is pulled near the end of the thick filament (C). Resting length represents the point where the overlap of the actin and myosin is similar to that seen in B. It also represents the length where muscle exerts its greatest active tension and where less stretch (A) puts parts of the actin moiety in contact with cross-bridges (antiparallel regions of D) which, on stimulation, resist rather than facilitate further shortening. Thus, at both A and C the tension which the muscle fiber exerts in response to stimulation (active tension) is less than at resting length (B). The transition from C to B to A may result from the elastic properties of the muscle (passive tension) or its contractile properties (active tension). The transition from A to B to C results from external forces (an antagonistic muscle or gravity) acting on the muscle. (*Modified from* Spiro, D., and Sonnenblick, E. H.: Comparison of the ultrastructural basis of the contractile process in heart and skeletal muscle. Circ. Res., 15:15, 1964)

to a different site. It is further speculated that during a single contraction each activated bridge goes through 50 to 100 cycles of attachment and detachment per second. It is this reaction between actin and myosin that is sometimes called the *active state*.

The active state is apparently the result of the intracellular release of Ca^{++}. It begins with the combination of a small number of myosin cross-bridges with receptor sites on the actin molecule and builds up as more actin and myosin molecules interact. Generally not all the actin and myosin molecules in a single muscle fiber are activated at once, but those that are activated

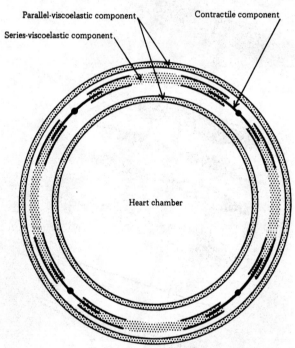

Fig. 5-4. A simplified diagram of the functional anatomy of the heart. A. V. Hill characterized muscle as a 3-component system: (1) contractile component, (2) series-elastic component, and (3) parallel-elastic component. This classification was based more on the functional than on the anatomical characteristics of the muscle but there are obvious structural equivalents. The tendon of skeletal muscle is, for example, in series with the contractile elements (actin and myosin) and the sarcolemma is parallel with them. On the basis of more recent studies, however, physiologists have emphasized the importance of the viscous properties in the above model. If we are to be complete, then, we should modify the original concept to consist of (1) a contractile component, (2) a series-viscoelastic component, and (3) a parallel viscoelastic component.

tend to stretch the myofilaments or connective tissue elements which lie in front of and behind them. In other words, they exert a force on the elements (myofilaments, tendons, etc.) with which they are in series. Contraction, in fact, has been characterized as a stretching of viscoelastic elements in series with the contractile elements (Fig. 5-4).

The tendon and actomyosin cross-bridges are part of the so-called *series-viscoelastic component*. Thus actomyosin has both a viscoelastic and a contractile function. In the contracting muscle fiber some of the actomyosin that has been activated by Ca^{++} stretches some of the actomyosin not activated by Ca^{++}. In the relaxed muscle, on the other hand, the actomyosin is not activated, but still helps to resist stretch by virtue of its viscoelastic properties. In the resting muscle actin and myosin are not readily separated or rearranged, and this resistance to separation constitutes an important factor in the series-viscoelastic characteristics of the muscle. Contributing to these viscoelastic characteristics is the sarcolemma, lying parallel to the actomyosin and characterized as the major part of the *parallel-viscoelastic component*. These two factors, the (1) series-viscoelastic component, and (2) parallel-viscoelastic component, like the rubber in a slingshot, are responsible for the resistance of the muscle to stretch, but unlike rubber have important viscous properties as well. There is internal slipping in muscle, for example, meaning the muscle does not return to its exact prestretched position when the distending force is removed unless there is also a muscle contraction.

MUSCLE TWITCH

The response of a muscle or single muscle fiber to an adequate stimulus is called a twitch (Fig. 5-5). The twitch is usually preceded by an impulse which is conducted along an axon to a group of motor end-plates. The axon terminal of each *motor end-plate* then releases the chemical transmitter acetylcholine, which in turn helps to initiate a self-propagating impulse in the associated muscle fiber. Events beginning with the arrival of the axon impulse at the end-plate and ending with a propagated action potential in the muscle fiber are referred to as neuromuscular transmission. The events following neuromuscular transmission can be di-

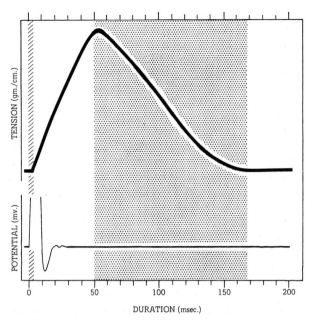

Fig. 5-5. Response of the tibialis anticus muscle to a single supermaximal stimulus delivered to the popliteal nerve of a cat. The muscle action potential (*lower tracing*) was recorded from 2 surface electrodes on the muscle. The tension (isometric) began to develop about 3 msec. after the onset of the action potential and far outlasted the electrical change. The periods of latency and relaxation have been shaded. (*Modified from* Creed, R. S., et al.: Reflex Activity of the Spinal Cord. P. 6. Oxford, Clarendon Press, 1932)

vided into four periods (latency, contraction, relaxation, recovery) which together constitute the twitch.

Period of Latency

There is an interval of from 2 to 6 msec. from the time the muscle or muscle fiber is stimulated until it develops tension. This is called the period of latency. It normally results from neuromuscular transmission and begins with an increased *permeability* of the muscle cell membrane to Na^+ and an associated *depolarization*, which is conducted at a rate of about 30 m. per second through the cell membrane. There is also a decrease in cell volume, an increase in pH, a release of heat, and an initial relaxation during this period.

The depolarization is followed by a movement of K^+ out of the cell, which results in a return to resting polarity (*repolarization*). The period of depolarization generally lasts throughout the period of latency and into the period of contraction. It is associated with an interval of reduced excitability called the *refractory period*. During the initial period of refraction (about 2 msec.) the muscle fiber is incapable of being stimulated electrically or through its motor neuron (absolute refractory period), but during the second stage (relative refractory period), superthreshold stimuli can initiate an additional contraction. In skeletal muscle the refractory period is sufficiently short (less than 10 msec.) and the period of contraction sufficiently prolonged (5-60 msec.) to permit a complete fusion of twitches (tetanus) in response to a high frequency of threshold stimuli.

Depolarization at the cell membrane is, by itself, incapable of initiating contraction. I have suggested that the significance of depolarization is that it brings about the intracellular release of Ca^{++}. Heilbrunn and Wiercynski in 1947, for example, showed that they could initiate a contraction by injecting $CaCl_2$ into the fiber with a micropipette, but that intracellular injection of $NaCl$, KCl, or $MgCl_2$ had no such action. Watanabe in 1958 demonstrated that two intracellular electrodes do not necessarily initiate a contraction even if a current greater than that of the action potential is maintained between them for the same duration as the action potential. These data have led many to conclude that the surface action potential is but the first step in initiating contraction.

A. V. Hill in 1949, on the other hand, objected to the view that a surface action potential would cause the release of an activator at the sarcolemma and that it would, in turn, diffuse throughout the muscle fiber and initiate contraction. Although this was a distinct possibility with certain slow muscles and certain small muscle fibers, he felt that it was impossible with most muscle for any activator to diffuse throughout the cell rapidly enough to initiate the twitch. The work of A. F. Huxley and coworkers (1955 through 1958) offered a ready answer to this dilemma; they demonstrated a continuation of the cell membrane into the fiber (T system) and, in addition, a closely associated intracellular system of tubules (sarcoplasmic reticulum). They also noted that there were sensitive spots at the surface of the cell membrane, which, when stimulated with a subthreshold stimulus, produced a localized contraction in part of the muscle fiber. The electrical changes at the sensitive spots on the cell membrane were localized to a few microns. These spots were the areas where the cell membrane invaginated to form the transverse tubule. In the frog they are near the Z line, but in the crab they are near the ends of the A band.

It has been suggested that depolarization in the transverse tubules results in a change in the charge across the sarcoplasmic reticulum. This may result from the movement of K^+ into the reticulum that causes the release of Ca^{++} from the reticulum. Calcium is also found in high concentration in mitochondria and enters the cell through the cell membrane during depolarization, so the reticulum is not the only potential source of Ca^{++} during activation.

The significance of the release of Ca^{++} is that Ca^{++} activates an actomyosin adenosine triphosphatase. One speculation is that each myosin bridge has a terminal ATP and that the Ca^{++} facilitates a connection between the ATP and actin. There follows a shortening of the extended bridge which brings the ATP in contact with adenosine triphosphatase (both actomyosin and myosin contain adenosine triphosphatase moieties), which hydrolyzes the ATP and, in so doing, breaks the bond between actin and myosin. During a single twitch this process might repeat itself 50 to 100 times in each (about 65) of the cross-bridges on the myosin molecule. In each case ATP is hydrolyzed to ADP and the ADP is converted back to ATP by a reaction with some other energy source, such as creatine phosphate (CrP). During muscle shortening (isotonic contraction), each cross-bridge would combine with a site on the actin which is progressively closer to the Z line (creeping phenomenon), in this way, pulling the actin closer to the center of the myosin. This has been characterized by Oosawa et al. as the "unidirectional propagation of a wave of association-dissociation between actin and myosin."

Some of this speculation is based on the study of muscle extracts. If muscle is stored for several days in a strong glycerol solution at low temperatures, the cell membranes break down and most of the soluble proteins and crystalloids are extracted. If this *glycerol-extracted muscle* is now placed in an isotonic salt solution containing K^+, Ca^{++}, and Mg^{++}, it will shorten on the addition of ATP. If the muscle is placed in a solution of K^+, Mg^{++}, and ATP, it will shorten on the addition of Ca^{++}.

During the period of latency the muscle also

shows a much-increased resistance to stretch, probably a result of the formation of cross-bonds between actin and myosin. This increase in the resistance to stretch is one of the signs used to study the *active state*. Eventually the series-elastic component is stretched sufficiently to result in tension being exerted on an external resistance. This marks the beginning of the period of contraction.

Period of Contraction

Contraction is an increase in muscle tension. It may be associated with either muscle lengthening or shortening or with no change in length (isometric contraction), and is the result of the chemical union of the myosin cross-bridges with actin. The amount of tension that develops is dependent upon a number of factors (Fig. 5-6). Each of these determines either (1) the number of effective actomyosin cross-bridges activated, (2) the rate at which the entire cell is activated, (3) the duration of the active state, or (4) the state of the series-elastic component.

One way to increase the number of effective actomyosin bridges activated is to have the muscle at its resting length prior to an isometric contraction. At resting length (Fig. 5-3) we have a relationship between the myosin and actin such that all the cross-bridges can find an effective receptor site on the actin. It is also possible to increase the number of actomyosin molecules and cross-bridges in a given cell using certain hormones which facilitate protein anabolism. Isometric exercise also facilitates the intracellular formation of actomyosin.

The rate and duration of activation may vary from one cell to the next. A muscle fiber from the internal rectus of the eye takes about 8 msec. for the period of latency and contraction combined (contraction time). Contraction time for a fiber from the gastrocnemius muscle, on the other hand, is about 30 msec., and that for a fiber from the soleus is about 110 msec. These times may be extended by cooling and other environmental changes. In Figure 5-7 we see a record of the isometric twitch tension developed at 25°C. and 35°C. Note that at the lower temperature the rate of tension development is slower and the duration of the twitch is longer. Nitrate ions also prolong the active state.

Period of Relaxation

Relaxation is a decrease in muscle tension associated with muscle lengthening or shortening or with no change in length. It probably results from an intracellular translocation of Ca^{++} that starts toward the end of contraction. The series-elastic elements of the muscle, however, continue to exert tension for a short time after the bridges have detached from the actin. We know that if we reduce the Ca^{++} concentration in actomyosin-ATP solutions to below 10^{-6} mols per liter, relaxing effects are produced. We know too that we can reduce Ca^{++} concentration in ATP solutions to 10^{-7} mols per liter by adding fragments of the internal membrane system (sarcoplasmic reticulum). It has been suggested that the reticulum picks up this Ca^{++} and deposits or precipitates it in its lumen.

Relaxation is also dependent upon the presence of inorganic pyrophosphate, ADP, or ATP. As muscle depletes its supply of ATP it eventually becomes rigid. This stiffness is similar to what one finds after death (*rigor mortis*) and to the state of glycerol-extracted muscle in an isotonic KCl solution without ATP. It is probably the result of the formation of tight cross-bridges

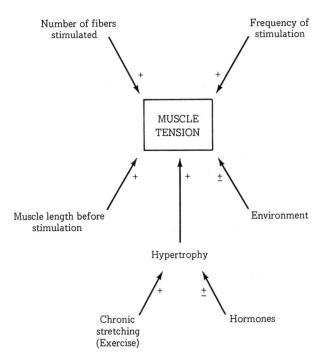

Fig. 5-6. Diagram of the factors which control skeletal muscle tension during an isometric contraction.

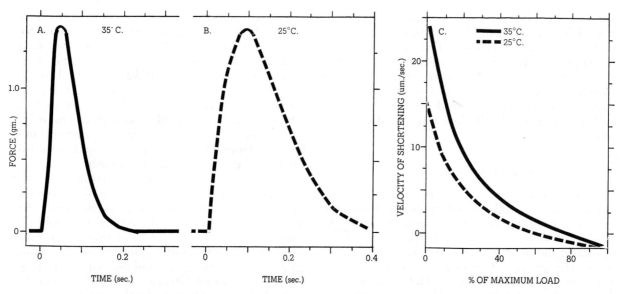

Fig. 5-7. The effect of temperature on the development of an isometric twitch (A and B) and a twitch associated with shortening (C) in the extensor digitorum longus muscle of the neonatal rat. Note that changing the temperature of the muscle from 35°C. to 25°C. results in a prolongation of the contraction time and a decrease in the rate of development of tension. In C we find that cooling also decreases the velocity of shortening per sarcomere at various loads (forces). (*Modified from* Close, R.: The relation between intrinsic speed of shortening and duration of the active state of muscle. J. Physiol., *180*:545, 1965)

between actin and myosin. Thus it seems that ATP serves a dual function. It acts as an energy source and it prevents rigor.

Period of Recovery

After relaxation is complete the muscle continues to produce more heat and use more oxygen than it did prior to stimulation. This is a period when the muscle continues to extrude the intracellular Na^+ and to bring in the extracellular K^+ that was exchanged during the action potential. The oxygen concentration of the sarcoplasm is being restored, nutritive supplies (adenosine triphosphate, creatine phosphate, glucose, glycogen, etc.) are being replenished, and waste products are being eliminated.

Summary

The events occurring during a single muscle twitch may be summarized as follows:
1. Neuromuscular Transmission
2. Excitation-Contraction Coupling
    A. Depolarization and conduction
        1. Na^+ influx (sarcolemma and transverse tubule)
        2. Cisternae of sarcoplasmic reticulum
    B. Intracellular release of an activator (Ca^{++}) into myofibrillar space
3. Development of the Active State
    A. Actin + Myosin-ATP → Actomyosin-ATP
    B. Shortening of actomyosin bonds
    C. Actomyosin-ATP → Actin + Myosin-ADP + P
    D. Myosin-ADP + CrP → Myosin-ATP + Cr
    E. Repetition of A through E
    F. Repolarization
        1. K^+ efflux
4. Deactivation
    A. Intracellular deposition of activator (Ca^{++})
    B. Formation of actin + myosin-ATP
    C. Relaxation
    D. Recovery
        1. Cr + P → CrP
        2. Lactic acid + O_2 → CO_2 + H_2O + glucose + glycogen

MUSCLE TETANUS

I have characterized the response of a muscle or muscle fiber to a single stimulus as a twitch. In Figure 5-8 we note that as we increase the frequency of stimulation of a skeletal muscle, we eventually reach a point at which there is a fusion of twitches. This is a point at which the force of contraction is greater than it was for a single twitch. This fusion of twitches in response to high-frequency stimulation of a single muscle fiber or a whole muscle is called a tetanus. It should not be confused with two neuromuscular disorders called tetanus and tetany. In both of these pathological situations we are dealing with a tetanic contraction, but in the case of the former, the tetanus is the response of the individual to the toxin of a bacillus, and in the case of the latter (tetany) to a change in environment (hypocalcemia, for example.)

The student should not confuse the role of Ca^{++} in excitability with its intracellular role. I have emphasized that its role in the muscle fiber is probably the activation of contraction. Its effect on the cell membrane, however, is to decrease its excitability. Thus a diminution in the intracellular release of Ca^{++} might decrease the amount of tension developed in response to stimulation, while a decrease in extracellular Ca^{++} could result in more stimuli arriving at the neuromuscular junction.

The point at which a complete fusion of twitches occurs varies from one muscle to the next. The internal rectus of the eye, for example, is classified as a *fast muscle* since it has a contraction time of only 8 msec. It requires about 350 stimuli per second to produce a complete tetanus in this muscle. A good example of a slow muscle is the soleus. It has a contraction time of about 110 msec. and a *critical fusion frequency* of 30 stimuli per second.

Increasing the rate of stimulation from 1 stimulus per second to the critical fusion frequency produces a 2- to 5-fold increase in the force of contraction over that found in the twitch. This relationship is sometimes expressed in terms of a tetanus-to-twitch ratio. At the point of complete tetanus in Figure 5-8, most of the contractile elements have been stimulated and the elastic structures have reached a fixed length. Increasing the frequency of stimulation above this level has little additional effect on either the force or the smoothness of contraction. For example, in rat gastrocnemius muscle the tetanus-to-twitch ratio has been reported by Schottelius et al. to be 2.93. In other words, they found that the gastrocnemius muscle exerted almost three times as much force during a complete tetanus as it did during a twitch. They also showed that the force of contraction at 75 stimuli per second was 2075 gm., and at 135 per second, 2085 gm. These forces were not significantly different from one another.

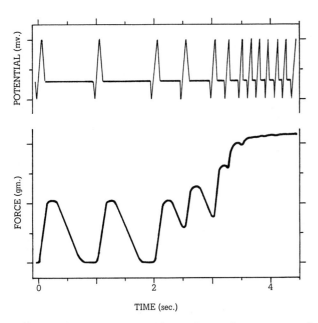

Fig. 5-8. Electrical and mechanical response of skeletal muscle to increasing frequencies of stimulation. At a frequency of 1 stimulus/sec. there is a series of twitches, at 2 stimuli/sec. there is an incomplete tetanus, and at 7 stimuli/sec. a complete tetanus.

CONTROL

Skeletal muscle, unlike cardiac and certain types of smooth muscle, is normally dependent upon the central nervous system for the initiation of contraction. The force with which a single muscle fiber responds to this stimulation is dependent on the frequency of stimulation and the conditions within and around the muscle fiber. If the nerve supply to muscle is cut peripheral to the spinal cord (*decentralization*), the muscle becomes *flaccid* (loses much of its tone). Although it no longer contracts in response to impulses originating in the spinal cord, it responds to direct electrical stimulation.

After several months to a year of denervation, the muscle fibers decrease markedly in volume and lose many of their myofibrils; eventually the entire myofiber disappears. During the first few weeks of decentralization the muscle periodically *fibrillates*. Fibrillation is an irregular, asynchronous twitching which is too weak to be seen through the skin and is best studied by electromyography. It is associated with action potentials of 10 to 200 μv. and 1 to 2 msec. duration. It is apparently the result of a supersensitivity which the decentralized muscle develops to circulating acetylcholine. In certain decentralized facial muscles, for example, stimulation of parasympathetic neurons to the salivary glands initiates skeletal-muscle contraction or twitching.

Fibrillation should not be confused with fasciculation. *Fasciculation* is a more synchronous, more forceful contraction. It results from the spontaneous discharge of motor neurons and is sufficiently forceful and synchronous to be seen through the skin. It occurs in amyotrophic lateral sclerosis and progressive muscular atrophy.

Apparently, then, the nerve supply to muscle serves not only to initiate contraction but also to maintain the muscle itself and to suppress its irritability. In the absence of a nerve supply the muscle cells become smaller (*atrophy*) and eventually disappear. This atrophy can be slowed by periodic electrical stimulation of the muscle, by massage, and by other types of physical therapy, though such measures merely retard the degenerative process. If nervous reinnervation eventually occurs and all the muscle cells have degenerated the body does not form new ones. If a few muscle fibers remain and are reinnervated, these slowly increase in size (*hypertrophy*). Muscle fibers, unlike blood cells, characteristically respond to an increased demand by hypertrophy rather than by an increase in the number of cells (*hyperplasia*).

Motor Unit

A single, somatic efferent neuron with the muscle fibers it innervates is called a motor unit. The cell body for most of these neurons lies in the ventral (anterior) horn of the gray matter of the spinal cord. Their axons generally help form mixed nerves and branch near their terminations. A single neuron may innervate as few as three muscle fibers in the case of the extrinsic muscles of the eye and as many as 150 in some of the leg and back muscles. The total number of motor units in a muscle is about 200 to 600. It is gen-erally agreed that each muscle fiber receives only one part of one neuron, but there have been a few demonstrations of single skeletal-muscle fibers which receive multiple innervation.

In most contractions only a small percentage of the total motor units to a muscle contract at any given time. During reflex, postural contractions between 5 and 25 stimuli per second pass over a single motor neuron. During maximum effort the *frequency per motor neuron* may go as high as 50 per second. In most motor units these frequencies are inadequate to produce a complete summation of twitches (complete tetanus or wave summation), but the muscle contraction is smooth and can be sustained. The mechanism for this is apparently an *asynchronous* firing of motor units. That is to say that the various motor units do not each fire at the same time. Thus there is a spatial summation of waves which is responsible for a smooth, sustained contraction.

I would also like to emphasize that most muscular efforts are associated with the simultaneous contraction of antagonistic muscles. If two antagonistic muscles contract with the same force, they serve the function of *fixation*. If one contracts with more force than the other, the more forceful serves as an *accelerator* and its antagonist as a *decelerator*.

The dampening action of antagonistic muscles, the well controlled asynchronous firing of synergistic motor units, and the total body coordination of all its muscles are the result of integration within the spinal cord and brain. This integration determines which motor neurons are stimulated or depressed, when they are stimulated, and with what frequency. Damage to part of the integrating system may result in an exaggerated to and fro movement called tremor, or to some of the more pronounced disturbances in coordination such as parkinsonism or hemiballismus.

All-or-none Law

Stimulation of the axon of a somatic efferent neuron with a single, threshold stimulus results in the axon's activating all the muscle fibers it innervates. Increasing the stimulus strength changes neither the response of the axon nor the response of the muscle. Apparently, in the case of the motor unit or skeletal-muscle fiber, the stimulus merely triggers a series of events at the cell membrane which results in a propagated nerve impulse or a propagated muscle action potential. That is to say that the response of a muscle fiber or axon bears the same relationship with its stimulus as the response of a gun to pressure on the trigger or the response of a fuse to heat. Providing the pressure on the trigger detonates the shell or the heat ignites the fuse, one does not change the response by increasing the stimulus strength. More pressure on the trigger does not change the velocity of the bullet, just as more heat does not change the reaction of the fuse.

The American physiologist *Bowditch* in 1871 expressed this relationship between stimulus and response in his now-famous *all-or-none* law: "An induction shock produces contraction or fails to do so according to its strength; if it does so at all it produces the greatest contraction that can be produced by any strength of stimulus in the condition of the muscle at the time." This law applies to the axon, muscle fiber, motor unit, and the entire heart. Apparently in the heart an action potential is conducted from one fiber to the next. In skeletal muscle the fibers are better insulated from one another and therefore have a higher electrical impedance.

Although there is an all-or-none relationship between the stimulus and the motor unit or skeletal-muscle fiber, we can modify the force of contraction by artificially by-passing the sarcolemma. This can be done experimentally, but apparently does not occur in the normal human body. One means of circumventing the sarcolemma is by the use of intracellular electrodes. Using these we can produce contraction in a limited area of the cell. Microelectrodes have also produced contraction in a single sarcomere of a muscle fiber by stimulating a selected point on the sarcolemma with a subthreshold stimulus. Apparently such a stimulus applied at a point where the transverse component (T system) joins the sarcolemma induces a highly localized intracellular conduction along the sarcoplasmic reticulum and an associated release of Ca^{++} within a single sarcomere, but no propagated potential. In short, only a local excitatory state is produced.

Fiber Length

A skeletal muscle, when removed from the body, acquires a length (*equilibrium length*) shorter than any at which it exists in the body (Figure 5-3). This shortening is not associated with an action potential or contraction, but rather is an expression of the elastic and viscous properties of the muscle. If the muscle is slowly

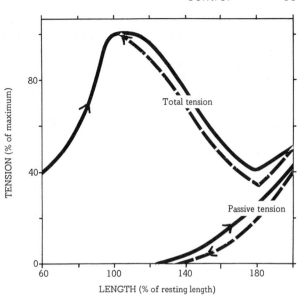

Fig. 5-9. Length–tension diagram for an isolated single muscle fiber stimulated by a volley of adequate stimuli. The solid lines are tensions recorded in an ascending order of lengths and the interrupted lines are tensions recorded in a descending order of lengths. The initial increase in total tension with distension has been credited to stretching the series-viscoelastic component and bringing the myosin cross-bridges into an optimal relationship with actin receptor sites. The decrease in total tension with further stretch may be due to pulling some of the active sites of actin away from the myosin moiety. (*Modified from* Ramsey, R. W., and Street, S. F.: The isometric length-tension diagram of isolated skeletal muscle fibers of the frog. J. Cell. Comp. Physiol., 15:21, 1940)

stretched, as in Figures 5-3, 5-9, and 5-10, eventually each muscle fiber will exert a tension. This is called *passive tension* and is a result of the muscle's viscoelastic properties. Prior to the point at which the muscle begins to exert passive tension, the muscle shows its greatest force in response to stimulation. If stretched beyond this point, it begins to show a decrease in force in response to stimulation (decreased active tension). It is believed that stretching a muscle beyond the point at which a passive tension occurs results in a separation of some of the active sites on the actin moiety from the myosin cross-bridges.

If, on the other hand, the muscle is allowed to approach equilibrium length, there is an overlap of actin such that some of the actin is in contact with bridge sites which are oppositely-directed.

Thus the bridges at one side of the midline of the myosin tend to pull the actin in a direction opposite to that in which the contralateral bridges are working. As the sarcomere length changes from 3.0 to 2.2 μ, the tension in response to stimulation increases. From 2.2 to 1.9 μ tension is maximum. From 1.9 to 1.6 tension decreases.

By subtracting the tension a muscle develops in response to stretch prior to stimulation (passive tension) from the total tension developed in response to stretch and stimulation, one obtains the active tension, or, *total tension minus passive tension equals active tension*. The active tension (sometimes called developed tension) is used as an indication of the behavior of the activated contractile component at various lengths.

In the normal human body, skeletal muscle usually has only a very limited range of lengths. This is a result of the lever systems that generally limit skeletal-muscle length. For all practical purposes in the healthy person, changing the length of most skeletal muscles prior to stimulation has little effect on the strength of the contraction, since, in the healthy person, most skeletal muscle functions at the stable upper plateau of the length-tension diagram of Figure 5-9. If, on the other hand, the tendon is cut, an entirely different situation may exist. The relative stability of skeletal-muscle length is in marked distinction to that of the heart, urinary bladder, and stomach, to mention but a few other muscular structures. Here, muscle-cell length, and therefore the force developed by a single muscle fiber or muscle, may vary tremendously.

Up to this point I have said little about the *viscoelastic* properties of skeletal muscle. I have emphasized in Figure 5-4 that some of the elastic elements lie parallel to the contractile elements and others in series with them. I have also indicated that actomyosin may serve both an important contractile and viscoelastic function. Figure 5-10 illustrates the response of the frog sartorius muscle to a load or force. Like a rubber band, this muscle increases in length (within limits) in response to a load. Unlike a rubber band, it does not snap back to its original length when the load is removed. Muscle, then, can be characterized as having *plastic* as well as elastic and contractile properties. These plastic properties are perhaps more pronounced in the case of the stomach, body bladders, and the uterus, which, within limits, can accommodate to distention without a maintained increase in intraluminal pressure. In Figure 5-10 the load on the frog sartorius muscle was increased at a

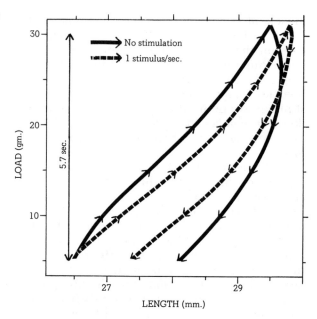

Fig. 5-10. Load-length diagram of a resting frog sartorius muscle loaded and unloaded at a rate of 4.4 Gm./sec. Note that progressive increases in load produced substantial increases in muscle length and that decreases in load produced less substantial decreases in length. These data are, in part, the basis for the observation that muscle has viscoelastic properties. (*Modified from* Alexander, R. S., and Johnson, P. D., Jr.: Muscle stretch and theories of contraction. Am. J. Physiol., 208:413, 1965)

rate of 4.4 gm. per second for 5.7 sec. and then decreased at the same rate for the same period of time. As you can see, the curve of lengthening is different from that of viscoelastic recoil. These curves can be modified by stimulating the muscle with a single, maximal stimulus every second. We see, then, that contraction can reset the length of the muscle.

Stretch has a chronic effect on muscle as well as the acute one discussed above. Cutting the *tendon* to a muscle, thus preventing it from being stretched, causes it to *atrophy*. Stretching a muscle frequently, as in exercise, causes it to *hypertrophy*. Muscles get bigger and stronger by forming more protein (actin and myosin) within the living cells at hand or by increasing other intracellular elements.

Environment

We can modify the response of a muscle to a given stimulus not only by changing its length prior to stimulation, but also by changing its environment. One of the functions of the circulatory system is to keep such changes to a minimum. It brings in O_2, glucose, lipids, buffers, and phagocytes, and carries away CO_2, lactate, acid, heat, and other products of metabolism. During severe exercise, however, the circulatory system may be unable to maintain a relatively constant environment.

Figure 5-11 illustrates how muscle, after a period of inactivity, may change its response to a series of stimuli of equal strength. In this example the second twitch contraction is stronger than the first, and the third is stronger yet. Henry Pickering Bowditch in 1871 described this relationship as a treppe or staircase effect. Later, the Nobel laureate Albert Szent-Györgyi defined "treppe" as a condition in which activity creates a situation favorable for activity. Treppe, in other words, is a frequency-related phenomenon. In Figure 5-11 the stimulation was changed from 1 stimulus per 10 min. to 1 stimulus per second. Note the progressive increase in shortening as the higher frequency is maintained (treppe). A further increase to 2 stimuli per second would eventually result in a twitch-tension greater than any at 1 stimulus per second. Eventually, however, one reaches a point of maximum twitch-tension, or the twitches fuse to form a tetanus.

Apparently the mechanism responsible for the direct relationship between the frequency of contraction and twitch-tension is the change in intra- and extracellular Ca^{++}, K^+, and Na^+ concentrations resulting from progressive increases in activity. These as well as other changes in the ionic environment change the response of a muscle to a single stimulus. They may (1) in-

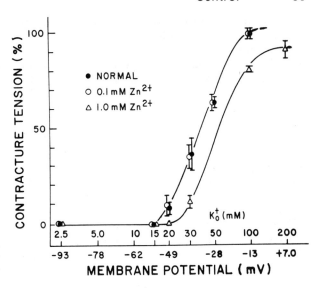

Fig. 5-12. Relationship between the external concentration of K^+ (K_O^+) and the tension developed by a frog toe muscle. Increases in the external concentration of K^+ depolarize the cell (make the transmembrane potential less negative). As K_O^+ increases, eventually a potential is reached (−50 mv.) which results in the muscle's developing tension. This potential is sometimes referred to as the *mechanical threshold*. Within limits a progressively greater depolarization produces a progressively greater muscle tension. Note that at −20 mv. we obtain *mechanical saturation*. At this potential a maximal tension is developed. Agents such as nitrates and caffein may lower the mechanical threshold from −50 to −65 mv. A 1-mM. concentration of Zn^{++}, on the other hand, raises the mechanical threshold and increases the duration of the spike potential. In other words we may, by changes in the environment of a cell, change (1) the character of its action potential or (2) the relationship between its transmembrane potential and its contractile response. (Sandow, A., et al.: Role of the action potential in excitation-contraction coupling. Fed. Proc., 24:1119, 1965)

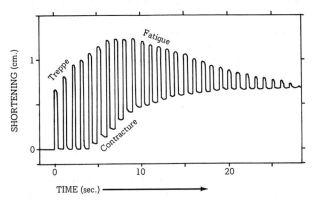

Fig. 5-11. Record from a frog gastrocnemius muscle stimulated once a second with a supermaximal stimulus.

crease the peak tension, (2) prolong the periods of contraction and relaxation, (3) increase the maximal rate of tension change during contraction or relaxation, and (4) decrease the period of latency. The mechanisms whereby various physiologic and pharmacologic agents modify these parameters include changes (1) in the character of the action potential, (2) in the relationship between action potential and contractile mechanism, and (3) in the contractile apparatus itself. For example, an increase in the duration of the action potential or a lowering of the mechanical threshold (see Figure 5-12) can increase contractility. In the intact individual, on the other hand, changes in the amplitude of the action potential of the skeletal muscle fiber are characteristically of little consequence, since they normally are not only well above the mechanical threshold for the fiber but above the point of mechanical saturation as well. For example, in Figure 5-12 the point of mechanical saturation is a transmembrane potential of −20 mv. Usually in skeletal muscle you expect to see the transmembrane potential change from a resting level of −80 mv. (well below the mechanical threshold) past −20 to +40 mv. (well above the point of mechanical saturation).

You will also note in Figure 5-11 that activity can create a situation unfavorable for activity. The muscle eventually relaxes less completely (*physiologic contracture*) and fails to contract with maintained vigor (*fatigue*).

MUSCLE PERFORMANCE (CONTRACTILITY)

We can characterize the contractile state in terms of the:
1. Force (dynes or dynes/cm.^{2}) developed
2. Degree of shortening (cm.)
3. Duration (sec.) of contraction
4. Rate of change of force (dynes/sec.)
5. Rate of shortening (cm./sec.)

All these quantities are an index of the degree of contraction of a muscle. It is also true, however, that these parameters interact with one another. For example, as we hang heavier and heavier loads on a muscle it eventually shortens increasingly less in response to stimulation. Since the load lifted is equivalent to the force exerted, we see in this example that an increase in force can be associated with a decrease in shortening. In Figure 5-7 we see that an increase in force (load lifted) may also be associated with a decrease in the velocity of shortening, and that no change in force or tension is sometimes associated with an increase in the duration of the twitch and a decrease in the rate of change of tension (sometimes called dP/dt). Some investigators have attempted to incorporate the above parameters into a single formula. If, for example, one studies an isometric contraction in a manner similar to that illustrated in Figure 5-7A and -B, he can measure the area under the twitch-tension curve. This area would be in units of gram-seconds. Quantities measured in this way are called *tension-time indices* or *impulses*.

Muscle contractility has also been characterized in terms of the:
1. Number of actomyosin cross-bridges activated
2. Synchrony of activation
3. Duration of activation

CHEMISTRY

Skeletal muscle in a lean person constitutes about 50 per cent of the body weight. Its metabolism (usually measured in terms of calories, ergs, or gram-centimeters) during exercise may increase as much as one hundred fold. This means that during increased activity skeletal muscle is capable of removing large quantities of nutrients from, and adding numerous wastes and heat to, its immediate environment and the entire circulatory system. The chemical reactions associated with skeletal-muscle metabolism are summarized in Figure 5-13.

Until 1930 it was emphasized that the formation of *lactic acid* supplied the energy for muscle contraction, but then Lundsguard reported that a muscle poisoned with *iodoacetate* and placed in an anaerobic environment did not form lactic acid, but could still contract "a hundred times." In 1932, A. V. Hill suggested that it was the breakdown of phosphocreatine, and later Szent-Györgyi suggested that it was the breakdown of *adenosine triphosphate* (ATP) that supplied the energy for contraction. Since then it has been shown that myosin and actomyosin can split ATP and that ATP added to glycerol-extracted muscle can produce shortening.

It is now generally held that ATP is the immediate metabolic unit of payment in contraction and that other sources of energy in the muscle serve to replenish ATP as it is broken down. Support for this concept comes from the study of *fluorodinitrobenzene*. It inhibits in vitro and in vivo the catabolic action of creatine phosphokinase on *creatine phosphate*. When muscle is poisoned with fluorodinitrobenzene, it continues to contract several times in response to stimulation. This occurs even though the muscle is no longer able to break down creatine phosphate. Thus creatine phosphate, oxygen, various lipids, glycogen, and so on seem to serve, as do savings accounts, to replenish our in-pocket currency (ATP). Just as many people have more money in the bank than they do in their pocket, so it is with muscle. There is, for

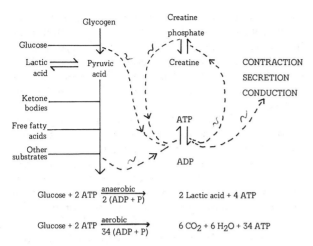

Fig. 5-13. Summary of various energy yielding (*interrupted lines*) reactions that occur in the body. The anaerobic conversion of muscle glycogen or blood glucose to lactate, or of creatine phosphate to creatine, can result in the formation of adenosine triphosphate (ATP) from adenosine diphosphate (ADP). The aerobic conversion of lactate, pyruvate, free fatty acids, or ketone bodies (derived from fats) to $CO_2 + H_2O$ also results in the formation of ATP. ATP is a readily used energy source for contraction, secretion, and conduction.

example, about five times more creatine phosphate in muscle than ATP.

In Figure 5-13 we note that ATP is produced by both anaerobic and aerobic reactions. The conversion of creatine phosphate to creatine, and of glucose or glycogen to pyruvate (Embden-Myerhof pathway), are some of the important anaerobic reactions. The conversion of free fatty acids, ketone bodies, or pyruvate to CO_2 and H_2O are some of the important aerobic reactions. Certain microorganisms are able to survive on the basis of anaerobic reactions alone, but multicellular organisms have far too high an energy requirement. They are ultimately dependent upon the more productive aerobic reactions. In the resting, well-oxygenated person, an important aerobic source of energy is the breakdown of lipid. Under conditions of a low oxygen concentration, the muscle becomes more dependent upon the conversion of glycogen and glucose to lactate (Fig. 5-14). Much of the lactate is picked up by the blood and carried to the liver and other organs, where it is converted back to glycogen or glucose or further broken down to CO_2 and H_2O.

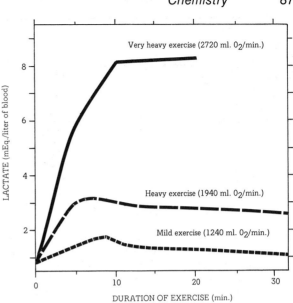

Fig. 5-14. Changes in arterial blood lactate concentration in 10 normal men during exercise on a bicycle ergometer. Prior to exercise the average oxygen consumption was 322 ml./min. The oxygen consumption at the end of the different work loads is indicated above. The oxygen debt accumulated during each of the above exercises was 5,878 ml., 3,552 ml., and 2,067 ml. of O_2 respectively. (*Modified from* Wasserman, K., et al.: Excess lactate concept and oxygen debt of exercise. J. Appl. Physiol., 20:1303, 1965)

Exercise

Some muscles contain large concentrations of mitochondria and the red pigment *myoglobin*. These red muscles, are capable of sustained activity even during the sluggish circulation of blood, and constitute the "dark meat" of the chicken and other animals. In man, all the muscles of the body contain at least some red-muscle fibers. The myoglobin in these fibers is similar to the hemoglobin found in the red blood cell. They both combine with and store oxygen. Hemoglobin acts as a reservoir of oxygen for the plasma and myoglobin for the mitochondria.

Other muscles, white or pale muscles, contain smaller quantities of myoglobin and mitochondria than red muscle, but do contain higher concentrations of actomyosin. They are stronger than red muscle, but are not capable of sustained activity. A good example of white muscle is the breast muscle of the chicken and the leg muscle of the frog. It is a very unusual frog that can jump

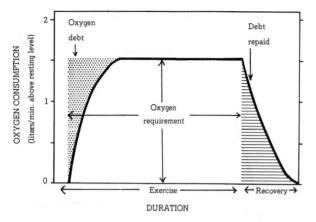

Fig. 5-15. Simplified diagram of oxygen consumption during and after exercise. Initially muscle depends primarily on anaerobic sources of energy (adenosine triphosphate, creatine phosphate, glycolysis, etc.). As these sources diminish, the muscle must either stop contracting or replenish them. The replenishment is accomplished by oxidative reactions (ketone bodies to $CO_2 + H_2O$, etc.), which continue even after the exercise is stopped. When muscle uses anaerobic reactions for energy or uses its own oxygen stores, it is said to create an *oxygen debt*. Apparently contraction, secretion, and conduction are all associated with the accumulation of such a debt.

more than a few times in succession. On the other hand the muscles of the whale and other diving mammals contain tremendous concentrations of mitochondria and myoglobin which probably make it possible for these air-breathing animals to swim for extended periods while submerged. In man, most of the muscles are mixed. The heart comes closest to being a pure red muscle. An athlete can generally run a 200-meter race without taking a breath, but must reach a steady state between oxygen intake and consumption if he is going to run much further.

Figure 5-15 is a schematic graph of the oxygen consumption of a subject during exercise. We see that initially much of the energy utilized is either from anaerobic sources or an oxygen reserve or some combination of the two. After exercise is over, oxygen consumption is maintained at an elevated level for an extended period of time. This apparently results in a restoration of the resting concentrations of ATP, creatine phosphate, and lactic acid, as well as of the oxygen reserves. In Figure 5-14 I have pictured the changes in blood lactate that occur during various grades of exercise. In the most strenuous of the exercises, an average oxygen debt of 5,878 ml. of oxygen was accumulated. That is to say that after the exercise was over the subjects' oxygen consumption remained elevated over the resting level for a period of time until 5,878 ml. of oxygen over and above the resting level had been utilized.

In Table 5-1 we see an interesting comparison of the speeds that variously-well-trained runners were able to maintain when racing different distances. As the distance to be run increases, the performance becomes markedly more dependent upon systems other than skeletal-muscle. Performance (speed) becomes limited not only by skeletal-muscle reserve, but by cardiac, respiratory, and hepatic reserve, as well as by all the other systems of the body—which must remain metabolically active if they are to survive.

Energy Expenditure

Muscle releases energy in two forms, as heat and as work. This energy expenditure can be expressed in terms of calories, ergs, gram-centimeters, foot-pounds, or kilowatt-hours (1 cal. = 4.18×10^{10} ergs = 4.6×10^7 gm.-cm. = 3.08×10^3 ft.-lb. = 1.17×10^{-3} kw.-hr.). The study of the energy expenditure of muscle has proven to be a very useful means of characterizing the response of muscle to stimulation. In Figure 5-16, for example, we note the energy expenditure of muscle under a variety of different conditions. We see that the muscle in these experiments releases less energy, and therefore needs less energy for isometric contractions than for contractions associated with shortening.

The increase in heat production during an isometric contraction (*isometric heat*) is associated with the development of the active state (*heat of activation*) and with internal muscle shortening. A number of ways have been developed to study the heat of activation in the absence of internal shortening. In one, the muscle is allowed to shorten maximally before stimulation. In the second, the muscle is distended before stimulation to such a degree that the actin and myosin can not interact on stimulation. Under such circumstances the heat released is about 40 per cent of the isometric heat.

The fact that a muscle releases more heat in a contraction associated with shortening than in one in which there is no shortening is called the *Fenn effect*. This additional amount of heat, the *heat of shortening,* is dependent on the amount

Event	Time (sec.)	Runner	Year	Speed (m.p.h.)
Men				
100 yards	9.1	Bob Hayes	1964	22.5
100 meters	10.0	Armin Hary	1960	22.4
220 yards	19.5	Tommie Smith	1966	23.1
200 meters	19.7	John Carlos	1968	22.7
400 meters	44.5	Tommie Smith	1967	20.1
440 yards	44.7	Curtis Mills	1969	20.1
800 meters	104.3	Peter G. Schnell	1962	17.2
880 yards	105.1	Peter G. Schnell	1962	17.1
1,000 meters	136.2	Juergen May	1965	16.4
1,500 meters	213.1	Jim Ryun	1967	15.7
1 mile	231.1	Jim Ryun	1967	15.6
2,000 meters	296.2	Michel Jazy	1966	15.1
3,000 meters	459.6	Kipchoge Keino	1965	14.6
2 miles	499.6	Ron Clarke	1968	14.4
3 miles	772.4	Ron Clarke	1965	14.0
5,000 meters	804.2	Kipchoge Keino	1965	13.9
6 miles	1607.0	Ron Clarke	1965	13.4
10,000 meters	1659.4	Ron Clarke	1965	13.5
15,000 meters	2694.6	Emil Zatopek	1951	12.5
10 miles	2797.8	Jerome Drayton	1970	12.9
20,000 meters	3562.8	Ron Clarke	1965	12.6
15 miles	4368.2	Ron Hill	1965	12.4
25,000 meters	4522.6	Ron Hill	1965	12.4
30,000 meters	5554.6	T. F. K. Johnston	1965	12.1
Women				
100 yards	10.0	Chi Cheng	1970	20.5
100 meters	11.1	Irena Kirszenstein	1965	20.2
220 yards	22.9	Mary Burvill	1964	19.7
200 meters	22.4	Chi Cheng	1970	20.0
400 meters	51.7	Nicole Duclos	1969	17.3
440 yards	51.0	Marilyn Neufville	1970	17.6
800 meters	120.5	Vera Nikolic	1968	14.9
880 yards	122.0	D. Willis	1962	14.8
1 mile	276.8	Marie Gommers	1969	13.0

Table 5-1. World track records for men and women. Note that as the distance increases, the maximum average speed decreases. Body reserves and metabolites are important limiting factors in this relationship.

of external shortening and the load or velocity of shortening. We see the relationship between load and velocity of shortening in Figures 5-7 and 5-16. It has also been noted that an additional amount of heat is released when the muscle lengthens during relaxation.

We see, then, that the amount of heat released during a twitch is dependent upon:
1. The activation process of the muscle
2. The internal work performed during a contraction
3. The tension developed (in the case of shortening this would equal the load)
4. The centimeters of shortening
5. The degree of lengthening during relaxation

In a tetanus the amount of heat released is also dependent upon the number of action potentials. Each action potential reinitiates the activation process or active state and thus increases the total amount of heat released. Thus, in a tetanus, once the muscle has completed its shortening, its heat release is solely the result of the activation process. The heat release associated with the maintenance of muscle tension or shortening or both is called the heat of maintenance and is approximately equal to the number of action potentials times the heat of activation.

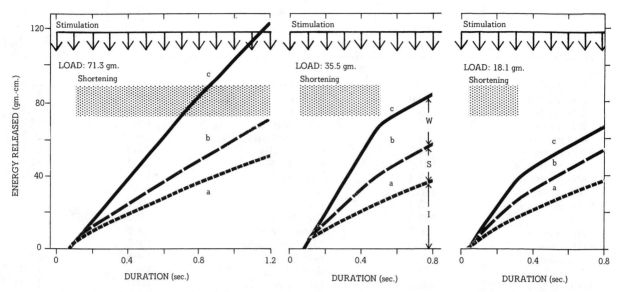

Fig. 5-16. Comparison of the energy expenditure in (a) an isometric tetanic contraction with that of (c) tetanic contractions in which loads of 71.3, 35.5, and 18.1 Gm. were lifted 7 mm. All experiments were on frog sartorii muscles at 0°C., a length of 3.1 cm., and an initial load of 2.0 Gm. Note that the muscle releases less heat during (a) an isometric contraction than during (b) a contraction associated with shortening. We also find that as the load is increased the total amount of work performed is increased and the velocity of shortening is diminished. At a load of 71.3 Gm. it took the muscle 1.1 sec. to shorten 7 mm., but at a load of 18.1 Gm. it took only 0.26 sec. to shorten 7 mm. The heat released during an isometric contraction is designated by the letter I and the extra heat released during shortening (the heat of shortening) by the letter S. During shortening energy is also released in the form of work (W). This can be calculated by multiplying the load times the distance moved. Thus, we see that lines b and c become parallel once shortening has stopped, but the muscle continues to release heat (maintenance heat) as long as stimulation is applied. (*Modified from* Hill, A. V.: The effect of load on the heat of shortening of muscle. Proc. Roy. Soc. Lond., *159*:300, 1964)

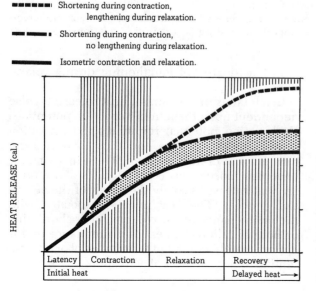

It has also proven rewarding to follow the time course of heat production during a twitch. The heat released during latency, contraction, and relaxation is called the *initial heat*, and the heat released after relaxation is called the *delayed heat* (see Fig. 5-17). The delayed heat is approximately equal to the initial heat and is

Fig. 5-17. Heat production in skeletal muscle during a twitch in which there is shortening during contraction and lengthening during relaxation, during a twitch in which there is shortening during contraction but no lengthening during relaxation, and during isometric contraction and relaxation. From data such as these it has been concluded that both muscle shortening during contraction and muscle lengthening during relaxation result in additional heat being released. The heat of shortening is represented by the dotted area.

markedly reduced in a hypoxic environment. The initial heat, on the other hand, is, at least for several twitches, little affected by hypoxia.

In summary, we have noted that skeletal muscle maintains posture and produces movement by means of a series of twitches or tetanic contractions. Each twitch is associated with an action potential and a series of reactions between actin, myosin, and ATP. These reactions are associated with a release of heat which is not immediately dependent upon the presence of oxygen. If work or shortening occurs, additional energy is required. After relaxation, the muscle remains for a period metabolically more active than before stimulation. Apparently this phase is dependent upon the presence of oxygen and results in the muscle's returning to a condition similar to that which existed before stimulation.

Chapter 6

GENERAL PRINCIPLES
THE ELECTROCARDIOGRAM

THE HEART
ELECTRICAL PROPERTIES

GENERAL PRINCIPLES

Intrinsic Control

The heart differs from a skeletal muscle such as the diaphragm in its dependence upon an extrinsic nerve supply. The diaphragm contracts about 16 times a minute in the resting person and stops contracting if its motor nerves are cut, while the heart contracts about 70 times a minute in the resting person and usually contracts more rapidly if its nerve supply is destroyed. In the diaphragm and other skeletal muscles, efferent neurons initiate a contraction; in the heart, neurons modify an intrinsic stimulation system. At the center of this system is the cardiac pacemaker cell. It, unlike any skeletal-muscle cell, has an unstable transmembrane potential, i.e., it depolarizes in the absence of an extrinsic stimulus. The frequency of its depolarization, however, can be modified by the autonomic nervous system and other factors (temperature, epinephrine, etc.).

Figure 6-1 compares the transmembrane potentials of two different types of cardiac cell—the pacemaker and the follower cell. In the *pacemaker cell* we note (1) a slow movement of the transmembrane potential toward zero (pacemaker prepotential) and (2) rapid depolarization (the beginning of an action potential). The pacemaker prepotential is apparently due to a slow leakage of Na^+ into the cytoplasm, which continues until a critical potential is reached. At this point a rapid depolarization occurs and is propagated throughout the heart (Fig. 6-4).

Normally the heart contains but one pacemaker. There are, however, other cells in the heart which have a pacemaker prepotential. These cells represent potential or reserve pacemakers. They do not normally pace the heart because they are stimulated by the action potential initiated by the dominant pacemaker before their prepotential reaches a critical firing level. If, on the other hand, the dominant pacemaker is depressed or destroyed or if the potential pacemaker becomes electrically isolated from the dominant pacemaker, it begins to fire on its own and occasionally to pace the contraction of other cells as well. Such a cell is known as an ectopic pacemaker. It should be emphasized, however, that not all the muscle fibers in the heart nor-

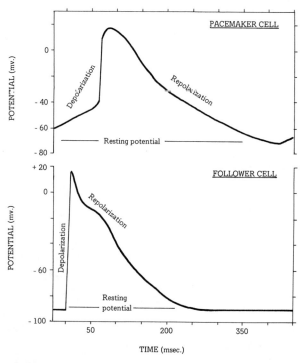

Fig. 6-1. Transmembrane potential of a pacemaker cell and a follower cell from the heart. These measurements were made possible by the development in 1946 by Graham and Gerard of hollow glass microelectrodes (1 μ in diameter) which could be inserted into a single cell.

mally have a pacemaker prepotential. Those that do not are called *follower cells* (Fig. 6-1), being dependent upon stimulation by a pacemaker.

Characteristic of both the cardiac pacemaker and follower cell is a delay in repolarization. The action potential lasts about 1 msec. in the neuron and about 10 msec. in the skeletal-muscle fiber, while it may last well over 100 msec. in cardiac-muscle fibers. This prolongation of the action potential is associated with a delay in the loss of intracellular K^+ and a lengthening of the *refractory period*.

Intercellular Conduction

An action potential in a skeletal-muscle fiber is characteristically conducted from its neuromuscular junction throughout the cell membrane of that particular muscle fiber, but there is no conduction from the muscle fiber to surrounding muscle fibers. There are some areas in the heart in which the electrical resistance between fibers is less than 5 ohm-cm.^{2}, and others in which the resistance is about 2,000 ohm-cm.^{2} The points of low resistance lie in the *intercalated disc* (Figs. 6-2 and 6-3); here an action potential can pass from one cell to the next. The heart also contains at least two groups of interconnecting cells—the atrial and ventricular groups. The two atria constitute at least one electrical continuum and the two ventricles another. Stimulation of any cell in the atrium can result in the conduction of an action potential throughout the atria, the same being true for the ventricles. Normally an action potential is initiated by a pacemaker cell in the sino-atrial node (S-A node). The potential spreads throughout the atria and is carried by means of atrioventricular conducting link to the ventricles (Fig. 6-4).

Implied in this presentation is that the heart has but one functional pacemaker. This is normally the case, but as I have already noted, other cells in the heart are capable of pacemaker activity. Usually the cell with the highest frequency of depolarization predominates; if the normal pacemaker is suppressed or if intercellular conduction is inhibited, however, additional pacemakers develop, each firing at a different frequency. While a single cardiac pacemaker produces a well-coordinated contraction of the heart associated with a good pumping action, many pacemakers firing independently of one another leads to a chaotic, incoordinate con-

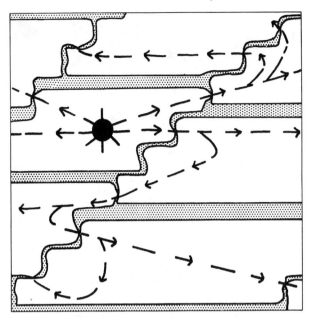

Fig. 6-2. A hypothesis on intercellular conduction in the heart. The pacemaker cell is designated by the large dot and is surrounded by follower cells. The action potential (*dashed lines*) originates from the pacemaker cell and passes from one cell to the next at points in the intercalated disc where 2 adjacent cell membranes come together. This is a point of low electrical resistance. Areas where the cell membranes (*solid lines*) are separated by intercellular material (*dotted areas*) are of higher resistance and do not normally support intercellular conduction. Note that the diagram shows multiple pathways from one cell to the next. Normally the refractory period is sufficiently prolonged to prevent the activation of any one fiber more than once in response to one pacemaker firing.

traction called *fibrillation*. In ventricular fibrillation little or no blood is pumped.

Pattern of Activation

The frequency (number of depolarizations/min.) at which the normal pacemaker (sino-atrial node) fires, as well as the velocity (meters/sec.) at which the action potentials are conducted, result from the intrinsic properties of the various cells of the heart. These intrinsic properties can be and are markedly modified by the autonomic nervous system, as well as by certain other extrinsic influences. The cardiac parasympathetic neurons of the vagus, for example, can so depress the S-A node that a different pacemaker will

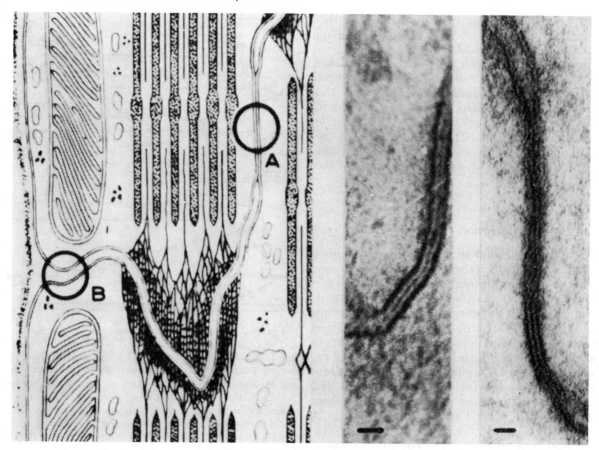

Fig. 6-3. Diagram and electron micrographs of the intercalated disc. The intercalated disc is a structure composed of 2 plasma membranes in the heart which come close to each other. At B we note an intercellular space in the intercalated disc of about 100 Å, but at A and in the right frame the cell membranes seem to fuse. This point of apparent fusion may represent an area of low intercellular resistance and, hence, an area of intercellular conduction. Calibration lines represent 200 Å. (*Reprinted with permission from* Taccardi, B., and Marchetti, G. (eds.): International Symposium on the Electrophysiology of the Heart. P. 20. New York, Pergamon Press, 1965)

originate the cardiac action potential—called *vagal escape*. Cardiac parasympathetic neurons can also slow conduction across the A-V node and in the atria. Cardiac sympathetic neurons generally increase the frequency of pacemaker discharge and hasten conduction throughout the heart.

Figure 6-5 illustrates the pattern of activation usually occurring in the heart. The action potential originates from an area (the S-A node) where the superior vena cava joins the right atrium, and is conducted throughout the right and left atria at a velocity of about 0.9 meter per second (Table 6-1). Near the end of right atrial activation the A-V junctional tissue starts to depolarize and conduct the action potential very slowly to a structure (the A-V node) lying in the interatrial septum above the tricuspid valve. It is activated when approximately two thirds of the two atria are depolarized. The *A-V node* conducts the impulse to the *common bundle*, which in turn carries it to the basal interventricular septum. It is here that the *bundle branches* are formed. The impulse is then carried down the bundle branches to the middle of the septum, where the *Purkinje fibers* are given off. They, in man and dog, extend throughout much of the ventricular endocardium.

General Principles

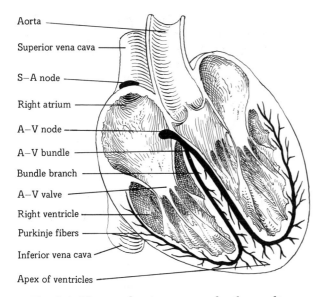

Fig. 6-4. The conduction system for the cardiac action potential. Normally the S-A node depolarizes first and then the rest of the atria. After a delay at the A-V junctional tissue, the action potential is conducted down the A-V node, to the A-V bundle, to the bundle branches, to the Purkinje fibers, and then to the right and left ventricles.

Structure	Conduction Rate (meters/sec.)
Atrium	0.9
A-V junctional tissue	0.05
A-V node	0.1
A-V bundle	2
Purkinje fibers	2
Ventricular endocardium	1
Ventricle	0.3

Table 6-1. Conduction velocities in cardiac tissue.

(see "References," Chapter 9). Briefly, however, there is an activation of the apical portion of the interventricular septum during the first 10 msec. This is from both the right and left endocardial surface to the middle, but predominantly from the left. Within 20 msec. the first action potential has reached the epicardium, located at the anterior apical surface. Within 60 msec., the entire right ventricle and most of the left ventricle have been activated. The only areas not depolarized are a small portion of muscle near the posterior lateral epicardium of the left ventricle, and a portion of the basal interventricular septum.

This activation pattern is characteristic of a normal man with a heart rate of 70 beats per minute, a right ventricle wall thickness of 4 mm., and a left ventricle wall thickness of 15 mm. In newborn infants both the right and left ven-

From this point the activation of the heart becomes a considerably more complex phenomenon. The interested student is referred to the international symposium on electrophysiology of the heart, edited by Taccardi and Marchetti

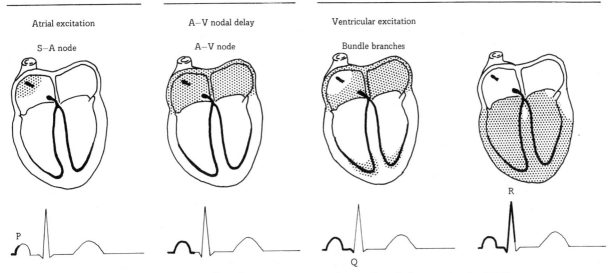

Fig. 6-5. Sequence of cardiac excitation and associated changes in the ECG.

tricular walls are approximately the same thickness and both the right and left ventricles are completely activated at approximately the same time. Increases in heart rate during exercise or anesthesia are generally associated with a more rapid activation of the atria and ventricles.

In this activation-pattern, the apex of the heart starts contracting before the base does, causing the projection of blood, very early in the cardiac cycle, toward the pulmonary and aortic valves. The specialized conducting system of the ventricles (A-V bundle and Purkinje fibers) serves two important functions. It speeds up ventricular activation and it imposes a pattern of activation on the ventricles. Of interest is the observation that the pattern of ventricular repolarization is different from that of depolarization.

Prolonged Inexcitability

In Figures 6-1 and 6-11 we note the transmembrane potentials in two different types of cardiac cell. You will note that each cell responds to a stimulus with a *period of depolarization* at least nine times longer than that for skeletal muscle. This means that each cardiac cell has a prolonged period of inexcitability which starts with depolarization and ends during repolarization. This interval of inexcitability is called the *absolute refractory period* and is followed by an interval of reduced excitability, termed the *relative refractory period*. These intervals are so long that they prevent the heart from being tetanized. The heart's responding to stimuli by means of a twitch serves it well in its function as a pump, since once the heart has contracted and ejected its contents it is important that it relax and receive more blood. The cardiac refractory period also prevents the spread of abnormally rapid rhythms. In atrial flutter, for example, the A-V node frequently conducts only every third action potential to the ventricles.

Figure 6-6 illustrates the importance of the refractory period to the contractile response of the heart. Note that for the greater part of ventricular contraction, the ventricles are refractory. Eventually a point in the cycle is reached at which the ventricle responds to stimulation, but this is now associated with a prolonged *latent period*. The duration of latency following stimulation seems, within limits, to vary inversely with the closeness of the stimulus to the absolute refractory period.

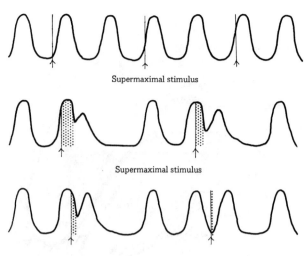

Fig. 6-6. Recordings of the contractions of a frog's ventricle. Arrows indicate the points at which a supermaximal stimulus was applied and dots indicate the period of latency. The ventricle is in an absolute refractory period during the initial phases of contraction. Stimulation during the relative refractory period is associated with a prolonged latent period. In no case did the extrinsic stimulus produce an extrasystole, but rather a premature systole followed by a compensatory pause.

The overall effect is to guarantee a certain amount of cardiac relaxation regardless of how often the stimuli are supplied. Also of interest is the fact that a single extrinsic stimulus does not necessarily produce a true extra contraction (extrasystole), but may bring about a *premature contraction* followed by a *compensatory pause* (Fig. 6-21, trigeminy).

Once the transmembrane potential has returned to normal the cardiac cell exhibits, at first, a period of supernormal excitability lasting about 0.1 sec.

THE ELECTROCARDIOGRAM

Techniques

Since the body fluids conduct electrical charge and since the spread of the action potentials throughout the atria and the ventricles are synchronous, it is possible to record by means of surface electrodes electrical potentials related to the cardiac action potentials (Fig. 6-5). Early in the century, when these techniques were

The Electrocardiogram

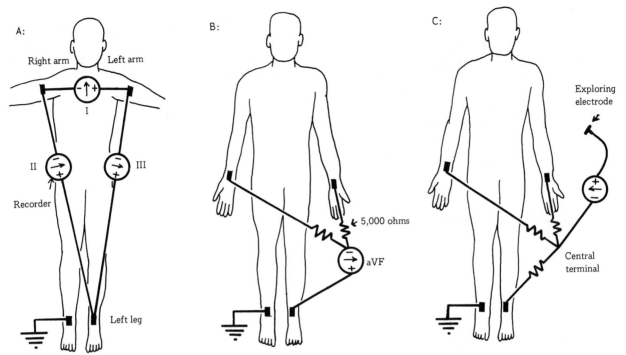

Fig. 6-7. Techniques for recording the electrocardiogram (ECG). (A) *Standard bipolar limb leads* I (left arm–right arm), II (left leg–right arm), and III (left leg–left arm). In this technique there is measurement of the voltage between 2 appendages. Each electrode goes to a pole on the galvanometer (recorder), so that a normal individual will produce positive P and R waves. (B) *Augmented unipolar limb leads.* The two arms and the left leg are used here. Two of these appendages are connected to the negative pole of the galvanometer. The third (the right arm for lead aVR, the left arm for lead aVL, and the left leg for lead aVF) is connected to the positive pole of the galvanometer. This technique sometimes yields negative P, R, and T waves in normal individuals. Lead aVF is shown above. (C) *Unipolar leads.* The exploratory electrode is placed at various chest positions (unipolar precordial leads), on the appendages (unipolar limb leads VR, VL, or VF), or elsewhere (the esophageal lead for example). The central terminal provides a fairly stable indifferent point.

being developed, an approach was used (bipolar standard limb leads I, II, and III) in which electrodes were placed on the four appendages (Fig. 6-7). The right leg electrode served to ground the patient and two other electrodes were sent to the two terminals of a recorder. In standard limb lead II, for example, the right arm electrode was attached to the negative pole and the left leg electrode to the positive pole of the recorder. Other techniques were later developed for studying the electrical output of the heart (Figs. 6-8 and 6-9).

Figure 6-9 shows the various electrocardiogram (ECG) patterns that might be obtained if 12 different recording techniques were used on a single healthy individual. Although the direction and amplitude of the various waves differ, there is a basic consistency throughout. In most cases there is a wave, arbitrarily called P, followed by an isoelectric line, a Q wave, an R wave, an S wave, an isoelectric line, a T wave, and finally another isoelectric line. This sequence is repeated each cardiac cycle. In some of the leads neither a Q wave nor an S wave may be visible, but the P, R and T waves are constant components of the normal ECG regardless of the technique employed.

Figure 6-10 introduces the terminology used in ECG studies. "Segment" refers to a period from the end of one wave to the beginning of another,

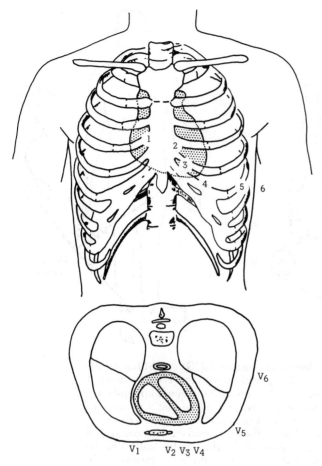

Fig. 6-8. Standard placement for the exploratory electrode in the precordial unipolar ECG.

approximately the same time as the action potential of the follower cell and ends about 40 msec. after the end of the period of rapid depolarization for this particular cell. The isoelectric line of the ECG extends for the plateau

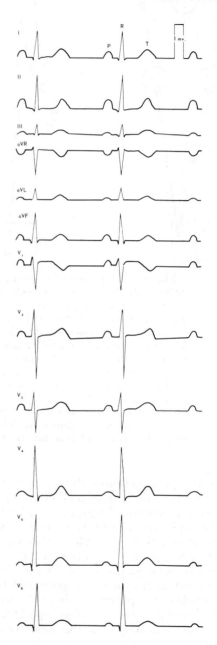

Fig. 6-9. A normal 12-lead ECG. An ECG is usually obtained at a paper speed of 25 mm./sec. (abscissa) and a sensitivity such that 1 mv. (ordinate) produces a 1 cm. vertical deflection.

while "interval" is usually more inclusive. Since the QRS complex seen in this figure is due to ventricular activation, and since a Q wave is not always present in all leads, the cardiologist frequently uses the letter R to designate the beginning of the QRS complex and may sometimes use the letter Q to designate the beginning of an RS complex not associated with a Q wave. In the following discussion I will adhere to the terminology of Figure 6-10.

Interpretation

Ventricular pattern. Figure 6-11 compares the transmembrane potential in a single ventricular follower cell with the potential recorded from lead V_3. In this particular example a P wave is not shown. Note that the QRS complex begins at

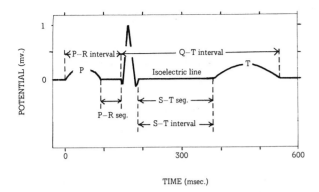

Fig. 6-10. Standard terminology for the ECG.

period of the intracellular potential and the T wave begins at approximately the same time as the period of rapid repolarization. We see that the QRS complex is not a simple uniphasic wave and that its voltage is considerably less than that

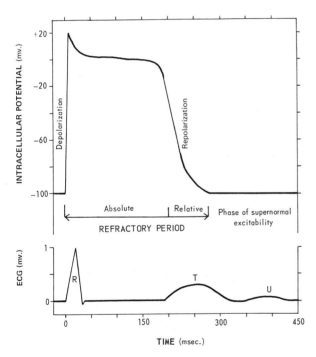

Fig. 6-11. Comparison of a transmembrane potential from a ventricular follower cell and the ECG pattern from lead V_3. Note that the QRS complex occurs at about the same time as ventricular depolarization and the T wave during ventricular repolarization. The period of depolarization in the heart is much more prolonged than in skeletal muscle due to a delayed increase in K^+ permeability, which results in a more prolonged refractory period for the heart.

of the transmembrane potential. In essence, though, the *QRS complex* does help us understand the depolarization process going on in the ventricles; its duration is a good estimate of the time it takes to activate the ventricles. Prolongation of this time may mean damage to the rapidly conducting A-V bundle or bundle branches, cardiac hypertrophy, or cardiac distension. The *T wave* occurs during ventricular repolarization. The *Q-T interval*, since it appears at a time when there is some degree of depolarization, also represents the refractory period or interval of reduced excitability for the ventricles. The T wave is sometimes followed by a *U wave* representing a short period of supernormal sensitivity (lowered threshold). The U wave is best seen in lead V_3 and is seldom noted in other leads.

The small amplitude of the R wave results from the considerable distance of the recording electrode from the source of current, and from the fact that the various cells are carrying the waves of depolarization in different directions. That we usually can characterize ventricular activation by a triphasic wave (QRS) is an indication of the rapidly changing activation-pattern in the ventricles. The Q wave in the standard limb leads is generally associated with the depolarization of a part of the apex and interventricular septum, the R wave with the depolarization of the major portion of the right and left ventricles, and the S wave with the depolarization of the base of the left ventricle and the base of the interventricular septum (Fig. 6-5). The *S-T segment* represents a period of little electrical change in the heart, i.e., the plateau between depolarization and repolarization.

Atrial pattern. In Figure 6-9 the *P waves* (waves of atrial depolarization) are shown as fairly simple unipolar waves. There is evidence, however, that with further amplification some notching is generally seen. The P wave is followed by the *P-R segment* or isoelectric line. This represents a period following complete atrial activation but during which the action potential is spreading through the A-V conduction system. Prior to this, during the latter third of the P wave, A-V conduction had been delayed at the A-V junctional tissue and the A-V node.

The period of *atrial repolarization* occurs during the QRS complex and usually cannot, therefore, be demonstrated in the ECG. It is frequently seen in cases in which the atria and ventricles have separate pacemakers or the atrial rate is unusually rapid. When noted, it is usually in a direction opposite to the P wave and is designated T_P or Ta.

Clinical Use

The ECG is used to study the sequence of activation in the heart, the time utilized for activation, and the overall voltage output of the heart. When any of these parameters changes markedly, it is usually an indication of pathology or irritation. In short, the ECG has proven an important diagnostic tool for the physician. The physiologist too, in studying abnormal patterns of activation, has gained considerable insight into the functioning of the normal heart.

Sequence. In Figure 6-12 we see a comparison between various abnormal and normal electrocardiograms. In the third pattern in the group we note the following abnormal sequence:

$$P, QRS, T, QRS, T$$

In addition, the T-R segment is markedly prolonged. This is an example of a periodic depression of the sinus (S-A node) rhythm and the subsequent development of a nodal (A-V node) pacemaker. In this particular example A-V nodal firing is associated with a QRS and T wave, but no obvious P wave.

Another abnormal sequence is seen in the example of second degree block:

$$P, QRS, T, P, P, QRS, T$$

In this case the A-V node is conducting only every other atrial wave to the ventricles. In some cases the failure of the A-V node to conduct impulses to the ventricles can be of great survival value. Note, for example, the case of a partial heart block (second degree) associated with atrial fibrillation. If most of these atrial waves were not stopped at the A-V node, the ventricular rate would be inefficient. Characteristic of atrial fibrillation is a marked ventricular arrhythmia. The fact that the A-V node generally maintains a relatively long refractory period makes it an important mechanism for preventing the spread of abnormally rapid atrial rhythms to the ventricles.

Duration. Table 6-2 contains some normal ranges of duration for various events in the ECG. In the example of first-degree block given in Figure 6-12 we note a normal sequence of events but a markedly prolonged P-R segment and P-R interval. This is usually indicative of A-V nodal depression. If this depression becomes more

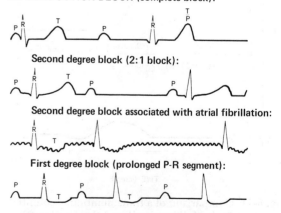

Fig. 6-12. Some abnormal changes in the sequence and duration of the ECG waves. The above patterns have been divided into the following categories: those associated solely with a sinus rhythm, those associated with an ectopic pacemaker or ectopic pacemakers, and those associated with an A-V conduction block. These designations show a considerable degree of overlap.

marked, it can lead to a second- or third-degree (complete) heart block. Another case in which knowing duration has diagnostic value is paroxysmal tachycardia. Here, following a normal ECG pattern, we note a sudden onset of high-amplitude, long-duration waves. The upright waves are the result of ventricular depolarization, and the inverted waves, of repolarization. The long duration of the upright waves is indicative of a slow rate of depolarization for the ventricles. This can be due to suppression of the A-V conducting system but is probably the result, here, of a ventricular pacemaker fairly distant from the A-V bundle and bundle branches.

A final example of the value of duration-analysis is complete heart block. Here we see an interval between adjacent P waves which is less than the interval between adjacent R waves. This is the result of an electrical isolation of the ventricles from the atria. The atria have a sinus (S-A node) rhythm and the ventricles have a separate pacemaker. The atria, as a result, may be contracting 90 times per minute while the ventricles are contracting only 30 or 40 times per minute. Characteristic of most potential pacemaker cells outside the S-A node is an intrinsic rate of depolarization of less than 40 beats per minute. So long as the S-A node sends impulses to these potential pacemakers at a frequency greater than their intrinsic rate of depolarization, they will not pace, but once the impulses to these cells are cut off, their potential as pacemakers is revealed. These conclusions are documented in Figure 6-13. In this particular example a dog

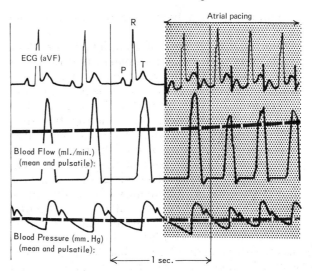

Fig. 6-13. Record from an unanesthetized dog before and during atrial pacing. The pacemaker discharge is seen just preceding each P wave. The heart rate before pacing was 126/min. When the dog was paced artificially at 155/min. the normal pacemaker cell apparently became a follower cell with an unstable membrane potential. Blood pressure was measured from the abdominal aorta and blood flow from the ascending aorta. (Courtesy of L. N. Cothran and R. S. Shepard)

had two electrodes sewn into the right atrium. After recovering from surgery he had a heart rate of 126 beats per minute. When 155 stimuli per minute were sent to the atrial electrodes the dog

ECG Event	Normal Duration (sec.)	Normal Amplitude* (mv.)	Associated Event in the Heart
P wave	<0.11	<0.3	Atrial depolarization
P-R interval	0.12-0.20		Atrial depolarization, conduction in A-V node, bundle, and bundle branches
P-R segment	<0.14		Conduction in A-V node, bundle, and bundle branches
Q wave	<0.04		Initial activation of ventricle
QRS wave	0.04-0.12	0.5-3.0	Ventricular depolarization
Q-T interval	0.35-0.43		Ventricular depolarization through ventricular repolarization (ventricular refractory period)
T wave	<0.4	<0.5	Ventricular repolarization (return of excitability)
U wave	0.1	<0.1	Supernormal excitability

* Values given in this column are for standard limb leads only.

Table 6-2. Analysis of the ECG.

responded with a heart rate of 155 beats per minute.

One of the dangers of complete heart block is that the ventricular pacemaker that develops fires so slowly that the resulting ventricular rate is inadequate to meet the body's needs. This condition is frequently treated by implanting electrodes in the ventricles and leading the electrodes to either an external electronic pacemaker or a subcutaneously implanted, battery-operated one. Figure 6-14 illustrates the surgical technique for implanting electrodes in the ventricle and a pacemaker subcutaneously in the abdomen. It also illustrates the effect of ventricular pacing on the ECG of a patient with A-V block.

Frequently the candidate for an electronic pacemaker is so debilitated that thoracic surgery would be fatal. In these cases a catheter with electrodes at the tip is placed in the jugular vein and led to the apex of the right ventricle. Fortunately, intravascular clotting has not been a major problem. A week or more later the patient is usually sufficiently strong for the permanent implantation of electrodes in the ventricle and the subcutaneous implantation of the pacemaker.

One to five years later the patient must have the batteries in his pacemaker replaced, but this is a rather simple procedure involving only local anesthesia. Originally, investigators tried leading the wires from the pacemaker electrodes to a pacemaker that hung around the patient's neck. This enabled the patient to increase his ventricular rate to 80 when he went bowling and to decrease it to 60 when he went to bed, but the associated puncture wound in the skin created a problem as far as infection was concerned and the control of rate was not judged to be worth the additional risk. More recently, investigators have been trying to control the rate by radio transmitters carried by the patient, and to coordinate atrial and ventricular contraction during ventricular pacing.

Analysis of the duration of ECG events can also be helpful in identifying the abnormal entry of an atrial action potential into the ventricles (Fig. 6-15). In this case (Wolff-Parkinson-White syndrome) the atrial action potential is not delayed appreciably before entry into the ventricles (note shortened or nonexistent P-R segment) and is not rapidly conducted throughout the ventricles at first (note low amplitude and small slope of delta wave). It has been suggested that cases such as this are due to a pathway in addition to that of the A-V bundle. Such a condition results in the ventricles' beginning their contraction before the atria finish contracting.

Voltage (the dipole). Activation of the atria involves the movement of numerous action potentials in many directions. The complexity of this movement can be appreciated by referring once again to Figure 6-2. The activation-pattern becomes even more compounded in the ventricles. Here we are dealing with a number of systems which conduct at different velocities. In addition, the thickness of the muscle mass means we must think in three dimensions rather than just two. In many phases of ventricular activation there is simultaneous movement of action potentials from right to left, from left to right, from endocardium to epicardium, from posterior to anterior, and from anterior to posterior. It has proven useful to characterize all the negative and positive charges in the heart at a particular instant in terms of a single *dipole*. A dipole is characterized as containing equal numbers of positive and negative charges separated by a small

Fig. 6-14. (A) Recording from a patient with an A-V block. Note that in one cardiac cycle there was a 1.9-sec. interval between consecutive R waves. (B) Surgical procedure for implanting a pacemaker and associated electrodes. (C) Recording from the above patient with A-V block after the implantation of a pacemaker. Note the pacemaker artifact. The prolonged duration of the R wave indicates that ventricular activation is now quite prolonged.

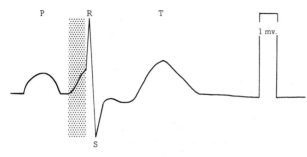

Fig. 6-15. Wolff-Parkinson-White (W-P-W) Syndrome. In this condition there is a slurring of the initial deflection of the QRS complex (delta wave) and frequently an associated shortening of the P-R segment and a prolongation of the QRS interval. Traditionally this has been explained on the basis that an abnormal pathway exists between the atria and ventricles and thus the usual delay of the atrial impulse at the A-V junctional tissue followed by rapid ventricular activation via the A-V bundle does not occur. Recently (Sherf and James, 1969), however, it has been proposed that there are 3 functional pathways connecting the S-A and A-V nodes (anterior, middle, and posterior internodal tracts) and that they are normally responsible for activating the A-V conducting system. It is hypothesized that in the W-P-W syndrome only the posterior internodal tract is functional. This tract bypasses the bulk of the A-V node to enter its lower reaches. The implications of this concept are extensive. They include the possibility of a sinus rhythm without associated P waves and aberrant ventricular conduction due to damage of parts of the atrial myocardium.

distance. I have drawn a dipole in Figure 6-16, and in Figures 6-18 through 6-20 have shown some methods for their computation. The dipole is represented as having a positively charged head and a negatively charged tail and results from a thin layer of positive and negative charges at the junction between resting and depolarized muscle. When all the cells in the heart exhibit a resting potential or when all are depolarized there is no dipole.

Voltage (the volume conductor). If we now put the dipole in a homogeneous volume conductor, we find that the potential, V, at any point in the conductor is proportional to the dipole moment, m (number of charges at each pole times the distance between poles), and the cosine of the angle (in Figure 6-16 the angle would be for recording point 2) made between the dipole and a line extending from the recording point to the center of the dipole, and inversely proportional to the square of the distance, r, from recording electrode to dipole. This relationship is expressed by the following formula:

$$V = \frac{m \cos \theta}{r^2}$$

From this formulation we see that the potential obtained at a point depends on the dipole itself and on the position of the exploratory electrode. If the electrode is near the head of the dipole we obtain a positive deflection on our recorder, and if it is near the tail, a negative deflection. It is also possible, theoretically, to find a point in the volume conductor at which the electrode is equally influenced by the head and the tail; at this point the electrode would conduct no current.

Characterizing the cardiac dipole at any instant in the cardiac cycle is infinitely more complex in the human body, though approximate descriptions have proven useful. It should be borne in mind, of course, that the body is not a homogeneous volume conductor. The *specific resistivity* for the lungs, for example, is about 2,000

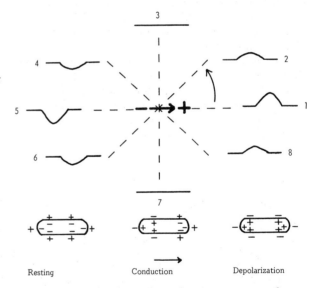

Fig. 6-16. (*Top*) The effect of a moving wave of positivity, followed by a wave of negativity (a dipole) on various electrodes placed in a volume conductor. The deflections produced are the result of attaching each exploratory electrode to the positive pole of a recorder. Note that the greatest positive deflection is recorded from electrode 1. A smaller amplitude is seen at electrode 2 and no displacement at electrode 3. Such recordings are used to define the movement of charge or charges. (*Bottom*) The movement of a wave of depolarization across a cardiac cell.

ohm-cm., for the blood about 160 ohm-cm., and for the heart as a whole about 400 ohm-cm.

Voltage (the cardiac dipole). The cardiac dipole studied by means of electrodes on the skin depends, in part, on the anatomic position of the heart and the activation pattern in the heart. In pregnancy and obesity the dipole may change because the abdominal contents push the apex further cephalad. In Figure 6-17, pregnancy might result in the apex moving toward an axis of 0°. Hence, one should bear in mind, in interpreting the dipole, the anatomical position of the heart. Normally in man the apex-to-base axis of the ventricle lies almost parallel to the diaphragm, and the right ventricle, anterior to the left.

In Figure 6-17 we note some of the patterns obtained when electrodes are placed at different points on the body. If standard lead I is defined as representing the horizontal axis (0° or 180°), then lead II will represent an axis of 60°, aVF 90°, lead III 120°, aVR −150° (or 210°), and aVL −30°. In this system, then, we are able to characterize the cardiac dipoles by means of studying potentials picked up from the surface of the skin at the two wrists and the left ankle (Fig. 6-7). In fact it would make no difference whether in lead aVF the potential were picked up from the right or the left ankle, since both appendages have their only contact with the torso at the crotch. The left leg lead is a crotch lead, in effect, the right arm lead a right shoulder lead, and the left arm lead a left shoulder lead. Connecting these three points (the two shoulders and the crotch)

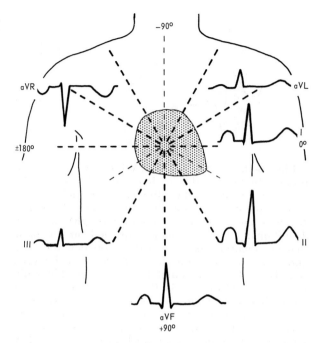

Fig. 6-17. The electrical axis of the heart.

by means of straight lines, one obtains a triangle surrounding the heart. We will refer to this imaginary triangle as *Einthoven's triangle* after the 1924 Nobel laureate who did so much in the development of ECG recording techniques.

In Figure 6-18 we see an illustration of how four different cardiac dipoles (P, Q, R, and S) might affect the voltage output from ECG leads

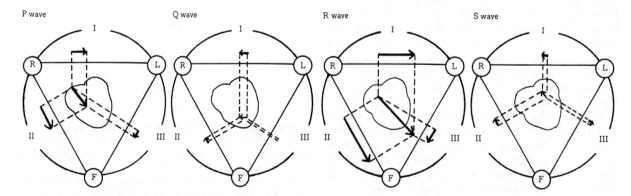

Fig. 6-18. The effect of cardiac vectors on the standard bipolar limb leads. The P, Q, R, and S waves each have been characterized as a single vector, which will be recorded differently in each of the bipolar limb leads. Note that, although the P and R waves have the same direction, the magnitude (represented by the length of the line) of the R wave is greater. This figure was derived from the data in Figure 6-17.

I, II, and III. We note that atrial activation produces a dipole moving toward the apex of the heart in such a way as to give maximum deflection in lead II and minimum deflection in lead III. The Q wave is a dipole of small moment moving from left to right in such a way as to give the most marked negative deflection (opposite to R-wave deflection) in lead I. Its moment, in other words, is less than that of the P wave and its direction quite different. The R wave is seen to have the greatest moment of all of the waves and is characterized as moving toward the apex and giving a maximum deflection in lead II. It is during the R wave that most of the action potentials are spreading from the endocardium of the right and left ventricles to the epicardial surfaces. The S wave, like the Q wave, is of low moment. It occurs during the activation of a part of the base of the left ventricle and interventricular septum.

In Figure 6-19 I have presented a technique frequently used to calculate the direction of dipole movement during ventricular activation. In this procedure a right to left movement of the

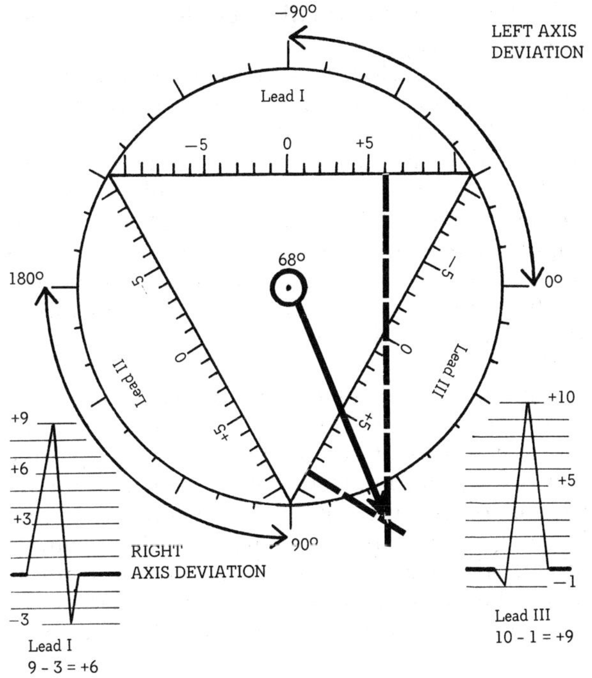

Fig. 6-19. Calculation of the mean electrical axis of the heart. In this example the mean electrical axis has been calculated from the QRS complex of lead I and III. The mean polarity for each of the QRS complexes was estimated by subtracting the amplitude of the negative deflection from that of the positive deflection. All measurements were made from the isoelectric line. The calculated mean polarity for each lead was then marked on the appropriate side of the triangle and a perpendicular line (*heavy dashed line*) drawn from each mark. Next a line was drawn (*heavy solid line*) from the center of the circle to the point where the 2 dashed lines intersected. The mean electrical axis (68°) could now be read at the point where the heavy solid line crosses the circle. An axis less than 0° is a left axis deviation and one greater than 90° a right axis deviation. It should be remembered that the mean electrical axis of the heart can vary in the absence of any change in the anatomical position of the heart.

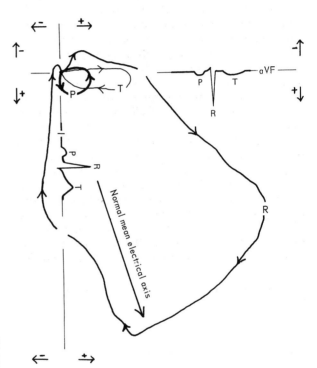

Fig. 6-20. Tracing from a frontal plane vectorcardiogram. In this record voltages from lead aVF controlled vertical deflections and voltages from lead I controlled horizontal deflections. The P wave loop (*smallest circle*) begins and ends at the isoelectric point (*large dot*). The QRS loop (*largest loop*) and T wave loop (*oblong loop*) also begin and end at the isoelectric point. In this pattern the Q wave is represented by a negative deflection above the horizontal isoelectric line and the S wave by a deflection to the left of the vertical isoelectric line. Superimposed on the vectorcardiogram are patterns from ECG leads I and aVF and an arrow indicating the major axis of the QRS loop. This is roughly the same as the mean electrical axis (+70°) that would be calculated by means of the Einthoven triangle.

dipole is represented as an axis of 0° and a left to right movement as an axis of 180°. A caudal movement represents an axis of 90°, and a cephalad movement, an axis of −90°. For mathematical purposes, Einthoven's triangle is assumed to be equilateral, the heart is assumed to be in the center of the triangle, and the body is assumed to be a homogeneous conductor. On the basis of these assumptions, many workers have computed what they call the *mean electrical axis* of the heart by using the data obtained from two of the three standard bipolar limb leads.

Voltage (the vectorcardiogram). I would emphasize that the methods discussed in characterizing the atrial dipole, the initial ventricular dipole, the major ventricular dipole, the third ventricular dipole, and the mean electrical axis, while not theoretically valid, have proven useful in diagnosis and usually yield adequate estimates. Some disadvantages of the methods we have discussed are that they emphasize but one or two points on a given ECG wave and that frequently the various leads used for the computation are not recorded simultaneously. These are serious considerations since inspiration, for example, can change the height of an R wave.

A better technique for studying the electrical axis of the heart is *vectorcardiography*. Here one usually uses an X-Y recorder or an oscilloscope. The voltage from lead I, for example, can be sent to that part of the recorder that governs the horizontal deflection,* and the voltage from lead aVF

* The horizontal axis in the scalar technique is used as a time base.

can be led to that part of the recorder that governs the vertical deflection. Time intervals can be indicated by decreasing the intensity of the electron beam at regular intervals (every 20 msec. for example). Figure 6-20 illustrates such a vectorcardiogram (VCG).

The advantages of the VCG are that it uses all the information from any two ECG leads; that changes in pattern from one cardiac cycle to the next are easily identified; and that the two leads being compared are simultaneously recorded. It is also possible, when using an oscilloscope, to record important high-frequency changes in the ECG that are not normally recorded on the direct writing recorder.

In studying electrical changes on the frontal plane, the investigator has lead I control the horizontal deflection and lead aVF the vertical. Electrical changes across the horizontal plane of the heart or through the sagittal plane are also frequently studied by placing electrodes at the appropriate points. Some vectorcardiographers have even developed three-dimensional vectorcardiograms. This has been done by means of special viewing devices or by varying the width of the trace.

Voltage (significance). In analyzing the ECG and VCG the investigator can learn much about the activation-pattern of the heart. An inverted wave, for example, frequently means that the direction of conduction has reversed. In Figure 6-12 we see an inverted P wave (in the example of a nodal pacemaker) which we interpret to mean that atrial conduction has changed from a movement toward the A-V node to one away

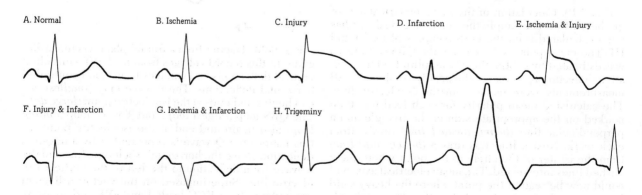

Fig. 6-21. Some abnormal voltage changes in the ECG. In the analysis of the ECG the investigator looks for inversion of the T wave (ischemia), elevation of the S-T segment (injury), prolongation of the Q wave, and changes in the amplitude of the R wave (bigeminy, trigeminy, quadrigeminy, etc.).

from it. Frequently the changes in conduction are less marked and therefore result only in a change in amplitude in the various leads. This is seen in Figure 6-21 in the example of trigeminy. The higher amplitude of the ectopic pattern results, in part, from a different pattern of activation.

In Figure 6-21 we also note an example (ischemia) of a change in the pattern of ventricular repolarization without an associated change in the pattern of depolarization. This is probably due to the fact that repolarization is more closely associated with the recovery process and therefore more dependent on coronary blood flow, as well as upon the presence of certain ions. The T waves usually increase in amplitude, for example, in hyperpotassemia, and are sometimes lower in *hypopotassemia* (Figure 6-22). Cardiac hyperemia is frequently associated with an increase in the amplitude of the T wave, and ischemia with an inversion (Fig. 6-21), a decrease in amplitude, or just a change in shape. The T wave is the most variable of the ECG waves, ranging in lead I from an amplitude of 0.05 to 0.55 mv.

Analysis of the shape of the various ECG and VCG waves or loops has also proven useful. Unilateral *cardiac hypertrophy* or *distention* due to hypertension, valvular stenosis, or valvular insufficiency can often be diagnosed on this basis. Figure 6-23, for example, illustrates how a peaked P wave can result from right atrial enlargement. In this example the enlargement resulted in prolongation of the period of activation

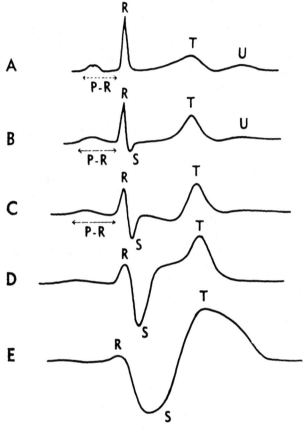

Fig. 6-22. The effect of increases in serum K concentration on the ECG. At A we see the ECG at a normal serum K level (5 mEq./liter). At B (7 mEq./liter) there is a decrease in the resting transmembrane potential of the cardiac cells, a slowing of cardiac conduction (prolonged QRS complex) and an increase in the amplitude of the T wave. By the time the concentration has doubled (10 mEq./liter) there is a disappearance of the P wave and an even more marked slowing of conduction. These symptoms can be treated by the administration of intravenous Ca^{++} or the reduction of the serum K level (Surawicz, B., and Lepeschkin, E.: The electrocardiogram in hyperpotassemia; *reprinted from* The Heart Bulletin, *10*:67, 1961; copyrighted by The Medical Arts Publishing Foundation, 1603 Oakdale Street, Houston, Texas 77004).

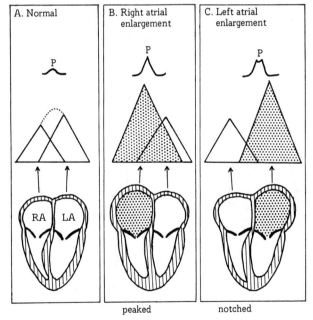

Fig. 6-23. The production of peaked and notched P waves by atrial enlargement. In A we see the normal electrical output of the right and left atria during activation. In B and C we note that changes in this output can result from right and left atrial enlargement. (*Modified from* Rushmer, R. F.: Cardiovascular Dynamics. 2nd ed., p. 283. Philadelphia, W. B. Saunders, 1961)

and an increase in the voltage output of the right atrium. Since the period of activation for the right atrium did not extend beyond the time necessary to activate the left atrium, the duration of the P wave did not change appreciably, but the sum of the dipoles associated with the right and left atrial activation did. In the same figure we note that enlargement of the left atrium produced not only an elevation of the P wave and a notching, but also a prolongation.

Analysis of the electrical axis of the heart is also useful in diagnosing unilateral enlargement. Normally the axis lies between 0 and 90° (Fig. 6-19). A deviation to the left (toward −90°) occurs in left cardiac distention or hypertrophy, and to the right in right cardiac distention or hypertrophy. If, however, the anatomical position of the heart is also abnormal, one change may cancel the other. In a person with a vertical heart associated with left ventricular hypertrophy, that is, the electrical axis might be normal. Figure 6-24 shows how unilateral ventricular hypertrophy may change a frontal plane VCG.

Etiology of abnormalities. One important cause of abnormal cardiac function is the development of an ectopic pacemaker (Fig. 6-25). This results from the failure of the normal pacemaker to dominate the other cells of the heart.

This dominance, in turn, is based on (1) the ability of the normal pacemaker to depolarize more frequently than the other cells of the body and (2) the ability of cardiac cells to conduct this depolarization throughout the heart. Unless both conditions obtain, an ectopic pacemaker may develop. If the blood pressure is increased, for example, there may be such a reflex depression of the S-A node that another pacemaker becomes dominant. On the other hand, scar tissue in the heart resulting from trauma, infection, or coronary occlusion; depression of the A-V conducting system by the vagus nerve; or generalized depression of intercellular conduction by drugs such as quinidine may separate potential pacemakers from the dominance of the S-A node and lead to the development of ectopic pacemakers. *Conduction block*, however, usually leads to the development of multiple pacemakers, whereas *sinus depression* leads to the development of a single ectopic locus.

Ectopic pacemakers also result from certain cells either becoming *hyperirritable* or depolarizing more frequently than the S-A node. Changes in the environment of the cell such as those occurring in cardiac ischemia or hypoxia, and disturbances in calcium and potassium levels can lead to the development of such hyperirritable foci — as can a number of hormones and drugs. Agents that decrease the refractory period of the heart may also facilitate the development of an ectopic pacemaker.

For a number of years physiologists have been

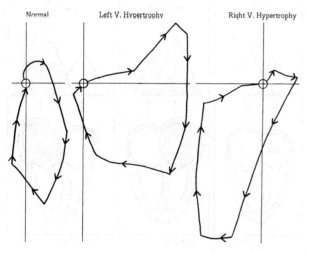

Fig. 6-24. The vectorcardiogram and ventricular hypertrophy. The above patterns represent frontal plane QRS loops in a normal individual, an individual with a left ventricular hypertrophy, and an individual with a right ventricular hypertrophy. (*Modified from* Kimura, N., and Toshima, H.: Essential difference between vectorcardiogram and electrocardiogram. Jap. Circ. J., 27:62, 1963)

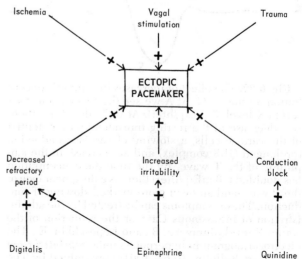

Fig. 6-25. Factors which may result in the development of an ectopic pacemaker.

discussing the mechanism or mechanisms involved in abnormal tachycardia, flutter, and fibrillation. Some feel that *circus movement,* or reentry, is an important etiological factor in at least some of these abnormal rhythms. They hypothesize that in certain arrhythmias the wave of depolarization for the heart does not disappear at the end of a single contractile cycle, but rather moves in a circle, stimulating groups of cells many times. Others emphasize that abnormal tachycardia (100-250 beats/min.) and flutter (250-350 beats/min.) are usually due to a single, rapidly firing pacemaker, while fibrillation, they feel, usually results from multiple ectopic pacemakers.

Diagnosing the abnormality in a person with chronic changes in his ECG is simpler than it is in one with periodic bouts of chest pain or dizziness. An increasingly popular diagnostic technique is the exercise ECG, in which electrodes are placed on the chest away from any skeletal muscle whose action potentials might distort the ECG; the subject is then asked to run on a moving treadmill, to pedal a bicycle, etc., while the ECG is being taken. Some investigators have even recorded ECG's on a single subject during a normal 24-hour day (Fig. 6-26). In this case the electrodes are sometimes attached to a small radio sending set around the belt and the signals picked up by a nearby tape recorder, frequently carried in a small briefcase or purse. The record is generally analyzed by a computer. Such techniques permit the study of heart function during both stress and relaxation.

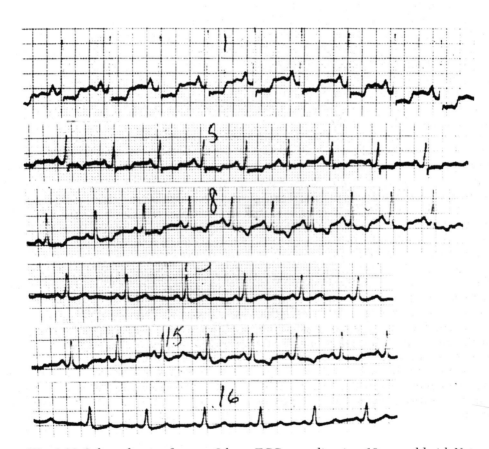

Fig. 6-26. Selected strips from an 8-hour ECG recording in a 19-year-old girl. Note the varying degrees of depression of the S-T interval and the T wave throughout the recording period. This woman complained of "tightness in the chest and palpitations" but prior to the 8-hour recording there were no clinical or laboratory data to indicate pathology. The woman was diagnosed to have "vasoregulatory aesthenia." (Aranaga, C. E., et al.: Eight-hour electrocardiogram: Technique and clinical application. Brit. Heart J., 29:350, 1967)

Chapter 7

REVIEW
CARDIAC PERFORMANCE
CARDIAC NUTRITION
SEMILUNAR VALVES
ATRIOVENTRICULAR (A-V) VALVES
CARDIAC CYCLE
COMPARISON OF THE RIGHT AND LEFT HEART
CARDIAC CATHETERIZATION
HEART SOUNDS
MURMURS

THE HEART—CONTRACTION

REVIEW

In Chapter 5 I described contraction as a series of reactions between actin and myosin, associated with the release of heat and the production of tension. We have noted that a single muscle fiber can change (1) the amount of force or tension it develops, (2) its degree of shortening, (3) its duration of contraction, (4) its rate of change of force, and (5) its velocity of shortening. These changes result from variations in the muscle fiber's environment, variations in its length prior to stimulation, and variations in the load it lifts or the resistance it must overcome. In the heart, such changes are also effected by the alteration of autonomic nervous system tone.

In Chapter 6 I emphasized some of the differences between skeletal muscle and heart muscle. The heart, unlike skeletal muscle, has an intrinsic stimulator or pacemaker, conducts action potentials from one cell to the next, and does not exhibit tetanic contractions. An area in the right atrium, the sino-atrial node, normally contains the cell which, due to the instability of its membrane potential, paces the entire heart. It is the periodic depolarization of this cell that results in the conduction of action potentials throughout the atria and to the junctional tissue of the atrioventricular node, which, after a delay, carries the stimulus to the A-V conducting system. We will be concerned in this chapter with the phenomena resulting from these waves of depolarization.

CARDIAC PERFORMANCE

The heart, like skeletal muscle, acts as a physiological transducer, converting nutritive energy into heat and tension. The tension (dynes/cm.) eventually results in the ejection of blood. We can see the relationship between intraventricular pressure and the ejection of blood (change in the volume of the lumen of the ventricle) in Figure 7-1. This curve can be divided into four segments:

1. Curve 1 represents a period of *ventricular filling*. Here the ventricles are increasing in size but there is little change in intraventricular pressure.

2. Next (Curve 2) there is a period (*isovolumic contraction*) in which the ventricles are contracting and rapidly increasing their intraluminal pressure, but not yet ejecting blood.

3. Curve 3 designates the period of *ejection*. During the ascending part of the curve the blood is being rapidly (see wide separation of 20-msec. markers) pumped to the arteries. During the descending part of the curve we have the period of reduced ejection (note smaller separation of 20-msec. markers).

4. Curve 4 represents the period of *isovolumic relaxation*. Here the intraventricular pressure is rapidly declining and the size of the ventricle is static. We note, then, that the ventricles produce their greatest pressure change (1) prior to ejection and (2) prior to filling.

The function of the ventricle is to pump blood from a low-pressure system (veins and atrium) to a high-pressure system (arteries) (Fig. 7-11). This function is facilitated by valves which prevent blood from passing from the ventricle into the atrium (atrioventricular valves) or from the arteries back into the ventricle (semilunar valves). The ventricle first creates pressure, which eventually forces open the semilunar valves, in turn producing flow into the arteries and an associated decrease in the volume of the ventricle. Though cardiac performance can be and sometimes is described in the same terms as skeletal-muscle performance (force, shortening, and duration), it is generally more con-

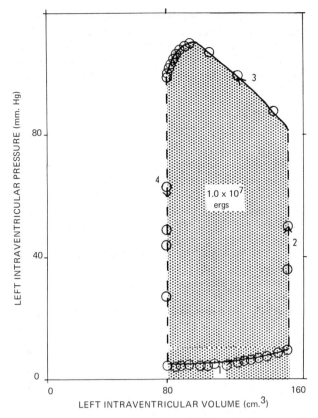

Fig. 7-1. Work of the left ventricle. The horizontal deflections of a recorder are controlled by size transducers around the surface of the left ventricle and interventricular septum. The vertical deflections are controlled by a pressure transducer in the left ventricle. The circles have been added to the deflections every 20 msec. In this example, the left ventricle pumped 80 cm.^{3} of blood from a pre-ejection intraluminal volume of 160 cm.^{3}. The total elevation in pressure in the ventricle was 105 mm. Hg. Prior to ventricular contraction, ventricular filling had been associated with a 5-mm. Hg increase in pressure. The work done by the heart (7500 mm. Hg-cm.^{3} or 1.0×10^7 ergs) is represented by the shaded area. The work done on the heart during filling is represented by the shaded area below line 1. This curve consists of 4 phases: (1) ventricular filling; (2) isovolumic ventricular contraction; (3) ventricular ejection; and (4) isovolumic relaxation.

venient to use terms of pressure, volume change, and duration.

The rate at which the heart develops pressure (dP/dt) is usually examined prior to ejection during isovolumic contraction. On the other hand, the acceleration of blood flow from the heart (dI/dt in ml./sec./sec.) is studied early during the period of ejection. By the time we reach the descending part of Curve 3 in Figure 7-1 (reduced ejection), deactivation of the contractile elements has begun and much of the flow is merely inertial. There is some evidence that about one third of the way through the period of ventricular contraction some parts of the ventricle have already begun to relax.

The more significant aspects of cardiac performance may be summarized as follows:
1. Intraventricular pressure
   a. Mean ejection pressure—Average pressure (in mm. Hg) in ventricle during the ejection of blood
   b. Peak systolic pressure—Greatest pressure (in mm. Hg) in the ventricle during a single cardiac cycle
   c. Peak rate of change of pressure—Greatest rate at which pressure is developed (dP/dt in mm. Hg/msec.) during isovolumic contraction
   d. Tension-time index—Average pressure developed by a ventricle during contraction times the duration of contraction (mm.Hg-sec.)
2. Ejection of blood or change in volume
   a. Stroke volume—The amount of blood ejected (ml.) by a ventricle in one cardiac cycle
   b. Stroke volume index—The stroke volume divided by the surface area of the skin (ml./meters^2) or the body weight (ml./kg.)
   c. Mean ejection rate—Stroke volume divided by the duration of ejection (ml./sec.)
   d. Peak rate of change of flow—The greatest rate at which flow is developed (dI/dt in ml./sec./sec.) during a cardiac cycle
   e. Cardiac output—The amount of blood pumped from a ventricle in one minute (liters/min.)
   f. Mean ejection velocity—The average speed of the blood (cm./sec.) as it passes the semilunar valves of a ventricle
   g. Peak rate of change of velocity—The greatest rate at which velocity is developed (dv/dt in cm./sec./sec.)
   h. Duration of ejection—The time (sec.) during which blood is leaving a ventricle during a single cardiac cycle
   i. Heart rate—The frequency of cardiac cycles (beats/min.) based on the interval represented by one cardiac cycle or the number of cardiac cycles in a period of 10 to 60 sec.
3. Energetics
   a. Energy expenditure
   b. Nutritional requirement
   c. Efficiency

Cardiac Work

The work performed by skeletal muscle (W in dyne-cm., or ergs) is generally calculated from the force exerted by the muscle (F in dynes) and the distance the muscle shortens (dL in cm.):

$$W = (F)(dL)$$

The work of a ventricle is usually calculated from the pressure it develops (P in dynes/cm.^{2}) and the decrease in volume it undergoes (dQ in cm.^{3}):

$$W = (P)(dQ)$$

In practice, the ventricle during a single contraction exerts not a single pressure but many different ones. If, however, we know the pressures developed by the ventricle (P_1, P_2, etc.) and the volume change associated with each pressure (dQ_1, dQ_2, etc.) we can calculate the work of the heart:

$$W = (P_1)(dQ_1) + (P_2)(dQ_2) + \text{etc.}$$

or

$$W = \int P(\text{mm. Hg}) \times dQ(\text{cm.}^3) \times \frac{1330(\text{dynes/cm.}^2)}{(\text{mm. Hg})}$$

This is equivalent to obtaining the area under the pressure-volume curve (shaded part of Fig. 7-1). Work may be computed for 1 cardiac cycle (*stroke work*) or for 1 min. (cardiac work or minute work). It can also be expressed in terms of the body weight or body surface area (*stroke work index or cardiac index*).

In the absence of the appropriate equipment the above calculations may be rather time-consuming; some investigators have therefore used the following shortcut:

$$W = (\overline{P})(dQ)$$

Here $\overline{P}$ is the average pressure during ejection and dQ the total volume of blood ejected from the ventricle.

The right ventricle pumps the same volume of blood as the left, but does so by producing approximately one fifth as much pressure. Thus the stroke work performed by the left and right ventricles is about 1.0×10^7 ergs plus 0.2×10^7 ergs, or a total of about 1.2×10^7 ergs.

Mechanical Efficiency

The heart, like any other muscle, has only a limited amount of energy available to it. The efficiency with which it uses this energy, as well as the amount available, are important limiting factors in exercise, and in coronary arteriosclerosis may determine whether the heart continues to function in the resting individual. Efficiency can be calculated using the following formula:

Mechanical efficiency (%) =
$$\frac{\text{work performed (ergs)} \times 100}{\text{total energy released (ergs)}}$$

The total energy released can be calculated on the assumption that it is equal to the total heat released (1 kilocalorie = 4.19×10^{10} ergs) plus the work performed. We can also estimate the total energy expenditure by first measuring the oxygen consumption of the heart and then determining the *caloric equivalent* for that amount of oxygen. On an average, when 1 liter of oxygen at standard temperature and pressure is burned in the heart, 4.86 kilocalories (kcal.) of energy are released. This value will vary with the nutrients which the heart catabolizes. The approximate value for 1 liter of oxygen when only carbohydrate is catabolized is 5.0 kcal. The caloric equivalent when only fat is catabolized is 4.7 and when only protein is catabolized is 4.5 kcal. per liter of oxygen. In the following formula the mechanical efficiency is calculated using 2.075 kg.-meters as the *energy equivalent* for 1 ml. of oxygen (1 kg.-meter = 2.34×10^{-3} kcal.):

Mechanical efficiency (%) =
$$\frac{\text{worked performed (kg.-m./min.)} \times 100}{\text{O}_2 \text{ consumption (ml./min.)} \times 2.075 \text{ (kg.-m./ml.O}_2\text{)}}$$

Usually the efficiency will not exceed 10 to 15 per cent.

The work performed by the heart will vary with the ventricular stroke volume, ejection pressure, and rate. You will note, for example, in Figure 7-2 that decreases in stroke work are produced by decreases in stroke volume (see mitral stenosis) and that increases in stroke work are produced by increases in stroke volume (see aortic and mitral insufficiency) and ejection pressure (see aortic stenosis).

During exercise the heart exhibits (1) an increase in the stroke work associated with an increase in the stroke volume, (2) an increase in the minute work associated with an increase in heart rate and stroke volume, and (3) a decrease in efficiency. In other words the heart is now expending more energy per liter of blood

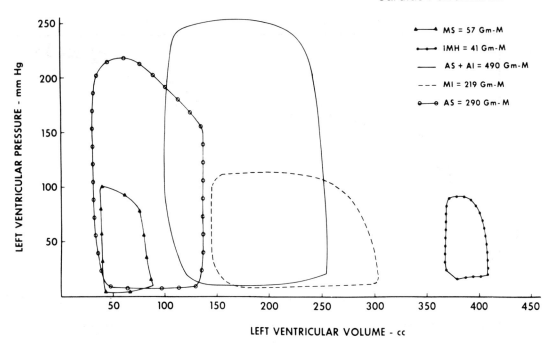

Fig. 7-2. Left ventricular pressure-volume curves from patients with mitral stenosis (MS), idiopathic myocardial hypertrophy (IMH), aortic stenosis and insufficiency (AS and AI), mitral insufficiency (MI), and aortic stenosis (AS). The volume of the left ventricle at different stages of a cardiac cycle was calculated from pictures obtained using the technique of biplane angiocardiography. Normally the heart performs about 100 gram-meters of work/stroke. Note that in mitral stenosis and idiopathic myocardial hypertrophy the stroke volume is reduced and, as a consequence, the stroke work is reduced. In mitral insufficiency, on the other hand, the ventricle changes markedly in size during systole due to the regurgitation of blood past the mitral valve and into the atrium and therefore the stroke work is greater than normal (219 Gm.-m.). Stroke work in aortic stenosis (290 Gm.-m.) is greater than normal because of the high pressures which the ventricle develops in this condition. (Dodge, H. T., et al.: Usefulness and limitations of radiographic methods for determining left ventricular volume. Am. J. Cardiol., 18:20, 1966)

pumped than at rest, but in the normal individual the extra amount of blood pumped is usually worth the extra cost. (Analogously, the driver of a car that gets 15 miles per gallon at 25 m.p.h. and 10 miles per gallon at 50 m.p.h. will nevertheless drive at a speed of 50 m.p.h. if he is in a hurry and if his car is in good condition.) Conditions that produce a more serious decrease in efficiency include flutter, fibrillation, and excessive cardiac distention (congestive heart failure).

Cardiac Power

Another method of studying the heart is by simultaneously recording the pressure in the left ventricle and the flow in the ascending aorta. Such an investigation is presented in Figure 7-3. If one also integrates the pressure developed by the left ventricle (P) with the flow in the ascending aorta during ejection (I) one can calculate the power (W/dt) developed by the ventricle during ejection:

$$(W/dt) = \int P(mm.\ Hg) \times I\ (cm.^3/sec.) \times \frac{1330\ (dynes/cm.^2)}{(mm.\ Hg)}$$

Power calculated in this way would be in ergs per second. If we now multiply the left ventricular power by the duration of ejection, we obtain the *work* performed by the left ventricle. Data ob-

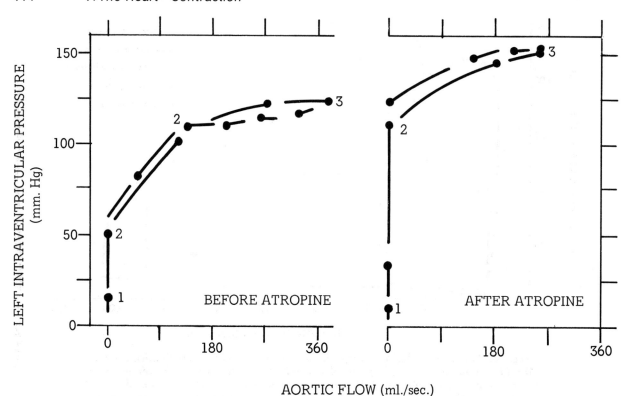

Fig. 7-3. Relationship between left intraventricular pressure and aortic flow (approximately equals cardiac output) in a conscious dog before (*left*) and after (*right*) the intravenous injection of atropine (0.25 mg./kgm.). Prior to the atropine the heart rate was 71/min. and the stroke work was 0.55×10^7 ergs. After atropine the heart rate rose to 196/min. and the stroke work decreased to 0.43×10^7 ergs. The period of isovolumic ventricular systole begins at point 1 and ends at 2. The period of ejection begins at 2 and reaches a maximum at 3. The patterns are interrupted every 20 msec. and are recorded from an electromagnetic flowmeter probe (60 mm. circumference) around the ascending aorta and a catheter in the left ventricle. The electrocardiogram was normal. (Courtesy of R. L. Green and R. S. Shepard)

tained in this way is comparable to that obtained by integrating intraventricular pressure with the change in ventricular volume.

CARDIAC NUTRITION

We have noted that during skeletal muscle contraction heat is released in response to (1) activation, (2) shortening, and (3) an increased load. In cardiac muscle we find that the oxygen consumption (ml. O_2/min.) is increased when the heart rate, stroke volume, or mean ejection pressure increases. Figure 7-4 shows the relationship between oxygen consumption (ordinate) and the product of heart rate and the pressure developed by a heart (HR × dP) that was neither ejecting nor receiving blood. Apparently, under conditions in which there is no marked shortening of the cardiac fibers, the energy requirement of the heart is well correlated with this product (HR × dP).

The degree to which the needs of the heart can be met is largely dependent upon (1) the vascular system of the heart and (2) changes in the efficiency of the heart. The fibers of atria and ventricles, unlike those of skeletal muscle, contract a minimum of about 60 times a minute. In the resting athlete this rate may be somewhat less. In strenuous exercise it is usually more than double. Yet in exercise, the heart, unlike skeletal muscle, does not appreciably increase its *arteriovenous oxygen difference*. In addition,

Cardiac Nutrition   115

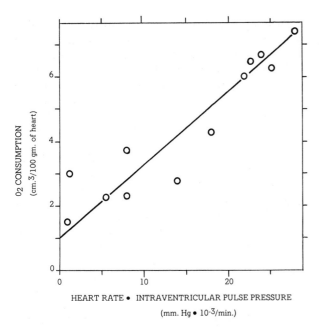

Fig. 7-4. Relationship between oxygen consumption and the product of heart rate and intraventricular pulse pressure in a dog whose left ventricle is contracting isovolumically. Note that even when the heart is not contracting (asystole) some O_2 is being consumed. During hypoxia the ratio of (oxygen consumption)/(H.R. · dP) decreases. It will also change in response to catecholamines (norepinephrine and epinephrine), Ca^{++}, and cardiac glycosides (digitalis and ouabain). (*Redrawn from* Katz, L. N.: Recent concepts of the performance of the heart. Circulation, *28*:127, 1963; by permission of the American Heart Association, Inc.)

pouchings—the aortic sinuses or sinuses of Valsalva. These sinuses prevent the occlusion of coronary flow when the aortic valves are open. Most of the coronary blood passes to the coronary capillaries. These have a lumen diameter of about 6 μ and a *permeability* greater than that of the capillaries of skeletal muscle. In the adult there is generally 1 coronary capillary for each muscle fiber, but in the newborn, where the muscle fibers are not so thick, the ratio is about 1 to 6.

The coronary capillaries drain into venules, which lead to veins roughly paralleling the coronary arteries. Most of these veins lead to the *coronary sinus*, which empties into the right atrium near the point of entrance of the inferior vena cava. A few smaller vessels, the anterior cardiac veins, empty directly into the right atrium, and a small amount of blood may be carried from the capillaries to either ventricle by way of the thebesian veins (venae cordis minimae).

It is also recognized that blood from the coronary arteries may pass to the vasum vasorum of the pulmonary arteries (extracardiac arteries) or directly back into the chambers of the heart by way of small connections between the coronary arterioles and the cardiac lumen (arterioluminal

its ability to accumulate an oxygen debt is very limited. All this adds up to the important fact that the heart is very susceptible to hypoxia (low concentrations of O_2) and extremely dependent upon a capillary blood supply which both meets its minute-to-minute needs at rest and can be rapidly elevated to meet increased metabolic needs. You can place a tourniquet on a limb for 30 min. without damage to its skeletal muscle, but if you cut off the blood supply to the heart for 10 min. you are in danger of damaging the heart (Table 17-2).

Anatomy of the Nutritional System

The heart receives its blood supply from the right and left coronary arteries, which originate at the beginning of the ascending aorta (Fig. 7-5). At this point the aorta contains three out-

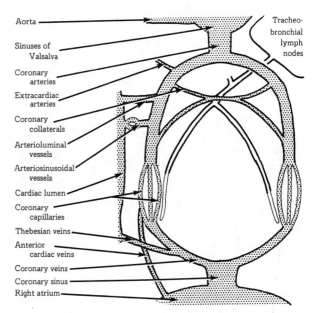

Fig. 7-5. Diagram of cardiac blood and lymph flow. The term cardiac lumen refers to either of the atria or ventricles. The lymph capillaries and ducts (*unshaded*) have not been labeled.

vessels). More numerous, however, are the arteriosinusoidal vessels. They too originate at the coronary arterioles, but form permeable sinusoids 50 to 250 μ in diameter in the endocardium and empty directly into the chambers of the heart. The heart also drains some of its intercellular fluid into *lymph ducts*. The major lymph duct of the heart passes under the arch of the aorta and into the tracheobronchial lymph nodes.

The anastomoses (coronary collaterals) between the right and left coronary arteries probably provide some protection against coronary ischemia (inadequate blood supply) due to a coronary blood clot or *coronary spasm*, but the heart, like the brain, is rather vulnerable to circulatory obstruction. Acute coronary occlusion in the human heart may lead to the death of the tissue (infarction) supplied by that portion of the vasculature. On this basis, the microcirculation of the heart is generally regarded as showing very little overlap.

In contrast, it is interesting to note the effect of a slow, progressive narrowing of a coronary artery. Under these circumstances, collateral circulation develops sufficiently to prevent cardiac damage. Increases in the intercoronary collateral channels have been reported in diseases associated with cardiac hypertrophy, valvular disease, chronic anemia, and narrowing of the coronary artery.

Another indication of the importance of oxygen to the heart is the large number of mitochondria found in its muscle fibers (Fig. 7-6). These mitochondria constitute an effective system for rapid oxidative phosphorylation.

Figure 7-7 illustrates variations in coronary flow during a single cardiac cycle.

You will note that during ventricular systole, coronary flow is minimal. In fact, the major portion of the left coronary artery may show a retrograde flow during ventricular systole, or in other words, when systolic pressure is at its greatest in

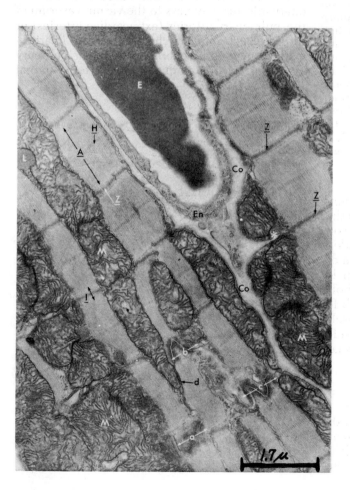

Fig. 7-6. An electron micrograph of 2 adjacent cardiac muscle cells from the bat. Note the capillary and its erythrocyte (E) as well as the connective tissue (Co) fibrils which separate these 2 cells. The endothelial (En) cells of the capillary contain numerous vesicles which may be important in transport of materials across the capillary endothelium. Note also the invagination of the sarcolemma at the Z line. It is here that the transverse system of tubules opens to the intercellular fluid. Note also the mitochondria (M) and associated lipid droplets (L). An intercalated disk is seen at a, b, and c. The distance between 2 adjacent Z lines in the myofibril is 1.7 μ. (Porter, K. R., and Bonneville, M. A.: An Introduction to the Fine Structure of Cells and Tissues. 3rd ed., plate 39. Philadelphia, Lea & Febiger, 1968)

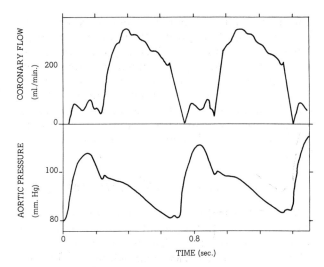

Fig. 7-7. Relationship between coronary flow and aortic pressure in the dog. Note that coronary flow, unlike flow elsewhere in the body, is minimal during peak aortic pressure. (*Modified from* Gregg, D. E., and Fisher, L. C.: Blood supply to the heart. *In* Hamilton, W. F. (ed.): Handbook of Physiology. Sec. 2 (Circulation), vol. II, p. 1551. Washington, American Physiological Society, 1963)

the ventricles, the coronary arteries have been compressed by the contracting ventricular musculature. In Fig. 7-7, the period of coronary ischemia lasts about one third of the cardiac cycle. It has been estimated that approximately 70 to 80 per cent of the coronary flow occurs during diastole.

Clinical Significance of the Nutritional System

Clinically, there has been a great deal of interest in coronary blood flow. Cardiac ischemia can result from coronary atherosclerosis, thyroid disease, hyperinsulinism, diabetes, adrenal dysfunction, gout, myocardial infarction, cardiac distension, coronary spasm, and hypertension. It is frequently associated with acute precordial pain (*angina pectoris*) similar in nature to that experienced in a forelimb when it is exercised after the blood supply has been cut off by a tourniquet. The pain probably results from the stimulation of pain fibers by accumulated metabolites or by a decrease in pH.

Cardiac ischemia can also produce any one of many abnormal cardiac rhythms, ranging from paroxysmal ventricular tachycardia to ventricular fibrillation. Treatment may consist merely in decreasing the myocardial oxygen demand, or in increasing coronary blood flow or the oxygen-carrying capacity of the blood. Increases in blood flow can sometimes be produced by coronary dilators such as the nitrites (isoamyl nitrite) and the nitrates (nitroglycerin). Decreasing the oxygen demand can be accomplished by restricting activity, eliminating such cardiac stimulants as caffein, lowering arterial blood pressure, and by the use of tranquilizers and cardiac depressants such as quinidine. Continued hypoxia may cause irreversible damage.

Cardiac Nutrients

The heart is capable of catabolizing glucose, lactate, pyruvate, esterified and nonesterified fatty acids, and to lesser extent acetate, ketone bodies, and amino acids. Soon after a meal or after the intravenous infusion of glucose the respiratory quotient (R.Q. = CO_2 produced/O_2 used) of the heart is about 0.9 and the main nutrients catabolized are glucose, lactate, and pyruvate. After an overnight fast, on the other hand, the R.Q. sinks to 0.8, and after a more prolonged fast reaches 0.7. Under these conditions primarily fatty acids are catabolized (Table 7-1). In its capacity to utilize fatty acids as a source of energy the heart resembles skeletal muscle. It differs from skeletal muscle, however, in its greater ability to withdraw lactate from the circulation and to use it as an energy source (p. 457).

Despite the considerably higher concentration of glucose in the blood (Table 15-1), the heart (under testing conditions) removes almost as much lactate from the blood as it does glucose. There is, as mentioned in Figure 5-14, a marked outpouring of lactate into the blood from active muscles during exercise. Hence the role of lactate in supplying energy to the heart is probably much more important during exercise than it is at rest.

SEMILUNAR VALVES

Two sets of valves separate the arteries from the ventricles (Fig. 7-8). On the right are the pulmonary valves and on the left the aortic valves. These are known collectively as the semilunar valves. They are open when the energy level of blood tending to leave the ventricle is greater than that entering it from the artery. Conversely,

	Carbohydrate(%)		Noncarbohydrate(%)
Glucose	17.90	Fatty acids	67.0
Lactate	16.46	Amino acids	5.6
Pyruvate	0.54	Ketones	4.3
Total	34.90	Total	76.9

Table 7-1. Contribution of various nutrients to myocardial oxygen usage. Note that under the conditions of the experiment the heart uses 67 per cent of its oxygen to metabolize fatty acids. The data were obtained from a subject in the postabsorptive state by sampling from catheters in the coronary sinus (coronary venous blood) and in a systemic artery. Coronary blood flow was determined by a nitrous oxide technique. You will note that the sum of the different nutrients accounts for more than 100 per cent of the simultaneously measured oxygen consumption. This is due to the storage of incompletely oxidized substances. The nutrients extracted from blood by the heart depend, in part, on their concentration in the blood and, in part, on the concentration of various hormones in the blood. The heart usually cannot extract glucose at concentrations less than 60 mg. per 100 ml. of plasma. This threshold is lowered in response to an injection of insulin and elevated in diabetes mellitus. The quantity of glucose extracted increases progressively as the plasma level increases from threshold up to 100 mg. per 100 ml. of plasma (maximum for the heart). The threshold for lactate is 1.6 mg. and for pyruvate 0.4 mg. per 100 ml. of plasma. (Bing, R. J.: Myocardial metabolism. Circulation, 12.637, 1955)

they close when the energy level toward the ventricle is greater than that from the ventricle. We are concerned here with three types of energy:

1. Pressure (potential) energy (E_p) — The product of the pressure (P) and the volume of blood moved by that pressure: E_p (ergs) = P(dynes/cm.^{2}) × Q (cm.^{3})

2. Gravitational (potential) energy (E_g) — The weight (w) of the blood between 2 points times the difference in height of the 2 points (h): E_g (ergs) = w(dynes) × h(cm.)

3. Kinetic energy (E_k) — One half the product of the mass of blood moved (m) and the square of its velocity (v): E_k (ergs) = ½ m(gm.) × v^2 (cm./sec.)^2

In the initial period of ventricular ejection the pressure in the left ventricle of a reclining person exceeds the pressure in the ascending aorta and the velocity of blood flow increases (Fig. 7-9), i.e., the blood acquires more kinetic energy. During the period of reduced ejection the pressure in the ascending aorta becomes greater than that in the ventricle, but the valves remain open because now the kinetic energy of the blood leaving the heart more than compensates for the negative pressure gradient. Eventually the kinetic energy becomes so low that the valves close.

ATRIOVENTRICULAR (A-V) VALVES

Two sets of valves also separate the atria from the ventricles. On the right is the *tricuspid valve* and on the left the *bicuspid* or *mitral valve*. These A-V valves are open when the energy level in the atrium exceeds that in the ventricle, and closed when the energy level in the ventricle exceeds that in the atrium. They generally open in response to ventricular relaxation and close near the end of atrial contraction. They are attached by means of collagenous fibers (the *chordae tendinae*) to the *papillary muscles* (Figs. 7-8 and 6-4). The papillary muscles are activated early during ventricular contraction and, by shortening, maintain tension on the A-V valves at a time when the intra-ventricular pressure is markedly higher than the intra-atrial pressure. Also at this point in the cycle the apex-to-base dimensions of the heart are decreasing.

The chordae tendinae and papillary muscles effectively prevent the flapping of the A-V valves into the atrium and the regurgitation of blood from the ventricle. Regurgitation into the atrium is most completely prevented if the start of atrial contraction precedes ventricular contraction by less than 147 msec. During atrial contraction blood is rapidly pushed past the A-V valves and into the ventricles. The kinetic energy which the blood acquires at this time apparently keeps the A-V valves open while the ventricular pressure acting on the valves is slowly rising. For a short period, blood flows against a pressure gradient. This may prevent regurgitation more effectively than the closing of the valves solely on the basis of ventricular contraction. Thus atrial contraction apparently serves not only to move blood rapidly into the ventricle, but also prevent regurgitation. If atrial contraction precedes ventricular contraction by too great an interval, the valves may close before the ventricles begin to contract. If the interval is too short, the periods of atrial and ventricular contraction overlap.

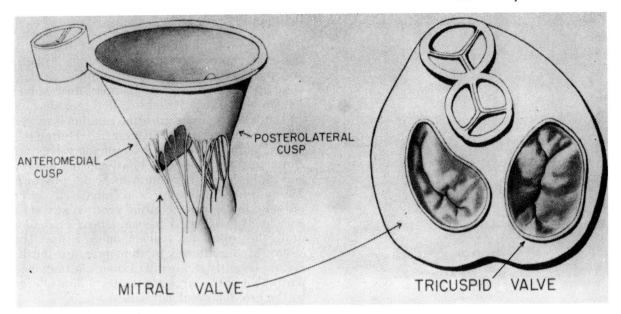

Fig. 7-8. The semilunar and atrioventricular valves. (A) Lateral view of semilunar (*left*) and atrioventricular (*right*) valves. (B) Superior surface of closed valves. These valves effectively prevent the backward flow of blood, but do not normally impede its forward movement. (*After* Spalteholz: *in* Rushmer, R. F.: Cardiovascular Dynamics. 2nd ed., p. 37. Philadelphia, W. B. Saunders, 1961)

CARDIAC CYCLE

A heart contracting 75 times a minute has a cycle of 0.8 sec.:

$$\frac{60 \text{ sec./min.}}{75 \text{ cycles/min.}} = 0.8 \text{ sec./cycle}$$

During this 0.8 sec. the atria contract (*atrial systole*), causing a small pressure increase in the atria, the veins leading to them, and the ventricles. It also results in a movement of blood to the ventricles and, since there is no effective veno-atrial valve, a movement of blood back to the veins. Atrial relaxation and recovery (known collectively as *atrial diastole*) follow and are associated with a movement of blood from the veins to the atria. Also during this 0.8 sec. there is a ventricular contraction (systole) which results in the movement of blood into the arteries and an increase in intra-arterial pressure. *Ventricular systole* is associated with marked compression of the coronary vessels and a reduction in coronary blood flow. Ventricular relaxation and recovery (*ventricular diastole*) result in a decrease in intraventricular pressure, an increase in coronary blood flow (Fig. 7-7), and a movement of the blood that has accumulated in the atria and veins during ventricular systole into the ventricle.

Some of the changes occurring during a single cardiac cycle are summarized in Figure 7-9 and discussed below. We will concern ourselves first with those changes involving the left heart.

Ventricular Systole (0.27 sec.)

Isovolumic contraction (0.05 sec.). The first part of ventricular systole usually begins with the closure of the A-V valves and ends with the opening of the semilunar valves. During it, there is a rapid increase in intraventricular pressure. Some muscle fibers shorten and others lengthen, but, since the A-V valves and semilunar valves are both closed, the volume of the ventricle is relatively constant (Fig. 7-10). Initially there is a slight increase in intra-atrial pressure (the *c* wave, due to the movement of the A-V valves toward the atrium), but it soon decreases (the *x* wave) as a result of atrial relaxation.

Ejection (0.22 sec.). When the pressure in the ventricle exceeds that in the aorta, the semilunar valves open and the blood flow from the ven-

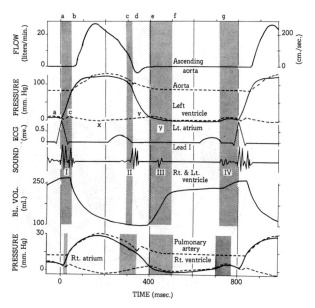

Fig. 7-9. The cardiac cycle in an individual with a heart rate of 75 beats/min. These tracings represent estimates of the cardiodynamics in a 75 kg. man. The upper tracing was obtained from a square wave electromagnetic flow meter transducer around the ascending aorta of an unanesthetized dog. The output of the transducer has been calibrated in terms of flow volume (ml./min.) and flow velocity (cm./sec.). In order to calculate the stroke volume (60 ml.) from this deflection, one multiplies the average flow during ejection (16,000 ml./min.) by the duration of ejection (0.0037 min.). Note that the cardiac cycle has been divided into 7 parts: (a) isovolumic ventricular systole; (b) systolic ejection; (c) protodiastole; (d) isovolumic ventricular relaxation; (e) rapid ventricular filling; (f) diastasis; and (g) atrial systole.

tricle rapidly changes from zero to a maximum value. This takes about 100 msec. and represents about 45 per cent of the period of ventricular ejection. When the rate of ejection reaches its maximum, pressure in the aorta starts to exceed that in the ventricle. Aortic pressure reaches its maximum about 80 msec. after the maximum rate of flow is obtained. The elevation in the aortic pressure curve is called the *anacrotic limb*.

As ejection continues, the rate of flow into the aorta decreases and there is a decrease in the aortic pressure (the *catacrotic limb*). During the period of ventricular ejection, the atria are filling and the ventricles getting smaller.

Ventricular Diastole (0.53 sec.)

Protodiastole (0.04 sec.). During the initial phase of ventricular relaxation, intraventricular and intra-aortic pressures fall rapidly, as does ventricular ejection, but the semilunar valves remain open. This period is called protodiastole and ends with the closure of the semilunar valves.

Isovolumic relaxation (0.08 sec.). During this period both the A-V and semilunar valves are closed and the ventricles relax, causing marked and rapid decrease in intraventricular pressure. The arterial distension and increased aortic pressure that occurred during ventricular systole are slowly dissipated to maintain arterial flow until the semilunar valves open again. The ability of the arteries to store pressure during systole is partly the result of their elasticity and partly that of the large resistance to flow offered by the *arterioles* and capillaries.

The closure of the semilunar valves is preceded by a sharp dip, the incisura or dicrotic notch, and followed by a wave, the *dicrotic wave*. In typhoid fever the dicrotic wave is sometimes so exaggerated that it is felt as a separate pulse.

Rapid filling (0.11 sec.). When the pressure in the ventricle sinks below that in the atrium, the A-V valves open. This begins the period of rapid ventricular filling. You will note in Figure 7-9 that the ventricles show their greatest increase in volume during this period. Prior to it, pressure has been building in the atria and veins. With the opening of the A-V valves the intra-atrial pressure declines (the *Y wave*) as a result of blood flow into the relaxing ventricle. It is important to realize that this period of filling is adequate at a heart rate of 75 beats per minute to maintain cardiac function in the absence of atrial systole. It has been estimated, for example, that atrial systole at a heart rate of 75 beats per minute is responsible for less than 20 per cent of ventricular filling. If this were not the case, the use of artificial ventricular pacemakers would not have proven as effective as it has in the treatment of complete A-V block. Atrial systole, on the other hand, becomes much more essential for ventricular filling when the heart rate is markedly elevated.

Diastasis (0.20 sec.). This is a period of relatively mild pressure and volume change in the heart. Frequently there is a slight increase in intra-atrial and intraventricular pressure during the latter part of this period due to the accumula-

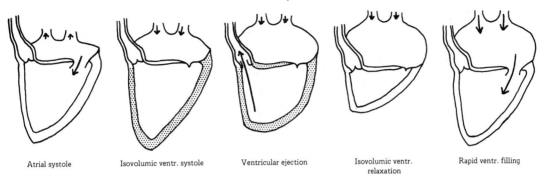

Atrial systole   Isovolumic ventr. systole   Ventricular ejection   Isovolumic ventr. relaxation   Rapid ventr. filling

Fig. 7-10. The heart valves during a cardiac cycle. In this diagram, only the left heart is represented. The dots indicate the portions of the heart that are contracting, and the arrows the direction of blood flow. Note that from the end of isovolumic ventricular relaxation through the period of atrial systole, the A-V valves are open. The semilunar valves are open only during the period of ventricular ejection and protodiastole.

tion of blood in the atria, ventricles, and veins. This period disappears at heart rates of 100 beats per minute.

Atrial systole (0.1 sec.). Atrial systole constitutes the last period of ventricular diastole. It results in a 5-mm. Hg increase in pressure in the atrium (*a wave*), a small increase in the volume of the ventricles, distension of the inferior vena cava, and an effective closure of the A-V valves. Of interest, however, is the fact that the superior vena cava decreases in size during atrial systole. Apparently at least part of this vein constricts during the contraction of the atrium. It is tempting to speculate that this is a mechanism for preventing increases in intracranial pressure.

COMPARISON OF THE RIGHT AND LEFT HEART

In Figure 7-11 I have characterized the heart as two major pumps in series with each other. Each pump (ventricle) has valves (the A-V valves) on its input side that prevent regurgitation from the pump, and valves on its output side (the semilunar valves) that prevent regurgitation into the pump. These pumps have two functions: (1) to receive blood from the atria, veins, and capillaries, thus preventing *congestion* and maintaining *flow* here, and (2) to send blood to the arteries, thus maintaining *pressure* and flow there. In cardiac arrest, the veins become congested and arterial pressure declines. In cardiac failure, venous congestion is also characteristic.

The atria are pumps of secondary importance (Fig. 8-12). In the individual with a heart rate of 75 beats per minute, atrial contraction produces less than 20 per cent of the flow into the ventricles. On the other hand, when the heart rate is more rapid they may be responsible for a considerably greater part of the ventricular inflow. They have no effective valves preventing regurgitation into the veins, but the character of their contraction is such that most of their blood is ejected toward the ventricles rather than toward the veins. The atria also serve as *reservoirs*. While the A-V valves are closed and the ventricle is expelling blood, the atria relax and receive blood from the veins. When the A-V valves open during ventricular diastole, much of the blood that has accumulated in the atria now passes rapidly into the ventricles.

There is yet a third important component of this system. The arteries offer very little resistance to flow, and therefore, convert little of the heart's energy output into heat. When, on the other hand, the blood in the systemic system (greater circulation) reaches the arterioles and capillaries, it encounters great resistance, and there is a marked decrease in pressure or a decrease in the energy level of the blood in these vessels. In the case of the arterioles, this re-

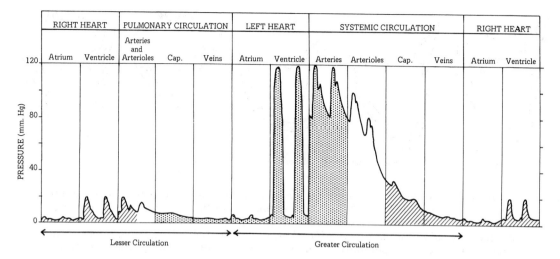

Fig. 7-11. Diagram illustrating the pressures and pressure variations in the greater and lesser circulations of a reclining subject. The lesser circulation is a low pressure system, and the greater circulation a high pressure system. In this presentation the right heart is characterized as primarily preventing congestion in the systemic veins and capillaries and providing pressure to the pulmonary arteries. The left heart is represented as preventing congestion in the pulmonary veins and capillaries and providing pressure to the systemic arteries. The systemic arterioles serve as a type of dam or high resistance system, maintaining a high pressure in the arteries and a lower pressure in the capillaries.

sistance is controlled primarily by the *sympathetic nervous system* and by certain *intrinsic mechanisms* of the arteriole itself. When the smooth muscle of the arteriole contracts, there is arteriolar vasoconstriction and an associated increase in *resistance*. The systemic arteriole thereby acts as a dam protecting the vein and capillary from distension and the pressures of the arterial system. It is also essential for the distension of the arteries and the maintenance of their pressures. After death, the arterioles dilate and arterial blood is shunted into the venous system by the highly elastic arteries. The fact that no blood is found in the arteries after death led many, prior to the work of William Harvey (1628), to believe that the function of the heart was simply to pump blood into and out of the veins.

In summary, then, the pump (heart) sends a fluid (blood) to highly elastic vessels (*arteries*), whose output is controlled by small, highly muscular vessels (arterioles) parallel to one another. This means that the amount of blood pressure in the distensible system of vessels (veins) depends upon the degree of constriction of the arterioles and the effectiveness of the heart as a pump.

The right ventricle is the weaker of the two ventricles. At birth, both the right and left hearts contain about the same amount of muscle, but the left heart encounters greater resistance to flow and therefore acquires considerably more actomyosin than the right heart. In the adult, the right ventricle in the resting individual shows a pressure range of 0 to 25 mm. Hg above atmospheric pressure, and the left ventricle of 0 to 120 mm. Hg above atmospheric pressure. In the right heart, the semilunar valves open at a pressure of about 7 mm. Hg, and in the left heart at about 80 mm. Hg. This means that the pressure range in the pulmonary artery of a resting individual is 25 to 7 mm. Hg (25/7), and in the aorta, 120 to 80 mm. Hg (120/80).

These relatively low pressures in the pulmonary artery (25/7) result in low pressures in the pulmonary capillaries (about 5 mm. Hg). In the systemic capillaries there is a wide range of pressures; while the average is about 35 mm. Hg, a pressure of 70 mm. Hg is characteristic for the glomerular capillaries of the kidney. Many systemic capillaries, on the other hand, periodically have such low pressures that they collapse. The *distribution* of the heart's output, then, depends to a very great degree on the state of constriction of the various systemic arterioles.

The low pressures in the pulmonary capillaries prevent the filtration of fluid into the alveoli (pulmonary congestion) and the high pressures in glomerular capillaries facilitate the filtration of fluid into the nephron (urine formation). When the pressure in the glomerular capillaries goes below a certain level, urine production stops and blood volume usually increases. When the pressure in the pulmonary capillaries increases, lung congestion may occur. Thus it is important that the left heart remove enough blood from the pulmonary system to prevent congestion there, and that the right heart not produce so much pressure that congestion results.

Figure 7-9 stresses some additional differences between the right and left heart. Note that right atrial systole begins about 20 msec. prior to left atrial systole. Note also that the period of isovolumic contraction begins earlier and persists longer in the left ventricle and that the pulmonary valves close about 28 msec. after the aortic valves. In short, the events in the right and left hearts occur in the same order but vary in degree, time of onset, and duration.

CARDIAC CATHETERIZATION

In 1929, Forssmann secretly passed a tube into his own heart. Later, in 1941, Cournand and Ranges demonstrated the safety and value of this catheterization procedure. Since that time, cardiac catheterization has become an increasingly common diagnostic tool. Catheters may be placed in the right heart or pulmonary artery of the unanesthetized individual by first entering a vein (the antecubital or saphenous veins, for example) with a needle, and then passing a tube through the vein and into the right heart (Fig. 7-12, 1). By this method one can determine the pressure in the venae cavae, the right atrium, the right ventricle, the pulmonary artery, or, if one passes the catheter as far as possible into the pulmonary artery ("wedge position"), the pressure in the pulmonary vein. This same catheter may be forced through the interatrial septum and used to measure pressure in the left heart (*transeptal approach*). A more common approach to the left heart (*retrograde aortic catheterization*) is through a systemic artery (brachial or right femoral artery, for example). Other techniques of left heart catheterization include passing a needle and catheter through the chest and into the left atrium or into the apex of the left ventricle. The latter two methods introduce the danger of blood entering the pericardial sac and compressing the heart (cardiac tamponade) either during catheterization or after the removal of the catheter.

The cardiac catheter has been used for many things. Some kinds have a small microphone at their tip for the study of heart sounds, some have electrodes for the study of the ECG or for cardiac pacing, some contain pressure transducers for the study of blood pressure, and most contain a lumen for the withdrawal of blood samples (to determine oxygen concentration) or the injection of radiopaque dyes during fluoroscopy. In Figure 7-12, 2, I have indicated the normal pressure and oxygen concentration found in various parts of the circulatory system. Diagrams 3 through 13 of Figure 7-12 show how these pressures may change in various cardiac abnormalities. Note that an interatrial or an interventricular septal defect (Fig. 7-12, 3, 4, and 5) may elevate pressure and oxygen concentration in the right heart. We also see that mitral stenosis and insufficiency (Fig. 7-12, 8 and 9) cause an increase in left atrial pressure which can result in a pressure increase as far back as the right ventricle. Pulmonary and aortic stenosis (Fig. 7-12, 6 and 11) produce a marked increase in pressure in the associated ventricle, and a decrease in the artery. Increases in pressure in an atrium, ventricle, vein, or artery may lead to cardiac distention and hypertrophy.

HEART SOUNDS

At the beginning of ventricular systole there is a sudden elevation in pressure that sets up vibrations in the heart (first heart sound); this can be heard by placing one's ear or a stethoscope to the chest. Later in the cardiac cycle, vibrations of the aorta and pulmonary artery (second heart sound) appear as the semilunar valves close (Fig. 7-9). These and related vibrations, since they are most often studied by the physician listening to them through a stethoscope, are usually called heart sounds. If, on the other hand, a microphone is placed in the heart or on the chest wall and the signal that is picked up is relayed to an oscilloscope, it is seen that most of the vibrations associated with what we call heart sounds lie outside the audible range (Fig. 7-13). In common practice all the low-frequency vibrations associated with the contraction and relaxation of the heart are called heart sounds, though many of them are not, strictly speaking, sounds at all.

The *first heart sound* (more properly, vibration) has been variously described as a "lub" (English), "frou" (anglicized French), "doop" (German), "r-rupp" (Turkish), and "htah" (Russian). It consists of a "mitral component" and a "tricuspid component." Usually the mitral valves

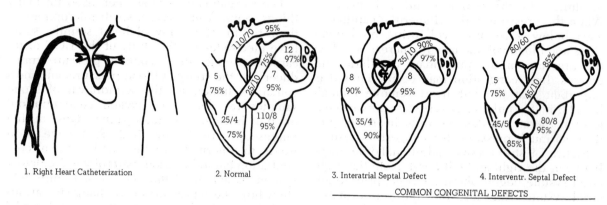

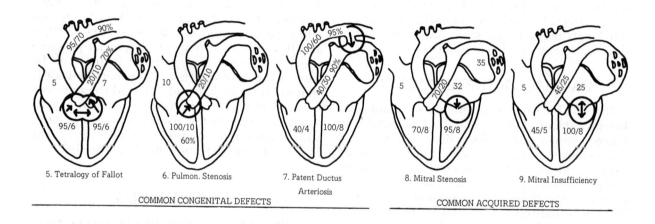

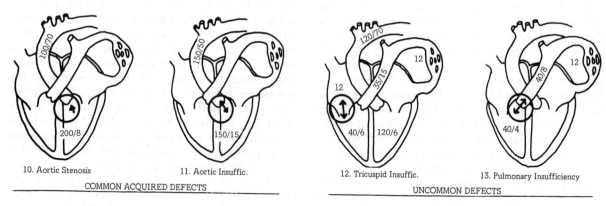

Fig. 7-12. Cardiac catheterization. (*Modified from* Luisada, A. A.: Cardiac catheterization. Hosp. Med., 2:22, 23, 1965)

Heart Sounds

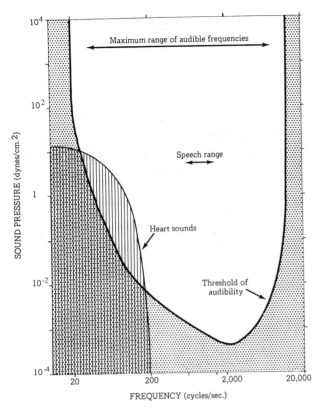

Fig. 7-13. Audibility of various frequencies. The heart sounds are characteristically of low frequency and must, therefore, be of a high intensity to be heard. Certain murmurs, on the other hand, reach frequencies of 1,000 cycles/sec. and can be perceived at a relatively slight energy level.

close and the vibrations in the left ventricle begin a few milliseconds before the closure of the tricuspid valves and the associated vibrations in the right ventricle. In most cases the vibrations in the right and left ventricles fuse to produce what appears to be a single first heart sound, but in some normal individuals the first heart sound is clearly split into a *mitral* and a *tricuspid component*. Each component is usually loudest just after the closure of its A-V valves and is best heard by placing a miniature microphone in the associated ventricle. A number of investigators have emphasized that the first heart sound is due primarily to the closure of the A-V valves, but our best evidence seems to indicate that it results from the rapid rise of pressure in the ventricles during isovolumic contraction. This rise, in turn, produces a vibration of the ventricular wall, interventricular septum, chordae tendinae, and A-V valves. The turbulent flow of blood during the first part of ventricular ejection may also contribute to the first sound. In conditions in which the intraventricular pressure does not rise rapidly, such as *mitral insufficiency* (failure of the mitral valve to close completely), the first heart sound is muffled. In conditions in which the activation of the right ventricle is delayed (right bundle branch block) the splitting of the first sound is exaggerated. And in conditions in which one of the ventricles is distended (aortic stenosis) the first heart sound may increase in loudness.

The *second heart sound* (or vibration) is also two-part. The first component is due to vibrations in the aorta, and the second to vibrations in the pulmonary artery. In expiration the second sound from the right heart usually fuses with that of the left, but in inspiration the second sound is heard in two parts—an early aortic followed by a later pulmonary sound. The second sound is usually characterized as a "dup" (English), "ti" (French), "teup" (German), and "ta" (Turkish and Russian). It is loudest after the closure of the semilunar valves and is best heard by placing a microphone in the ascending aorta. This sound is much louder in the ascending aorta than in the left ventricle, and louder in the pulmonary artery than in the right ventricle. After the ventricles stop contracting, there is a flow of blood from the arteries toward the ventricles. This contributes not only to the dicrotic pressure wave, but also to vibrations in the blood column, the arterial wall, and the semilunar valves which produce the second sound.

A *third sound* is most frequently heard in small children and is usually associated with the period of rapid ventricular filling. Luisada and MacCanon reported some instances in which it occurred while the A-V valves were closed, and concluded that the transition in the muscle from "active relaxation to passive distension" may contribute to the vibrations in the ventricle responsible for the third sound.

A *fourth sound* is also sometimes heard when there is an elevation in left intraventricular pressure at the end of diastole. This is associated with atrial systole.

It is usually impractical to place a small intravascular microphone in the right heart and pulmonary artery and study the vibrations there, or to perform a similar experiment on the left heart. It is practical, however, to place microphones on on the chest wall. Figure 7-14 shows chest posi-

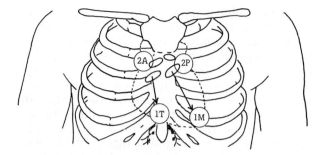

Fig. 7-14. Areas on the chest where the sounds associated with the closure of the aortic (2A), pulmonary (2P), mitral (1M), and tricuspid (1T) valves are loudest.

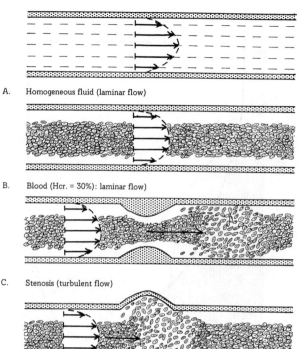

Fig. 7-15. Laminar and turbulent flow. Water (A) or blood (B) flowing through a straight tube or vessel without branches will tend to flow more rapidly at the center of the tube than at the wall. In blood vessels with a small radius the rapidly moving axial stream consists of a higher concentration of red blood cells than the peripheral stream of plasma. In blood vessels with a large radius such as those seen above, the red blood cells are fairly evenly dispersed at normal hematocrits. If, on the other hand, the hematocrit is markedly reduced as above, an axial stream will appear as in B. If, however, the velocity of flow is sufficiently increased as in stenosis or exercise (C), the axial stream will be disrupted by a nonlaminar (turbulent) flow. Turbulence may also result from an increase in the radius of a vessel or part of a vessel, as in a saccular or fusiform aneurysm (D) or a branching of a vessel. Here, high velocity flow is set up perpendicular to the axial stream. Turbulence results in a loss of kinetic energy by the stream and a sound (murmur or bruit).

tions at which the first and second sounds for the right and left heart are best heard. Note that the first sound is best heard near the apex of the ventricle, and the second sound near the semilunar valves.

MURMURS

If we inject dye into the center of a stream of blood or water flowing slowly in a long, straight, glass cylinder, we note that the dye stays in the center of the stream (*axial stream*). Under such circumstances we have a *laminar flow* (Fig. 7-15), i.e., a movement of a fluid in which the numerous layers of the fluid do not mix. In the case of blood and water, the most central layers move most rapidly, while the peripheral layer is stationary. Implied here is that in laminar flow the cylinder or vessel offers no resistance. The resistance is strictly between layers (laminae) of the fluid, i.e., it is internal. As we increase the velocity of the axial stream, we eventually reach a point—the *critical velocity*—at which the dye moves from the axial stream and colors the entire solution. It is at this point that laminar flow changes to *turbulent flow*. In turbulence the tube is filled with eddies, i.e. there is flow at right angles to the original axial stream. When the kinetic energy of the axial stream ($1/2\ mv^2$) reaches a certain level, some of its energy is dissipated in disordered (nonlaminar) flow.

In the resting, healthy adult, the blood in most of the arteries (the vessels of greatest flow velocity) moves in a laminar (streamlined) fashion, but there are many situations leading to turbulence and associated sounds, called *murmurs* and *bruits*. These sounds are sometimes noted during exercise when the blood in the arteries is circulating at a much greater velocity than in the individual at rest.

Use of Murmurs in Determining Arterial Pressure

Turbulence is also intentionally induced to determine arterial blood pressure. In this technique a cuff is placed on an appendage and pumped up to a pressure well in excess of the suspected systolic arterial pressure. The pressure in the cuff is then slowly reduced, a stethoscope is placed over an artery just peripheral to the cuff, and the investigator notes the pressure at which he hears (1) the first sound (a faint, clear tapping called the first Korotkow sound). As pressure is decreased, the sound gradually increases in intensity and is usually followed by (2) an abrupt muffling of the sound, and later (3) disappearance of the sound. The appearance of the sound occurs at the cuff pressure which lets blood pass the constriction in each cardiac cycle. At this pressure (approximately the arterial systolic pressure) the constriction of the arteries permits a rapidly accelerating stream of blood to pass the constriction for a few milliseconds of each cardiac cycle. This apparently produces an abrupt displacement of the arterial wall and surrounding tissues distal to the compression, which is heard as the first Korotkow sound. The increase in the intensity of the sound as the cuff pressure is further reduced is due in part to the turbulence (compression murmur) that occurs distal to the cuff as progressively more blood passes the constriction. The abrupt muffling is apparently due to a reduction of the vibrations set up in the artery and tissues distal to the cuff at that pressure which permits blood to pass the cuff throughout the entire cardiac cycle. The muffled sound ceases when turbulence disappears.

In the past there has been considerable disagreement on whether to use the muffling or the disappearance of the sound as the index of diastolic pressure. The American Heart Association in 1967 recommended that the onset of muffling "should be regarded as the best index of diastolic pressure." It should be borne in mind, however, that this is just an index and not the actual diastolic pressure. Studies with arterial catheters have shown that on an average the muffling occurs at a pressure 7 to 10 mm. Hg higher than the true diastolic pressure. Because of this many investigators prefer to record the pressure at which the sound appears, the pressure at which there is muffling, and the pressure at which the sound disappears. A typical set of values for a resting person might be 130/82/74 (systolic/first diastolic/second diastolic sound). In cases where the disappearance and softening occur at the same time the pressures might be reported as 130/82/82, for example. It is interesting in this regard that when one uses the cuff method for estimating arterial pressure after strenuous exercise, one sometimes obtains values such as 140/70/30 or even finds that the sound does not disappear at all. This is due to the increased velocity of flow which occurs during and after exercise. A similar situation may exist when the hematocrit is abnormally low. In arteriosclerosis, on the other hand, the pressures obtained by the cuff method may be considerably higher than the true systolic and diastolic pressures. In other words, the cuff pressures at which the Korotkow sounds are heard are dependent not only upon the pressures in the arteries but also upon the characteristics of the blood (its velocity and viscosity) and the characteristics of the blood vessels (their resistance to compression).

Use of Murmurs in Diagnosis

The recognition of murmers in certain areas of the body has also proven useful in diagnosis. We find, for example, that a constriction of channels may occur in response to certain pathological conditions as well as to a blood pressure cuff. Aortic stenosis is one such condition. Here the valvular orifice between the left ventricle and aorta is constricted to such an extent that the heart, in maintaining an adequate flow into the aorta during ejection, produces a high-velocity jet of blood:

$$\text{Flow velocity (cm./sec.)} = \frac{\text{volume flow (cm.}^3\text{/sec.)}}{\text{cross-sectional area of vessel (cm.}^2\text{)}}$$

Sounds (turbulence) may also occur where there is marked distention of a blood vessel (saccular and fusiform aneurysms). In these areas the average velocity of flow is reduced, but since the number of layers (laminae) of fluid is increased, the axial stream moves at a greater velocity and turbulence results. One also finds high-velocity flow, and hence sounds, in areas of abnormal A-V shunts and in patent ductus arteriosus.

Another cause of turbulence is a decrease in the viscosity of the fluid under study. With blood this results from a decrease in the concentration of the red blood cells (a decreased hematocrit).

By definition a decrease in viscosity is a decrease in the internal resistance of the fluid, i.e., there is less resistance between layers. All other factors being equal, the velocity of the axial stream increases as the viscosity of the blood decreases.

Sir Osborne Reynolds in 1883 reported a method for determining approximately when turbulence would occur. He first measured the average velocity ($\bar{v}$ in cm./sec.) of a fluid, its density (ρ in gm./cm.^{3}), its viscosity (η in poises), and the radius of the tube through which it moved. From these data he determined a number, R (*Reynolds' number*):

$$R = \frac{\bar{v} \times \rho \times r}{\eta}$$

If R exceeds 970 ± 80 (mean ± standard deviation), turbulence may be expected.

Chapter 8

DETERMINATION OF
CARDIAC OUTPUT
STROKE VOLUME
HEART RATE
CARDIAC MASSAGE

THE HEART—CARDIAC OUTPUT

The cardiac output is the volume of blood pumped from the left ventricle per minute. It is equal to the heart rate times the stroke volume of the left ventricle. An individual with a heart rate of 70 beats per minute and a stroke volume of 80 ml. per beat would have a cardiac output of:

Cardiac output = (70 beats/min.)(80 ml./beat) = 5,600 ml./min.

If this individual had a surface area of 1.75 m.², he would have a *cardiac index* of:

$$\text{Cardiac index} = \frac{5.6 \text{ l./min.}}{1.75 \text{ m.}^2} = 3.2 \text{ l./min./m.}^2$$

The cardiac output is controlled by the heart rate and the stroke volume (Fig. 8-1), which in turn are controlled by peripheral resistance, venous return of blood, the autonomic nervous system, the endocrine system, the external environment of the heart, and various intrinsic

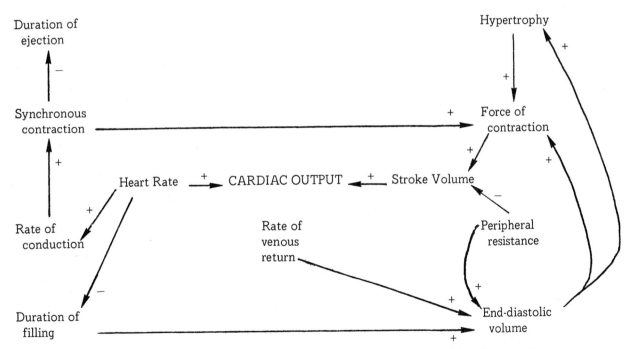

Fig. 8-1. Control of cardiac output. This chart shows the relationships (+→, positive relationship; −→, negative relationship) that exist between heart rate and stroke volume in a heart-lung preparation. The nervous system, endocrine system, and the environment of the heart can also change the heart rate and stroke volume. Sympathetic nerves can, for example, increase the rate of conduction of the action potential, increase the force of contraction, increase the peripheral resistance, and increase the heart rate.

cardiac factors (amount of actomyosin, relationship between actin and myosin, oxygen concentration, etc.).

Cardiac outputs up to 35 liters per minute and stroke volumes up to 180 ml. have been recorded during maximum activity, but these are probably exceptional cases. In Figure 8-2 I have presented some graphs showing how cardiac output, stroke volume, heart rate, and arterial pressure may change in healthy males during different levels of exercise on a bicycle ergometer. The oxygen consumption in these studies is used as an index of the amount of work performed. Note that as the oxygen consumption (amount of worked performed) increases, arterial pulse pressure, heart rate, stroke volume, and cardiac output also increase. Since blood is a delivery system for oxygen, one would expect increased oxygen consumption to be associated with increased cardiac output. The mechanisms for the elevation of cardiac output in these studies are an increase in stroke volume and heart rate. The increase in pulse pressure is probably, in this case, the result of a more forceful contraction of the heart associated with a decrease in arterial resistance to flow.

DETERMINATION OF CARDIAC OUTPUT

The techniques most commonly used for determining cardiac output in clinical situations are the Fick and the indicator-dilution methods. In the former, one measures the concentration of O_2 in a systemic and a pulmonary artery and the oxygen consumption (ml. of O_2/min.) of the subject. These calculations can also be performed on the basis of CO_2 concentrations and CO_2 production. In the indicator-dilution tech-

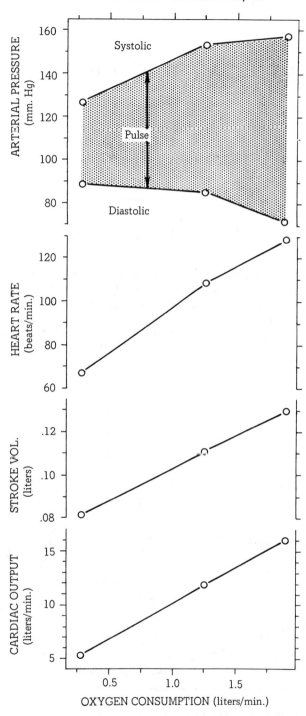

Fig. 8-2. Response of the healthy male to exercise. The averages given above are based on the responses of 3 to 7 men ranging in age from 20 to 73 years to varying degrees of work on a bicycle ergometer. Arterial pressure was obtained by auscultation from the cubital area, heart rate from an ECG, and oxygen con-

sumption by analysis of the expired air. Cardiac output (C.O.) was calculated by determining the body's CO_2 production in ml./min. (V_{CO_2}), the concentration of CO_2 in the arterial blood in ml. of CO_2/ml. of blood (C_a), and the concentration of CO_2 in venous blood (C_v):

$$C.O. = \frac{V_{CO_2}}{C_v - C_a}$$

Note that the maximum work done on the bicycle ergometer resulted in a 6.7-fold increase in oxygen consumption, in a 3.2-fold increase in cardiac output, in a 1.6-fold increase in stroke volume, in a 1.9-fold increase in heart rate, and in a 2.3-fold increase in pulse pressure over that found before exercise. (*Data from* Klausen, K.: Comparison of CO_2 rebreathing and acetylene methods for cardiac output. J. Appl. Physiol., 20:765, 1965)

nique an indicator is rapidly injected into a systemic vein near the right heart and the change in its concentration in a systemic artery near the left heart is measured. If the indicator is radioactive the change in its arterial concentration can be measured by means of a Geiger counter. If it is a dye such as cardiogreen (indocyanine green) its concentration in arterial blood can be studied by means of a cuvette densitometer.

The Fick Principle

The Fick principle is that the amount of a substance taken up by an organ (or released by an organ) per unit time is equal to the arterial concentration of that substance minus the venous concentration (venous minus arterial in the case of a substance released by the organ) times the blood flow through the organ. In the case of the lung the following formula would apply:

$$\text{Pulmonary blood flow (ml. blood/min.)} = \frac{\text{oxygen consumption of the body (ml. } O_2/\text{min.)}}{\text{arteriovenous oxygen difference (ml. } O_2/\text{ml. blood)}}$$

The pulmonary blood flow so obtained is approximately equal to the cardiac output.

In a typical study on the adult, an oxygen consumption of 250 ml. per minute, an arterial O_2 concentration of 0.19 ml. per milliliter of blood, and a venous O_2 concentration of 0.14 ml. per milliliter of blood might be found. Substituting these values in the above equation we obtain the following cardiac output:

$$\text{Cardiac output} = \frac{250 \text{ ml. } O_2/\text{min.}}{0.190 - 0.140 \text{ ml. } O_2/\text{ml. blood}} = 5{,}000 \text{ ml. blood/min.}$$

Since the arterial concentration of O_2 is the same for any systemic artery, determining the arterial concentration is a relatively simple matter. The venous concentration of O_2, however, varies from one systemic vein to the next. Therefore the venous sample is generally obtained from a catheter placed in either the pulmonary artery or right ventricle. By obtaining the sample in this way the investigator gets an average O_2 concentration for venous blood. He lets the right atrium serve as a mixing chamber in other words.

The Stewart-Hamilton Technique

In the Stewart-Hamilton technique a known amount of indicator is rapidly injected into a systemic vein. At the same time, arterial blood is withdrawn through a cuvette-densitometer and the changes in the optical density of the blood are determined. If the entire dye curve could be recorded before any dye circulated through the heart a second time, the cardiac output could be obtained from the following calculation:

$$\text{Cardiac output} = \frac{\text{quantity of dye injected}}{\text{average concentration under dye curve} \times \text{duration of curve}}$$

Unfortunately dye does circulate through the heart a second time before the dye curve is complete. It is therefore essential in determining cardiac output by this method to extrapolate the curve that would occur if there were no recirculation, and to use the extrapolated curve in calculating both the average arterial concentration of the dye and the duration of the curve. The extrapolation technique is simple, involving a plotting on semilog paper (log of concentration v. time) of the descending slope of the dye curve. In this semilog plot the first part of the descending slope is found to be a straight line. Eventually the recorded concentration moves away from the straight line on the semilog paper; it is at this point that recirculation is assumed to occur. The straight line is then extended to the bottom of the semilog paper and the extrapolated points are transferred to the original optical density recording. The average height of the extrapolated curve in milligrams of dye per milliliter of blood, and its duration are then determined. If, for example, 5 mg. of cardiogreen produced an extrapolated curve with an average height of 0.0016 mg. dye per milliliter of blood, and a duration of 35 sec., the calculated cardiac output would be:

$$\text{Cardiac output} = \frac{(5 \text{ mg. dye})(60 \text{ sec./min.})}{(0.0016 \text{ mg. dye/ml. blood})(35 \text{ sec.})} = 5{,}360 \text{ ml. blood/min.}$$

STROKE VOLUME

The ventricle does not normally receive additional blood during ejection, since the A-V valves are closed at this time. Thus its output during each cycle depends on the amount of *diastolic filling* and the completeness of ejection. The amount of blood in a ventricle at the beginning of systole (*end-diastolic volume*) and the amount of blood there at the end of systole (*end-systolic volume*) varies with the amount and type

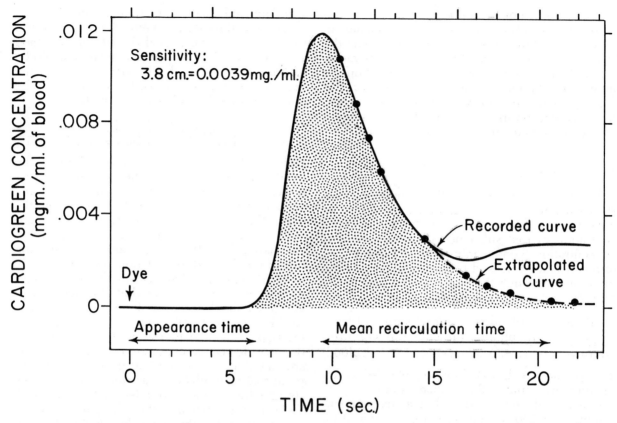

Fig. 8-3. Cardiogreen dye-dilution curve from a normal conscious dog. The dye (2.5 mg. of cardiogreen) was injected rapidly into the jugular vein at 0 time and blood samples were obtained from the thoracic aorta through a cuvette densitometer by means of a motorized syringe that withdrew 28 ml. of arterial blood per minute. The dye first appeared in the densitometer 6 seconds after the injection (6 sec. = appearance time). Recirculation of the dye began 9 seconds later and the peak of the recirculation curve was noted at 20.4 seconds. The time from the peak of the first curve to the peak of the second curve is called the mean recirculation time (11.2 sec.). The cardiac output was calculated as follows:

$$C.O. = \frac{2.5 \text{ mg.} \times 60 \text{ sec./min.}}{(0.0037 \text{ mg./ml.}) \times (15.6 \text{ sec.})} = 2{,}600 \text{ ml./min.}$$

of nervous tone and the body position. The ventricle's size may also change in response to certain drugs (Fig. 8-4).

In Figure 8-5 I have estimated the ability of the left ventricle in an average adult to change its blood volume. Usually the resting stroke volume averages about 80 ml. (Curve 1) and the amount of blood left in the ventricle at the end of systole (*functional residual capacity*) is about 60 ml. If, on the other hand, the ventricle contracts as forcefully as it is able (Curve 4), the amount of blood in the ventricle at the end of systole is reduced to 25 ml. (*residual volume*).

We see, then, that the heart can increase its stroke volume by as much as 35 ml. (*systolic reserve volume*) by increasing its force of contraction. We also note in the diagram that the end-diastolic volume of a ventricle may be increased by as much as 40 ml. (*diastolic reserve volume*) by increasing the amount of filling during ventricular diastole (Curve 2). In total, a normal resting stroke volume of 80 ml. can, in the average individual, be increased to about 175 ml. if he increases the venous return of blood to his ventricle and decreases his heart's end-systolic volume.

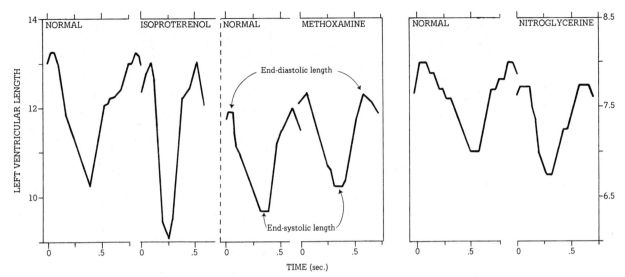

Fig. 8-4. Effect of isoproterenol (2.5 μg./min.), methoxamine (1.0 mg./min.), and nitroglycerine (0.6 mg.) on the length of the ventricles. These studies were performed on reclining, unanesthetized patients who had received radiopaque clips near the base and apex of the right and left ventricles during cardiac surgery. Two weeks to 16 months after implantation the distance between clips throughout the cardiac cycle was studied by means of cineradiography (30 frames/sec.). No attempt was made to measure the distance between clips in absolute units such as centimeters. The data from 3 studies are presented above. The authors conclude: (1) that beta adrenergic stimulation with isoproterenol consistently decreases "the end-systolic dimensions of both ventricles"; (2) that alpha adrenergic stimulation with methoxamine increases "ventricular end-diastolic dimensions" and "left ventricular end-systolic dimensions"; and (3) that nitroglycerine decreases both "end-diastolic and end-systolic dimensions." (*Redrawn from* Harrison, D. C., et al.: Studies on cardiac dimensions in intact unanesthetized man. Circulation, 29:188, 190, 1964; and Williams, J. F., et al.: Studies on cardiac dimensions in intact unanesthetized man. Circulation, 32:768, 1965; by permission of the American Heart Association, Inc.)

The ability of the heart to change its stroke volume is in marked contrast to that of the lungs. The lumen of the heart, unlike the alveolus of the lungs, is almost completely surrounded by actomyosin (Fig. 5-4). The significance of this is that if the heart should be distended past a certain point, actin and myosin would no longer be able to interact. Since the pump for the lungs (the diaphragm) is not a part of the lungs, this is not a problem here.

I should reemphasize, however, that Figure 8-5 applies only to the average, normal, adult heart. If over a period of many months the heart is exposed to distending forces, its volumes increase beyond the limits indicated. This may occur in chronic hypertension, in aortic stenosis, or in response to weeks of strenuous exercise. Cardiac enlargement is generally associated with the laying down of more actomyosin (*hypertrophy*) and an increase in the size of both the ventricular wall and its lumen (*distension*).

Cardiac distension associated with hypertrophy may create some important problems. One may be in the coronary circulatory system. It has been suggested by a number of authors that in the adult, an increase in the amount of muscle in the heart is not associated with a comparable increase in the number of coronary capillaries. The second problem is that the distended heart must develop more tension in order to produce the same amount of pressure in its lumen as is produced in the lumen of a nondistended heart. As the cross-sectional area (cm.^{2}) of the ventricle increases, in other words, the amount of force (dynes) and tension (dynes/cm.) necessary to produce a given pressure (dynes/cm.^{2}) also in-

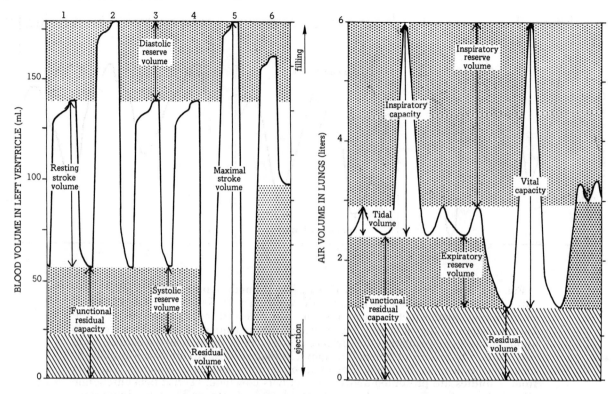

Fig. 8-5. Comparison of the ability of the average adult heart and lungs to change their volume. Note that the muscular heart can only double its resting stroke volume, whereas the highly distensible lungs can increase their resting tidal volume by about 10 times. In the case of the heart one can increase its stroke volume by a more complete diastolic filling (2), or a more complete ejection (4), or a combination of both (5). Pattern 6 represents the condition in a reclining individual. (*In part from the data of* Levinson, G. E., et al.: Studies of cardiopulmonary blood volume. Circulation, 33:347-356, 1966; Levinson, G. E., et al.: Studies of cardiopulmonary blood volume-measurement of left ventricular volume by dye dilution. Circulation, 35:1038-1048, 1967; and Dodge, H. T., et al.: Usefulness and limitations of radiographic methods for determining left ventricular volume. Am. J. Cardiol., 18:10-24, 1966)

creases. *Laplace*, in 1821, defined this relationship between tension (T), pressure (P), and radius (r) for a sphere and a cylinder as follows:

Sphere—2T(dynes/cm.) = P(dynes/cm.^{2}) × r(cm.)
Cylinder—T(dynes/cm.) = P(dynes/cm.^{2}) × r(cm.)

The heart, of course, is neither a sphere nor a cylinder, but as with them, the amount of tension needed to produce a given pressure increases as the radius increases.

Heterometric Autoregulation

In the chapter on skeletal muscle contraction we noted that as one slowly stretches a muscle the thin, intracellular myofilaments are pulled toward the ends of thick myofilaments (myosin). This change in the relationship between actin and myosin modifies the amount of force the muscle exerts before (passive tension) and in response to (active tension) stimulation. The length-tension diagram for skeletal muscle is the result of these changing relationships between actin and myosin and is reviewed in the first graph in Figure 8-6.

The length-tension diagram for the heart is presented in the second graph. Note that here too, within limits, the passive, active, and total tension increase as the muscle is stretched. There are, however, some differences between these two types of muscle. Note that the heart, unlike skeletal muscle, starts to develop a marked

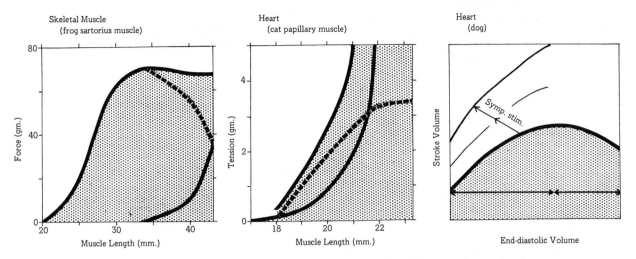

Fig. 8-6. The length-tension diagram for skeletal and cardiac muscle. Within limits skeletal muscle and the heart will increase their active tension (*interrupted lines*), passive tension (*solid line*), and total tension (*shaded*) in response to an increase in muscle length. In the case of the heart, this results in cardiac distention producing an increased stroke volume. When distention is so severe as to produce a decreased stroke volume, or no change in stroke volume, the heart is said to be in decompensation. Sympathetic stimulation results in the heart's producing a greater stroke volume at each fiber length. This is sometimes referred to as a shift of the Starling curve to the left.

amount of passive tension well before maximum active tension is reached. The heart is apparently more resistant to stretch than is skeletal muscle. In the example in Figure 8-6 this is probably the result of the greater quantities of *collagen* and elastic tissue in the heart (papillary muscle) than in skeletal muscle. Another important factor in decreasing the entire heart's *compliance* is the presence of a rigid, fibrous pericardium around it. The importance of this resistance to distension is best realized if you remember that the heart is a series of actin and myosin filaments surrounding a lumen (Fig. 5-4). If the lumen becomes too large, the actin separates from the myosin and therefore cannot interact with it to produce tension.

The last graph in Figure 8-6 shows the relationship between end-diastolic volume and stroke volume. As the end-diastolic volume increases, the heart must exert more tension in order to produce the same amount of pressure (law of Laplace) and stroke volume. Distention apparently causes a large enough increase in tension that stroke volume is not only maintained but is increased.

The response of the heart to distention is sometimes referred to as heterometric autoregulation. The cardiac muscle fiber, that is, responds to an increase in length by an increase in tension, which is dependent upon the characteristics of the muscle rather than upon associated nerves or hormones. Otto Frank, in 1895, noted the length-tension relationship in the heart, and E. H. Starling, using a heart-lung preparation, later reported on the effect of changes in arterial pressure or venous return on heart size. Starling, in 1915, stated what is now frequently called *Starling's law of the heart* or the *Frank-Starling mechanism*: "The energy of contraction is a function of the length of the muscle fiber."

The Frank-Starling mechanism was first clearly demonstrated on a *heart-lung preparation*. In this preparation, the viscera of the chest are removed from the body and blood from a glass reservoir passes to the right heart, which pumps the blood through the lungs or an oxygenating chamber and back to the left heart; here the blood is pumped through the aorta and into the coronary arteries or directly back to the reservoir. If the blood is pumped through the lungs, they are in turn controlled by a mechanical ventilator. In this preparation it is possible to increase the *pressure load* in the heart by constricting the tubing leading from the aorta (aortic stenosis) to the reservoir. It is also possible to increase the *volume load* in the heart by elevating the venous reservoir (increasing venous return). Within limits, both an increase in arterial pressure load (elevated peripheral resistance) and an increase in volume load (increased venous return), by increasing the end-diastolic volume of the heart, increase the tension developed by the heart. Since the heart-lung preparation is isolated from the body, it is of course not influenced by the central nervous system or the endocrine system.

When there is an increase in *peripheral resistance* (an increase in arterial pressure, for example), there is also a reduction in stroke volume for a few beats, but as the heart is distended it contracts more forcefully and its stroke volume soon returns to normal. It is also possible to increase the stroke volume without an increase in end-systolic volume. This frequently occurs when there is an increase in *venous return*. Here distension occurs at the beginning of systole but not necessarily at the end. Increases in venous return may occur when the period of cardiac filling is increased (a decrease in heart rate); when venous tone is increased; or when we change from a standing to a recumbent position. It is also changed as a result of contraction of skeletal muscle (milking action).

It was once felt that running would so increase venous return as to distend the heart, but studies on heart size have shown that the heart in most cases gets smaller during exercise. These observations and others like it have led many to speculate that under most circumstances the nervous system is the predominant means for the rapid control of cardiac output and that the Starling mechanism is important primarily when the central or peripheral nervous system is depressed (anesthetization or atropinization) or is for some other reason unable to meet the body's needs. It has been shown, for example, that an individual increases his heart rate when he is told to "get set" prior to a race. In this way the nervous system seems to increase cardiac output even before there is a need for increased cardiac output.

Starling also pointed out that if cardiac dilation "goes on sufficiently long the dilatation must pass the optimum length of muscle fiber and the muscle then has to contract at such a mechanical disadvantage that the heart fails altogether. With the failure of the 'prime mover' all other mechanisms of the body stop work and the animal is dead." Such hearts are said to be in decompensation or failure (Table 8-1). Further distension may so separate the thick and thin contractile filaments that cardiac action potentials are no longer followed by tension (cardiac arrest). The *failing heart* can sometimes be successfully treated with agents that increase cardiac force (Ca^{++} and digitalis) or agents that decrease arterial pressure (isoproterenol). The heart in decompensation should be protected from marked increases in venous return, increases in peripheral resistance, or other factors (caffeine for example) that might increase its metabolic requirements. This is to say that the heart in decompensation has lost most of its reserve (ability to increase its energy output or efficiency) and should be protected from any stress.

Patient	End-diastolic Vol. (cm^3)	End-systolic Vol. (cm^3)	Stroke Vol. (cm^3)	Ejection Fraction
1	305	262	43	0.14
2	149	80	69	0.46
3	352	301	51	0.14
4	325	258	67	0.21
5	407	367	40	0.10
6	293	241	52	0.18
7	155	96	59	0.38
8	261	177	85	0.33
9	364	272	92	0.25
10	297	266	31	0.10
11	165	86	79	0.48
12	237	154	83	0.35
13	226	177	49	0.21
Normal	140	60	80	0.57

Table 8-1. Blood volume in the left ventricle in patients with primary cardiopathy or arteriosclerosis (patient No. 12 only). (Data from Dodge, H. T., Sandler, H., Baxley, W. A., and Hawley, R. R.: Usefulness and limitations of radiographic methods for determining left ventricular volume. Am. J. Cardiol., 18:10-24, 1966)

Homeometric Autoregulation

There are intrinsic factors other than changes in cardiac fiber length that may modify stroke volume. The cardiac cell, for example, may deplete its own oxygen supply, increase its CO_2 concentration, decrease its pH, elevate its intracellular Na^+ concentration, decrease its intracellular K^+ concentration, elevate its intracellular

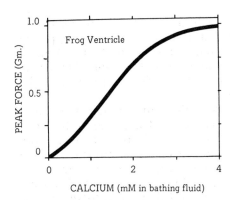
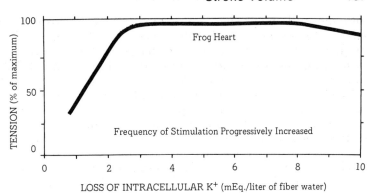

Fig. 8-7. The relation between Ca^{++} and K^+ concentration and the development of tension by the frog heart. In the first series of experiments a strip of ventricle was placed in progressively higher concentrations of Ca^{++} and its peak isometric tension measured. In the second group of experiments, the frequency of stimulation to a heart was increased and the amount of K^+ lost to the bathing fluid measured. Note that as more K^+ is lost by the cell the tension produced by the heart increases. (*Modified from* Niedergerke, R.: The rate of action of calcium ions on the contraction of the heart. J. Physiol., *138*:508, 1957; and Hajdu, S.: Mechanism of staircase and contracture in ventricular muscle. Am. J. Physiol., *174*:373, 1953)

Ca^{++} concentration, and increase its temperature. All of these changes can result from elevated metabolic activity or an increase in heart rate. In Figure 8-7 we note that, within limits, one can also increase the tension that a ventricle develops by increasing the amount of Ca^{++} in its external environment.

Sodium. Sodium, calcium, and potassium apparently all play roles in modifying (1) the transmembrane potential and (2) the contractile reaction. In the case of Na^+ we have an ion which exists in the interstitial fluid at a concentration of approximately 145 mEq. per liter and in the cytoplasm of the cell at a concentration of about 14 mEq. per liter (Fig. 1-4). Progressively decreasing the concentration of extracellular Na^+ to about 10 to 20 per cent of normal results in (1) a decrease in the amplitude of the action potential (Fig. 3-3) and (2) an increase in the force of the heart's contraction. There is also a decrease in the rate of depolarization of pacemaker and follower cells, leading to a slowing of the heart rate and a decrease in the velocity of conduction of action potentials in the heart (most pronounced at the atrioventricular node). If these markedly reduced concentrations of Na^+ are maintained, a cardiac standstill in diastole eventually results.

Such changes in extracellular Na^+ concentration are undoubtedly more severe than any occurring in either a healthy or diseased individual, but they do emphasize the importance of Na^+ in producing depolarization. They also suggest that Na^+ in some way depresses contractility, though the mechanism whereby this happens is not clear. Some investigators have suggested that Na^+ competes with Ca^{++} for storage sites in the cell membrane and in its associated transverse tubules. This would explain the increase in cardiac contractility when the extracellular Na^+ concentration is 10 per cent of normal. With an increased sequestering of Ca^{++} in response to hyponatremia, that is, more Ca^{++} could be released into the cytoplasm in response to stimulation. It is the intracellular release of Ca^{++} from its storage depots and the movement of Ca^{++} into the cell, you will remember, that initiate the contractile reaction.

Several explanations have been offered for the movement of Na^+ out of the cell membranes and the transverse tubules, causing increased cardiac contractility. One of these is an increase in heart rate (Fig. 8-8). It has been suggested that within certain limits, increasing the heart rate induces greater movement of Na^+ from the cell membrane into the cytoplasm, and that the resultant increase in membrane Ca^{++} causes the increased force accompanying increased contractility. It has also been suggested that cardiac glycosides such as digitalis and ouabain cause increases in contractility by depressing the Na^+ pump and thereby inhibiting the movement of Na^+ back into the cell membrane. Consistent with this view is the observation that the glycosides also produce a less negative resting charge in the cytoplasm, due apparently to an increase in the intracellular concentration of Na+.

Calcium. Calcium ions, like Na+, play a number of roles in muscle. At cytoplasmic Ca++ concentrations below 10^{-7} molar the muscle remains relaxed, but, within limits, an increase in the intracellular Ca++ concentration above this level produces increasingly forceful contraction. On the other hand, decreases in the extracellular Ca++ concentration below the resting level of 4.9 mEq. per liter not only decrease the force of contraction but also tend to (1) make the membrane more irritable and (2) decrease its resting transmembrane potential (make the cytoplasm less negative). The increased irritability is due only in part to the depolarization, since it can occur in the absence of a depolarization. In addition, there is a more frequent depolarization of the cardiac pacemaker (increased heart rate) and a tendency to develop ectopic pacemakers and ventricular fibrillation. In some cases a decrease in Ca++ concentration leads to an increase in electrical activity associated with a cessation of contractile activity, while increased extracellular Ca++ concentration leads to increased strength of contraction and an elevated threshold for activation (producing cardiac slowing). If the Ca++ concentration is increased enough, a sustained contraction of the heart results. This is called Ca^{++} *rigor*. The most common clinical disturbance caused by hypercalcemia is a tetany (intermittent, involuntary, painful skeletal muscle contractions) that may be sufficiently severe to produce asphyxiation due to laryngospasm.

Potassium. Potassium ions are normally found in the interstitial fluid at a concentration of 5 mEq. per liter. Increases in their concentration result in a number of changes opposite to those produced by increased Ca++ concentration. For this reason K+ infusions are often used in treating hypercalcemia and digitalis intoxication. An increase in the extracellular K+ concentration, for example, initially (1) decreases the resting transmembrane potential (Fig. 3-5) and increases the irritability of the cell, and (2) decreases the force of contraction. In addition, it decreases the conduction velocity and rate of rise of the propagated action potential. The effects of these changes on the ECG are seen in Figure 6-22. Apparently as we move toward an extracellular concentration of K+ six times the normal value, we reach a point at which the conduction is so depressed that numerous areas become electrically isolated from the rest of the heart and develop their own pacemaker. Eventually the heart begins to fibrillate or reaches a standstill. Although the initial response to progressive hyperkalemia is depolarization, which increases irritability, the depolarization sometimes becomes so severe that the membrane goes below

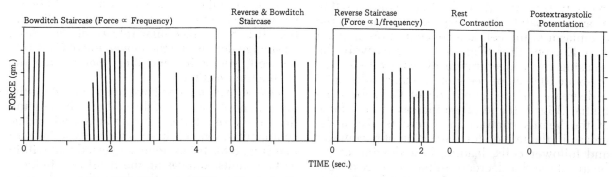

Fig. 8-8. The relationship between frequency of stimulation and cardiac tension. The Bowditch staircase or treppe phenomenon is one in which there is a direct relationship between frequency of stimulation and force of contraction. The reverse staircase is a situation in which there is an inverse relationship between frequency and force. During the periods of activation, contraction, relaxation, and recovery, the intracellular and extracellular environments are being modified. It is these various changes that are responsible for the changes in tension seen above. (*Redrawn from* Hajdu, S., and Leonard, E.: The cellular basis of cardiac glycoside action. Pharmacol. Rev., *11*:185, 1959; copyright © 1959, The Williams and Wilkins Company, Baltimore)

the critical firing level. When this occurs there is a loss of excitability and the propagated action potential disappears. Hypokalemia produces less marked changes in the heart. Here there is hyperpolarization of the cardiac cells associated with a decrease in excitability. Hypokalemia most commonly causes a skeletal muscle weakness that may be so severe as to produce apnea.

Bowditch staircase. In Figure 8-8 we have a demonstration of how activity or lack of it may change cardiac force. Note that with increases in the frequency of contraction there may be a progressive increase in the force of contraction (Bowditch staircase). This change in force results at least in part from environmental changes produced by elevated metabolic activity. We also note in Figure 8-8 that increases in rate may be associated with decreases in force. One mechanism for this may be a decrease in the cardiac filling time and a resultant decrease in end-diastolic volume (Starling's law). There may also be a shift away from an optimal environmental concentration of metabolites either inside or outside the cell.

In Figure 8-9 we note that increases in the rate of contraction are also associated with decreases in the Q-T interval. As the heart rate increases, that is, the period of ventricular depolarization is decreased. These changes can be demonstrated in both the heart-lung preparation and the normal individual.

Hypertrophy

A third way in which the cardiac cell regulates itself is through protein anabolism and catabolism. It apparently responds to stresses such as arterial hypertension, aortic and pulmonary stenosis, aortic and pulmonary insufficiency, and exercise by a laying down of additional actomyosin in its cytoplasm (hypertrophy). Decreases in activity, on the other hand, cause a decrease in its actomyosin (atrophy). The response of the heart to the stresses mentioned above, though considerably less rapid than that associated with heterometric and homeometric autoregulation, frequently is an effective mechanism for slowly increasing cardiac force. In arterial hypertension, aortic and pulmonary stenosis, and aortic and pulmonary insufficiency, the hypertrophy in many cases produces a sufficient increase in cardiac force to maintain a fairly normal stroke volume under conditions which would otherwise

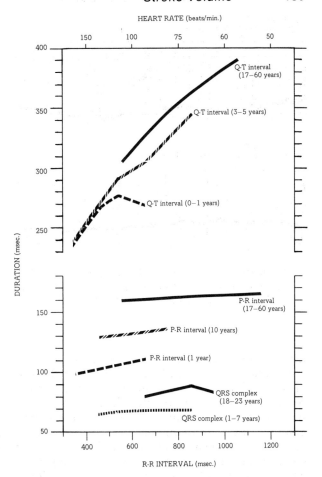

Fig. 8-9. The effect of heart rate and age on the electrocardiogram. Data based on 12,604 cases reported in the literature. (*Data supplied by* Eugene Lepeschkin, M.D.)

decrease stroke volume. In the case of the athlete, on the other hand, the hypertrophy produces an increase in the cardiac reserve and the resting stroke volume. The resting cardiac output, on the other hand, is kept constant by an associated decrease in heart rate (Fig. 8-10).

Extrinsic Control

I have summarized in Figures 8-1 and 8-11 some of the factors important in the control of cardiac force and stroke volume. In these diagrams I have emphasized that elevation in the venous return of blood to the heart, or elevation in arterial resistance to flow, may increase the end-diastolic volume of the heart and, through

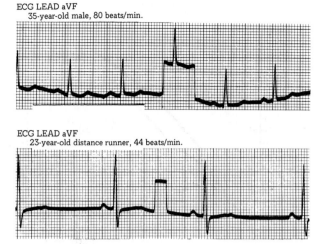

Fig. 8-10. Comparison of the ECG's of an office worker and an athlete. Both subjects were supine and rested 5 minutes prior to the recording.

this distension, increase the heart's force of contraction (Frank-Starling mechanism). I have also noted that changes in the internal or external environment of the heart sometimes modify its performance. These environmental changes result from increases in the metabolic activity of the heart or changes in its ion content (such as occur when the heart rate is elevated). They also result from increases in skeletal-muscle metabolism, abnormal nutrition, breath-holding, hyperventilation, or kidney malfunction. Finally, there are many hormones which affect cardiac metabolism and, as a consequence, stroke volume. For years there has been a controversy over which of the many factors plays the predominant role in cardiac control. Nor is it an easy question to answer. It is important to realize, however, that we have a tremendous array of potentially important overlapping controls.

The autonomic nerves, exhibiting reflex responses within several seconds after stimulation, markedly modify cardiac function. The value of the autonomic nervous system to the heart seems to lie in the speed with which it can change cardiac performance; cardiac nerves are often the first relief mechanisms in stress (exercise, hemorrhage, etc.). When autonomic changes are inadequate, some of the other controlling mechanisms come into play (cardiac distension, accumulation of metabolites, etc.).

Parasympathetic nerves. The preganglionic parasympathetic fibers to the heart originate in the medulla oblongata and pass down the right and left cervical vagi. Both vagi give off fibers to all parts of the two atria. Unilateral section or block of a cervical vagus nerve has little effect on cardiac function, but bilateral block or section in a conscious subject results in a 50 to 100 per cent increase in the heart rate in man and dog (though not in all mammals). One apparently very important function of the vagi is to decrease the frequency of depolarization of the S-A node and, as a result, to decrease the rate of contraction of the entire heart. This *negative chronotropic* (cardiac slowing) *action* of the vagi is a result of a continuous firing of parasympathetic neurons to the S-A node. *Parasympathetic tone* (activity) increases when arterial pressure increases, thus slowing the heart, and decreases during exercise or when arterial pressure is lowered, thus speeding the heart. These changes in heart rate may, in turn, bring about changes in stroke volume.

Some parasympathetic fibers from the right and left cervical vagi also go to the A-V node. Stimulation of these may further delay conduction of the atrial impulse to the ventricles (first-degree block) or prevent conduction altogether (second- or third-degree block). First-, second-, or third-degree *block* may occur in response to a reflex increase in vagal parasympathetic tone due to an elevation of arterial pressure. In Figure 8-12 we have an example of the importance of delay at the A-V junctional tissue. Note that as the P-R interval changes from 120 msec. to 0 msec., the systolic pressure in the left ventricle and aorta also diminish. A similar decrease in pressure occurs if we increase the P-R interval above 120 msec.

A third effect of parasympathetic stimulation is a direct *negative inotropic action* on the atria, i.e., decreased force of contraction by the atrial

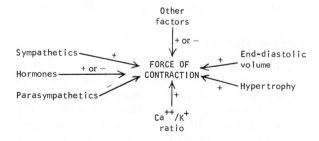

Fig. 8-11. Control of cardiac contractile force.

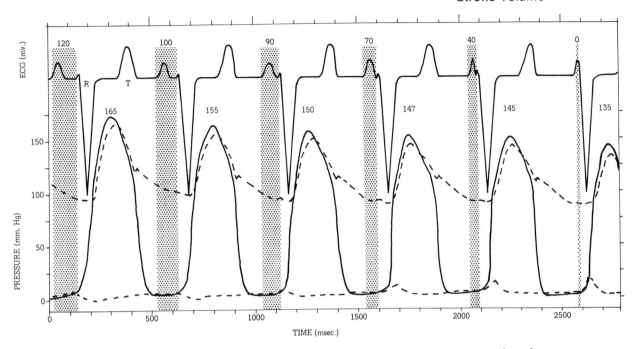

Fig. 8-12. Importance of the P-R interval. In these experiments pacemaker electrodes were placed on an atrium and ventricle and the interval between atrial and ventricular activation varied (synchronous pacing). The shaded areas represent the duration of the P-R interval. The parameters recorded are the ECG, aortic pressure (*interrupted line*), left intraventricular pressure (*solid line*), and left atrial pressure (*interrupted line*). Note the progressive decrease in the aortic and intraventricular pressure as the P-R interval decreases and the appearance of the cannon wave in the atrial pressure tracing. (*Modified from* Stephenson, S. E., and Brockman, S. K.: P wave synchrony. Ann. N. Y. Acad. Sci., *11*:911, 1964)

fibers due to vagal stimulation. This is not an indirect action through a change in heart rate, since it has been shown to occur during atrial pacing. Some have characterized it as a shift of the Starling curve (Fig. 8-6) to the right.

There is also physiological evidence that vagal stimulation may have a direct negative inotropic action on the ventricles. Though there is disagreement as to the importance of this, the vagi unquestionably exert an important indirect tonic influence on the ventricles through their control of atrial rate, force, and conduction.

Sympathetic nerves. Sympathetic nerves innervate all parts of the atria and ventricles. They have a *positive chronotropic action* on the S-A node but exert considerably less tone here in resting man and dog than do the vagal parasympathetic fibers. They can increase the heart rate during exercise or hypotension, but have little or no effect in lowering the heart rate during hypertension.

Cardiac sympathetic fibers have a direct *positive dromotropic action* (i.e., they increase the rate of conduction of the action potentials throughout the heart) and a direct *positive inotropic action* throughout the heart. One basis for this observation is in Table 8-2. In this study the dog's S-A node was destroyed and his heart rate was maintained at 116 beats per minute by atrial pacing. Under these conditions stimulation of a sympathetic ganglion (stellate ganglion) did not change the heart rate, but did decrease the duration of atrial systole; the duration of ventricular systole; the duration of ventricular isovolumic contraction; and the duration of ventricular relaxation. Since the total duration of the cycle was kept constant by atrial pacing, the period of rapid ventricular filling plus diastasis was increased.

Apparently the resting sympathetic tone to the heart is rather small, but sympathetic tone may increase markedly during exercise, excitement,

	Control	Stellate Stimulation	Change (%)
Duration of atrial systole (msec.)	120	87	−28
Duration of ventricular systole (msec.)	270	207	−23
Duration of ventricular diastole (msec.)	250	313	+25
Duration of isovolumic contraction (msec.)	105	55	−48
Relaxation time (msec.)	93	75	−19

Table 8-2. Changes in the cardiac cycle due to stellate ganglion stimulation in a dog with a heart rate maintained at 116 beats per minute by atrial pacing. The durations are estimated from pressure recordings. (Sarnoff, S. J., and Mitchell, J. H.: The control of the function of the heart. *In* Hamilton, W. F. (ed.): Handbook of Physiology. Sec. 1 (Circulation), vol. I, p. 512. Washington, American Physiological Society, 1962)

or hypotension. When this happens, increased sympathetic tone to the heart may be associated with increased sympathetic tone to the adrenal medulla and an associated release of *epinephrine* into the blood. Epinephrine, like the cardiac sympathetic fibers, has a direct positive chronotropic (Gk. *chronos*, time; plus *tropos*, a turning), dromotropic (Gk. *dromos*, run), and inotropic (Gk. *ino*, fiber) action on the heart. It may also, like the sympathetic fibers, increase the irritability of the heart (i.e., lower the heart's threshold to stimulation and lead to the development of ectopic pacemakers).

HEART RATE

The heart rate can be calculated on the basis of a number of cardiac cycles or on the basis of a single cardiac cycle. In Figure 8-13, for example, we note a single cardiac cycle of 2 sec. On the basis of this one cycle we can calculate the heart rate represented by its interval:

$$\text{Heart rate} = \frac{60 \text{ sec./min.}}{2 \text{ sec./cycle}} = 30 \text{ cycles/min.}$$

The preceding interval, however, is only 0.76 sec., which represents a heart rate of 79 beats per minute. This beat-to-beat variation in heart rate is characteristic of dogs (Fig. 8-13) and children and is called either *sinus arrhythmia* or *inspiratory tachycardia*, since it is due to reflex decreases in vagal inhibition of the S-A node during inspiration.

Not only can the heart rate vary from beat to beat, but also from person to person. In Figure 8-10 we saw the ECG's of a reclining distance runner and a reclining adult who leads a relatively sedentary life. The runner's heart rate is lower and his stroke volume higher than that of an average adult. He, as a result of his strenuous training program, has developed a highly muscular (hypertrophic) and somewhat distended heart. For most of us, however, our heart rate ranges between 60 and 200 beats per minute. During sleep we approach a rate of 60, and during intense exercise, one of 200 beats per minute. Heart rates between 250 and 350 beats per minute (flutter) and in excess of 350 (fibrillation) are indicative of pathology. They may result from multiple ectopic pacemakers (Fig. 6-25) or circus movement, or from Ischemia, scar tissue, and certain drugs.

Marked cardiac slowing (*bradycardia*) may also result from ischemia, scar tissue, and certain drugs. A drug that increases parasympathetic nervous tone to the S-A node can slow the ventricles, as can one that prevents A-V conduction. In the latter case, the ventricles beat more slowly than the atria. There is also a condition in which the ventricles beat more rapidly than the atria (*A-V dissociation*).

The importance to the individual of a change in heart rate is not at all clear. Obviously if an increased rate is associated with no change in stroke volume, it will produce an increase in cardiac output, but an increase in rate produces a decrease in the ventricular filling time (Table 8-3, duration of diastole), which may result in decreased stroke volume. On the other hand, increases in rate may produce an increase in the force of contraction (staircase phenomenon shown in Fig. 8-8). It is interesting, then, to review some of the work done on cardiac pacing. Miller et al. (Fig. 8-14) reported that in ventricular pacing (in anesthetized dogs), stroke volume begins to decrease at heart rates in excess of 30 to 70 beats per minute, and that cardiac outputs increase with increasing rates only up to 90 to 180 beats per minute. In other words, some dogs showed a

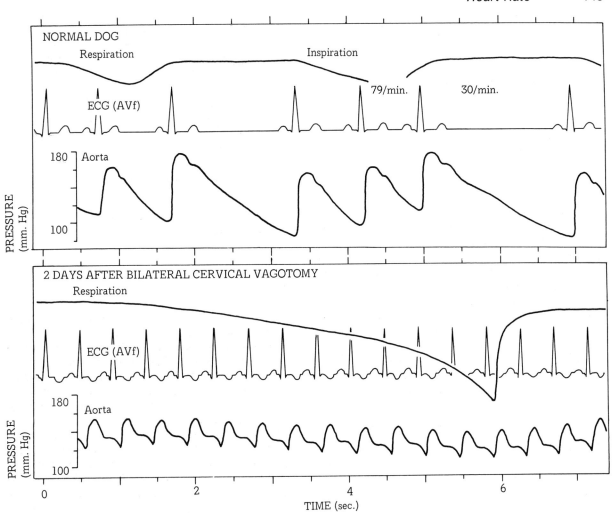

Fig. 8-13. Record from an unanesthetized dog before and after bilateral cervical vagotomy. Note that prior to vagotomy the dog had an increase in heart rate during inspiration (inspiratory tachycardia or sinus arrhythmia), which disappeared when the vagi were cut. Note also that section of the vagi was followed by an elevation in heart rate, a decrease in pulse pressure, and a prolongation of inspiration. (*Modified from* Shepard, R. S., and Whitty, A. J.: Bilateral cervical vagotomy: A long-term study on the unanesthetized dog. Am. J. Physiol., *206*:268, 1964)

decrease in stroke volume at rates greater than 30, while others showed no decrease until the rate exceeded 70.

Ross et al. paced 17 patients with an atrial pacemaker at rates ranging from 50 to 190 beats per minute and found no consistent increase in cardiac output with an increasing rate. It seems, then, that an increase in heart rate is not by itself a completely effective means of increasing cardiac output; to be effective it must be associated with an increase in venous return. An increase in heart rate might, however, be considered a mechanism for preventing *cardiac distension* and *venous congestion*.

Parasympathetic Control

The parasympathetic nerves to the heart usually control heart rate by two important means: (1) depression of the S-A node, and (2)

	Heart Rate	
	75/min.	200/min.
Duration of cardiac cycle (msec.)	800	300 (−62.5%)
Duration of ventricular systole (msec.)	270	160 (−40.7%)
Q-T interval (msec.)	270	160 (−40.7%)
Duration of absolute refractory period (msec.)	270	160 (−40.7%)
Duration of relative refractory period (msec.)	50	20 (−60.0%)
Duration of diastole (msec.)	530	140 (−73.5%)

Table 8-3. Changes in the cardiac cycle due to cardiac speeding.

depression of the A-V node (Fig. 8-15). Under most conditions the vagi are constantly depressing the rate at which the pacemaker fires, and delaying conduction across the A-V node. Under some conditions (e.g., marked increase in arterial pressure) vagal tone may so increase as to completely depress the S-A node or prevent conduction of some of the atrial potentials to the ventricle (heart block). When there is complete sinus depression an ectopic pacemaker usually develops (*vagal escape*). A decrease in vagal tone occurs in response to many common drugs (anesthetics, barbiturates, atropine, etc.) and to decreases in arterial pressure. This results in an increase in heart rate and a decrease in the P-R segment of the ECG.

Sympathetic Control

The cardiac sympathetic nerves can stimulate the S-A node to fire more rapidly. They, unlike the parasympathetic fibers, do not seem to exert a very important influence on the sinus pacemaker in the resting man or dog. However they can markedly increase the heart rate in response to exercise or to a decrease in arterial pressure. They also exert a much more important direct influence on the ventricles than do the parasympathetic fibers, and can therefore increase the rate of firing of ventricular pacemakers (when present), as well as of the S-A node. It is important to bear in mind that, although the sympathetic and parasympathetic fibers are antagonistic in some respects, the sympathetic fibers play a unique role in the case of the ventricle (Table 8-2).

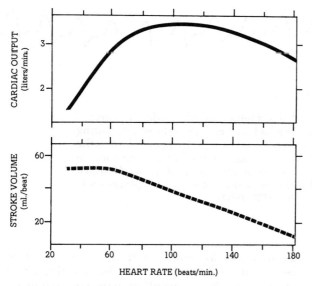

Fig. 8-14. Relationship between heart rate, stroke volume, and cardiac output in 6 anesthetized dogs with A-V block and a ventricular pacemaker. The dogs' ventricles were paced at rates ranging from 30 to 240 stimuli/min. (*Redrawn from* Miller, D. E., *et al.*: Effect of ventricular rate on the cardiac output in the dog with chronic heart block. Circ. Res., 10:659, 1962; by permission of the American Heart Association, Inc.)

Other Control Mechanisms

Figure 8-16 shows a comparison of the response to exercise of a normal dog and one in whom all the nerves to and from the heart have been severed. Both before and after cardiac decentralization, the dog responds to various grades of exercise on a treadmill with similar increases in cardiac output and oxygen consumption. Before decentralization the increase in cardiac output is associated with a marked increase in heart rate but very small changes in stroke volume and heart size. After decentralization the increase in heart rate during exercise is reduced but there is a marked increase in stroke volume and heart size. Apparently there are mechanisms for controlling cardiac output in the absence of nerves. One of these is the increased venous return of blood that occurs during exercise. In the absence of an associated increase in heart rate this causes cardiac distension and resultant increases in the force of contraction and stroke volume (Starling mechanism). Other mechanisms for increasing cardiac output are the release of the hormone epinephrine from the adrenal medulla, changes in the concentration of Ca^{++}, K^+, O_2, and CO_2, and changes in temperature in the heart.

CARDIAC MASSAGE

The function of the heart is to pump blood. Asphyxiation, electrical shock, many drugs, and coronary occlusion can cause cessation of this pumping action by initiating ventricular fibrillation or arrest. When a patient's heart stops pump-

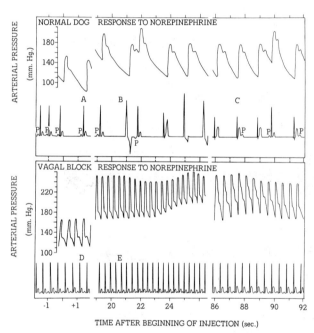

Fig. 8-15. Importance of the cervical vagi to the unanesthetized dog's response to norepinephrine (3 μgm./kg. in 30 sec.). Before the cervical vagi were blocked, the dog showed a normal ECG pattern at A, but at B (20 sec. after the injection of norepinephrine) the arterial pressure had risen, the S-A node was depressed, and an ectopic pacemaker had developed in the ventricle (vagal escape). At C, a P wave is seen, but does not produce ventricular activation (A-V dissociation) because of vagal depression of the A-V node. After blocking the vagi with xylocaine the dog responded to the same dose of norepinephrine with a more marked increase in arterial pressure, and cardiac speeding. The ECG patterns at D and E are similar. Apparently reflex slowing of the heart in response to arterial hypertension is capable of preventing a marked elevation of arterial pressure.

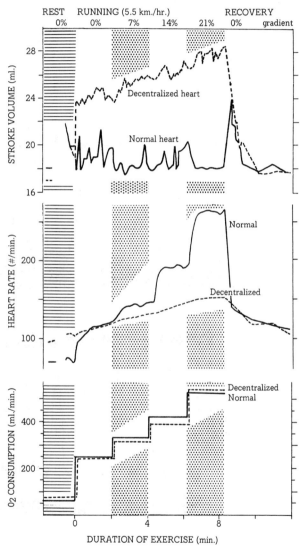

Fig. 8-16. Response of the dog before (*solid line*) and after (*interrupted line*) cardiac decentralization to exercise on a treadmill. The treadmill speed was 5.5 km./hr. and the slope was changed from 0% to 21%. The exercise was maintained for 8 min. (*Modified from* Donald, D. E., and Shepherd, J. T.: Sustained capacity for exercise in dogs after complete cardiac denervation. Am. J. Cardiol., *14*:856, 1964)

ing blood and he stops breathing he is said to be *clinically dead*. If the patient's brain is at 37°C. (body temperature) it will begin to show irreversible changes within 4 to 6 min. Now the patient is said to be *biologically dead*. At this point it is still possible to start the patient's heart beating again and to restore normal body function to those parts of the body not dependent upon the brain; after 4 to 6 min. of cardiac arrest the surviving patient remains a "vegetable."

It is imperative if the heart stops pumping adequate amounts of blood that some means be rapidly initiated to restore a well-oxygenated blood supply to the brain. The recommended method is cardiac massage and artificial respiration until the heart starts beating effectively and the individual starts breathing. If the chest is open the physician massages the heart by squeezing and releasing about 60 times a minute. If the chest is closed, the recommended method is to push down on the caudal portion of the body of the sternum 60 times a minute. If the massage is effective, a pulse is felt (Fig. 8-17) and the pupils of the eye are generally not dilated. Once an effective circulation has been restored to the brain and to the heart itself, a fibrillating heart

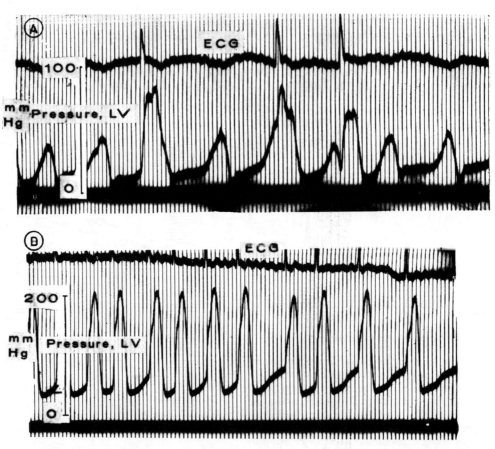

Fig. 8-17. Cardiac massage, the electrocardiogram, and ventricular pressure in a 45-year-old female with severe mitral stenosis. At A external heart compression is being administered. Note that when compression occurs at the same time as the R wave (pulse 3) that a higher pressure is obtained (80 mm. Hg) than when compression occurs in the absence of an R wave (pulse 2, 45 mm. Hg) or when an R wave occurs in the absence of compression (pulse 7, 60 mm. Hg). At B we note the pressure pattern after the intracardiac injection of 0.3 mg. of epinephrine and after further cardiac massage. The patient recovered and complained only of a "sore breast bone." (Elam, J. O.: Development of cardiac resuscitation. Intern. Anesthesiol. Clin., 2:118, 119, 1963)

can be defibrillated by means of electrical shocks alone or in combination with certain drugs. Since coronary ischemia produces fibrillation, massage and artificial respiration should generally precede any attempt at defibrillation. The American Heart Association, American National Red Cross, Industrial Medical Association, and United States Public Health Service have, to date, been reluctant to teach this technique to laymen as extensively as they have artificial respiration because of the difficulty in recognizing when it is appropriate to give massage and because of the danger during massage of (1) producing damage to the heart and liver, (2) initiating internal bleeding, and (3) fracturing ribs and puncturing the lungs.

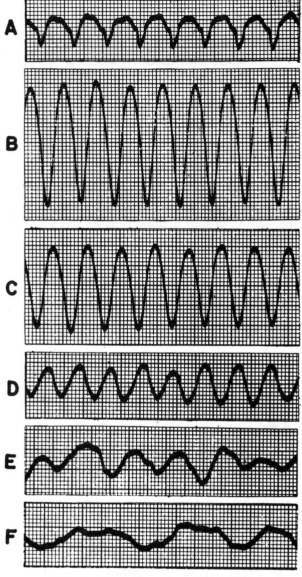

Fig. 8-18. The dying heart (ECG lead II). The records shown above were taken about 1 minute apart and show a transition from ventricular tachycardia (A), through ventricular flutter (B and C), through an intermediate stage between flutter and fibrillation (D), to ventricular fibrillation (E and F). (Marriott, H. J. L.: Practical Electrocardiography. 4th ed., p. 103. Baltimore, Williams & Wilkins, 1968; copyright © 1968, The Williams and Wilkins Company, Baltimore)

Chapter 9

ELECTRICAL PROPERTIES
ACTIVATION-CONTRACTION COUPLING
MULTI-UNIT SMOOTH MUSCLE
SINGLE-UNIT SMOOTH MUSCLE
CONTROL

SMOOTH MUSCLE

In the preceding four chapters I have emphasized that contraction in striated muscle involves the sliding of thin filaments (actin) over a thick filament (myosin). In the case of mammalian smooth muscle, however, no one has yet been able to demonstrate the presence of thick and thin filaments side by side (Fig. 9-1). This has led many to suggest that smooth muscle does not contain important amounts of myosin and that contraction here is not the result of sliding. Others have suggested that in smooth muscle thin filaments slide over adjacent thin filaments.

Smooth muscle also differs from the two types of striated muscle in that it is not usually the major element in the organ in which it is found; hence its properties may be considerably masked by associated epithelial and connective tissue or by other types of muscle. In a sense it is the great "et cetera classification" of contractile structure. One type of smooth muscle may differ from another as much as it differs from skeletal or cardiac muscle. Despite this marked variability, however, a few generalizations can be made.

ELECTRICAL PROPERTIES

All living muscle cells at rest apparently contain an *electric potential* across their cell membrane. This potential is maintained by the cell membrane as a result of its limited permeability to charged particles and its active transport systems (usually called "pumps"). In the case of skeletal muscle, the transmembrane potential is sufficiently high (about 80 mv.) and stable that the only physiological stimulus capable of destroying it is a release of highly concentrated acetylcholine at the neuromuscular junction. Cardiac muscle presents a somewhat different picture. Though most of its cells (follower cells) have stable potentials similar to those in skeletal muscle, a few of them (pacemaker cells), have less stable potentials. As positively charged ions leak into these cells, a critical potential is eventually reached and the cell depolarizes. This depolarization, unlike that in skeletal muscle, is conducted from one cell to the next.

The various smooth muscles of the body exhibit all the characteristics discussed above (stable transmembrane potential, pacemaker potential, intercellular conduction) plus some uniquely their own. Various types of transmembrane potentials reported for smooth muscle are shown in Figure 9-2. The first is similar to that seen in skeletal muscle, the second to a cardiac follower cell, and the eighth is from a pacemaker cell without the prolonged period of depolarization characteristic of cardiac pacemakers.

Patterns 4 through 7 are from smooth-muscle cells whose resting transmembrane potentials vary periodically. These fluctuations (called slow waves) represent variations in the excitability of the cell. They are not considered action potentials. In these drawings an elevation of the wave represents a reduction in the polarity and the threshold of excitability. When the wave is lowered we have *hyperpolarization* and, as a consequence, decreased *excitability*. In hyperpolarization, that is, it requires a stronger stimulus to excite. As can be seen from the illustration, these *slow waves* may be associated with volleys of spike potentials, an occasional spike, or no action potentials at all.

In Figure 9-3 we note that hyperpolarization increases the amplitude of the spike potential and that hypopolarization decreases it. If hyperpolarization is sufficiently increased, the spike potentials disappear altogether; and if polarity is reduced enough, intercellular conduction may cease (i.e., the spike potential of one cell will not be sufficient to depolarize a neighboring cell).

A. Skeletal muscle (frog's leg):

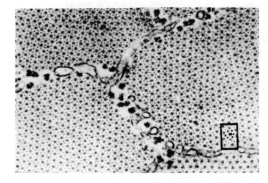

B. Cardiac muscle (cat's papillary muscle):

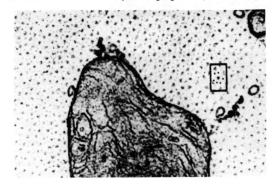

C. Smooth muscle (mouse epididymus):

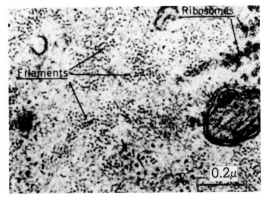

Fig. 9-1. Comparison of transverse sections through (A) skeletal muscle, (B) cardiac muscle, and (C) smooth muscle. Note in skeletal and cardiac muscle that each thick filament (myosin) is surrounded by an array of thin filaments (actin). The rectangles in these electron micrographs contain areas which have been touched up. In smooth muscle, on the other hand, we find irregular groupings of thin filaments (long arrows) and scattered thick filaments (short arrows) which do not seem to have a constant relationship with the thin filaments. It may be that the contractile mechanism hypothesized for striated muscle (sliding of thin filaments over thick filaments) is unimportant in most smooth muscle. (A: Huxley, H. E.: The mechanism of muscular contraction. Sci. Am., *213*:21, 1965; B, C: Fawcett, D. W.: An Atlas of Fine Structure. The Cell: Its Organelles and Inclusions. Pp. 239, 243. Philadelphia, W. B. Saunders, 1966)

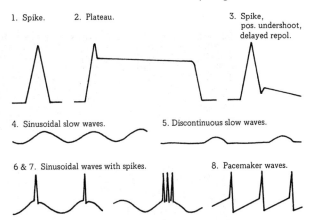

Fig. 9-2. Types of transmembrane potentials found in smooth muscle. (1) spike type; (2) plateau type; (3) spike followed by positive undershoot and delayed repolarization; (4) sinusoidal slow waves not associated with contraction; (5) regular discontinuous slow waves not associated with contraction; (6 and 7) sinusoidal slow waves with spikes; (8) pacemaker slow waves with associated spikes. (*Redrawn from* Burnstock, G., et al.: Electrophysiology of smooth muscle. Physiol. Rev., *43*:497, 1963)

Normally the transmembrane and spike potentials are lower in smooth than in striated muscle. The taenia coli of the large intestine sometimes have transmembrane potentials as high as 70 mv. and spike potentials of 60 mv., but on the other hand the pregnant guinea pig uterus may have resting potentials of less than 35 mv. The duration of the spike potential in smooth muscle varies from 5 msec. in the vas deferens to 500 msec. in the dog ureter.

ACTIVATION-CONTRACTION COUPLING

The action potential in all muscles initiates an increase in the intracellular concentration of Ca^{++}. In striated muscle the sarcoplasmic reticulum usually constitutes an important mecha-

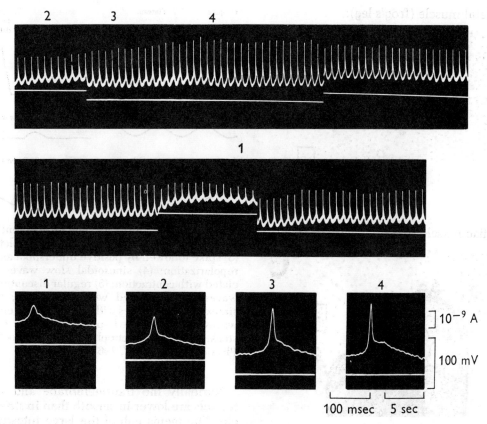

Fig. 9-3. Effects of hyperpolarization and depolarization on the spontaneous electrical activity of the taenia coli of the guinea pig. Electrical recording was from a single intracellular microelectrode. Note that normal electrical activity at 2 includes sinusoidal slow waves and associated spikes. Hyperpolarization (3 and 4) resulted in an increase in the spike potential and rate of change of potential. Incomplete depolarization (1) resulted in a decrease in the spike potential and the rate of change of potential. Lower records represent a faster time base. (Kuriyama, H., and Tomita, T.: The responses of single smooth muscle cells of guinea pig taenia coli to intracellularly applied currents, and their effect on the spontaneous electrical activity. J. Physiol., *178:* 276, 1965)

nism for Ca^{++} release and pickup. In smooth muscle, the reticulum is poorly developed and most of the Ca^{++} must come (1) from other sources in the cell, (2) from the cell membrane, and (3) through the membrane from the external environment. Apparently it is the accumulation of Ca^{++} around the myofilament that initiates contraction, and the associated hydrolysis of ATP and creatine phosphate that brings about the production of lactic acid, an increase in oxygen consumption, and the production and maintenance of tension in all muscles. In Figure 9-4 we see a demonstration of the dependence of the contractile mechanism of the carotid artery on Ca^{++}. You will note that the Ca^{++} concentration-tension curves for the carotid artery and psoas muscle are superimposed on each other. Figure 9-5 illustrates the effect of epinephrine on tension, creatine phosphate concentration, and ATP concentration. Beviz et al. found that early in the response to epinephrine the concentration of ATP and creatine phosphate consistently decreased, and that the concentration of lactic acid increased. ATP concentration returned to normal levels while the muscle was still exerting tension.

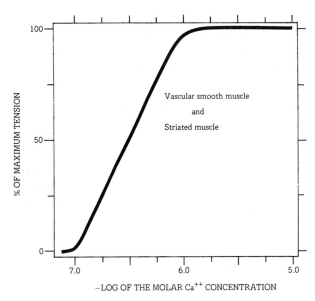

Fig. 9-4. Effect of Ca^{++} concentration on the tension developed by individual hog psoas (skeletal) and carotid artery (smooth) muscle. The fibers were extracted in 50% glycerol for 1 to 3 months prior to the study. The psoas fibers were teased to a diameter of 50 to 100 μ and developed a maximum tension of 1,000 Gm./cm.^{2} on the addition of Ca^{++}. The fibers from the media of the carotid were teased to a thickness of 150 to 300 μ and developed a maximum tension of 100 Gm./cm.^{2} The curves developed for the 2 different muscles were identical. (*Redrawn from* Filo, R. S., et al.: Glycerinated skeletal muscle: Calcium and magnesium dependence. Science, *147*:1582, 1965; copyright March, 1965 by the American Association for the Advancement of Science)

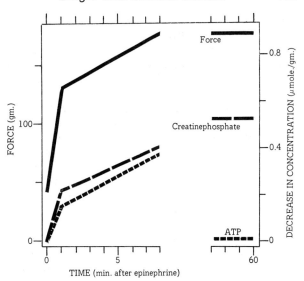

Fig. 9-5. Influence of epinephrine (10^{-6} moles/liter every 15 min.) on isolated bovine mesenteric arteries. Prior to the epinephrine, ATP was at a concentration of 0.91, and creatine phosphate (CP) at 0.66 μmoles/Gm. of wet weight. After 1 min. there were significant decreases in ATP and CrP (*see graph*) and increases in lactic acid (+1.0 μmoles/Gm.). After 1 hour the tension was maximal and the muscle had a normal concentration of ATP, but had lost 80% of its CrP. (*Modified from* Beviz, A., et al.: Hydrolysis of adenosinetriphosphate and creatinephosphate on isometric contraction of vascular smooth muscle. Acta Physiol. Scand, 65:270, 1965)

MULTI-UNIT SMOOTH MUSCLE

The ciliary muscle (intrinsic muscle of the eye), nictitating membrane and iris, pilomotor muscle (skin), and the muscles of some of the larger blood vessels are classified as multi-unit smooth muscle. Functionally these structures resemble skeletal muscle. They do not exhibit intercellular conduction and their contraction is usually initiated by stimulation of a motor nerve rather than by the firing of an intrinsic pacemaker. They differ from skeletal muscle, however, in that their motor nerve fibers (autonomic neurons) synapse between the central nervous system and the muscle fiber. The postsynaptic neurons do not generally form well-defined terminal neuromuscular junctions with the muscle fibers they innervate; rather they release their neurohormone at spots along the neuron where the enveloping neurilemma is absent and where there are numerous synaptic vesicles. This results in more diffuse innervation with a considerably greater overlap in control than is found in skeletal muscle, with its well-defined motor units. The neurotransmitter released near smooth muscle usually persists longer, too, and may cause a *series* of action potentials rather than the one found in skeletal muscle.

SINGLE-UNIT SMOOTH MUSCLE

A second type of smooth muscle, single-unit muscle (visceral muscle), includes most visceral muscle and resembles the heart more than it does skeletal muscle in function. This category includes the muscle of the gastrointestinal tract, the ureter, the uterus, and some arteries and veins. These structures continue to show well-coordinated rhythmic contractions when removed from the body, due to the firing of one or more intrinsic *pacemakers*. They, like the heart, conduct impulses from one cell to the next. It has been suggested that this *intercellular conduction* results from the proximity of one cell membrane to the next (Fig. 9-6) or from the fusion of one cell membrane with another (Fig. 6-3).

In the ureter the pacemaker lies in or near the renal pelvis and may initiate a contraction that travels throughout the ureter to the urinary bladder. In the intestine, on the other hand, there seem to be many functioning pacemakers, all initiating contractions that travel only a small distance and then disappear. Also, the *intrinsic nerve plexi* play an important role in the co-ordination of contraction in the intestine. A segment of intestine removed from the body exhibits peristalsis (a wave of contraction moving toward the anal end of the intestine); though this can be eliminated by drugs or procedures that depress or destroy the intrinsic nerve fibers, periodic, localized contractions throughout a segment of intestine will persist.

In some structures intrinsic pacemakers fire only under certain conditions. In the uterus under estrogen domination, for example, functioning pacemaker zones can be found in the ovarian end of some specimens and in the cervical part of others. In the absence of estrogens or in progesterone domination the uterine pacemakers disappear.

It is also important to remember that many smooth muscles capable of intrinsic pacemaker

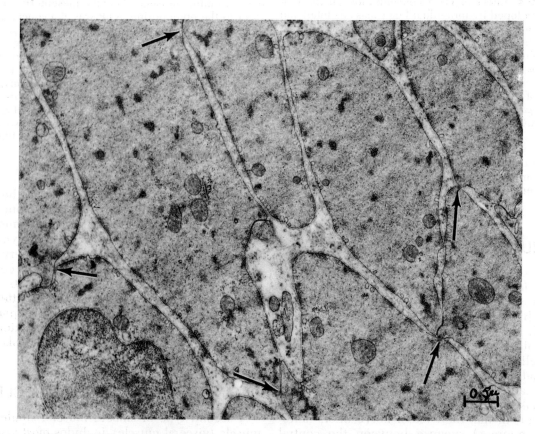

Fig. 9-6. Electromicrograph from the smooth muscle of the circular layer of the small intestine of a dog. Arrows indicate points of close apposition of adjacent cell membranes. These are areas where intercellular conduction may occur. (Dewey, M. M., and Barr, L.: A study of the structure and distribution of the nexus. J. Cell. Biol., 23: 555, 1964)

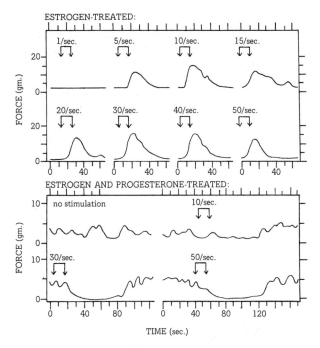

Fig. 9-7. Response of the immature, isolated rabbit uterus to stimulation of its hypogastric nerves. One group of rabbits received 70 μg. of estradiol-17β every other day for 4 days (estrogen-treated), and a second group received 70 μg. of estradiol-17β every other day for 4 days and, then, 5 mg. of progesterone daily for 5 days (estrogen- and progesterone-treated). Stimulation of the hypogastric nerves (5–50 stimuli/sec.) to both untreated and estrogen-treated uteri produced contractions, but stimulation of the nerves to the estrogen- and progesterone-treated uteri inhibited the spontaneous contractions characteristic of this preparation. (Redrawn from Miller, M. D., and Marshall, J. M.: Uterine response to nerve stimulation: Relation to hormonal status and catecholamines. Am. J. Physiol., 209:860, 861, 1965)

activity under certain conditions may, under different conditions, have contractions initiated by the stimulation of motor nerves. In Figure 9-7, for example, we note that isolated uteri from estrogen-treated, immature rabbits show no intrinsic rhythmicity, but do contract in response to stimulation of sympathetic neurons in their hypogastric nerves. If, on the other hand, a regime of estrogen and progesterone treatment is initiated, the uterus demonstrates intrinsic contractions which are inhibited by the stimulation of sympathetic neurons in the hypogastric nerves.

The urinary bladder, vas deferens, and retractor penis also contain smooth muscles which are dominated by a pacemaker under certain circumstances, and by the central nervous system under other circumstances. The urinary bladder, for example, has a great variety of specialized receptors which send impulses to the central nervous system (free nerve endings, Krause end-bulbs, Golgi-Mazzoni and genital corpuscles, and Ruffini endings). These receptors make possible the sensations of pain, temperature, and pressure in the bladder, and perform a role similar to that of the proprioceptive system in skeletal muscle and its associated tendons and aponeuroses. Impulses in a single afferent fiber from the bladder of a cat have been recorded and found to increase in frequency as the intravesicular pressure rises, and to disappear after the bladder is emptied. Vesicoconstrictor and vesicorelaxer areas (which change the tone of detrusor muscle in the urinary bladder) have been located in the sacral cord, medulla oblongata, pons, and midbrain. Other areas associated with micturition have been reported in the septum pellucidum, amygdala, and cerebral cortex. In short, control and integration of bladder function involve an extensive central nervous representation. It is interesting to note that the urinary bladder, unlike multi-unit or skeletal muscle, exhibits automatic responses when removed from the body. If, for example, the ureter of a rabbit is filled with warm saline, the bladder will contract in the absence of an extrinsic nerve supply. Destruction of the intramural ganglia, however, abolishes this response.

CONTROL

The force of contraction of smooth muscle, like that of striated muscle, is determined by the number and frequency of contractile reactions initiated, and by the synchrony of these reactions. These, in turn, depend on the energy available to the muscle, the number of contractile molecules present, the relationship between these molecules, the amount of Ca^{++} available to the muscle cell, and the rate at which the Ca^{++} is released to the myofilaments.

Transmembrane Potential

We have at our disposal essentially two ways of controlling muscle function. One is through modification of the electrical potential across the cell membrane, and the other is through modification of the contractile apparatus. The major way—possibly the only way—the body can initiate contraction is through depolarization of the muscle cell membrane. In the case of skeletal muscle this results from the stimulation of somatic efferent neurons and the subsequent release of acetylcholine at the neuromuscular junction. In the case of the ciliary muscle of the eye (smooth muscle), depolarization results from the stimulation of parasympathetic neurons and their release of acetylcholine. Depolarization of the radial muscle of the eye, on the other hand, is initiated by adrenergic sympathetic fibers.

In skeletal muscle the stimulation of somatic efferent fibers initiates contraction. The physi-ological inhibition of the contraction of skeletal muscle occurs only at the level of the central nervous system. In some smooth muscles, however, *inhibition* is an important peripheral phenomenon. We have seen in Figure 9-7 how, under certain circumstances, the stimulation of sympathetic fibers to the uterus can depress its intrinsic contractions. Norepinephrine and epinephrine have been shown to inhibit the intrinsic tone of the isolated coronary artery as well.

I have indicated that nerve stimulation of skeletal muscle causes the release of acetylcholine, which in turn completely depolarizes the cell membrane. This acetylcholine is rapidly destroyed, and while the muscle is contracting, the membrane repolarizes. In smooth muscle,

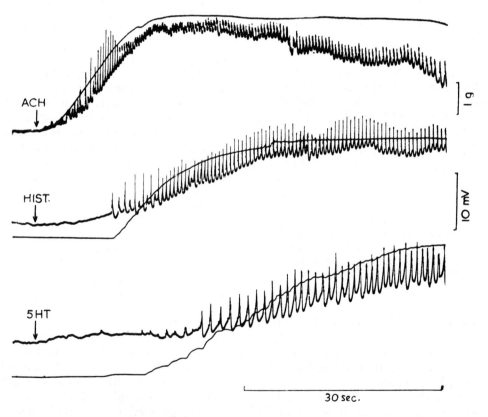

Fig. 9-8. Mechanical (*smooth line*) and electrical (*spiked line*) response of isolated ganglion-free muscle from the taenia coli of the guinea pig to 10^{-6} Gm./ml. of acetylcholine, 10^{-7} Gm./ml. of histamine, and 5×10^{-8} Gm./ml. of 5-hydroxytryptamine. Note that each of these agents increases muscle tension and the frequency of the spike potentials and decreases the membrane potential. A similar effect results from stretching this muscle. (Bülbring, E., and Burnstock, G.: Membrane potential changes associated with tachyphylaxis and potentiation of the response to stimulating drugs in smooth muscle. Brit. J. Pharmacol., *15*:614, 1960)

Control 155

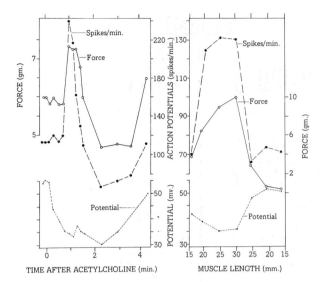

Fig. 9-9. Effect of acetylcholine (1 μg.) and stretch on the cell membrane potential, frequency of action potentials, and the force exerted by the taenia coli muscle of the guinea pig. Note that acetylcholine produced a prolonged period in which the membrane potential was reduced, the frequency of spike potentials increased, and the muscle force increased. Muscle stretch produced less marked changes in the frequency of the spikes, but a more forceful contraction. (Redrawn from Bülbring, E.: Correlation between membrane potential, spike discharge and tension in smooth muscle. J. Physiol., *128*:207, 211, 1955)

however, a stimulus (nerve impulse, serotonin, histamine, hormone, stretch, hypoxia, hypercapnea, Ca^{++}, Na^+, K^+, H^+) frequently acts by shifting the transmembrane potential. In Figure 9-8, for example, we note that the infusion of acetylcholine, histamine, or 5-hydroxytryptamine around an isolated strip of taenia coli from the intestine both decreases the membrane potential (seen as an elevation in the base line) and increases the rate at which the intrinsic pacemaker fires. This action of acetylcholine has also been demonstrated in the nictitating membrane, iris, muscularis mucosa of the esophagus, stomach, longitudinal and circular muscle of the intestine, ureter, urinary bladder, vas deferens, uterus, and amnion. In Figure 9-9 we note that stretching muscle from the colon from an initial length of 15 mm. to 20, 25, or 30 mm. also lowers the resting membrane potential and increases the frequency of the spikes. Decreasing the length from 30 to 15 mm. has the opposite effect.

We see, then, that an important response of smooth muscle to different types of stimulation is a modification of the *transmembrane potential. Cold, stretch, acetylcholine, histamine, 5-hydroxytryptamine, parasympathetic stimulation,* an *increase in extracellular K^+*, and a *decrease in extracellular Ca^{++}* are all capable of lowering this potential, thereby increasing *excitability* or initiating an action potential. When lowering the potential does not itself initiate contraction, it may allow an extrinsic stimulus to do so. During reduced polarity an intrinsic pacemaker takes less time to reach a critical transmembrane potential, so that the frequency of depolarization increases.

Hyperpolarization and an associated decrease in excitability and the rate of pacemaker firing occur in the intestine in response to epinephrine or sympathetic stimulation. The uterus, on the other hand, shows greater sensitivity to the reproductive hormones. In an ovarectomized rabbit, the uteri have a transmembrane potential of only 35 mv. and show little or no spontaneous activity. After *estrogen* treatment the potential increases to 46 mv., but spontaneous potentials also appear. Thus we see that an increase in polarity is not always associated with a decrease in the number of spike potentials; if a maintained decrease in polarity is sufficiently marked, any further depolarization occurring in response to stimulation may be too weak to produce a *propagated* or conducted *potential*. Progesterone treatment after the estrogen priming mentioned above increases the resting potential from 46 mv. to 54 mv. and reduces the frequency of spike potentials.

After estrogen treatment or late in pregnancy the posterior pituitary hormone *oxytocin* lowers the resting membrane potential slightly, elevates excitability, and increases the frequency of contraction in the uterus. In Figure 9-10 we note some of the changes that take place in a human uterus during pregnancy. During the second and third trimester of pregnancy the estrogen and progesterone content of the blood increase markedly. As pregnancy progresses during the third trimester, the frequency and force of the myometrial (uterine) contractions also increase, as do the excitability of the uterus and its responsiveness to oxytocin. After parturition the amount of estrogen and progesterone in the blood and the number of uterine contractions are reduced.

Our understanding of the mechanisms underlying changes in the resting membrane potential is still highly speculative. It has been suggested,

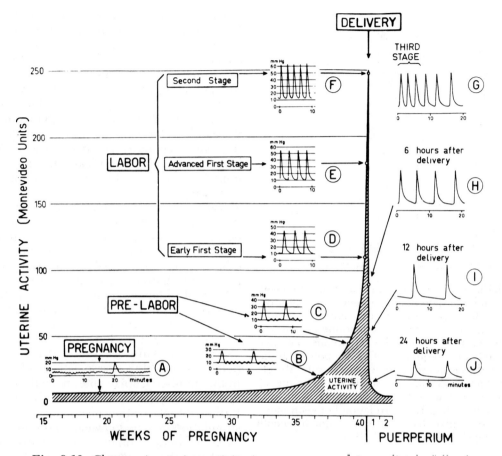

Fig. 9-10. Changes in uterine activity in pregnancy and immediately following parturition. (Annals of the New York Academy of Sciences, vol. 75, Caldeyro-Barcia, R., et al. Fig 1, p. 814; copyright by New York Academy of Sciences, 1959; reprinted by permission)

however, that agents decrease the transmembrane potential by increasing the permeability of the cell membrane to Na+ or by interfering with the pumps which extrude positively charged ions from the cell. Conversely, an increase in the transmembrane potential results from agents that decrease permeability to Na+, decrease intracellular K+, or facilitate the pumps. Acetylcholine is said to increase the permeability of the membrane to Na+ (depolarization). Epinephrine, heat, and factors which elevate metabolic rate hyperpolarize the membrane, it has been suggested, by increasing the activity of the sodium pump. Lack of glucose also reduces polarity. Calcium ion not only is an essential part of the excitation-contraction coupling system, but also can hyperpolarize the membrane by decreasing its permeability to Na+. Thus an increase in Ca++ can both depress excitability (membrane effect) and increase the force of contraction in response to an adequate stimulus (effect on contractile mechanisms).

Contractile Mechanism

One limiting factor in any contraction is the amount of contractile protein present in the cell. There are in the body a number of hormones that control the amount of protein produced by a muscle. In the uterus, for example, more protein is laid down if estrogen is present in the blood than when it is absent or in low concentrations.

Another limiting factor is the amount of Ca++ released around the myofilaments. There have recently been some suggestions that certain hormones increase the force of contraction in smooth

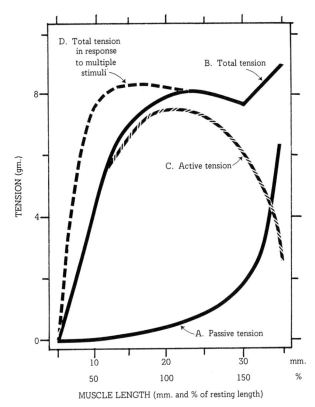

Fig. 9-11. Length-tension diagram for a strip of taenia coli in which the intrinsic rhythmicity has been depressed by 10^{-6} Gm./ml. of epinephrine. Curve A represents the tension developed in response to stretch (passive tension), curve B the tension in response to stretch followed by a single electrical stimulus (total tension), curve C the difference between A and B (active tension), and curve D the total tension produced by multiple stimuli. Each strip studied was 20 mm. long before it was removed from the colon. (*Redrawn from* Mashima, H., and Yoshida, T.: Effect of length on the development of tension in guinea-pig's taenia coli. Jap. J. Physiol., *15*:468, 1965)

muscle by increasing the influx of Ca^{++} into the cell during depolarization and by facilitating the release of Ca^{++} from certain sites where it is bound or deposited.

Still another means of modifying the response of a muscle to its action potential is stretch. In Figure 9-11 we see the results obtained from a study of the taenia coli muscle. In these experiments the intrinsic rhythmicity of this muscle was depressed by hyperpolarizing its cells with epinephrine. The muscle was then stretched or allowed to shorten and the tension it developed at the various lengths before stimulation (passive tension) and after the application of a single stimulus (total tension) was recorded. The length-tension diagram obtained by this technique is similar to that obtained for skeletal muscle. It is tempting, then, to suggest that stretching smooth muscle pulls actin filaments away from the center of the myosin molecule (sliding-filament theory) and that this results, at first, in more actin receptor sites becoming available to the myosin cross-bonds. This concept was discussed more fully in the chapter on skeletal muscle. The problem with this explanaton is that to date it has not been established that the actin and myosin of smooth muscle exist in separate filaments.

REFERENCES

Åberg, A. K. G., and Axelsson, J.: Some mechanical aspects of an intestinal smooth muscle. Acta Physiol. Scand., *64*:15-27, 1965.

Abildskov, J. A., et al.: Symposium on the present status of electrocardiography. Am. J. Cardiol., *14*: 285-436, 1964.

Ahlborg, B., Bergström, J., Ekelund, L.-G., and Hultman, E.: Muscle glycogen and muscle electrolytes during prolonged physical exercise. Acta Physiol. Scand., *70*:129-142, 1967.

Alexander, R. S.: Contractile mechanics of venous smooth muscle. Am. J. Physiol., *212*:852-858, 1967.

Alexander, R. S., and Johnsion, P. D., Jr.: Muscle stretch and theories of contraction. Am. J. Physiol., *208*:412-416, 1965.

American Heart Association: Recommendations for human blood pressure determination by sphygmomanometers. New York, 1967.

Aranaga, C. E., Mower, M. M., Staewan, W. S., and Tabatznik, B.: Eight-hour electrocardiogram: Technique and clinical application. Brit. Heart J., *29*: 345-351, 1967.

Bar, U., and Blanchaer, M. C.: Glycogen and CO_2 production from glucose and lactate by red and white skeletal muscle. Am. J. Physiol., *209*:905-909, 1965.

Bellet, S.: Clinical Disorders of the Heart Beat. Philadelphia, Lea & Febiger, 1963.

Bellet, S., and Muller, O. F.: The electrocardiogram during exercise. Circulation, *22*:477-487, 1965.

Benchimol, A.: Cardiac functions during electrical stimulation of the heart. Effect of exercise and drugs in patients with permanent pacemakers. Am. J. Cardiol., *17*:27-42, 1966.

Berne, R. M., and Levy, M. N.: Heart. Ann. Rev. Physiol., *26*:153-186, 1964.

Beviz, A., Lundholm, L., Mohme-Lundholm, E., and Vamos, N.: Hydrolysis of adenosine triphosphate and creatine phosphate on isometric contraction of vascular smooth muscle. Acta Physiol. Scand., 65: 268-272, 1965.

Bing, R. J.: Myocardial metabolism. Circulation, 12: 635-647, 1955.

Bohr, D. F.: Electrolytes and smooth muscle contraction. Pharmacol. Rev., 16:85-111, 1964.

Brady, A. J.: Active state in cardiac muscle. Physiol. Rev., 48:570-600, 1968.

Braunwald, E.: The control of ventricular function in man. Brit. Heart J., 27:1-16, 1965.

Braunwald, E.: Heart. Ann. Rev. Physiol., 28:227-266, 1966.

Braunwald, N. S., Gay, W. A., Jr., Morrow, A. G., and Braunwald, E.: Sustained, paired electrical stimuli. Am. J. Cardiol., 14:385-393, 1964.

Bruce, T. A., and Chapman, C. B.: Left ventricular residual volume in the intact and denervated dog heart. Circ. Res., 17:379-385, 1965.

Bülbring, E.: Correlation between membrane potential, spike discharge and tension in smooth muscle. J. Physiol., 128:200-221, 1955.

Bülbring, E., and Burnstock, G.: Membrane potential changes associated with tachyphylaxis and potentiation of the response to stimulating drugs in smooth muscle. Brit. J. Pharmacol., 15:611-624, 1960.

Burnstock, G., and Holman, M. E.: Smooth muscle—Autonomic nerve transmission. Ann. Rev. Physiol., 25:61-85, 1963.

Burnstock, G., Holman, M. E., and Prosser, C. L.: Electrophysiology of smooth muscle. Physiol. Rev., 43:482-527, 1963.

Caldeyro-Barcia, R., and Poseiro, J. J.: Oxytocin and contractility of the pregnant human uterus. Ann. N. Y. Acad. Sci., 75:813-830, 1959.

Cannon, W. B., and Rosenblueth, A.: The Supersensitivity of Denervated Structures. New York, Macmillan, 1949.

Chardack, W. M., Gage, A. A., and Greatbatch, W.: Correction of complete heart block by a self-contained and subcutaneously implanted pacemaker. J. Thorac. Cardiovasc. Surg., 42:814-830, 1961.

Close, R.: The relation between intrinsic speed of shortening and duration of the active state of muscle. J. Physiol., 180:542-559, 1965.

Comroe, J. H.: Physiology of Respiration. Chicago, Year Book Publishers, 1965.

Constantin, L. L., Franzini-Armstrong, C., and Podolsky, R. J.: Localization of calcium-accumulating structures in striated muscle fibers. Science, 147: 158-159, 1965.

Creed, R. S., Denny-Brown, D., Eccles, J. C., Liddell, E. G. T., and Sherrington, C. S.: Reflex Activity of the Spinal Cord. Oxford, Clarendon Press, 1932.

Davila, J. C.: Symposium: Measurement of left ventricular volume. Am. J. Cardiol., 18:1-42, 1966.

Dewey, M. M., and Barr, L.: A study of the structure and distribution of the nexus. J. Cell Biol., 23:553-585, 1964.

Dill, D. B.: Symposium on work and the heart. Am. J. Cardiol., 14:729-889, 1964.

Dodge, H. T., Sandler, H., Baxley, W. A., and Hawley, R. R.: Usefulness and limitations of radiographic methods for determining left ventricular volume. Am. J. Cardiol., 18:10-24, 1966.

Donald, D. E., and Shepherd, J. T.: Sustained capacity for exercise in dogs after complete cardiac denervation. Am. J. Cardiol., 14:853-859, 1964.

Ebashi, S., Oosawa, F., Sekine, T., and Tonomura, Y.: Molecular biology of muscular contraction. Tokyo, Igaku Shoin, 1965.

Eichna, L. W.: Proceedings of a symposium on vascular smooth muscle. Physiol. Rev. (Suppl.), 5:365, 1962.

Elam, J. O.: Respiratory and circulatory resuscitation. In Fenn, W. O., and Rahn, H. (eds.): Handbook of Physiology. Sec. 3 (Respiration), vol. II, pp. 1265-1312. Washington, American Physiological Society, 1965.

Ernst, E.: Biophysics of the striated muscle. Budapest, Akadémiai Kiodó, 1963.

Evans, J. R.: Structure and Function of Heart Muscle. New York, American Heart Association, 1964.

Evans, W.: Which leads shall we take? Am. Heart J., 64:143-148, 1962.

Fawcett, D. W.: An Atlas of Fine Structure. The Cell: Its Organelles and Inclusions. Philadelphia, W. B. Saunders, 1966.

Filo, R. S., Bohr, D. F., and Ruegg, J. C.: Glycerinated skeletal muscle: Calcium and magnesium dependence. Science, 147:1581-1583, 1965.

Fishman, A. P., and Richards, D. W. (eds.): Circulation of the Blood. Men and Ideas. New York, Oxford University Press, 1964.

Fletcher, G. F., Hurst, J. W., and Schlont, R. C.: Electrocardiographic changes in severe hypokalemia. A reappraisal. Am. J. Cardiol., 20:628-631, 1967.

Fry, D. L.: Myocardial mechanics: Tension-velocity-length relationships of heart muscle. Circ. Res., 14: 73-85, 1964.

Garb, S.: The effects of potassium, ammonium, calcium, strontium, and magnesium on the electrocardiogram and myogram of normal heart. J. Pharmacol. Exp. Ther., 101:317-326, 1951.

Gergely, J.: Biochemistry of Muscle Contraction. Boston, Little, Brown & Co., 1964

Gergely, J., et al.: The relaxing factor of muscle. Fed. Proc., 23:885-939, 1964.

Gibbs, C. L., and Ricchiuti, N. V.: Activation heat in muscle: Method for determination. Science, 147: 162-163, 1965.

Glenn, W. W. L.: Cardiac pacemakers. Ann. N. Y. Acad. Sci., 111:813-1122, 1964.

Gregg, D. E., and Fisher, L. C.: Blood supply to the heart. In Hamilton, W. F. (ed.): Handbook of Physiology. Sec. 2 (Circulation), vol. II, pp. 1517-1584. Washington, American Physiological Society, 1963.

Hajdu, S.: Mechanism of staircase and contracture in ventricular muscle. Am. J. Physiol., *174*:371-380, 1953.

Hajdu, S., and Leonard, E.: The cellular basis of cardiac glycoside action. Pharmacol. Rev., *11*:173-209, 1959.

Ham, A. W., and Leeson, T. S.: Histology. 4th ed. Philadelphia, J. B. Lippincott, 1961.

Hamilton, W. F. (ed.) Handbook of Physiology. Sec. 2 (Circulation), vol. I-III. Washington, American Physiological Society, 1962, 1963, 1965.

Harrison, D. C., Glick, G., Goldblatt, A., and Braunwald, E.: Studies on cardiac dimensions in intact unanesthetized man. Circulation, *29*:186-194, 1964.

Hawthorne, E. W.: Heart. Ann. Rev. Physiol., *27*:351-394, 1965.

Hecht, H.: Normal and abnormal transmembrane potentials of the spontaneously beating heart. Ann. N. Y. Acad. Sci., *65*:700-730, 1957.

Heymans, C., and Neil, E.: Reflexogenic Areas of the Cardiovascular System. London, J. & A. Churchill, 1958.

Hider, C. F., Taylor, D. E. M., and Wade, J. D.: The sequence of contraction of the left and right ventricles of the dog. Quart. J. Exp. Physiol., *50*:456-465, 1965.

Hill, A. V.: The effect of load on the heat of shortening of muscle. Proc. Roy. Soc. Lond., *159*:297-318, 1963-64.

Hukuhara, T., Nakayama, S., and Fukuda, H.: On the problem whether the intestinal motility is of a neurogenic or myogenic nature. Jap. J. Physiol., *15*:512-522, 1965.

Huxley, A. F.: Muscle. Ann. Rev. Physiol., *26*:131-152, 1964.

Huxley, H. E.: The mechanism of muscular contraction. Sci. Am., *213*:18-27, 1965.

Hurwitz, L., Joiner, P. D., Von Hagen, S.: Calcium pools utilized for contraction in smooth muscle. Am. J. Physiol., *213*:1299-1304, 1967.

Irisawa, H., Morio, M., and Seyama, I.: The presence of notches on the normal P wave. Jap. J. Physiol., *15*:17-27, 1965.

Johansson, B., and Bohr, D. F.: Rhythmic activity in smooth muscle from small subcutaneous arteries. Am. J. Physiol., *210*:801-806, 1966.

Johansson, B., Jonsson, O., Axelsson, J., and Wahlström, B.: Electrical and mechanical characteristics of vascular smooth muscle response to norepinephrine and isoproterenol. Circ. Res., *21*:619-634, 1967.

Johansson, B., and Ljung, B.: Spread of excitation in the smooth muscle of the rat portal vein. Acta Physiol. Scand., *70*:312-322, 1967.

Johnson, R. A., and Blake, T. M.: Lymphatics of the heart. Circulation, *33*:137-142, 1966.

Kahn, A. J., and Sandow, A.: Effects of bromide, nitrate, and iodide on responses of skeletal muscle. Ann. N. Y. Acad. Sci., *62*:139-175, 1955.

Katz, A. M.: The descending limb of the Starling curve and the failing heart. Circulation, *32*:871-875, 1965.

Katz, L. N.: Recent concepts of the performance of the heart. Circulation, 28:117-135, 1963.

Kimura, N., and Toshima, H.: Essential difference between vectorcardiogram and electrocardiogram. Jap. Circ. J., 27:61-67, 1963.

Klausen, K.: Comparison of CO_2 rebreathing and acetylene methods for cardiac output. J. Appl. Physiol., *20*:763-766, 1965.

Kosan, R. L., and Burton, A. C.: Oxygen consumption of arterial smooth muscle as a function of active tone and passive stretch. Circ. Res., *18*:79-88, 1966.

Kuriyama, H., Osa, T., and Toida, N.: Membrane properties of the smooth muscle of guinea-pig ureter. J. Physiol., *191*:225-238, 1967.

Kuriyama, H., Osa, T., and Toida, N.: Electrophysiological study of the intestinal smooth muscle of the guinea-pig. J. Physiol., *191*:239-256, 1967.

Kuriyama, H., Osa, T., and Toida, N.: Nervous factors influencing the membrane activity of intestinal smooth muscle. J. Physiol., *191*:257-270, 1967.

Kuriyama, H., and Tomita, T.: The responses of single smooth muscle cells of guinea pig taenia coli to intracellularly applied currents, and their effect on the spontaneous electrical activity. J. Physiol., *178*:270-289, 1965.

Kuru, M.: Nervous control of micturition. Physiol. Rev., *45*:425-494, 1965.

Langer, G. A.: Ion fluxes in cardiac excitation and contraction and their relationship to cardiac contractility. Physiol. Rev., *48*:708-757, 1968.

Lepeschkin, E.: Das Elektrokardiogramm. Dresden, Leipzig, Verlag von Theodor Steinkopff, 1942.

Levinson, G. E., Frank, M. J., Nadimi, M., and Braunstein, M.: Studies of cardiopulmonary blood volume-measurement of left ventricular volume by dye dilution. Circulation, *35*:1038-1048, 1967.

Levinson, G. E., Pacifico, A. D., and Frank, M. J.: Studies of cardiopulmonary blood volume. Circulation, *33*:347-356, 1966.

Lister, J. W., Delman, A. J., Stein, E., Grunwald, R., and Robinson, G.: The dominant pacemaker of the human heart: Antegrade and retrograde activation of the heart. Circulation, *35*:22-31, 1967.

Luisada, A. A.: Cardiac catheterization. Hosp. Med., *2*:21-26, 1965.

Luisada, A. A., and MacCanon, D. M.: Functional basis of heart sounds. Am. J. Cardiol., *16*:631-633, 1965.

Lundholm, L., and Mohme-Lundholm, E.: Energetics of isometric and isotonic contraction in isolated vascular smooth muscle under anaerobic conditions. Acta Physiol. Scand., *64*:275-282, 1965.

Marriott, H. J. L.: Practical Electrocardiography. 4th ed. Baltimore, Williams & Wilkins, 1968.

⸺: Ways and means of conduction. Dis. Chest, *55*:93-94, 1969.

Mashima, H., and Yoshida, T.: Effect of length on the development of tension in guinea-pig's taenia coli. Jap. J. Physiol., *15*:463-477, 1965.

McCutcheon, E. P., Baker, D. W., and Wiederhielm, C. A.: Frequency spectrum changes of Korotkoff sounds with muffling. Med. Res. Engin., 8:30-33, 1969.

McCutcheon, E. P., and Rushmer, R. F.: Korotkoff sounds. An experimental critique. Circ. Res., 20:149-161, 1967.

Miller, D. E., Gleason, W. L., Whalen, R. E., Morris, J. J., and McIntosh, H. D.: Effect of ventricular rate on the cardiac output in the dog with chronic heart block. Circ. Res., 10:658-663, 1962.

Miller, M. D., and Marshall, J. M.: Uterine response to nerve stimulation: Relation to hormonal status and catecholamines. Am. J. Physiol., 209:859-865, 1965.

Mommaerts, W. F., and Langer, G. A.: Fundamental concepts of cardiac dynamics and energetics. Ann. Rev. Med., 14:261-296, 1963.

Nakajima, A., and Horn, L.: Electrical activity of single vascular smooth muscle fibers. Am. J. Physiol., 213:25-30, 1967.

Nayler, W. G.: Calcium exchange in cardiac muscle: A basic mechanism of drug action. Am. Heart J., 73:379-394, 1967.

Niedergerke, R.: The rate of action of calcium ions on the contraction of the heart. J. Physiol., 138:506-515, 1957.

Peachey, L. D.: Transverse tubules in excitation-contraction coupling. Fed. Proc., 24:1124-1134, 1965.

Petelenz, T.: Extracoronary blood supply of the sinuatrial node. Cardiologia, 47:57-67, 1965.

Podolsky, R. J., et al.: Excitation-contraction coupling in striated muscle. Fed. Proc., 24:1112-1152, 1965.

Porter, K. R., and Bonneville, M. A.: An Introduction to the Fine Structure of Cells and Tissues. 3rd. ed. Philadelphia, Lea & Febiger, 1968.

Prec, O., Katz, L. N., Sennett, L., Rosenman, R. H., Fishman, A. P., and Hwang, W.: Determination of kinetic energy of the heart in man. Am. J. Physiol., 159:483-491, 1949.

Ralston, H. J., Inman, V. I., Strait, L. A., and Shaffrath, M. D.: Mechanics of human isolated voluntary muscle. Am. J. Physiol., 151:612-620, 1947.

Ramsey, R. W., and Street, S. F.: The isometric length-tension diagram of isolated skeletal muscle fibers of the frog. J. Cell. Comp. Physiol., 15:11-34, 1940.

Randall, W. C. (ed.): Nervous Control of the Heart. Baltimore, Williams & Wilkins, 1965.

Remington, J. W., et al.: Mechanical aspects of cardiac muscle. Fed. Proc., 21:954-1005, 1962.

Richardson, K. C.: The fine structure of autonomic nerve endings in smooth muscle of the rat vas deferens. J. Anat. (London), 96:427-442, 1962.

Rosenbluth, J.: Smooth muscle: An ultrastructural basis for the dynamics of its contraction. Science, 148:1337-1339, 1965.

Ross, J., Jr., Linhart, J. W., and Braunwald, E.: Effects of changing heart rate in man by electrical stimulation of the right atrium. Circulation, 32:549-558, 1965.

Rushmer, R. F.: Cardiovascular Dynamics. 2nd ed. Philadelphia, W. B. Saunders, 1961.

Rushmer, R. F.: Initial ventricular impulse. A potential key to cardiac evaluation. Circulation, 39:268-283, 1964.

Rushmer, R. F., Van Citters, R. L., and Franklin, D. L.: Some axioms, popular notions, and misconceptions regarding cardiovascular control. Circulation, 27:118-141, 1963.

Sandow, A.: Excitation-contraction coupling in muscular response. Yale J. Biol. Med., 25:176-201, 1952.

Sandow, A., and Preiser, H.: Muscular contraction as regulated by the action potential. Science, 146:1470-1472, 1964.

Scher, A. M.: The sequence of ventricular excitation. Am. J. Cardiol., 14:287-293, 1964.

Schottelius, B. A., Thomson, J. D., and Hines, H. M.: Mechanical properties of skeletal muscle. Am. J. Physiol., 180:191-194, 1955.

Selzer, A.: The Heart: Its Function in Health and Disease. University of California Press, 1966.

Shepard, R. S., and Whitty, A. J.: Bilateral cervical vagotomy: A long-term study on the unanesthetized dog. Am. J. Physiol., 206:265-269, 1964.

Sherf, L., and James, T. N.: A new electrocardiographic concept: Synchronized sinoventricular conduction. Dis. Chest, 55:127-140, 1969.

Sonnenblick, E. H.: Series elastic and contractile elements in heart muscle: Changes in muscle length. Am. J. Physiol., 207:1330-1338, 1964.

Sonnenblick, E. H.: Instantaneous force-velocity-length determinants in the contraction of heart muscle. Circ. Res., 16:441-451, 1965.

Sonnenblick, E. H., Spiro, D., and Spotnitz, H. M.: The ultrastructural basis of Starling's Law of the Heart. The role of the sarcomere in determining ventricular size and stroke volume. Am. Heart J., 68:336-346, 1964.

Spencer, M. P., and Greiss, F. C.: Dynamics of ventricular ejection. Circ. Res., 10:274-279, 1962.

Spiro, D., and Sonnenblick, E. H.: Comparison of the ultrastructural basis of the contractile process in heart and skeletal muscle. Circ. Res., 15:14-35, 1964.

Stephenson, S. E., and Brockman, S. K.: P wave synchrony. Ann. N. Y. Acad. Sci., 111:907-914, 1964.

Surawicz, B.: Electrolytes and the electrocardiogram. Am. J. Cardiol., 13:656-662, 1963.

Taber, R. E., Stoye, E., Ludovico, R., Green, E. R., and Gahagan, T.: Treatment of congenital and acquired heart block with an implantable pacemaker. Circulation, 29:182-185, 1964.

Taccardi, B., and Marchetti, G. (eds.): International Symposium on the Electrophysiology of the Heart. New York, Pergamon Press, 1965.

Tomita, T.: Current spread in the smooth muscle of the guinea-pig vas deferens. J. Physiol., 189:163-176, 1967.

Warner-Chilcott Laboratories. A programmed course in electrocardiography. Thirty-six installments, 1964.

Wasserman, K., Burton, G. G., and Van Kessel, A. L.: Excess lactate concept and oxygen debt of exercise. J. Appl. Physiol., *20*:1299-1306, 1965.

Weiss, G. B.: Homogeneity of extracellular space measurement in smooth muscle. Am. J. Physiol., *210*:771-776, 1966.

Wells, J. B.: Comparison of mechanical properties between slow and fast mammalian muscles. J. Physiol., *178*:252-269, 1965.

White, P. D.: The closed-chest method of cardiopulmonary resuscitation—Revised statement. Circulation, *31*:641-645, 1965.

Wilkie, D. R.: Muscle. Ann. Rev. Physiol., *28*:17-38, 1966.

Williams, J. F., Glick, G., and Braunwald, E.: Studies on cardiac dimensions in intact unanesthetized man. Circulation, *32*:767-771, 1965.

Zuberbuhler, R. C., and Bohr, D. E.: Responses of coronary smooth muscle to catecholamines. Circ. Res., *16*:431-440, 1965.

Part IV

CIRCULATION

Every living cell in the body needs nutrients and produces waste products. In man and other multicellular organisms diffusion of these substances into and out of the cell is not by itself an adequate transport mechanism; we need the heart and circulation to carry nutrients to and wastes from the cell. The heart and brain, of all the body's organs, are most dependent upon this circulation. If the heart stops pumping blood for 4 to 6 min., irreversible damage to the brain occurs. If blood flow to the heart is stopped for 10 min., the heart begins to fibrillate. If blood flow to the skin or to skeletal muscle stops for several hours, necrosis occurs.

Blood constitutes about 8 per cent of the body weight of a lean person. It carries *nutrients* from the lungs (oxygen), alimentary tract (water, carbohydrates, amino acids, lipids, minerals, vitamins, etc.), liver (glucose, vitamins, etc.), skin (lipids, vitamins, etc.), fat depots (lipids, etc.), and, in fact, from every cell in the body (lactic acid, etc.). It delivers its *wastes* to the lungs (carbon dioxide), kidneys (urea, water, minerals, acid, etc.), liver (lactic acid, bilirubin, etc.), alimentary tract (iron, etc.), skin (heat, etc.), heart (lactic acid), and also to every cell in the body.

The blood, then, is one means the body has of maintaining *homeostasis* (a constant internal and external environment for the cell). Its plasma *proteins* are important in keeping a fairly constant plasma pH (buffer function); in preventing excess fluid from leaving the blood vessels (coagulation and colloid osmotic pressure); and in transporting various molecules (hormones, drugs, etc.). Its *paltelets* contain procoagulants, which facilitate or initiate the formation of fibrin threads, and serotonin, which produces vasoconstriction. The platelets also adhere to fibrin and in this way plug up circulatory leaks. *Red cells* constitute about 40 per cent of the blood and are important both in transporting oxygen and carbon dioxide and in maintaining a constant pH, while the *white cells* destroy microorganisms and defend the body against foreign proteins. The water in blood carries many of the minerals, nutrients, and wastes which are either needed or produced by the cell.

The organ responsible for blood pressure and flow is the *heart*. The highly elastic *arteries* store some of the energy produced by the heart during systole and release it during diastole, thus affording a fairly constant blood flow to the tissues of the body during diastole. Distribution of the blood is accomplished by the *arterioles*. When an arteriole increases its muscular tone, blood flow to the area it serves is decreased; when it relaxes, blood flow is increased. It is possible to increase blood flow to all parts of the body at once by increasing cardiac output. It is also possible to maintain a constant cardiac output and increase blood flow to one part of the body (local arteriolar dilation) by decreasing blood flow to another part (local arteriolar constriction). This complex system is controlled by both *sympathetic* and *parasympathetic nerves* to the heart and sympathetic fibers to the arterioles. The smooth muscle of the arterioles also responds to *stretch, hypoxia, hypercapnea, hyperkalemia,* and circulating *hormones* by a change in its contractile state.

Eventually the blood passes to connecting links (A-V anastomoses and capillaries) between the arteries and veins. Some of the *A-V anastomoses* are in the skin, where they carry large quantities of blood rapidly from the arteries to the veins. Their function here is to rapidly transport large amounts of body heat to the skin, where it can be radiated to the outside world. The *capillaries*, on the other hand, are just large enough to allow a single red cell to pass slowly through. Their wall is one cell thick and generally permeable to molecules smaller than albumin. It is here that blood gives up some of its nutrients to the perivascular space and takes on some of the wastes from it.

The *veins* carry blood back to the heart. Their pressure, compared to that in adjacent arteries, is small, and they are readily distensible. (Rapidly injecting large quantities of fluid into a vein scarcely increases its pressure.) This quality makes the veins an important pressure-volume buffer mechanism. The veins prevent increases in blood volume (due to kidney damage or excessive fluid intake) from markedly increasing the arterial pressure and, as a consequence, increasing the work of the heart. They also serve as an important blood reservoir. In a normal person about 65 per cent of systemic blood is in the systemic veins. After a decrease in the volume of circulating blood due to hemorrhage, kidney malfunction (diuresis), or water deprivation, the veins constrict, thus making a greater percentage of the blood available to arteries and capillaries.

Some veins lead into vessels which are one cell thick and highly permeable, but which have larger diameters than do capillaries. These structures (*sinuses* and *sinusoids*) are found in (1) the liver, where they expose the hepatic cells to the nutrient-rich blood from the digestive tract, (2) bone marrow, (3) the spleen, and (4) the dura mater, where they allow the constituents of the cerebrospinal fluid to enter the blood.

In addition to the capillaries, sinuses, and sinusoids there is another system of highly permeable vessels, the *lymph capillaries*. These lie adjacent to most blood capillaries but are generally more permeable. They too carry parts of the interstitial fluid away from the cell, and also convey lipids from the intestinal tract. Their advantage over the blood capillary is greater permeability, enabling them to carry back to the blood large molecules or substances that escaped from the blood capillary during a period of increased permeability or damage. The lymph capillaries lead to *lymph ducts*, which carry the lymph to *lymph nodes*. The lymph nodes add lymphocytes and monocytes to the lymph and serve to phagocytose and destroy any microorganisms or foreign proteins in the lymph. Eventually the lymph ducts carry their contents to a vein.

Chapter 10

FLUID DYNAMICS
HEMODYNAMICS

ARTERIES, ARTERIOVENOUS ANASTOMOSES, AND VEINS

The veins, arteries, and A-V anastomoses are vessels leading to and from the heart. They can be characterized in terms of their size, their muscle tone, their elasticity, and their compliance. Each of these factors may vary from moment to moment. The characteristics of blood contained in the arteries and veins also changes from moment to moment. In short, blood flow in the body is a very complex phenomenon not easily studied. Let us then begin this section with a review of some of the principles of flow in an artificial system.

FLUID DYNAMICS

The Fluid

A fluid (gas or liquid) is a substance that tends to conform to the shape of its container. As is not the case with solids, one layer of a fluid readily slides over an adjacent one. If a particle denser than the fluid is placed in it, the particle will settle (or sediment) to the bottom of the container, i.e., it will exert a *shearing force* on the fluid. The speed with which the particle moves to the bottom of the container is proportional to the shearing force it exerts, and inversely proportional to the lack of internal slipperiness of the fluid (*viscosity*). The ease with which one layer of a fluid slides over an adjacent layer (proportional to 1/viscosity) varies from one fluid to the next. Oil, for example, has less internal slipperiness than does water and is therefore said to be more viscous. Some fluids (blood, for example) have different viscosities at different flow rates. Others obey the Newtonian laws (Newtonian fluids), that is, they do not change their viscosity with a changing rate of flow.

The Tube

In Figure 10-1 I have illustrated some of the factors that modify the flow (I, in cm.³/sec.) of water through rigid, cylindrical vessels. Note that as the horizontal tube gets farther (L, distance in cm.) from the pressure (P_1, in dynes/cm.²) source (water tower or pump), its contained pressure (P_2) decreases. The pressure in the tower, that is, helps overcome resistance in the system. Note too that part of the pressure in the water tower is dissipated in imparting velocity (v, in cm./sec.) to the fluid (m, fluid mass in gm.). The energy lost by the tower in this way equals $\frac{1}{2}mv^2$, where m is calculated from the volume of fluid pumped (Q, in cm.³) times the density (ρ, in gm./cm.³) of the fluid:

$$m = Q \times \rho \qquad (1)$$

Most of the energy in these examples (Fig. 10-1) is lost in overcoming the resistance of the system. The energy lost between points a and d, for example, is equal to the pressure difference (dP, in dynes/cm.²) between these two points, times the volume of fluid moved (Q, in cm.³). In examples A and B the energy output (E in dyne-cm., or ergs) of the water tower can be calculated from the mass of fluid moved, the square of the velocity imparted to the mass, the pressure head (dP), and the volume of fluid moved:

$$E = \tfrac{1}{2} mv^2 + P_1 Q \qquad (2)$$

Up to this point we have been considering energy dissipation in horizontal tubes; in water flowing uphill, additional energy is needed to overcome gravity. In Figure 10-1C the energy of the water tower is used as follows:

$$E = \tfrac{1}{2} mv^2 + P_1 Q + wh \qquad (3)$$

That is to say that an additional amount of work must be done to lift the fluid (w, weight of fluid in dynes) to a particular height (h in cm.). Note that the height is calculated only from the heights at w and z ($h_z - h_w = h$). The fact that the

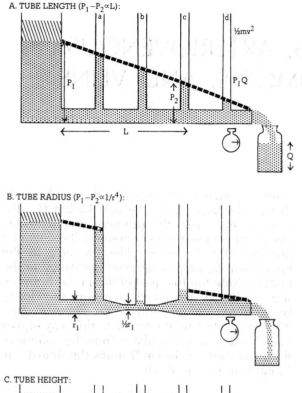

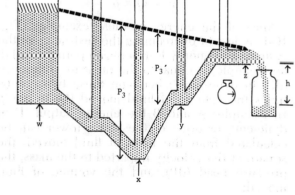

Fig. 10-1. Effect of tube length (L), radius (r), and height (h) on fluid dynamics (pressure (P), mass of fluid collected (m), velocity of fluid in the tube (v), and fluid volume collected (Q)). In each diagram the energy source is represented by a water tower at the left. Part of the energy (*upper shaded area*) in each tower is dissipated in imparting a velocity to the blood (1/2 mv²) and part (*lower shaded area*) in overcoming resistance (P_1Q) or resistance plus gravity ($P_1Q + mh$). In A the radius of the conducting tube is constant and therefore the pressure drop between any 2 points ($P_1 - P_2$) will be proportional to the length of the tube between those points. In B the tube is constricted and less of the energy of the tower is dissipated in imparting a velocity to the blood and more is dissipated in

water column goes first downhill and then uphill is not material to the *wh* calculation.

It is also important to note in Figure 10-1 that in all three examples the energy output of the water towers is equal. The constriction of the tube at B and the elevation of the output spout at C, however, modify the distribution of that energy. In B, more of the energy of the tower is used to overcome resistance (P_1Q is increased) and less to produce flow (1/2 mv² is decreased). In other words, the volume of fluid moved (Q) and the mass of fluid moved (m = Q × ρ) have decreased and the pressure head (P_1) has increased. The constriction has produced an increase in velocity at the constriction, but a decrease elsewhere:

$$\text{Velocity} = \frac{\text{flow}}{\text{cross-sectional area of vessel}} = \frac{I}{\pi r^2} \quad (4)$$

$$v_a \text{ cm./sec.} = \frac{I \text{ cm.}^3/\text{sec.}}{\pi (r_1 \text{ cm.})^2} = \text{velocity at } r_1;$$

$$v_b \text{ cm./sec.} = \frac{I \text{ cm.}^3/\text{sec.}}{\pi (1/2\, r_1 \text{ cm.})^2} =$$

velocity at constriction;

$$v_b = 4 v_a$$

Note too that the increased velocity at the constriction results in a decrease in the lateral pressure there. This relationship between pressure and velocity was noted by the Swiss mathematician Daniel Bernoulli (1700-1782). He stated what has come to be known as Bernoulli's principle: When the velocity of flow increases, more of the pressure head is converted to kinetic energy and the lateral pressure decreases.

Resistance

The French physician Jean L. M. *Poiseuille* (1799-1869) defined the factors that regulate flow (I, in ml./sec.) as follows:

$$I = (P_1 - P_2) \frac{(\pi r^4)}{8L\eta} \quad (5)$$

overcoming resistance. Therefore there is a decrease in blood flow. Note the large pressure drop across the constricted area. This drop is inversely proportional to the fourth power of the radius. In C, unlike A and B, part of the energy in the water tower is dissipated to overcome gravity (mh). Therefore, as in B, the flow is reduced.

That is to say that flow between two points in a horizontal tube is directly proportional to the pressure difference between those two points ($P_1 - P_2$) and the fourth power of the radius (r^4) of the tube, and inversely proportional to the length of the tube (L) and the viscosity of the fluid (η) in the tube. If the *pressure* is always measured in terms of the same horizontal reference point, the effect of gravity may be disregarded. For example, one could use the Poiseuille formula to calculate flow (I) at point x in Figure 10-1C if one knew the pressure exerted at w by the water column, and the pressure, P_3', at x, as well as $(\pi r^4)/(8L\eta)$. That is to say if the investigator keeps his pressure-measuring device (pressure transducer) at the same height at all times he need not correct for gravity in the Poiseuille equation. If the pressure transducer is placed at the same height as the conduit, however, a correction factor must be included. In measuring pressures in man the pressure transducer is generally maintained at the level of the heart.

The Poiseuille formula may be further simplified to read:

$$I = \frac{(P_1 - P_2)}{\Omega} \quad (6)$$

Here Ω (omega) is the resistance in the system:

$$\Omega = \frac{8L\eta}{\pi r^4} \quad (7)$$

Note the similarity between equation (6) and that for Ohm's law:

$$I = \frac{V}{\Omega} \quad (8)$$

In the study of electricity I represents the current (amperes); V, the electrical potential (volts); and Ω the electrical resistance (ohms).

In fluid dynamics the resistance is frequently reported in units of dyne-sec. per cm.[5] or mm. Hg-sec. per cm.[3] (or PRU, peripheral resistance units). If, for example, a system exhibited a flow of 100 ml. per second and a pressure head of 50 mm. Hg, its resistance would be:

$$\Omega = \frac{50 \text{ mm. Hg}}{100 \text{ ml./sec.}} = 0.5 \text{ PRU}$$

If this system were in series (Fig. 10-2B) with two similar resistances the total resistance would be 1.5 PRU:

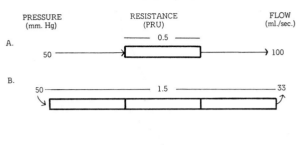
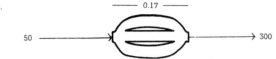

Fig. 10-2. Relationship between pressure drop across a system, resistance of the system, and flow. In part B, 3 resistances are in series with one another. In part C the 3 resistances are in parallel.

$$\Omega_T = \Omega_1 + \Omega_2 + \Omega_3 = \quad (9)$$

$$0.5 + 0.5 + 0.5 = 1.5 \text{ PRU}$$

If these resistances were parallel with one another (Fig. 10-2C), the total resistance would be 0.17 PRU:

$$\frac{1}{\Omega_T} = \frac{1}{\Omega_1} + \frac{1}{\Omega_2} + \frac{1}{\Omega_3} = \quad (10)$$

$$\frac{1}{0.5} + \frac{1}{0.5} + \frac{1}{0.5} = \frac{3}{0.5}$$

$$\Omega_T = 0.17 \text{ PRU}$$

Turbulence

In formulas (5) and (6) we emphasized the direct relationship between the pressure head and the flow. That is to say that as the pressure head ($P_1 - P_2$) increases, (due to the more forceful contraction of a pump or to a higher water tower, for example), the flow (I) is increased. This relationship is seen in Figure 10-3. However, we note that the relationship between pressure and flow is linear only up to a point; beyond it, an increase in pressure may be associated with a decrease in flow or with a less marked increase than was seen at a lower pressure (i.e., some of the energy of the system is wasted in nonlaminar—turbulent—movement). In Chapter 5 under murmurs we discussed turbu-

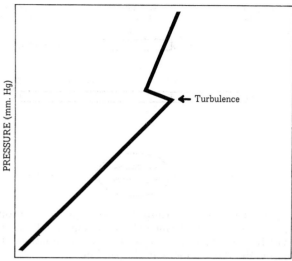

Fig. 10-3. Effects of turbulence on the relationship between flow and pressure head. As the flow increases, eventually a critical velocity is reached at which turbulence occurs. When there is turbulence the resistance ($(P_1 - P_2)/I$) increases and some of the energy in the system is used to produce this nonlaminar flow. In other words, the relationship between laminar flow and pressure is changed.

lence and noted Reynolds' finding that when R in the following formula exceeded 1000 (970 ± 80), turbulence occurred:

$$R = \frac{v \times \rho \times r}{\eta} \qquad (11)$$

That is to say that increases in the velocity of the fluid (v, in cm./sec.), the density of the fluid (ρ, in gm./cm.^{3}), the radius of the tube (r in cm.) or decrease in the viscosity of the fluid (η in Poiseuille's formula) produce nonlaminar flow if these changes are sufficiently marked.

HEMODYNAMICS

The flow of blood through the vascular system does not lend itself to the exact type of analysis used by Poiseuille in the characterization of the flow of water through rigid horizontal tubes. Blood vessels are conical (not cylindrical), elastic (not rigid) structures which branch and anastomose and contain a heterogeneous fluid (blood) whose viscosity (internal friction) changes with the shape of the vessel, the velocity of blood flow, and the concentration of red cells. In addition, flow and pressure in parts of the vascular system are not constant as in the water tower systems shown in Figure 10-1. However it will, from time to time, prove useful to compare blood flow in the circulatory system to water flow in rigid cylindrical tubes.

Characteristics of Blood Vessels

The blood vessels of the body can be divided into three major systems: a distributing system (arteries, arterioles, and arteriovenous anastomoses), a *diffusion system* (capillaries, sinuses, and sinusoids), and a collecting system (venules and veins). In this chapter we will be concerned primarily with the distributing and collecting systems. These systems accommodate changes in blood volume brought about by the pumping action of the heart, by hemorrhage, or by a greater intake of fluids (drinking, transfusion, etc.). They must maintain pressure and they must be sufficiently strong to withstand the stresses placed on their walls. Many are also responsible for regulating the distribution of the cardiac output.

Anatomy. The distributing vessels include the *arteries, arterioles*, and *A-V anastomoses*. The arteries characteristically have a ratio of wall thickness to lumen radius (w:r) of about 1:5, the arterioles of 3:4, and the A-V anastomoses of 5:1. The muscular arterioles and A-V anastomoses control the amount of blood leaving the distribu-

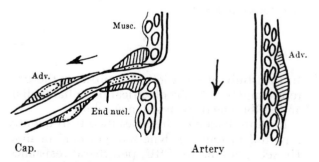

Fig. 10-4. Camera-lucida drawing of a precapillary sphincter. The sphincter is that part of the vessel where we see the last smooth muscle cell (Musc.) before the vessel becomes a capillary (Cap.). Also seen in the drawing is an adventitial cell (Adv.), an arrow indicating the direction of flow, and the nucleus for the endothelial lining of the vessels (End. nucl.). (Sandison, J. C.: Contraction of blood vessels and observations on the circulation in the transparent chamber of the rabbit ear. Anat. Rec., 54:116, 1932)

tion (arterial) system—the arterioles letting blood pass to the capillaries, and the anastomoses letting it pass directly to the veins. Under most circumstances over 99 per cent of the cardiac output passes through the arterioles into capillaries. That part of the vessel between the arteriole and the capillary, where we see the last smooth muscle cell, is called the *precapillary sphincter* (Fig. 10-4).

The character of the diffusing system varies from one area to the next. In the mesentery it consists of groups of branching and anastomosing channels with walls one cell thick. Zweifach has characterized this system as containing several *thoroughfare channels* (also called a-v bridges and preferential vessels) and their branches ("*true capillaries*"). The beginning of the channel (called a metarteriole) contains discontinuous muscle cells wrapped around the endothelium at widely separated points. At some points, where the thoroughfare channel leads to a "true capillary," precapillary sphincters have been noted. Zweifach believes that the thoroughfare channels are continuously open and pass fluid into the intercellular space. Its branches periodically open and close and are thought to pick up fluid from the intercellular space.

In the skin and intestinal wall, on the other hand, there seems to be no extensive network of branching and anastomosing capillaries. Instead, the majority of capillaries originate directly from the arterial wall or from arterioles. There is considerable evidence that the thoroughfare vessels are not typical constituents of any capillary net other than that in the mesentery. They have not been found in the subcutaneous tissue of the bat, the rabbit's ear, the urinary bladder of the frog, or the conjunctiva of the human eye.

The first part of the collecting system is the *venule*. It begins in the postcapillary area where the vascular endothelium is first associated with smooth muscle. The venule leads into the veins, which, unlike the venules, have a double layer of circular and longitudinal muscle.

Resistance to stress. The force per unit area of wall thickness (stress, σ, in dynes/cm.^{2}) exerted on a blood vessel is proportional to the pressure (P, in dynes/cm.^{2}) in that vessel and the radius (r in cm.) of the lumen of the vessel, and inversely proportional to the thickness of the wall (w in cm.) of the vessel:

$$\sigma = \frac{P \times r}{w} \quad (12)$$

Since the tension (T) exerted on the wall of a blood vessel is equal to the product of the pressure in the vessel and the radius of the lumen of the vessel (law of Laplace) we can substitute in the above formula as follows:

$$\sigma = \frac{T}{w} \quad (13)$$

In Table 10-1 we note that the greatest tension exerted on a vessel wall occurs in the aorta. This

Vessel	Radius of Lumen (μ)	Wall Thickness (μ)	Pressure (mm. Hg)	Tension (dynes/cm.)	Elastic Fibers	Collagen Fibers	Smooth Muscle
Aorta	1.3×10^4	0.2×10^4	100	170,000	+6	+4	+3
Arteries	0.4×10^4	0.1×10^4	90	48,000	+4	+1	+5
Arterioles	4.0×10	3.0×10	60	320	+2	+2	+4
A-V Anastomoses	1.0×10	5.0×10	50	67	0	0	+5
Capillaries	4.0	1.0	30	16	0	0	0
Venules	1.0×10	0.2×10	20	27	0	+1	0
Veins	5.0×10^3	0.5×10^3	15	1,000	+1	+1	+2
Vena Cava	1.6×10^4	0.15×10^4	10	21,000	+2	+4	+3

Table 10-1. Structural and functional characteristics of the blood vessels in man. The vessels are characterized for a young reclining man in terms of their approximate radius, wall thickness, transmural pressure, the tension produced on the vessel by this pressure, and the amount (in arbitrary units) of elastic fibers, collagen fibers, and muscle. The tension was obtained by multiplying the radius (in cm.) by the pressure (in dynes/cm.^{2}) (1 mm. Hg = 1.33×10^3 dynes/cm.^{2}). Arteriovenous anastomoses vary in total diameter from 55 μ in the nailbed to 300 μ in the dog's tongue. (Modified from Burton, A.C.: Physiology and Biophysics of the Circulation. Chicago, Year Book Publishers, 1965; and Liebow, A.: Situations which lead to changes in vascular patterns. *In* Hamilton, W. F. (ed.): Handbook of Physiology. Sec. 2 (Circulation), vol. II, pp. 1251-1276. Washington, American Physiological Society, 1963)

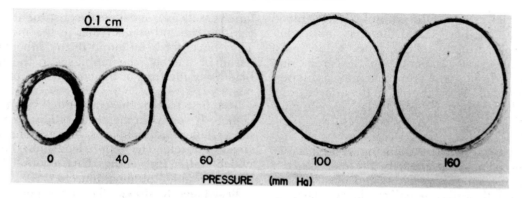

Fig. 10-5. Cross sections of rabbit aortas fixed at various distending pressures. (Wolinsky, H., and Glagov, S.: Structural basis for the static mechanical properties of the aortic media. Circ. Res., *14*:402, 1964; by permission of the American Heart Association, Inc.)

is also the point in the vasculature at which the wall is the thickest and the *collagen* fibers are most numerous. Collagen is approximately 25 times as strong as elastic tissue but 15 times less distensible. Thus we see that although collagen is poorly adapted to increases in size, it gives a strength to the vessel that prevents "blowout" or rupture.

Distensibility. Distensibility (D) is the capacity of a structure to respond to large increases in its relative volume (dQ/Q) with only small changes in pressure (dP).

$$D = \frac{(dQ/Q)}{dP} \quad (14)$$

That is to say if a 10 per cent increase in volume produces a pressure rise of 2 mm. Hg, the distensibility of that vessel is 5 per cent per millimeter Hg.

Sometimes a system is characterized in terms of the pressure change (dP) produced by a particular volume change rather than a relative volume change. This is called *compliance* (c) or *capacitance:*

$$c = \frac{dQ}{dP} \quad (15)$$

In Figure 10-5 and 10-6 we note the response of a segment of the thoracic aorta to distention. Note in Figure 10-6 that initially an increase in the radius of the aorta is associated with a linear increase in tension, but that eventually a small increase in radius produces a more marked increase in tension (reduced compliance). If this distention were carried far enough the vessel would rupture.

If we now decrease the radius of the vessel by removing fluid from it, we note a decrease in tension. A perfectly *elastic* structure would show the same length-tension diagram in response to distention as during a decrease in distention. A perfectly elastic structure would also show the same tension at a constant length regardless of how long that length were maintained. Elas-

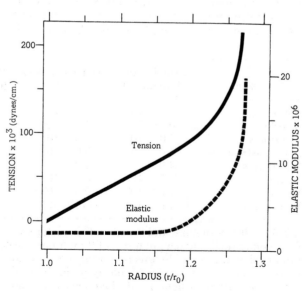

Fig. 10-6. Length-tension relationship in a segment of the thoracic aorta of the dog. (*Modified from* McDonald, D. A.: Blood Flow in Arteries. P. 164. London, Edward Arnold (Publishers), 1960)

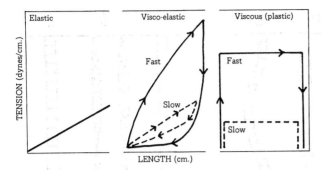

Fig. 10-7. Comparison of the length-tension relationships of an elastic, visco-elastic, and viscous material. In an elastic structure, within limits, the tension produced is directly proportional to the amount of distention. In a visco-elastic substance the tension is proportional to both the degree of lengthening and the speed of lengthening. In addition the curve of distention and shortening inscribe a *hysteresis loop*. In a viscous substance the tension is proportional to the speed of distention and independent of the degree of lengthening. When lengthening stops, tension returns to zero.

ticity, in other words, is the capacity of a substance to return to its prestretched shape without the loss of energy. Biological systems are not perfectly elastic, but rather *viscoelastic*, i.e., they tend to return to their prestretched state but show a progressive loss of tension if stretch is maintained. Figure 10-7 contains an illustration of the length-tension relationships of both a perfectly elastic and a viscoelastic substance.

The English investigator Robert Hooke (1635-1703) was the first to note that, within limits, the distance an elastic body is stretched is directly porportional to the force acting on it. If one exceeds these limits the shape of the body is permanently changed. We note, however, that if we plot the stress (σ, in dynes/cm.^{2}) exerted on various structures (see equation (13) on the ordinate, and relative change in length $((L - L_0)/(L_0))$ on the abscissa, different substances exhibit different slopes. The slope is a measure of the elastic modulus, or Young's modulus (Y):

$$Y = \frac{\sigma}{(L - L_0/L_0)} \qquad (16)$$

The higher the elastic modulus, the higher the stress resulting from a given distention.

You will note in Figure 10-6 that the elastic modulus of the aorta, remains fairly constant, within limits, but that when these limits are exceeded it increases markedly with further distention. This is probably due to the fact that the initial distention serves to stretch the more distensible elastic fibers and the viscoelastic smooth muscle cells, but also serves to uncoil or unfold the less distensible collagen fibers (Fig. 10-8). Once the collagen fibers have uncoiled, their high elastic modulus is expressed and, as a result, a further increase in radius is associated with a more pronounced increase in tension. This arrangement is fairly common throughout the vascular system. Within limits, it permits a vessel to receive a large volume of blood without markedly increasing its intraluminal pressure. Once the pressure starts to decrease, the elastic fibers release their stored energy and thus maintain blood flow. The collagen fibers act as a "jacket," which is brought into action only when the fibers are straightened by an unusually great increase in blood pressure. During the usual variations in pressure, they are not sufficiently strained to limit distention. Thus they serve to prevent "blowout" during high blood pressure. The function of the collagen fibers of the blood vessels is very similar to that of the pericardium of the heart. Both prevent overdistention.

Muscle tone. The smooth muscle in the blood vessels is of two types—tension muscle and ring muscle. *Tension muscle* (Fig. 10-9) is attached to elastic fibers and membranes, using them as skeletal muscle uses a tendon. They are in highest concentration in the aorta and decrease in number as we pass to the arterioles. They appear again, in small concentration, in the veins. When they contract they increase the tension exerted on the blood by the elastic fibers and in this way increase blood pressure.

Ring muscle is most highly concentrated in the so-called *muscular vessels* (arterioles and arteriovenous anastomoses), but is also an important component of the veins. Generally the veins contain fewer elastic fibers than their neighboring arteries and are therefore more dependent on the relaxation of smooth muscle for their distensibility. Those veins which are subject to higher intravascular pressures are generally those with a greater concentration of smooth muscle. Thus we find that there is a greater concentration of smooth muscle in the veins of the leg, where gravity increases the intravascular pressure in the erect individual, than in the veins of the thorax, abdomen, or neck.

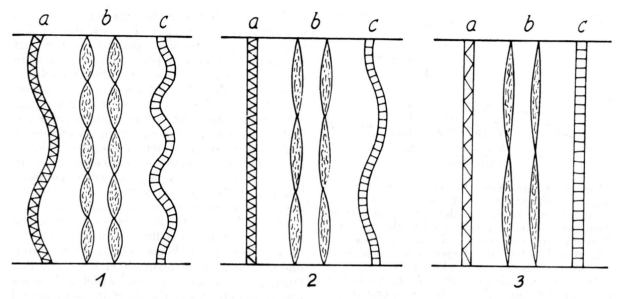

Fig. 10-8. Schematic diagram of changes in elastic fibers (a), smooth muscle (b), and collagen fibers (c) during distention (1 → 2 → 3) of a blood vessel. Note that the collagen fibers are more coiled than the elastic fibers and, therefore, the initial phases of distention result in an expression of the elastic characteristics of the elastic fibers while the collagen fibers are uncoiling. It is only during the more marked distention seen at 3 that the high elastic modulus and strength of the collagen fibers become important. (Bader, H.: The anatomy and physiology of the vascular wall. *In* Hamilton, W. F. (ed.): Handbook of Physiology. Sec. 2 (Circulation), vol. II, p. 879. Washington, American Physiological Society, 1963)

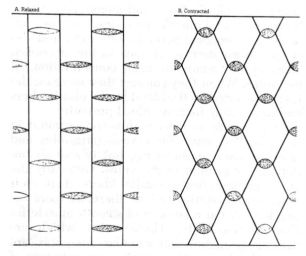

Fig. 10-9. Model for the relationship between muscle fibers, elastic fibers, and membranes in tension muscles. This type of arrangement exists in elastic arteries. (Kapal, E., and Bader, H.: Ein Modell für die Wirkungsweise der glatten Muskulatur in der Aortenwand. Z. Biol., *110*:237, 1958)

The arterioles and A-V anastomoses, though sometimes having smaller amounts of smooth muscle than the arteries, characteristically have the greatest concentration of muscle found in any of the blood vessels. They, unlike the arteries and veins, are sufficiently strong to close their lumens completely by increasing their smooth-muscle tone. When a precapillary sphincter (Fig. 10-4) or *arteriole* constricts, it decreases the pressure and flow in the vascular beds it serves. Generalized arteriolar constriction, on the other hand, is one of the body's major means of increasing arterial pressure, and generalized arteriolar dilation is one of the major means of decreasing arterial pressure. During exercise, for example, blood flow to skeletal muscle may increase up to twentyfold, but arterial pressure remains fairly constant (Fig. 8-2). This is probably due to the fact that associated with marked arteriolar dilation in skeletal muscle is increased vasoconstriction elsewhere in the body and an increase in the energy output of the heart (increased cardiac output).

The muscular vessels collapse in response to a decrease in their intravascular pressure. The pressure at which they close off completely during a given degree of arteriolar tone has been called by Alan Burton their *critical closing pressure*. When there is marked vasoconstriction the critical closing pressure is high. With reduced vasoconstriction, the vessel remains open at this pressure. We find, for example, that a man placed in a hot environment shows full vasodilation in the fingers and a critical closing pressure there as low as 10 mm. Hg. The vessels, in other words, remain open at pressures greater than 10 mm. Hg. In the cold, marked vasoconstriction and closure occur as soon as the pressure gets below 60 mm. Hg. There is also evidence that urine production ceases at arterial pressures less than 70 mm. Hg. (100 mm. Hg. being normal) due to the cessation of renal blood flow.

Gangrene may, in some cases, result from interference by an embolus with the transfer of pressure to areas where the critical closing pressure is high. It has even been suggested that irreversible *shock* may result from the prolonged shutdown of certain vascular beds in response to hypotension or excessive arteriolar vasoconstriction. Such a shutdown can result in necrosis and the release of toxins into the circulation. Thus we see that the arterial pressure is important not only in that it pushes blood to low-pressure areas (capillaries and veins), but also in that it keeps many vascular beds *patent* (open); it is important to maintain both the arterial pressure and the cardiac output. An adequate *cardiac output* alone is no guarantee that all the tissues of the body are receiving an adequate *blood supply*.

The *arteriovenous anastomoses*, unlike the arterioles and precapillary sphincters, send their blood to the veins without a preliminary exchange with the extravascular fluids. They are widely distributed in the body, being found in the skin, subcutaneous tissues, erectile tissues, bone, cartilage, mucous membranes, lungs, and possibly in the kidneys. They may be straight or coiled (Fig. 10-10), and usually contain a thick muscular wall on the arterial side and a thinner, funnel-shaped widening on the venous end. A narrow, intermediate portion of the vessel seems to be the part most concerned with constriction. Evidence of their existence includes photographs such as that seen in Figure 10-10, and the fact that glass beads many times the size of a capillary, when injected into an artery, appear in the venous blood. Spheres with a diameter

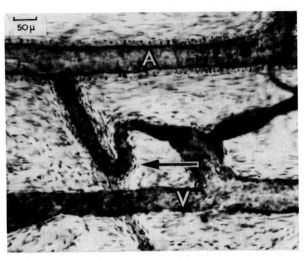

Fig. 10-10. An arteriovenous anastomosis in the human ear. The specimen was injected with Berlin blue, stained with hematoxylin, and cleared. The arrow points to the terminal portion of an intercalated segment, at the end of which there appears a thickening suggestive of the accumulation of muscle or epitheloid cells. (Prichard, M. M. L., and Daniel, P. M.: Arterio-venous anastomoses in the human ear. J. Anat., 90:309-317, Plate 1, 1956; by permission of the Cambridge University Press)

as large as 420 μ, for example, have been reported to pass from the arteries to the veins of the lung, and spheres as large as 90 to 440 μ have passed from renal arteries to renal veins. Though most investigators believe these particular observations are not representative of the normal person, they are convinced that beads up to 40 μ can pass through most A-V anastomoses.

The functional significance of the A-V anastomoses is far from settled. It has been suggested that, under most circumstances, less than 1 per cent of the cardiac output passes through these usually closed passages. Probably their best-understood action is in the skin. When a person becomes overheated, A-V anastomoses of the skin and subcutaneous tissues open up and in this way increase blood flow through the skin, thereby facilitating the radiation of body heat from the skin. An increase in skin temperature not associated with a reddening of the skin, then, is the result of the increase in flow in these thick-walled shunts. In this regard it is interesting to note that in 1840 Julius Robert Mayer reported that venous blood tends to become "arterialized" in individuals living in the tropics. The concentration of venous oxygen, that is, increases when the A-V anastomoses open up. Sometimes the clinician who wishes to obtain a sample of arterial blood but does not wish to go into an artery for it will have the patient immerse his forearm in warm water for about 15 min. He can then obtain a sample from the cubital vein which is usually representative of arterial blood.

The importance of these shunts in organs other than the skin has not been established. It has been suggested, however, that they are important in temperature regulation wherever they are found. In the respiratory tract they may serve to warm the inspired air, and in this way protect the alveoli from irritation. In metabolically active areas they may carry away excess heat. There is also the possibility they are concerned with pressure regulation. An effective means of lowering arterial blood pressure might well be an opening up of the A-V anastomoses.

Before leaving the A-V anastomosis let me also point out that *abnormal* connections between arteries and veins may also occur. These connections, unlike the normal ones we have been discussing, generally lack the muscle and control systems so characteristic of the normal ones. Abnormal shunts may be congenital (inherited) or acquired. Abnormal anastomoses have been reported between the pulmonary arteries and veins, which have resulted in diminution of the concentration of oxygen in the arterial blood and associated cyanosis. Abnormal connections between systemic arteries and veins, on the other hand, initially cause a fall in arterial pressure, an engorgement of the veins, and an increase in heart rate and cardiac output. Eventually the blood volume is increased and the heart becomes enlarged. In certain cases of liver damage the portal veins become engorged with blood and the surgeon intentionally forms an anastomosis between the portal vein and vena cava (portal caval shunt).

Growth of blood vessels. Antyllus, 1700 years ago, was perhaps the first to note that tying off an artery to a limb in the treatment of an aneurysm does not necessarily result in the loss of that limb. Fifteen hundred years later John Hunter noted that if he ligated the major nutritive artery to a stag's antler, growth of the antler proceeded normally and that a prodigious growth of new vessels eventually developed. He concluded that "vessels go where they are needed."

Apparently as the cells of the body enlarge (hypertrophy) and multiply (hyperplasia) or as the metabolic activity of an area increases, there is also an increased need for nutrients. On an acute basis these changes in need are met, for the most part, by (1) an increase in flow brought about by the dilation of blood vessels and (2) a change in the extraction of nutrients from the blood (that is, a change in the *arteriovenous difference*). Most tissues, for example, remove only about one third of the oxygen passing through their capillaries. They can, in times of need, increase this quantity more than twofold. Even in the resting individual, however, the heart leaves only about 4 to 7 per cent of the oxygen in the venous blood. This high A-V oxygen difference in the heart means that it, more than most organs, is dependent upon arteriolar dilation in times of increased metabolic need.

In chronic abnormalities the nutritive needs of an area are sometimes met by increases or decreases in the number of blood vessels and in the size of the existent vessels; under certain circumstances one can either develop a *collateral circulation* or decrease the number of blood vessels in an area in response to a pro-

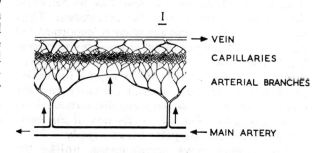

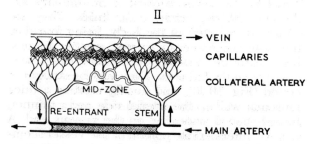

Fig. 10-11. Diagram to illustrate one concept of how a collateral circulation may develop. In this diagram arterial occlusion is shown to distend an adjacent vascular network. (Longland, C. J.: The collateral circulation of the limb. Ann. Roy. Coll. Surg. Eng., *13*:162, 1953)

longed change in the need of that area. Liebow defined collateral circulation as "blood flow that pursues a channel or system of vessels which is alternative to or develops in substitution for a major vascular pathway."

The definition states, then, that we can (1) develop a collateral circulation by expanding the preexisting arteries, capillaries, veins, or all three, or (2) grow more vessels. Figure 10-11 is one example of how a preexisting pathway might be expanded. Here we note that the circulation in a part of an artery is stopped by ligation, spasm, or perhaps an embolus. In response to this, the network of small vessels leading from the artery is expanded and is eventually able to meet the needs of the area as well as they were met before the obstruction. In this process the lumen of the vessels that form the collateral circulation becomes larger, and smooth muscle cells begin to hypertrophy and multiply. In this process the vessels become more tortuous; some capillaries may develop into arteries or, in the case of obstruction of a vein, some of the capillaries develop into veins. It is interesting to note, in this regard, that some workers have removed a segment of the external jugular vein of a dog and used it to connect two ends of the carotid artery. In time the size of its lumen decreases, the thickness of its wall increases two to three times and it, for all intents and purposes, becomes an artery.

One of the yet-unanswered questions in physiology is what makes a collateral circulation develop. Most hypotheses revolve around four basic possibilities or combinations of these. It has been suggested that (1) chemical changes such as hypoxia, (2) mechanical changes such as an increase in pressure, (3) changes in sympathetic tone, or (4) changes in the concentration of the hormones that control anabolism and catabolism can initiate or facilitate the development of the collateral circulation. It has been reported, for example, that cortisone inhibits the development of a collateral circulation and the proliferation of connective tissue. On the other hand, Stefani in 1886 noted that the axillary artery of the salamander could be ligated with no untoward effects. If he denervated the extremity and then ligated, however, gangrene developed. Denervation alone did not result in gangrene.

One of the crucial aspects of the development of collateral circulation is the rapidity with which it becomes effective in meeting the needs of an area. Rosenthal and Guyton showed in a dog that 1 hour after the ligation of the femoral artery, preexisting collaterals opened and returned the flow to 70 per cent of normal. In the case of the coronary arteries it may take 3 to 4 weeks to get a comparable change in collateral flow. Some investigators have tied the main pulmonary artery to 1 lung and have reported that the development of a collateral blood supply may continue for 18 months. The rapid development of a collateral blood supply characteristic of femoral or carotid occlusion is readily understood. With the exception of the aortic and pulmonary valves the arteries have no valves to hinder the movement of their blood. Both the right and left common carotid arteries and the right and left vertebral arteries communicate in the medulla of the brain at the circle of Willis. Thus the right vertebral has the potential, once its lumen size is increased, of sending a large volume of blood to the right and left brain. It has the capacity, in other words, to take over the function of the common carotid artery.

In the case of the microcirculation of the heart we have a potentially dangerous situation. Here there are generally a very limited number of preexisting collateral circuits. In the heart, as well as in many other areas, new vessels must be produced, and in these areas the organ is much better able to cope with a slowly developing occlusion (atherosclerosis, for example) than with one of rapid onset.

Ligation of the main pulmonary artery to one lung has been reported in the dog and in man. It results over a period of 18 months in the development of a flow from the bronchial arteries to the alveoli of the affected area. Prior to ligation bronchial flows of about 25 cm.^{3} per minute per square meter of body surface were found. After 18 months the flow had increased to 1,000 cm.^{3} per minute per square meter. This particular ligation, then, resulted in a left-to-left shunt (left heart to aorta to bronchial arteries to pulmonary capillaries to pulmonary vein to left heart). Occlusion of a pulmonary vein, on the other hand, results in an opening up of short, narrow, precapillary vessels to the bronchial veins.

Deterioration of blood vessels. In the preceding section I have emphasized that the vessels of the circulatory system are not a static group of tubes. They can, in response to changing situations, dilate, constrict, and increase or decrease their number. It is becoming increasingly clear that they also change their character with age. By the time an individual has reached middle age a progressive deterioration of *elastic fibers* throughout the body has begun. In the case of the eyes this becomes evident in the progressive failure of the process of accommodation and an increasing need for bifocal or trifocal glasses to compensate for this. In the case of the blood vessels, elastic fibers become frayed and fragmented and are replaced by *collagen*. *Smooth muscle* cells begin to degenerate and the arteries become progressively *distended* until the age of about 60, at which time the arteries have almost completely lost their elasticity and ability to distend. At the age of 20, about 0.4 per cent of the human aorta is *calcium*, but by 80 this has increased to about 7 per cent. In some cases the calcification process is much accelerated. Any abnormal thickening and hardening of the walls of arteries is called *arteriosclerosis*.

The thinner walls of the veins represent a different problem. In young people the *valves* in the veins of the appendages effectively prevent retrograde flow. In the erect individual, however, we must add to the normally low pressures in the veins of the leg, thigh, and rectum the pressure produced by the column of blood above these points (Fig. 10-12).

One means of keeping venous pressure low is by the contraction of the skeletal muscles adjacent to or around the veins. Apparently in a horizontal position the energy for the return of all the blood to the right heart comes mainly from the energy of the left heart, but in the erect position the energy for venous return in the head, neck, and trunk comes primarily from the left heart, whereas the energy for the return of blood from the limbs comes primarily from the contraction and relaxation of skeletal muscles (the so-called *muscle pumps*). The pulsation of adjacent arteries, the peristalsis of the intestine, the

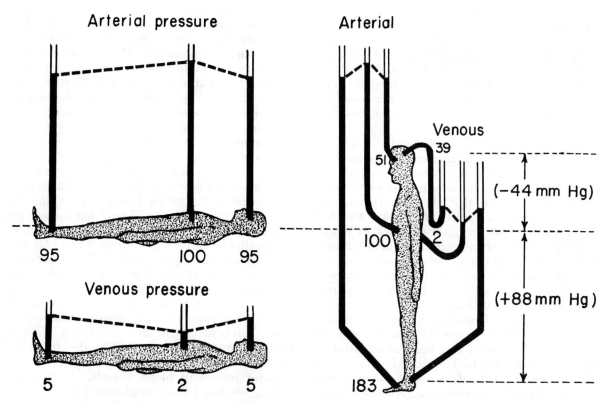

Fig. 10-12. Arterial and venous pressures in the horizontal and motionless erect subject. The pressures are given in mm. Hg, reference point being the level of the right atrium (pressure transducers were at the level of the right atrium). (From Physiology and Biophysics of the Circulation by A. C. Burton. P. 97. Copyright © 1965, Year Book Medical Publishers, Inc. Used by permission)

Hemodynamics 177

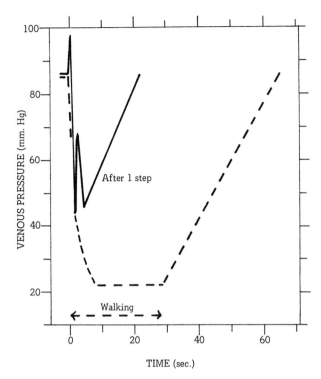

Fig. 10-13. Change in venous pressure in the dorsum of the foot during walking. The solid line shows the result of taking one step and the interrupted line the result of walking for 30 sec. (*Redrawn from* Pollack, A. A., and Wood, E. H.: Venous pressure in the saphenous vein at the ankle in man during exercise and changes in posture. J. Appl. Physiol., *1*:650, 1949)

buoyancy effect of the abdominal contents, and the respiratory movements all assist in pumping (or milking) the blood in the veins back to the heart, thus reducing pressure and distention in the veins of the lower extremities. Undoubtedly one of the most important factors in reducing venous pressure in the lower extremities, however, is the contraction of skeletal muscle which occurs during walking and running (Fig. 10-13). When the pressure in the superficial veins or those around the anus (hemorrhoidal veins) is allowed to get excessive the valves become incompetent and the distention may become irreversible (*varicose veins* or *hemorrhoids*). This may result from prolonged standing in one position or an increased resistance to venous return (pregnancy or other factors that may compress the inferior vena cava).

Age is another cause of venous deterioration. There are an average of 13.6 valves in the greater saphenous vein in the child, and 10.7 in the adult. By the age of 70 about 70 per cent of the valves in the veins have atrophied and there is extensive backflow into the lower extremities when one stands erect.

Atherosclerosis is still another type of blood vessel deterioration. In this condition lipid accumulates in the intima of the wall, leading to a decrease in lumen size, an accumulation of collagen fibers, calcification, and thrombosis (formation of a blood clot). It was at one time thought to be a disease of old age, but studies of accident victims and soldiers killed in the Korean war (average age 23) have shown that most young adults also have extensive atherosclerotic plaques in their arteries. These plaques are most often found in (1) the coronary arteries, (2) the brachiocephalic, common carotid, basilar, and cerebral arteries, as well as the arteries of the circle of Willis, (3) the abdominal aorta, and (4) the iliac and popliteal arteries. The renal arteries, on the other hand, are usually devoid of such plaques.

The danger of atherosclerosis is that these plaques tend to interfere with blood flow. Parts of the plaque may even break off and occlude flow farther downstream. When atherosclerosis prevents or markedly decreases flow to an area, death or malfunction of that area may result. In the coronary circulation, this may cause *heart attack,* angina pectoris (chest pain), or some change in the ECG. A common result of cerebral ischemia (inadequate flow) is a *stroke,* and gangrene can result from iliac or popliteal ischemia.

The death rate in the United States from heart attack is now 500,000 per year, and from stroke, 200,000 per year. It is estimated that at least 5 per cent of the adult males in this country show some signs of a heart disorder and that the basic disease responsible for most of these deaths and disorders is atherosclerosis. It is interesting, in this regard, to look at the charts in Figure 10-14. Here we note that the incidence of coronary heart disease is higher in men who smoke, are obese, are hypertensive, have a high blood cholesterol level, have an ECG abnormality, or have parents with a coronary condition than in men who do not show these symptoms. Interestingly enough, women show a lower incidence than men. It is suggested that their estrogens partially protect them from atherosclerosis. It has been shown, for example, that ovarectomized women have a high incidence of atherosclerosis and that males can be protected

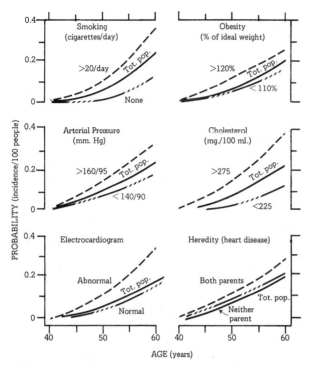

Fig. 10-14. Probability of developing coronary heart disease in men after age 40. The study was conducted over a 10-year period on a group of men who were from 40 to 54 years of age at the onset. The indices used for coronary heart disease were autopsy findings after sudden death, myocardial infarction with survival, angina pectoris (intense chest pain), or a history of an abnormal ECG in response to exercise. It was noted that men who smoked, were obese, had hypertension, a hypercholesterolemia, an abnormal ECG, or a parent who had had a heart attack were themselves more prone to develop coronary heart disease than those without this background. The symbols used are > (greater than), < (less than), and Tot. pop. (total population). (Redrawn from Doyle, J. T.: Etiology of coronary disease: Risk factors influencing coronary disease. Mod. Conc. Cardiovasc. Dis., 35:2, 1966)

from atherosclerosis by the injection of estrogens. These hormones do, however, have a feminizing effect on the male.

The exact pathogenesis of atherosclerosis is not known. Its incidence has been increased by increasing the amount of fats, *cholesterol,* and meat in the diets of animals. It has also been noted that in the U. S. the average serum cholesterol level is 240 mg. per 100 ml. of serum, whereas in Yugoslavia it is 191 mg. per 100 ml. In the U. S. the incidence of myocardial infarction is 11.8 per 1,000 men, and in Yugoslavia, 2.1 per 1,000. These and other data have led many to suggest that lipids move from the bloodstream into the arterial wall and that the higher the cholesterol level the greater the infiltration. It is further suggested that high arterial pressure accelerates this process by making the wall more permeable. Or possibly atherosclerosis is secondary to an increased *coagulability* of the blood. We know that anticoagulants can reduce the incidence of atherosclerosis and that a clot formed in a vessel lumen may cling to the wall and subsequently become endothelialized, organized, and infiltrated with lipid. A third theory on the cause of atherosclerosis is that an injury occurs in the vessel wall and that this results in a breakdown in the intercellular cement and a subsequent infiltration of the area with lipid. This *intramural injury,* it is hypothesized, results in a greater synthesis of fat.

Characteristics of the Blood

Blood is a suspension of formed elements (cells and cell fragments) in plasma (an aqueous solution of electrolytes and nonelectrolytes). The major formed element, the red blood cell (erythrocyte), constitutes about 45 per cent of the blood volume. Its concentration does, however, change in response to certain physiological (hypoxia or hemorrhage) or pathological (aplastic anemia or polycythemia) conditions. Erythrocyte concentrations (called *hematocrits*) in excess of 60 per cent cause deformation of the erythrocyte, though hematocrits as high as 70 per cent have been reported in polycythemia vera.

Laminar flow. Blood flow, like the flow of a Newtonian fluid (see "Fluid Dynamics," at the beginning of this chapter), can be characterized as either laminar or turbulent. Although laminar flow is the rule in the body, turbulence does normally occur at the branches of certain blood vessels and when the velocity of the blood is markedly increased (as in strenuous exercise). As I have pointed out, each of the layers in a bloodstream moves at a different velocity, the layer next to the endothelium having a velocity of zero and that in the *axial stream* having the maximum velocity. For the most part the erythrocytes in the arteries and veins are oriented parallel to the walls of the vessel, but a small amount of microturbulence (movement between layers) has been reported. Generally the layers of blood next to the endothelium (plasma layer) have a hematocrit lower than that of the axial stream (Fig. 10-15). When one views the living blood vessel under a microscope, however, the differences in hematocrit from layer to layer appear much exaggerated due to the differences in velocity of the various layers. For example, the axial stream may have a velocity twice of another layer, but the same hematocrit. This means that twice as many erythrocytes per second pass a point on the axial stream as pass the more peripheral point, although the hematocrits are the same. These observations are based on the analysis of rapid-exposure photographs of the blood stream in a living person and the study of quick-frozen preparations. It is now possible by using liquid propane (−170°C.) to quick-freeze an artery or vein in less than 0.1 sec.

The differences in hematocrit in the axial stream and in the plasma layer have led to some interesting though unproven speculation. It has been suggested, for example, that certain blood vessels obtain blood primarily from the axial stream (high hematocrit), while others obtain it primarily from the plasma layer (low hematocrit). It may be that such a sampling system is of little importance in the normal individual, but of great importance where the hematocrit is low and where, as a consequence, the plasma layer is increased in size.

Viscosity. The observation that blood contains a number of layers of fluid, each flowing at a different velocity, helps us understand the importance of the internal friction (viscosity) of each of these layers on its neighbor. A slowly moving lamina provides a drag on a rapidly moving lamina. The physiologist generally characterizes this internal drag or friction in terms of viscosity or relative viscosity, the commonly used unit being the *poise*. A liquid of 1 poise, with a velocity gradient of 1 cm. per second per centimeter, would exert a force between layers of 1 dyne per square centimeter of contact. Water at room temperature has a viscosity of about 1 centipoise.

With blood, viscosity is generally reported as *relative viscosity*. In determining this, the investigator observes the time (t_w) it takes water to fall from one point to another in a standard glass tube (Ostwald viscometer), and the time (t_b) it takes a sample of blood to fall the same distance, in the same viscometer, at the same temperature:

$$\text{Relative viscosity of blood} = \frac{t_b}{t_w}$$

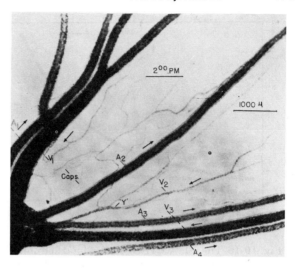

Fig. 10-15. Photomicrographs (1/2000 sec. exposure) from the vertically mounted mesentery of a frog anesthetized with urethane. In the first picture we note a normal circulation. Characteristic of such a circulation is the absence of clumps of red cells, a fairly homogeneous axial stream of cells, a peripheral plasma layer, and a state of vascular tone. The second picture was taken 18 hrs. 40 min. later (1 hr. 50 min. before the heart stopped). Note that veins V_1, V_2, and V_3 are wider (decreased tone) and that many of the red cells have formed large masses and settled. This has resulted in low red cell counts in many of the arteries (A_1, A_2, and A_3). Plasma layers are indicated by a P and vessel endothelium by an E. (Knisely, M. H., Warner, L., and Harding, F.: Antemortem settling. Angiology, *11*:554, 560, 1960; copyright by Angiology Research Foundation, 1960. Reprinted by permission)

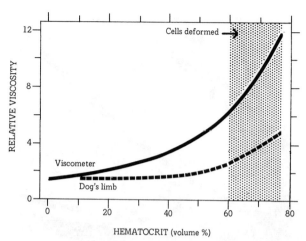

Fig. 10-16. The relative viscosity of blood at different hematocrits. In these experiments the relative viscosity of blood was determined with a viscometer of a diameter greater than 1 mm. and with the vascular bed of the hind limb of a dog. At hematocrits greater than 60%, the erythrocytes become deformed. (*Modified from* Whittaker, S. R. F., and Winton, F. R.: The apparent viscosity of blood flowing in the isolated hindlimb of the dog, and its variation with corpuscular concentrations. J. Physiol. (London), 78:358, 1933)

The relative viscosity of blood obtained from the Ostwald viscometer is about 3 to 4 at a normal hematocrit of 45 to 50 per cent. As you can see from Figure 10-16, however, it varies with the concentration of erythrocytes. Whittaker and Winton have pointed out that "normal blood" may exhibit a viscosity "of anything between 2 and 100 times that of water, depending on the design of the apparatus in which it is measured." In Figure 10-16 we note, for example, that blood at a hematocrit of 50 per cent has a relative viscosity of 2 in a dog's limb and 4.6 in an Ostwald viscometer.

Mean blood flow. A 72-kg. male has approximately 5,550 ml. of blood (7.7% of his body weight). When he is reclining, approximately 85 per cent of this is in the systemic circulation and 15 per cent in the heart and pulmonary system (Levinson, et al., 1966). The distribution of the 85 per cent in the systemic circulation is shown in Figure 10-17. Note that 65 per cent of the systemic blood is in the venous system. It is also interesting to note that as blood passes from the ascending aorta, where its average velocity is 28 cm. per second, to the smaller arteries, to

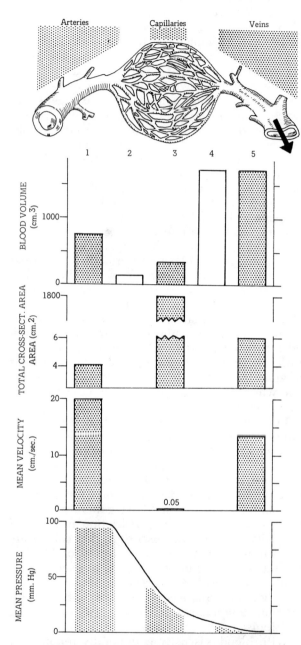

Fig. 10-17. Flow and pressure in the systemic circulation. In this presentation the normal blood volume in the systemic (1) arteries, (2) arterioles, (3) capillaries, (4) venules, small veins, sinuses, sinusoids, and (5) large veins is estimated, as is the total cross-sectional area. Note that although the pressure in the veins is less than that in the capillaries the velocity of flow in the veins is greater than in the capillaries.

the arterioles, to the capillaries, its velocity decreases. This is due to the expanding nature of the vascular system. In the case of the ascending aorta, approximately 5,500 ml. of blood pass through it each min. (92 ml. per second). This means that the average cross-sectional area of the ascending aorta is 3.3 cm.²:

$$A_x = \text{Cross-sectional area} = \frac{\text{flow (cm.}^3\text{/sec.)}}{\text{velocity (cm./sec.)}}$$

$$A_x \text{ (aorta)} = \frac{92 \text{ cm.}^3\text{/sec.}}{28 \text{ cm./sec.}} = 3.3 \text{ cm.}^2$$

By similar means one can calculate the cross-sectional area of the capillaries and veins parallel to each other, assuming almost all the cardiac output passes through veins and capillaries.

$$A_x \text{ (capillaries)} = \frac{92 \text{ cm.}^3\text{/sec.}}{0.05 \text{ cm./sec.}} = 1,800 \text{ cm.}^2$$

$$A_x \text{ (veins)} = \frac{92 \text{ cm.}^3\text{/sec.}}{14 \text{ cm./sec.}} = 6.6 \text{ cm.}^2$$

If we assume that there are 330 cm.³ (Fig. 10-17) of blood in the capillaries, and that the average capillary has a radius (r) of 4 μ and a length (l) of 1,000 μ, each capillary would occupy the following volume ($Q_{cap.}$):

$$Q_{cap.} = \pi r^2 \times l = (3.14)(4 \ \mu)^2 (1,000 \ \mu) = 50,000 \ \mu$$

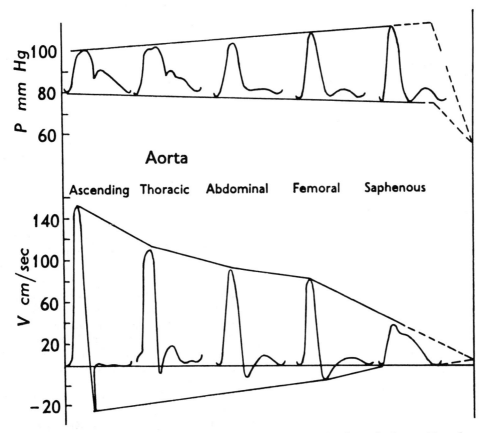

Fig. 10-18. Flow and pressure waves at different distances from the heart. Note that as the pressure wave gets further from the heart in a normal healthy individual, although the mean pressure is slowly decreasing, the pulse pressure is increasing. This is analogous to waves at the seashore. As the wave approaches the shore (that is, as the water becomes shallower), the wave's amplitude increases until the wave breaks. Past the proximal arteriole (*dotted line*) both the mean and pulsatile pressure fall. Note that the fluctuations in the flow velocity decrease as the blood gets further from the heart. (McDonald, D. A.: Blood Flow in Arteries. P. 271. London, Edward Arnold (Publishers), 1960)

There would, therefore, be the following total number of capillaries ($n_{cap.}$):

$$n_{cap.} = \frac{330 \text{ cm.}^3 \text{ blood}}{50 \times 10^{-9} \text{ cm.}^3 \text{ blood/cap.}} = 6.6 \times 10^9 \text{ cap.}$$

Unfortunately many of these data are based on assumptions which may not be accurate, but they do emphasize an important aspect of the systemic circulation, that as the blood passes from the aorta to its branches it enters an expanding vascular bed; this, in turn, results in a progressive decrease in the velocity of flow. By the time an erythrocyte reaches a capillary it is traveling at a speed (0.05 cm./sec.) such that it will remain in the capillary for about 2 sec., or (0.1 cm. long)/(0.05 cm. per second). Apparently this is enough time for the erythrocyte to give off significant amounts of oxygen to the intercellular fluid and take on large quantities of carbon dioxide. Thus the low velocity of flow in the capillary serves it well in its diffusion function, while the markedly higher velocities in the arterial and venous systems serve them well in their transport function.

Pulsatile blood flow and pressure. Up to this point we have, for the most part, been characterizing the circulatory system in terms of average or mean parameters. We have described the ascending aorta as having a pressure of 100 mm. Hg, a flow volume of 5.5 liters per minute, and a flow velocity of 1,800 cm. per minute (30 cm./sec.). It must be remembered that these specific pressures and flows exist in the aorta for only a few milliseconds of each cardiac cycle; averages do not tell the whole story. Actually, the pressure in the aorta of a resting individual usually varies between 120 mm. Hg (systolic pressure) and 80 mm. Hg (diastolic pressure), and the flow between +22 liters per minute (6,600 cm./min.) and 0 liters per minute (0 cm./min.). These moment-to-moment variations are not only of physiological importance, but also of diagnostic value, since they can be used to estimate the state of the heart and blood vessels.

In Figure 10-18 we note that as the blood passes from the heart toward the periphery, the pulsatile flow pattern becomes dampened, but that the pulse pressure (systolic pressure minus diastolic pressure) becomes exaggerated. In the capillaries, however, both the flow and the pressure pulses are so reduced that the flow is generally characterized as continuous and the pressure as nonpulsatile. The arterioles, with their relatively high resistance to flow, convert

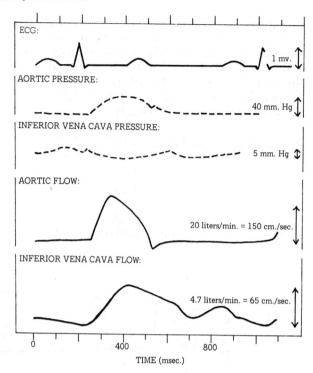

Fig. 10-19. Comparison of pressure and flow in the aorta and vena cava. Note that the flow patterns are expressed in terms of both volume flow and velocity flow. This is possible by keeping the cross-sectional area of the vessel constant.

the pulsatile pressure and flow patterns of the arteries to the nonpulsatile patterns of the capillaries. As we pass toward the large veins of the body we once again note pulsatile variations in flow and pressure (Fig. 10-19). The greatest flow in the venae cavae usually occurs soon after the relaxation of the atrium, and the greatest pressure during atrial contraction. The pulse pressures in the veins are not sufficiently great that they can be detected by palpation, but in cases where they are exaggerated (congestive heart failure, for example) the jugular vein can frequently be seen to pulsate.

The arterial *pulse pressure* (dP), on the other hand, can be readily palpated and characterized as weak, normal, or strong. Its strength is related to (1) cardiac stroke volume (Q), (2) arterial capacitance (c), and (3) arterial runoff during ventricular ejection. In fact, some authors have used the following formula to approximate the stroke volume:

$$Q \text{ (ml.)} = dP \text{ (mm. Hg)} \times c \text{ (ml./mm. Hg)}$$

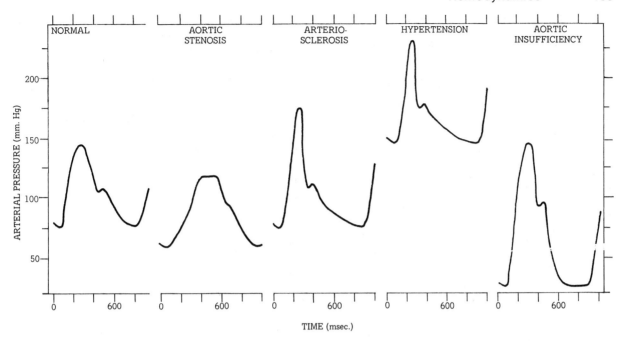

Fig. 10-20. Use of the pulse pressure of the brachial artery in diagnosis. The magnitude of the pulse pressure as well as its rate of rise and decline can be an important diagnostic aid.

This formula involves a number of assumptions, e.g., that drainage from the arterial system during ventricular ejection is unimportant. Hamilton (1962), on the basis of empirical relationships between stroke volume and the brachial pulse pressure, states the following approximation: "At arterial pressures below 120 mm. Hg, 1 mm. pulse pressure = 1 ml. stroke volume per square meter (stroke index)." At higher pressures and in arteriosclerosis, the figure for the stroke index is less due to the lower compliance of the arteries.

The pulse pressure is also elevated in arteriosclerosis, hypertension, and aortic insufficiency (Fig. 10-20), as well as when the stroke volume is increased. In arteriosclerosis the compliance of the arteries has decreased, causing a more rapid increase in pressure (dP/dt) during the first part of ventricular ejection, followed by a more rapid decrease in pressure. The character of the pulse wave in hypertension is similar to that in arteriosclerosis. In both cases there is a loss of compliance, but in hypertension the loss is due to the distention of the arteries by the high blood pressure.

Another characteristic that hypertension and arteriosclerosis have in common is an increase

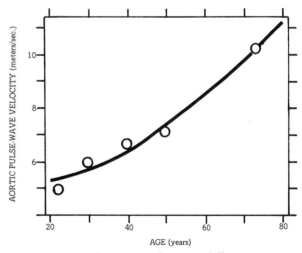

Fig. 10-21. Pulse-wave velocity at different ages. As one becomes older the mean arterial pressure increases, the cross-sectional area of the arteries increases, elastic fibers disappear from the arteries, and the concentration of calcium and collagen in the arteries increases. All of these modifications result in a decreased distensibility (or capacitance) in the arterial system and, as a consequence, changes in the character of the pulse pressure (shape, amplitude, velocity). (*Redrawn from* Medical Physics by O. Glasser. Copyright © 1950, Year Book Medical Publishers, Inc. Used by permission)

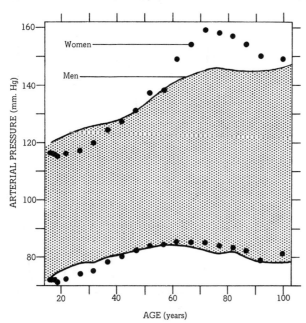

Fig. 10-22. Arterial pressures of men (*solid lines*) and women (*circles*) at different ages. The data for ages 16 through 64 were published in 1950 and were collected from 7,722 healthy men and 7,984 healthy women. The data for ages 65 through 100 were collected from 2,998 healthy males and 2,759 healthy females. The only 10-year groups containing less than 300 subjects were males (25 subjects) and females (28 subjects) between the ages of 95 and 106. These are characterized above in the 100-year-old column. Note that for males the pulse pressure (*shaded area*) at age 16 is about 45 mm. Hg, at 45 about 48 mm. Hg, at 60 about 57 mm. Hg, and at 90 about 67 mm. Hg. This marked increase in pulse pressure after age 45 is the result of arteriosclerosis. Note, too, that at age 45 the systolic pressure of the female becomes the same as that for the male and eventually exceeds the male value. Age 45 is the average age for the onset of menopause. (*Data from* Master, A. M., Dublin, L. I., and Marks, H. H.: The normal blood pressure range and its clinical implications. JAMA, *143*:1464-1470, 1950; and Master, A. M., Lasser, R. P., and Jaffe, H. L.: Blood pressure in apparently healthy aged 65 to 106 years. Proc. Soc. Exp. Biol. Med., *94*:463-467, 1957)

in the velocity of transmission of the pressure wave—also a result of the decreased compliance of the arteries. You will note in Figure 10-21 that the velocity of the pressure wave increases with age. This is but one of the many signs of the hypertension (Fig. 10-22) and arteriosclerosis that develop as the individual grows older.

In aortic insufficiency (Fig. 10-20), the aortic valves allow blood to move back into the ventricle during ventricular diastole. As a result there is a rapid decrease in arterial pressure after the dicrotic notch and a decrease in diastolic pressure not associated with a comparable decrease in systolic pressure. Or, the pulse pressure increases as a result of a decrease in diastolic pressure. This is in marked contrast to the condition of aortic stenosis, in which constriction of the aortic valve prevents the rapid transfer of blood and pressure to the systemic arteries. As a consequence the rate of rise of arterial pressure is reduced, as is the pulse pressure.

The *diastolic pressure* is dependent on (1) the rate of outflow from the arterial system and (2) the time interval between ventricular contractions. The rate of outflow can be increased, and hence the diastolic pressure decreased, by aortic insufficiency or arteriolar dilation. The level of diastolic pressure in the normal individual at any particular heart rate is sometimes used, therefore, as a rough indication of the peripheral resistance offered by the arterioles.

In the *pulmonary circulation* the arterioles and capillaries do not offer the resistance to flow that they do in the systemic system. As a result, the pulse wave from the right ventricle seems to play a more important role in the development of pressure and flow in the pulmonary vein than does the pulse wave from the left ventricle in the development of pressure and flow in the venae cavae. Compare, for example, Figures 10-23 (pulmonary circulation) and 10-19 (systemic circulation). In the venae cavae one can correlate flow and pressure almost totally with events in the right atrium (a wave, atrial systole; c wave, closure of A-V valves; x wave, relaxation of the atrium; v wave, filling (stretch) of atrium with blood; y wave, opening of A-V valves). In the pulmonary veins, however, a distinct ventricular pressure wave is conducted through the pulmonary capillaries and superimposed on the changes in pressure emanating from the left atrium. In the case of the venae cavae the major flow into the right atrium occurs when the pressure in the venae cavae is decreasing (x wave), whereas in the pulmonary vein the major flow into the atrium frequently begins during an increase in venous pressure.

Fig. 10-23. Pressure and flow patterns in the pulmonary circulation of the unanesthetized dog. The interval between 1 and 2 is the transmission time for the pulse wave from the pulmonary artery to the pulmonary capillaries (0.09 sec.) and between 2 and 3 is the transmission time from pulmonary capillaries to veins (0.03 sec.). The pulse pressure in the pulmonary artery (PA) is about 25 mm. Hg (40−15), and in the pulmonary vein (PV) about 3 mm. Hg (10−7). Note that in the pulmonary vein, flow (Q) is continuous and pulsatile and, unlike the inferior vena cava (see Fig. 10-19), shows its greatest flow during its greatest pressure. From data such as these it is concluded that in the pulmonary system, unlike the systemic circulation, the major determinant of venous flow is the pressure wave transmitted from the right ventricle. PV_w: pressure recorded from a catheter passed (wedged) toward the pulmonary capillaries from a pulmonary vein. (Morkin, E., et al.: Pattern of blood flow in the pulmonary veins of the dog. J. Appl. Physiol., *20*:1123, 1965)

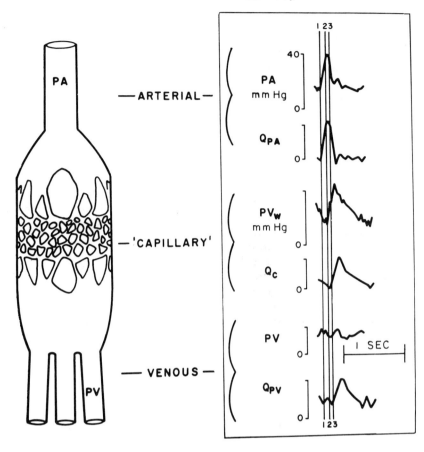

Chapter 11

CARDIAC OUTPUT
DISTRIBUTION OF
CARDIAC OUTPUT
BLOOD VOLUME
CENTRAL NERVOUS
SYSTEM

CIRCULATORY CONTROL

One function of the circulatory system is to supply an adequate amount of blood to all the cells of the body. This is accomplished by regulation of the *cardiac output*, regulation of the *distribution* of that output, and regulation of *venous return*. Most vessels in the body also serve as reservoirs in that they contain more blood than is necessary to perform their usual tasks. These *blood reservoirs* serve to meet the extra demands made on the circulatory system. The arterial reservoir, for example, permits increased blood flow to the capillaries when the arterioles dilate. The venous reservoir permits an increased cardiac output when the heart rate and force increase and compensates for losses of blood due to hemorrhage or dehydration.

In addition, man can usually sustain large increases in blood volume without a dangerous stress on either the heart or its vessels. The major *blood reservoir* and *volume buffering* functions of the circulatory system are served by the systemic veins and the pulmonary system. Apparently these structures respond to a decrease in blood volume with an increase in *vascular tone*, and to an increase in blood volume with a decrease in vascular tone.

The *kidney* too plays an important part in the body's response to changes in blood volume. In response to increases in blood volume it can increase urine production (Fig. 21-12), and in response to decreases in blood volume, can decrease urine production.

Distribution of the cardiac output is to a great degree dependent upon the maintenance of adequate blood pressure. We have in the preceding chapter emphasized the fact that the arteries of the kidneys generally have a high critical closing pressure, which is to say that at *mean arterial pressures* of less than 70 mm. Hg the small vessels of the kidney collapse. It is important to realize that in the erect individual large arterial pressures below the heart must be maintained in order to facilitate an adequate blood supply to the brain. Arterial pressure is largely dependent on two structures: (1) the heart, which moves blood into the arteries, and (2) the resistance vessels, which restrict blood flow from the arteries. Constriction of the resistance vessels initially decreases blood flow out of the arterial system, resulting in distention of the arteries; as a consequence, arterial pressure (and strain on the heart) increases progressively—until the flow from the arterial system returns to normal. The greatest resistance to flow is offered by the arterioles, precapillary sphincters, A-V anastomoses, capillaries, and venules. Of these structures the arterioles and precapillary sphincters seem to be the most capable of varying their diameters in response to nervous stimulation, hormones, and local factors, and as a consequence are apparently most important in the control of peripheral resistance. The A-V anastomoses are also capable of varying their diameters and therefore the systemic arterial pressure, but seem to be more involved with temperature regulation than with the regulation of blood pressure.

We should emphasize the importance of arterioles and precapillary sphincters in the control of *blood flow* to the capillaries. *Generalized dilation* can, by lowering arterial pressure, decrease blood flow to the capillaries, but *localized dilation* may have little or no effect on arterial pressure and may thus increase blood flow to the area served by the dilated vessel. If dilation is widespread, however, it must be associated either with constriction elsewhere or with an increase in cardiac output if arterial pressure is to be maintained. Our aim in this chapter will be to see how the body integrates these various parameters into an effective transport system.

CARDIAC OUTPUT

In Chapter 8 ("Cardiac Output") we emphasized that the heart can increase its cardiac output by increasing its heart rate or its stroke volume. Both heart rate and stroke volume are in turn controlled by cardiac sympathetic and parasympathetic fibers. Under conditions in which these neurons are depressed (anesthesia) or are unable to maintain homeostasis (a constant internal environment), hormones (epinephrine, etc.), local metabolites (H^+, CO_2, K^+, or hypoxia), or cardiac distention may also become important in increasing cardiac output. The significance of increases in cardiac output is that they tend to shift large quantities of blood from the venous reservoirs to the arterial system. This in turn results in arterial distention, an increase in

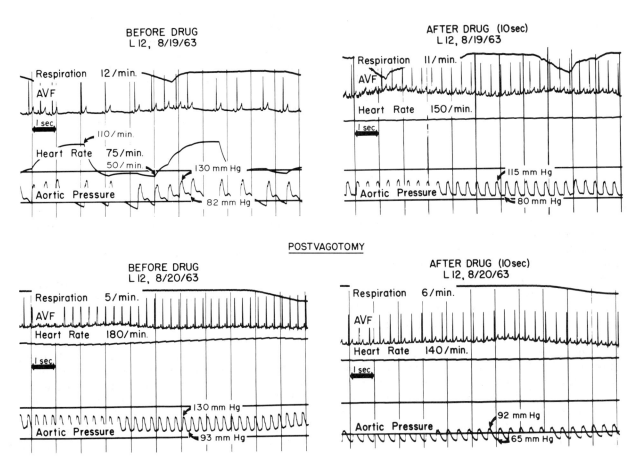

Fig. 11-1. Response of the unanesthetized dog to a ganglionic blocking agent (hexamethonium chloride, 5 mg./kg.) before and after bilateral cervical vagotomy. Note that the normal dog responds to a loss of central autonomic control with an increase in heart rate and a decrease in systolic aortic pressure. After bilateral section of the cervical vagi the dog responds to ganglionic blockade with a decrease in heart rate as well as in systolic and diastolic pressure. Apparently in the resting dog the parasympathetic nervous system exerts the predominant role in the control of heart rate. Note also that in the absence of cardiac parasympathetic fibers, the speeding of the heart which occurs during inspiration (inspiratory tachycardia) disappears. The decrease in aortic pressure which occurs in response to ganglionic blockade results from a decrease in sympathetic tone to the blood vessels. (Courtesy of J. S. Life and R. S. Shepard)

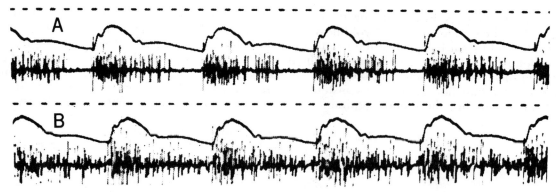

Fig. 11-2. The importance of pressure and epinephrine to the stimulation of carotid sinus afferent fibers. In A we see a record of the pressure in the lingual artery and the electrical potentials in the baroreceptor neurons. In B we see the change in impulse traffic 4 minutes after the local application of 0.25 ml. of epinephrine (1 μg./ml.) to the carotid sinus. In both A and B the mean pressure is 145 mm. Hg. Section of the baroreceptor afferent neurons results in a rapid increase in arterial pressure to 300 mm. Hg and a marked tachycardia. (Heymans, C. J. F., and Folkow, B.: Vasomotor control and the regulation of blood pressure. *In* Fishman, A. P., and Richards, D. W. (eds.): Circulation of the Blood. Men and Ideas. P. 465. New York, Oxford University Press, 1964)

arterial pressure, and an increase in blood flow (volume and velocity) through the capillaries.

In a resting person, the heart rate is approximately 70 beats per minute. If all the nerves to and from the heart are blocked or destroyed, the heart rate in man and dog increases to about 140 beats per minute (Fig. 11-1). If just the parasympathetic fibers to the heart are blocked or destroyed, however, the heart rate will go to about 160 beats per minute. In short, the heart rate in a resting person is constantly being slowed by the nervous system. It is the high degree of parasympathetic tone to the sino-atrial node that is responsible for this slowing. This tone is partly the result of the intrinsic rhythmical properties (membrane instability) of neurons in the central nervous system, and partly that of nervous reflexes involving the cardiac parasympathetic fibers of the vagus nerve.

Mechanoreceptor Reflexes

Located in the walls of certain blood vessels throughout the body are sensory neurons which, when stretched, send impulses to the central nervous system. This group of neurons and associated structures probably should be called a stretch receptor system, but is more generally referred to as a baroreceptor or pressoreceptor system since it is usually the pressure in the lumen of the blood vessel that produces the stretch stimulating the neurons. One type of baroreceptor fires under almost all conditions. When the pressure increases, its rate of firing increases; when the pressure decreases, its rate of firing decreases (Fig. 11-2). A second type of baroreceptor fires only when the pressure exceeds resting levels.

The carotid sinus and aortic arch baroreceptor reflex. The Greeks over 2,000 years ago produced a strange type of anesthesia by compressing an artery in the neck, named the carotid artery from the Greek "karos," meaning a deep sleep. Later, John Czermak (1828-1873) used carotid pressure to slow the heart, and Etienne Marey (1830-1904) stated what has come to be called Marey's law of the heart: "When the blood pressure increases, the heart rate decreases, and when the blood pressure decreases the heart rate increases." Then, in 1929, Corneille Heymans (born 1892) performed a cross-circulation experiment in two dogs. In this experiment he perfused the isolated carotid sinus of one dog (the recipient) with blood from another dog (the donor). When he raised the arterial pressure in the donor dog by the injection of epinephrine (adrenaline) he noted that the blood pressure and heart rate in the recipient decreased. In 1938 he received the Nobel Prize for this work. From his and subsequent studies we conclude that an increase in pressure in either the carotid sinus or aortic arch brings about vasodilation and

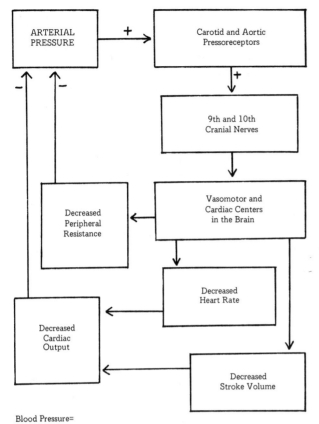

Blood Pressure= Peripheral Resistance x Cardiac Output

Fig. 11-3. The carotid sinus, aortic arch baroreceptor reflex. In man the pressures in the large arteries of the chest and neck stimulate pressoreceptors which by reflex slow the heart, decrease its force (or stroke volume), and dilate the arterioles (decrease peripheral resistance). This reflex is intensified by increases in pressure up to 200 mm. Hg and is depressed by decreases in pressure down to 30 mm. Hg. The significance of the reflex is that it tends to keep the blood pressure in the head, neck, and chest constant and, therefore, to maintain a fairly constant blood flow to the brain. Thus, decreases in cerebral arterial pressure which might occur when we change from a horizontal to a vertical position are prevented by a reflex increase in cardiac output and peripheral resistance. This reflex does not, on the other hand, usually keep the arterial pressure constant in arteries below the heart.

a reflex decrease in heart rate, both of which decrease arterial pressure.

I have in Figure 11-3 summarized the role played by the pressoreceptors of the carotid sinus and aortic arch in circulatory control. They can be characterized as the guardians of pressure homeostasis in the head, neck, and chest. It should be emphasized, however, that they are primarily concerned with the maintenance of pressure in the areas in which they are found. Take, for example, an individual changing from a reclining to a standing position. In the absence of the pressoreceptors or the efferent limb of the reflex they control (the autonomic nervous system can be blocked by ganlionic blocking agents such as hexamethonium chloride), when a subject stands, the arterial pressure in the abdomen and legs increases, the pressure in the head and neck decreases, and brain ischemia (inadequate blood flow) develops, possibly leading to dizziness or fainting. In the presence of this pressoreceptor reflex the individual, on standing, increases his cardiac output, constricts a number of noncerebral arterioles, and in this manner maintains a constant arterial pressure and flow in the head and neck. The pressure in the abdomen and leg, on the other hand, increases. We see in the baroreceptor reflex, then, an explanation of the anesthesia (brain ischemia) produced by the Greeks, and of the dizziness occasionally experienced by someone fastening a tight collar. Apparently this reflex becomes less effective after prolonged bed rest or weightlessness. Patients on returning to a vertical position after days of confinement to bed, and astronauts, after returning to the earth's gravitational field, frequently experience dizziness.

The impulses from the aortic arch pressoreceptors are carried in the right and left tenth cranial nerves (vagus) to the brain. Those of the carotid sinus are carried in the right and left nerve of Hering (carotid sinus nerve) to the right and left ninth cranial nerves (glossopharyngeal) and then to the brain. These impulses in the brain facilitate a high degree of parasympathetic tone to the heart and a depression of sympathetic cardiac and vasoconstrictor tone. It is probable that in man and dog (but not in all animals) the cardiac sympathetic fibers play a minor role in the control of the heart during rest or restricted activity. In response to increases in arterial pressure there would be an increase in *vagal tone* to the heart (decreased heart rate) and a decrease in *sympathetic tone* to the arterioles (decreased peripheral resistance). It has also been suggested that there might be an increase in cholinergic sympathetic tone to the blood vessels. In response to decreases in arterial pressure, on the other hand, there would be a decrease in vagal tone to the heart (increased heart rate), an increase in sympathetic tone to the heart (positive chronotropic, inotropic, and dromotropic action), and an increase in adrenergic sympathetic tone to the arterioles (increased peripheral resistance). Destruction of the sensory neurons from the carotid sinus and aortic arch results in the conversion of an arterial pressure of 120/80 to 300/200 mm. Hg.

Bilgutay and Lillehei recently suggested that certain types of hypertension are due to sclerosis of pressoreceptor areas or to a resetting of their susceptibility to arterial pressure, and reported on two hypertensive patients (260/160 mm. Hg) in whom they implanted electrodes on the carotid sinuses. These electrodes were led to a variable-rate implanted stimulator. Using this technique, investigators were able to lower the systolic pressure to below 140 mm. Hg. It is not yet clear, however, whether the technique will be effective over prolonged implantation.

Other mechanoreceptor reflexes. A number of other mechano- or baroreceptor reflexes have been suggested other than those associated with the carotid sinus and aortic arch. Their importance to the body is not well understood and is still debated. It has been demonstrated, however, that under certain conditions increases in pressure in the atria, left ventricle, pulmonary arteries, lungs, pulmonary capillaries, or pulmonary veins can produce reflex bradycardia, arteriolar dilation, and decreased venous tone that results in decreased venous return of blood to the heart, decreased cardiac output, and a decrease in both arterial and venous pressures. Possibly such a reflex is important in preventing venous congestion in the systemic and pulmonary systems and an associated edema (pulmonary edema, ascites, general edema).

It is to be noted that all the mechanoreceptor reflexes mentioned above depress the cardiovascular system (decrease heart rate and vasoconstriction). *Bainbridge* in 1915 suggested a facilitatory one. He reported that under special conditions (anesthesia, slow heart rate) a marked increase in venous return produced cardiac acceleration. To date, however, a receptor mechanism for this response has not been demonstrated, and many deny that such a reflex exists. Brainbridge believed that an increase in venous return elicited the stimulation of vagal afferent fibers from the great veins and right atrium, a reflex decrease in cardiac parasympathetic tone, and an increase in sympathetic tone to the heart. Others have suggested that the afferents for this reflex (the Bainbridge reflex) may originate from the pulmonary vessels, while still others believe the increase in heart rate is a nonreflex response of the heart itself to increased venous return. The Bainbridge effect has in any case stimulated much writing and discussion by physiologists in the last fifty years.

Specialized receptors believed to respond to increases in pulse pressure have been reported in the thoracic aorta, and others have been noted in the small peripheral veins. These are thought to respond to increases in pressure by stimulating afferent fibers to bring about a local arteriolar constriction. Receptors in the mesenteries may regulate the local distribution of flow there. In addition, it has been noted that distention of the left atrium brings about a diuresis dependent on a functioning vagus nerve. It is suggested that there are volume receptors (stretch receptors) in the left atrium which reflexly depress the release of antidiuretic hormone from the pituitary gland and thus increase the amount of urine formed by the kidneys. These receptors are visualized as controlling blood volume.

Before leaving the mechanoreceptors, I would emphasize that they are influenced by factors other than pressure and distention. As we have noted in Figure 11-2, for example, epinephrine modifies the output of the carotid sinus afferent fibers in the absence of any change in blood pressure. The drug veratrine has also been shown to have this effect on the mechanoreceptors of the left ventricle (Bezold-Jarisch reflex). When veratrine is injected into the left coronary artery in dosages having no effect when injected into a systemic vein, there is reflex slowing of the heart, hypotension, and apnea that can be prevented by section of the vagus nerves. That is to say that the vagi seem to carry the sensory impulses for this reflex (*coronary chemoreflex* or *Bezold-Jarisch reflex*) from the heart. It has also been suggested that in patients with myocardial infarcts, substances are released by the damaged tissue that initiated this reflex, producing the hypotension commonly associated with these infarcts.

DISTRIBUTION OF CARDIAC OUTPUT

The amount of blood that flows through a particular capillary system depends not only on (1) the amount of blood pumped by the heart and (2) the pressure difference between the heart and the capillaries, but also on (3) the relative resistance to flow offered by the vessels leading into the capillary system. If the afferent vessels dilate and all other vessels stay the same, flow to this area increases; but if all the vessels dilate at once, flow to this area may not change. In Figure 11-4 we note the distribution of the cardiac output at rest. We see that approximately

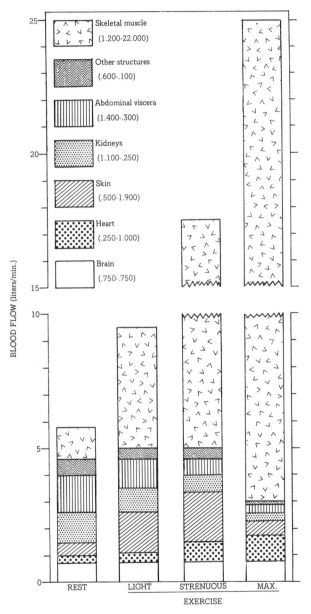

Fig. 11-4. Distribution of blood flow at rest and during exercise. The numbers in parentheses represent the maximum range in flow that occurs during rest and during strenuous or maximal exercise. (*Data from* Chapman, C. B., and Mitchell, J. H.: The physiology of exercise. Sci. Am., *212*:91, 1965)

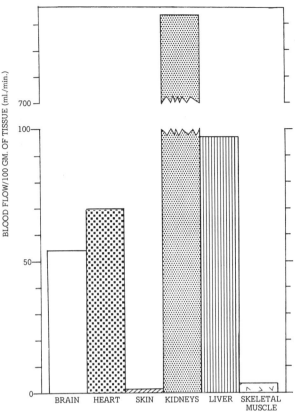

Fig. 11-5. Blood flow to different structures in the resting individual.

24 per cent of the cardiac output goes to the abdominal viscera, 21 per cent to skeletal muscle, 19 per cent to the kidneys, 13 per cent to the brain, 9 per cent to the skin, and 3 per cent to the heart.

In Figure 11-5 we get a somewhat different perspective. Here we note that when flow is studied on the basis of the weight of the structures served, the kidneys get a disproportionately large blood supply, and skeletal muscle and the skin get a disproportionately low flow. These relationships are not surprising when one remembers the significance of these flows. Flow to the kidneys serves not only to supply the metabolic needs of this organ but also to assist the entire organism in eliminating both waste products and substances that exist in excess (water, Na^+, K^+, H^+, Cl^-, Ca^{++}, etc.). Flow to the liver, too, is concerned with the removal of certain wastes (bilirubin) from the body, as well as the addition to or extraction of certain nutrients (blood sugar, vitamins, etc.) from the blood. In the case of the heart we are dealing with a muscle which is considerably more active metabolically in the resting individual than is skeletal muscle. Probably the most active skeletal muscle in a resting individual is the diaphragm, which contracts about 16 times a minute. The skin is the least metabolically active of all the structures listed. This, at least in part, is due to the fact that the skin consists of an outer layer of dead cells or cell fragments and an inner layer rich in intercellular material.

In Figure 11-4 we also note that the distribution of blood flow may change markedly. For example, during strenuous exercise the flow to skeletal muscle may increase tenfold. Such an increase in our example is associated with a threefold rise in cardiac output. Apparently there has been a change in the relative resistances throughout the vascular system, since although the cardiac output has increased, there is a decrease in flow to the abdominal viscera (−52%) and kidneys (−46%). The flow to the brain remains unchanged. The increases to the heart are needed to meet the increased metabolic needs of the heart associated with increases in cardiac output. Increases in flow to the skin are needed to dissipate the increased heat production resulting from the contractions of the skeletal muscles. It is interesting to note, however, that in the most strenuous exercise of which one is capable (maximal exercise), blood flow to the skin begins to decrease even though heat production continues to be elevated.

Nervous Control

It is apparent that the fine degree of vasoconstrictor and vasodilator control discussed above is to a great extent the result of central nervous system activity. It is important to bear in mind, however, that the importance of the nervous system varies tremendously from one area to another. We find very few vasoconstrictor fibers to the brain and heart, for example, but large numbers to skeletal muscle, the skin, the splanchnic area, and the kidneys. Blocking the nerves to the heart and brain produces little or no increase in blood flow, but blocking nerves to skeletal muscle can produce a fivefold increase in blood flow. Apparently, then, there is a considerable degree of nervous tone to the blood vessels of skeletal muscle in the resting individual.

In Figure 11-6 I have summarized the maximum changes in blood flow that can result from stimulation of vasoconstrictor and vasodilator fibers to different areas. Note that vasoconstrictor fibers have a pronounced effect on the kidney, skin, skeletal muscle, and abdominal viscera. Note also that vasodilator fibers seem most important in skeletal muscle.

We pointed out in Chapter 4 ("The Autonomic Nervous System") that smooth muscle may be innervated by either sympathetic or parasympathetic nerve fibers or by both sympathetic and parasympathetic fibers, or have no innervation at all. The placenta is probably the only vascular system in the human body that does not receive sympathetic neurons. *Parasympathetic fibers*, on the other hand, seem to be directly concerned with the control of vascular tone only in the salivary glands, the tongue, and parts of the genital organs. In these structures they produce vasodilation.

The *sympathetic fibers* can, in skeletal muscle, produce either vasoconstriction or vasodilation. It was once thought that sympathetic vasodilation was due only to a decrease in sympathetic tone; now it is generally recognized that there are nerve fibers originating at the thoracic or lumbar spinal cord which, when stimulated, produce dilation. These sympathetic neurons apparently stimulate postganglionic fibers which release acetylcholine at their endings. The vasodilator action of these cholinergic postganglionic sympathetic fibers is not disturbed by agents that block the action of norepinephrine or epinephrine. Neither the *beta blocking agent* dichloroisoproterenol nor the *alpha blocking agent* phenoxybenzamine interferes with the action of the cholinergic sympathetic fibers.

You will remember from our discussion of the autonomic nervous system the hypothesis that *alpha receptors* are receptor sites on vascular smooth muscle which are stimulated by *epinephrine* and *norepinephrine* to produce vasoconstriction, and *beta receptors* are sites which are stimulated by epinephrine and isoproterenol to produce vasodilation. Norepinephrine is a precursor of epinephrine and is released in highest concentration at postganglionic sympathetic endings. Epinephrine is found at postganglionic sympathetic endings but is most concentrated in the secretions of the adrenal medulla. In Figure 11-7 we note that epinephrine has a dual effect on the vasculature of skeletal muscle and certain of the abdominal viscera. Initially it produces a decrease in flow—probably due to the

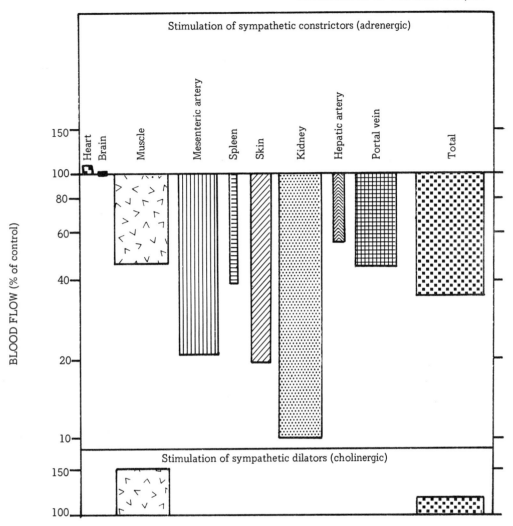

Fig. 11-6. Maximum response of various vascular beds to the stimulation of sympathetic constrictor nerves and sympathetic dilator nerves. The width of each bar is approximately proportional to the percentage of the cardiac output received by each vascular bed. (*Redrawn from* Green, H. D.: Ganglionic and adrenergic blocking agents. Minn. Med., *41*:244, 1958; and Green, H. D., and Kepchar, J. H.: Control of peripheral resistance in major systemic vascular beds. Physiol. Rev., *39*:676, 1959)

stimulation of alpha receptors (vasoconstriction). Later, it produces an increase in flow (vasodilation) probably due to the stimulation of beta receptors. In the skin, kidney, hepatic artery, and portal vein there is merely a decrease in flow in response to epinephrine. Apparently these organs have few, if any, beta receptors. On the other hand, epinephrine is believed to have little direct effect on either the cerebral or coronary vessels. It does, however, produce an increased metabolic activity of cardiac muscle fibers that is followed by an increased coronary flow. This is consistent with the view that blood flow to the heart and brain is controlled almost exclusively by their need for nutrients, and that their blood vessels are little affected by either sympathetic stimulation or the injection of epinephrine, but are responsive to hypercapnea and hypoxia.

Figure 11-8 should be compared with Figure 11-7. It shows the response to epinephrine after alpha-receptor blockage with phenoxybenzamine. Note that this blocking agent prevents

194   11/Circulatory Control

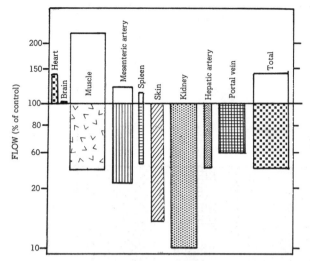

Fig. 11-7. Response of various vascular beds to an intra-arterial injection of 1-10 μg. of epinephrine. The shaded bars represent initial responses and the clear bars secondary responses. The width of each bar is approximately proportional to the percentage of the cardiac output received by each vascular bed. (*Redrawn from* Green, H. D.: Ganglionic and adrenergic blocking agents. Minn. Med., *41*:243, 1958)

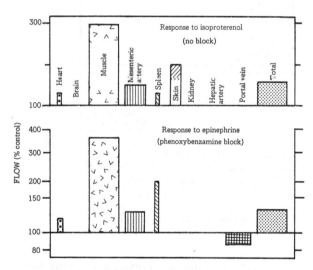

Fig. 11-8. Response of various vascular beds to the intra-arterial injection of 1 μg. of epinephrine after alpha blockade with phenoxybenzamine. Note the similarity between these responses and those seen to the intra-arterial injection of 1 μg. of isoproterenol. The width of each bar is approximately proportional to the percentage of the cardiac output received by each vascular bed. (*Redrawn from* Green, H. D.: Physiologic action of ganglionic and adrenergic blocking agents on peripheral circulation. Minn. Med., *41*: 243, 1958)

many of the vasoconstrictor actions of epinephrine and results in epinephrine's producing an effect similar to that of isoproterenol. Information such as this has led to the classification of isoproterenol as a beta adrenergic agent.

Chemoreceptor reflexes. There are bodies in the area of the carotid sinuses and aortic arch that are sensitive to changes in the chemical nature of the blood. The carotid bodies lie cephalad to the carotid sinuses and are composed of epithelioid cells in close contact with sinusoidal blood vessels. They receive their blood from the occipital or pharyngeal artery and drain into the internal jugular vein. Their blood flow (2,000 ml./100 gm. tissue/min.) and sensory innervation are great. Their oxygen usage (9 ml./min./100 gm.) is three times that of brain tissue. Epithelioid tissue similar to that in the carotid bodies is also found on the anterior surface of the aortic arch and at the roots of the right and left subclavian arteries.

In Figure 11-9 we see the effect of a 10-per cent decrease in the oxygen content of the inspired air on the frequency of afferent action

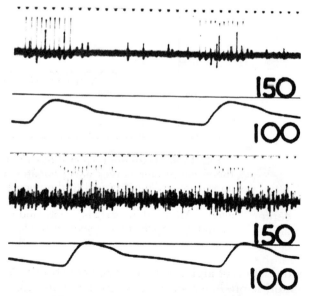

Fig. 11-9. Action potentials in the peripheral portion of a cut carotid sinus nerve (sensory impulses) in a cat while breathing room air (*upper tracings*) and while breathing 10% oxygen (*lower tracings*). Arterial pressure was obtained from the femoral artery. A 100 and 150 mm. Hg reference line is shown. (Neil, E. *In* Heymans, C., and Neil, E.: Reflexogenic Areas of the Cardiovascular System. P. 139. London, J. & A. Churchill, 1958)

potentials from the carotid sinus nerve. Note that this hypoxia increases the number of sensory impulses from the carotid sinus area. It has been demonstrated that the carotid and aortic bodies are sensitive to decreases in the partial pressure of oxygen in the plasma of the blood (P_{O_2}), but not to the total amount of oxygen in the blood. That is to say we can stimulate the carotid and aortic bodies by decreasing the P_{O_2} of the plasma but not by decreasing the number of red blood cells (anemia), which carry most of the blood's oxygen, or by decreasing the amount of oxygen carried by the red blood cell (CO poisoning). The chemoreceptors are also stimulated by (1) poisons that interfere with the utilization of O_2 (sodium cyanide), (2) increases in the partial pressure of CO_2 in the plasma (P_{CO_2}), and (3) increases in the acidity of the plasma.

In Figure 11-10 I have summarized the response of the chemoreceptors to graded concentrations of oxygen in the inspired air. Normally the inspired air has a P_{O_2} of about 150 mm. Hg. We note that as we decrease the P_{O_2} from 160 to 80 mm. Hg, more and more sensory fibers fire. At a P_{O_2} of 80 mm. Hg the threshold has been reached for all the chemoreceptor afferent fibers, but as the P_{O_2} is lowered still more, the frequency of impulses in each of these afferent fibers increases up to a maximum of about 12 per second.

The significance of these receptors is that they increase respiratory rate and depth in response to hypoxia, hypercapnea (increased P_{CO_2}), acidity, or increased temperature. Their importance in the control of the cardiovascular system, however, is probably less marked and less well understood, and differs from one species to another. In Figure 11-11, for example, we see some evidence that in the dog stimulation of the carotid chemoreceptors by a KCN-induced hypoxia produces bradycardia, and stimulation of the aortic chemoreceptors produces tachycardia and an increase in arterial pressure. In part D of this figure we also note that KCN produces hypotension when these receptors are denervated. Apparently then, hypoxia has the direct effect upon the arterioles of producing vasodilation, though this is usually masked by the reflex stimulation of the arterial chemoreceptors by hypoxia.

During mild exercise or in a resting, awake person, the peripheral chemoreceptors probably do not play an important role in the control of arterial pressure. In cases of cyanide poisoning, breath-holding, or conditions that cause a

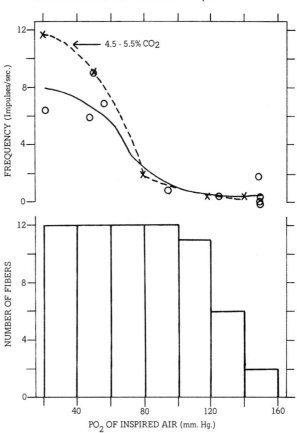

Fig. 11-10. Effect of lowering the P_{O_2} of the inspired air on the sensory impulses in the aortic nerve. In the upper illustration we note that as we lower the P_{O_2} the frequency of impulses in the nerve fiber being studied increases. We also note that adding CO_2 to the inspired air increases the frequency at certain P_{O_2}'s. In the lower graph we note that, within limits, the number of fibers stimulated also increases as the P_{O_2} decreases. (*Data from* Paintal, A. S., and Riley, R. L.: Response of aortic chemoreceptors. J. Appl. Physiol., *21*:543-548, 1966)

moderate degree of hypoxia in the arterial blood, reflex vasoconstriction occurs. The chemoreceptors are also thought to be responsible for Traube-Herring (Mayer) waves—periodic fluctuations in arterial pressure not associated with respiratory movements. These sometimes occur after a severe hemorrhage. In such cases there is probably periodic hypoxia which develops in the chemoreceptors due to ischemia, resulting in a reflex increase in arterial pressure due to vasoconstriction. The increased pressure stimulates the carotid and aortic pressoreceptors, which reflexly bring the pressure down again. This fluctu-

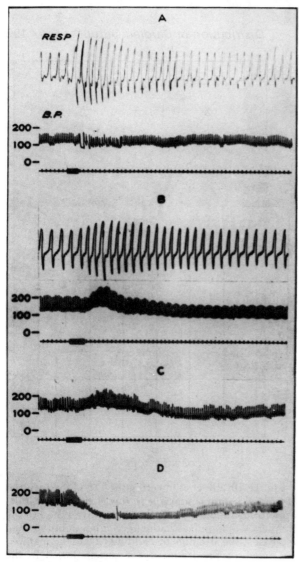

Fig. 11-11. Action of KCN on the carotid and aortic bodies of a dog under chlorolose anesthesia. In part A, 50 μg. of KCN was injected into the left common carotid artery. Hyperpnea and bradycardia (see B.P., blood pressure record) resulted. In part B the carotid sinuses and bodies were decentralized and 250 μg. of KCN was injected into the ascending aorta. Hyperpnea, tachycardia, and an increase in arterial pressure resulted. In parts C and D, the ventilation was maintained constant by artificial respiration. In C the chest was opened and the procedures performed in part B were repeated. Tachycardia and hypertension were produced by the KCN injection into the ascending aorta. In D, the dog was prepared the same as C, but the cervical vagi were cut (that is, the aortic bodies were also decentralized) and a decrease in pressure occurred in response to KCN. (Comroe, J. H.: The peripheral chemoreceptors. In Fenn, W. O., and Rahn, H. (eds.): Handbook of Physiology. Sec. 3 (Respiration), vol. I, p. 574. Washington, American Physiological Society, 1964)

ation between the firing of presso- and chemoreceptors results in fluctuations of the arterial pressure.

It has recently been demonstrated that during deep sleep in the cat, arterial oxygen concentrations fall and the afferent fibers from chemoreceptors carry more impulses. In the absence of the chemoreceptor function, severe hypotension results. These and other data are consistent with the view that when parts of the brain are depressed, as in deep sleep or by anesthesia, the chemoreceptors provide an important stimulus to the central nervous system to increase respiratory rate and depth, as well as arterial pressure. It is in part through the peripheral chemoreceptors, then, that we protect the brain from severe hypoxia.

Central chemoreceptors. Hypoxia has not only an indirect facilitatory action on the vasomotor and cardio-inhibitory centers of the medulla through the peripheral chemoreceptors, but also, within limits, a direct facilitatory action. For example, in cases of abnormal increase in the cerebrospinal fluid pressure around the medulla, the blood vessels supplying the medulla are compressed and local hypoxia develops. This hypoxia facilitates activity in both the vasomotor and cardio-inhibitory centers. As a result, there is generalized vasoconstriction in the viscera, skin, and skeletal muscle which increases arterial pressure and alleviates compression of the medullary vessels, thus also preventing the associated medullary ischemia. There is also bradycardia. Local medullary hypercapnea, on the other hand, seems to facilitate only vasomotor-center activity.

Mechanoreceptor reflexes. We have discussed the mechanoreceptors in an earlier section of the chapter. There we noted that increases in arterial pressure increase the stimulation of aortic and carotid pressoreceptors, which reflexly decrease cardiac output by decreasing heart rate, and produce reflex vasodilation in susceptible vessels. This decrease in cardiac output and peripheral resistance tends to lower the arterial pressure. Decreases in arterial pressure also act through these receptors to increase cardiac output and peripheral resistance, and in this way to bring arterial pressure back to normal levels.

Triple response. If one places a rubber band around the forearm, pulls it back, and then releases it, a localized red line appears due to capillary dilation; this becomes a generalized red flare, due to arteriolar dilation; and finally, if the rubber band did sufficient tissue damage, there is a local swelling or wheal due to an efflux of fluid from the blood. Capillary dilation is apparently a direct response of the capillaries to the stimulus, and the wheal the result of the release of a local vasodilator (histamine) from the damaged cells. The red flare, however, is dependent upon the presence of sensory nerve fibers. It does not occur if the peripheral nerves have been sectioned and given time to degenerate. It does occur, however, it the stimulus is applied after nerve section but before the nerve has degenerated. Apparently it results from the stimulation of sensory nerve fibers which give off collateral branches to the arterioles. Although this response does not necessarily involve the central nervous system, it is sometimes called an *axon reflex*. In the case of the skin, this hyperemia in response to irritation may play a role in the skin's defense against infection and in the repair of damaged tissue. It has not, however, been demonstrated elsewhere in the body.

Vasoactive Substances

A number of substances found in the body are capable of producing either vasoconstriction or vasodilation. One of these is *epinephrine*, produced in the adrenal medulla and released in response to the stimulation of certain preganglionic sympathetic neurons. The adrenal medulla is the only endocrine gland in the body innervated by preganglionic neurons. The epinephrine is released into the blood, acting on the sino-atrial node to make it fire more rapidly (positive chronotropic action), and on the individual muscle fibers of the heart to make them conduct more rapidly (positive dromotropic action) and contract more forcefully (positive inotropic action). These effects of epinephrine on the heart, then, tend to markedly increase the cardiac output (Fig. 11-12). In Figure 11-7, we outlined the response of the various vascular beds of the body to epinephrine. We noted the dual action of epinephrine on the flow in skeletal muscle and some of the abdominal viscera, and its vasoconstrictor action in the skin, kidney, hepatic artery, and portal vein. In summary, epinephrine increases cardiac output and changes the distribution of the cardiac output. Under most circumstances it is less important in the control of the cardiovascular system than are the nerves to the heart and blood vessels, but when these nerves

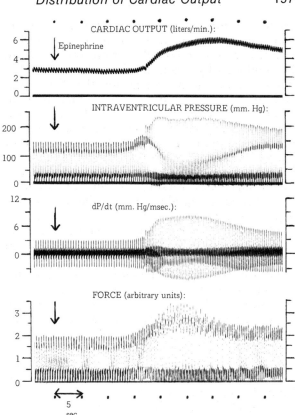

Fig. 11-12. Response of a pentobarbital-anesthetized dog to epinephrine (0.5 µg./kg., I.V.). The force was measured from a 2.3-cm. segment of the left ventricle by means of a Walton-Brodie transducer. The transducer keeps the end-diastolic length of this segment constant. Apparently the increased cardiac output in response to epinephrine is due to (1) an increase in cardiac contractility (+ inotropic action of epinephrine) and (2) a decrease in peripheral resistance ($beta_1$ receptor stimulation by epinephrine). After $beta_E$ blockage with AY-21,011 (5 mg./kg., I.V.) the positive inotropic action of epinephrine is prevented and its effect on the cardiac output is markedly reduced. The AY-21,011 is an experimental drug supplied by the Ayerst Laboratories. (Courtesy of R. Green and R. S. Shepard)

do not meet the body's needs, epinephrine may be released. Thus it serves as an auxiliary system when nervous responses are inadequate.

A number of other vasoactive agents have been reported. These, though of pharmacologic and pathologic interest, have not been shown to play an important role in the normal person. One of these is *renin*, secreted by the juxtaglomerular cells on the afferent renal artery. In 1934 Goldblatt et al. reported that severe diastolic hypertension could be produced by placing partially constricting clamps on the main renal arteries. Later, Page showed that hypertension could also be produced by wrapping the kidneys in cellophane. It was at first suggested that these procedures led to a renal ischemia which brought about the release of renin; now there is good evidence that renal ischemia alone does not lead to the release of renin, but rather to some other change in the kidney. Perhaps here too there is a baroceptor mechanism. Once the renin reaches the plasma it acts on a substrate, alpha-2-globulin, by splitting a peptide bond to liberate the decapeptide angiotensin I (Fig. 11-13). This is relatively inactive and is split in blood to form the octapeptide, angiotensin II. It is the angiotensin II that produces vasoconstriction. It has also been suggested that angiotensin II stimulates the adrenal cortex to release aldosterone, which acts on the kidneys to increase the reabsorption of Na^+ from the glomerular filtrate (urine) back into the blood. This is more fully discussed in Chapter 21 under renal hypertension.

Another vasopressor agent of doubtful physiologic action is *vasopressin*. It can be extracted from the posterior pituitary, concentrated, and then injected. Under these circumstances it produces vasoconstriction.

There is, however, a substance found in the blood platelet which may be important locally. This substance, *serotonin* (5-hydroxytryptamine), is released by the platelets when they are broken down during the formation of a blood clot. The local vasoconstriction which they produce probably helps prevent bleeding.

Finally, there are a number of vasodilating agents found throughout the body which are of great clinical concern. One substance, histamine, is released by damaged cells. It, like heparin, is found in particularly high concentration in the mast cells. When released, it produces arteriolar dilation, venous constriction, and, if in high enough concentrations, contraction of the smooth muscle of the respiratory tract. It is the local release of histamine that produces swelling (edema) after tissue damage. A massive release of histamine, on the other hand, sometimes occurs in response to certain foreign proteins in the blood. This may lead to constriction of the respiratory tract (asthma) due to muscular contraction and edema, or to such a marked decrease in arterial pressure as to result in death (anaphylactic shock). Also of interest is the suggestion by some investigators that certain vasodilator nerve fibers (histaminergic fibers), when stimulated, cause the release of histamine.

Some of the other vasoactive substances released by damaged cells are adenosine, lysolecithin, and *bradykinin*. Bradykinin is of interest in that it is released by certain glands (sweat glands, salivary glands, and the exocrine portion of the pancreas) when they are stimulated by autonomic neurons. Thus secretion may be associated with vasodilation. Bradykinin is similar to histamine in that it increases capillary permeability and stimulates the movement of leukocytes to injured areas.

Autoregulation

In addition to the autonomic nervous system, the sensory neurons of the axon reflex, the endocrine system, histamine, serotonin, vasopressin, renin, and bradykinin, there are a number of agents released by most metabolically active cells which also may modify vascular tone. Some of these *metabolites* (CO_2 and H^+) have a general effect on the body through their action on peripheral chemoreceptors and the brain, as well as some effects on the areas in which they are produced. Others (K^+, pyruvate, acetate, citrate, etc.) under most circumstances play only a local role. The significance of these metabolites is that when their concentration in a small area increases, they have the potential to produce local vasodilation. If their concentrations change sufficiently, in other words, they constitute a local feedback mechanism that operates in the absence or presence of the previously mentioned factors to maintain an adequate local blood supply. It is not yet clear what role autoregulation

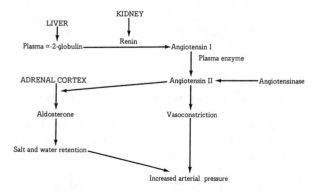

Fig. 11-13. The renin-angiotensin theory of hypertension.

plays in the cardiovascular dynamics of the resting person. Some believe that the autonomic nervous system is primarily responsible for maintaining adequate cardiac output, arterial pressure, and body temperature, and that autoregulation is the primary factor in controlling the distribution of the cardiac output. Others regard autoregulation merely as an important reserve mechanism, i.e., they feel it is masked and overpowered by the centrally directed neurohormonal mechanisms. They emphasize, for example, the observation that in moderate exercise the CO_2 concentration in the arterial blood frequently goes down rather than up, and cite other observations (Fig. 11-14) indicating that the nervous system overcompensates for stressful, and for that matter expected, situations ("get-set" response).

On the other hand there are blood vessels on which autonomic neurons and epinephrine have little or no direct effect. These consist primarily of the blood vessels of the heart and brain. It has also been suggested that the renal arteries respond to increases in arterial pressure by an increase in arteriolar tone that does not involve neurons. The arteries of the skin and brain, on the other hand, are more sensitive to changes in CO_2 concentration than are other vessels of the body. This has led to the suggestion that locally produced CO_2 in these areas is important in producing vasodilation and a local increase in blood flow.

One condition over which the autonomic nervous system may have little control is the occlusion of blood flow that occurs when a tourniquet is placed on a limb. A situation similar to this is the occlusion of blood flow when one sits or lies in one position for a prolonged period. During the occlusion, metabolites accumulate and the blood pressure in the affected area decreases. As a result, vascular tone may also decrease, and when blood flow resumes there is less resistance than before occlusion. This increase in blood flow following ischemia is called *reactive hyperemia* and has been demonstrated in skeletal muscle, the heart (Fig. 11-15), the kidney, the brain, the intestine, and the liver, but not in cutaneous tissue.

Myogenic hypothesis. Two of the major theories advanced to explain reactive hyperemia are (1) the myogenic, and (2) the metabolic hypotheses. The myogenic concept was first proposed by Bayliss in 1902. According to Bayliss, the arterioles respond to stretch by an increase in their smooth-muscle tone, which is independent of any nerves. The smooth muscles, in other words, are thought to depolarize with greater frequency as they are stretched, and with less frequency as stretch decreases.

It is further suggested that such an autoregulatory pattern of control is important in certain vascular fields to prevent an overdistension or underdistension of the fields and to help maintain an adequate blood flow. In Figure 11-16, for example, we see a plot of the relationship between flow and perfusion pressure. Note that as we change from a pressure of 40 to 80 mm. Hg, the flow increases by 3 ml. per minute and the peripheral resistance increases by 46 per cent. Apparently this increase in peripheral resistance is due to an increase in arteriolar tone. When the pressure increases from 80 to 120 mm. Hg, the flow is raised by an additional 19 ml. per minute and the peripheral resistance decreases by 45 per cent. Under these circumstances, blood pressure apparently results in the exertion of more tension on the muscular vessels than the smooth muscles can overcome. As a result the vessels are distended and peripheral resistance decreases. A similar situation has been shown to exist in the normal and denervated kidney. Here too, increases in the perfusion pressure produce

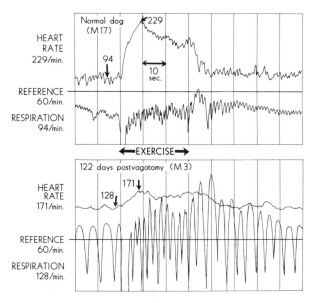

Fig. 11-14. Changes in the dog's heart rate during exercise (3 m.p.h. at a 5% slope for 30 sec.) before and after bilateral cervical vagotomy. Note that in the normal dog there is a rapid increase in heart rate at the beginning of exercise and that as the exercise continues the heart rate stabilizes at a lower value.

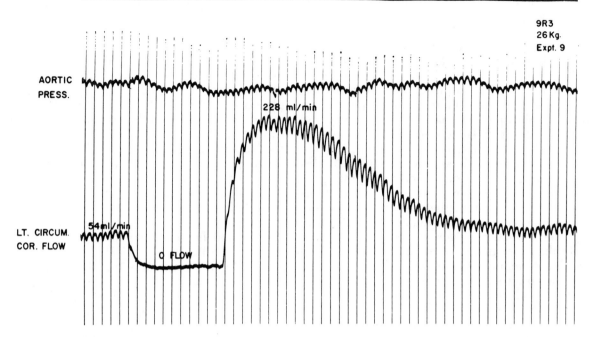

Fig. 11-15. Reactive hyperemia in the heart. In these experiments the coronary artery was occluded for 10 sec. and then the occlusion was released. Note that the subsequent increase in flow (hyperemia) more than compensates for the loss of flow. (Olsson, R. A.: Kinetics of myocardial reactive hyperemia blood flow in the unanesthetized dog. Circ. Res. (Suppl. 1), *14* and *15*:I-83, 1964; by permission of the American Heart Association, Inc.)

an increase in peripheral resistance — within limits. In certain types of reactive hyperemia, on the other hand, we may be dealing with the response of vessels to a decrease in perfusion pressure. This results in a decrease in vascular tone such that when the perfusion pressure rapidly returns to its previous level there is an initially marked increase in flow (hyperemia), which is maintained until vascular tone returns to normal.

Hypoxia. Another cause of reactive hyperemia are the changes in the environment of the arteriole resulting from the metabolic activity of surrounding cells. These changes include hypoxia, hypercapnea, acidity, hyperkalemia, increases in lactate, and so on. Part of the case for hypoxia as a change inducing vasodilation is presented in Figure 11-17. Note that if the partial pressure of the O_2 in the blood perfusing the arterioles and precapillary sphincters of skeletal and cardiac muscle drops from 100 mm. Hg to 70 mm. Hg (Fig. 11-17), these vessels dilate and in this way decrease the resistance to (increase the conductance of) flow. Hypoxia, on the other hand, has little effect on resistance in the renal vessels.

The question that arises in situations other than reactive hyperemia is whether sufficient changes develop in the PO_2 around the arterioles and precapillary sphincters of the heart and skeletal muscle to produce important changes in blood flow. Some believe this to be the most important mechanism regulating the distribution of flow to these areas. Others emphasize that at rest and during exercise the PO_2 of the blood perfusing the arterioles remains at about 100 mm. Hg, and that flow is regulated by changes in sympathetic tone to these vessels. They point out that in those few conditions in which the arterial PO_2 does decrease, the aortic chemoreceptors are activated to produce generalized vasoconstriction. In summary, then, both local hypoxia and changes in sympathetic tone are generally acknowledged to be vasodilators. The

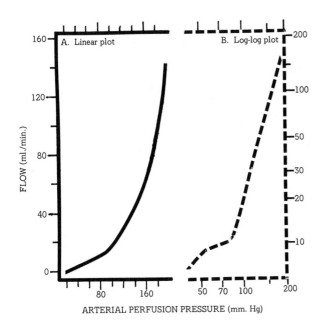

Fig. 11-16. Blood flow in a dog's skeletal muscle at different perfusion pressures. Part A is a linear plot of the data and Part B a plot on log-log paper. That part of the curve between 58 and 82 mm. Hg is probably the result of a vasoconstriction in response to the rising pressure, since it is associated with an increase in peripheral resistance. (*Redrawn from* Green, H. D., and Rapela, C. E.: Blood flow in passive vascular beds. Circ. Res. (Suppl. 1), *14* and *15*:I-15, 1964; by permission of the American Heart Association, Inc.)

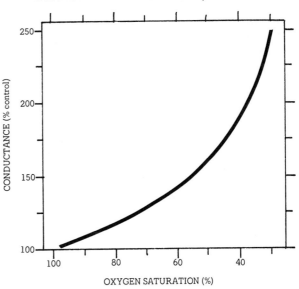

Fig. 11-17. Effect of O_2 saturation on vascular conductance (conductance equals 1/resistance). (*Redrawn from* Guyton, A. C., et al.: Evidence for tissue oxygen demand as the major factor causing autoregulation. Circ. Res. (Suppl. 1), *14* and *15*:I-61, 1964; by permission of the American Heart Association, Inc.)

question that remains is to what extent these variables influence the control of blood flow under various conditions of rest and stress.

Hypercapnea and acidity. The partial pressure of CO_2 in the arterial blood is 40 mm. Hg, and in the venous blood, 46 mm. Hg. The pH of venous blood is 0.02 below that of arterial blood (pH 7.4). That is to say that as the blood passes into the capillaries it picks up the acid-forming gas, CO_2, most of which combines with H_2O to form H^+ and HCO_3^-. It has been shown that a 30 to 100 per cent increase in hindlimb and intestinal blood flow can be produced by exposing arterial blood to an atmosphere of 2 to 10 per cent CO_2. Large increases in coronary flow have also been demonstrated when the heart is exposed to increases in CO_2, but these can be prevented if the pH is kept constant. Increases in P_{CO_2} in the absence of pH changes, on the other hand, seem to be quite effective in increasing renal flow. Lactic, pyruvic, acetic, citric, nitric, hydrochloric, and carbonic acid have all been used to lower the pH of arterial blood to an area. A 0.5 decrease in the pH has been shown to markedly increase both coronary and forelimb flow. We note that in the case of a limb, then, increases in flow may result from either a decrease in pH, an increase in P_{CO_2}, or a combination. In the kidney, on the other hand, increases in P_{CO_2} are far more effective than a decrease in pH in increasing renal blood flow, while in the heart an increase in coronary flow is more readily produced by changes in pH.

Release of potassium. It has been demonstrated that with increased metabolic activity, skeletal muscle releases increasing amounts of K^+ into the intercellular fluid and blood. This observation has led some investigators to conclude that the release of K^+ by active cells is one of a number of important local mechanisms which increase blood flow to skeletal muscle. Potassium ions have been shown to increase vasodilation in skeletal muscle, the heart, the intestine, and the kidney. In the limb, increases in K^+ of as little as 1 to 3 mEq. per liter have been shown to increase blood flow there. In addition, reduction of K⁺ concentration in the limb and kidney have been followed by vasoconstriction.

Release of other substances. Adenosine triphosphate (ATP), adenosine diphosphate (ADP), adenosine monophosphate (AMP), and adenosine are all potent vasodilators. The breakdown products of adenosine have been reported in the coronary blood during hypoxia, and AMP has been reported in the renal blood after renal occlusion. Although ATP does not freely cross the intact cell membrane, it has been reported in the venous effluent of the rabbit's ear after stimulation of the skin. On the basis of these observations, it has been suggested that these substances may be important vasodilators under some conditions.

BLOOD VOLUME

I have emphasized the importance of cardiac output and its distribution in the circulatory system. A third important area is the control of blood volume. In a 24-year-old person there is an average blood volume of 5,548 ± 601 ml. (mean ± standard error), and in a 41-year-old person, one of 5,505 ± 612 ml. In the 24-year-old (72.5 ± 9.7 kg.), this represents about 77.2 ± 8.8 ml. per kilogram, and in the 41-year-old (79.6 ± 9.8 kg.), about 69.7 ± 8.3 ml. per kilogram.

I have, in Table 11-1, estimated the distribution of this blood volume at the end of ventricular diastole. Note that over 50 per cent of the blood volume lies in the systemic veins and less than 20 per cent in the systemic arteries. There is no important *blood storage* in stagnant backwaters like the spleen in man, as there is in cats and dogs. The dog, for example, responds to (1) hemorrhage, (2) the stimulation of sympathetic nerves, or (3) the release of epinephrine or norepinephrine, by contracting his spleen and releasing to the general circulation blood with a higher than normal concentration of red blood cells.

Change in Distribution

In man the mobilization of blood from stagnant backwaters is unimportant, since all the blood is circulating. Exceptions to this general rule sometimes have serious repercussions however. A *stagnant stream* may result from tourniquets or such related phenomena as sitting or lying in one position for a prolonged period, or an increase in intra-abdominal pressure due to pregnancy or ascites (fluid in abdominal cavity). A semi-stagnant reservoir may also develop in a heart chamber as a result of cardiac distention (Table 8-1). One serious aspect of these stagnant areas is that should some of the formed elements of the blood release a procoagulant, a large *blood clot* might develop. In the presence of a rapidly moving blood stream the procoagulant is washed away and diluted to a point where it is ineffective. Should a small clot form in a rapidly moving stream it is moved to a small blood vessel where it becomes trapped. The pulmonary capillaries serve as an effective *filtering system* keeping clots from the systemic arteries and thus protecting the highly vulnerable coronary and cerebral circulations. Small clots in systemic and pulmonary capillaries and arterioles generally do not interfere sufficiently with circulation to cause serious tissue damage and, as I have mentioned, are eventually compensated for by the development of a collateral circulation. Large clots, on the other hand, may disrupt flow to a large area and rapidly cause irreversible tissue damage.

Most vessels in the body do, however, serve as reservoirs in that they are distended with more blood than is necessary. This *blood reservoir* in the arteries and veins serves to meet the extra demands constantly occuring in the body.

In the case of the arteries in a reclining individual, there is an ejection of blood from the ventricles about every 0.8 sec. which is initially associated with a distention of the arteries and

	Per Cent	Quantity (ml.)
Heart (end-diastolic volume)	8	442
Pulmonary vessels	7	389
Large systemic veins	31	1,720
Small systemic veins, venules, and venous sinuses	31	1,720
Systemic arteries	18	1,000
Systemic capillaries	5	277
Total	100	5,548

Table 11-1. An estimate of the distribution of the blood volume. (In part from the data of Levinson, G.E., Pacifico, A.D., and Frank, M.J.: Studies of cardiopulmonary blood volume. Circulation, 33:347-356, 1966; and Chien, S., et al.: Blood volume and age: Repeated measurements on normal men after 17 years. J. Appl. Physiol., 21:583-588, 1966)

an increase in flow to the capillaries. As ejection continues (period of reduced ejection), we reach a point at which flow from the arteries exceeds that into them. This continues until the next period of rapid ventricular ejection. Under most circumstances there is sufficient blood in the arteries to maintain capillary flow until the next cardiac cycle. You will also remember that the greatest flow through the coronary vessels occurs during ventricular diastole rather than systole (Fig. 7-7). The importance, then, of a small reservoir of blood in the arterial system is that it helps maintain capillary blood flow during ventricular diastole. It is important to realize, however, that the arterial system is, by comparison with the venous system, a small reservoir which, in the young adult, can maintain flow to the capillaries for only a few seconds after the last ventricular contraction.

The ability of an older person to maintain flow between ventricular contractions is much reduced as a result of decreased elasticity in the arteries. In Figure 11-18, for example, you will note that a change in pressure from 75 to 125 mm. Hg increases the volume of an isolated aorta by 95 per cent in a 22-year-old and by 15 per cent in a 75-year-old. That is to say that as an individual ages the arteries become less distensible and less able to store pressure. As a result, capillary flow becomes progressively more pulsatile and less continuous. The elderly person is also less capable of supporting a slow heart rate.

The systemic and pulmonary veins constitute our main *blood reservoir* and *pressure buffering system*. Increases in blood volume due to drinking and transfusion result in a far greater increase in the amount of blood in the veins than in the arteries. We find that an extra 40 ml. of blood added to the arterial system raises arterial pressure about 40 mm. Hg, but in the venous system only about 0.2 mm. Hg. The venous pressure and volume buffering mechanism is not equally developed in all organs, however, being most effective in the lungs and liver.

The great complicance of the veins also creates problems during changes in *posture*. In a reclining person the legs and thighs contain about 14 per cent of the total blood volume (750 ml.). In a person standing freely this is increased by 140 ml., and in a person standing half a minute after working, the increase in blood volume over that in the horizontal position is about 660 ml. If a healthy man stands erect after running almost to exhaustion, the pooling of the blood in the

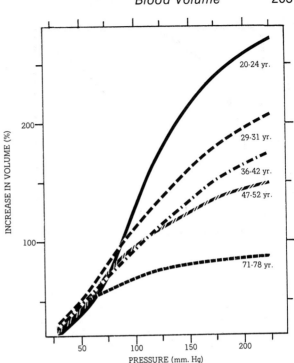

Fig. 11-18. Effect of age on the distensibility of isolated human thoracic aortas. The decrease in elasticity in the arteries of elderly individuals results in their decreased ability to distend in response to increases in pressure. (*Redrawn from* Hallock, P., and Benson, J. C.: Studies on elastic properties of human isolated aorta. J. Clin. Invest., *16*:597, 1937)

veins of the legs and thighs and the associated vasodilation in the skin and skeletal muscle may so reduce flow to the brain as to produce syncope (fainting).

One of the first responses to a decrease in circulatory blood volume due, for example, to hemorrhage is a reduction of arterial pressure and a reflex increase in heart rate and arteriolar constriction. The increase in heart rate serves to rapidly shift part of the blood from the venous reservoir to the arterial system. The arteriolar vasoconstriction decreases the outflow of blood from the arterial system until the pressure in that system has risen sufficiently to return the arterial outflow to normal. Although an increase in heart rate and an associated increase in arteriolar resistance are the usual means of rapidly shifting part of the blood volume to the arterial side, individuals with a constant heart rate (subjects with implanted pacemakers) must rely on increases in stroke volume. Increases in cardiac

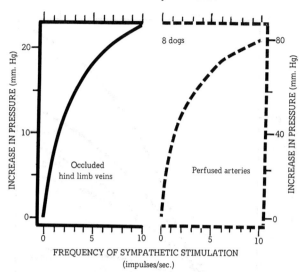

Fig. 11-19. Effect of stimulation of the lumbar sympathetic chain of 8 dogs on the pressure developed in (1) occluded hind limb veins (capacity vessels), and (2) the arteries of a perfused (constant flow) hind limb (resistance vessels). Changes in the pressure in the carotid sinus had little effect on the hind limb veins. (*Redrawn from* Browse, N. L., Donald, D. E., and Shepherd, J. T.: Role of the veins in the carotid sinus reflex. Am. J. Physiol., *210*:1427, 1966)

output due to increases in heart rate generally result in a smaller end diastolic cardiac volume than increases in stroke volume.

It has also been suggested that decreases in blood volume produce venoconstriction. In Figure 11-19, for example, we note that sympathetic stimulation (though the lumbar sympathetic chain) produces an increase in pressure in the saphenous vein, presumably due to an increase in smooth-muscle tone in the saphenous vein. To date, however, it has been hard to delineate the mechanisms producing venoconstriction or dilation. It has been noted that venoconstriction is frequently associated with an increase in cardiac output, but one of the important reflexes which increase cardiac output, the arterial baroreceptor reflex, has little or no effect on venous tone. Yet we know that decreases in total blood volume are followed by a marked decrease in pulmonary blood volume and liver size, as well as a generalized movement of blood away from all veins.

Muscular exercise, strong emotional stimuli, and cold showers produce increases in cardiac output, decreases in arterial resistance, and venoconstriction in the limbs. These same effects can be produced by the beta adrenergic stimulator isoproterenol, and by epinephrine. Norepinephrine, on the other hand, constricts both the arteries (resistance vessels) and the veins (capacitance vessels).

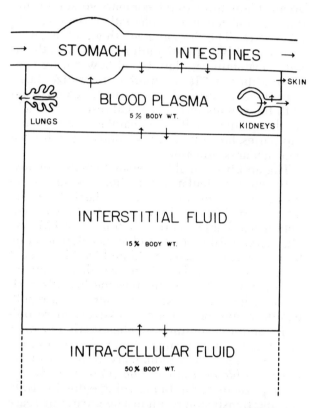

Fig. 11-20. Body fluid compartments. The extracellular fluid constitutes about 20 per cent of the body weight and the intracellular fluid about 40 per cent of the body weight. An increase in the blood volume due to drinking water (*see arrows*) will eventually result in an increase in the interstitial and intracellular fluid volume as well. An excessive loss of fluid to the expired air (evaporation) or through vomiting, diarrhea, bleeding, sweating, or urine formation will eventually lead to a decrease in fluid volume throughout the body. We see, then, that the various fluids are in equilibrium with one another and that a decrease in volume of one compartment may affect the volume of another. It is also true that a change in the osmotic pressure in one compartment affects the osmotic pressure of the others. (Gamble, J. L.: Chemical Anatomy, Physiology, and Pathology of Extracellular Fluid. 6th ed., p. 3. Cambridge, Harvard University Press, 1954)

Extravascular Fluid

Decreases in blood volume sometimes produce decreases in the capillary blood pressure, resulting in a movement of fluid from the inter- and intracellular spaces into the vascular system (Fig. 11-20). If this movement is extensive, it slowly produces tissue dehydration and dilution of the blood (a decrease in the concentration of red blood cells). In the dog, the dilution may be more than compensated for by a contraction of the spleen, which sends blood rich in red cells into the circulatory system. In man there is a progressive increase in red cell formation in hemopoietic tissues and frequently an increase in the number of immature red cells in the circulation, but seldom an increase in red cell concentration.

Fluid Intake and Excretion

Under most circumstances changes in blood volume or osmotic pressure are eventually compensated for by changes in the intake or excretion of water and electrolytes. Important among the hormones controlling water excretion is the *antidiuretic hormone* (ADH), which literally prevents diuresis (or the production of a large volume of urine). Its role in the maintenance of a fairly constant blood volume is outlined in Figure 11-21. ADH is released from the posterior pituitary gland under most conditions, its concentration in the blood being greatest when there is a decrease in blood volume or an increase in the plasma osmotic pressure. Its main target organ is the kidney, where it serves to increase the absorption of water into the blood from the urine. In the absence of ADH, as much as 30 liters of water a day are lost in the urine. These mechanisms are discussed more thoroughly on pages 390 through 396.

Another mechanism which can increase blood volume by decreasing urine volume is the release of the hormone *aldosterone* by the adrenal cortex. Aldosterone is carried by the blood to the kidneys, where it stimulates the movement of sodium from the urine into the intercellular space and from there into the blood. Or, it serves as an anti-natriuretic agent, thus creating an environment around portions of the nephron and in the blood which facilitates the withdrawal of water from the urine and its movement into the intercellular space of the kidney, from which it passes

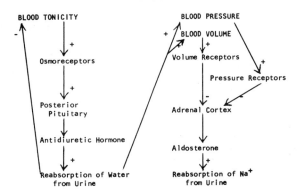

Fig. 11-21. A theory on the control of blood tonicity and blood volume. In this scheme the osmotic pressure of the blood is visualized as stimulating the release of an antidiuretic hormone. The blood pressure and/or blood volume is characterized as inhibiting the release of aldosterone (the "anti-natriuretic" hormone).

into the blood. These mechanisms are more thoroughly discussed on pages 385 through 387.

In summary, then, it has been demonstrated that we respond to dehydration or blood hypertonicity by an increase in the release of ADH, which, through its effect on the permeability of parts of the kidney apparatus, increases the movement of water back into the blood from the urine (Fig. 11-21). It is further demonstrated that we respond to hypotonicity by a reduction in the release of ADH and, as a consequence, an increase in the excretion of water from the body. Decreases in blood volume or pressure not associated with a change in the tonicity of the blood probably result in an increase in aldosterone secretion and an increase in the movement of sodium from the urine into the blood. This movement of sodium out of the urine causes water to move from the urine back into the blood, and from the intracellular to the extracellular environment. An increase in blood volume or pressure may stimulate volume and pressure receptors to inhibit aldosterone secretion.

Gilmore and Daggett (Table 11-2) studied the response of eight anesthetized dogs to increases in blood volume (blood volume having been increased by 3 per cent of the body weight) and found that the maximum urine production occurred 74 min. after the increase in blood volume. It had, at this time, changed from 0.13 ml. per minute to 2.20 ml. per minute. At the same time, sodium excretion had increased from 18 mEq. per minute to 334 mEq. per minute. Denervation

	Before Denervation		After Denervation	
	Control	Diuresis	Control	Diuresis
Urine volume (ml./min.)	0.13 ± 0.06	2.20 ± 1.43	0.11 ± 0.08	0.86 ± 0.75
Sodium excretion (µEq./min.)	18 ± 12.5	334 ± 168	19 ± 30.6	156 ± 128
Time to maximum diuresis (min.)		74 ± 34		104 ± 43
Time to maximum natriuresis (min.)		76 ± 37		121 ± 51
Osmolal excretion (µosmols/min.)	188 ± 92	1,074 ± 264	165 ± 112	576 ± 286

Table 11-2. Effect of an acute intravascular volume increase on renal function before and after complete denervation of the heart. Before denervation, 36 per cent of the volume of the infused dextran solution was recovered as urine in 3 hr. After denervation, 23 per cent was recovered in 3.5 hr. Diuresis figures are for period when it was maximal. All values are expressed as means ± their standard errors. (Gilmore, J.P., and Daggett, W.M.: Response of the chronic cardiac denervated dog to acute volume expansion. Am. J. Physiol., *210*:509-512, 1966)

of the hearts of these dogs prolonged and reduced the response of their kidneys to the hypervolemia.

It is hypothesized that there are sensory fibers from the heart which inhibit the release of aldosterone.

Increases in the osmotic pressure of the blood apparently stimulate osmoreceptors as well, causing the sensation of thirst (also experienced in response to decreases in blood volume or pressure not associated with changes in the osmotic pressure).

CENTRAL NERVOUS SYSTEM

We have, up to this point, concentrated on the role of peripheral mechanisms in the control of the cardiovascular system. We have noted the ability of different areas to release such potentially vasoactive agents as renin, histamine, and epinephrine into the blood. We have emphasized the importance of the kidney in regulating blood volume. We have discussed some of the intrinsic properties of the various vascular organs, the ability of blood vessels to increase their muscular tone in response to stretch, the ability of the heart to increase its force of contraction in response to distention, and the response of the various circulatory organs to local metabolites.

We have also considered some of the sensory inputs acting on the central nervous system. We have noted the uneven distribution of autonomic neurons throughout the body. We have observed the predominant role of parasympathetic fibers in the control of heart rate, and of sympathetic fibers in the control of heart force and vasoconstriction. In addition, we have considered the multiple actions of sympathetic fibers upon arterioles. These fibers have been shown to maintain vasoconstriction in certain arterioles by releasing norepinephrine near the smooth muscle; they can increase vasoconstriction by increasing the number of times they release norepinephrine per second, or decrease it by decreasing their rate of firing. In addition to this type of postganglionic sympathetic fiber there is the type that releases acetylcholine. These cholinergic fibers are usually found in skeletal muscle and produce vasodilation. There is yet another type of postganglionic vasodilator fiber, thought to cause the release of a histamine-like substance.

Figure 11-22 summarizes some of the preceding information. It emphasizes some of the stimuli that activate *afferent fibers* to carry impulses into the *central internuncial system*. The presso- and chemoreceptors, as well as the visceral (pain, hunger, pressure) and cutaneous fibers (pain, heat, cold, pressure, touch) and the special senses (vision, hearing, smell) are also indicated. One should add to this the afferent fibers from skeletal muscle (i.e., from the muscle spindle and Golgi tendon organ).

We note in Figure 11-22 that much of the body's sensory information (proprioceptive, visceral, and cutaneous fibers) goes directly to the spinal cord, where it either inhibits or facilitates

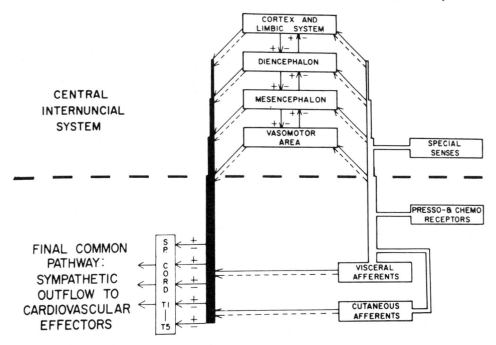

Fig. 11-22. Diagram of possible nervous organization for the regulation of cardiovascular function. Solid lines and plus signs represent excitatory connections. Interrupted lines and minus signs represent inhibitory connections. For simplicity the parasympathetic outflow from the central nervous system has been omitted. (Peiss, C. N.: Supramedullary cardiovascular regulation. *In* Price, H. L., and Cohen, P. J. (eds.): Symposium on the Effects of Anesthetics on the Circulation. P. 37. Springfield, Charles C Thomas, 1964)

sympathetic outflow. Some of this information may also ascend or descend in the cord and modify parasympathetic outflow. Other information (from carotid and aortic chemo- and pressoreceptors, and special senses) enters the brain directly, where it inhibits or facilitates parasympathetic outflow and the internuncial outflow to the sympathetic and parasympathetic fibers leaving the spinal cord.

Spinal Cord

In the intermediolateral horn of the gray matter of T-1 through L-2 lie the cell bodies of the preganglionic sympathetic neurons. Impinging on each neuron are multiple facilitatory and inhibitory internuncial fibers from higher and lower levels of the central nervous system, from both the ipsilateral and contralateral sides of the cord, and from ipsilateral sensory neurons. For some time it was believed that impulses from the medulla of the brain controlled the output of these preganglionic fibers. In 1906, for example, Sherrington published the record shown in Figure 11-23 and concluded:

When in the dog complete *transection of the spinal cord* through the eighth cervical segment is practiced, a severe fall in the general arterial *blood pressure* ensues, and vasomotor reflexes cannot be elicited. But in the course of some days this is largely recovered from, and after some weeks the blood pressure will, with the animal in the horizontal position, often be practically normal. When the animal is then anesthetized and curarized, artificial respiration being maintained, it is usually easy to obtain good and often very large vasomotor reflexes, on stimulation of the central ends of divided afferent or mixed nerves, for instance, of the internal saphenous nerve, the blood pressure rising 50 mm. and more. These reflexes upon the vascular musculature are purely spinal, since the cord has been divided just headward of the thoracic region.

Many interpreted the work of Sherrington to mean that blood pressure is controlled predominantly by the brain and that when the sympathetic nerves are separated from the brain by transection of the cord, *sympathetic tone* dis-

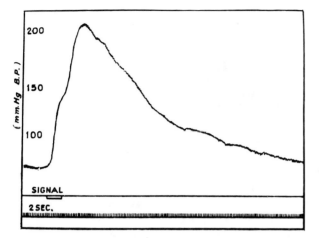

Fig. 11-23. Spinal vasomotor reflex in a dog under chloroform and curare anesthesia 300 days after spinal transection at C-8. The signal indicates the period of electrical stimulation of the central end of a digital nerve of the hind limb. (Sherrington, C. S.: Integrative Action of the Nervous System. P. 242. New Haven, Yale University Press, 1906)

appears and, as a result, blood pressure falls. This was not to deny that there are spinal cord reflexes which can increase the blood pressure, but only to indicate that in the resting dog with a fully functioning central nervous system they play a minor role in the control of blood pressure.

Recently, however, it was demonstrated that spinal cord transection need not result in an acute and long-lasting depression of vascular reflexes. In 1934, Hermann et al. showed that slow removal of the medulla over a period of 2 hours resulted in no appreciable fall in blood pressure. He also reported that interruption of the cord by cocaine produced only a small decrease in blood pressure. Peiss has concluded that: "The inherent capabilities of *spinal sympathetic centers* to mediate vasomotor reflexes and to support vasomotor tone is not necessarily dependent upon a flow of impulses from higher levels of the central nervous system, nor do the spinal centers require long periods of time to slowly develop these capabilities when higher influences are removed."

Medulla

The medulla is the part of the brain stem between the spinal cord and pons. It gives origin to the ninth through twelfth cranial nerves. Since the ninth nerve (glossopharyngeal) transmits sensory impulses from the carotid sinus and the carotid body, and the tenth nerve (vagus) transmits sensory impulses from the aortic arch, the pulmonary arteries and veins, and the heart and carries parasympathetic impulses to the heart, it has long been thought to contain reflex centers controlling the heart and blood vessels.

The importance of the medulla to the cardiovascular system was demonstrated by Carl Ludwig almost one hundred years ago when he performed serial transections of the brain stem. Transecting first the more rostral portions of the brain stem and then the more caudal, he found that blood pressure dropped most severely when he cut at the rostral portion of the medulla. Points were eventually located on the floor of the fourth ventricle of the medulla that produced vasoconstriction and dilation when stimulated. The Webers localized a vagal center in the medulla. From these and other data came the concept that there are in the dorsolateral reticular formation of the rostral two thirds of the medulla a *vasoconstrictor-cardioaccelerator center* and a *vasodepressor-cardioinhibitory center*. Stimulation of one center was visualized as inhibiting the other. In addition, there is a considerable, but not complete, crossover of fibers on the right and left sides. That is, the right vasomotor center is visualized as controlling blood vessels on both the right and left sides of the body.

Recently, however, physiologists have begun to question the emphasis placed on these areas. It is generally recognized that they receive sensory input from the pressorreceptors, and that they send impulses to the heart which have both negative and positive chronotropic, dromotropic, and inotropic action. It is also felt that they send impulses to the blood vessels that increase and decrease vasoconstrictor tone, but that not all important cardiovascular reflexes are concentrated in the medullary centers. The essence of a center is that it is an area of integration containing many synapses. We now know that many of the afferent fibers from the baroreceptors do not end in the medulla, but send impulses further up the brain stem and to the cerebrum (pons, midbrain, diencephalon). We know too that some descending fibers from the cerebral cortex and hypothalamus (part of the diencephalon) have a profound influence on the cardiovascular system and yet pass through the medulla without synapsing. Livingston has expressed these relationships as follows: "The nervous system appears to be made up less of independent linear pathways

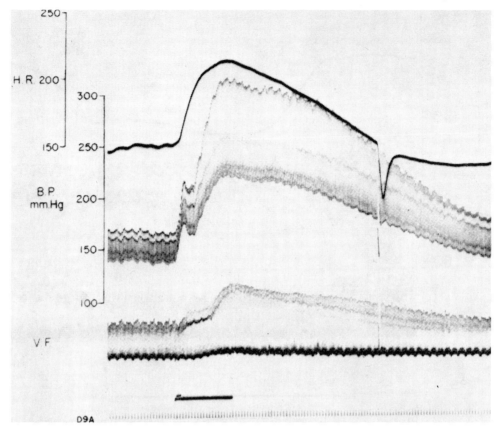

Fig. 11-24. Cardiovascular response of a vagotomized dog under chlorolose anesthesia to 20 seconds of stimulation (*horizontal line*) of the posterior hypothalamus. Note that the heart rate (*upper thick line*) increases from 150 to 210 per minute, the diastolic, systolic and pulse pressures increase, and the ventricular force (V.F.) increases. The solid bar represents the period of stimulation. The lower time lines are 1 second apart. (Peiss, C. N.: Sympathetic control of the heart. *In* Luisada, A. A. (ed.): Cardiovascular Functions. P. 317. New York, McGraw-Hill, 1962; copyright 1962, used with permission of McGraw-Hill Book Company)

than of mutually interdependent loop circuits which stitch together the various parts of the brain into a functional whole."

The Hypothalamus and Other Parts of the Brain Stem

There is now evidence that important areas of synapse exist in the pons, hypothalamus (part of the diencephalon), and mesencephalic tectum, as well as in the medulla, spinal cord, and cerebral cortex. These areas have to do with the integration of cardiovascular reflexes. Transection of the *pons* and pontine hemorrhage have been shown to produce increases in arterial pressure, while transection of the *midbrain* results in a decrease in blood pressure of about 30 mm. Hg. Ablation of the hypothalamus also lowers blood pressure. Stimulation of a *posterior part of the hypothalamus* (Fig. 11-24), on the other hand, produces cardiovascular changes similar to those occurring during exercise (increased heart rate and ventricular force), and stimulation of an *anterior area* (Fig. 11-25) has the opposite effect.

It appears that the hypothalmus exerts a tonic influence on the blood vessels through its action on the sympathetic nerves, and a tonic influence on the heart through its action on parasympathetic fibers. Like the medulla, it receives impulses from peripheral presso- and chemoreceptors and is important in both the body's response to changes in blood pressure and the concentra-

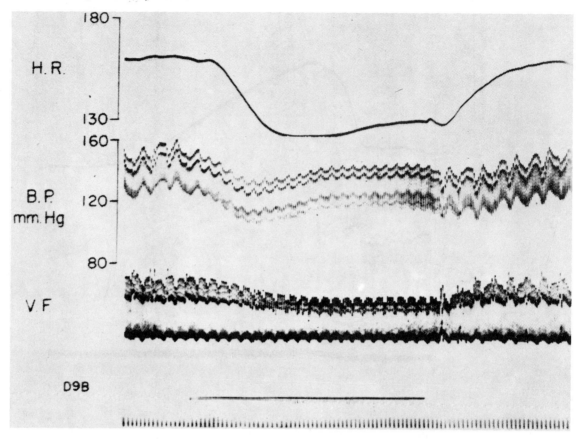

Fig. 11-25. Cardiovascular response of the same dog studied in Figure 11-24 to stimulation of the preoptic region of the hypothalamus. Note that the heart rate, blood pressure, and ventricular force all decreased. (Peiss, C. N.: Sympathetic control of the heart. *In* Luisada, A. A. (ed.); Cardiovascular Functions. P. 319. New York, McGraw-Hill, 1962; copyright 1962, used with permission of McGraw-Hill Book Company)

tion of metabolites in the blood. It also receives fibers from the cerebral cortex, and can apparently change the cardiovascular state of the individual to meet the changing needs of skeletal muscle. Just prior to running, for example, impulses originating in the cerebral cortex initiate a complex of skeletal muscle contractions and relaxations. Apparently some of these impulses produce stimulation and inhibition in the hypothalamus which simultaneously integrate cardiovascular and skeletal muscle changes.

This type of thinking in physiology is quite different from that of 15 years ago. At that time it was generally held that in exercise the cardiovascular system *responds to the changes which the skeletal muscles initiate* (hypoxia, hypercapnea, acidity). In my view, the *cardiovascular and skeletal muscle systems respond simultaneously during exercise*. This is sometimes referred to as the "get set" response. When a person preparing to run a race is told to get on his mark and get set, his heart rate sometimes increases either before he begins to run or the instant he starts. Increased heart rate, in other words, sometimes occurs before the skeletal muscles have produced systemic acidosis, hypoxia, or hypercapnea. In fact, the cardiovascular system generally over-responds at first. As the race continues, the heart rate and cardiac output decrease (Fig. 11-14).

Probably the main reason many investigators failed to realize the importance of the hypothalamus in integrating cardiovascular changes is that they studied the anesthetized animal, *i.e.*, a preparation in which the hypothalamus had been depressed (Fig. 11-26).

There is good evidence, then, that the hypo-

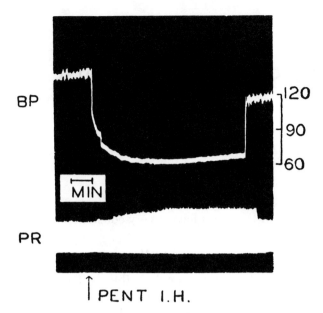

Fig. 11-26. Response of a 3-kg. cat to the injection of pentothal (0.04 ml., 4%) into the posterior hypothalamus. Note that the blood pressure decreased and the cardiac interval increased for a period of 25 minutes. PR represents the cardiac interval (1/pulse rate). (Redgate, E. S., and Gellhorn, E.: The tonic effects of the posterior hypothalamus on blood pressure and pulse rate as disclosed by the action of intrahypothalamically injected drugs. Arch. Int. Pharmacodyn., *105*:196, 1956)

thalamus contains some of the major centers of cardiovascular regulation. It apparently receives impulses from the cerebral cortex which tend to integrate cardiovascular changes with changes in skeletal muscle. When these areas in the hypothalamus are bilaterally destroyed, the cardiovascular response normally associated with exercise does not occur. In addition, there are areas in the hypothalamus concerned with *heat regulation*, control of the release of the *antidiuretic hormone*, control of the release of other pituitary hormones, production of the reactions we characterize as *fear* and *rage*, and with *hunger*, appetite, and satiety. In short, the hypothalamus seems to contain centers which regulate every aspect of glandular, smooth muscle, and cardiac muscle function. Put another way, it seems to be the main center for the integration of autonomic nervous system activity.

The Cerebral Cortex

The chief action of the cerebral cortex on the autonomic system is probably exerted through neurons that descend from the motor cortex in the pyramidal and extrapyramidal tracts and synapse in the hypothalamus. Their messages are then relayed to synapses in the mesencephalic tectum, where they synapse once again, and finally to the intermediolateral horn of the gray matter of the thoracic and lumbar cord, where they synapse with preganglionic sympathetic neurons. It is felt that this pathway passes through the medulla but is not interrupted there. It is also felt that this pathway, unlike those that synapse in the medulla, results in the stimulation of cholinergic postganglionic sympathetic neurons rather than adrenergic ones. Such an arrangement might result in a cortically induced skeletal muscle contraction associated with vasodilation in that muscle.

Chapter 12

DISTRIBUTION
FLOW
STRUCTURE
PERMEABILITY
FILTRATION
THE LYMPHATIC SYSTEM
ARTERIOVENOUS DIFFERENCES

CAPILLARIES AND SINUSOIDS

There are in the body a number of vessels that allow fluid to move into and out of their lumens. These include the blood capillaries, sinusoids, venules, and lymphatic capillaries. The physiologist calls these vessels *semipermeable; i.e.,* some but not all substances can penetrate their *endothelial-cell* barrier and *basement membrane.* The term semipermeable as I shall use it does not imply the presence of pores or holes in the endothelial cell and its basement membrane, or between cells. While true that the permeability of many capillaries and sinusoids is due to pores, many capillaries apparently have none; clearly their semipermeability stems from some other property.

DISTRIBUTION

The blood and lymphatic capillaries are found in most parts of the body. The former has an averge diameter of about 8 μ and the latter of about 14 μ. Blood sinusoids, on the other hand, are usually larger and do not have such a general distribution. They are found primarily in the liver, bone marrow, and spleen. The lymph nodes contain what we might call a lymph sinusoid.

The blood capillaries vary in concentration from one area to the next. They are most concentrated in glands, sparsely concentrated in the cornea of the eye, and completely absent in the lens of the eye and the normal cartilaginous plate of long bones. It has been estimated that there are approximately 300 capillaries per square millimeter of transverse section of skeletal muscle. Of these, only about 60 are patent at any one time in a resting person; most are collapsed and devoid of blood. Flow is quite variable, however, one capillary being patent one minute and collapsed the next. It is also not uncommon in a muscle capillary for the direction of flow to reverse periodically, though the capillaries of the brain have a very constant flow.

FLOW

Blood flows more slowly in the capillaries than it does anywhere else in the body (0.8 to 2.0 mm./sec.). This is 200 times faster than lymph flows in lymph ducts, however. Many blood capillaries are so small that a red blood cell is deformed in passing through it. Hence one sees in the smaller capillaries many deformed or bent red cells moving toward a venule, their fronts convex and their arteriolar ends concave. During illness the red cells sometimes stick together to form what is called a *rouleau,* or what Knisely calls "sludge." Frequently these red cell groups can be disaggregated by the intravenous administration of a low-molecular-weight dextran (40,000 to 70,000). The dextran is thought to reduce the electrostatic bonding forces between the red cells, and to dilute the blood. Unfortunately, large amounts of dextran sometimes produce an aggregation of blood platelets (thrombocytes).

Sickle cell anemia is another condition producing a change in the red cell that can interfere with blood flow in the capillaries. The sickle cell is an abnormal red cell, somewhat rigid and tending to interlock with other sickle cells. Thus it may plug up the capillaries and interfere with flow.

STRUCTURE

Sinusoids

Figure 12-1 is an electron micrograph of a liver sinusoid. Characteristic of all sinusoids (Table 12-1) is the presence of a distinct intercellular space through which water and electrolytes pass freely. Note that some of these spaces are more than 1 μ in diameter—sufficiently large, that is,

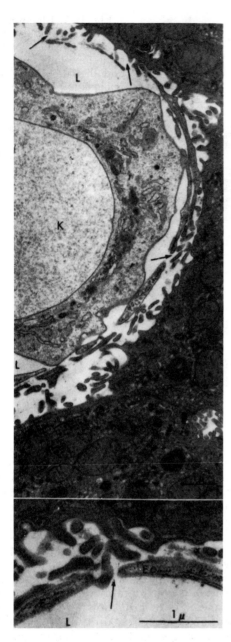

Fig. 12-1. Electron micrograph of a liver sinusoid. In the upper illustration we see a large Kupffer cell (K) in the lumen (L) of a sinusoid. In the lower illustration we note the protrusion of a microvillus from a liver parenchymal cell into the intercellular space (*arrows*) of 2 endothelial cells (E). Note that there is very little basement membrane (x) and that it, like the endothelial cells, is discontinuous. (Majno, G.: Ultrastructure of the vascular membrane. *In* Hamilton, W. F. (ed.): Handbook of Physiology. Sec. 2 (Circulation), vol. III, p. 2320. Washington, American Physiological Society, 1965)

to permit the free movement of proteins and other plasma constituents. Even leukocytes and erythrocytes sometimes pass through the intercellular space.

Also characteristic of the sinusoid is the absence of a continuous sheath (*basement membrane*) around the endothelial cells. In the sinusoids of the *spleen*, the discontinuous basement membrane sends irregular processes into the endothelial cells and is also attached to reticular cells, which send processes through the basement membrane and through the endothelial cell into the lumen of the sinusoid. Thus the endothelial cells receive support from both the basement membrane and the surrounding cells. The *liver* endothelial cells seem to receive most of their support from the parenchymal microvilli which pass between and through the endothelial cells. The endothelial cells of *bone marrow*, on the other hand, appear to receive no appreciable local reinforcement or support. There is neither a basement membrane with its supporting fibers, nor supporting reticular cells (as in the spleen), nor parenchymal villi (as in the liver).

The sinusoids contain channels called fenestrae (L. *fenestra*, window) which permit the free movement of fluid and small molecules between the vascular and extravascular compartments. We note too in Figure 12-1 that the endothelial cells can be quite thin. Some liver sinusoidal cells have been reported to be as thin as 0.02 μ. Thus we note four important factors contributing to the permeability of the sinusoid: (1) the intercellular space, (2) the absence of a continuous basement membrane, (3) the presence of intracellular fenestrations, and (4) the small width of the endothelial cell.

Capillaries

There are as many different types of capillaries as there are organs in the body. All blood capillaries, however, have certain characteristics in common. One of these is the presence of a continuous basement membrane, which serves as a point of attachment for the endothelial cells and as an important barrier between the blood and extravascular fluid. It is usually about 0.05 μ thick and is a protein-carbohydrate complex which is in some cases mostly collagen. It seems to be permeable to water, electrolytes, O_2, CO_2, N_2, lactic acid, glucose, amino acids, and various phagocytic cells.

The basement membrane also surrounds a cell,

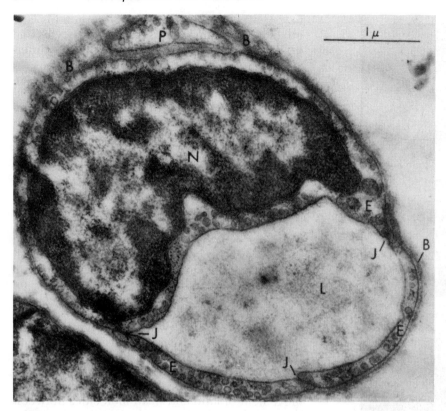

Fig. 12-2. Electron micrograph of a capillary from rat skeletal muscle. Note the tight junctions (J) between endothelial cells (E) and the nucleus (N) of one endothelial cell projecting into the lumen (L) of the capillary. Note, too, the basement membrane (B) and the pericyte (P) which it encloses. (Palade, G. E.: Blood capillaries of the heart and other organs. Circulation, 24:371, 1961; by permission of the American Heart Association, Inc.)

the *pericyte* (once called the Rouget cell), which lies adjacent to the capillary (Fig. 12-2). The pericyte is believed to be phagocytic. We have already noted the projection of pseudopodia from the extravascular space into the lumen of the sinusoid; it is interesting to observe that this same phenomenon appears in capillaries. Intraluminal pseudopodia from perivascular cells have been noted in the capillaries of the normal corpus luteum, the pituitary, and the adrenal cortex. Intravenous injection of carbon black in the rat apparently accelerates the formation of these pseudopodia in the adrenal cortex, while the injection of trypan blue accelerates their formation in the pituitary. The proportion of phagocytic protrusions coming from the highly branched pericyte, and not from cells lying outside the basement membrane, is still a matter of speculation.

Another characteristic of the capillary is that its endothelial cells are generally flat (or low) except where the nucleus lies. In the capillaries of striated muscle, for example, most of the endothelial cells are about $0.2\ \mu$ wide. The capillary with the highest cells, the dermal capillary, has a thickness of about $4\ \mu$.

Fenestrated capillaries. The capillaries found in various glands of the body and in such structures as the choroid plexus of the ventricles of the brain, the ciliary body of the eye, the glomerulus of the kidney, the gallbladder, and the vagina contain intracellular fenestrations (Fig. 12-3) between 0.02 and $0.17\ \mu$ wide. There are about 30 per square micron, covering approximately 35 per cent of the capillary surface. Most fenestrations contain a single diaphragm (Fig. 12-4) which is considerably thinner than the cell membrane. These are known as closed fenestrae, but apparently remain quite permeable. It has been suggested that the diaphragms open and close periodically.

Unfenestrated capillaries with a continuous basement membrane. The capillaries found in skeletal, cardiac, and most smooth muscle, the nervous system, and the skin do not have intracellular pores or an open intercellular space. The obvious question is how fluid passes through the endothelial barrier and the basement membrane. To date there is no completely satisfactory answer. It is known, however, that the endothelium of these capillaries contains a tremendous number of vesicles (Figs. 12-2 and 12-5).

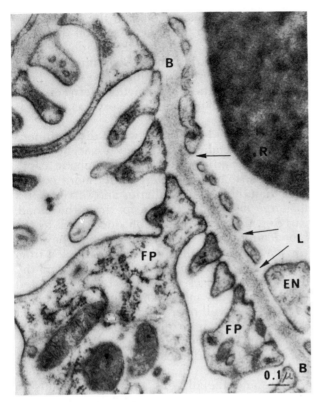

Fig. 12-3. Electron micrograph of the fenestrae (*arrows*) of the endothelium (EN) of a renal glomerular capillary of a rat. Note the red blood cell (R) in the lumen (L) of the capillary and the continuous basement membrane (B). FP represents foot processes (pedicels) from the visceral epithelium. These fenestrae are larger (diameter, 0.05 - 0.1 μ) and more numerous than those seen in Figure 12-4. They are not bridged by a diaphragm and contain a thicker basement membrane. (Farquhar, M. G.: Fine structure and function in capillaries of the anterior pituitary gland. Angiology, *12*: 279, 1961; copyright by Angiology Research Foundation, 1961; reprinted by permission)

In the rat myocardium, for example, the vesicles all have a fairly uniform diameter of 0.06 to 0.07 μ and represent about one third of the cell volume. They have frequently been called pinocytotic (Gk. *pineo*, to drink; plus *kytos*, a hollow) vesicles because of the general belief that they result from invagination of a cell membrane and the subsequent closing off of a vesicle filled with extracellular fluid in the cytoplasm. For this reason it has been suggested that the vesicles constitute a transport system across the thin endothelial cell, though it is generally agreed that such a system could not account for all or even most of the movement between intravascular and extravascular spaces.

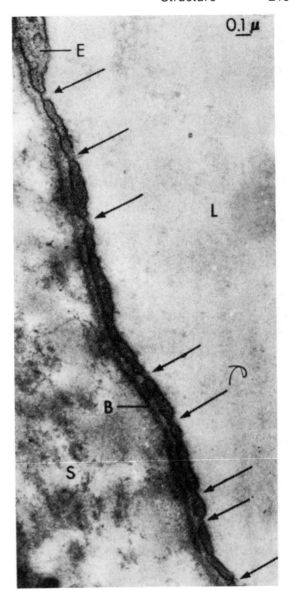

Fig. 12-4. Electron micrograph of the closed fenestrae of a capillary in the ciliary body of the eye of a rabbit. The arrows point to the fenestrae and their respective diaphragms. Lying between the extravascular space (S) and the lumen (L) of the capillary are the basement membrane (B) and the endothelial cells (E). (Pappas, G. D., and Tennyson, V. M.: An electron microscopic study of the passage of colloidal particles from the blood vessels of the ciliary processes and choroid plexus of the rabbit. J. Cell Biol., *15*:228, 1962)

In Figure 12-5 we see the typical *intercellular junction* of an unfenestrated capillary. It sometimes contains a flap projecting into the lumen of the capillary, significant because it increases

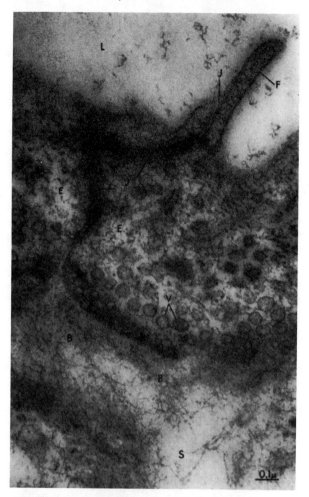

Fig. 12-5. Electron micrograph of a capillary in the ventricle of a rat. Projecting into the lumen (L) of the capillary is an endothelial flap (F). These are generally located at the point of junction (J) between 2 endothelial cells (E). Note that the 2 adjacent cell membranes come together and seem to form a tight seal. Characteristic of unfenestrated capillary endothelium is a high concentration of vesicles. Characteristic of all blood capillaries is a continuous basement membrane (B). Also note at S the fibrillar structure of the basement membrane. (Courtesy of Palade, G. E. *In* Majno, G.: Ultrastructure of the vascular membrane. *In* Hamilton, W. F. (ed.): Handbook of Physiology. Sec. 2 (Circulation), vol. III, p. 2300. Washington, American Physiological Society, 1965)

the surface area of endothelium exposed to blood. One cell characteristically interdigitates with the next. The intercellular gap at most points is about 0.01 μ (100 A), and the cell membrane, about 0.0025 μ thick. We find, however, that at a point near the lumen the two plasma membranes come together to seal off the intercellular space. In other words, we do not have in blood capillaries the discontinuous endothelium characteristic of sinusoids. It has been suggested that the tight junction between cells is not impermeable and may serve as a narrow filter.

Unfenestrated capillaries with a discontinuous basement membrane. Closely associated with the blood capillaries are the lymphatic capillaries. These are characteristically more permeable than the blood capillaries and serve to pick up *large molecules* that escape from the blood, carrying them via the lymph ducts back to the blood circulatory system (Fig. 12-6). They also carry water and other dissolved particles from the extravascular spaces, and nutrients from the digestive tract, back to the blood. It is estimated that during a day more than 50 per cent of blood proteins and well over 100 per cent of the blood plasma are lost from the blood stream and returned to it by the lymphatic system. Structurally, the lymphatic capillaries represent an interesting transition between the sinusoid, with its discontinuous endothelial cells and discontinuous or absent basement membrane, and the blood capillaries, with their continuous endothelial cells and basement membranes. The lymph capillaries have a continuous endothelial layer but a discontinuous basement membrane. With some exceptions, the intercellular gap is similar to that in blood capillaries. One of these exceptions is the lymph capillaries in the mouse diaphragm, which sometimes have gaps as large as 0.2 μ. The endothelial cells of the lymph capillary, like those of skeletal muscle, contain vesicles.

PERMEABILITY

I have in Table 12-1 summarized some of the structural information determining the permeability of the sinusoids and capillaries. The sinusoids are understandably the most permeable vessels in the body. They contain (1) a large intercellular space, (2) open intracellular fenestrae, and (3) little or no basement membrane. The lymphatic capillaries, on the other hand, are less permeable in that they contain neither intracellular fenestrations nor large intercellular spaces.

The blood capillaries show many variations, some having intracellular fenestrations and others none. They, unlike the sinusoids and

	Continuous								Fenestrated											Discontinuous		
	Striated Muscle	Myocardium	Central Nervous System	Sm. Muscle (digest & repro. syst.)	Lung	Subcut. & Adipose Tissue	Dermis	Placenta	Endocrine Glands	Synovial Membrane	Renal Glomerulus	R. Peritubular Capillary	Ciliary Body	Choroid Plexus	Exocrine Pancreas	Salivary Glands	Gallbladder	Intestinal Villus	Vagina	Sinusoids (liver & bone marrow)	Sinusoids (spleen)	Lymphatic Capillary
Endothelium																						
Continuous	x	x	x	x	x	x	x	x														x
Intracellular fenestrations																						
Closed									x	x		x		x	x	x	x	x	x	x	x	
Open									x	x			x							x	x	
Discontinuous																				x	x	
Low cells	x	x	x	x			x	x	x	x	x	x	x	x	x	x	x	x	x	x		x
High cells					x																x	
Basement Membrane																						
Continuous	x	x	x	x	x	x	x	x	x	x	x	x	x	x	x	x	x	x	x			
Discontinuous or absent																				x	x	x
Pericytes																						
Present	x	x	x	x	x	x	x	x	x	x	x	x	x	x	x	x	x	x	x			
Absent																						x

Table 12-1. Characterization of capillaries and sinusoids. (Majno, G.: Ultrastructure of the vascular membrane. *In* Hamilton, W.F. (ed.): Handbook of Physiology. Sec. 2 (Circulation), vol. III, pp. 2293-2375. Washington, American Physiological Society, 1965)

lymphatic capillaries, all have an intact basement membrane. I have indicated that the presence of fenestrae facilitates a certain amount of movement across the endothelium, but we have not adequately explained movement across a basement membrane or unfenestrated endothelial cell, which obviously occurs.

The Movement of Small Molecules

Small molecules such as H_2O, CO_2, and so on probably move into and out of the endothelial cell and basement membrane readily without the help of fenestrae or intercellular spaces. Larger water-soluble molecules (Table 12-2) such as glucose, however, pass the capillary barrier less rapidly and may be more dependent for their movement upon the character of the intercellular junction, the transport systems in the cell, etc. In Table 12-2 we see a comparison of the size of various lipid-insoluble particles with the ease with which they penetrate the capillary barrier (specific permeability). These data are consistent with the concept that these

Substance	Molecular Weight	Approx. Molecular Radius	Specific Permeability
H_2O	18	1.5	28
NaCl	58	2.3	15
Urea	60	2.6	14
Glucose	180	3.7	6
Sucrose	342	4.8	4
Raffinose	504	5.7	3
Inulin	5,500	12-15	0.3
Myoglobin	17,000	19	0.1
Serum Albumin	67,000	36	0.001

Table 12-2. Permeability of mammalian muscle capillaries to lipid-insoluble molecules. The specific permeability is given in terms of mols per second per square centimeter of membrane per mols per milliliter of concentration difference, multiplied by 10^5. The molecular radius is in angstroms (1 A = 10^{-4} μ). (Landis, E.M., and Pappenheimer, J.R.: Exchange of substances through the capillary walls. *In* Hamilton, W. F. (ed.): Handbook of Physiology. Sec. 1 (Circulation), vol. II, pp. 961-1034. Washington, American Physiological Society, 1963)

molecules pass through the barrier primarily by a passive process.

Lipid-soluble particles (O_2, CO_2, barbiturates), on the other hand, seem to penetrate the barrier more readily than lipid-insoluble particles (NaCl, glucose, H_2O, etc.) of comparable size. This may be due to the fact that they move through the cell more readily. Thus they can use all the mechanisms the lipid-insoluble particles can, with the advantage of being better able to dissolve in parts of the cell membrane. You will remember that the cell membrane is composed of lipid and protein.

Pinocytosis and the Large Molecule

Up to this point we have been concerned primarily with the movement of large quantities of fluid and small dissolved particles across a leaky barrier. It should not be forgotten, however, that small quantities of large molecules may also pass this barrier, even when the barrier is a continuous, unfenestrated layer of endothelium with a continuous basement membrane. For example, we find that 2 min. to 1 hour after the injection of gold (molecular radius 0.003 to 0.0250 μ) or ferritin (0.01 μ), these particles appear in (1) vesicles of the endothelial cell, (2) the basement membrane, and (3) the pericapillary spaces. In the case of larger particles, only an occasional vesicle is labeled and the basement membrane and perivascular space remain uncontaminated. Apparently particles with a radius less than 0.03 μ that cannot diffuse past the endothelial cell can be transported through it by the process of endothelial microphagocytosis (formation of pinocytotic vesicles). This process could not be expected to move the large quantities of material that diffusion and filtration do, but it probably is important in the transport of significant quantities of material which cannot readily pass through a capillary barrier.

Diapedesis

Particles and cells too large to be moved by filtration or pinocytosis through a continuous endothelium do, nevertheless, enter and leave the circulatory system. This occurs through sinusoids and lymph capillaries, whose barriers are incomplete, and through capillaries with a continuous endothelium and basement membrane. Apparently a number of the cells of the body can (1) migrate through all types of capillary barriers, (2) phagocytose microorganisms or other large particles, and (3) return through the capillary. This movement of a cell through the capillary wall is called diapedesis (Gk. *dia*, through; plus *pedesis*, leap). The cell capable of this phenomenon apparently contains enzymes that can change solids (gels) into liquids (sols) and then back into solids. Many body cells, in other words, have the machinery to facilitate their flowing through a temporarily fluid cell membrane or intercellular space.

Many white blood cells, for example, pass out of the blood through the capillary barrier during infection and phagocytose the infectious agents in the perivascular space. The pericyte and endothelial cell, as well as other perivascular cells, are also capable of phagocytosis. Some investigators have suggested that the pericyte not only projects pseudopodia into the blood but may also, under some circumstances, detach itself from the basement membrane, migrate into the blood, and change into a plasma cell. It is in the liver, spleen, and bone marrow, however, that most phagocytosis occurs. Here perivascular cells send pseudopodia through the intercellular endothelial space to remove worn-out blood cells and other material. Here too, because of the incompleteness of the barrier, blood cells migrate into and out of the perivascular space with greater frequency than anywhere else in the body.

Control of Permeability

The sinusoids and capillaries seem particularly well-adapted for their function. It has been estimated that in an active muscle no fiber is more than 12 μ from a capillary. The shortness of this distance means that the movement of small particles from and toward the pericapillary space should be a simple matter. The question which arises is whether movement of O_2, CO_2, electrolytes, water, nutrients, and wastes through the capillary is an important limiting factor. Most of the evidence indicates that it is not. It has been noted, for example, that when heavy water is injected into the arterial blood perfusing a forearm, nearly all of it is lost to the tissue during a single transit. About 30 per cent in the case of Na^+, 50 per cent in that of thiocyanate, and 40 per cent in that of glutathione, is lost to the intercellular tissues in the first passage through the capillaries. The permeability of the capillary, that is, does not seriously limit the quantity of water, urea, Na^+, Cl^-, CO_2, O_2, and H^+ that moves into and out of the capillary. The amount, type, and pressure of blood that *flows* to the capillary, however, are important factors.

Apparently nothing is gained by increasing capillary permeability except in the specialized case of diapedesis. The disadvantage of general increased permeability is that more proteins escape from the plasma. It is hardly surprising, therefore, that capillary permeability is generally quite stable. Permeability does not change appreciably in response to changes in pH from 4.0 to 8.0, or to changes in CO_2 concentration, but may increase in severe hypoxia (occlusion of blood supply to a rabbit's ear for 2 hours), tissue damage, and *anaphylactic shock*. It has been suggested that in tissue damage and anaphylactic shock, increased permeability is due to the release of histamine and kinins (bradykinin, etc.) into the intercellular space. Haddy (1960), on the other hand, suggests that histamine probably has no direct effect on the "permeability of the capillary membrane" but may act by producing a local arteriolar dilation and small-vein constriction. He states that the resultant rise in capillary pressure would "increase the surface area available for filtration both by distending capillaries already open and by opening those that are closed. Further, it might increase the size of pores in beds in which pores exist.

FILTRATION

There is in the capillaries an interchange of water, electrolytes, gases, and organic products between the blood and intercellular space. This is in part the result of diffusion (the tendency of particles to disperse or to be in motion), and in part that of filtration (forcing a fluid through a permeable membrane or barrier). Starling, at the beginning of this century, suggested that filtration is determined by two forces, the hydrostatic and osmotic pressures inside and outside the capillary.

Hydrostatic Pressure

Given a constant osmotic pressure, greater volumes of fluid move out of the capillary as we increase the blood pressure in the capillary lumen (*capillary hydrostatic pressure*) or decrease the pressure in the perivascular space (*tissue hydrostatic pressure*). Filtration, that is, depends partly on the *effective hydrostatic pressure* — capillary hydrostatic pressure minus tissue hydrostatic pressure.

In Table 12-3 I have indicated some of the variations that occur in the hydrostatic pressures throughout the body. In the case of the glomerular capillaries of the kidney, very high hydrostatic pressure is maintained. These capillaries,

Capillary Bed	Hydrostatic Pressure (mm. Hg)		
	Capillary		Tissue
	Arterial End	Venous End	
Kidney			
Glomerular	65		20
Peritubular	17	10	10
Retina			20
Hand			
40 cm. below heart	45	33	
Heart level	32	12	3
30 cm. above heart	23	10	
Local heating, intradermal histamine, inflammation, or reactive hypermia	49-60		
Lung	16	6	0

Table 12-3. Capillary and tissue hydrostatic pressures in the body. (In part from Landis, E.M., and Pappenheimer, J.R.: Exchange of substances through the capillary walls. *In* Hamilton, W.F. (ed.): Handbook of Physiology. Sec. 1 (Circulation), vol. II, pp. 961-1034. Washington, American Physiological Society, 1963)

because of their high effective hydrostatic pressures, deliver to the lumens of the nephrons approximately 190 liters of ultrafiltrate each day. A second capillary net in series with the glomerular capillaries, the peritubular capillaries, has a much smaller effective hydrostatic pressure, which is in part responsible for the reabsorption into the blood of almost as much fluid (189 liters/day) as the glomerular capillaries lose. The difference between what the glomerular capillaries lose and what the peritubular capillaries gain is the urine volume. Thus in one organ we have some capillaries adapted for filtration and others for reabsorption.

In the blood capillary system of the skin and skeletal muscle we find that the amount of fluid reabsorbed into the total system almost equals the amount filtered from the blood. The excess filtered fluid is normally carried back to the blood by the lymphatic system. When this does not happen, the tissues accumulate water and become *edematous*. If more is reabsorbed than filtered, the tissues become *dehydrated*. Note too in Table 12-3 that the higher the capillary in relation to the heart, the lower the capillary pressure. We find that any of a number of procedures that dilate arterioles (local heating, etc.) locally also increase capillary pressure.

In the case of the pulmonary capillaries we are dealing with a low-pressure system which, like the peritubular capillaries, is organized to facilitate absorption from the perivascular space. This is particularly important in the case of the lungs, since a high capillary hydrostatic pressure can lead to pulmonary edema (excess liquid in the alveoli). This, in turn, interferes with the diffusion of O_2 into the blood from the alveolar air.

Osmotic Pressure

Increasing the concentration of dissolved particles (particles/ml.) in the blood impedes the filtration of fluid out of the blood, since these particles create a pressure (capillary osmotic pressure) in opposition to the capillary hydrostatic pressure. Similarly, particles dissolved in the perivascular space exert an osmotic pressure in opposition to the tissue hydrostatic pressure. The average osmotic pressure of the blood plasma is about 7.6 atmospheres, or 5,776 mm. Hg. The average osmotic pressure in the perivascular space of skeletal muscle is 5,756 mm. Hg. We note a resulting *effective osmotic pressure* in this particular system of −20 mm. Hg (5,756 − 5,776). We shall in our discussion use a minus (−) to indicate pressure resulting in the movement of fluid into the capillary, and a plus (+) for pressure causing outward movement.

The cause of the difference in the osmotic pres-

Plasma Protein	Concentration (gm./100 ml.)	% of Total Protein	Molecular Weight	Approx. Osmotic Pressure (mm. Hg)
Whole plasma	7.0	100		25
Albumin	3.6	51	69,000	16.4
γ-Globulins	0.7	11	156,000	0.9
Fibrinogen	0.3	4	340,000	0.2
α-Lipoprotein	0.28	4	160,000-400,000	0.2
β-Lipoprotein	0.25	3.8	2×10^6	
β_1-Metal combining protein	0.2	3	90,000	0.7
β_2-Globulins	0.2	3	(150,000)	0.4 (?)
β_1 Lipid poor euglobulin	0.13	2	(150,000)	0.2 (?)
α_1-Acid glycoprotein	0.03	0.4	45,000	0.2 (?)
Remaining known components	0.4	5		<1.0
Total	6.0	87		20 (approx.)
Unidentified	1	13		5

Table 12-4. Some protein components of human plasma. (Landis, E.M., and Pappenheimer, J.R.: Exchange of substances through the capillary walls. *In* Hamilton, W. F. (ed.): Handbook of Physiology. Sec. 1 (Circulation), vol. II, pp. 961-1034. Washington, American Physiological Society, 1963)

sures of plasma and perivascular fluid is apparently the higher concentration of proteins in the plasma (Table 12-4) than in the perivascular fluid. Since all but the large molecules (colloids) pass readily through the endothelial barrier, we shall concern ourselves only with their osmotic properties. The colloid osmotic pressure of plasma is about 25 mm. Hg, and that of the perivascular fluid of skeletal muscle, about 5 mm. Hg. Thus, using the data in Table 12-3 on the hand at heart level, our calculation of the filtration pressure at the arteriolar end of a capillary is as follows:

Capillary hydrostatic pressure	+32 mm. Hg
Tissue hydrostatic pressure	−3 mm. Hg
Effective Hydrostatic Pressure	+29 mm. Hg
Capillary colloid osmotic pressure	−25 mm. Hg
Tissue colloid osmotic pressure	+5 mm. Hg
Effective Colloid Osmotic Pressure	−20 mm. Hg
Filtration Pressure	+9 mm. Hg.

Given the above conditions, that is, fluid tends to leave the capillary at its arteriolar end. On the venular end, all the above conditions prevail, except that the capillary hydrostatic pressure has decreased to 12 mm. Hg. This is a decrease of 20 mm. Hg and means that the filtration pressure will now be −11 mm. Hg (+9, −20). In this example, then, fluid would be leaving the capillary at the arteriolar end and entering at the venular end. Such conditions lead to a good exchange between the blood and the extravascular space. This is not the only situation that may prevail however; local dilation of an *arteriole* or an increase in *venous pressure* (Table 12-3, hand 40 cm. below heart) can lead to a sufficiently marked increase in pressure throughout the capillary that filtration will occur at all parts of the capillary, while a decrease in venous pressure (hand 30 cm. above heart) or a local vasoconstriction may lead to the reabsorption of fluid into the blood at all points along the capillary. It is also possible for one capillary to lose fluid throughout its length while another is reabsorbing it.

Control of Filtration

From our discussion we see that a major means of regulating capillary filtration is by controlling the pressure in the capillary. This is primarily the function of the arteriole leading into the capillary network. This arteriole, by dilating, brings the capillary pressure almost up to that in the arteries, and by constricting, decreases the pressure to that in the veins. In the glomerular capillaries of the kidney we have the interesting situation of there being an arteriole on either side of the capillary; here the afferent arteriole can constrict, thereby *decreasing* capillary hydrostatic pressure, or the efferent arteriole can constrict, thus *increasing* capillary hydrostatic pressure.

There are other factors, of course, important in regulating filtration. Capillary function is dependent upon relatively high arterial pressure and relatively low venous pressure. These pressures can change within certain limits without seriously affecting filtration, but excessive increases in venous pressure, as in heart failure, or arterial hypertension, as in arteriosclerosis, can sufficiently increase capillary hydrostatic pressure to cause the accumulation of excessive quantities of fluid in the extravascular spaces (edema). In most extravascular spaces the viscoelastic properties (Fig. 10-7) of the tissues are such that when fluid accumulates, the tissue hydrostatic pressure goes up for a period and tends to restrict further filtration.

Another cause of edema (Table 12-5) is a decrease in the capillary colloid osmotic pressure. This may result from malnutrition. When there is inadequate caloric intake and an inadequate reserve of carbohydrates and lipids, either the production of the plasma proteins by the liver decreases or the destruction of plasma proteins by the body increases. In either or both cases

---

1. Increased capillary hydrostatic pressure
    a. Hypertension
        (1) Arteriosclerosis
        (2) Kidney disorders (nephritic syndromes)
        (3) Endocrine disturbances
    b. Venous congestion
        (1) Cardiac failure
        (2) Venous occlusion (thrombosis, tumors, pregnancy)
2. Decreased capillary colloid osmotic pressure
    a. Starvation
    b. Liver malfunction
    c. Kidney disorders (nephrotic syndromes)
3. Increased tissue colloid osmotic pressure
    a. Capillary hypoxia
    b. Histamine release
    c. Capillary damage
4. Malfunction of the lymphatic system
    a. Obstruction (elephantiasis)
    b. Surgical removal of lymph nodes

Table 12-5. Some causes of edema.

more fluid moves to the extravascular space because of a change in the effective colloid osmotic pressure. In extreme cases of starvation, body weight may increase due to the accumulation of water throughout the body, and the abdomen may become distended with water (ascites). The capillary colloid osmotic pressure may also change in cases of liver damage or when kidney damage is such that albuminurea develops.

Changes in tissue colloid osmotic pressure may also cause edema. Such pressure changes can result from an increase in the permeability of the capillaries due to the release of histamine in the body, the development of severe local hypoxia, or capillary damage.

THE LYMPHATIC SYSTEM

There are lymph capillaries throughout the body (Fig. 12-6), carrying their contents to lymph ducts and lymph nodes. The lymph ducts contain *valves* that prevent the backward flow of lymph and resemble the veins structurally. Lymph flow, however, is more dependent upon forces (milking action of muscle) external to the system, e.g., rhythmic contraction of the intestines, squeezing by skeletal muscle contraction, and changes in pressure during breathing. This results in a much slower movement of fluid in the lymph vessels than occurs in the veins.

Lymph ducts converge at the lymph node—an important proliferative area for some of the blood cells, a source of antibodies, an area of phagocytosis, and the point at which some of the lymph is reabsorbed into the blood.

One function of the lymphatic system is to transport fluid, electrolytes, plasma proteins, etc. (Table 12-6) from the extravascular compartment to the blood. In persons who fail to develop a normal lymphatic system or whose lymph vessels become clogged by a parasite (elephantiasis), marked edema develops in the affected areas (Fig. 12-7) from the accumulation of fluid and colloids in the extravascular spaces. It has been shown in the dog, for example, that a *volume of lymph* equivalent to the animal's total plasma volume is carried in the thoracic duct back to the blood each day. This volume can be increased up to twelvefold by infusing 2,000 ml. of blood (the dog's blood volume) into a dog's vein.

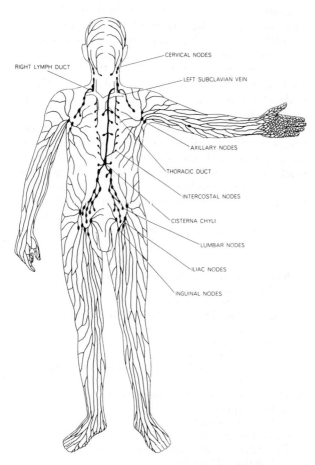

Fig. 12-6. The lymphatic system. The diagram illustrates some of the larger superficial (*thin lines*) and deep (*thick lines*) vessels of the lymphatic system. (Mayerson, H. S.: The lymphatic system. Sci. Am. 208(6):81, 1963; copyright © June 1963 by Scientific American, Inc. All rights reserved)

	Plasma	Lymph
Protein (%)	6.9	2.6
Sugar (mg./100 ml.)	128	124
Non-protein N (mg./100 ml.)	27.2	27
Urea (mg./100 ml.)	22	23.5
Amino acids (mg. N/100 ml.)	4.9	4.8
Calcium (mg./100 ml.)	10.4	9.2
Chloride (mg./100 ml.)	392	418

Table 12-6. Chemical composition of the plasma and thoracic duct lymph of the dog. It should be noted that the protein content of the lymph varies from 1.7 in the lymphatics of the leg and 3.7 in the lymphatics of the heart and lung to 4.8 in the lymphatics of the liver. (Drinker, C. F., and Yoffey, J. M.: Lymphatics, Lymph, and Lymphoid Tissue: Their Physiological and Clinical Significance. Cambridge, Harvard University Press, 1941)

The Lymphatic System   223

the entire amount is returned to the blood by the lymphatic vessels (Fig. 12-8).

The lymphatic system also functions in transport from the intestinal tract. It was the intestinal lymphatic vessels (lacteals), in fact, that were first discovered. Gasparo Aselli, in 1622, noted these normally inconspicuous structures in the mesentery of a dog that had recently eaten. The lacteals had absorbed fat from the intestine and as a result had a milky white appearance (L. *lac*, milk). We now know that the lacteals are the major pathway carrying long-chain fats such as *stearic* and *palmitic acids* and the principal steroid of the body, *cholesterol*, to the blood (thoracic duct to subclavian vein). Most other nutrients are picked up by the capillaries and carried to the liver in the portal vein.

Lymph too is probably the means by which some of the hormones reach the blood. Hista-

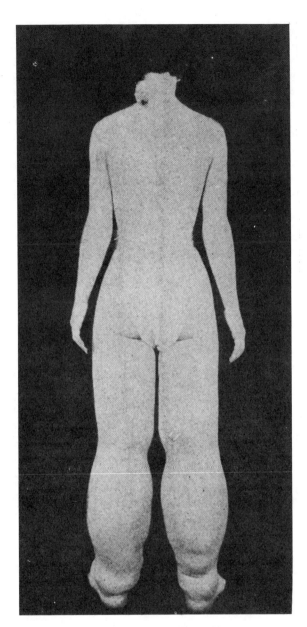

Fig. 12-7. Bilateral edema (elephantiasis) in a 19-year-old girl. This condition was of a 6-year duration in the right leg and of a 4½-year duration in the left. It is caused by an occlusion of the lymphatic flow by the parasite *Filaria*. (Drinker, C. F., and Yoffey, J. M.: Lymphatics, Lymph, and Lymphoid Tissue; Their Physiological and Clinical Significance. P. 292. Cambridge, Harvard University Press, 1941)

The amount of *plasma protein* lost to the extravascular space each day by man is estimated to be in excess of 50 per cent of the amount contained in the blood. Under normal circumstances

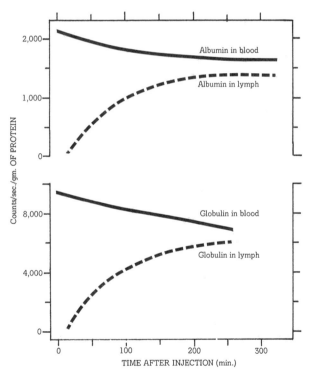

Fig. 12-8. Fate of radioactive albumin (*upper curves*) and globulin (*lower curves*) injected into a vein. Note that within 100 minutes after injection substantial amounts of protein had disappeared from the blood (*solid curves*) and had appeared in the thoracic duct (*interrupted curves*). (*Redrawn from* Mayerson, H. S.: The lymphatic system. Sci. Am., *208*:85, 1963; copyright © June 1963 by Scientific American, Inc. All rights reserved)

mine and renin have been shown to be more highly concentrated in some lymph than in the blood. Recently Kolmen produced hypertension in a dog by removing one kidney and restricting the blood supply to the other. When he diverted the flow in the thoracic duct to the intestinal lumen he found that the hypertension decreased.

ARTERIOVENOUS DIFFERENCES

During exercise the body increases cardiac output and modifies its distribution. We also find that in many tissues the amount of oxygen per milliliter of blood flow used is greater during exercise than at rest. This is illustrated in Figure 12-9. Here we note that as exercise becomes more strenuous, the heart rate and cardiac output rise and the difference in oxygen content between arteries and veins (A-V O_2 difference) increases. This indicates that in at least some tissues in the resting person, the capillary blood has an *oxygen reserve* which can be called upon.

I have, in Figure 12-10, indicated the proportion of oxygen removed by the various tissues in the resting person. Normally there are 19 ml. of oxygen in the arteries per 100 ml. of blood and 14.3 ml. in the veins. This represents an A-V O_2 difference of 4.7 ml. of oxygen. Someone with

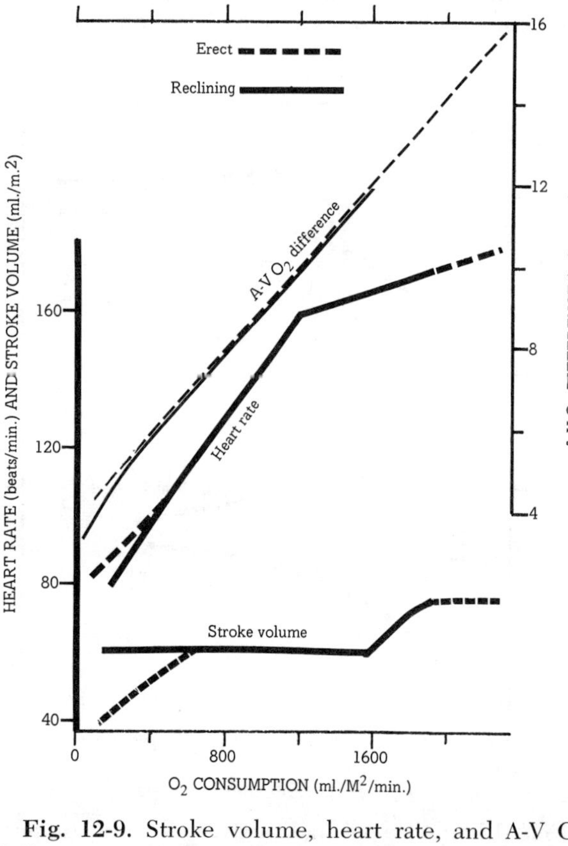

Fig. 12-9. Stroke volume, heart rate, and A-V O_2 difference as functions of oxygen consumption during work in the erect and reclining positions. Note that the A-V O_2 difference changes from 4 ml. of oxygen to 15 ml. of oxygen per 100 ml. of blood as the severity of the exercise increases. (*Redrawn from* Rushmer, R. F.: Postural effects on the baselines of ventricular performance. Circulation, *20*:902, 1959; by permission of the American Heart Association, Inc.)

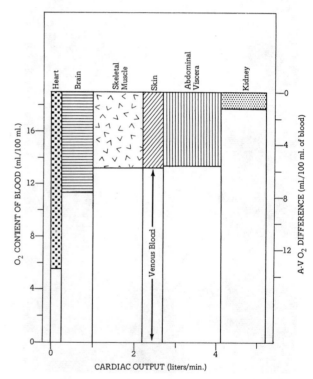

Fig. 12-10. A-V O_2 differences in the resting individual. The width of each bar is an indication of the flow (ml./min.) to the various tissues (I) and the height of each lower bar a representation of the oxygen content of the venous blood from each tissue. The height of the upper bars represents the respective A-V oxygen differences (dO_2) and the area of each upper bar the oxygen used by the tissue per minute. (O^2):

$$O_2 \text{ (ml./min.)} = I \text{ (ml. of blood/min.)} \times dO_2 \text{ (ml./ml. of blood)}.$$

Note that although 19 ml. of oxygen/100 ml. of blood are presented to each tissue, there are wide variations in the oxygen content of the venous blood.

a cardiac output of 5,800 ml. blood per minute, then, consumes 270 ml. oxygen per minute:

$$(5{,}800 \text{ ml. blood/min.})(4.7 \text{ ml. } O_2/100 \text{ ml. blood}) = 273.6 \text{ ml. } O_2/\text{min.}$$

It is important to note in Figure 12-10 that not all tissues have the same A-V O_2 difference. The kidneys, for example, normally extract only about 6 per cent of the oxygen passing through them, whereas the heart extracts about 70 per cent. In other words, the flow of blood to the kidney is out of proportion to its need for oxygen. The heart, on the other hand, is probably more dependent on an increase in flow during times of increased oxygen demand than any other organ in the body.

Chapter 13

CEREBROSPINAL FLUID
INTRAOCULAR FLUID
ENDOLYMPH AND
PERILYMPH — THE
LABYRINTHINE FLUIDS
AMNIONIC FLUID

EXTRAVASCULAR FLUID SYSTEMS

In the preceding chapters we have been concerned primarily with the circulation of blood in the arteries, capillaries, sinusoids, veins, and the heart, and the contribution of the lymphatic system to the circulation. In this section we shall consider some other fluid systems found in the body.

CEREBROSPINAL FLUID

The cerebrospinal fluid is a solution formed, for the most part, by the highly vascular *choroid plexi* in the chambers of the brain (ventricles, and ducts of the brain). A small part of it passes into the central canal of the spinal cord (Fig. 13-1), but most leaves the chambers of the brain in the area of the fourth ventricle (in the medulla). For most persons there are three major points at which the fluid leaves the brain — the medial foramen (of Magendie) and the two lateral foramina (of Luschka). In animals below the anthropoid ape the medial foramen is absent, and in about 20 per cent of apparently normal human subjects the lateral foramina are absent.

After the fluid moves through the medial and lateral foramina and into the cisterna magna below the cerebellum, it passes through the subarachnoid space that surrounds the cord and brain. The subarachnoid space is a channel of variable size formed by two membranes (or meninges). The central membrane is the pia mater, which is thin and adheres to the spinal cord and brain on their surface and in their fissures and sulci. It, unlike the other meninges, also follows blood vessels into the substance of the central nervous system. The periphery of the subarachnoid space is formed by the arachnoid membrane. This membrane sends out numerous fibers, the arachnoid trabeculi (arranged like a spider's web), to the pia. These, though fairly weak, help stabilize the cord and brain in the vertebral canal and cranium. Areas in which the *subarachnoid spaces* become large

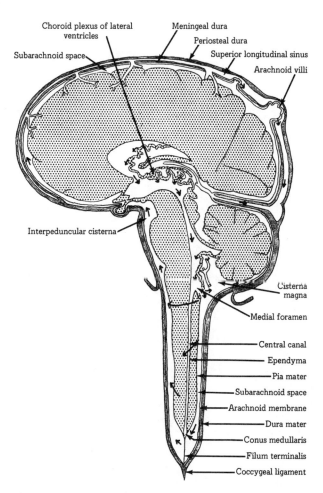

Fig. 13-1. The formation and absorption of cerebrospinal fluid. Most of the cerebrospinal fluid is formed by the choroid plexi, which release their secretions into the ventricles and ducts of the brain. Most of this fluid in man passes from the chambers of the brain to the subarachnoid spaces and cisternae through the lateral and medial foramina of the medulla. Most of the cerebrospinal fluid re-enters the blood through and the arachnoid villi that project into the superior longitudinal (sagittal) dural sinus.

are called *cisternae* (cisterna magna or cerebellomedullaris, cisterna superior above the cerebellum, cisterna chiamatis at the optic chiasma, cisterna interpeduncularis at the midbrain, and cisterna pontis below the pons).

The arachnoid membrane is closely applied to the dura mater, which in turn is attached to the vertebral column by the posterior longitudinal ligament, the coccygeal ligament, and some cervical strands. The cephalad portion of the dura mater is attached to the borders of the foramen magnum and is an inseparable part of the cranial periosteum. The dura also contains a meningeal portion in the cranial cavity. This provides support by partitioning the brain between the cerebellum and cerebral hemispheres (tentorium cerebelli) and between the right and left cerebral hemispheres (falx cerebri).

We also find in the cranium that sinuses separate the periosteal layer of dura from the meningeal layer of dura at certain points. These dural sinuses are lined by endothelium and receive blood from the veins of the skull, meninges, and brain. They drain into the ophthalmic, jugular, and vertebral veins. Invaginating the venous dural sinuses (primarily the superior longitudinal sinus) are finger-like projections (Fig. 13-2), the arachnoid villi. It is through these villi that most of the cerebrospinal fluid enters the blood.

Function

The brain and cord weigh about 1,500 gm. outside the body. While in the body they float in the cerebrospinal fluid (CSF), and are *buoyed up* by it to such an extent that their weight is reduced to about 50 gm. Under these circumstances the various supporting structures in the brain and cord (dentate ligaments, arachnoid trabeculi, etc.) are strong enough to prevent the movement of the brain into the boney border of the foramen magnum or the movement of any part of the cord or brain into a bone. In an air encephalogram, as much fluid as possible is replaced by air in the cord to provide a good x-ray contrast. This causes a traction on nerve roots and is so painful that sedation is usually required. The procedure is followed by a headache, intensified by the slightest amount of jarring.

The CSF, being incompressible, also protects the central nervous system from changes in velocity such as occur when one stops or accelerates rapidly. When, for example, one is hit on the head by a brick, the hard cranium is pushed toward the brain but the brain, by virtue of its inertia, tends

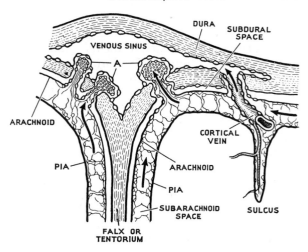

Fig. 13-2. Reabsorption of cerebrospinal fluid at the superior longitudinal sinus (venous sinus). A, arachnoid villi. (Keele, C. A., and Neil, E.: Samson Wright's Applied Physiology. 11th ed., p. 48. New York, Oxford University Press, 1965)

to remain stationary. The CSF provides an important cushion or *buffer zone* between the hard cranium or vertebral column and the fragile central nervous system.

Besides its buoyancy and shock-absorber action, the CSF has been implicated as an important *transport system*. Approximately 700 ml. of CSF are formed and reabsorbed into the blood each day. As the fluid passes through and around the central nervous system it loses glucose and phosphate and picks up urea and creatinine. The quantity of CSF circulating through the brain and cord (0.0005 liters/min.) is, of course, rather small compared to the 1,200 liters of blood per day that pass to the central nervous system; however it may, like the lymphatic flow, prove to be an important transport system.

It should also be noted that CSF provides an *environment* for the central nervous system different from that provided by its capillaries. The concentrations of Ca^{++} and K^+, for example, are lower in CSF than in blood plasma. Investigators who have tried to increase the concentration of Ca^{++} in the CSF by injections into the cisterna magna or lateral ventricles have found that sleep or unconsciousness results. Small increases in K^+ produce increases in the frequency, and decreases in the amplitude, of the electroencephalogram. Thus it may be that CSF serves an important function in the control of the environment in certain areas of the central nervous system.

Formation

Most of the CSF is formed by the *tela choroidea*. This consists of a highly vascularized pia and a close associated layer of choroidal epithelium which lies between the pia and the CSF. The *choroidal epithelium* is a modified continuation of the lining (formed by *ependymal cells*) of the ventricles of the brain.

Part of the evidence that the tela choroidea of the ventricles produces CSF is its highly vascular structure. It has also been found in the dog that if the cerebral aqueduct (between the third and fourth ventricles) is plugged, a hydrocephalus (Fig. 13-3) results. If, on the other hand, the choroid plexus of one lateral ventricle is removed and both interventricular foramina (ducts which connect the lateral ventricles with the third ventricle) are blocked, the ventricle without a choroid plexus collapses and the contralateral ventricle dilates.

In Table 13-1 I have compared the composition of CSF to that of blood plasma. It is of interest that, although there are from 6,500 to 7,500 mg. of *protein* per 100 ml. plasma, there are only about 25 mg. of protein per 100 ml. CSF. The differences between plasma and CSF are not, however, limited to the large molecules. Apparently there is an *active transport* of Na^+ and Cl^- into the ventricles of the brain that results in the formation of *hypertonic* CSF. The hypertonicity of the secretion of the tela choroidea is probably responsible for the movement of most of the water from the blood into the CSF. This active transport process can be blocked by the carbonic anhydrase inhibitor *acetazolamide*.

Note also in Table 13-1 that there is a lower K^+ concentration in CSF than in plasma. It has been suggested that the same system in the tela choroidea that moves Na^+ into the CSF moves K^+ out of it. I do not wish to imply, of course, that all concentration differences between the plasma and CSF are due to active transport. The low concentration of urea in CSF probably results from the low permeability of the secretory cells to this substance. The lower concentration of phosphate and glucose is probably the result of their uptake by nervous tissue.

We find as we trace CSF from its source at the tela choroidea that the concentration of urea in it increases from 60 to 81 per cent of that of plasma, and that the concentration of creatinine changes from about 5 to 17 per cent. The importance of these observations is still not well understood, but one should remember that the blood capillaries of the central nervous system are not so readily permeable to many substances as are the capillaries in other vascular beds. The capillaries of the central nervous system, like those of skeletal muscle (Table 12-1), are

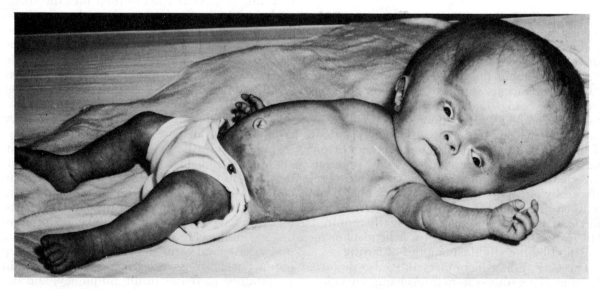

Fig. 13-3. Hydrocephalus in a 4-month-old infant. (Dekaban, A.: Neurology of Infancy. P. 227. Baltimore, Williams & Wilkins, 1959; copyright © 1959, The Williams and Wilkins Company, Baltimore)

Substance	Cerebrospinal Fluid		Aqueous Humor		Dialyzed Plasma	
Na^{+1}	1.03	± 0.005	0.96	± 0.01	0.945	± 0.003
K^+	0.52	± 0.04	0.955	± 0.02	0.96	± 0.005
Mg^{++}	0.80	± 0.05	0.78	± 0.04*	0.80	± 0.02
Ca^{++}	0.33	± 0.01	0.58	± 0.005	0.65	± 0.02
HCO_3^-	0.97	± 0.04	1.26	± 0.03	(1.04)	
H_2CO_3	1.61		1.29		(1.00)	
Cl^-	1.21	± 0.007	1.015	± 0.01	1.04	± 0.006
Br^-	0.715	± 0.02	0.98	± 0.15	0.96	± 0.01
I^-	0.004	− 0.04	0.32	± 0.51	0.85	± 0.01
CNS^-	0.06	± 0.21	0.46	± 1.24	0.59	
Phosphate	0.34	± 0.05	0.58	± 0.05		
Glucose	0.64	± 0.02	0.86	± 0.25	0.97	± 0.01
Urea	0.81	± 0.02	0.87	± 0.02	(1.00)	
Ascorbic acid	1.55		18.5		(1.00)	
Protein	0.0036		0.0019			
pH	7.27		7.48		7.46	

* In man this ratio is 1.13 to 1.30.

Table 13-1. Distribution of various substances in the cerebrospinal fluid, aqueous humor of the eye, and plasma dialyzate of the rabbit. All data are expressed, where known, in terms of the ratio ± the standard error between the concentration of each substance in the indicated fluid and that in blood plasma. Values in parentheses are estimates. (Davson, H.: Intracranial and intraocular fluids. *In* Magoun, H.W. (ed.): Handbook of Physiology. Sec. 1 (Neurophysiology), vol. III, pp. 1761-1788. Washington, American Physiological Society, 1960)

unfenestrated and have continuous basement membranes. They, unlike the capillaries of skeletal muscle, have a covering of neuroglial processes (Fig. 13-4) over 85 per cent of their surface. Perhaps this additional covering accounts for their reduced permeability.

In Figure 13-5 we have a comparison between the uptake of various substances from the blood by skeletal muscle and by the brain. Note that chloride, urea, and inulin all move more rapidly into skeletal muscle than into the brain. Observations such as these have led to the conclusion that there is a *blood-brain barrier* that is much more pronounced than the capillary barriers found elsewhere in the body. Apparently there is also an important brain-blood barrier. There are a number of substances, in other words, that do not move readily from the blood to the interstitial fluid of the brain or from the interstitial fluid of the brain to the blood. These substances move more freely from the interstitial fluid of the brain to CSF.

Reabsorption

The major site for the reabsorption of CSF is the microscopic arachnoid villi (Fig. 13-2) that project into the dural sinuses. At these points the

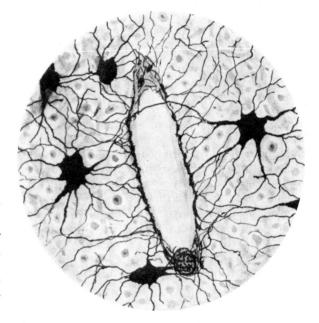

Fig. 13-4. Membrane around a cerebral capillary. Note that the fibrous processes of astrocytes (large dark cells) form a barrier around the cerebral capillary. (Glees, P.: Neuroglia Morphology and Function. P. 25. Oxford, Blackwell Scientific Publications, 1955)

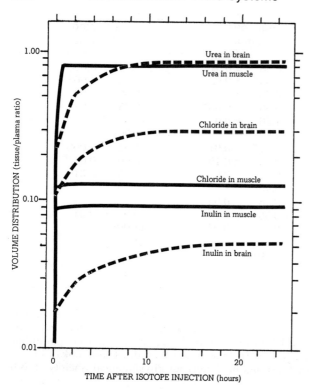

Fig. 13-5. The blood-brain barrier. Comparison of the uptake of the isotopes of urea, chloride, and inulin from the blood by the cerebral cortex and skeletal muscle of the rat. Volume distribution is in terms of the tissue/plasma ratio. Note the slower uptake of urea and inulin by the brain. (Bain, W. A.: Quart. J. Exp. Physiol., 22:269-274, 1932)

CSF is separated from the blood in the dural sinuses by two layers of cells, the arachnoid layer and the endothelium of the sinus itself. Evidence of the importance of these areas in the reabsorption of CSF rests primarily on a number of studies in which dyes were injected into the subarachnoid space. Prussian blue and colored gelatin solutions have both been shown to reenter the circulation primarily through the arachnoid villi. Very little of these dyes were seen in the epidural *lymphatic system* and there are no lymphatics in either the meninges or the brain.

The large-*molecular-weight* substances such as albumin and inulin move from the CSF to the blood almost exclusively by way of the *arachnoid villi*. A very small percentage of the smaller molecules such as sucrose and creatinine pass through the pia and neuroglial cells to the capillaries of the brain and spinal cord.

Here they are moved very slowly into the circulation (brain-blood barrier). In the case of *lipid-soluble* substances (O_2, CO_2, thiourea and ethyl alcohol), movement from CSF into the capillaries of the central nervous system is much more rapid. Although the escape of small lipid-soluble molecules from CSF to these capillaries is significant, most of the molecules get back into the blood via the arachnoid villi.

Pressure

A person lying quietly on a bed has a CSF pressure throughout the central nervous system of from 70 to 160 mm. H_2O (5 to 12 mm. Hg). Under most circumstances this pressure increases when the pressure in the vessels leading to or from the brain increases, and decreases when the arterial or venous pressure decreases. Apparently, then, the CSF pressure tends to maintain a fairly constant *transmural* (cerebral blood pressure minus CSF pressure) pressure. Or, an increase in arterial pressure associated with an increase in CSF pressure limits the *distention* of an artery, and a decrease in arterial pressure associated with a decrease in CSF pressure helps keep that artery *patent*.

If we occlude the jugular veins by pressure on the surface of the neck, CSF pressure increases throughout the system in 10 to 12 sec. Releasing the pressure produces an equally rapid decline in CSF pressure. If occlusion of the jugular veins does not produce a prompt increase in CSF pressure (*Queckenstedt-Stookey sign*), there is a block between the arachnoid villi and the point of measurement.

Changes in CSF pressure also occur in response to changes in the pressure in the superior vena cava and aorta. Sneezing, coughing (Fig. 13-6), or straining cause changes in CSF pressure. Prolonged vigorous expirations against a closed glottis (Valsalva maneuver or straining during defecation or urination) produce an elevation in arterial, superior vena caval, and CSF pressure. The increase in CSF pressure is in part due to the transmission of pressure to the superior sagittal sinus and serves, as I have said, to prevent rupture (or overdistention) of the cerebral vessels.

Fortunately, there are few situations in which an increase or decrease in cerebral vessel pressure is not associated with a similar change in CSF pressure. When a pilot pulls out of a dive, for example, much of his blood and cerebro-

Fig. 13-6. Changes in arterial pressure (A), CSF pressure (B), and transmural cerebral artery pressure (D) in response to a cough. Line C indicates the time in seconds. (Hamilton, W. F., et al.: Physiologic relationships between intrathoracic, intraspinal and arterial pressures. JAMA, *107*: 855, 1936)

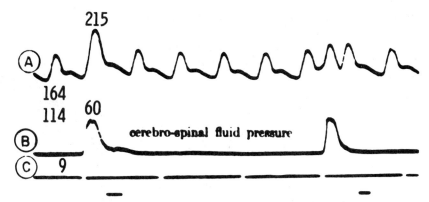
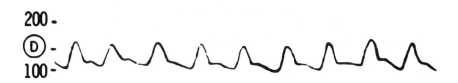

spinal fluid may be forced caudally; both the CSF pressure in and around the brain and the pressure in the cerebral vessels fall, but the normal transmural (across-the-wall) pressures and vessel diameters are maintained. The incompressibility of CSF and the rigidity of the bones that surround it prevent excessive caudal movement of the cord.

When one changes from a reclining to a *standing* or sitting position the CSF pressure at the cisterna magna changes from about 110 to 0 mm. H_2O, and the pressure in the lumbar region changes from 110 to about 450 mm. H_2O. At the same time there is a decrease in the jugular vein pressure of 7 mm. Hg (95 mm. H_2O) and a decrease in cerebral artery pressure of about 25 mm. Hg (Fig. 10-12). The fact that when one stands there is a simultaneous decrease in pressure both in and around the blood vessels tends to keep the vessels patent. This mechanism does not, on the other hand, completely prevent a decrease in flow to the brain. As I have pointed out, the carotid and aortic mechanoreceptors simultaneously produce a reflex increase in arterial pressure, which serves to return cerebral flow to normal. In the absence of such a reflex, dizziness and fainting occur. Generally it takes only 8 sec. of complete brain ischemia to produce unconsciousness.

Besides gravity, CSF pressure is dependent on the amount of CSF formed by the tela choroidea and the amount reabsorbed at the arachnoid villi. Formation of CSF seems to be relatively independent of CSF pressure. We have mentioned, for example, that CSF continues to be formed even after there has been a block in its pathway to the subarachnoid space and an associated increase in pressure in the lateral ventricles. Its reabsorption, however, seems to be much more dependent on the effective hydrostatic pressure (CSF pressure minus pressure in the dural sinus) at the point of reabsorption. When CSF pressure is markedly higher than that in the venous sinus, CSF moves readily through the highly permeable membranes into the sinus. When sinus pressure is higher than CSF pressure, the villi projecting into the sinus collapse. Thus these villi function as a *valve*, permitting fluid to move into the dural sinus when CSF pressure exceeds sinus pressure, and impeding the movement of plasma into the CSF when sinus pressure exceeds CSF pressure.

In 1903, Cushing described five patients with arterial hypertension associated with rapidly rising intracranial pressure. Since then, it has been shown in the unanesthetized dog that an elevation in intracranial pressure produces a significant increase in arterial pressure. Some have suggested that this is due to another type of mechanoreceptor reflex, while others feel that the increase in intracranial pressure (CSF pressure) so restricts the cerebral blood flow as to produce brain asphyxia (increased CO_2 and decreased O_2). The brain responds to this asphyxia by causing an increase in arterial blood pressure throughout the body.

The latter suggestion is consistent with one put forward by Roy and Sherrington in 1890. They stated that:

> Anemia of the central nervous system excites vasoconstrictor nerves with the result that owing to constriction of the vessels of the digestive, urinary and other systems, the arterial blood pressure rises, causing an increased flow of blood through the cerebrospinal vessels.

They went on to make the following original and important observation:

> We conclude that the chemical products of cerebral metabolism contained in the lymph which bathes the walls of the arterioles of the brain can cause variations of the calibre of the cerebral vessels: that in this reaction the brain possesses an intrinsic mechanism by which its vascular supply can be varied locally in correspondence with local variation of functional activity.

INTRAOCULAR FLUID

In many respects the eye is like the central nervous system. Embryologically, it develops as an outgrowth of the brain. Structurally there are many similarities. It has a thick outer membrane, the sclera, which resembles and is continuous with the dura of the brain (Fig. 13-7). A vascular choroid membrane in the ciliary body, like the choroid plexi of the ventricles, secretes a low-protein fluid, the *aqueous humor*, into the posterior chamber. This fluid passes over the lens, through the pupil, and into the anterior chamber, and is carried by the canal of Schlemm to various veins.

Between the lens and the retina lies the *vitreous body*. It is a dilute gel containing collagen-like proteins, mucopolysaccharides, hyaluronic acid, and water (over 98 per cent). It contains a lower concentration of phosphate and glucose than the aqueous humor.

Function

Briefly, the aqueous humor (1) transmits and refracts light, (2) maintains corneal curvature (by pressure), (3) maintains a fairly stable pH (through its buffers), and (4) transports nutrients and wastes. The vitreous body is less well understood but is also important in the transmission

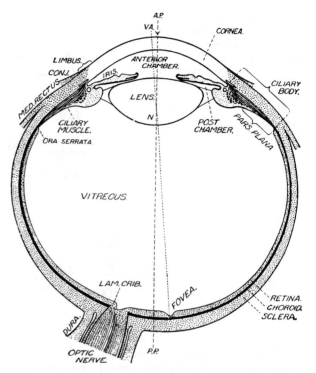

Fig. 13-7. Structure of the human eye. Intraocular fluid is formed by a continuation of the vascular choroid coat into the ciliary body. It is secreted into the posterior chamber, flows around the lens and through the pupil into the anterior chamber, and is carried by the canal of Schlemm in the corneoscleral junction to the veins. The vitreous body is a dilute gel (over 98% water) on the posterior side of the lens. Fluid from the capillaries of the eye diffuse through it. (*Modified from* Salzmann, M.: *in* Last, R. J.: Eugene Wolff's Anatomy of the Eye and Orbit. London, H. K. Lewis and Co., Ltd. 6th ed., p. 30. 1968)

of light, support, the maintenance of pH, and transport.

We shall postpone our discussion of light-transmission until we consider sensation, but I would like to emphasize the importance of substantial intraocular pressure in maintaining the curvature of the cornea—and therefore its refractory properties. Normally, the pressure in the anterior and posterior chambers is about 20 mm. Hg (as opposed to an average CSF pressure of about 8 mm. Hg).

The transport function of the intraocular fluid is probably more important to the eye than that of CSF is to the central nervous system. Although the aqueous humor has a turnover of only about 1.3 per cent per minute, at least one part of the eye, the lens, is apparently quite dependent on it for nutrients and for the removal of waste products. As can be seen in Table 13-1 (page 229), the composition of the aqueous humor of the eye differs in many ways from that of CSF. The lens contains no blood vessels and is probably dependent upon the aqueous humor for glucose, phosphate, and amino acids. It also continually produces lactic acid, which the aqueous humor buffers and carries away.

The vitreous humor may serve a nutritive and buffer function for the retina. Fluids from capillaries apparently diffuse through it rather freely.

Formation

About 6 ml. of aqueous humor are formed each day. This results from active transport in an extension of the choroid membrane over the ciliary muscle (Fig. 13-8). This vascular choroid contains processes which extend into the posterior chamber of the eye. The fluid forming the aqueous humor first passes through the capillaries into the intercellular space, and then through the lining of epithelial cells between the intercellular space and the posterior chamber of the eye.

Thus there seems to be a barrier between the blood and the aqueous humor. We can determine the effectiveness of this barrier—the blood-aqueous barrier—by injecting various substances into the blood, maintaining their blood concentration, and after a specified interval testing their concentration in the aqueous humor. It is common practice to designate the permeability of the barrier by a constant, k. A high k value indicates a high degree of permeability, and a low value a low degree of permeability. In Table 13-2 I have listed the k values for a number of substances. Note that in the case of the aqueous humor there is a large range in the k values. This range is considerably smaller for skeletal muscle capillaries. The broader range for aqueous humor is probably due primarily to the layer of epithelial cells between the posterior chamber and the capillaries of the ciliary body.

You will note in Table 13-2 that the lipid-soluble substances methyl thiourea, ethyl thiourea, propyl thiourea, and ethyl alcohol permeate the barrier much more rapidly (have a higher k) than does the non-lipid-soluble substance sucrose. Apparently the lipid-soluble

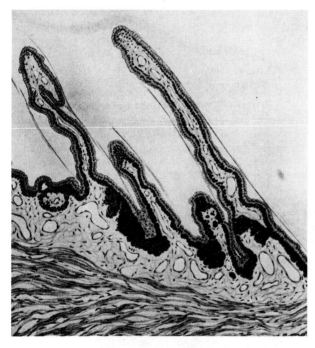

Fig. 13-8. Human ciliary body. In this picture we note that the ciliary body consists of a layer of smooth muscle with a vascular covering containing processes projecting into the posterior chamber. Each process contains a tangle of capillaries and is covered by a double layer of epithelium. (Reproduced by permission from Eugene Wolff's Anatomy of the Eye and Orbit. London, H. K. Lewis and Co., Ltd. 6th ed., p. 68. 1968)

Substance	k_{in}* (1/min.)
Sucrose	0.00072
Creatinine	0.0022
Thiourea	0.0082
Methyl thiourea	0.0115
Ethyl thiourea	0.0144
Propyl thiourea	0.0220
Ethyl alcohol	0.0500
Sodium	0.0080
Bromide	0.0104
Thiocyanate	0.0130
Potassium	0.0104

* k_{in} is a transfer constant in the equation $dC_{Aq}/dt = k_{in}C_{Pl} - k_{out}C_{Aq}$. C_{Aq} and C_{Pl} are the concentrations in aqueous humor and plasma, respectively.

Table 13-2. Permeability of the blood-aqueous barrier to various substances in the rabbit. (Davson, H.: Intracranial and intraocular fluids. *In* Magoun, H.W. (ed.): Handbook of Physiology. Sec. 1 (Neurophysiology), vol. III, p. 1771. Washington, American Physiological Society, 1960)

particles mentioned, as well as O_2, not only pass freely from the capillary, but also pass relatively freely through the cell membranes of the epithelial layer.

It has been suggested that the capillaries surrounding the posterior and anterior chambers of the eye contribute to the aqueous humor just as the capillaries surrounding the vitreous body contribute to it. The iris, for example, contains capillaries and does not have a closely packed epithelial layer on its anterior surface, but certain substances do not pass so readily from the blood to aqueous humor as they do from blood to most other extravascular spaces. Perhaps the capillaries of the iris are too few, or their surrounding connective tissue too dense, to permit the movement of large quantities of fluid into the anterior or posterior chambers.

Reabsorption

The aqueous humor is reabsorbed from the anterior chamber into the blood at the corneoscleral junction. Here a channel, the canal of Schlemm, leads into the intrascleral, episcleral, and conjunctival veins. The intraoptic fluid pressure is about 20 mm. Hg and the intravenous pressure here about 12, so that we have an effective hydrostatic pressure of about 8 mm. Hg moving the aqueous humor toward the blood. Colored solutions in the aqueous humor have been found to move rapidly into the veins around the canal of Schlemm, even when the intraocular pressure is well below normal. Dyes have not, however, been observed to move into the vessels of the iris or other potential accessory channels.

If the drainage of aqueous humor through the

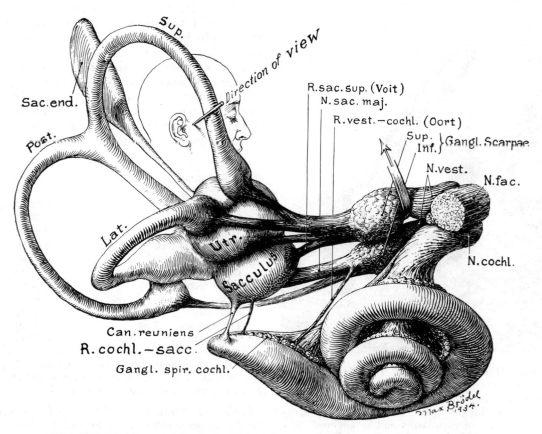

Fig. 13-9. The human labyrinth. Within the osseous labyrinth of the temporal bone is the membrane-lined, endolymph-containing, membranous labyrinth shown above. (Hardy, M.: Observations on the innervation of the macula sacculi in man. Anat. Rec., 59:412, 1935)

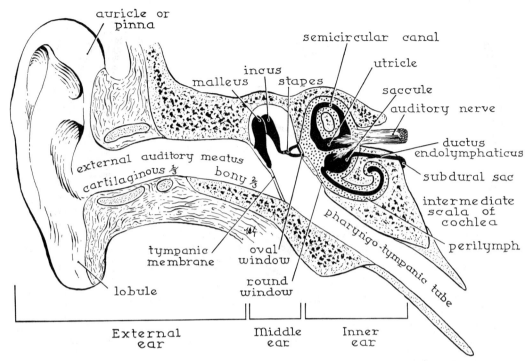

Fig. 3-10. Diagram of the ear. Note that the inner ear consists of 2 channels. One (osseous labyrinth) containing perilymph (*stippled*) and a second (membranous labyrinth) containing endolymph (*solid*). (Ham, A. W.: Histology. 6th ed., p. 1003. Philadelphia, J. B. Lippincott, 1969; *redrawn and modified from* Addison, W. H. F.: Piersol's Normal Histology, 15th ed. Philadelphia, J. B. Lippincott, 1932)

canal of Schlemm is interfered with, the intraocular pressure increases. This condition is called *glaucoma* (Gk., opacity of the crystaline lens) and can lead to damage of the fibers of the optic nerve and blindness.

ENDOLYMPH AND PERILYMPH—THE LABYRINTHINE FLUIDS

Lying in the temporal bone is a maze of interconnecting cavities filled with liquid and containing specialized receptors concerned with hearing and equilibrium. Collectively, the enclosed structures are called the inner ear or labyrinthine (Gk. *labyrinthos*, a maze) apparatus. They include the cochlea (which contains specialized receptors for sound), the utricle (gravity, acceleration, etc.) the saccule (poorly understood), the semicircular canals (acceleration, etc.), and associated structures (Fig. 13-9). Lying in these osseous channels is a second series of membrane-lined canals. The more peripheral channel is the *osseous labyrinth* and contains a fluid called perilymph. The *membranous labyrinth* is surrounded by this perilymph but contains a markedly different fluid, endolymph (Fig. 13-10).

Perilymph

The perilymph contains concentrations of K^+, Na^+, and Cl^- similar to those of CSF, but more than twice as much protein (Table 13-3). It is in direct communication with the *subarachnoid space* of the brain by a channel (cochlear aqueduct or perilymph duct) running from the cochlea through the temporal bone.

In the cochlea (Fig. 13-11), perilymph lies in the *scala vestibuli* above the cochlear duct and the *scala tympani* below it. These two perilymph channels are connected at the apex of the cochlea by the *helicotrema*. Sound waves set up vibrations of the tympanic membrane, which in turn moves the ossicles (maleus, incus, and stapes) of the middle ear. The *stapes* is inserted

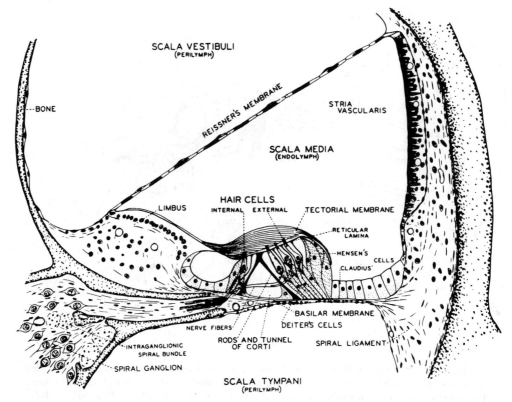

Fig. 13-11. Cross section of the human cochlea. The 2 outer chambers (scala vestibuli and scala tympani) contain perilymph and the central large chamber (scala media) contains endolymph. The fluid in the spiral ligament and tunnel of Corti (the clear space through which the nerve fibers are seen to pass) also resembles perilymph. (*Reprinted from* Davis, H., *in* Rosenblith, W. A. (ed.), Sensory Communication, by permission of The MIT Press, Cambridge, Massachusetts. P. 121, 1961)

on the *oval window* and its movements produce movements of the oval window and the perilymph on the medial side of the window. These vibrations are conducted in the perilymph of the scala vestibuli, through the helicotrema, to the perilymph of the scala tympani, and finally to the round window, which, like the oval window, separates the perilymph from the middle ear. As they pass through the perilymph, the waves also set up vibrations in the endolymph of the cochlear duct (scala media) and in the organ of Corti. This, in turn, produces impulses in the cochlear nerve.

The perilymph of the cochlea, besides conducting sound waves to the specialized receptors of the cochlear duct, also serves to nourish the cells of the organ of Corti. At present we do not know the extent to which the various cells of the labyrinth depend on the perilymph for their nutrition; nor is it clear how rapid the turnover of perilymph is. It is well recognized, however, that the basilar membrane is permeable to the perilymph and that the intercellular fluid of the spiral ligament resembles perilymph much more than it resembles endolymph. The high concentration of K^+ in the endolymph makes it a poor fluid for bathing most excitable cells.

Endolymph

Endolymph is found in the membranous labyrinths (cochlear duct, utricle, saccule, semicircular canals, and endolymphatic duct). Its high concentration of K^+ resembles that found in intracellular fluids, but its low concentration of proteins resembles that found in CSF. The marked difference of endolymph from CSF and perilymph is shown in Table 13-3.

	Cerebrospinal Fluid	Perilymph	Endolymph
Potassium (mEq./liter)	4.2 ± 0.15*	4.8 ± 0.4	144.4 ± 4.0
Sodium (mEq./liter)	152.0 ± 1.8	150.3 ± 2.1	15.8 ± 1.6
Chloride (mEq./liter)	122.4 ± 1.0	121.5 ± 1.2	107.1 ± 1.4
Protein (mg./100 ml.)	21 ± 2	50 ± 5	15 ± 2

* Values expressed as means ± standard error.

Table 13-3. Composition of spinal fluid, perilymph, and endolymph. Endolymph was collected from the utricle. Cochlear endolymph gave similar but less consistent results. (Davis, H.: Excitation of auditory receptors. *In* Magoun, H. W. (ed.): Handbook of Physiology. Sec. 1 (Neurophysiology), vol. I, pp. 565-584. Washington, American Physiological Society, 1962)

The difference in its concentration from that of other fluids could only be accomplished by an active process. The secretion of endolymph into the membranous labyrinth probably occurs at the secretory epithelium lying between the cochlear duct and the spiral ligament. It has been suggested that secretion also occurs at the endolymphatic saccule, which lies in the epidural space. Reabsorption probably occurs at the same points.

Another important characteristic of the endolymph is its charge. In the cochlea, the endolymph carries a charge 80 mv. greater than that in the perilymph of the scala vestibuli and scala tympani, and 80 mv. greater than that in the spiral ligament and extracochlear tissues in general. Since the interior of the hair cells of the organ of Corti have a charge of -70 mv., the resting transmembrane potential of these cells is 150 mv. $(80 - (-70))$. It has been suggested that this high transmembrane potential is an important mechanism in the response of the hair cells to sound waves. Possibly the motion of the tectorial and basilar membranes produced by the sound waves distorts the hair cells in such a manner as to increase their permeability and thus decrease their transmembrane potential. This might result in a partial depolarization of the hair cell (formation of a receptor potential), which generates either an action potential or the release of a chemical mediator that fires a nerve impulse to the brain.

The high *endocochlear potential* disappears after the injection of the metabolic poison cyanide into the scala tympani or scala media or after a few minutes of hypoxia; it is probably dependent, therefore, on a constantly functioning metabolic process. Interestingly enough, the extracochlear endolymph does not show the same high potential (+80 mv.). The endolymph of the utricle, for example, seldom has a potential greater than +5 mv. This is approximately the potential existing between the perilymph at the round and oval windows and the perilymph at the helicotrema.

The endolymph in the semicircular canals, the utricle, and the saccule, like that surrounding the organ of Corti, serves to buoy up various specialized receptors containing hair cells, as well as to bend the hair cells. In the case of the organ of Corti, the bending occurs in response to sound waves, and in the other cases, in response to acceleration and changes in the body in relation to a gravitational force. Each of these structures, then, is a physiological transducer converting one form of energy into a nerve impulse.

AMNIONIC FLUID

Another important fluid system is that associated with the developing fetus in a pregnant woman. If the oocyte (egg) released by the ovary is *fertilized* within 12 to 24 hours it will migrate down the oviduct and, about 4½ days after ovulation, *implant* itself in the uterus. By the beginning of the second month of pregnancy, a well-developed *chorionic cavity* and associated fluid are present (Fig. 13-12); by the end of the

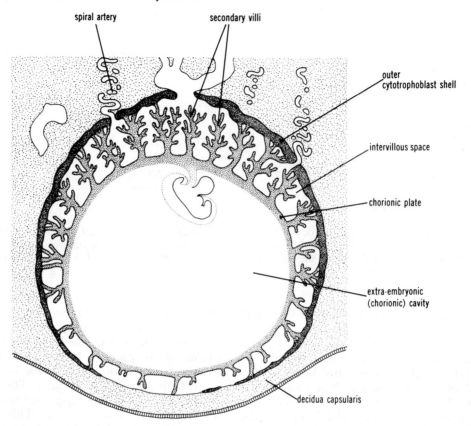

Fig. 13-12. Diagram of the human embryo at the beginning of the second month of development. Note the well formed vascular villi dipping into the maternal lake of blood fed by the spiral artery. Note that at the abembryonic pole these villi are less numerous and more poorly developed. (Langman, J.: Medical Embryology. P. 62. Baltimore, Williams & Wilkins, 1963; *Modified after* von Ortmann. Copyright © 1963, The Williams and Wilkins Company, Baltimore)

third month, a large *amnionic cavity* and associated fluid predominate (Fig. 13-13). As the pregnancy progresses, the volume of fluid around the embryo increases from approximately 30 ml. at 10 weeks to 300 ml. at 20 weeks, to 600 ml. at 30 weeks, and finally to 800 ml. at 40 weeks. After the second month of pregnancy this fluid is primarily amnionic.

Formation and Reabsorption

The source of fetal nutrients, besides the chorionic and amnionic fluids, is the maternal blood in the placenta. Early in pregnancy the maternal vessels apparently erode and produce small pools of blood in the *intervillous space* (Fig. 13-14). These pools are supplied by numerous maternal arteries and lead back into many maternal veins.

It is felt that most of the nutrients pass from the placental pool of blood to the fetus via the *umbilical veins*, while most of the wastes leave the fetus through the *umbilical artery* (rather than by means of diffusion between fetus and amnionic fluid and between amnionic fluid and placenta). Several layers of tissue separate the fetus' umbilical capillaries from the blood in the intervillous space however. This *placental barrier* consists of the endothelium of the fetal capillary, its basement membrane, a connective tissue space containing collagen fibrils, a second basement membrane, and a syncytial layer which forms the maternal microvilli projecting into the intervillous space. This barrier prevents

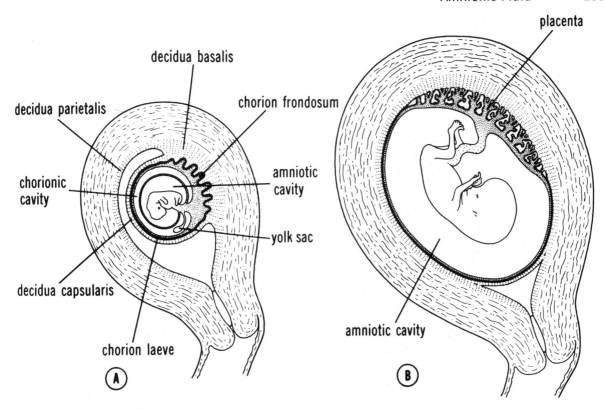

Fig. 13-13. Diagram of the human embryo in the uterus at the end of the second month (A) and at the end of the third month (B) of pregnancy. At the end of the second month, the villi at the abembryonic pole have disappeared and a yolk sac is still found in the chorionic cavity. By the end of the third month the amnion and chorion have fused and the chorion laeve and decidual parietalis have fused to practically eliminate the uterine cavity. (Langman, J.: Medical Embryology. P. 63. Baltimore, Williams & Wilkins, 1963; *modified after* Stark; copyright © 1963, The Williams and Wilkins Company, Baltimore)

the free movement of proteins and cells between mother and fetus.

As pregnancy progresses, the *uterine flow* increases from about 50 ml. per minute at the tenth week to 500 ml. per minute at term (Fig. 13-15). At term the uterus and fetus together use about 24 ml. of oxygen per minute. This results in a uterine *A-V oxygen difference* of about 4.7 ml. of oxygen per 100 ml. of blood flow, which is similar to the maternal A-V oxygen difference between the aorta and vena cava in the third trimester of pregnancy (reported to be between 4.4 and 4.8 ml. of oxygen per 100 ml. of blood).

The rate at which new amnionic fluid is formed and old fluid is reabsorbed at term is approximately 600 ml. per hour. This represents a turnover of approximately 75 per cent of the amnionic fluid each hour. The pressure in the amnion is about 10 mm. Hg, while that in the intervillous space is about 15 mm. Hg. During uterine contractions, however, the intra-amnionic pressure may increase to many times this value.

Function

Probably one of the major functions of the amnionic fluid, like the cerebrospinal fluid, is to *buoy up* (reduce the weight of) the structures it bathes and to act as a *shock absorber*. It may also receive wastes from and supply nutrients to some of the more superficial fetal areas, such as the skin. Amnionic fluid, like cerebrospinal fluid

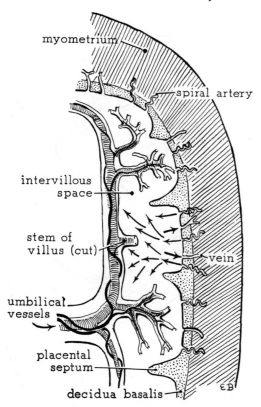

Fig. 13-14. Circulation of fetal (*left*) and maternal (*right*) blood. The maternal blood enters the intervillous space via the spiral arteries and there comes in contact with the placental barrier which surrounds the fetal capillaries. The fetal capillaries give up waste products to the intervillous blood and take on nutrients. The fetal blood then passes back to the fetus via the 2 umbilical veins (*dark*) and the maternal blood returns to the mother via her veins. (Ham, A. W.: Histology. 6th ed., p. 913. Philadelphia, J. B. Lippincott, 1969)

and plasma but unlike endolymph, has as its major cation Na$^+$ (Table 13-4). Its concentrations of K$^+$, Ca^{++}, sugar, and protein are markedly lower than those of plasma. It has not been established whether these concentration differences are due primarily to the placental barrier, to an active transport system, to the utilization of certain substances by the fetus, or to some other factor.

The fact that the pregnant uterus and its contents are an important source of *hormones* presents some interesting problems. Apparently chorionic gonadotropin and "growth hormone-prolactin" are produced by a part of the de-

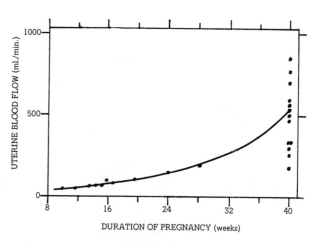

Fig. 13-15. Estimates of uterine blood flow in pregnancy. (*Redrawn from* Hytten, F. E., and Leitch, I.: The Physiology of Human Pregnancy. P. 73. Oxford, Blackwell Scientific Publications, 1964)

veloping placenta (trophoblast), but are secreted only into the maternal blood. Steroid hormones, on the other hand, such as the *estrogens*, are synthesized by the fetal placental unit and are found in higher concentrations in the amnionic fluid (63 μg./100 ml.) and umbilical cord blood

	Amnionic Fluid	CSF	Plasma
Sodium (mEq./liter)	125	152	148
Potassium (mEq./liter)	4	4	8
Calcium (mg./100 ml.)	0+	3	10
Sugar (mg./100 ml.)	30	64	100
Protein (mg./100 ml.)	250	21	7,200
Estriol (μg./100 ml.)	63		15
Unconjugated	(2)		(2)
Sulfate	(9)		(3)
Glucosiduronate	(32)		(4)
Sulfoglucosiduronate	(9)		(6)

Table 13-4. Composition of amnionic fluid, cerebrospinal fluid, and blood plasma. (Data on amnionic fluid from Hytten, F.E., and Leitch, I.: The Physiology of Human Pregnancy. Oxford, Blackwell Scientific Publications, 1964; and Levitz, M.: Conjugation and transfer of fetal-placental steroid hormones. J. Clin. Endocr., 26:773-777, 1966)

(110 μg./100 ml.) than in the maternal plasma (15 μg./100 ml.). Apparently most of the steroids in the amnion and fetus are *stored* in a conjugated form (Table 13-4). The process of *conjugation* is probably the chief means by which the fetus is protected from these potent hormones. Levitz has suggested that these steroids move into and out of the amnionic fluid by way of the intervillous space, the fetus, and the chorion (Fig. 13-16).

Estrogens (estradiol, estrone, etc.) infused into the umbilical vein or amnionic fluid are found conjugated in the amnionic fluid and are eventually found in the maternal urine as well. It is suggested that conjugation of these estrogens by the fetus, amnion, and chorion is an important means of inactivation and storage. It is further suggested that the fetal circulation and chorion are important means for transporting estrogens to the maternal circulation. The chorionic membranes are probably negatively charged lipoprotein. They would, therefore, attract the lipid-soluble, unconjugated steroids and repel the negatively charged, water-soluble conjugates. Thus they are well-adapted to serve the dual function of conjugation and transport to the maternal circulation. The placental and fetal membranes also contain sulfatases which hydrolyze sulfates, thus facilitating the transfer of estrogens to the mother. The highest concentration of unconjugated estriol is in the intervillous blood.

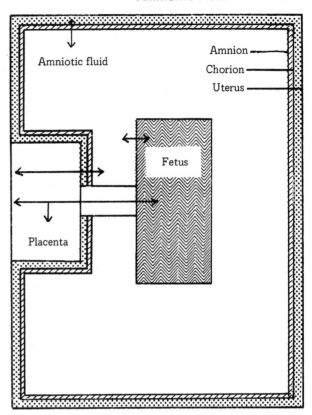

Fig. 13-16. Suggested routes of transfer for steroids. (*Redrawn from* Levitz, M.: Conjugation and transfer of fetal-placental steroid hormones. J. Clin. Endocr. Metab., 26:775, 1966)

Chapter 14

HEMOSTASIS

CLOT FORMATION
FORMATION OF A PLATELET PLUG
RESPONSE OF THE BLOOD VESSELS
TESTS FOR HEMOSTATIC FUNCTION

Blood is a suspension of cells and cell fragments in plasma—usually flowing freely through its arterial and venous channels. In order for it to perform its function as the major transporter of nutrients and waste products it is essential that (1) some of the vessels in each area of the body remain patent (open), and (2) the blood remain fluid and at roughly the viscosity of water. Also essential are mechanisms restricting loss of blood from the vessels (hemostatic mechanisms). Unfortunately, these may also seriously restrict blood flow through the arteries and veins.

When a blood vessel ruptures, the amount of blood lost from the vessel depends in part on (1) the characteristics of the vessel, (2) the characteristics of the blood, (3) the pressure in the vessel, and (4) the characteristics of the perivascular space. If the rupture occurs in a tissue that restricts extravascular fluid movement, the lost blood builds up an interstitial pressure, thereby producing a hematoma which impedes further hemorrhage. If the injury occurs at the surface of the skin or in a lumen or body cavity, however, the hemorrhage may be considerably more serious.

In addition to increased interstitial pressure, there are three mechanisms, essentially, that prevent hemorrhaging. These include (1) vasoconstriction, (2) platelet (a cell fragment in the blood) aggregation, and (3) the formation of a blood clot. In clot formation, liquid plasma (a sol) changes into a semisolid (a gel), called a clot. It, like the platelet aggregate, sometimes forms a plug which adheres to damaged tissue and in this way both prevents further hemorrhaging and provides a tissue into which fibroblasts can migrate during the slow process of repair following hemostasis.

Each of the three mechanisms is important for hemostasis. We find, however, that patients can survive fairly well conditions in which either the clotting mechanism or platelet function is severely impaired. For example, agents such as dicoumarol, which suppress the clotting mechanisms, have been used for many months in patients without serious hemorrhage resulting. Platelet-depleting procedures such as total body radiation and the injection of ^{32}P have also been used experimentally in animals without a high incidence of hemorrhagic problems. In conditions in which both the platelet and clotting functions are impaired, however, the danger of serious hemorrhaging is markedly increased.

CLOT FORMATION

Normally the blood contained in the arteries, capillaries, veins, and heart is fluid. It may, however, during prolonged stasis or in response to platelet disruption, cell damage, or tissue damage, change into a semisolid similar in consistency to gelatin (a clot). Plasma generally contains a number of factors important in clot formation, as well as fibrinogen—the precursor of the fibrin threads which form the skeleton of clots (Fig. 14-1). Apparently what usually activates this plasma system is the release of platelet factor 3 from the platelets or of thromboplastin from a cell or tissue, or both. If, for example, we carefully withdraw some blood into a syringe, it will, after a period of 10 to 15 min., change into a semisolid mass (a blood clot). This phenomenon can be accelerated by (1) agitating the blood sample, (2) increasing its exposed surface area, or (3) slightly increasing its temperature. It can be decelerated by (1) coating the syringe with paraffin or silicone, (2) cooling the blood, or (3) both.

Activation of Prothrombin

One of the first steps in clot formation is producing a certain quantity of *thrombin* from *prothrombin*. It is quite likely that minute amounts of thrombin are constantly being pro-

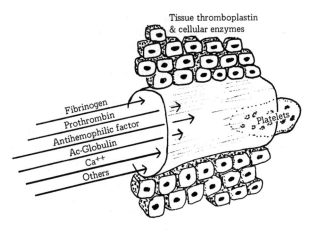

Fig. 14-1. Factors important in the formation of a blood clot. Either thromboplastin or platelet factor 3 or a combination of these can induce the intrinsic factors in the plasma (fibrinogen, prothrombin, antihemophilic factor, plasma accelerator globulin, and Ca++) to form insoluble fibrin threads. (Courtesy of W. H. Seegers)

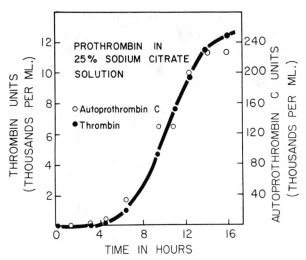

Fig. 14-2. The *in vitro* activation of prothrombin to autoprothrombin C and thrombin in the presence of 25% sodium citrate. In this experiment Seegers et al. showed that the prothrombin complex contained both the thrombin and autoprothrombin C moieties. The citrate served merely as a mild cleavage-producing agent. They extended this concept by showing that intermediates of prothrombin activation may result in the formation of a number of other procoagulants as well as anticoagulants. (Marciniak, E., and Seegers, W. H.: Autoprothrombin C: a second enzyme from prothrombin. Canad., J. Biochem. Physiol., 40:602, 1962)

duced from prothrombin, and that this thrombin is destroyed by antithrombin. During cell, tissue, or platelet disruption, however, the procoagulants (accelerators of clot formation) released speed up the formation of thrombin to such an extent that its concentration in the blood rises. This rise is in part responsible for the conversion of the fibrinogen in plasma to insoluble fibrin threads. Once these threads have been formed, platelets and red blood cells adhere to them and water and electrolytes become trapped in them. This complex of materials makes up the clot.

In this chapter we shall use the term prothrombin to refer to a complex or aggregate with a molecular weight of 68,900 or some multiple thereof (Fig. 14-3). Prothrombin is formed in the liver and normally exists in the plasma at a concentration of approximately 20 mg. per 100 ml. When placed in a test tube containing 25 per cent sodium citrate (Fig. 14-2), it forms *autoprothrombin C* (factor Xa, molecular weight 22,000 (?)) and thrombin (molecular weight 33,000). Apparently the citrate serves to break down some of the weak bonds (probably hydrogen bonds) which hold the thrombin and autoprothrombin C precursors in the prothrombin aggregate. Seegers et al. have suggested that essentially the same thing occurs in the body during clot formation. That is, there is a degradation of prothrombin to autoprothrombin III (factor X), autoprothrombin C (factor Xa), prethrombin, and thrombin, as well as to a number of other procoagulants. Their concept is diagrammed in Figure 14-3.

Plasma-platelet activation. In response to (1) exposure to the collagen of open wounds, (2) endotoxin, (3) thrombin, or (4) disruption, the platelets release a procoagulant called platelet factor 3. It has been suggested that this factor, acting with the procoagulants of plasma, facilitates the formation of thrombin from the prothrombin complex (autoprothrombin III + prethrombin and others) in the following manner:

$$\begin{array}{c} \text{Autoprothrombin III (X)} \\ \Big| \begin{array}{l} \text{Platelet factor 3} \\ \text{Ca}^{++} \text{ (IV)} \\ \text{Antihemophilic factor (VIII)} \end{array} \\ \downarrow \\ \text{Autoprothrombin C (Xa)} \\ \text{(small amount)} \end{array} \quad (1)$$

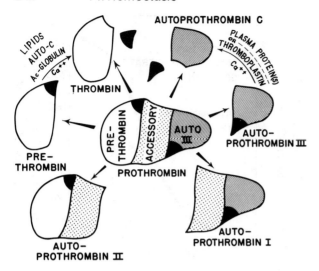

Fig. 14-3. Components of the prothrombin complex. (Courtesy of W. H. Seegers)

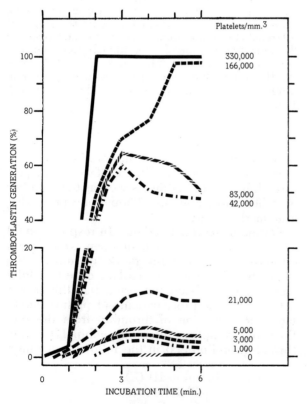

Fig. 14-4. The effect of platelet concentration in plasma on the generation of "plasma thromboplastin" (probably the procoagulant generated is mainly autoprothrombin C). (*Modified from* Miale, J. B.: Laboratory Medicine—Hematology. 2nd ed., p. 715. St. Louis, C. V. Mosby, 1962)

$$\text{Prethrombin} \xrightarrow[\substack{\text{Autoprothrombin C (Xa)} \\ \text{Platelet factor 3} \\ \text{Ca}^{++} \text{ (IV)} \\ \text{Ac-globulin (V)}}]{} \text{Thrombin} \quad (2)$$

The thrombin formed now acts on prothrombin to catalyze the formation of larger quantities of prethrombin than were originally present:

$$\text{Prothrombin (II)} \xrightarrow{\text{Thrombin}} \text{Prethrombin} \quad (3)$$

The thrombin produced in these reactions not only catalyzes the conversion of fibrinogen to fibrin, but also activates inhibitor source material in the plasma to neutralize antihemophilic factor:

$$\text{Antihemophilic Factor (VIII)} \xrightarrow{\text{Inhibitor}} \text{Inactive Antihemophilic Factor}$$

$$\text{Inhibitor source material} \xrightarrow{\text{Thrombin}} \quad (4)$$

As the thrombin concentration builds up, in other words, one factor that facilitates its production is removed and thrombin production decreases. By such a negative feedback we prevent the complete depletion of prothrombin.

Plasma-tissue-platelet activation. When there is disruption of the perivascular tissue and an associated release of tissue juice or thromboplastin into the blood, thrombin formation is accelerated even more than when there is merely a disruption of platelets. With the release of tissue extract into the blood, greater quantities of platelet factor 3 (see reactions 1 and 2) are released into the plasma. In addition, the thromboplastin of the tissues and the intrinsic factors of the plasma act to increase the production of thrombin and autoprothrombin C:

$$\text{Autoprothrombin III (X)} \xrightarrow[\text{Ca}^{++} \text{ (IV)}]{\text{Tissue extract (III)}} \text{Autoprothrombin C (Xa)} \quad (5)$$

$$\text{Prethrombin} \xrightarrow[\text{Ac-globulin (V)}]{\text{Autoprothrombin C (Xa)}} \text{Thrombin} \quad (6)$$

Tissue extract thromboplastin, in other words, catalyzes the production of a procoagulant (autoprothrombin C) that is produced only in small quantities by the catalytic action of platelet factors (reaction 1), and therefore increases the total quantity of thrombin produced.

Summary of current theories on the formation of thrombin. Prior to this decade our understanding of the processes that lead to the formation of thrombin were clouded by nonuniformity of terminology and perspective. In recent years, however, much of the confusion has disappeared and many of the debates have been more academic than clinical.

One sign of this increased understanding is the data presented in Table 14-1. The roman numerals used here are those recommended by a majority vote of the *International Committee on Nomenclature of Blood Clotting Factors*. One of the academic problems on which there is still no general agreement is whether the precursor of thrombin and the precursor of autoprothrombin C (X) are released by the liver as a loosely bound prothrombin complex or as separate entities. There is also evidence that factors VII (autoprothrombin I in Fig. 14-3) and IX (autoprothrombin II in Fig. 14-3) have the properties of autoprothrombin C. It is generally agreed, however, that a single agent, *coumarin*—by acting on the liver—reduces the concentration of prothrombin and factors VII, IX, and X in the blood. It is also accepted by many that factors I through V and VII through XIII are important in hemostasis. "Deficiencies" in each of these factors have been reported in patients, and have been shown to be associated with bleeding disorders. The term deficiency has been applied in such pathologic conditions. However, it should be kept in mind that a molecular aberration, as well as a decrease in concentration, may cause the "deficiency." In the case of hypocalcemia, progressive decreases in Ca^{++} concentration sufficient to produce a bleeding disorder first produce tetany and cardiac standstill.

The clinically useful facts concerning the formation of thrombin from a prothrombin complex (or a prothrombin molecule, depending on one's bias) are that this process can be markedly accelerated by (1) the release of lipoprotein from the platelets (platelet factor 3), (2) the release of a lipid-protein complex (thromboplastin) from cells and the perivascular tissues, or (3) a combination of the two. Both the platelet factor and the tissue factor function with the plasma factors to produce thrombin. The factors which activate the plasma factors are separated from each other by a barrier (plasma membrane of the platelets, erythrocytes, or endothelial cells). Upon disruption of the barrier a series of events begins which leads to thrombin production. The rate at which thrombin is produced varies with the character of the procoagulants and anticoagulants present or produced. For example, persons with low concentrations of the procoagulant antihemophilic factor have a tendency to bleed excessively in response to certain stresses, even though they have normal concentrations of the other procoagulants.

Another area of contention is the sequence of the various activation processes. I have presented in this chapter an approach which is consistent with the evidence on hand. Others, however, have proposed a more rigid, stepwise progression of events, in which each of the clotting factors listed in Table 14-1, except fibrinogen, Ca^{++}, tissue thromboplastin, and factor VII, is an enzyme which is converted to its active form by another enzyme. This perspective is frequently called the "waterfall" or "cascade" concept and is summarized in Figure 14-5.

Conversion of Fibrinogen

The next step in the formation of a clot is the conversion of the fibrinogen normally found in plasma to insoluble fibrin threads (Fig. 14-6). Fibrinogen is a molecular dimer with a molecu-

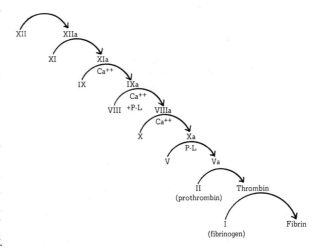

Fig. 14-5. The "waterfall" or "cascade" sequence for reactions leading to the formation of fibrin. This is a sequence of reactions once proposed to explain the relationship between procoagulants. It is currently believed incorrect by a number of investigators. P-L, phospholipid.

Factor	Probable Synonyms	Function
I	Fibrinogen	Forms fibrin-s and -i
II	Prothrombin	Forms thrombin and enzymes that facilitate thrombin formation
III	Tissue thromboplastin	Catalyzes the formation of autoprothrombin C (Xa)
IV	Ca^{++}	Cofactor in the formation of thrombin and fibrin-i
V	Ac-globulin Plasma accelerator globulin Proaccelerin Prothrombin accelerator Labile factor	Facilitates thrombin formation
VI	Obsolete	
VII	Serum prothrombin conversion accelerator (SPCA) Proconvertin Stable factor Stable prothrombin conversion factor Autoprothrombin I Co-thromboplastin	Facilitates the formation of autoprothrombin C (Xa)
VIII	Antihemophilic factor (AHF) Antihemophilic globulin (AHG) Platelet cofactor I Thromboplastinogen Antihemophilic factor A	Facilitates the formation of autoprothrombin C (Xa)
IX	Plasma thromboplastin component (PTC) Christmas factor Antihemophilic factor B Antihemophilic globulin B Platelet cofactor II Autoprothrombin II	Facilitates the formation of autoprothrombin C (Xa)
X	Stuart-Prower factor Stuart factor Prower factor Prephase accelerator (PPA)* Plasma thromboplastin* Coagulation product I* Autoprothrombin III Autoprothrombin C*	Catalyzes the formation of thrombin from prethrombin
XI	Plasma thromboplastin antecedent (PTA) Antihemophilic factor C	Possibly facilitates the formation of factor IX and/or factor XII
XII	Hageman factor Surface factor Contact factor Clot-promoting factor	Possibly facilitates the formation of factor XI
XIII	Fibrin-stabilizing factor (FSF) Plasma transglutaminase Laki-Lorand factor Fibrin-stabilizing enzyme Fibrinase	Catalyzes the formation of fibrin-i

* All of these are probably activated factor X (Xa).

Table 14-1. A list of some of the factors thought to be important in clot formation. Not included in this list, but also important in facilitating thrombin formation, is platelet factor 3. Each list of synonyms begins with the terms most commonly used. The comments on the function of each factor are for the most part superficial, dividing clot formation into five categories, namely, those that:

1. Facilitate or directly catalyze the formation of autoprothrombin C
2. Facilitate or directly catalyze the formation of thrombin
3. Form thrombin
4. Catalyze the formation of fibrin-s or fibrin-i
5. Form fibrin-s and fibrin-i

Some prefer to further specify, in the case of category 2, whether the factor acts by (a) facilitating the release of platelet factors or (b) facilitating the release of tissue thromboplastin.

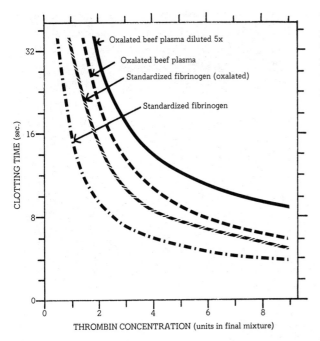

Fig. 14-6. The effect of thrombin concentration on the clotting time (time to form visible fibrin threads). Preparation 1 differs from the others in that it contains Ca^{++}. Ca^{++} has been removed from the other preparations by the addition of potassium oxalate ($K_2C_2O_4$). (*Redrawn from* Seegers, W. H., and Smith, H. P.: Factors which influence the activity of purified thrombin. Am. J. Physiol., *137*:349, 1942)

lar weight of 330,000. In the presence of thrombin, fibrinogen loses two pairs of peptides and some small carbohydrate fragments:

$$\text{Fibrinogen} \xrightarrow{\text{Thrombin}} \text{Fibrin monomer} + \text{Peptides} + \text{Carbohydrate} \quad (7)$$

The loss of one type of peptide (peptide A, molecular weight 1,900) from fibrinogen results in the formation of a monomer (fibrin) with a molecular weight slightly less than that of fibrinogen. Several such fibrin monomers then undergo end-to-end polymerization:

$$\text{Fibrin monomer} \rightarrow \text{Fibrin polymer} \quad (8)$$

The loss of another peptide (peptide B, molecular weight 2,500), on the other hand, produces fibrin monomers that undergo side-to-side polymerization. With these two systems of polymerization, then, the plasma can construct extensive networks of fibrin threads. These threads are highly organized, and appear under an electron microscope as light and dark bands of a characteristic periodicity.

Fibrin formation as outlined above will occur in a test tube containing only an aqueous solution of fibrinogen and thrombin. The fibrin formed in such a mixture, however, is soluble in a 5 molar urea solution (fibrin-s). The fibrin formed when one adds thrombin to plasma, on the other hand, is not soluble in a 5 molar urea solution (fibrin-i). Apparently in plasma, thrombin activates a substance with a molecular weight of 330,000 (inactive plasma transglutaminase, or factor XIII) which helps stabilize the fibrin network:

$$\text{Inactive Plasma Transglutaminase (XIII)} \downarrow \text{Thrombin}$$
$$\text{Active Plasma Transglutaminase (XIIIa)} \quad (9)$$

$$\text{Fibrin-s} \xrightarrow{\text{Active plasma transglutaminase}} \text{Fibrin-i} + NH_3 \quad (10)$$

Fibrinolysis

Once the clot has been formed and has served to prevent hemorrhage, various processes begin to bring about its dissolution (Fig. 14-7). One of these is the conversion of profibrinolysin (plasminogen) to a proteolytic enzyme with a great affinity for fibrin, fibrinogen, and other proteins and peptides (*fibrinolysin*, or plasmin). High concentrations of fibrinolysin can result in clot dissolution in a matter of seconds and an associated release of anticoagulants. The activation of profibrinolysin occurs in response to tissue enzymes (cytofibrinokinase), certain bacterial enzymes (streptokinase and staphylokinase), and certain substances found in urine (urokinase). Both the cytofibrinokinase and urokinase occur naturally in man. The others are used for the treatment of certain clotting problems. Apparently there is also an antifibrinolysin in the blood which tends to counter the action of fibrinolysin.

Leukocytes are strongly attracted to fibrin and seem to migrate into clots. The eosinophil serves as a source of profibrinolysin and the neutrophil phagocytizes fibrin (Fig. 15-10, p. 268). Reticuloendothelial cells phagocytose both fibrin and fibrin-split products, as well as thrombin and autoprothrombin C.

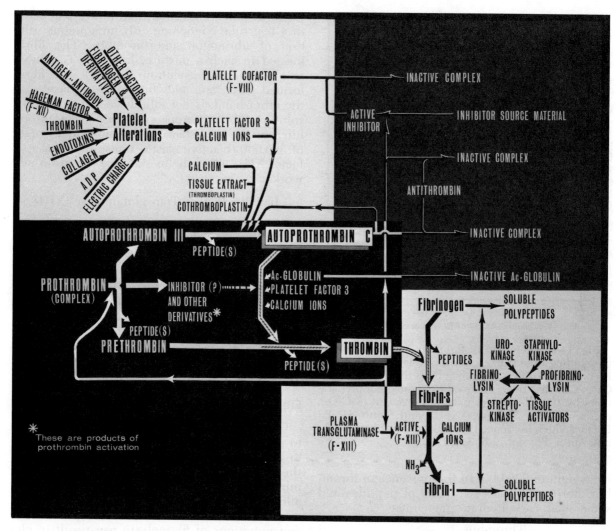

Fig. 14-7. A summary of some of the mechanisms which (1) initiate the breakdown of prothrombin (*upper left*), (2) are associated with the formation of thrombin (*lower left*), (3) are associated with the formation and dissolution of fibrin-i (*lower right*), and (4) are associated with the neutralization of procoagulants (*upper right*). For simplicity the role of leukocytes and macrophages in clot dissolution and the role of heparin in the inactivation of procoagulants have been omitted. (Courtesy of W. H. Seegers and G. Murano)

Anticoagulants

For every bleeding disorder a physician treats, there are probably a hundred cases of serious thrombosis. There is, therefore, a great deal of medical interest in anticoagulant therapy. The anticoagulants can be divided into three general categories: (1) those that remove Ca^{++} from solution, (2) those that interfere with liver function, and (3) those that depress the action of organic procoagulants.

Removal of Ca^{++}. In this category we have (1) sodium citrate and ethylenediaminetetraacetate (EDTA), which chelate Ca^{++}, and (2) sodium and postassium oxalate, which form microcrystalline precipitates with Ca^{++}. These anticoagulants are used for the storage of blood outside the body. They are not used for the prolongation of clotting time in the body, because, as I have mentioned, concentrations sufficient to affect the clotting system also produce tetany and ventricular standstill.

Depression of prothrombin production. The anticoagulant most commonly used for extended periods in vivo was originally derived from spoiled sweet clover. It was noted that cattle and swine that fed on improperly cured sweet clover frequently died from a hemorrhagic condition. In 1940 Link et al. isolated and identified the toxic agent in sweet clover as bishydroxycoumarin (dicoumarol). The following year it was used by Butt, Allen, and Bollman and Bingham, Meyer, and Dohle for the treatment of patients in whom excessive thrombosis was a problem.

Numerous coumarin and indandione derivatives are now used for prolonging the clotting time. They all, apparently, act on the liver to competetively inhibit the utilization of vitamin K (named the "Koagulations Vitamin" by Dam in 1935) in the production of prothrombin (Fig. 14-8). Or in other words, they serve to decrease the plasma concentration of the prothrombin complex. On the first day of Dicumarol therapy, for example, a patient might receive an oral dose of 300 mg., on the second day, 200 mg., and on subsequent days, a dose sufficient to maintain one fifth the normal concentration of prothrombin in the plasma. (This will vary from one patient to the next.) As a rule, the one-stage prothrombin time is used to estimate the concentration of prothrombin. In this test a standard solution of Ca^{++} and thromboplastin is added to citrated or oxalated plasma from the patient, and the time for clot formation is determined (prothrombin time). Slow clot formation is usually associated with a decrease in prothrombin concentration.

Heparin. Heparin is a heat-stable, water-soluble, sulfated mucopolysaccharide that is produced in the body and combines with proteins to alter their physical and biological properties. It is an excellent anticoagulant but has not been demonstrated in the blood of normal persons. Clinically, it is frequently given intravenously every 4 hours or intramuscularly (200 mg.) every 12 hours. Many insist, however, that to be continuously effective it must be given by intravenous drip. Dicumarol, on the other hand, can be taken orally and acts longer, but does not have maximum effect for 72 hours. Heparin in the plasma works in at least two ways. It (1) inhibits the conversion of fibrinogen to fibrin (i.e., is an antithrombin), and (2) in large concentrations inhibits the production of thrombin. Excess heparin can be countered by a stoichiometric titration of protamine sulfate. The action of 1 mg. of heparin, that is, can be countered by 1 mg. of protamine sulfate.

FORMATION OF A PLATELET PLUG

Normally the platelets flow freely, suspended in the blood. If there is damage to the intima of a blood vessel, however, such that collagen comes in contact with the blood, a few platelets (1) adhere to the denuded collagen, (2) degranulate, and (3) release a number of substances into their external environment. One of these substances is *adenosine diphosphate* (ADP), which facilitates the formation of a platelet aggregate. This aggregate can serve as a plug to help prevent hemorrhage, or may, when it occurs in small vessels, completely occlude the lumen of the vessel.

Stabilization of the Plug

During degranulation of the platelets that form the platelet plug there is also a release of platelet factor 3. In a rapidly moving blood stream it may be washed away and diluted in the blood, so that it does not initiate the formation of either a thrombus (stationary clot) or an embolus (a circu-

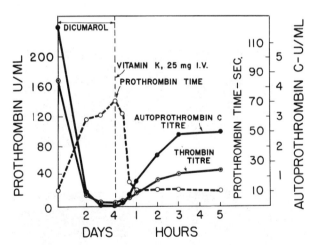

Fig. 14-8. Response of the dog to Dicumarol and vitamin K. Note that the quantity of autoprothrombin C (in units/ml.) and thrombin (in units/ml.) that the dog could produce when his plasma was activated was decreased during Dicumarol therapy. This, as well as the increased prothrombin time, is apparently the result of the depressed production of prothrombin initiated by the Dicumarol. Intravenous vitamin K quickly neutralized the effect of Dicumarol. Thus the effect of Dicumarol on the liver is a reversible one. (Reno, R. S., and Seegers, W. H.: Two-stage procedure for the quantitative determination of autoprothrombin III concentration and some applications. Thromb. Diath. Haemorrh., *18*:205, 1967)

lating clot or other foreign solid). In a vessel in which there is either partial or complete stasis, however, a substantial concentration of platelet factor 3, and therefore of thrombin and fibrin as well, may result. The thrombin not only catalyzes the formation of fibrin but may also consolidate the platelet plug. Prior to the appearance of thrombin the movement of platelets into and out of the plug is common; after its appearance, platelet cohesion is practically irreversible.

Clot Retraction

In the material above I have talked about a series of events that may lead to hemostasis:

1. Platelet adhesion to collagen
2. Platelet degranulation associated with the release of ADP, platelet factor 3, and platelet lipids
3. Platelet cohesion in response to ADP
4. Platelet consolidation in response to thrombin
5. Production of a platelet-fibrin plug

The final step in this sequence is a process called clot retraction. In this process the platelet-fibrin clot diminishes over a period of several hours and extrudes a clear fluid called *serum* (Fig. 14-9). This serum is similar to plasma except that it (1) contains no fibrinogen, (2) contains substances released by the platelets during degranulation, and (3) contains markedly different concentrations of the procoagulants. It contains serotonin, ADP, ATP, K^+, platelet factor 3, autoprothrombin II (IX), autoprothrombin III (X), fibrinogen-split products, autoprothrombin I, and a small amount of antithrombin. It has, in addition, neutralized autoprothrombin C (Xa), some of the autoprothrombin I, thrombin, antihemophilic factor (VIII), Ac-globulin, and tissue thromboplastin. Serum is also produced by stirring a wooden paddle in citrated plasma to which Ca^{++} has been added. The fibrin formed adheres to the paddle, and what remains when the paddle is removed is called serum. This serum, of course, contains lower concentrations of the platelet factors.

The degree of clot retraction that occurs decreases proportionately as the concentration of platelets is decreased below 100,000 per cubic millimeters of blood. It is, on the other hand, inversely related to fibrinogen concentration, and can be modified by changes in temperature, cell volume, pH, thrombin concentration, Ca^{++} concentration, and the means and duration of storage. Clot retraction will not occur in the absence of platelets or in blood stored at 4°C. for more than five days. Apparently as the platelet-

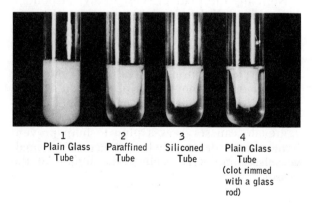

Fig. 14-9. Clot retraction in test tubes with different surfaces. Each tube contained a plasma, platelet, Ca^{++}, and thrombin mixture and was maintained at 37°C. for 2 hours. Note the retracted clot in tubes 2, 3, and 4 surrounded by the clear serum. (Ballerini, G., and Seegers, W. H.: A description of clot retraction as a visual experience. Thromb. Diath. Haemorrh., 3:147-164, 1959)

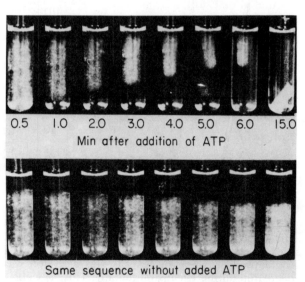

Fig. 14-10. Response of a protein from human blood platelets to ATP. In the upper series pictures were taken of the protein 0.5, 1, 2, 3, 4, 5, 6, and 15 min. after the addition of 10^{-3} M. ATP. In the lower series the study was repeated without the addition of ATP. (Lüscher, E. F.: Retraction activity of the platelets. *From* Bettex-Galland, M.: Unpublished data, 1960. *In* Johnson, S. A., et al. (eds.): Blood Platelets. P. 449. Boston, Little, Brown & Co., 1961)

fibrin clot accumulates thrombin, a concentration is reached which causes the contraction of a platelet factor, *thrombosthenin*. This protein, like actomyosin, needs adenosine triphosphate (ATP) for its shortening (Fig. 14-10). The importance of this retraction, however, is not well understood. The force exerted during retraction is small and is not sufficient to pull the cut ends of a wound together. It may, on the other hand, serve to open up a vessel previously occluded by the clot. In some cases, that is, the clot may occlude the vessel lumen, thus allowing the accumulation of thrombin and other procoagulants in the area. Once the thrombin concentration has built up, the clot retracts. This relieves any ischemia that may have developed and allows leukocytes and cells capable of forming fibroblasts into the area for the next process—dissolution of the clot and repair of the tissue.

RESPONSE OF THE BLOOD VESSELS

Unfortunately little is known concerning the role of blood vessels in hemostasis. We know that a number of hemorrhagic conditions exist in which both the clotting and platelet systems are normal. For example, vitamin C is required for the synthesis of hyaluronic acid found both in the intercellular cement of the endothelial lining and the pericapillary ground substance. Absence of vitamin C in the diet therefore causes scurvy, a condition in which there is bleeding into the skin and from the mucous membranes.

The smooth muscle of the blood vessel also is important in the prevention of hemorrhage. Following the transverse section of an arteriole, for example, a sufficiently intense constriction of the vessel occurs so that bleeding is either stopped or markedly decreased. In addition, the platelets that tend to accumulate at an area of vascular damage will during degranulation release an agent, *serotonin*, which also produces a local vasoconstriction. This will further decrease the bleeding and, in so doing, facilitate the local accumulation of procoagulants and the growth of a platelet-fibrin plug.

TESTS FOR HEMOSTATIC FUNCTION

It is beyond the scope of this introductory text to cover completely the various tests for hemostatic function, but it is important at this stage that the student have some appreciation of the distinctions between a bleeding disorder due to a defect in the clotting system, for example, and one due to a defective platelet or vascular system.

Coagulation Time

In determining coagulation time, a clotting system is activated, a stopwatch is started, and the time at which a clot is formed is noted. Such studies are performed in a test tube under standard conditions. The results one obtains depend on the type of test tube used (in terms of surface area exposed to the blood, for example), the temperature of the solution, the amount of agitation during the test, and numerous other factors. The variables are great and it is difficult to give a set of normal values for every hospital, but with standardized techniques, reproduction can be obtained.

Whole-blood clotting time. In the Lee-White clotting time, approximately 5 ml. of blood are drawn into a syringe from a vein, and 1 ml. is placed in unsiliconized tubes marked 1, 2, and 3. The stopwatch is started the moment blood is withdrawn into the syringe. The tubes are placed in a water bath at 37°C. and tube 1 is tilted every 30 sec. until a clot is noted. The technician then switches to tube 2, tilting it every 30 sec. until a clot is noted, and finally to tube 3. The time required for a clot to form in tube 3, to the nearest 30 sec., is the whole-blood clotting time. The value usually obtained using this technique is between 10 and 15 min. A value of less than 7 min. generally indicates poor technique.

Clotting time for recalcified plasma. Another technique is to (1) mix the blood drawn from a vein with citrate or oxalate in order to remove the Ca^{++}, (2) centrifuge it, and (3) extract the supernatant plasma. Equal portions of plasma are then added to three test tubes, a standard $CaCl_2$ solution is added to each, and the test is conducted as above. A clot normally forms in 80 to 120 sec.

One-stage prothrombin time of Quick. In this test standard quantities of $CaCl_2$ and thromboplastin are added to citrated or oxalated plasma and the clotting time noted. The solution added to the plasma is such as to provide an optimal concentration of Ca^{++} and an excess of thromboplastin. By performing the experiment in this way, early workers hoped to obtain a clot formation time directly proportional to the prothrombin concentration. This does not happen, however, since the speed of the reactions also depends on the presence of other procoagulants

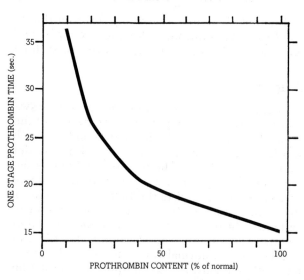

Fig. 14-11. Estimation of the prothrombin content of plasma in terms of per cent of normal (abscissa) from a one-stage prothrombin time (ordinate).

and on the concentration of fibrinogen. This latter problem can be eliminated by a *two-stage prothrombin assay*. Here the activation is done in two stages. In the first, thromboplastin and Ca^{++} are added to defibrinated, citrated plasma. In the second stage, a standard quantity of fibrinogen is added and the time for the formation of a clot is determined. The one-stage method normally yields a clotting time of 12 to 15 sec. (Fig. 14-11).

Bleeding Time

The bleeding time is primarily a test of the hemostatic capacity of the small blood vessels. In one test, the Duke method, the earlobe is punctured with a sharp No. 11 Bard-Parker blade. The stopwatch is started when the first drop of blood is noted, and the wound is blotted with filter paper every 30 sec. The point at which the filter paper no longer stains red is taken as the bleeding time. Ivy developed a modification of this method in which he placed a sphygmomanometer cuff around the arm and inflated it to 40 mm. Hg. This resulted in a pressure of 40 mm. Hg in the capillaries and probably placed the hemostatic mechanisms under somewhat of a stress. A wound was then made below the cuff and the procedure used in the Duke method followed. Using the Duke method one normally obtains a bleeding time of 1 to 3 min.; using the Ivy method, it is generally 2 to 4 min.

Tests for Capillary Fragility

There are two general categories of tests for capillary fragility—the positive pressure test and the negative pressure test. In the latter, a suction cup is used on the skin and the number of petechiae (minute hemorrhagic spots) that result are noted. In the former, a blood pressure cuff is put on the arm and the pressure in it placed halfway between systolic and diastolic arterial pressure. The cuff remains at this pressure for 5 min. and is them removed. Normally after the cuff is removed, few if any petechiae are noted. In cases of increased capillary fragility, however, petechiae develop on the antecubital area, the volar aspect of the wrist, and the dorsum of the hand. The test is graded from 1 to 4+, depending on the number of petechiae.

Summary

Bleeding disorders can be divided into three broad categories: coagulopathies (coagulation dysfunction), thrombocytopathies (platelet dysfunction), and angiopathies (blood vessel dysfunction). The inherited coagulopathies listed in Table 14-2, with the exceptions of fibrin stabilizing factor (factor XIII) "deficiency" and factor VII "deficiency," are all associated with prolongation of the Lee-White whole-blood clotting time. In each of these, however, there is usually a normal bleeding time as well. In afibrinogenemia and in prothrombin and factor V, VII (autoprothrombin I), and X (autoprothrombin III) "deficiencies," the prothrombin time is prolonged. In factor VIII (AHG), IX (autoprothrombin II), XI, and XII "deficiencies," however, prothrombin time is normal; apparently these factors are not important in prothrombin activation when, as in the prothrombin time procedure, excess thromboplastin is used as the activator. Afibrinogenemia can be distinguished from the other coagulopathies by determining the amount of fibrinogen.

Prolonged bleeding times with normal whole-blood clotting and prothrombin times are usually associated with the thrombocytopathies. A platelet count enables the exclusion of thrombocytopenia. In von Willebrand's syndrome (an inherited thrombocytopathy), clot retraction is normal, but in hereditary hemorrhagic thrombasthenia, clot retraction is absent or grossly prolonged. Prolonged bleeding times may also be found in certain types of angiopathies, such as scurvy.

Coagulopathies
  Inherited
    Hypofibrinogenemia (factor I)
    Dysfibrinogenemia (factor I)
    Hypoprothrombinemia (factor II)
    Parahemophilia (factor V)
    Factor VII "deficiency" (factor VII)
    Hemophilia A (factor VIII)
    Hemophilia B (Christmas disease) (factor IX)
    Stewart-Prower factor "deficiency" (factor X)
    Plasma thromboplastin antecedent "deficiency" (factor XI)
    Hageman trait (factor XII)
    Fibrin stabilizing factor "deficiency" (factor XIII)
    Combined "deficiencies"
  Acquired (affecting clotting and fibrinolytic system)
    Abruptio placentae
    Dead fetus syndrome
    Amniotic fluid embolism
  Acquired (affecting predominantly the clotting system)
    Generalized Shwartzman equivalents
    Thrombotic thrombocytopenic purpura
    Purpura fulminans
    Hemolytic-uremic syndrome (Gasser syndrome)
    Shock
    Giant hemangioma (Kasabach-Merrit syndrome)
  Acquired (impaired procoagulant production)
    Liver disorders
    Vitamin K malabsorption
    Coumarin and indanedione therapy
  Acquired (circulating anticoagulants)
    Immuno-antibodies against coagulation components
    Heparin therapy
    Thrombolytic therapy
  Acquired (dysproteinemias)
    Macroglobulinemia Waldenstroem
    Cryoglobulinemia
    Multiple myeloma
    Hypergammaglobulinemia
Thrombocytopathies
  Inherited
    Hereditary congenital thrombocytopenia
    Hereditary constitutional thrombocytopathy (von Willebrand's syndrome)
    Hereditary hemorrhagic thrombasthenia
    Connective tissue diseases
    Ehlers-Danlos syndrome
    Osteogenesis imperfecta
    Pseudoxanthoma elasticum
  Acquired
    Thrombocytopenias
    Thrombocytopathy
    Thrombocythemia
Angiopathies
  Inherited
    Hereditary hemorrhagic telangiectasia (Osler)
    Hereditary familial purpura simplex (Davis)
  Acquired
    Scurvy
    Purpura senilis
    Anaphylactoid purpura (Schoenlein-Henoch)

Table 14-2. A list of bleeding disorders.

Chapter 15

SEDIMENTATION RATE
HEMATOCRIT
PLASMA
TRANSFUSION
RESISTANCE TO INFECTION
FORMED ELEMENTS

BLOOD

Blood forms the transport system responsible for carrying gases, acids and bases, nutrients and wastes, hormones, water and electrolytes, heat, and cells that help defend us against infection. If carefully drawn into a syringe containing an anticoagulant (sodium citrate) and then put in a standard sedimentation tube (2.5-mm. bore, 100 mm. in length), it will slowly separate into a transparent, amber *upper layer* (plasma) and a *dense, red lower layer* (red cells). Between these two layers is a thin, white *buffy coat* (white cells and platelets). If the sedimentation tube is left vertical for 1 hr., blood from a normal, healthy male will form an upper layer of amber fluid 0 to 6 mm. thick (94 to 100 mm. of red fluid), while blood from a healthy female will form a layer of amber supernatant 0 to 9 mm. thick. For a woman, that is, a *sedimentation rate* of 0 to 9 mm. per hour is normal.

SEDIMENTATION RATE

The sedimentation rate for a red blood cell depends on its radius (r), its thickness (c), its density (d_1), the density of the fluid in which it is suspended (d_2), the viscosity of the fluid (η), and the gravitational force (g):

$$\text{Sedimentation rate} = \frac{2rc(d_1 - d_2)g}{\eta k} = \frac{2rc(d_1 - d_2)g}{7.65}$$

The specific gravity of the plasma is generally about 1.03, and that of the red blood cell, 1.10. When the concentration of plasma proteins is elevated, however, the red blood cells—once they are removed from the body—either clump or arrange themselves like a stack of coins (in a *rouleau*). In either case the blood now contains a number of masses larger than a single red cell which sediment more rapidly. An increase in the concentration of plasma proteins, in other words, increases the sedimentation of blood outside the body.

The protein most effective in increasing the sedimentation rate is fibrinogen. Others, in order of importance, are alpha-2 globulin, gamma globulin, beta globulin, and alpha-1 globulin. Albumin, on the other hand, has very little effect. Some of the clinical conditions in which the sedimentation rate is elevated include (1) acute and chronic infections, (2) acute localized infections, (3) myocardial infection, (4) malignant tumors associated with necrosis, (5) hyper- and hypothyroidism, (6) lead and arsenic intoxication, (7) nephrosis, (8) internal hemorrhage, (9) pregnancy after the third month, (10) menstruation, and (11) tuberculosis. In conditions in which the shape of the red blood cell makes rouleau formation impossible, the sedimentation rate is generally low. This occurs in sickle cell anemia and in hemolytic anemias in which the red blood cell becomes spheroid. High sedimentation rates are also usually lacking in (1) thyphoid and undulant fever, (2) acute allergies, (3) uncomplicated virus diseases, and (4) infectious mononucleosis.

HEMATOCRIT

The separation of blood into its three components can be accelerated and exaggerated by centrifugation (Fig. 15-1). When blood is centrifuged the percentage of the total volume occupied by the packed red cells is called the hematocrit. Healthy men usually have an average hematocrit of 45 per cent (38 to 54%), and women, one of 40 per cent (36 to 46%). In *polycythemia* the hematocrit may reach 60 to 70 per cent. Under these conditions blood flow is impeded. In anemia the hematocrit may go as low as 25 per cent. In *leukemia* the number of white cells may increase as much as thirty fold, and, as a consequence, the volume occupied by the buffy coat may also increase greatly.

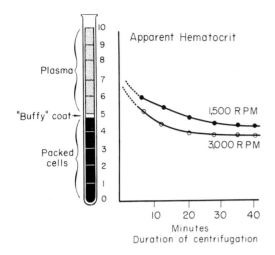

Fig. 15-1. Measurement of the hematocrit. The hematocrit obtained depends on the technique used and the sample being studied. (After Millar, W. G., 1925. From Physiology and Biophysics of the Circulation by A. C. Burton. P. 27. Copyright © 1965, Year Book Medical Publishers, Inc.; used by permission)

One of the important factors in obtaining a hematocrit is to carefully withdraw the blood sample. Frequently a syringe with a nonwettable surface (siliconized syringe) is used so as to protect the cells from its foreign surface. Careless handling damages the cell membranes of the red cells, causing the release of the red pigment hemoglobin into the plasma. This gives the plasma a color ranging from amber (small amount of hemoglobin) to red (large amount of hemoglobin).

Another important factor is the technique used in centrifugation. Blood is generally centrifuged at 1,500 g (or 1,500 times 980 cm./sec.²) for 15 min. If the speed is increased too much above 1,500 g, some of the cells are destroyed, hemoglobin is released into the plasma, and the apparent hematocrit decreases. It should be borne in mind, however, that regardless of what technique is used to determine the hematocrit, there is always some *plasma trapped* between the red cells. Normally this amounts to less than 5 per cent. The clinician does not usually correct the hematocrit for this, since he is interested only in an approximation, though in certain dye and radioisotope studies in which the tracer is limited to the plasma the investigator can determine the amount of plasma in the packed red cell mass. He can also, on the other hand, estimate this value and make an approximate correction. Corrections of this sort are important in studies on blood volume in which one determines the plasma volume and, using a corrected hematocrit, calculates the total blood volume.

PLASMA

The major constituents of plasma are listed in Table 15-1. You will note that the *plasma proteins* constitute about 7 per cent, the *inorganic substances* about 1 per cent, and the nonprotein organic substances about 1 per cent of the total plasma. An additional 90 per cent is water. The inorganic substances Na^+, Ca^{++}, K^+, Cl^-, HCO_3^-, $H_2PO_4^-$, and HPO_4^{--} are of particular interest, since they are the major charged particles of the plasma. Normally plasma contains 155 mEq. of anions per liter. You will note that the cation Na^+ is present at a concentration of 142 mEq. per liter, and the anions Cl^- and HCO_3^-, at 132 mEq. per liter.

Osmotic Pressure

The electrolytes listed above are also responsible for most of the 5,776 mm. Hg plasma osmotic pressure. Though the plasma proteins too carry a negative charge and contribute to the osmotic pressure, their contribution is of the order of only 25 mm. Hg. This osmotic pressure (called colloid osmotic pressure or oncotic pressure) is important, however, because of the size of the proteins responsible for it. Unlike the electrolytes, they do not pass readily into the perivascular space, and therefore tend to maintain a higher osmotic pressure in the blood than in the perivascular space. This, in turn, results in less water escaping from the blood than would occur if there were no proteins (colloids) dissolved in the plasma.

The importance of oncotic pressure is clear from the case of a brother and sister who had plasma albumin concentrations 1,000 to 500 times below normal (analbuminemia). They apparently responded to their lack of albumin by doubling their plasma globulin concentration, but this was not sufficient to bring their plasma oncotic pressure to normal levels. As a result, the reabsorption of water by the peritubular capillaries from the nephrons of the kidney was reduced. Normally 99 per cent of the 120 ml. of water per hour which enter the nephrons of the kidney is reabsorbed into the blood. In these analbuminemic persons only 95 per cent was re-

	mg./100 ml.	mEq./liter
Acetoacetate + acetone	1	
Aldosterone	0.000006	10
Amino acid nitrogen	4	
Ammonia	0.055	
Ascorbic acid	1	
Bicarbonate		29
Bilirubin	0.3	
Calcium	10	5
Ceruloplasmin	32	
Chloride	360	103
Cholesterol	180	
Copper	0.180	
Cortisol (17-hydroxycorticoids)	0.012	
Creatinine	1	
Glucose	100	
Iodine (protein-bound)	0.006	
Iron	0.11	
Lactic acid	10	
Lipids	550	
Magnesium	2	2
Nonprotein nitrogen	25	
Phenylalanine	1	
Phosphates ($H_2PO_4^-$ and HPO_4^{--})	4	2
Phospholipid	170	2
Potassium	19	4
Protein		
Total	7,000	
Albumin	4,000	
Globulin	3,000	
Fibrinogen	400	
Pyruvic acid	1.5	
Sodium	330	142
Urea	30	
Uric acid	3	

Table 15-1. Major constituents of plasma (see Table 31-1 for the plasma concentration of the hormones). The concentration of charged particles can be expressed in mg./100 ml. or mEq./liter. To convert from the former to the latter a constant (k) is computed for the particle in question:

$$\text{mEq./liter} = (\text{mg./100 ml.}) \cdot k$$

This constant is calculated as follows:

$$k = \frac{(\text{valence}) \cdot (\text{dissociation factor}) \cdot (10)}{(\text{molecular weight})}$$

For Na⁺ the constant is 0.43:

$$k_{Na^+} = \frac{(1) \cdot (1) \cdot (10)}{23}$$

In the case of calcium about half is bound to protein, and in the case of the phosphates at a pH of 7.4 about 80% are bivalent and 20% monovalent.

absorbed. Had this gone unchecked, polyuria and dehydration would have resulted, but they have apparently survived their defect by means of renal vasoconstriction, which markedly decreases blood flow to the kidneys and, therefore, water loss from the body.

Plasma Proteins

I have compared the average dimensions of some of the plasma proteins to sodium, chloride, glucose, and hemoglobin in Figure 15-2. The size of the plasma proteins is such that they do

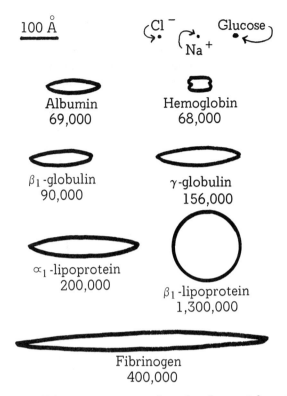

Fig. 15-2. Dimensions and molecular weights of hemoglobin and some of the plasma proteins. (*Modified from* Oncley. *In* Harper, H. A.: Review of Physiological Chemistry. 12th ed., p. 200. Los Altos, Lange Medical Publications, 1969)

not readily move out of the blood to the extracellular space, while the size of the other substances enables them to move rather readily into and out of the blood.

The protein hemoglobin is normally trapped inside red blood cells. When a cell is damaged, hemoglobin is released into the plasma and can then move through the glomerular capillaries into the urine. Albumin, on the other hand, is normally found only in trace amounts in the urine. The main difference between the albumin (molecular weight 69,000) and hemoglobin (molecular weight 68,000) molecules is their length. It has been suggested that while blood is flowing, the ellipsoid plasma proteins orient their long axis in the direction of flow. Thus they, unlike the more spherical hemoglobin molecules, do not readily move out of the blood. *Stasis*, on the other hand, accelerates the movement of plasma proteins from the blood. Under normal conditions, in other words, plasma proteins do not readily move into the perivascular space, and are responsible for the osmotic pressures being higher in the blood than in the interstitial fluid.

The plasma proteins can be divided, by paper electrophoresis or ultracentrifugation, into albumin, α_1-globulin, α_2-globulin, β_1-globulin, β_2-globlin, γ-globulin, fibrinogen, and the lipoproteins. Using our more advanced techniques, the albumin, globulin, and lipoprotein fractions can be still further subdivided. Of particular interest in this regard are some trace proteins which migrate with the globulins. Included in this group is prothrombin, found at a concentration of about 10 mg. per 100 ml. of plasma. Antibodies are found in the gamma-globulin fraction and are manufactured chiefly in the lymph nodes, the intestine, and the spleen. Albumin, fibrinogen, and prothrombin are produced by the *liver*.

Plasma protein levels decrease (hypoproteinemia) in certain intestinal diseases, such as sprue, in the late stages of starvation, and in kwashiorkor (malignant malnutrition). It should be emphasized, however, that the plasma proteins when catabolized yield only about 1,400 kilocalories (Cal.). This is not even enough energy to meet a man's metabolic needs for one day. Normally the *albumin-globulin* (A-G) *ratio* is about 2:1, but in severe liver damage or kidney disease (nephrosis) the plasma albumin may drop to below 3.5 gm. per 100 ml. of plasma and the globulin may increase to above 3.0 gm. per 100 ml. This decrease in the A-G ratio to 1:1 in the case of liver damage is due primarily to the failure of the liver to replenish the albumin normally lost from the body. In nephrosis the primary problem is excessive loss of albumin in the urine rather than merely its replenishment. Congenital protein deficiencies have also been reported.

Buffering capacity. The plasma proteins, besides accounting for the osmotic pressure's being higher in the blood than in the tissue spaces, are also responsible, by virtue of their amphoteric nature, for about one sixth of the *buffering capacity* of the blood. In the previous chapter I stressed their importance in blood *coagulation*; later in this chapter I shall consider their importance (as antibodies) in the body's defense reactions.

Protein complexes. We should also note that some of the regularly occurring plasma proteins react with certain substances added to or normally present in the blood. In some cases the reaction results in destruction or permanent change of the reactants, and in others it means only that the substance is loosely bound to the protein and will be released when the plasma concentration of that substance is low. Some of the substances bound by the plasma proteins are listed in Figure 15-3. Note that a number of dyes (Evans blue, Congo red, etc.), minerals (Ca, Fe, etc.) nutrients (lipids, vitamins, etc.), waste products (bilirubin, uric acid, etc.), hor-

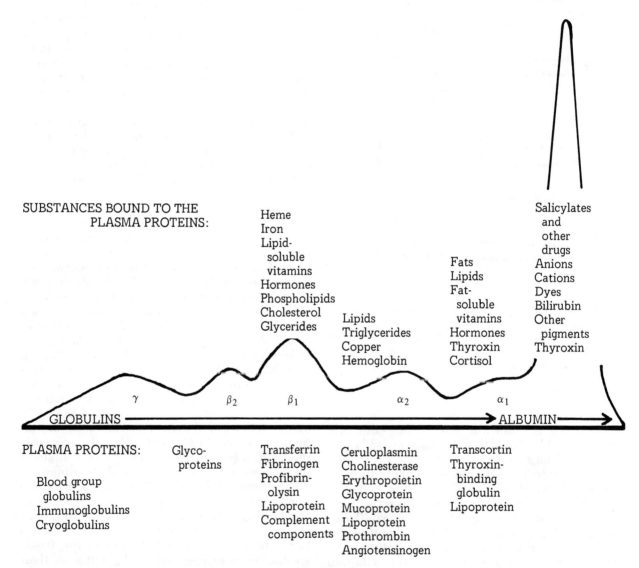

Fig. 15-3. Survey of the substances bound to plasma proteins. The plasma proteins can be subdivided into 6 fractions (albumin, alpha$_1$, alpha$_2$, beta$_1$, beta$_2$, and gamma globulin) by electrophoresis. The albumin fraction contains the highest concentration of proteins, offers the greatest surface, and has the most varied binding capacity. In the globulin fractions, however, are contained the antibodies, the procoagulants (prothrombin and fibrinogen), and proteins important in carrying iron, copper, and numerous nutrients. (*Modified from* Bennhold, H.: Transport function of the serum proteins: Historical review and report on recent investigations on the transport of dyestuffs and of iron. *In* Desgrez, P., and De Traverse, P. M. (eds.): Transport Function of Plasma Proteins. West-European Symposia on Clinical Chemistry. Vol. 5, p. 6. Amsterdam, Elsevier Publishing Co., 1966)

mones (steroids, thyroxin, etc.), and drugs (barbiturates, digitoxin, streptomycin, etc.) combine with the proteins. Since the plasma proteins tend to stay in the blood, their binding with hormones, drugs, and so on tends to keep these substances in the circulation and out of the kidney tubules. Thus the plasma, by virtue of its containing such proteins, acts as an effective reservoir of many agents. (See Table 31-2 for a listing of the half-lives of some of the hormones.)

One example of its importance as a reservoir is a case of atransferritinemia reported in the literature. In this report a 6-year-old girl was found to have a *transferrin* deficiency in the plasma. This resulted in an iron-binding capacity of her serum of only 15 µg. per 100 ml., and a disappearance time for iron of 5 min. Normally, the serum-binding capacity is 330 µg. per 100 ml. and the disappearance time, between 70 and 140 min. In addition, this case showed extensive iron deposits in the liver associated with cirrhosis and a grayish-brown appearance of the skin.

TRANSFUSION

Blood is usually transfused into a patient to supply something the patient needs. Transfusions of packed red blood cells have been used to treat anemia; transfusions of platelet concentrates, to treat thrombocytopenia; and transfusions of plasma and plasma fractions, to treat certain bleeding disorders. Transfusions are also performed to increase the patient's blood volume and blood pressure. An adult can normally lose 500 ml. of blood without its seriously affecting the cardiovascular system, but more severe blood losses can result in decreased blood pressure and due to the loss of red blood cells, serious decreases in the ability of the blood to carry oxygen.

Hemorrhaging is best treated by the infusion of whole blood but, when this is not available, can be treated by the injection of plasma or plasma substitutes. In Figure 15-4 we note a comparison of the effectiveness of whole blood, water, and saline in the treatment of a hemorrhage that had caused a fall in arterial pressure to 30 mm. Hg for 15 min. Note that the best result was obtained when the dog's own blood was reinfused (II). It is interesting to note, however, that the slow injection of water over a period of 50 min. (III), the rapid injection of water over a period of 7 min. (IV), the injection of 50 ml. of saline per kilogram of body weight in 7 min. (V),

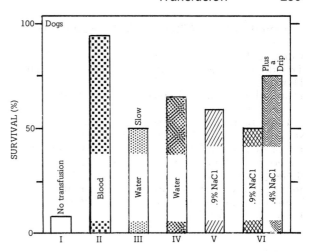

Fig. 15-4. The effect of transfusions of whole blood (II), NaCl solutions (V and VI), and water (III and IV) on the survival of dogs after a hemorrhage that produced an arterial pressure of 30 mm. Hg for 15 min. All dogs that survived were alive, active, and able to pass urine 2 days after the procedure. Group I, no transfusions after hemorrhage. Group II, whole blood transfusion. Group III, water slowly (50 min.) infused. Group IV, water rapidly (7 min.) transfused. Group V, 0.9% saline rapidly (7 min.) transfused. Group VI, 0.9% or 0.45% NaCl solutions rapidly transfused and followed by a continuous drip to maintain arterial pressure. The authors feel that the advantage of the 0.45% saline solution, if real, is due to the expansion of the red blood cell without an associated damage to the membrane. (*Modified from* Coleman, F., et al.: A hypotonic solution for emergency blood substitution. Surgery, 60:393, 1966)

and a constant infusion of 0.45 per cent saline or 0.9 per cent saline (VI) were all more effective in preventing death than no replacement of the lost volume (V).

One of the problems in transfusion is obtaining the appropriate blood or blood fraction. The first *blood bank* was established by Fantus at Cook County General Hospital in 1936. Since that time blood banks have spread to every major city, simplifying transfusion problems and facilitating such procedures as cardiac surgery using the heart-lung machine. Initially, 4 per cent sodium citrate was added to the blood to prevent clotting and facilitate storage, but this blood could be safely stored only for less than a week. It was later found that adding citric acid to whole blood to reduce its pH, and dextrose as an additional source of nutrients for the red blood cell, enabled safe storage for about 21 days

Ingredients	ACD Formula A	ACD Formula B
Citric acid	8.0 gm.	4.8 gm.
Trisodium citrate	22.0 gm.	13.2 gm.
Dextrose	24.5 gm.	14.7 gm.
Distilled water to	1,000.0 ml.	1,000.0 ml.

Table 15-2. Blood preservatives. Blood can be stored in blood banks up to 21 days at 4°C. when 15 ml. of ACD formula A or 25 ml. of ACD formula B are added to 100 ml. of whole blood. During storage there is progressive deterioration of the blood. (From Johnson, S.A., and Greenwalt, T.J.: Coagulation and Transfusion in Clinical Medicine. P. 35. Boston, Little, Brown & Co., 1965)

at 4°C. The most common solution in current use for the storage of whole blood is called ACD solution, discussed in Table 15-2.

ACD solution has not, however, eliminated all the problems involved in the storage of whole blood. As can be seen in Figure 15-5, there is progressive deterioration of procoagulant activity in blood stored at 4°C. and a disappearance of platelets as storage continues. The cell found in highest concentration is the red blood cell. It has a QO_2 (oxygen consumption expressed in milliliters of O_2 per grams of tissue per hour) comparable to that of other cells if one calculates the QO_2 in terms of the total weight of the cell minus the weight of its hemoglobin. Apparently it is the metabolic activity of the red cell that is responsible for many of the problems in blood storage.

You will note in Table 15-3 that blood dextrose disappears and lactic acid accumulates during storage. It is also important to bear in mind that under normal conditions the red cell has a life span of only about 120 days; hence it is not surprising that more and more of the intracellular components of the older erythrocytes (K^+ and hemoglobin) appear in the plasma as storage continues. The presence of K^+ or ammonia in the plasma can be particularly serious in persons with cardiac decompensation and kidney or liver malfunction, since the liver and kidney keep the K^+ and ammonia concentrations in the blood at safe levels, and since increases in the K^+ concentration modify the electrical characteristics of the myocardium. Whole blood is considered suitable for transfusion if 70 per cent of its erythrocytes are present in the recipient 24 hrs. after transfusion. It has been shown that a person with a normal liver can tolerate up to 10 pints of ACD blood in an hour if each 2 pints of donor blood is supplemented with 10 ml. of a 10 per cent solution of calcium gluconate. The gluconate acts as a source of Ca^{++} to replace that removed by the citrate in the ACD blood. The primary reason for adding Ca^{++} is to protect the myocardium from hypocalcemia, which may lead to fibrillation or other abnormal cardiac electrical activity.

The advantage of using whole blood or plasma in transfusion is that both contain the proteins

	Concentration in Plasma				
	Days Stored				
	0	7	14	21	28
Dextrose (mg./100 ml.)	350	300	245	210	190
Lactic acid (mg./100 ml.)	20	70	120	140	150
Inorganic phosphate (mg./100 ml.)	1.8	4.5	6.6	9.0	9.5
pH	7.00	6.85	6.77	6.68	6.65
Hemoglobin (mg./100 ml.)	0-10	25	50	100	150
Sodium (mEq./liter)	150	148	145	142	140
Potassium (mEq./liter)	3.5	12	24	32	40
Ammonia (μg./100 ml.)	50	260	470	680	-------

Table 15-3. Changes in ACD blood during storage at 4°C. Apparently the cells in the blood deteriorate and release their intracellular elements (inorganic phosphate, hemoglobin, potassium) to the plasma. The cells also continue to use dextrose and produce lactic acid and ammonia during storage. (From Strumia, M.M., Crosby, W.H., Greenwalt, T.J., Gibson, J.G., and Krevans, J.R.: General Principles of Blood Transfusions. Philadelphia, J.B. Lippincott, 1963)

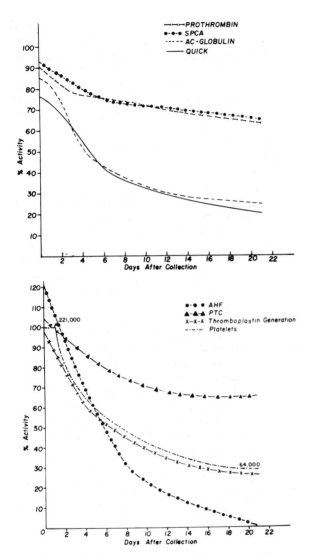

Fig. 15-5. Changes in the level of coagulation factors during storage of ACD whole blood in vacuum bottles. SPCA, combined factor VII and X activity; Quick, prothrombin time plotted as per cent of normal from dilution curve. Note that the platelet count dropped from 221,000 to 64,000/mm.³ of blood during storage. (Annals of The New York Academy of Sciences, vol. 115, Goldstein, R., et al. Figs. 1a and 1b, p. 427. Copyright by the New York Academy of Sciences, 1964; reprinted by permission)

which do not readily pass out of the blood. *Isotonic saline* (0.9% NaCl), on the other hand, contains nothing that is restricted to blood, so that infusing it only temporarily increases blood volume (though it may be considerably better than distilled water). When *water* is rapidly injected into the blood it tends to produce a hypotonic environment around the blood cells. This results in the movement of water into the cell, which in turn expands the cell, increases its permeability, and causes the diffusion of intracellular material into the plasma. Normally there are 17 mg. of K^+ in each 100 ml. of plasma, and 420 mg. per 100 ml. in each red blood cell. Levels of K^+ as high as 90 mg. per 100 ml. of plasma have been reported clinically in uremia and may produce serious cardiovascular changes.

In cases of severe dehydration investigators inject isotonic *dextrose* (5% glucose solution) into the blood, rather than water. As the dextrose is slowly metabolized, the water is taken up by the tissues and the tonicity of the blood is kept fairly constant. For hemorrhaging, when neither whole blood nor plasma is available, a 6 per cent solution of *dextran* in isotonic NaCl is preferable to saline alone. Dextran is a high-molecular-weight (average 75,000), water-soluble glucose polymer, which, like the plasma proteins, does not readily pass from the blood. When injected in a concentration of 12 per cent in distilled water it effectively reduces edema caused by nephrosis.

Blood Types

With the exception of identical twins, each person has a different hereditary background and, as a consequence, forms some proteins and polysaccharides that differ from those of his relatives. Frequently when microorganisms invade the body, when tissues from another person are transplanted into the body, or when a foreign protein or polysaccharide passes into the body through the nasal mucosa or intestinal mucosa, a person reacts to the invaders (*antigens*) by producing *antibodies*.

When blood is transfused from donor to recipient, the recipient sometimes reacts to foreign molecules in the donor's red blood cells by forming antibodies. Fortunately, however, by the time the antibody titer (concentration) has reached a level sufficient to destroy the foreign molecules, the recipient has removed the foreign cells by other processes. Problems sometimes arise when a second transfusion is attempted using the same donor blood. Under these circumstances the antibodies may already be in sufficiently high concentration to be effective against the donor's cells.

In the reaction discussed above a person forms antibodies in response to foreign molecules. Immunity in this case is *acquired* following an exposure. There is another type of response to foreign substances (*agglutinogens*), however, which is due to the presence of an *inherited*, antibody-like substance (*agglutinin*) in the plasma.

There are so many different blood types at present that a person can be almost as readily identified by his blood type as by his fingerprints. Thus when a transfusion is contemplated, the safest procedure is to mix the donor's blood and the recipient's serum or plasma on a slide and see if the donor's red cells agglutinate. If they do, the transfusion is unsafe. We do not generally worry about the donor's plasma agglutinating the recipient's cells because the agglutinogens and antibodies in the donor's plasma get so diluted by the recipient's blood as to be ineffective. If more than 500 ml. of blood is to be transfused into a single person, however, it is wise to see if the donor's plasma causes the recipient's cells to agglutinate.

In emergency situations, one may not have the time or facilities for a *cross-match*. Under these circumstances, plasma or a plasma-expander is sometimes infused. It has also proven helpful to classify a person according to his blood type. The simplest classification in common use is the A, B, O, AB, Rh-positive, and Rh-negative one.

A and B antigens. In 1900, Landsteiner described three blood groups: A, B, and AB (Table 15-4). Group A contains antigen A; group B, antigen B; and group AB, both antigen A and B. Two years later, a group O was described by DeCastello and Sturli. It contained neither antigen A nor antigen B. It was found that if type A red blood cells were placed in the plasma from either a type B or O person, they would agglutinate. On the basis of this and similar observations with the other blood types, it was established that type A blood also contains an anti-B (frequently called β) agglutinin in its plasma; type B blood, an anti-A (frequently called α) agglutinin; type O, both an α and a β agglutinin; and type AB, neither an α nor a β agglutinin. On this basis, type O was called the universal donor, since its red cells were agglutinated by neither α nor β agglutinins, and type AB, the universal recipient, since it contained neither the α nor β agglutinins.

Between the years 1911 and 1924, Mendel's laws of heredity, first presented in 1865, were shown to apply to the blood groups. It was demonstrated that there are three allelic genes—A, B, and O. A person of blood group A was characterized as having a genotype of AA or AO, a person with a phenotype of B had a genotype of BB or BO, a type AB phenotype had a genotype of AB, and a type O phenotype was an OO genotype.

By use of the ABO classification, deaths from transfusion reactions were kept down to 2 per 1,000 transfusions. Later however, it was found that type A blood could be further divided into A_1 (sometimes called A) and A_2. This meant that the type AB group could be subdivided into A_1B and A_2B. It also became apparent that type O blood contained a very weak antigen. In rare cases, persons repeatedly transfused with O blood develop an anti-O antibody. More common among the other blood groups noted since then are the M, N, S, and P groups, but undoubtedly the most important group discovered since Landsteiner's original work at the beginning of this century was the Rh factor reported by Landsteiner and Wiener in 1941.

The Rh factor. In 1940, Landsteiner and Wiener injected the red cells of Rhesus *monkeys* into rabbits and guinea pigs. The animals that received these red cells developed antibodies which agglutinated not only Rhesus red cells, but also the red cells of about 85 per cent of

Blood Group	Genotype	Plasma Agglutinin	Incidence in U.S. (%)		
			Caucasian	Negro	Oriental
O	OO	α, β	45	48	36
A	AA, AO	β	41	27	28
B	BB, BO	α	10	21	23
AB	AB	—	4	4	13

Table 15-4. The ABO blood groups.

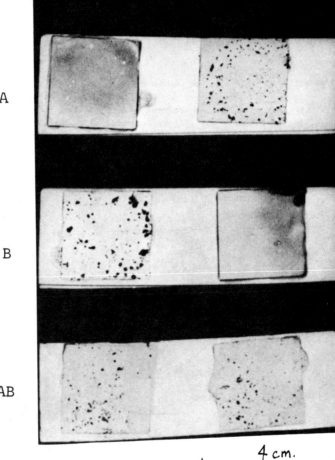

Fig. 15-6. Red cell agglutination. In this test one drop of a saline suspension of red blood cells (RBC concentration, 1–2%) is added to one drop of a known antiserum. If, for example, type B erythrocytes are added to an anti-B serum, a clumping of erythrocytes will be noted within 3 to 5 minutes. (Wiener, A. S.: Blood Groups and Transfusion. 3rd ed., p. 15. Springfield, Charles C Thomas, 1943)

all white persons. The investigators hypothesized that the antibodies from the Rhesus-sensitized rabbit and guinea pig were reacting against a factor (the Rh factor) on the membrane of most human red cells. It was later established that many humans (called *Rh-negative*) whose red cells were not agglutinated by the anti-Rhesus factor also formed antibodies against blood from persons (called *Rh-positive*) whose red cells contained the Rhesus factor. Thus in 1941 an explanation was put forward for the 0.2 per cent death rate among recipients of transfused blood. After this period, with a knowledge of the ABO and Rh groups, the physician could transfuse blood with much greater safety.

The Rh system, unlike the ABO system, is one

in which the individual produces antibodies in response to an antigen. In about 50 per cent of transfusions from an Rh-positive donor to an Rh-negative recipient, no sensitization occurs. In almost all persons, however, sensitivity results if transfusion is repeated several times.

Another problem stemming from Rh differences concerns pregnancy. It has been calculated that 1 out of 42 Rh-negative women in their second pregnancy, and 1 out of 12 in their fifth, deliver children with hemolytic disorders (erythroblastosis fetalis). Rh-positive women may also deliver erythroblastotic children, but the incidence is much lower. The usual cause of erythroblastosis fetalis is the development by an Rh-negative mother of antibodies against the red cells of her Rh-positive fetus. This condition is unusual in a first child, but with subsequent Rh-positive children the mother's antibody titer (concentration) may increase. When it exceeds a certain level, a sample of her amnionic fluid is frequently tested for the concentration of anti-Rh factor.

If the concentration of Rh antibodies in the fetus is excessive (Fig. 15-7), hemoglobin and other intracellular elements will enter the blood stream and, if nothing is done, cause the death of the fetus. The hemolysis occurring in this condition results in jaundice, edema, and enlargement of the spleen and liver. Since the blood-

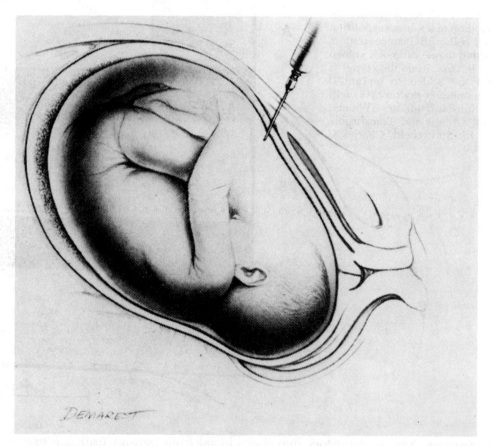

Fig. 15-7. A technique for sampling the amnionic fluid in a case of erythroblastosis fetalis. In a case of a Rh⁻ mother and a Rh⁺ fetus, where the mother's antibody titer is high, it is frequently useful to check the antibody content of the amnionic fluid. If the amnionic antibody titer is high also, a premature delivery is initiated. If the concentration in the amnion is low, then pregnancy should not be terminated. (Freda, V. J.: The Rh problem in obstetrics and a new concept of its management using amniocentesis and spectrophotometric scanning of amniotic fluid. Am. J. Obstet. Gynec., 92:345, 1965)

brain barrier is not fully developed, there may also be signs of brain damage (convulsions, respiratory difficulties, spasticity, mental retardation, or a combination of these).

The usual treatment for erythroblastosis fetalis due to the reaction of an Rh-negative mother to an Rh-positive fetus is to remove the fetus and perform an exchange transfusion on it. A femoral artery and vein or the umbilical artery and vein of the child are used for this procedure. Blood is removed from the fetus by the artery and replaced with an equal amount of Rh-negative blood of the appropriate ABO type. Removing blood from the baby decreases the concentration of the anti-Rh factors which it received from the mother through the placental barrier. It also eliminates the baby's Rh-positive red cells, many of which are coated with antibodies that will eventually cause hemolysis. Since it is impossible to eliminate all the antibodies from the infant's blood, the baby is reinfused with Rh-negative red cells rather than Rh-positive cells.

An alternate procedure, sometimes used in the case of fetuses too premature to be delivered, is to insert a needle in the mother's abdomen and the fetus' abdomen and infuse Rh-negative blood into the fetus' abdominal cavity.

Rh and Hr subgroups. The Rh system, like the ABO system, was soon found to be much more complex in humans than was originally thought. Fisher, for example, initially recognized three factors (C, D, and E) to which a donor might become sensitized. Wiener gave different names to these factors (Rh_o, rh', rh"). The Rh-positive person was characterized as having the D factor (Rh_o), and the Rh-negative person, as lacking it. Both the Rh-negative and Rh-positive subjects, on the other hand, might also have (1) both the C (rh') and E (rh") factors, (2) one of these factors, or (3) neither of these factors. As can be seen from Table 15-5, in time the number of factors found in the blood became greater and the entire subject grew more complicated.

One complication arose when an Rh-positive

Classification			Rate of Occurrence in Caucasians (%)
Rh-Hr (Wiener)	CDE (Fisher)	"Neutral"	
Individual factors			
Rh_o	D	Rh 1	85
rh'	C	Rh 2	70
rh"	E	Rh 3	30
hr'	c	Rh 4	80
hr"	e	Rh 5	97
hr	f, ce	Rh 6	64
rh_i	Ce	Rh 7	69
$rh^w 1$	C^w	Rh 8	2
rh^x	C^x	Rh 9	0.03
hr^V	V, ce^s	Rh 10	<1
$rh^w 2$	E^w	Rh 11	Very rare
rh^G	G	Rh 12	87
Combinations			
Rh_o, hr', hr", hr	cDe		2.7
Rh_o, rh', hr"	CDe		41
Rh_o, rh', rh^w, hr"	C^wDe		2
Rh_o, rh", hr'	cDE		15
Rh_o, rh', rh"	CDE		0.2
hr', hr", hr	cde		38
rh', hr"	Cde		0.6
rh', rh^w, hr"	C^wde		0.005
rh', hr'	cdE		0.5
rh', rh"	CdE		0.01

Table 15-5. Rh terminology. Three types of classification are presented. Although on the surface the Fisher classification is much simpler than that of Wiener, many investigators now suggest that the Wiener approach or some modification of it be used. (From Wintrobe, M.M.: Clinical Hematology. 5th ed. Philadelphia, Lea & Febiger, 1967)

mother gave birth to an Rh-negative infant with erythroblastosis fetalis. The mother was found to have an agglutinin in her blood which reacted to all Rh-negative and some Rh-positive blood. Just as all Rh-positive blood types contain an Rh_o (D) factor, in other words, so all Rh-negative ones contain the antigen called by Wiener the Hr factor.

A number of transfusion reactions still cannot be explained using the classifications A, B, AB, O, Rh-positive and Rh-negative, though a vast majority can be prevented through use of this system. Initially the subdivisions of the Rh factors presented by Fisher (C, D, and E) were most popular, undoubtedly because of their simplicity, but at present there is a great deal of evidence favoring Wiener's (Rh_o, rh', etc.). For a more complete discussion of this the reader is referred to Miale (1967) and Wintrobe (1967) (see references at end of this section). In closing, let me remind the reader that in fact there is no such thing as a universal donor or a universal recipient. It can be said, however, that certain blood types are more safely infused into an untyped recipient than others.

RESISTANCE TO INFECTION

Microorganisms are on our skin, in our large intestine, in the air we breathe, and in the food we eat. As a consequence they sometimes gain entry to the intercellular space and, if not destroyed, cause cellular damage and eventually death. We have two main lines of defense against these invaders. Either the body's phagocytic cells surround the microorganisms and incorporate them into the cytoplasm, where they are exposed to various proteolytic and lipolytic enzymes, or, certain cells form antibodies which enter the plasma and inactivate or destroy the invaders. This second form (antibody formation) requires several days and might be considered an auxiliary mechanism for use when phagocytosis alone fails to destroy the microorganisms.

Phagocytosis

In 1924, Aschoff described a defense system in the body consisting of cells scattered throughout the connective tissue. This mechanism, which he called the *reticuloendothelial system* (RES),

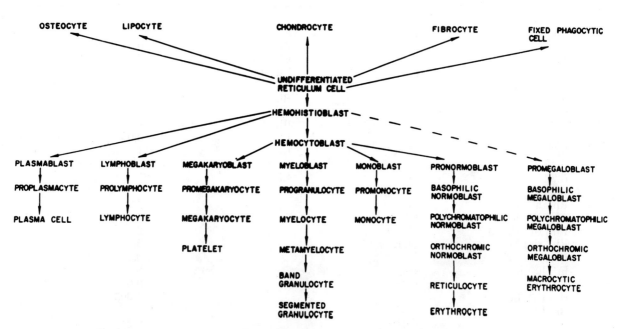

Fig. 15-8. Normal origin of the formed elements of the blood according to the neounitarian theory of Downey. Under some conditions lymphocytes may also form other cell types. It should be emphasized that this concept is not accepted by all investigators. Some believe, for example, that progranulocytes and monoblasts are not derived from the same stem cell (hemocytoblast or reticulum cell). (Miale, J. B.: Laboratory Medicine—Hematology. 3rd ed., p. 2. St. Louis, C. V. Mosby, 1967)

comprises a number of large cells (macrophages and reticuloendothelial cells) held in position by *reticular fibers*. In certain blood and lymph channels these cells may help form the *endothelium*. Both the macrophages and reticuloendothelial cells are phagocytic. The reticulum cell may, in addition, be responsible for the production of other cell types (Fig. 15-8), some of which are phagocytic (neutrophils, for example), and some of which are not (erythrocytes).

Organs containing large concentrations of *macrophages* include the *spleen, liver* (Fig. 12-1), *bone marrow*, and *lymph nodes*. In the past, it was common to classify the phagocytic cells found in those areas as *fixed* or *wandering*, but now it is recognized that some "fixed" cells become mobile and some wandering cells become "fixed." The macrophages probably remove microorganisms and aged blood cells, antigen-antibody complexes, and inert particles such as carbon, colloidal metal, and certain dyes (bromosulfophthalein).

The red blood cell *survives* about 120 days in the blood stream; one population of lymphocytes, about 2 days; another group of lymphocytes, between 100 and 200 days; and the granulocytes (neutrophils, basophils, eosinophils), between 2 and 14 days. It is not clear how the macrophages know which blood cells to remove, but one suggestion is that as the cells age or deteriorate, changes occur in their surface membranes. Possibly such changes permit contact with the macrophage, which in turn initiates phagocytosis (Fig. 15-9).

Chemotaxis. Certain cells are thought to release agents which attract phagocytes to them. Perhaps in many cases the breakdown of a cell membrane causes the release of a *chemotactic agent* which attracts the phagocyte. Chemotaxis may also be an important phenomenon in areas of local infection. A number of cells (mast cells, basophils, eosinophils), when damaged, are thought to release histamine or various kinins, which in turn are thought to attract both macrophages and microphages (neutrophils) to the area. The blood-borne phagocytic cells pass through or between the capillary endothelial cells by a process called *diapedesis*. Many or possibly most of the phagocytes get back to the blood through the lymphatic channels.

Fibrinolysis. Another function of the phagocyte may be to dissolve blood clots (Fig. 15-10) in the vascular system. Clots have been shown to attract microphages (neutrophils), which phagocytose the *fibrin threads* and in this way help

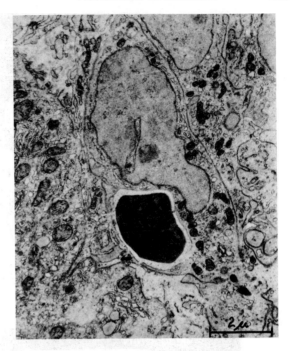

Fig. 15-9. Erythrophagocytosis by a macrophage. The erythrocyte is seen lodged in a fluid-filled vacuole in the cytoplasm of the histiocyte. Note the abundance of mitochondria and lysosomes in the histiocyte. (Halpern, B.: The reticuloendothelial system and immunity. Triangle, the Sandoz Journal of Medical Science, 6:176, 1964)

break up the clot and maintain flow. Phagocytosis may also be important in wound healing. A cut can initiate bleeding and the eventual formation of a clot. The *clot* is initially red, due to the presence of large numbers of red blood cells. In time the red cells are removed or destroyed and the clot becomes less red. The fibrin and blood cells disappear and are replaced by stabler cells and a different type of intercellular material (collagen, elastin, etc.).

Antibodies

Antibodies are *globulins* produced by vertebrates in response to various foreign molecules (antigens). Most foreign proteins, some bacterial polysaccharides, foreign deoxyribonucleic acid, and dextran have all been shown to cause antibody formation. These substances, that is, are *antigenic*. Antibodies produced in response to a specific antigen combine with the antigen. If the antigen is part of a bacterium or some other foreign cell, the reaction may destroy, inactivate,

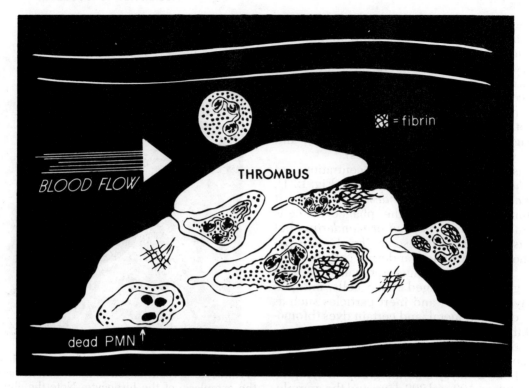

Fig. 15-10. A schematic representation of thrombolysis. It is suggested that an intravascular clot becomes infiltrated with neutrophils (PMN), which *phagocytose* some of the fibrin threads and possibly release a *fibrinolytic principle* from the intracellular granules. This process tends to decrease the size of a clot by digesting (lysing) parts of it and causing parts to be carried away by the blood stream. (Barnhart, M. I.: Importance of neutrophilic leukocytes in the resolution of fibrin. Fed. Proc., 24:852, 1965)

or in some other way facilitate the removal of the cell from the body.

Figure 15-11 illustrates how antibodies work in concert with other defense systems (phagocytosis, for example) in the removal of foreign cells. In these experiments pigeon red cells were injected into mice, producing a concentration of $10^{5.3}$ foreign cells per milliliter of mouse blood. It took the mouse about 28 min. to reduce this concentration to $10^{4.5}$. If the mice are immunized against pigeon red cells beforehand, however, it takes only about 7 min. to reduce the foreign red cells to a concentration of $10^{4.5}$, and only 11 min. to reduce the concentration to $10^{4.0}$.

Cross-reactivity. Antibodies do not usually react against a whole antigen molecule, but rather against one of its chemical groups. Hence there is cross-reactivity when two different molecules have an antigenic group in common. For example, an antibody formed against iodinated ovalbumin will precipitate both iodinated ovalbumin (the *homologous antigen*) and iodinated bovine serum albumin (the *heterologous antigen*).

Antibody Formation. Fetal and neonatal animals have little or no ability to form antibodies, but there is a progressive increase in the potential for antibody formation after birth. It has been reported, for example, that for each molecule of diphtheria toxoid injected, more than 10^6 molecules of antibody are formed. This potential eventually reaches a maximum and, with further aging, declines (Fig. 15-12). Evidence of this decline is seen in the greater success of tumor growth and the improved survival of skin grafts with advancing age.

Apparently just before or soon after birth the *thymus* stimulates the development of potential antibody-forming cells in the spleen, bone marrow, lymph nodes, Peyer's patches in the bowel, and various other organs. Part of the evidence for this is the observation that animals thymectomized soon after birth show a striking immunologic deficit. Two cells strongly implicated

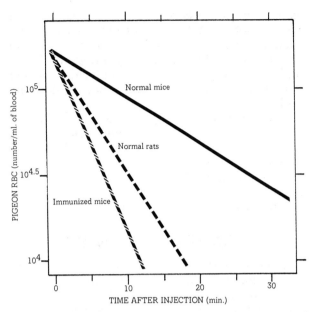

Fig. 15-11. Rate of elimination of identical doses of pigeon red cells in (a) normal mice, (b) normal rats, and (c) mice immunized against pigeon red cells. Apparently the rat has natural defenses against the foreign cells that are more efficient than those of the mouse, but note that immunization markedly facilitates the removal of the foreign cells. (*Redrawn from* Halpern, B.: The reticuloendothelial system and immunity. Triangle, the Sandoz Journal of Medical Science, 6:177, 1964)

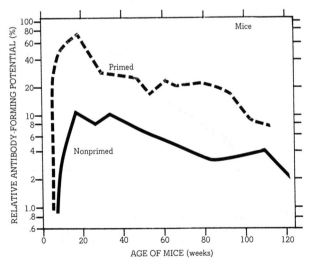

Fig. 15-12. Antibody-forming potential/spleen of primed and nonprimed mice at different ages. The highest individual titer is the 100% reference point. Each point represents 15 to 59 samples. (*Redrawn from* Albright, J. F., and Makinodan, T.: Growth and senescence of antibody-forming cells. J. Cell. Physiol., 67(Supp. 1):191, 1966)

in antibody production are the *plasma cell* and lymphocyte. Astrid Fagraeus, for example, noted that 2 days or so after the intravenous injection of an antigen, plasmablasts (Fig. 15-8) began to appear in the spleen. Plasma cells are also found to increase in number at the site of infection.

It has been suggested that plasmablasts respond to an antigen in a manner similar to that indicated in Figure 15-13. The initial response is liferation. The first cell division takes about 10 hrs. and each successive one progressively longer. As the cells divide, the concentration of ribosomes in the cytoplasm increases. These ribosomes begin to produce antibodies on about the third day (fifth series of cell divisions), and by the fifth day (eighth series of divisions) are releasing large quantities of them. The biological *half-life* of isotopically labeled antibody is about 13 days in adult humans.

As antibody production continues, the characteristics of the antibody change. Initially a molecule with a weight of 1 million and a sedimentation constant in the ultracentrifuge of 19 Svedberg units (S) is formed, but as time passes fewer of these molecules are produce and more having an approximate molecular weight of 160,000 and a sedimentation constant of 7 S are formed.

Also of interest is the observation that the initial response of an organism to an antigen prepares him for a second infestation. In rabbits, for example, the injection of bovine serum albumin or bovine gamma globulin causes antibody formation in about 5 days, and maximum titer between the eighth and twelfth days. The animal shows a negative precipitin test 25 to 30 days after initial exposure. This is not to say that all the antibodies have disappeared, but only that their concentration has so decreased that they cannot be demonstrated by the *precipitin test*. A more sensitive test, the *passive hemagglutination technique*, indicates that antibody production continues for many months after a single injection of bovine serum albumin. It has also been shown in the horse that if one uses a sensitive skin test, a single injection of alum-precipitated diphtheria toxin causes antibody production for about 9 months.

If the second injection of an antigen occurs while antibodies to that antigen are still present, there is prompt formation of antigen-antibody complexes. Whereas it took 8 to 12 days to reach a maximum titer initially, it now takes only about 5 to 9. The quick response to a second injection is apparently due to the starting level of antibody-producing cells.

Control of antibody formation. It has been noted in the rabbit that if an antigen is injected shortly after birth the animal not only produces an antibody to that antigen, but also responds less or not at all to that antigen when reinjected in later life. It appears, then, that the rabbit acquires an *immunologic tolerance*. Perhaps the same mechanism is responsible for the fact that we do not react to most of our own proteins, though we can form antibodies against some of the molecules in our own thyroid gland, brain, and lens. (Apparently these molecules do not usually come in contact with antibody-forming cells.)

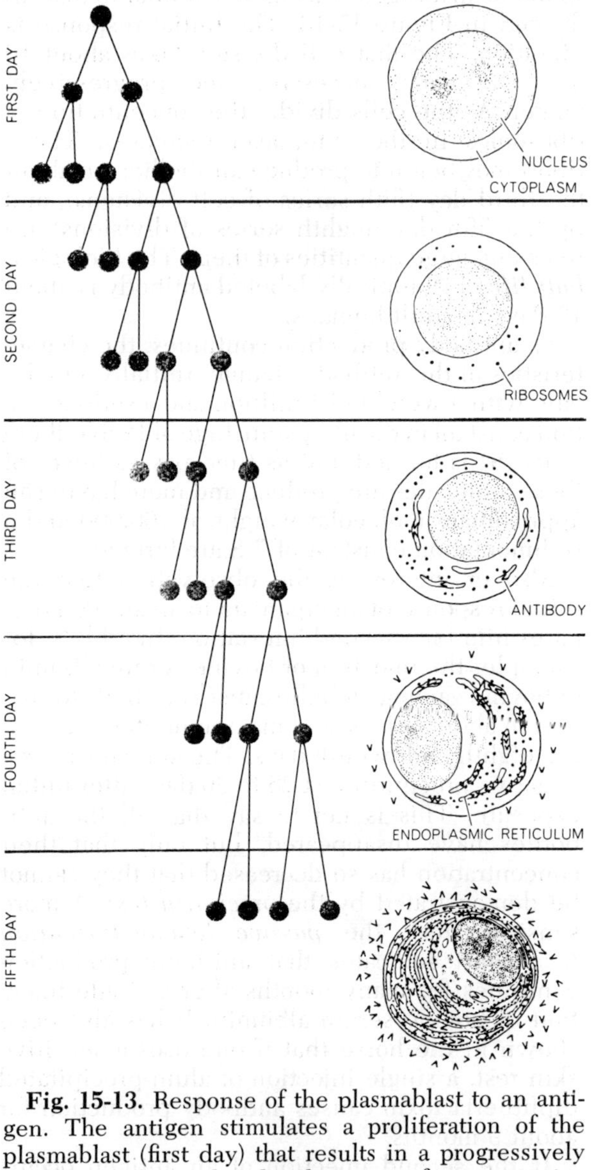

Another way the body controls antibody production is through the hormone *cortisone* (from the adrenal cortex), which decreases the formation of antibodies in response to an antigen. Irradiation and nitrogen mustards have also been used to suppress antibody production in organ transplants (e.g. heart or kidney).

FORMED ELEMENTS

Blood consists of formed elements (cells and cell fragments) plus the plasma in which they are suspended (Table 15-6). The formed elements usually carry a negative *surface charge*. Part of the evidence for this is the observation that at a pH of 7.4 they migrate toward the positive pole (anode) of an electrophoretic cell (Fig. 15-14). The rate of migration in plasma under specified conditions of voltage is about 1.0 μ per second for the erythrocyte, 0.6 for the

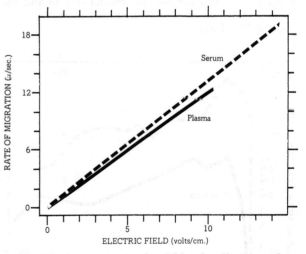

Fig. 15-13. Response of the plasmablast to an antigen. The antigen stimulates a proliferation of the plasmablast (first day) that results in a progressively smaller nucleus, the development of an extensive endoplasmic reticulum, increased numbers of ribosomes, and eventually (third, fourth, and fifth days) the formation of antibodies. (*From* Nossal, G. J. V.: How cells make antibodies. Sci. Am., *211*:109, 1964; copyright © December 1964 by Scientific American, Inc. All rights reserved)

Fig. 15-14. Migration of red blood cells in an electric field. Migration is seen to be toward the anode. The rate of migration is proportional to the voltage applied (in volts/cm. between electrodes). Some of these studies were performed in plasma and some in serum. (*Redrawn from* Abramson, H. A.: Electrophoresis of blood cells. *In* Sawyer, P. N. (ed.): Biophysical Mechanisms in Vascular Homeostasis and Intravascular Thrombosis. P. 5. New York, Appleton-Century-Crofts, 1965)

lymphocyte, 0.5 for the polymorphonuclear leukocyte, and 0.45 for the platelet. The negative charge at the surface of these structures may serve not only to decrease contact between the various formed elements during flow, but also to minimize contact with the intima of the vessel (also negatively charged at its surface).

Production

The formed elements are first seen in blood islands of the yolk sac during the first month of embryonic development (Fig. 15-15). The more peripheral cells probably form the primitive endothelium of the developing vessels, while the central cells may become the *primitive blood cells*. By the fifth month of pregnancy a placental circulation has been established and erythrocytes (red cells) are present in the blood at a concentration of about 2.9 million per cubic millimeter. Leukocytes are present at a concentration of about 12,000 per cubic millimeter. During this period the *liver* is the primary hemopoietic organ of the body, but afterwards the bone marrow and lymph nodes become increasingly more important, and the liver less so, as blood-forming organs.

Bone marrow becomes the prime source of blood cells after birth, but the lymph nodes continue producing lymphocytes and monocytes. Initially all the bones of the body contain the hemopoietic red marrow. Progressive infiltration of this marrow by fat occurs after age five, however, until, at age twenty, about half the marrow is infiltrated with fat and half with active hemopoietic cells (Fig. 15-16). At this time the red marrow lies primarily in the vertebrae, sternum, and ribs, though small quantities may also be found in the skull, innominate bone, and proximal epiphyses of the femur and humerus. Under certain conditions the highly labile *yellow* (fatty) *marrow* may be replaced by red marrow, and the spleen, lymph nodes, and liver may start producing erythrocytes, leukocytes, and platelets.

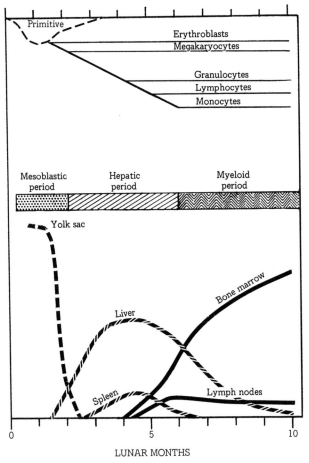

Fig. 15-15. Stages of hemopoiesis in the embryo and fetus. Primitive blood cells are formed initially in the yolk sac (mesoblastic period). Later (hepatic period) the liver, spleen, and thymus (not shown because of the small number of cells it produces) start producing less primitive cells (erythroblasts, erythrocytes, megakaryocytes, granulocytes, lymphocytes, and monocytes). Eventually (myeloid period) the bone marrow and lymph nodes take over the erythropoietic functions of the body. (*Modified from* Wintrobe, M. M.: Clinical Hematology. 6th ed., p. 2. Philadelphia, Lea & Febiger, 1967)

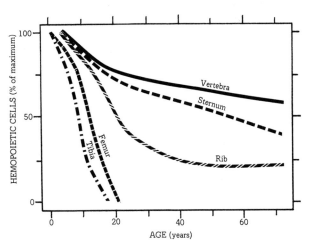

Fig. 15-16. The relative amount of hemopoietic cells in the marrow of different bones at various ages. (*Redrawn from* Wintrobe, M. M.: Clinical Hematology. 6th ed., p. 17. Philadelphia, Lea & Febiger, 1967)

	Before Hemorrhage	After Hemorrhage
Blood		
Erythrocytes	4,950,000	2,680,000
Total Leukocytes	4,180	7,240
Lymphocytes	3,060	4,670
Neutrophils	1,060	2,060
Eosinophils	20	190
Marrow		
Erythroid cells	420,000	820,000
Granulocytes	720,000	580,000
Neutrophils	540,000	300,000
Promyelocytes	17,000	6,000
Myeloblasts	32,000	8,000

Table 15-6. Cell counts (cells/mm.^{3}) in guinea pigs before and 3 days after an acute hemorrhage of 36 per cent of the blood volume. These data are consistent with the view that the guinea pig has a significant reservoir of leukocytes in the marrow which he releases into the circulation in response to hemorrhage. There is little evidence, on the other hand, of a large reservoir of erythrocytes; the guinea pig responds to hemorrhage with increased erythrocyte production. (Calculated from the data of Harris, P.F., Harris, R.S., and Kugler, J.H.: Studies of the leukocyte compartment in guinea-pig bone marrow after acute haemorrhage and severe hypoxia: Evidence for a common stem-cell. Brit. J. Haemat., 12:419-432, 1966)

Control of Hemopoiesis

Red bone marrow produces erythrocytes (is erythropoietic, that is), leukocytes (leukopoietic), and platelets (thrombocytopoietic). Infection results in the inhibition of *erythropoiesis* and stimulation of leukopoiesis. Hemorrhaging, on the other hand, has been shown to cause the release of stored leukocytes into the blood and increased erythropoiesis (Table 15-6). *Hypoxia* stimulates erythropoiesis and inhibits leukopoiesis. The mechanisms responsible for these actions are not clear, but if we assume that all or most of the blood cells come from a single cell type, the reticuloendothelial cell (Fig. 15-8), then we might speculate that increased red cell production results in part from increased hemopoietic activity and in part from decreased leukopoietic activity (formation of pronormoblasts at the expense of progranulocyte production).

The means by which hypoxia increases hemopoietic activity are numerous. There is evidence that renal hypoxia results in the kidney's releasing a *renal erythropoietic factor,* which in plasma forms erythropoietin. This substance, in turn, increases the production and release of red blood cells. Hypoxia may also act on the hypothalamus or on certain endocrine glands to cause increased hemopoietic activity. Androgens, estrogens, thyroxin, cortisol, growth hormone, prolactin, angiotensin, and adrenocorticotrophic hormone (ACTH) have all been shown to influence hemopoiesis. *Adrenocorticotrophic hormone* (ACTH), for example, increases the number of circulating neutrophils, decreases the number of lymphocytes, and decreases the number of eosinophils. The neutrophil response occurs in the absence of an adrenal cortex, but both the lymphocyte and eosinophil responses apparently result from the action of ACTH on the adrenal cortex. *Growth hormone* increases the number of erythroblasts in the bone marrow, and androgen-treated animals show an increase in erythropoiesis-like activity in their plasma.

	Average Cells/mm.^{3}	Range Cells/mm.^{3}	Per cent of Total Leukocytes
Erythrocytes			
Men	5,000,000	$4.5\text{-}6.0 \times 10^6$	
Women	4,500,000	$4.3\text{-}5.5 \times 10^6$	
Leukocytes	9,000	5,000-10,000	100
Granulocytes			
Neutrophils	(5,400)	(3,000-6,000)	50-70
Eosinophils	(270)	(150-300)	1-3
Basophils	(60)	(0-100)	0-1
Agranulocytes			
Lymphocytes	(2,730)	(1,000-4,000)	20-40
Monocytes	(540)	(200-800)	4-8
Platelets	300,000	$2\text{-}5 \times 10^5$	

Table 15-7. Normal values for the formed elements in blood.

Leukocytes

The leukocytes (Table 15-7) are part of a broad system which removes from the body many unwanted cells, cell fragments, and particles. This *reticuloendothelial system*, as it is called, includes all phagocytic cells, most of which are in the blood, liver, spleen, lymph nodes, lung, and gastrointestinal tract. It consists of macrophages or histiocytes, which are not characteristically in the circulating blood, and microphages (neutrophils), which are found in the circulating blood. Blood also contains other leukocytes which probably (1) produce and carry antibodies, (2) release other important substances, such as enzymes, and (3) take up toxins. Unfortunately much of what we say about the leukocytes is speculation based on far too little experimental evidence.

Neutrophils. The neutrophil is a cell that can move at a speed of about 19 μ per minute and migrate through a capillary or sinusoid endothelium. It appears to be actively phagocytic while in the blood stream—phagocytosing bacteria, small insoluble particles, and fibrin. It has also been suggested that it functions as a *trephocyte* (a cell that supplies nutrients to other cells) (Gk. *trephein*, to nourish). Some have reported that it breaks off parts of its cytoplasm and that these fragments are used by other cells as a source of nutrients and enzymes. The neutrophil also binds *vitamin B₁₂* and releases it to the plasma when the plasma concentration of this vitamin is low.

Basophils. Little is known of the function of basophils. It has been estimated that they contain about half the normal blood histamine and large quantities of heparin, and increase in number during healing. The *histamine* and *heparin* are thought to be contained in the metachromatic granules of the basophil, and to be released in response to basophil damage. It has been suggested that the *histamine* (1) acts on the blood vessels to increase blood and intercellular fluid flow in the areas in which it is released and (2) acts on other blood cells to attract them (chemotaxis) to the area of basophil damage. The heparin that is released may help prevent clotting in areas where tissue damage has resulted in the release of procoagulants (tissue thromboplastin, for example). Many have suggested that the basophil is similar in function to the *mast cell*, a connective tissue cell that can also release histamine and heparin.

Eosinophils. Eosinophils are normally more abundant in tissue fluid than in the blood and are most concentrated in the epithelium of the bowel, respiratory tract, and skin. Their concentration increases in response to the injection of a foreign protein; during the decomposition of body protein; and during allergic reactions. They are less motile than the neutrophil and appear to function in the process of detoxification. One of the most important substances acted upon by the eosinophil is the antigen-antibody complex, which is probably phagocytosed by the eosinophil. The eosinophil may also be an important source of profibrinolysin.

Lymphocytes. There is growing evidence that the lymphocyte as well as the plasma cell is capable of producing antibodies. McGregor has

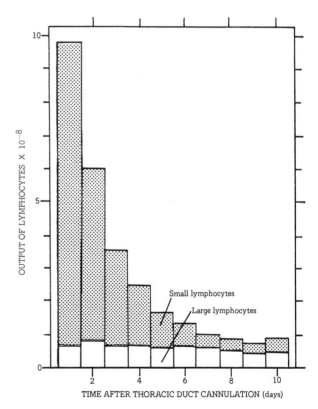

Fig. 15-17. Average output of small and large lymphocytes from the thoracic duct of 10 normal, adult Lewis rats on consecutive days after cannulation of the thoracic duct. Note that as the rats continued to lose lymph through their thoracic duct cannula the concentration of small lymphocytes in the lymph decreased. (*Modified from* McGregor, D. D.: Studies by thoracic duct drainage of the functions and potentialities of the lymphocyte. Fed. Proc., 25:1714, 1966)

noted, for example, that "rats can be depleted of small lymphocytes by cannulating the thoracic duct" and allowing lymph and associated cells to drain from the animal for several days (Fig. 15-17). Because of this depletion of small lymphocytes, the rats fail to show either an immune response to a first dose of tetanus toxoid or "the usual 'secondary' antibody response when challenged with tetanus toxoid 3 weeks later."

A number of other lymphocyte functions have been suggested but are much less widely accepted. It has been hypothesized, for example, that the lymphocyte is a totipotential cell, i.e., capable of forming a stem cell that can produce monocytes, megakaryocytes, granulocytes, and erythrocytes. Some have suggested that is is also capable of phagocytosis.

Monocytes. The monocyte, like the neutrophil, is motile, is capable of becoming phagocytic, and contains nucleases, proteinases, and carbohydrases. Unlike the neutrophil, it is rich in lipase. This makes it particularly effective in breaking down the lipoid capsule of the bacteria which cause tuberculosis (*Bacillus tuberculosis*) and leprosy (*Mycobacterium leprae*). The neutrophils and monocytes rapidly accumulate in an area of inflammation, the neutrophils predominating initially. The granulocytes disintegrate within about 3 days, after which the enlarged monocytes, the lymphocytes, and the hypertrophied lymphocytes predominate. Many believe the monocyte is a young cell that reaches its potential only after it leaves the blood stream.

Platelets. The blood platelets (thrombocytes) are cytoplasmic fragments split from megakaryocytes in the bone marrow. Their function in supplying procoagulants and a clot retraction factor, their role as plugs between fibrin threads, and their capacity to aggregate have been discussed in the preceding chapter.

Chapter 16

STRUCTURE
FRAGILITY
TRANSPORT OF GASES
OXYGEN TRANSPORT
CARBON DIOXIDE
TRANSPORT
BUFFERING ACTION

THE ERYTHROCYTE

Erythrocytes represent about 40 per cent of the blood volume and, as a consequence, are important in determining its viscosity and flow characteristics (Fig. 10-16). During its development, the mammalian erythrocyte loses its nucleus and accumulates the iron-containing protein *hemoglobin*, which constitutes approximately 25 per cent of the volume (3.4×10^8 molecules of hemoglobin per red blood cell) of the mature erythrocyte and is responsible for the transport of about 99 per cent of the O_2 found in whole blood (Table 16-1).

Red blood cells are also important in the transport of CO_2. Approximately 5 per cent of the CO_2 that enters the venous blood combines with hemoglobin (carbaminohemoglobin) and about 90 per cent of it is changed to H_2CO_3 and HCO_3^- by an enzyme, *carbonic anhydrase* (CA), found inside the erythrocyte. The intracellular conversion of CO_2 to HCO_3^- is as follows:

$$CO_2 + H_2O \overset{CA}{\rightleftharpoons} H_2CO_3 \rightleftharpoons H^+ + HCO_3^-$$

The carbonic anhydrase apparently accelerates both the rate of bicarbonate formation from CO_2 (systemic capillaries) and the formation of CO_2 from bicarbonate (pulmonary capillaries) by about 13,000 times.

STRUCTURE

The erythrocyte is well suited structurally to its role in the transport of O_2 and CO_2. Normally it has a diameter of about 8.5 μ and a thickness ranging between 1.0 and 2.4 μ (Table 16-2). For this reason erythrocytes must pass single file through the capillaries (Fig. 16-1). Their biconcave shape gives them a much larger surface for diffusion (163 μ^2) than would a spherical one (95 μ^2). It also means that no part of the cytoplasm is more than 1.2 μ from its cell membrane. A sphere of the same volume would have a radius of 2.75 μ, so that diffusion would take considerably longer to occur throughout the cell. The importance of these factors lies in the fact that an erythrocyte is in a capillary only about

	Arterial Blood (ml./100 ml.)	Venous Blood (ml./100 ml.)
Total O_2	20.3	15.5
(Dissolved O_2)	(0.3)	(0.12)
(O_2 combined with Hb)	(20)	(15.4)
(O_2 capacity of Hb)	(20.6)	(20.6)
Total CO_2	49.1	53.1
(Dissolved CO_2)	(2.4)	(2.7)
(Carbamino CO_2)	(2.2)	(3.3)
(CO_2 carried as HCO_3^-)	(44.5)	(47.1)
pH of plasma	7.40	7.376

Table 16-1. Transport of O_2 and CO_2 in the blood. The data are expressed in milliliters of O_2 or CO_2 per 100 ml. of whole blood. Note that most of the O_2 is transported as oxyhemoglobin, and most of the CO_2 as bicarbonate. The erythrocyte contains all the hemoglobin in the blood and most of the enzyme carbonic anhydrase, which is responsible for the formation of bicarbonate from CO_2 and water. Note too that venous blood is more acidic than arterial blood.

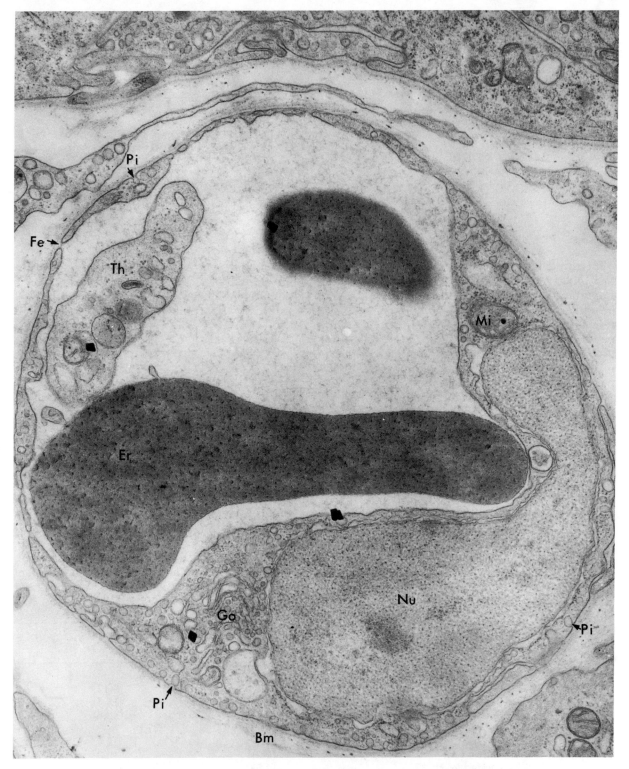

Fig. 16-1. Electron micrograph of a normal erythrocyte in a capillary of the gastric mucosa. Note the basement membrane (Bm), nucleus (Nu), Golgi complex (Go), mitochondria (Mi), pinocytotic vesicles (Pi), and fenestrations (Fe) of the capillary and its enclosed erythrocyte (Er) and platelet or thrombocyte (Th). The cytoplasm of the erythrocyte is relatively homogeneous, whereas that of the platelet contains an endoplasmic reticulum, mitrochondria, and large dense granules. (Rhodin, J. A. G.: An Atlas of Ultrastructure. P. 53. Philadelphia, W. B. Saunders, 1963)

Diameter	8.5 ± 0.41 μ
Maximum thickness	2.4 ± 0.13 μ
Minimum thickness	1.0 ± 0.08 μ
Area	163 μ^2
Volume	87 μ^3

Table 16-2. The dimensions (mean ± standard deviation) of normal erythrocytes. The maximum diameter is often given as 7.5 μ, but this is thought to be due to a shrinkage artifact. (From Burton, A.C.: Physiology and Biophysics of the Circulation. P. 33. Chicago, Year Book Publishers, 1965)

1 sec. and must, during this time, release or take up large quantities of O_2, CO_2, and HCO_3^-.

Another factor in the movement of O_2 and CO_2 into and out of the cell is the character of the cell membrane. In man this membrane, like that of other cells, is apparently associated with active transport systems that maintain differences in concentration between the intracellular and extracellular electrolytes (Table 16-3). Not all species, however, show the same concentration differences. Dogs and cats, for example, have more intracellular Na^+ than K^+.

Although Na^+ and K^+ are actively transported into and out of the erythrocyte in man, there is good evidence that O_2 and CO_2 move primarily by diffusion. Both gases are fat-soluble and move into the cell when their concentration in the plasma increases, and move out of the cell when their concentration in the plasma decreases. The importance of the *fat-solubility* of these gases is related to the chemical nature of the cell membrane. We know that this membrane consists of proteins, mucopolysaccharides, and lipids, but we do not know how these are arranged. Three suggestions concerning the structural characteristics of the membrane are presented in Figure 16-2. It is recognized in each that a good part of the membrane is lipid; undoubtedly this in itself is important in the diffusion of the lipid-soluble O_2 and CO_2 across the membrane.

FRAGILITY

Erythrocytes placed in water (1) swell, (2) become spherical instead of biconcave, and (3) lose their hemoglobin and other intracellular contents to their external environment. The loss of hemoglobin (hemolysis) following the formation of a red cell sphere apparently results from the increased permeability of the erythrocyte to hemoglobin, rather than from a rupture of the cell membrane (since the cells can regain their original biconcave shape after losing the hemoglobin). What remains of the erythrocyte after hemolysis is called a *red cell ghost*. The normal erythrocyte, however, does not release hemoglobin to the environment in all hypotonic solutions. Usually a small percentage of erythrocytes hemolyze in solutions having half the tonicity of plasma (0.45% saline, for example), and most of them hemolyze in solutions two fifths the tonicity of plasma (Fig. 16-3).

In disease the body's erythrocytes may become more fragile, i.e., they may hemolyze outside the body at higher environmental tonicities and be destroyed inside the body at a greater rate. This occurs in hemolytic *jaundice* and paroxysmal *nocturnal hemoglobinuria*. The theory has been advanced that in the latter condition a defective erythrocyte releases hemoglobin during sleep, when the pH of the blood tends to fall. The hemolytic phenomenon may be temporarily reduced by administering $NaHCO_3$ at bedtime. Normally, however, the plasma contains only 2 to 5 mg. of hemoglobin per 100 ml., and hemoglobin appears in the urine only after a concentration of over 135 mg.

Ions	Serum (mM./l.)	Erythrocytes (mM./l.)	Ratio of Erythrocyte to Serum	Ratio of Erythrocyte to Other Cells
Na^+	135	18	1:7	1.5:1
K^+	5	81	16:1	1:1.9
Cl^-	104	52	1:2	13:1
HCO_3^-	26	19	1:1.3	2.4:1

Table 16-3. Concentration of various ions in the serum and erythrocyte. Note that a human erythrocyte differs from most other cells in having a much higher intracellular concentration of Cl^-.

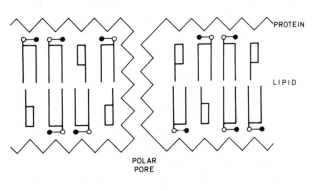

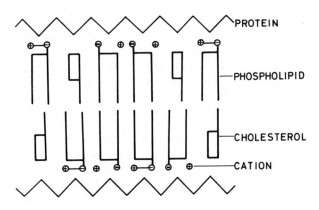

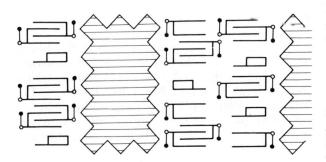

Fig. 16-2. Three hypotheses on the structure of the cell membrane. In the upper (Danielli, 1958) and middle (Booy and Bungenberg de Jong, 1956) illustrations the membrane is characterized as a bimolecular lipid compartment sandwiched between 2 layers of protein. In the lower illustration (Parpart and Ballentine, 1952) the cell membrane is pictured as a mosaic of protein (lined areas) and lipid. (van Deenen, L. L. M., and de Gier, J.: Chemical composition and metabolism of lipids in red cells of various animal species. *In* Bishop, C., and Surgenor, D. M. (eds.): The Red Blood Cell. Pp. 247, 248, 276. New York, Academic Press, 1964)

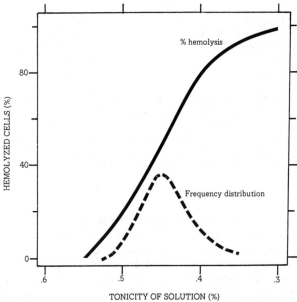

Fig. 16-3. Per cent hemolysis of normal erythrocytes at different tonicities. A frequency distribution curve is also shown. Note that about half of the cells will hemolyze in 0.45% saline. (*Redrawn from* Ponder, E.: Hemolysis and Related Phenomena. P. 102. New York, Grune & Stratton, 1948)

hemoglobin per 100 ml. plasma has been reached. In cases of rapid blood destruction, on the other hand, hemoglobin levels may reach 500 mg. hemoglobin per 100 ml. plasma and lead to a severe hypotension, the intravascular deposition of fibrin, renal shutdown, uremia, and death.

TRANSPORT OF GASES

When blood reaches the pulmonary capillaries it comes close to the air sacs (alveoli) of the lungs. At this point the gases in the alveoli and the gases dissolved in the plasma tend to reach an equilibrium. The amount of O_2, CO_2, N_2, and so on that diffuses out of the alveoli and dissolves in the plasma depends in part on the *per cent concentration* of each gas in the alveolus, and in part on the total *pressure* (usually atmospheric pressure) exerted on the mixture of alveolar gases. Dalton studied the relationship between gases in a mixture and concluded that:

Each gas in a mixture of gases behaves as if it alone occupied the total volume and exerts a pressure (its partial pressure) independently of the other gases present. The sum of the partial pressures of each gas is equal to the total pressure exerted by the mixture.

A practical application of *Dalton's law* is presented in Table 16-4. Here we note that if alveolar air contains 13.3 per cent O_2 and is exposed to a pressure of 760 mm. Hg, it exerts a partial pressure (P_{O_2}) of 101 mm. Hg:

$$P_{O_2} = (760 \text{ mm. Hg})(0.133) = 101 \text{ mm. Hg}$$

These same calculations can be made for CO_2 ($P_{CO_2} = 40$ mm. Hg), for N_2 ($P_{N_2} = 572$ mm. Hg), and the water vapor which the inhaled air acquires ($P_{H_2O} = 47$ mm. Hg). For the most part these four alveolar gases account for the total alveolar pressure:

$$\text{Alveolar pressure} = P_{O_2} + P_{CO_2} + P_{N_2} + P_{H_2O} = 101 + 40 + 572 + 47 = 760 \text{ mm. Hg}$$

The partial pressure of a gas in a mixture of gases is an important determinant of how much of that gas becomes dissolved in a liquid (*Henry's law*). Thus if there is either an increase in the concentration of a gas in the alveolus or an increase in the pressure exerted upon it, the concentration of that gas in the plasma increases. If, for example, we wish to increase the amount of O_2 carried by someone's plasma, we could have him breathe 100 per cent O_2 or we could place him in a hyperbaric (high-pressure) chamber.

OXYGEN TRANSPORT

The amount of oxygen that passes from the blood to the perivascular tissues or from the perivascular tissues to the blood depends on (1) the difference in the P_{O_2} between the blood and perivascular spaces, (2) the diffusion characteristics of the tissues, (3) the oxygen-carrying capacity of the blood, and (4) the speed with which the blood is traveling. In a resting person blood remains in the pulmonary and systemic capillaries for about 0.6 to 1 sec. This is normally enough time for the O_2 in the blood and in the perivascular space to reach an equilibrium.

Oxyhemoglobin Dissociation Curve

In Table 16-4 we note that the P_{O_2} of blood is less than the P_{N_2}. This is due to the higher percentage of N_2 in the alveolar air. On the other hand, the total amount of O_2 in the blood is many times that of N_2, since hemoglobin rapidly combines with and thus removes O_2 from solution as the P_{O_2} of plasma increases. The amount of O_2 carried by blood leaving the lungs is dependent upon the P_{O_2} in the alveolus and the amount of hemoglobin that passes through the alveolar capillaries. At a plasma P_{O_2} of 100 mm. Hg, normal blood at a resting flow rate carries 20 ml. of O_2 per 100 ml. of blood, and at a P_{O_2} of 40 mm. Hg, 15 ml. per 100 ml. of blood.

The normal relationship between the P_{O_2} of plasma and the degree of O_2 saturation of hemoglobin is shown in Figure 16-4. Note that at the usual P_{O_2} of arterial blood (100 mm. Hg), hemoglobin is 98 per cent saturated with O_2, and at the average P_{O_2} of venous blood (40 mm. Hg), 75 per cent saturated. In passing through the systemic capillaries, that is, each 100 ml. of blood has given up an average of 4.68 ml. of O_2 and contains an average *reserve* of 15.12 ml. of O_2. Some of this reserve is utilized by the tissues if their P_{O_2} is reduced still further. The P_{O_2} in the perivascular fluid of inactive skeletal muscle, for example, is about 40 mm. Hg (Fig. 16-4) but can go below 5 mm. Hg during maximum exertion. In the heart of a resting person the P_{O_2} around the coronary capillaries is about 22 mm. Hg.

Several aspects of the S-shaped curve in Figure 16-4 should be emphasized. One is its relative

Gas	Concentration	Partial Pressure		
	Alveolar Air (%)	Alveolar Air	Pulm. Art. Blood	Pulm. Vein Blood
O_2	13.3	101 mm. Hg	40 mm. Hg	100 mm. Hg
CO_2	5.3	40 mm. Hg	46 mm. Hg	40 mm. Hg
N_2	75.2	572 mm. Hg	572 mm. Hg	572 mm. Hg
H_2O vapor	6.2	47 mm. Hg		

Table 16-4. Partial pressures of various gases in the lung alveolus and blood at an atmospheric pressure of 760 mm. Hg.

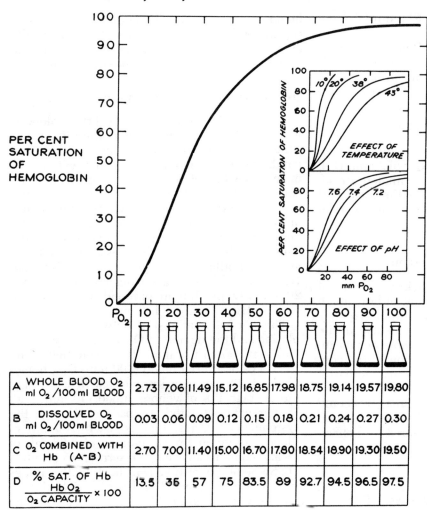

Fig. 16-4. Oxyhemoglobin dissociation curves for hemoglobin A. The large graph represents the situation in blood at a temperature of 38°C. and a pH of 7.40. The inserts indicate how the curve is modified by changes in temperature and pH. Fetal hemoglobin (hemoglobin F) and abnormal hemoglobins (B through E and G through H) do not exhibit this same curve. (*From* Physiology of Respiration by J. H. Comroe, Jr. P. 161. Copyright © 1965, Year Book Medical Publishers, Inc.; used by permission)

flatness at a P_{O_2} above 70 mm. Hg. This means that we do not under normal conditions appreciably increase the amount of O_2 carried by the hemoglobin when we inhale 100 per cent oxygen instead of 20 per cent oxygen, or when we hyperventilate (increase the amount of air inhaled per minute). And in fact, the hemoglobin does not become 100 per cent saturated until we reach a P_{O_2} of about 150 mm. Hg. The flatness of the curve also means that we can go well above sea level without significantly reducing the amount of O_2 in the blood. At sea level the P_{O_2} in the alveolus is approximately 100 mm. Hg, whereas at an altitude of 10,000 ft. it is about 70 mm. Hg.

Another important aspect of the hemoglobin dissociation curve is the steepness of its slope between a P_{O_2} of 10 and 40 mm. Hg. This is the P_{O_2} in the perivascular space of metabolically active tissues. The low P_{O_2} in these tissues facilitates the release of large quantities of O_2 by the hemoglobin. The high P_{O_2} in the lungs facilitates the near saturation of hemoglobin with O_2.

Modification of the Dissociation Curve

You will also note in Figure 16-4 that the amount of oxygen released by hemoglobin is dependent upon the temperature and pH of the hemoglobin environment. Decreases in *temperature* shift the curve to the left, increases, to the right, which is to say that decreases in temperature result in the hemoglobin becoming saturated at a lower P_{O_2}. At 10°C., for example, hemoglobin saturated with O_2 gives up practically no O_2 when exposed to a P_{O_2} of 25 mm. Hg. At 38°C., this same hemoglobin gives up 10 ml. of O_2 per 100 ml. of blood at a P_{O_2} of 25 mm. Hg.

Increases in pH (decreases in acidity) tend to shift the hemoglobin dissociation curve to the left, while decreases shift it to the right (*Bohr effect*). This is because the hemoglobin molecule (HHb) has less of an affinity for O_2 than does the hemoglobin ion (Hb^-):

$$HbO_2^- + H^+ = HHb + O_2$$

At a pH of 7.4 and a P_{O_2} of 30 mm. Hg, hemoglobin is 57 per cent saturated with O_2, but at a pH of 7.2 and a P_{O_2} of 30 mm. Hg, it is only 45 per cent saturated. Thus in tissues which become more metabolically active (Fig. 16-9), and as a consequence more acid, more O_2 is released by the hemoglobin. It should be noted, however, that changes in pH do not have a pronounced effect upon oxygen uptake at the P_{O_2} generally found in the lung.

In Figure 16-5 we note that increases in P_{CO_2}, like increases in acidity, shift the dissociation curve to the right. This is partly because CO_2 increases the acidity of the blood, but also because it combines with hemoglobin to form carbaminohemoglobin ($HbCO_2^-$). Since carbaminohemoglobin has much less affinity for O_2 than does hemoglobin (HHb) or partially oxygenated hemoglobin, an increase in P_{CO_2} tends to decrease the amount of oxyhemoglobin ($HHbO_2$):

$$HHbO_2 + CO_2 \rightleftharpoons HbCO_2^- + O_2 + H^+$$

Another condition affecting the dissociation curve is the presence of CO in the alveolar air. You will note in Figure 16-6 that hemoglobin has a much greater affinity for CO than for O_2. In the absence of O_2 a P_{CO} of 0.5 mm. Hg ties up about 98 per cent of the oxygen-binding sites on the hemoglobin molecule. When the P_{CO} is 0.5 mm. Hg and the P_{O_2} 100 mm. Hg, CO and O_2 each

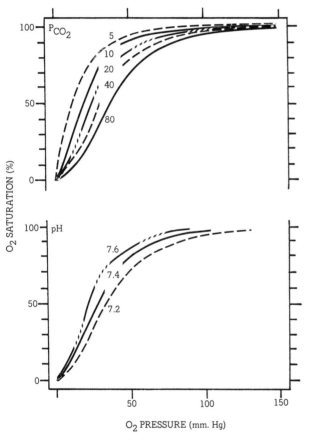

Fig. 16-5. Effect of the P_{CO_2} and the pH on the oxyhemoglobin dissociation curve. (*Redrawn from* Roughton, F. J. W.: Transport of oxygen and carbon dioxide. *In* Fenn, W. O., and Rahn, H. (eds.): Handbook of Physiology. Sec. 3 (Respiration), vol. I, pp. 776. Washington, American Physiological Society, 1964; *top*, from data of Bohr, C., et al.: Skand. Arch. Physiol., 16:402-412, 1904)

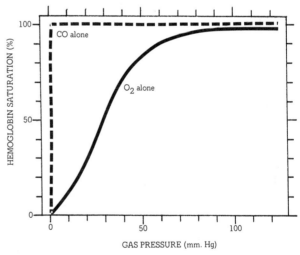

Fig. 16-6. Comparison of the carbon monoxide-hemoglobin dissociation and oxyhemoglobin dissociation curves. Maximum saturation of hemoglobin with O_2 occurs at a P_{O_2} above 120 mm. Hg and maximum saturation with CO occurs at a P_{CO} below 1 mm. Hg. Since both CO and O_2 compete for the same sites on the hemoglobin molecule, a small amount of CO may cause death due to hypoxia (anemic hypoxia). (*Redrawn from* The Lung, Second Edition, by J. H. Comroe, Jr., et al. P. 120. Copyright © 1962, Year Book Medical Publishers, Inc.; used by permission)

combine with about half the oxygen-binding sites. A concentration of 0.1 per cent CO (P_{CO} = 0.7 mm. Hg) in the air so decreases the O_2-carrying capacity of the blood that it is considered lethal; one of 0.005 per cent in a factory during an 8-hour day is regarded as dangerous. It is also interesting to note that the CO concentrations to which car passengers are exposed on crowded submerged expressways reportedly range from 0.0015 to 0.012 per cent.

It has been noted that drugs such as nitrites, aniline, sulfonamides, acetanilide, pamaquine, primaquine, and phenylhydrazine oxidize the iron in hemoglobin:

$$Fe^{++} \rightarrow Fe^{+++}$$

This oxidation process changes the hemoglobin to *methemoglobin*. It should not be confused with the process of oxygenation—a reaction involving the addition of O_2 to hemoglobin. Small amounts of methemoglobin are also formed in healthy persons, but are rapidly converted back to hemoglobin by the DPNH and *methemoglobin reductase* found in red cells. Since methemoglobin does not give up O_2 so readily as hemoglobin, a severe case of methemoglobinemia—like CO poisoning—produces severe tissue hypoxia.

CARBON DIOXIDE TRANSPORT

The amount of CO_2 dissolved in the plasma is dependent upon the partial pressure of CO_2 (P_{CO_2}) in the perivascular spaces of the alveoli and systemic capillaries (Table 16-4). The P_{CO_2} of the plasma, in turn, determines the amount of CO_2 carried in the blood as H_2CO_3, as HCO_3^-, and as carbamino compounds. A list of the amounts of CO_2 carried in each of these forms is presented in Table 16-1. Figure 16-7 shows a comparison between the oxygen dissociation and carbon dioxide dissociation curves for whole blood. Note that the CO_2 curve, unlike the O_2 curve, does not become flat at a partial pressure above 70 mm. Hg. It is not S-shaped, in other words, since blood does not approach CO_2 saturation at a P_{CO_2} of 70 mm. Hg.

Carbaminohemoglobin

Carbon dioxide in the blood combines with the free amino groups of proteins and amino acids to form carbamino compounds:

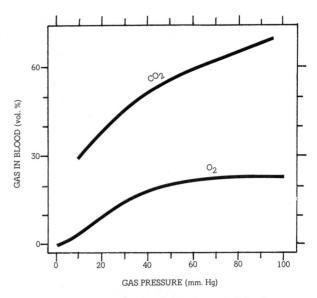

Fig. 16-7. Comparison of the O_2 and CO_2 dissociation curves for whole blood. Note that the curve for CO_2 does not have a flat portion. (*Redrawn from* The Lung, Second Edition, by J. H. Comroe, Jr., et al. P. 154. Copyright © 1962, Year Book Medical Publishers, Inc.; used by permission)

$$R-NH_2 + CO_2 \rightleftharpoons R-NHCOO^- + H^+$$

The blood protein most important for carrying CO_2 in this form is hemoglobin:

$$HbNH_2 + CO_2 \rightleftharpoons HbNHCOO^- + H^+$$

The amount of CO_2 carried by hemoglobin as carbaminohemoglobin is dependent upon both the P_{CO_2} and the P_{O_2} of the blood (Fig. 16-8). The importance of the P_{O_2} lies in the fact that oxyhemoglobin has less of an affinity for CO_2 than does unoxygenated hemoglobin. Hence in the systemic capillaries, where hemoglobin tends to give up O_2, it also tends to combine with CO_2.

Bicarbonate

In venous blood a total of 17 mM. per liter of CO_2 is carried in the plasma, and in the red cell, 6.2 mM. per liter. In both the plasma and the erythrocytes most of this is in the form of HCO_3^-. The cells surrounding the systemic capillaries (Fig. 16-9) are constantly forming CO_2. This CO_2 tends to diffuse throughout the intercellular spaces and to create a P_{CO_2} greater than that found in arterial blood. Thus when arterial blood

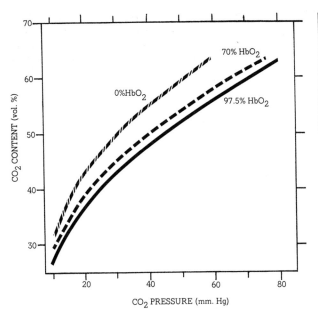

Fig. 16-8. Relationship between the P_{CO_2} and the carbon dioxide content of blood. Note that at any P_{CO_2} the carbon dioxide content of the blood goes down as the the concentration of oxyhemoglobin goes up. (*Redrawn from* The Lung, Second Edition, by J. H. Comroe, Jr., et al. P. 154. Copyright © 1962, Year Book Medical Publishers, Inc.; used by permission)

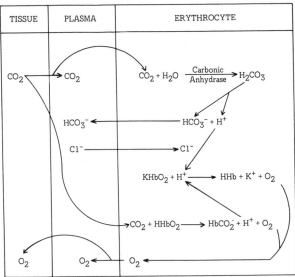

Fig. 16-9. A summary of the major chemical changes which occur in the erythrocyte and plasma of blood entering a systemic capillary. Note that CO_2 diffuses into the plasma and erythrocyte and O_2 diffuses out. In the pulmonary capillary the opposite occurs and all of the above changes are reversed.

enters the capillaries, CO_2 diffuses into the plasma and erythrocytes. During this diffusion a small amount of CO_2 combines with water to form H_2CO_3 and $H^+ + HCO_3^-$, but it is in the erythrocytes that most of the bicarbonate is formed. The erythrocytes contain the enzyme carbonic anhydrase in quantities sufficient to convert CO_2 rapidly into bicarbonate:

$$CO_2 + H_2O \underset{\text{anhydrase}}{\overset{\text{carbonic}}{\rightleftharpoons}} H_2CO_3 \rightleftharpoons H^+ + HCO_3^-$$

Most of the HCO_3^- formed during this reaction diffuses out of the erythrocyte through the cell membrane. As the HCO_3^- diffuses out, a negatively charged ion tends to diffuse in to replace it. Since the main anion of plasma is Cl^-, this is generally the ion that replaces the HCO_3^- (Table 16-3). This exchange of anions has come to be known as the "chloride shift." Cations do not move through the cell membrane of the human erythrocyte so freely as do Cl^- and HCO_3^-.

You will note in Figure 16-9 that the H^+ resulting from the formation HCO_3^- from CO_2 and H_2O is removed from solution by hemoglobin:

$$H^+ + KHbO_2 \rightleftharpoons HHb + K^+ + O_2$$

This helps prevent excessive lowering of the pH when CO_2 is added to the blood, and facilitates the release of O_2 by hemoglobin.

A small percentage of CO_2, as I have mentioned, also combines with hemoglobin. This too facilitates the release of O_2 by the hemoglobin. Thus in the systemic capillaries the addition of CO_2 facilitates the release of O_2 by hemoglobin, and in the pulmonary capillaries the loss of CO_2 facilitates the uptake of O_2 by hemoglobin. We have also noted that the loss of O_2 by the blood facilitates the uptake of CO_2, and that the uptake of O_2 by the blood facilitates the loss of CO_2 by the blood. We have a system, in other words, that is well suited for carrying large amounts of O_2 when the P_{CO_2} is low, and large quantities of CO_2 when the P_{O_2} is low.

BUFFERING ACTION

The blood contains a number of systems capable of maintaining relatively constant H^+ and OH^- concentrations. It has ways, that is, of maintaining a relatively constant pH even when

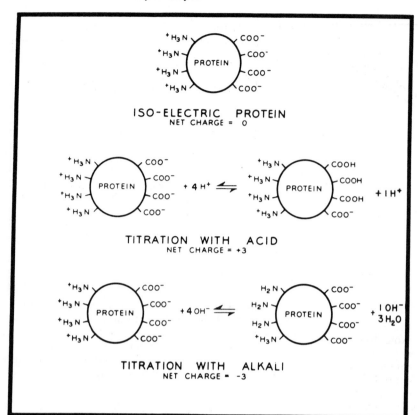

Fig. 16-10. Diagram of the buffering action of a protein. (Davenport, H. W.: The ABC of Acid-Base Chemistry. 4th ed., p. 13. Chicago, University of Chicago Press, 1958)

acidic or basic materials are added to it. The three systems responsible for this buffering action are (1) the $HPO_4^{--}/H_2PO_4^-$ system (least important), (2) the HCO_3^-/CO_2 system, and (3) the Pr^-/HPr (protein) system. When H^+ is added to the blood, for example, it combines with HPO_4^{--} or HCO_3^-, and protein anions to produce $H_2PO_4^-$, H_2CO_3, CO_2, and protein. When OH^- is added to the blood, $H_2PO_4^-$, H_2CO_3, and protein give up an H^+ to combine with the OH^-. We have already discussed the HCO_3^-/CO_2 system extensively, as well as the importance of the erythrocyte to it. In this section we shall be concerned primarily with the plasma proteins and the hemoglobin.

Figure 16-10 is a schematic diagram of how a protein responds to changes in pH. Proteins are amphoteric (Gk., *amphoteroi*, both). They are capable, in other words, both of donating H^+ when the pH is elevated, and of combining with H^+ when the pH is lowered. Hemoglobin is particularly important in this regard for two reasons. First, it has a much greater buffering capacity than the other blood proteins, and second, it is found in a much higher concentration than the other blood proteins:

Hemoglobin — 150 gm. per liter of blood
Plasma proteins — 38.5 gm. per liter of blood

If, for example, we change the pH of the blood from 7.5 to 6.5, hemoglobin and the plasma proteins tend to counter the change by combining with H^+ in the following manner:

Hemoglobin reaction 0.183 mEq. of H^+ per gram of hemoglobin (27.5 mEq. of H^+ per liter of blood)

Plasma protein reaction 0.11 mEq. of H^+ per gram of plasma protein (4.24 mEq. of H^+ per liter of blood)

When the pH is changed from 7.5 to 6.5, in other words, blood proteins combine with 31.74 mEq. of H^+ — hemoglobin being responsible for about 87 per cent (27.5/31.74) of this buffering action.

REFERENCES

Ahlquist, R. P.: A study of the adrenotropic receptors. Am. J. Physiol., 153:586-600, 1948.

Albright, J. F., and Makinodan, T.: Growth and senescence of antibody-forming cells. J. Cell. Physiol., 67(Supp. 1):185-206, 1966.

Aoki, V. S., and Brody, M. J.: Medullary control of vascular resistance and electrophysiological analysis. Circ. Res. (Suppl. 1), 18:73-85, 1966.

Bader, H.: The anatomy and physiology of the vascular wall. In Hamilton, W. F. (ed.): Handbook of Physiology. Sec. 2 (Circulation), vol. II, pp. 865-889. Washington, American Physiological Society, 1963.

Ballerini, G., and Seegers, W. H.: A description of clot retraction as a visual experience. Thromb. Diath. Haemorrh., 3:147-164, 1959.

Barnhart, M. I.: Importance of neutrophilic leukocytes in the resolution of fibrin. Fed. Proc., 24:846-853, 1965.

Bennhold, H.: Transport function of the serum protein. In Desgrez, P., and De Traverse, P. M. (eds.): Transport Function of Plasma Proteins. Pp. 1-12. New York, Elsevier Publishing Co., 1966.

Bilgutay, A. M., and Lillehei, C. W.: Surgical treatment of hypertension with reference to baropacing. Am. J. Cardiol., 17:663-667, 1966.

Bishop, C., and Surgenor, D. M.: The Red Blood Cell. New York, Academic Press, 1964.

Boggs, D. R.: Hemostatic regulatory mechanisms of hematopoiesis. Ann. Rev. Physiol., 28:39-56, 1966.

Browse, N. L., Donald, D. E., Shepherd, J. T.: Role of the veins in the carotid sinus reflex. Am. J. Physiol., 210:1424-1434, 1966.

Browse, N. L., and Shepherd, J. T.: Response of veins of canine limb to aortic and carotid chemoreceptor stimulation. Am. J. Physiol., 210:1435-1441, 1966.

Burton, A. C.: Physiology and Biophysics of the Circulation. Chicago, Year Book Publishers, 1965.

―――: Physical principles of circulatory phenomena: The physical equilibria of the heart and blood vessels. In Hamilton, W. F. (ed.): Handbook of Physiology. Sec. 2 (Circulation), vol. I, pp. 85-106. Washington, American Physiological Society, 1962.

Castleman, B., and Kibbee, B. U.: Case records of the Massachusetts General Hospital. New Eng. J. Med., 269:97-101, 1963.

Chapman, C. B., and Mitchell, J. H.: The physiology of exercise. Sci. Am., 212:88-96, 1965.

Chien, S., et al.: Blood volume and age: Repeated measurements on normal men after 17 years. J. Appl. Physiol., 21:583-588, 1966.

Clark, E. R., and Clark, E. L.: Caliber changes in minute blood vessels observed in the living mammal. Am. J. Anat., 73:215-250, 1943.

Cline, M. J.: Metabolism of the circulating leukocyte. Physiol. Rev., 45:674-720, 1965.

Coleman, F., Crowell, J. W., Smith, E. S., and Wilson, R. M.: A hypotonic solution for emergency blood substitution. Surgery, 60:392-394, 1966.

Comroe, J. H.: The peripheral chemoreceptors. In Fenn, W. O., and Rahn, H. (eds.): Handbook of Physiology. Sec. 3 (Respiration), vol. I, pp. 557-583. Washington, American Physiological Society, 1964.

―――: Physiology of Respiration. Chicago, Year Book Publishers, 1965.

Comroe, J. H., Forster, R. E., Dubois, A. B., Briscoe, W. A., and Carlsen, E.: The Lung. Chicago, Year Book Publishers, 1955.

Davenport, H. W.: The ABC of Acid-Base Chemistry. 4th ed. Chicago, University of Chicago Press, 1958.

Davis, H.: Excitation of auditory receptors. In Magoun, H. W. (ed.): Handbook of Physiology. Sec. 1 (Neurophysiology), vol. 1, pp. 565-584. Washington, American Physiological Society, 1962.

―――: Peripheral coding of auditory information. In Rosenblith, W. A. (ed.): Sensory Communication. Pp. 119-141. Cambridge, MIT Press, 1961.

Davson, H.: Intracranial and intraocular fluids. In Magoun, H. W. (ed.): Handbook of Physiology. Sec. 1 (Neurophysiology), vol. III, pp. 1761-1786. Washington, American Physiological Society, 1960.

Dekaban, A.: Neurology of Infancy. Baltimore, Williams & Wilkins, 1959.

Desgrez, P., and De Traverse, P. M. (eds.): Transport Function of Plasma Proteins. New York, Elsevier Publishing Co., 1966.

Doyle, J. T.: Etiology of coronary disease: Risk factors influencing coronary disease. Mod. Conc. Cardiovasc. Dis., 35:1-6, 1966.

Drinker, C. F., and Yoffey, J. M.: Lymphatics, Lymph, and Lymphoid Tissue: Their Physiological and Clinical Significance. Cambridge, Harvard University Press, 1941.

Elam, J. O.: Respiratory and circulatory resuscitation. In Fenn, W. O., and Rahn, H. (eds.): Handbook of Physiology. Sec. 3 (Respiration), vol. II, pp. 1265-1312. Washington, American Physiological Society, 1965.

Freda, V. J.: Recent obstetrical advances in the Rh problem. Bull. N. Y. Acad. Med., 42:474-505, 1966.

Frye, R. L., and Braunwald, E.: Studies on Starling's law of the heart. I. The circulatory response to acute hypervolemia and its modification by ganglionic blockade. J. Clin. Invest., 39:1043-1050, 1960.

Gamble, J. L.: Chemical Anatomy, Physiology, and Pathology of Extracellular Fluid. 6th ed. Cambridge, Harvard University Press, 1954.

Gilmore, J. P., and Daggett, W. M.: Response of the chronic cardiac denervated dog to acute volume expansion. Am. J. Physiol., 210:509-512, 1966.

Glasser, O. (ed.): Medical Physics. Vol. II, p. 190. Chicago, Year Book Publishers, 1950.

Glees, P.: Neuroglia Morphology and Function. Oxford, Blackwell Scientific Publications 1955.

Goldstein, R., Bunker, J. R., and McGovern, J. J.: The effect of storage of whole blood and anticoagulants upon certain coagulation factors. Ann. N. Y. Acad. Sci., *115*:422-442, 1964.

Green, H. D., and Kepchar, J. H.: Control of peripheral resistance in major systemic vascular beds. Physiol. Rev., *39*:617-686, 1959.

Green, H. D., and Rapela, C. E.: Blood flow in passive vascular beds. Circ. Res. (Suppl. 1), *14* and *15*: 11-16, 1964.

Green, H. D., Rapela, C. E., and Conrad, M. C.: Resistance (conductance) and capacitance phenomena in terminal vascular beds. *In* Hamilton, W. F. (ed.): Handbook of Physiology. Sec. 2 (Circulation), vol. II, pp. 935-960. Washington, American Physiological Society, 1963.

Guazzi, M., Baccelli, G., and Zanchetti, A.: Carotid body chemoreceptors: Physiologic role in buffering fall in blood pressure during sleep. Science, *153*: 206-208, 1966.

Guyton, A. C., et al.: Evidence for tissue oxygen demand as the major factor causing autoregulation. Circ. Res. (Suppl. 1), *14* and *15*:60-69, 1964.

Haddy, F. J.: Effect of histamine on small and large vessel pressures in the dog foreleg. Am. J. Physiol., *198*:161-168, 1960.

———: Role of chemicals in local regulation of vascular resistance. Circ. Res. (Suppl. 1), *18*:14-22, 1966.

Haddy, F. J., and Scott, J. B.: Metabolically linked vasoactive chemicals in local regulation of blood flow. Physiol. Rev., *48*:688-707, 1968.

Halpern, B.: The reticuloendothelial system and immunity. Triangle, *6*:174-181, 1964.

Hamilton, W. F.: The arterial pulse. *In* Luisada, A. (ed.): Cardiovascular Functions. Pp. 132-138. New York, McGraw-Hill, 1962.

Hammond, E.: Smoking in relation to mortality and morbidity. Findings in first thirty four months of follow-up in a prospective study started in 1959. J. Nat. Cancer Inst., *32*:1161-1188, 1964.

Hardy, M.: Observations on the innervation of the macula sacculi in man. Anat. Rec., *59*:403-418, 1935.

Harper, H. H.: Review of Physiological Chemistry. 12th ed. Los Altos, Lange Medical Publications, 1969.

Harris, P. F., Harris, R. S., and Kugler, J. H.: Studies of the leukocyte compartment in guinea-pig bone marrow after acute haemorrhage and severe hypoxia: Evidence for a common stem-cell. Brit. J. Haemat., *12*:419-432, 1966.

Harris, T. N., et al.: Recent advances on the biology and function of the lymphocyte. Fed. Proc., *25*: 1711-1741, 1966.

Haurowitz, F.: Antibody formation. Physiol. Rev., *45*: 1-47, 1965.

Henry, R. L.: Leukocytes and thrombosis. Thromb. Diath. Haemorrh., *13*:35-46, 1965.

Henry, R. L., and Steiman, R. H.: Mechanisms of hemostasis. Microvasc. Res. J., *1*:68-82, 1968.

Heymans, C., and Neil, E.: Reflexogenic Areas of the Cardiovascular System. London, J. & A. Churchill, 1958.

Hytten, F. E., and Leitch, I.: The Physiology of Human Pregnancy. Oxford, Blackwell Scientific publications, 1964.

Jaques, L. B.: The pharmacology of heparin and heparinoids. Progr. Med. Chem., *5*:139-198, 1967.

Jeffords, J. V., and Knisely, M. H.: Concerning the geometric shapes of arteries and arterioles. Angiology, *7*:105-136, 1956.

Johnson, S. A., and Greenwalt, T. J.: Coagulation and Transfusion in Clinical Medicine. Boston, Little, Brown & Co., 1965.

Jones, R. D., and Berne, R. M.: Evidence for a metabolic mechanism in autoregulation of blood flow in skeletal muscle. Circ. Res., *17*:540-554, 1965.

Keele, C. A., and Neil, E.: Sampson Wright's Applied Physiology. 11th ed. New York, Oxford University Press, 1965.

Knisely, M. H.: Intravascular erythrocyte aggregation (blood sludge). *In* Hamilton, W. F. (ed.): Handbook of Physiology. Sec. 2 (Circulation), vol. III, pp. 2249-2292. Washington, American Physiological Society, 1965.

Knisely, M. H., Warner, L., and Harding, F.: Antemortem settling. Angiology, *11*:535-588, 1960.

Korner, P. I.: The effect of section of the carotid sinus and aortic nerves on the cardiac output of the rabbit. J. Physiol., *180*:266-278, 1965.

Landis, E. M., and Pappenheimer, J. R.: Exchange of substances through the capillary walls. *In* Magoun, H. W. (ed.): Handbook of Physiology. Sec. 1 (Neurophysiology), vol. II, pp. 961-1034. Washington, American Physiological Society, 1963.

Langman, J.: Medical Embryology. Baltimore, Williams & Wilkins, 1963.

Laragh, J. H., Sealey, J. E., and Sommers, S. C.: Patterns of adrenal secretion and urinary excretion of aldosterone and plasma renin activity in normal and hypertensive subjects. Circ. Res. (Suppl. 1), *18* and *19*:158-174, 1966.

Levinson, G. E., Pacifico, A. D., and Frank, M. J.: Studies of cardiopulmonary blood volume. Circulation, *33*:347-356, 1966.

Levitz, M.: Conjugation and transfer of fetal-placental steroid hormones. J. Clin. Endocr., *26*:773-777, 1966.

Liebow, A.: Situations which lead to changes in vascular patterns. *In* Hamilton, W. F. (ed.): Handbook of Physiology. Sec. 2 (Circulation), vol. II, pp. 1251-1276. Washington, American Physiological Society, 1963.

Livingston, R. B.: Central control of receptors and sensory transmission systems. *In* Magoun, H. W. (ed.): Handbook of Physiology. Sec. 1 (Neurophysiology), vol. I. Washington, American Physiological Society, 1959.

Lloys, T. C., Jr.: Influence of blood pH on hypoxic pulmonary vasoconstriction. J. Appl. Physiol., *21*: 358-364, 1966.

Longland, C. J.: The collateral circulation of the limb. Ann. Roy. Coll. Surg. Eng., 13:161-176, 1953.

Luisada, A. (ed.): Cardiovascular Functions. New York, McGraw-Hill, 1962.

Lüscher, E. F.: Retraction activity of the platelets. In Johnson, S. A., Monto, R. W., Rebuck, J. W., and Horn, R. C., Jr. (eds.): Blood Platelets. Pp. 445-454. Boston, Little, Brown & Co., 1961.

Majno, G.: Ultrastructure of the vascular membrane. In Hamilton, W. F. (ed.): Handbook of Physiology. Sec. 2 (Circulation), vol. III, pp. 2293-2375. Washington, American Physiological Society, 1965.

Mammen, E. F.: Die Aktivierung des Prothrombins. Thromb. Diath. Haemorrh., 28:15-36, 1968.

Master, A. M., Dublin, L. I., and Marks, H. H.: The normal blood pressure range and its clinical implications. JAMA, 143:1464-1470, 1950.

Master, A. M., Lasser, R. P., and Jaffe, H. L.: Blood pressure in apparently healthy aged 65 to 106 years. Proc. Soc. Exp. Biol. Med., 94:463-467, 1957.

Mayerson, H. S.: The lymphatic system. Sci. Am., 208:80-90, 1963.

———: Blood volume and its regulation. Ann. Rev. Physiol., 27:307-322, 1965.

McDonald, D. A.: Blood Flow in Arteries. London, Edward Arnold (Publishers), 1960.

McGregor, D. M.: Studies by thoracic duct drainage of the functions and potentialities of the lymphocyte. Fed. Proc., 25:1713-1719, 1966.

Metcalf, D., and Brumby, M.: The role of the thymus in the ontogeny of the immune system. J. Cell. Physiol., 67:149-168, 1966.

Miale, J. B.: Laboratory Medicine—hematology. 3rd ed. St. Louis, C. V. Mosby, 1967.

Morkin, E., et al.: Pattern of blood flow in the pulmonary veins of the dog. J. Appl. Physiol., 20:1118-1128, 1965.

Nossal, G. J. V.: How cells make antibodies. Sci. Am., 211:106-115, 1964.

Olsson, R. A.: Kinetics of myocardial reactive hyperemia blood flow in the unanesthetized dog. Circ. Res. (Suppl. 1), 14 and 15:81-85, 1964.

Oparil, S., et al.: Role of renin in acute postural homeostasis. Circulation, 41:89-95, 1970.

Page, I. H., and McCubbin, J. W.: The physiology of arterial hypertension. In Hamilton, W. F. (ed.): Handbook of Physiology. Sec. 2 (Circulation), vol. III, pp. 2163-2208. Washington, American Physiological Society, 1965.

Paintal, A. S., and Riley, R. L.: Response of aortic chemoreceptors. J. Appl. Physiol., 21:543-548, 1966.

Palade, E. G.: Blood capillaries of the heart and other organs. Circulation, 24:368-384, 1961.

Peiss, C. N.: Supramedullary cardiovascular regulation. In Price, H. L., and Cohen, P. J. (eds.): Symposium on the Effects of Anesthetics on the Circulation. Pp. 32-43. Springfield, Charles C Thomas, 1964.

———: Sympathetic control of the heart. In Luisada, A. A. (ed.): Cardiovascular Functions. Pp. 314-323. New York, McGraw-Hill, 1962.

———: Concepts of cardiovascular regulation: Past, present and future. In Randall, W. C. (ed.): Nervous Control of the Heart. Pp. 154-197. Baltimore, Williams & Wilkins, 1965.

Phibbs, R. H.: Distribution of leukocytes in blood flowing through arteries. Am. J. Physiol., 210:919-925, 1966.

Pollack, A. A., and Wood, E. H.: Venous pressure in the saphenous vein at the ankle in man during exercise and changes in posture. J. Appl. Physiol., 1:649-662, 1949.

Prichard, M. M. L., and Daniel, P. M.: Arterio-venous anastomoses in the human ear. J. Anat., 90:309-317, 1956.

Randall, W. C.: Nervous Control of the Heart. Baltimore, Williams & Wilkins, 1965.

Rasmussen, A. T.: Outlines of Neuro-anatomy. 3rd ed. Pp. 45, 47. Dubuque, Iowa, William C. Brown, 1943.

Recommendations for Human Blood Pressure Determination by Sphygmomanometers. New York, American Heart Association, 1967.

Reynolds, S. R. M.: Maternal blood flow in the uterus and placenta. In Hamilton, W. F. (ed.): Handbook of Physiology. Sec. 2 (Circulation), vol. II, pp. 1585-1618. American Physiological Society, 1963.

Rhodin, J. A. G.: An atlas of Ultrastructure. P. 53. Philadelphia, W. B. Saunders, 1963.

Richardson, D. W., et al.: Role of hypocapnia in the circulatory responses to acute hypoxia in man. J. Appl. Physiol., 21:22-26, 1966.

Robinson, B. F., Epstein, S. E., Beiser, G. D., and Braunwald, E.: Control of heart rate by the autonomic nervous system: Studies in man on the interrelation between baroreceptor mechanisms and exercise. Circ. Res., 19:400-411, 1966.

Robinson, B. R., Epstein, S. E., Kahler, R. L., and Braunwald, E.: Circulatory effects of acute expansion of blood volume: Studies during maximal exercise and at rest. Circ. Res., 19:26-32, 1966.

Rosenthal, S. L., and Guyton, A. C.: Hemodynamics of collateral vasodilation in anesthetized dogs. Circ. Res., 23:239-248, 1968.

Roughton, F. J. W.: Transport of oxygen and carbon dioxide. In Fenn, W. O., and Rahn, H. (eds.): Handbook of Physiology. Sec. 3 (Respiration), vol. I, Pp. 767-825. Washington, American Physiological Society, 1964.

Rovick, A. A., and Randall, W. C.: Systemic circulation. Ann. Rev. Physiol., 29:225-258, 1967.

Ruch, T. C., and Patton, H. D.: Physiology and Biophysics. Philadelphia, W. B. Saunders, 1965.

Rushmer, R. F.: Postural effects on the baselines of ventricular performance. Circulation, 20:897-905, 1959.

———: Cardiovascular Dynamics. 2nd ed. Philadelphia, W. B. Saunders, 1961.

Rushmer, R. F., Van Citters, R. L., and Franklin, D. L.:

Some axioms, popular notions, and misconceptions regarding cardiovascular control. Circulation, 27: 118-141, 1963.

Sandison, J. C.: Contraction of blood vessels and observations on the circulation in the transparent chamber of the rabbit ear. Anat. Rec., 54:105-127, 1932.

Sawyer, P. N. (ed.): Biophysical Mechanisms in Vascular Homeostasis and Intravascular Thrombosis. New York, Appleton-Century-Crofts, 1965.

Scholer, J. F., and Code, C. F.: Rate of absorption of water from stomach and small bowel of human beings. Gastroenterology, 27:568-577, 1954.

Seegers, W. H.: Prothrombin in Enzymology, Thrombosis and Hemophilia. Springfield, Charles C Thomas, 1967.

———: Physiology of blood coagulation: Advances, problems, and present status. Pflueger. Arch., 299: 226-246, 1968.

———: Blood clotting mechanisms: Three basic reactions. Ann. Rev. Physiol., 31:269-294, 1969.

Seegers, W. H., and Smith, H. P.: Factors which influence the activity of purified thrombin. Am. J. Physiol., 137:348-354, 1942.

Selye, H.: The Mast Cells. Washington, Butterworth & Co., 1965.

Sharpey-Schafer, E. P.: Effect of respiratory acts on the circulation. In Hamilton, W. F. (ed.): Handbook of Physiology. Sec. 2 (Circulation), vol. III, pp. 1875-1886. Washington, American Physiological Society, 1965.

Shepherd, J. T.: Role of the veins in the circulation. Circulation, 33:484-491, 1966.

Sherrington, C. S.: The Integrative Action of the Nervous System. New Haven, Yale University Press, 1906.

Sonnenberg, H., and Pearce, J. W.: Renal response to measured blood volume expansion in differently hydrated dogs. Am. J. Physiol., 203:344-352, 1962.

Spaet, T. H., and Zucker, M. B.: Mechanism of platelet plug formation and role of adenosine diphosphate. Am. J. Physiol., 206:1267-1274, 1964.

Spain, D. M.: Atherosclerosis. Sci. Am., 15:48-59, 1966.

Staples, P. J., Gery, I., and Waksman, B. H.: Role of the thymus in tolerance. III. Tolerance to bovine gamma globulin after direct injection of antigen into the shielded thymus of irradiated rats. J. Exp. Med., 124:127-140, 1966.

Stern, S., and Braun, K.: Effect of chemoreceptor stimulation on the pulmonary veins. Am. J. Physiol., 210:535-539, 1966.

Sticker, E. M.: Extracellular fluid volume and thirst. Am. J. Physiol., 211:232-238, 1966.

Strumia, M. M., Crosby, W. H., Greenwalt, T. J., Gibson, J. G., and Krevans, J. R.: General Principles of Blood Transfusions. Philadelphia, J. B. Lippincott, 1963.

Tschirgi, R. D.: Chemical environment of the central nervous system. In Magoun, H. W. (ed.): Handbook of Physiology. Sec. 1 (Neurophysiology), vol. III, pp. 1865-1889. Washington, American Physiological Society, 1960.

van Deenen, L. L. M., and de Gier, J.: Chemical composition and metabolism of lipids in red cells of various animal species. In Bishop, C., and Surgenor, D. M. (eds.): The Red Blood Cell. Pp. 243-307. New York, Academic Press, 1964.

Whitby, L. E. H., and Britton, C. J. C.: Disorders of the Blood. 8th ed. London, J. & A. Churchill, 1957.

Whittaker, S. R. F., and Winton, F. R.: The apparent viscosity of blood flowing in the isolated hindlimb of the dog, and its variation with corpuscular concentrations. J. Physiol. (London), 78:339-369, 1933.

Wiedeman, M. P.: Patterns of the arteriovenous pathways. In Hamilton, W. F. (ed.): Handbook of Physiology. Sec. 2 (Circulation), vol. II, pp. 891-933. Washington, American Physiological Society, 1963.

Wiener, A. S.: Blood Groups and Transfusion. 3rd ed. Springfield, Charles C Thomas, 1943.

Wintrobe, M. M.: Clinical Hematology. 6th ed. Philadelphia, Lea & Febiger, 1967.

Wolinsky, H., and Glagov, S.: Structural basis for the static mechanical properties of the aortic media. Circ. Res., 14:400-413, 1964.

Zanjani, E. D., Contrera, J. F., Cooper, G. W., Gordon, A. S., and Wong, K. K.: Renal erythropoietic factor: Role of ions and vasoactive agents in erythropoietin formation. Science, 156:1367-1368, 1967.

Zweifach, B. W.: The microcirculation of the blood. Sci. Am., 200:54-60, 1959.

Part V

NUTRIENTS AND WASTE PRODUCTS

The cells of the body, if they are to survive, need fuel for their energy-consuming reactions. Part of this energy comes from nutrients stored in the cell, and part from the nutrients which enter the cell from the intercellular fluid. The intercellular fluid, in turn, is supplied by the blood and lymph, which carry organic compounds from the digestive tract, glucose from the liver, lipids from the adipose tissue, O_2 from the lungs, and lactic acid from skeletal muscle.

The ultimate source of these nutrients is the food we take into the alimentary tract. This must first be broken down physically and chemically to a size enabling ready movement through the membranes of the tract, capillaries, and body cells. Once this digested food passes from the alimentary tract to the bloodstream it is available to most of the cells of the body. Some of these cells utilize nutrients for growth, repair, and certain energy-requiring reactions. Some change it into forms that can be stored, such as fat and glycogen. The glycogen reserves of the liver are used in maintaining a constant blood sugar level, whereas those of the muscle appear to serve as a local source of glucose, used in the anaerobic, energy-yielding formation of lactic acid. Much of this lactic acid is released into the blood, where it is available to the heart and liver. The liver changes it back to glucose, and the heart in the presence of O_2 breaks it down into CO_2, H_2O, and energy. The fat depots also serve as a potential source of energy-rich compounds which can be released into the blood.

In order for the body to utilize fats, lactic acid, and many other compounds for energy, its cells must have large amounts of O_2 in their cytoplasm. Anaerobic catabolism in man can only temporarily meet the needs of his cells. Oxygen is delivered to the blood at the terminal portions of the respiratory tract. During inspiration, fresh air is pulled into the respiratory channels, and during expiration some of the stale air is pushed out. This results in there being a reservoir of O_2 continuously available to the blood. The adequacy of the O_2 supply to the cells, however, is dependent not only upon (1) the ventilation of the respiratory tract and (2) its diffusion characteristics, but also upon (3) the blood flow, (4) the hemoglobin concentration in the blood, (5) the adequacy of the cells' enzyme systems, and (6) the O_2 requirements of each cell.

The net result of each cell's chemical processes is that its internal environment is markedly changed. In performing its metabolic reactions, the body produces heat, urea, creatinine, uric acid, H^+, and CO_2, and may take up or lose excess amounts of water, acids, bases, or certain electrolytes. While some cells are growing and reproducing, others are aging and dying. This results in the formation and release of a number of additional substances (bilirubin, K^+, etc.). If a person is to survive, homeostasis must be maintained.

The respiratory system, urinary system, digestive system, and skin and mucous membranes are among the main contributors to homeostasis. In respiration we not only bring in O_2 to replace that which is used by the body, but also expel excess CO_2. The kidneys excrete urea, creatinine, excess water, and excesses of most of the minerals. The digestive tract is responsible for eliminating excesses of most of the heavy metals, as well as the bile pigments formed by the liver, bone marrow, and spleen during the breakdown of hemoglobin. Excess heat resulting from metabolism is carried by the blood to the skin and mucous membranes, where it is lost by the processes of radiation, convection, conduction, and evaporation.

Many of the excretory processes mentioned above also serve to maintain an acid-base balance in the body. For example, the addition of CO_2 to the blood in the systemic capillaries lowers the pH of the blood from 7.4 to 7.2. Our respiratory systems, by eliminating CO_2, prevent acidosis. The kidneys are also important in the control of acid-base balance, since they excrete H^+ in the urine in the form of NH_4Cl, NaH_2PO_4, and H^+. They can also increase the H^+ concentration in the body by excreting $NaHCO_3$, Na_2HPO_4, and OH^-. In a normal person, kidneys excrete 40 to 80 mEq. of nonvolatile acid a day, and the lungs, 13,000 mEq. of volatile acid a day.

Chapter 17

CELLULAR RESPIRATION
INSPIRATION
EXPIRATION
VENTILATION
DIFFUSION
AGING

RESPIRATION

Respiration in man is associated with (1) the movement of O_2 into and CO_2 out of the air sacs of the lungs, (2) the diffusion of these gases across the membranes that separate the alveolar lumen from that of the capillary, (3) the transport of O_2 and CO_2 by the blood, (4) the diffusion of these gases between the blood and the cytoplasm of the cell, and (5) the utilization and storage of O_2 and the elimination of the products of oxidation by the cell.

CELLULAR RESPIRATION

Every living cell in the body derives its energy from both *anaerobic* and *aerobic* reactions (Table 17-1). One important series of anaerobic reactions is the breakdown of glucose or glycogen to lactic acid and energy. The cells obtain this glucose from their own reserves (Table 17-2) and from the blood. Some of the *lactic acid* formed from glucose catabolism is released into the blood and used by the liver to produce more blood sugar, some is picked up by the heart and oxidized to produce CO_2, H_2O, and energy, and some is oxidized by the cells which produce it.

Should ventilation of the lungs stop, the cells would become dependent upon the body's *oxygen reserves* and upon anaerobic reactions for the energy needed to survive. Most of the body's oxygen is stored in the lungs and in the hemoglobin of the erythrocytes (Fig. 17-1). Only very small amounts are found combined with the myoglobin of skeletal and cardiac muscle, and even less is dissolved in the cytoplasm, intercellular fluid, and plasma. There is in the myoglobin of the heart, for example, only enough stored O_2 to provide energy for one cardiac contraction.

During hypoxia the cells continue to use O_2 at a fairly constant rate until their PO_2 diminishes to below 1 mm. Hg. At this time they become more dependent on the breakdown of glycogen and glucose to lactic acid and energy. Since these reactions result in the release of between one eleventh and one sixteenth of the energy that comes from the breakdown of glycogen and glucose in the presence of O_2, we find that the glycogen stored in the tissue rapidly diminishes once the O_2 reserves have been depleted (Fig. 17-2). This eventually results in the cessation of tissue function and irreversible tissue damage if ventilation of the lungs does not begin.

The organs in the body most susceptible to hypoxia are the brain and heart. If one cuts off the blood supply to the brain for 0.3 min., the

Structure	Mass (kg.)	O_2 Used (liters/day)	CO_2 Produced (liters/day)	R. Q.	Blood Flow (liters/day)
Brain	1.4	67	67	1.00	1,080
Heart	0.3	42	36	0.85	324
Kidney	0.3	25	20	0.80	1,814
Digestive Organs	2.6	73	51	0.70	2,160
Skeletal muscle	31.0	71	57	0.85	1,205
Skin	3.6	17	14	0.85	663
Other Structures	24.0	72	58	0.85	547
Whole Body	63.2	367	303	0.83	7,793

Table 17-1. A 24-hour metabolic study. R. Q., CO_2 produced/O_2 consumed. (From Mountcastle, V. B.: Medical Physiology. Vol. I. St. Louis, C. V. Mosby, 1968)

	O_2 Consumed $\left(\dfrac{\text{ml./min.}}{\text{100 gm.}}\right)$	Venous P_{O_2} (mm. Hg)	Blood Flow $\left(\dfrac{\text{ml./min.}}{\text{100 gm.}}\right)$	Glycogen Stores (mM./liter)	Minutes after Anoxia Functioning	Revivable
Brain (awake)	3.3	31	54	3	cortex 0.3 medulla 4.0	5 15
Brain (coma)	1.7					
Heart (at work)	8.0	19	70	29	5	10
Heart (at rest)	1.0				10	
Liver	2.0	40	54	292		50
Kidney	6.0	50	430	2		60
Skeletal muscle						
At work	7.9	12				
At rest	0.2	31	2.7	19	120	480

Table 17-2. Some of the factors determining the susceptibility of various structures to anoxia. Note that the brain, with its low glycogen reserve, is the organ most susceptible to hypoxia, and that skeletal muscle at rest, with its higher glycogen content and lower O_2 consumption, is among the organs least sensitive to anoxia. O_2 consumption, milliliters of O_2 consumed per minute/100 gm. of tissue; venous P_{O_2}, partial pressure of O_2 in veins of tissue; blood flow, milliliters of blood per minute/100 gm. of tissue; glycogen, millimols of glycogen per liter of tissue. (From Tenney, S.M., and Lamb, T.W.: Physiological consequences of hypoventilation and hyperventilation. *In* Fenn, W.O., and Rahn, H. (eds.): Handbook of Physiology. Sec. 3 (Respiration), vol. II, pp. 979-1010. Washington, American Physiological Society, 1965)

cerebral cortex ceases functioning normally, and after 5 min. of ischemia is irreversibly damaged. The heart continues to pump blood for only about 5 min. after its coronary vessels have been clamped, and suffers irreversible damage 5 min. later. During open heart surgery when there is little or no blood in the ventricles, the heart has a lower metabolic rate and can survive considerably longer periods of ischemia.

The survival time of any tissue during hypoxia is apparently dependent upon (1) its metabolic activity (sometimes measured in terms of O_2 consumption) and (2) its stores of glycogen, O_2, and other nutrients (Table 17-2). In cases of hypoxia due to too little O_2 in the inspired air (hypoxic hypoxia), survival time also depends upon (3) the distribution of the body's oxygen reserves. During such hypoxia there is a greater increase in the amount of blood going to the heart and brain than in that going to the kidney, for example. Skeletal muscle at rest, with its low level of metabolism (Table 17-2) and high concentration of glycogen, can withstand up to 8 hrs. of complete ischemia, whereas the brain, with its higher resting O_2 requirements and lower glycogen stores, can withstand only about 5 min. of complete ischemia.

INSPIRATION

Inspiration is the movement of air through the respiratory tracts toward the alveolar sacs (Fig. 17-3). The first part of the respiratory tract is a system of branching airways that might be classified as *conductive*. The terminal portion of the tract consists of the closely associated alveoli and alveolar capillary network. This is the area in which most of the diffusion of O_2 and CO_2 into and out of the blood occurs, and is called the *respiratory* zone. Lying between the respiratory and conductive portions of the airway is the *transitory* zone (Fig. 17-4).

The Upper Airway

The inspired air, as it passes through the conductive zone of the airway, is modified in three ways. Its *temperature* is brought closer to that of the body, *moisture* is added, and suspended *particulate matter* is removed. The nasal cavity seems to be highly important in all of these processes. For example, dry air at a temperature of −10°C. when inspired, passes through the nasal cavity, is warmed, humidified, and filtered, and enters the nasopharynx at a temperature of 25°C.

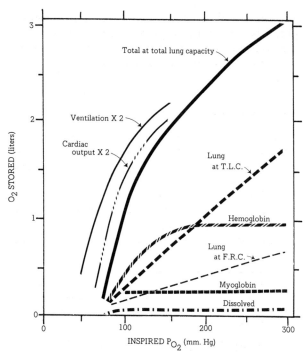

Fig. 17-1. Major O₂ stores in the body. Note that the amount of O$_2$ stored in the body is dependent upon the PO_2 of the inspired air. Under normal conditions the PO_2 is about 150 mm. Hg (760 mm. Hg × 0.20). After a normal expiration there is about 0.9 liter of O$_2$ stored in the hemoglobin of the body, 0.25 liter in the lungs (F.R.C., functional residual capacity), 0.2 liter in myoglobin, and 0.4 liter dissolved in the body fluids. Also shown is the total amount of O$_2$ that will be stored in the body after a maximum inspiration and the effect on this total of doubling the cardiac output or doubling the ventilation. (Redrawn from Tenney, S. M., and Lamb, T. W.: Physiological consequences of hypoventilation and hyperventilation. In Fenn, W. O., and Rahn, H. (eds.): Handbook of Physiology. Sec. 3 (Respiration), vol. II, p. 982. Washington, American Physiological Society, 1965)

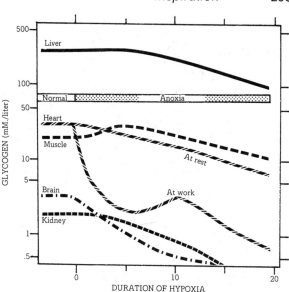

Fig. 17-2. Rate of depletion of tissue glycogen in response to anoxia. The heart was studied while pumping blood (At Work) and during rapid exsanguination (At Rest). The brain was studied during occlusion of its blood vessels. (Redrawn from Tenney, S. M., and Lamb, T. W.: Physiological consequences of hypoventilation and hyperventilation. In Fenn, W. O., and Rahn, H. (eds.): Handbook of Physiology. Sec. 3 (Respiration), vol. II, p. 986. Washington, American Physiological Society, 1965)

and a humidity of 100 per cent. On the other hand, air at a temperature of 55°C. is cooled to 37°C. in passing through the nasal cavity.

The amount of water vapor added to the inspired air may be considerable. For example, 1 cubic meter of air at 0°C. and 100 per cent humidity contains 4.85 gm. of water. The same air after it passes through the nasal cavity has a temperature of 30°C. and contains 30.4 gm. of water vapor (100% humidity).

Much of the removal of particulate matter occurs in the nasal cavity. Some particles dissolve in the mucous fluid covering the respiratory membranes. It has been estimated that approximately 0.9 gm. of mucus per square meter of respiratory epithelial surface is secreted every 10 min. in the humans. This mucus is carried continuously from the nasal cavity and tracheobroncial tree toward the hypopharynx, where it is swallowed by means of cilia projecting into the airway lumen. There are about 150 to 300 *cilia* on the surface of every cell, each about 6 μ long and in nearly continuous motion. It has been estimated that a cilium travels the equivalent of 1 mile a week.

Inhaled air does not always, of course, pass through the nasal cavity before entering the trachea. When it does not, changes in the temperature and moisture content of the air and removal of particulate matter occur in the trachea, bronchi, and so on. The extent to which the lower airway can substitute for the nasal cavity is not well understood. Apparently the nasal cavity, because of its large surface and *turbulent* air flow, is a considerably more effective mixing chamber than the lower airway with its more

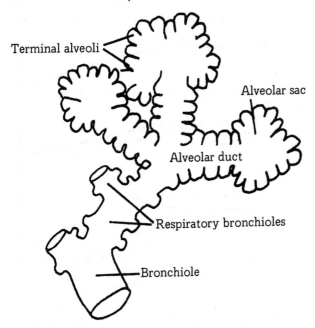

Fig. 17-3. Terminal portion of the respiratory tree. (*Redrawn from* Krahl, V. E. *In* Gray, P. (ed.): Encyclopedia of the Biological Sciences. P. 516. Van Nostrand Reinhold Co., 1970; copyright © 1970 by Litton Educational Publishing, Inc.; by permission of Reinhold Publishing Corp.)

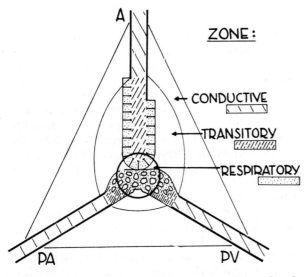

Fig. 17-4. Organization of the airways (A) and pulmonary vessels (PA and PV) into conductive, transitory, and respiratory portions. (Weibel, E. R.: Morphometrics of the lung. *In* Fenn, W. O., and Rahn, H. (eds.): Handbook of Physiology. Sec. 3 (Respiration), vol. I, P. 286. Washington, American Physiological Society, 1964)

laminar air flow. Two of the major problems in patients with total laryngectomies (who breathe through an opening in the neck) are (1) respiratory infection and (2) discomfort in dry atmospheres.

The Lower Airway

After the inspired air has been filtered, humidified, and brought to the proper temperature in the upper airways, it passes into the alveoli, where it increases the alveolar P_{O_2} and decreases the alveolar P_{CO_2} (Fig. 17-5). The degree to which the alveolar P_{O_2} is elevated, and the P_{CO_2} lowered, during inspiration normally depends on the depth of the inspiration, which in turn

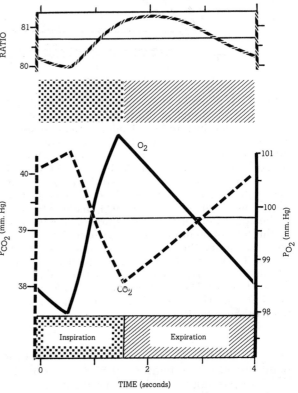

Fig. 17-5. Changes in the alveolus during one breathing cycle. Note that at 0.5 second after the onset of inspiration there begins an increase in the alveolar P_{O_2} and a decrease in P_{CO_2}. These changes are reversed at the beginning of expiration. Respiratory exchange ratio equals volume of CO_2/volume of O_2. (*Modified from* Weibel, E. R.: Morphometry of the Human Lung. P. 139. New York, Academic Press, 1963; and Staub, N. C.: The interdependence of pulmonary structure and function, Anesthesiology, 24:833, 1963)

depends on the force of contraction of the inspiratory muscles.

Muscles of Inspiration

Inspiration in a resting person results from contraction of the diaphragm and external intercostal muscles. These are skeletal muscle, and contract in response to impulses from the phrenic nerves, which innervate the diaphragm, and the intercostal nerves, which innervate the intercostal muscles. The *diaphragm* is a large, dome-shaped sheet of muscle separating the thoracic cavity from the abdominal cavity in mammals (Fig. 17-6). When maximally stimulated it descends as much as 10 cm. in the adult, thereby decreasing intrathroacic pressure and increasing intra-abdominal pressure. The increase in intra-abdominal pressure will be great if the abdominal wall is tense due to obesity, muscle tone, or the presence of fluid in the abdominal cavity (ascites). In deep anesthesia, the abdominal muscles, intercostal muscles, and accessory muscles of respiration are usually paralyzed. This causes pronounced distention of the abdomen during inspiration, characterized as *abdominal breathing*.

The *external intercostal muscles* also contract during a normal inspiration, and tend to elevate the ribs and increase the ventrodorsal dimensions of the chest. Their contraction also adds stability to the intercostal space and thus keeps it from being sucked in during inspiration.

A person increases the depth of his inspiration by stimulating more motor units in the diaphragm and in the external intercostal muscles, and by stimulating each motor unit at a higher frequency (Fig. 17-7). Normally, an adult inspires about 6 liters of air per minute. When his ventilation exceeds 50 liters per minute, as in exercise, the scalene and sternomastoid muscles too begin to contract during inspiration. When ventilation exceeds 100 liters per minute the posterior neck, trapezius, and back muscles begin to assist. Muscles other than the diaphragm and intercostals are called the *accessory muscles of inspiration,* since they do not normally participate in quiet breathing. In abnormal conditions such as emphysema they become considerably more important, and after complete paralysis of the diaphragm they help keep ventilation normal.

All the muscles mentioned above increase the size of the thoracic cavity. Another group of muscles aids in inspiration not by changing the size of the thoracic cavity, but by lowering the resistance to air flow. These include the mylohyoid, digastric, cheek, laryngeal, tongue, and posterior neck muscles, and the alae nasi, the platysma, and the levator palata.

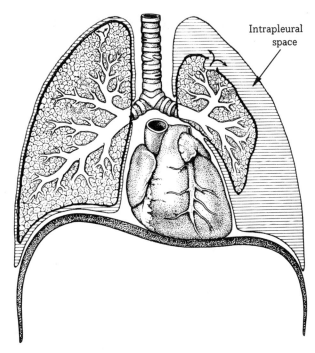

Fig. 17-6. Diaphragm, heart, and lungs. Normally, contraction of the diaphragm increases the craniocaudal dimensions of the chest and in so doing decreases the pressure in the intrapleural space and the alveoli. It is the decrease in pressure in the alveoli which causes inspiration. Should the intrapleural space become filled with air due to the rupture of a portion of the lung or the chest wall, contraction of the diaphragm will be markedly less effective in producing alveolar ventilation. (Adler, R. H.: Spontaneous pneumothorax. Hosp. Med., *1*:3, 1965)

Pressures Within the Thorax

The thoracic cavity is lined by a serous membrane, the parietal pleura, which is both continuous with and adjacent to a second membrane, the visceral pleura. The visceral pleura covers the lungs and is separated from the adjacent parietal pleura by a thin layer of serous (watery) liquid in the *intrapleural space*. This liquid serves as an effective *lubricant,* allowing the membranes of the lung to slide freely over those of the thorax during inspiration and expiration.

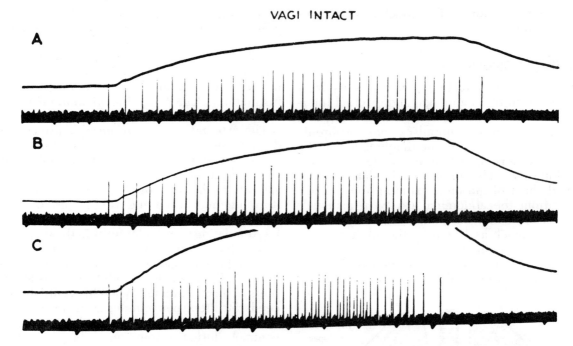

Fig. 17-7. Motor impulses (*lower line of each tracing*) in a part of the phrenic nerve of a cat at different depths of inspiration (*upper line of each tracing*). In A a single neuron is active. In B the depth of inspiration was increased by having the animal breathe a CO_2-rich gas. Note the increase in the frequency of firing of the neuron. In C the depth has been increased even more and there is now a firing of a second neuron of smaller spike potential. (Pitts, R. F.: The function of components of the respiratory complex. J. Neurophysiol., 5:409, 1942)

The pressure in the intrapleural space is about 3 cm. H_2O below atmospheric pressure, due to the forces exerted on the intrapleural liquid by the thorax, abdomen, and lungs. The lungs normally exist in a stretched state and, because of their elastic properties, tend to pull their attached visceral pleura inward. The thorax and abdomen tend to pull the parietal pleura away from the lungs. The result of these opposing forces is an intrapleural pressure below atmospheric pressure. The presence in the intrapleural space of a nondistensible fluid, the serous liquid, rather than distensible air, helps prevent the collapse of the lungs.

The elastic forces exerted on the intrapleural space are characterized in Figure 17-8. In part A of this figure the position of the chest (C) and lungs (L) are shown in a case in which the intrapleural pressure became equal to atmospheric pressure as a result of air entering the intrapleural space. This may happen after rupture of a lung (Fig. 17-6) or after an incision or wound through the skin at an intercostal space. In either case air gets into the intrapleural space and may result in the separation of the two pleurae and an associated (1) decrease in the volume of the lungs, (2) increase in the volume of the chest, and (3) descent of the diaphragm. Under normal conditions (see B, C, D, and E) the lungs are stretched. During a forceful or maximum expiratory effort (B) the chest is compressed and the diaphragm is pushed cephalad by the accessory expiratory muscles, but even under these conditions the lungs remain stretched (see lower arrow).

At the end of a normal inspiration (C) the lungs are considerably more distended and the chest less compressed than at the end of a maximum expiration. As the inspirations become deeper (D and E), the accessory muscles of inspiration prevent, and eventually reverse, the normal compression of the chest. In part E we note that the expansion of the chest is now greater than that found during pneumothorax (see A).

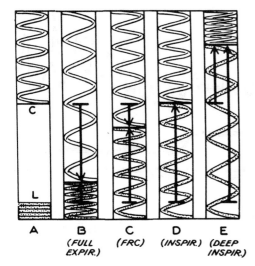

Fig. 17-8. Diagram of the elastic forces acting on the intrapleural space. At A air is allowed to enter the intrapleural space, and the lungs (L) and chest (C) recoil. At B we note the degree of lung distention (*lower arrow*) and chest compression (*upper arrow*) after a maximum expiration. At C, D, and E we see the results of increasingly more forceful inspirations. Note that as we pass from B (maximum expiration) to E (maximum inspiration) the lungs become progressively more stretched. This is due to the increasingly more forceful contractions of the diaphragm and the accessory muscles of inspiration. (*From* Physiology of Respiration by J. H. Comroe, Jr. Copyright © 1965, Year Book Medical Publishers, Inc.; used by permission)

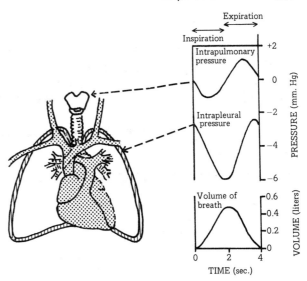

Fig. 17-9. Intrapleural and intrapulmonary pressure during respiration. These pressures will vary with body position and the site at which they are recorded. (*Data from* Perkins, J. F.: Respiration. *In* Encyclopedia Britannica. Vol. XIX, p. 218. Chicago, Encyclopedia Britannica, Inc., 1965)

Figure 17-9 shows a diagram of the average changes in intrapleural pressure that occur during quiet breathing. Note that although the average intrapleural pressure varies during quiet breathing, it does not reach atmospheric pressure (0 mm. Hg). This is not the case for the pressures recorded in the upper respiratory passages (bronchi, nasal cavity, etc.). During inspiration the intrapleural pressure decreases and brings down the pressure in the respiratory passages (intrapulmonary pressure). As the intrapleural pressure continues to decrease, air enters the lungs, and by the time the intrapleural pressure has reached its lowest point the intrapulmonary pressure has returned to 0 mm. Hg. When intrapleural pressure is minimum and intrapulmonary pressure is equal to that of the atmosphere, lung volume is maximum.

Intrapleural pressure can be measured by passing a needle between two ribs, injecting a fluid into the intrapleural space, and measuring the pressure in the expanded space. However there is some danger in this procedure of puncturing a lung with the needle. A more convenient method of estimating intrapleural pressure is by means of an esophageal tube with a balloon at its end. The tube is swallowed and its end placed in the lower intrathoracic esophagus. Pressures recorded by this technique are called *intrathoracic pressures*. *Intrapulmonary pressures* are measured by means of a tube in one of the patent respiratory passages.

EXPIRATION

In the resting person, expiration is usually a *passive* phenomenon. That is to say, it stems not from muscle contraction, but rather from muscle relaxation. In quiet inspiration the diaphragm contracts and in so doing stretches some of the body's tissues. In expiration, the diaphragm relaxes and the previously stretched tissues recoil. It is this recoiling which results in the expulsion of air from the lungs. At the end of a passive expiration the chest is said to be in the equilibrium position. *Active* expiration occurs in the average adult when the minute volume of air breathed begins to exceed 40 liters per minute,

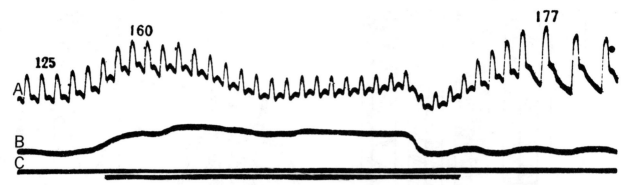

Fig. 17-10. The Valsalva maneuver and arterial pressure. A, arterial pressure (systolic pressure in mm. Hg shown); B, intrathoracic pressure; C, reference line and signal line indicating the beginning and end of the expiration against a closed glottis. Note that the beginning of the maneuver is associated with an increase in arterial pressure. As it is continued there is a decrease in arterial pressure due to an interference in the venous return of blood to the heart. At the cessation of the maneuver the venous return increases, the arterial pressure increases, and there is a reflex slowing of the heart. (Hamilton, W. F., et al.: Physiologic relationships between intrathoracic, intraspinal and arterial pressures. JAMA, *107*:854, 1936)

or during yelling, coughing, sneezing, or straining. It is produced by (1) the contraction of *abdominal muscles*, which pull the rib cage caudad and contract against the abdominal viscera, pushing the diaphragm cephalad, and (2) the contraction of the *internal intercostal muscles*, which depress the ribs and stiffen the intercostal space. The abdominal muscles includes the internal and external oblique muscles and the transversus abdominus, and are innervated by the lower six thoracic and first lumbar segments of the spinal cord.

During a forceful expiration the *intrathoracic pressure* may increase markedly. Trumpet players may reach an intrathoracic pressure of 160 mm. Hg on an arpeggio and 80 mm. Hg on a sustained note, whereas spontaneous coughing has been shown to elicit intrathroacic pressures in excess of 100 mm. Hg. In defecating, urinating, or lifting heavy loads one frequently takes a deep breath, followed by forceful expiration against a closed glottis. This is usually referred to as the *Valsalva maneuver*. It may result in an increase in intrathoracic and intra-abdominal pressure to as high as 200 mm. Hg.

In Figures 17-10 and 17-11 we have an illustration of the response of an individual to a Valsalva maneuver in which the intrathoracic pressure increased to 40 mm. Hg. The initial response (Fig. 17-10) is an increase in arterial pressure (tracing A) not associated with a change in transmural pressure, pulse pressure, or heart rate. The second response is seen as the Valsalva maneuver continues. Here the arterial pressure and aortic flow decrease due to the interference of the high intra-abdominal and intrathoracic pressures with venous return. The third change occurs as the Valsalva maneuver is discontinued. Here the intrathoracic and arterial pressures decrease. This is followed by increased venous return which results in an increase in cardiac output and arterial pressure. There is also a reflex decrease in heart rate due to the stimulation of the aortic and carotid pressoreceptors.

VENTILATION

The major functions of inspiration and expiration are (1) the delivery of O_2 to, and (2) the removal of CO_2 from, the alveoli. In other words, respiration serves to ventilate the alveoli. In a resting, 70-kg. adult, about 6,000 ml. of air are inspired per minute:

Respiratory minute volume =
(12 respirations/min.)(500 ml./respiration) =
6,000 ml./min.

Of this 6,000 ml. of air inspired per minute, only approximately 4,200 ml. per minute (70%) reach

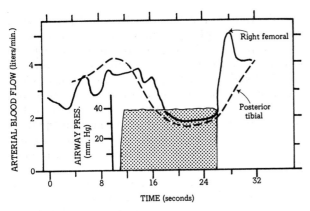

Fig. 17-11. The Valsalva maneuver (*stippled area*) and arterial flow in a healthy 28-year-old man. In experiments on 8 men flow decreased to 35 per cent of the control during forced expiration and increased to 19 per cent above control at the cessation of the expiration. Flow was determined by a dye dilution technique in which samples of blood were obtained from the femoral and posterior tibial arteries. (*Modified from* Fox, I. J., et al.: Effects of the Valsalva maneuver on blood flow in the thoracic aorta in man. J. Appl. Physiol., 21:1558, 1966)

the alveoli and associated respiratory zones (Fig. 17-4):

Alveolar ventilation =
(respiratory rate)(tidal volume − dead air space) =
(12/min.)(500 − 150 ml.) = 4,200 ml./min.

The dead air space represents the approximate volume of inspired air that remains in the conductive and part of the transitory respiratory passages at the end of inspiration. During increased metabolic activity both the respiratory rate and respiratory depth increase; hence both the respiratory minute volume and the alveolar ventilation are also elevated (Fig. 17-12). In this way increased needs of the body for (1) O_2 and (2) the elimination of CO_2 are met.

Vital Capacity

The vital capacity is the maximum amount of air that can be expired after a maximum inspiration. In a normal erect adult this is about 4,800 ml. per expiration, but it varies with (1) body position, (2) the size and shape of the chest, (3) the strength of the respiratory muscles, (4) the resistance to air flow, and (5) the compliance and elasticity of the system. Another definition of vital capacity is the resting tidal volume (about

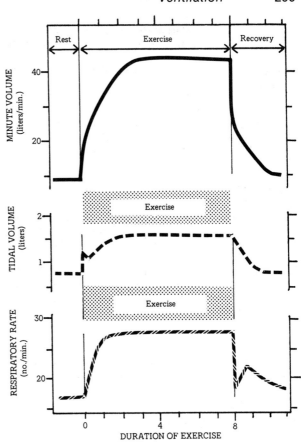

Fig. 17-12. The effect of exercise on minute volume, tidal volume, and respiratory rate. Note that the increase in minute volume is maintained for a number of minutes after the exercise is over (see O_2 debt). (*Redrawn from* Dejours, P.: Control of respiration in muscular exercise. *In* Fenn, W. O., and Rahn, H. (eds.): Handbook of Physiology. Sec 3 (Respiration), vol. I, pp. 631-648. Washington, American Physiological Society, 1964)

500 ml.), plus the inspiratory reserve volume (about 2,500 ml.), plus the expiratory reserve volume (about 1,600 ml.):

500 ml. + 2,500 ml. + 1,600 ml. = 4,600 ml.

Tidal volume. The tidal volume is the quantity of air inspired during a single respiratory cycle. The inspired air has *water vapor* added to it (Table 17-3), which results in a lowering of its P_{O_2} and P_{N_2} by the time it reaches the trachea. The P_{O_2} of the inspired air is lowered still further when it is mixed with alveolar air. The expired air is a mixture of the alveolar (low P_{O_2}) and tracheal (high P_{O_2}) gases and therefore has an

	Partial Pressure (mm. Hg)				
	P_{O_2}	P_{CO_2}	P_{N_2}	P_{H_2O}	Total
Dry air	159.1	0.3	600.6	0	760
Inspired tracheal air	149.2	0.3	563.5	47	760
Expired gas	116	28	569	47	760
Alveolar gas	100	40	573	47	760
Arterial blood	95	40	573		
Venous blood	40	46	573		
Tissues	<30	>50	573		

Table 17-3. Partial pressure of gases in the body. All values are approximations for a person at sea level (760 mm. Hg). In Denver, Colorado, the atmospheric pressure is about 640 mm. Hg and the P_{O_2}, P_{CO_2}, and P_{N_2} in the inspired air are less than shown above. Note that the dilution of the inspired air with water vapor lowers its P_{O_2} and P_{N_2}, and that the mixing of alveolar gas with the air of the dead space (conductive pathway) results in a higher P_{O_2} and a lower P_{CO_2} for expired air than for alveolar air.

average P_{O_2} that is higher, and an average P_{CO_2} that is lower, than what is found in the alveoli.

In an adult at rest the tidal volume is about 500 ml. of air. The last 150 ml. (100 to 200 ml.) inspired occupies primarily the conducting airway (nose, mouth, pharynx, trachea, bronchi, and bronchioles) and therefore does not serve to bring fresh air to respiratory structures (the alveoli, etc.). In expiration the first 150 ml. of air expired is primarily from the conducting airway and does not remove very much CO_2 from the body. In short, the first part of the respiratory passages (conducting system) in the adult represents a *dead space* of from 100 to 200 ml. In order to effectively ventilate the alveoli it is essential that each inspiration move more air than is represented by the volume of the dead space. If this does not happen, inspiration results only in moving stale air (air low in O_2 and high in CO_2) from the conducting system into the alveoli, and expiration results in moving stale air from the alveoli back to the conducting system. Thus if respirations are sufficiently shallow (as in panting), a person can develop fatal hypoxia regardless of the respiratory rate.

Tidal volume is a particularly important consideration when artificial respiration is being administered. Before 1951, the Schafer back-pressure method was recommended by most United States rescue agencies as an emergency technique for *artificial respiration* when hospital facilities were not available. Later, back–pressure-arm–lift and chest–pressure-arm–lift methods became the recommended techniques. Finally, in 1958, the National Academy of Sciences-National Research Council recommended expired-air methods (mouth-to-mouth and mouth-to-nose, for example) for emergency artificial respiration (Fig. 17-13). The advantages of the expired-air methods lie in the fact that they (1) result in a greater movement of air per inspiration (Table 17-4), (2) facilitate the maintenance of a patent airway, and (3) are safer.

Reserve volumes. After the usual inspiration, a resting person's lungs are capable of still greater expansion in response to greater contraction of the inspiratory muscles. The additional amount of air that can be inspired after a normal inspiration is called the *inspiratory reserve volume* (Fig. 17-14). After a normal expiration, an amount of air (the *functional residual*

Weight of Subject (kg.)	Average Tidal Volume of Air (ml.)						
	Rocking (prone)		Rocking (supine)		Schafer (BP)	Silvester (CPAL)	Expired Air
	60°	90°	60°	90°			
64	340	580	240	380	340	400	900
83	725	850	570	635	530	650	1,030

Table 17-4. Methods of artificial respiration. In the rocking method the subject is placed first with his head and then with his feet down. In the Schafer method, air is forced out of the lungs by pressure on the back (BP, back-pressure). In the Silvester technique, air is forced out of the lungs by pressure on the chest, and air is pulled into the lungs by lifting the arms (CPAL, chest-pressure–arm-lift). You will note that a much greater tidal volume results when air is expired into the subject through his mouth or nose or both. The expired-air technique also results in a higher arterial P_{O_2} and a lower arterial P_{CO_2} than do other methods. (From Elam, J.O.: Respiratory and circulatory resuscitation. *In* Fenn, W.O., and Rahn, H. (eds.): Handbook of Physiology. Sec. 3 (Respiration), vol. I, p. 1278. Washington, American Physiological Society, 1965)

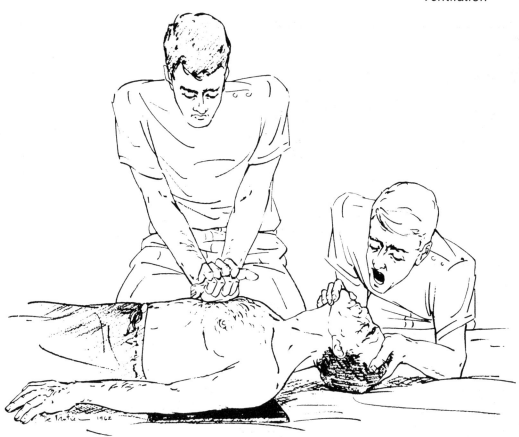

Fig. 17-13. Cardiorespiratory resuscitation. A cessation of breathing and a disappearance of the carotid pulse due to ventricular asystole or fibrillation can frequently be successfully treated by external cardiac massage and mouth-to-nose respiration. Note that the subject's head is extended to maintain the patency of the airway. (Elam, J. O.: Development of cardiac resuscitation. Int. Anesth. Clin., 2:107, 1963)

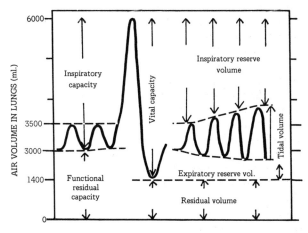

Fig 17-14. Subdivisions of lung volume. (*Modified from* Pappenheimer, J. R., et al.: Standardization of definitions and symbols in respiratory physiology. Fed. Proc., 9:602, 1950)

capacity) still remains in the lung. A certain amount of this can be expired (the *expiratory reserve volume*) by increasing one's expiratory effort. The air left in the lungs after maximum expiration is the *residual volume*. It is important in that it prevents the collapse of the alveoli and acts as a reservoir of O_2 on which the pulmonary blood can draw until the next inspiration. When the duct leading to an alveolus becomes occluded with mucus, etc., the blood eventually carries away most of the alveolar air and the alveolus collapses. This condition is called *atelectasis* (Gk. *ateles*, incomplete; plus *ektasis*, extension).

One of the problems in having a subject breathe pure O_2 instead of air is that he is more susceptible to atelectasis, since at one atmosphere and 100 per cent O_2 the rate at which the pulmonary blood depletes the alveolus of its gas is

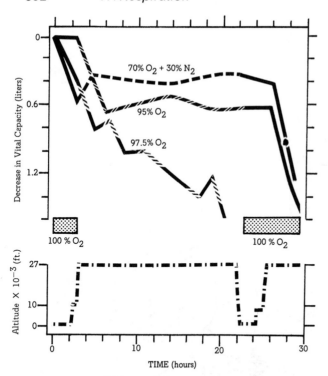

Fig. 17-15. Decreases in vital capacity in a 38-year-old man as a function of altitude (0–27 km., 0–27,000 ft. above sea level) and composition of inspired gas. The subject was chosen from a group of healthy men who had a tendency to develop atelectasis while breathing 100 per cent O_2 at an altitude of 27,000 ft. (258 mm Hg = 5 P.S.I.). Note that breathing 100 per cent O_2 and 97.5 per cent O_2 at 27,000 ft. led to a progressive decrease in vital capacity. These decreases and x-ray evidence were consistent with the view that at high O_2 concentrations this subject developed atelectasis. (*Redrawn from* Turaids, T., et al.: Absorptional atelectasis breathing oxygen at simulated altitude: prevention using inert gas. Aerospace Med., 38:190, 1967)

60 to 80 times faster than the rate of depletion when air at one atmosphere of pressure is breathed (Fig. 17-15). An advantage of having an *inert gas* such as N_2 in the inspired air, then, is that it prevents atelectasis when a respiratory tract is occluded. The quantity of air represented by the residual volume, the vital capacity, etc., varies with the state of health of the person (the presence or absence of atelectasis, for example), his age, his sex, and his size. In Figure 17-16 I have indicated how these volumes vary with the height of normal boys and girls, and in Table 17-7, how they vary with age.

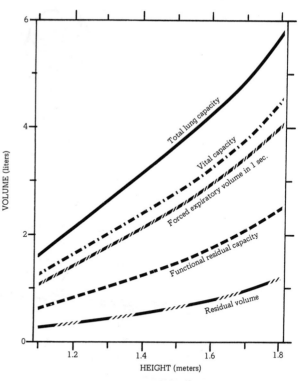

Fig. 17-16. Respiratory volumes in normal children in relation to height. (*Redrawn from* Cotes, J. E.: Lung Function. P. 318. Oxford, Blackwell Scientific Publications, 1965)

Pulmonary Dynamics

We are concerned in our study of the respiratory system with the movement of a gaseous fluid. For the most part the general principles set down in Chapter 10 ("Arteries, Arteriovenous Anastomoses, and Veins") concerning (1) fluid dynamics and (2) hemodynamics also apply to the movement of air in the respiratory passages. Briefly, these principles are as follows:

1. *Flow* of blood and air in the body is characteristically *laminar*, but may become *turbulent* (nonlaminar) in response to an increase in mean velocity, an increase in the density of the fluid, a decrease in its viscosity, or an increase in the radius of the vessel through which it passes.

2. Flow (I) results from pressure differences $(P_1 - P_2)$.

3. Flow is inversely related to the *resistance* (Ω) of the system:

$$I = \frac{P_1 - P_2}{\Omega}$$

4. Resistance is directly related to the length of the tube (L) through which the fluid passes and the viscosity (η) of the fluid, and inversely related to the radius (r) of the tube. In a system with a fluid of constant viscosity and a cylindrical, nondistensible, nonbranching tube of uniform radius, the resistance would be:

$$\Omega = \frac{8 L \eta}{\pi r^4}$$

5. *Parallel* resistances (Ω_1, Ω_2, Ω_3) produce a total resistance (Ω_T) which can be calculated as follows:

$$\frac{1}{\Omega_T} = \frac{1}{\Omega_1} + \frac{1}{\Omega_2} + \frac{1}{\Omega_3}$$

6. Resistances in *series* can be calculated as follows:

$$\Omega_T = \Omega_1 + \Omega_2 + \Omega_3$$

7. The volume of fluid added to a lung or vessel (dQ) is dependent upon the compliance (c) of that lung or vessel and the pressure change (dP) produced by the driving force (heart, diaphragm, etc.):

$$dQ = (c)(dP)$$

8. As a lung or vessel is distended, a point is eventually reached at which the compliance decreases markedly (Fig. 17-17). Attempts to distend a structure much beyond this point may result in rupture.

9. Distention of a vessel or lung stretches elastic elements in these structures. When the distending force or pressure is removed, the tension developed in the elastic elements during distention is used to return the structure toward its previous unstretched size.

10. A perfectly elastic structure develops a tension directly related to its distention, but not to either the speed or duration of distention.

11. Most organs in the body (lungs, skeletal muscle, vessels) are characterized as viscoelastic, since the tension they develop in response to distention is related to the speed and duration of distention as well as to the degree.

Compliance. I have defined compliance (c) as the ratio between a volume change (dQ) and a pressure change (dP):

$$c = \frac{dQ}{dP}$$

In young healthy men, the total compliance of

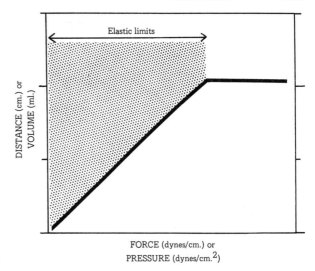

Fig. 17-17. Relationship between force and distance moved or pressure and volume displaced in an elastic body such as a spring or lung. The initial slope of the line (volume displaced/pressure change) represents the compliance of the spring, or normal lung within its elastic limits. (*Redrawn from* Physiology of Respiration by J. H. Comroe, Jr. Copyright © 1965, Year Book Medical Publishers, Inc.; used by permission)

the respiratory system is about 0.1 liter per centimeter of H_2O. This is the result of a compliance of 0.2 liter per centimeter of H_2O for both the pulmonary system and the thoracic cage:

$$\frac{1}{\text{Total compliance}} = \frac{1}{\text{pulmonary compliance}} + \frac{1}{\text{thoracic compliance}} = \frac{1}{0.1 \text{ liter/cm. } H_2O} = \frac{1}{0.2 \text{ liter/cm. } H_2O} + \frac{1}{0.2 \text{ liter/cm. } H_2O}$$

Compliance, then, is a measure of the ease with which a structure can be distended. In the case of the thoracic cage it depends on (1) the muscles that make up the thoracic cage, (2) the shape of the thoracic cavity, and (3) the structural characteristics of the connective tissue and bone of the cavity. For example, compliance is decreased in (1) diseases associated with skeletal-muscle rigidity, (2) conditions in which the diaphragm is pushed cephalad (obesity, abdominal tumor, pregnancy) and (3) scleroderma, kyphosis, and scoliosis.

Compliance in the lungs depends on (1) the residual volume, (2) the elastic characteristics of the lung tissue itself, and (3) the fluids found

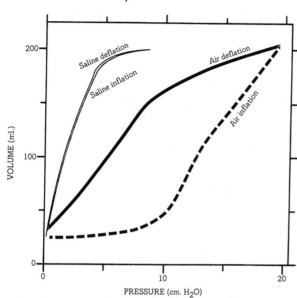

Fig. 17-18. Volume-pressure curves (compliance) from an isolated cat lung. Interrupted line represents points obtained during positive pressure filling of the collapsed lungs with air. Thick solid line represents points obtained during emptying of lungs. At each point, 30 seconds were allowed for the lung to reach equilibrium. Note the difference between air and saline inflation. Also note (*interrupted line*) that during air inflation initially the compliance is low. Eventually a pressure is reached (opening pressure) where the surface tension holding the alveoli closed is overcome and the compliance increased. (*Redrawn from* Radford, E. P., Jr.: Static mechanical properties of mammalian lungs. *In* Remington, J. W. (ed.): Tissue Elasticity. P. 185. Washington, American Physiological Society, 1957)

in the alveoli. If all the air in the alveoli is lost, for example, and the liquid-covered linings of the two sides of the alveoli adhere, the compliance of the lung is markedly diminished. That is to say, a relatively large pressure change is needed to overcome the intermolecular forces that result in the alveolar lining's being attracted to the lining liquid, which is attracted, in turn, to additional lining liquid, which is attracted to another alveolar lining. This is seen in Figure 17-18. Note that if we increase the pressure in a collapsed lung (interrupted line), there is very little volume change (low compliance) initially. Eventually, however, a pressure (opening pressure) is reached that produces an air-liquid interface in the alveoli. If this air-liquid interface can be maintained throughout the lung (uninterrupted line), the initial distention of the lung can be accomplished with a much smaller pressure change. It is also interesting to note in Figure 17-18 that a lung filled with liquid (saline) is much more compliant in response to a saline injection than a lung filled with air is in response to an air injection. That is to say, the liquid-air interface in the normal lung exerts a pressure due to its surface tension, which tends to resist lung distention.

We have discussed some of the problems associated with atelectasis (the collapse of alveoli) in the section above on *reserve volumes*. It may result from either (1) the occlusion of an airway or (2) a pneumothorax. It is characteristic of the fetus, which must, after birth, develop an intrapleural pressure of about -20 cm. H_2O to open up its alveoli to air. Under normal conditions all the alveoli, in children and adults, are patent (not atelectolic), and only the upper part (uninterrupted line in Fig. 17-18) of the air inflation curve applies for both inspiration and expiration.

Compliance varies not only with the degree of atelectasis, but also, in direct proportion, with lung volume. The compliance of the lung of a 2-day-old child (2.5 kg.) is approximately 5 ml. per centimeter H_2O, and that of a 70-kg.-adult (20 to 60 years old) is about 165. The reason for these differences is primarily the smaller *functional residual capacity* of the child (70 ml., as opposed to the adult's 2,700). Compliance increases as lung volume increases, in other words. For example, if one injects 1,000 ml. of air through the trachea and into the lungs, and the pressure increases by 5 cm. H_2O, then:

$$\text{Compliance of the 2 lungs} = \frac{1,000 \text{ ml.}}{5 \text{ cm. } H_2O} = 200 \text{ ml./cm. } H_2O$$

If half that volume goes to each lung, then:

$$\text{Compliance of 1 lung} = \frac{500 \text{ ml.}}{5 \text{ cm. } H_2O} = 100 \text{ ml./cm. } H_2O$$

To eliminate the variable of functional residual capacity, the term *specific compliance* has been introduced:

$$\text{Specific compliance} = \frac{\text{compliance}}{\text{lung volume}}$$

Compliance, as I have mentioned, is also dependent upon a third factor, the *intercellular fibers*. It decreases when the concentration of material not readily distended increases in the intercellular spaces. This may happen as a result of the laying down of collagenous tissue, or of a distention of the organ such that the slack in already-present collagen is eliminated by the distention itself (Figs. 10-6 and 10-8). Compliance increases when material that is not readily stretched is removed or made slack.

Changes in compliance. There are a number of conditions associated with changes in specific compliance. In *aging*, for example, the following three symptoms are noted: (1) an increase in residual volume and functional residual capacity, (2) no change in compliance, and (3) a decrease in the specific compliance of the lungs. At age forty-five, one's residual lung volume is about 1.5 liters, and at age seventy-five, about 1.9 liters (Fig. 17-19). Between the ages of forty-five and seventy-five there is not only an increase in the residual volume of about 13 ml. per year, but also a progressive increase in the amount of collagenous tissue in the lungs.

In severe diffuse alveolar *fibrosis*, on the other hand, the following symptoms are noted: (1) a decrease in the residual volume, (2) a decrease in compliance, and (3) a decrease in specific compliance. Pulmonary compliance in this condition may go as low as 10 ml. per centimeter H_2O, compared to a normal value of 165 ml. per centimeter H_2O.

In emphysema (Fig. 17-20), a narrowing of the respiratory channels results in distention of the alveoli which becomes so great that the elasticity of the lungs is impaired. In cases of emphysema in which damage to elastic fibers is not associated with an infiltration by collagen, the following symptoms appear: (1) an increase in the residual volume, (2) an increase in compliance, and (3) an increase in specific compliance.

Measurement of compliance. Compliance is measured under static conditions. We can readily determine pulmonary compliance in an anesthetized person with an open chest by injecting a known volume of air into a tracheal cannula and observing the pressure change that results in the cannula after the pressure has stabilized. It is simply a matter of repeating this procedure several times with different volumes of air and plotting the data in a manner similar to that in Figure 17-17. The slope of the line obtained gives the compliance.

In the unanesthetized person, a tube with a balloon at its end is usually swallowed and placed in the thoracic esophagus. Both the quantity of air inspired and the pressure change in the esophageal balloon are noted after several seconds of sustained inspiration. These studies are performed at several different depths of inspiration, and from them an estimate of pulmonary compliance is calculated.

Recoil. The inspiratory muscles cause distention of the lungs, which is associated with a *stretching* of elastic fibers and with increases in both the *surface of the alveoli* and the surface of the liquid lining the alveoli. The contraction of the inspiratory muscles, that is, imparts a potential energy to the lungs which is released during relaxation of the inspiratory muscles and is responsible for all or part of the air moved during expiration. We have discussed elastic tissue in Chapter 10 ("Arteries, Arteriovenous Anastomoses, and Veins"). In Figure 10-6 we noted that as we distend a vessel within its elastic limits, the vessel exerts progressively more tension on its contents. When the distending force is removed, the tension in the vessel is used to expel the contents of the vessel.

One of the important factors responsible for expelling air during passive expiration is the *surface tension* (Fig. 17-21) exerted by the liquid

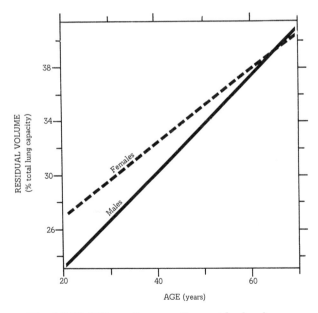

Fig. 17-19. Effect of age on the residual volume expressed in terms of per cent of total lung capacity. As the adult becomes older his total lung capacity decreases and his residual volume increases. At age 25, the vital capacity is 5.25 liters and at age 75 it is 3.20 liters. (*Redrawn from* Cotes, J. E.: Lung Function. P. 346. Oxford, Blackwell Scientific Publications, 1965)

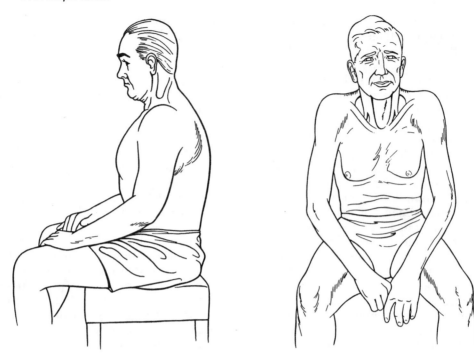

Fig. 17-20. Emphysema. Characteristic of emphysema is an increase in the functional residual volume of the lungs, an increase in the anteroposterior dimensions of the chest ("barrel chest"), and a greater dependence during inspiration on the contraction of muscles that elevate the shoulder girdle (accessory muscles of inspiration). (Barach, A. L.: Pulmonary emphysema. Hosp. Med., 2:13, 1965)

covering the alveolar surface. This liquid contains molecules that attract one another more than they are attracted by the air in the alveoli (Fig. 17-23), which results in the molecules in the middle of the solution being equally attracted in all directions while the molecules at the air-liquid and liquid-alveolar interfaces are attracted toward the center of the liquid. Since the alveolar liquid tends to form a sphere, the attraction of the water molecules away from their interfaces creates a surface tension which tends to shrink the surface to its smallest possible area, and therefore to increase the pressure and decrease the volume in the alveolus. In other words, the surface film on the alveolus contributes to the pressure of the air it surrounds (intrapulmonary pressure) in a manner similar to that in which the stretched rubber of a balloon or the liquid of a soap bubble contributes to the pressure of the air it encompasses.

In Figure 17-22 we note that the liquid lining the alveoli differs from water in two important ways. (1) It exerts less surface tension, and (2) its surface tension (dynes/cm.) increases as its area increases. Or, as the circumference of the alveolus increases, there is a progressive increase in surface tension and, as a consequence, an increased resistance to further distention. Similarly, we find that as the circumference of the alveolus decreases, there is a decrease in surface tension. This, then, tends to prevent the alveolus from collapsing.

There are in the lungs millions of alveoli of different radii. The situation might be compared to a forked tube (trachea and bronchi) leading to two soap bubbles (alveoli) of different radii (r). The pressure (P) in each bubble (intrapulmonary pressure) would be the same, but if the surface tension (T) exerted by each bubble were the same, small bubbles would collapse (atelectasis) and send their air into the large bubbles. These observations are consistent with the *law of Laplace* for a sphere, discussed in the section on stroke volume in Chapter 8:

$$T = \frac{Pr}{2}$$

In other words, if the surface tension exceeds

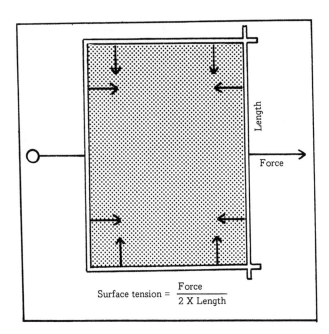

Fig. 17-21. The Maxwell frame can be used for measuring surface tension. (*Redrawn from* Clements, J. A., and Tierney, D. F.: Alveolar instability associated with altered surface tension. *In* Fenn, W. O., and Rahn, H. (eds.).: Handbook of Physiology. Sec. 3 (Respiration), vol. II, p. 1566. Washington, American Physiological Society, 1965)

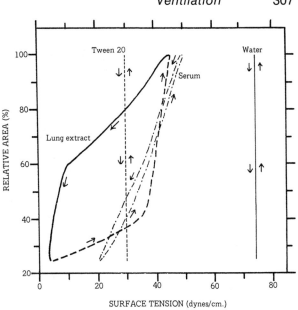

Fig. 17-22. Surface tension–area relationships of water, serum, lung extract, and Tween 20 (polyoxyethylene sorbitan monolaurate). (*Redrawn from* Clements, J. A., and Tierney, D. F.: Alveolar instability associated with altered surface tension. *In* Fenn, W. O., and Rahn, H. (eds.): Handbook of Physiology. Sec. 3 (Respiration), vol. II, p. 1568. Washington, American Physiological Society, 1965)

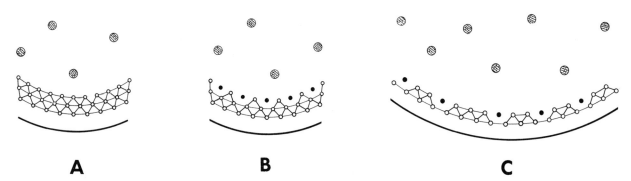

Fig. 17-23. Surface tension in a sphere. At A we note the intermolecular attractions in water (open circles are water molecules). At the surface the water and air (hatched circles are air molecules) form an interface. Note that the intermolecular attractive forces (*lines*) are such as to pull the water molecules at the air–liquid interface toward the center of the water. In other words, these forces result in a surface tension. At B we see the effect of adding surfactant (*solid circles*) to the water. Since the surfactant molecules are not attracted to water molecules as strongly as other water molecules they tend to accumulate at the air–liquid interface, where they lower the surface tension of the liquid. At C the volume of the liquid is the same as at B, but its surface area is increased. This results in a lower concentration of surfactant at the surface and, as a consequence, a greater surface tension (dynes/cm.) than is found at B.

(Pr)/2 the alveolus will collapse. In the body, collapse of the smaller alveoli is prevented by the production of a lining liquid which has a surface tension lower than that of water and which decreases its surface tension as the alveolus becomes smaller.

The factor in the lining liquid which gives it the characteristics mentioned above is a detergent-like agent, or *surfactant*. The surfactant (black molecules in Fig. 17-23) is not so strongly attracted to the molecules of the liquid as are the other molecules, and therefore accumulates at the surface where it decreases the surface tension. When the surface area of the alveolus increases, as in inspiration, water molecules in the liquid move up into the surface layer and decrease the concentration of the surfactant, thereby increasing the surface tension. Thus, at the end of expiration (when the alveoli are small) the surface tension is about 4 dynes per centimeter, and at the end of inspiration may reach 50 dynes per centimeter.

The importance of surfactant to our survival has been demonstrated in a number of cases in which its concentration was low. It has been shown that the surface tension of lung extracts is abnormally high in (1) babies who have died from "respiratory distress of the newborn" (hyaline membrane disease) (Fig. 17-24), (2) patients who have died after open heart surgery in which an artificial heart-lung machine was used, (3) patients with congestive collapse of the lungs after occlusion of one pulmonary artery, and (4) patients who have breathed 100 per cent O_2 for a prolonged period.

Resistance. Resistance (Ω) in the respiratory system is calculated from the air flow through the trachea (I) and the difference between the pressure at the nasal or oral cavity and that in the alveoli ($P_1 - P_2$):

$$\Omega = \frac{P_1 - P_2}{I}$$

It is the reciprocal of conductance:

$$\text{Conductance} = \frac{I}{\Omega}$$

In a normal, healthy person breathing 15 times a minute, the major resistance which must be overcome during inspiration is associated with the expansion of the lungs and chest (elastic recoil). The second most important resistance (about half the elastic recoil resistance) results

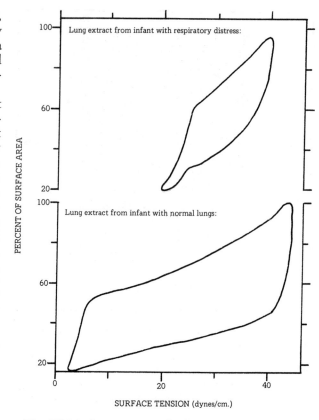

Fig. 17-24. Comparison of saline extracts of normal and atelectatic infants' lungs. The extract from the normal lung gives a minimum surface tension of 2 dynes/cm., whereas that from the infant with respiratory distress gives a minimum tension only down to 20 dynes/cm. (*Redrawn from* Clements, J. A.: Surface phenomena in relation to pulmonary function. Physiologist, 5:19, 1962)

from the character of the airway (Table 17-5), its length, and its radius. The trachea has a large diameter and incomplete *cartilaginous rings* which prevent it from collapsing completely, but further down the tract the cartilage is absent and the diameter of the lumen decreases.

The absence of cartilage creates no problem in a young healthy person, since on inspiration the intrathoracic pressure is more negative than the intrapulmonary pressure, and on expiration, the elastic recoil of the alveoli keeps the intrapulmonary pressure greater than the intrathoracic pressure. In other words, a positive transmural pressure across the bronchioles during both inspiration and expiration maintains their patency.

Classification and Name	Numerical Order	Number of Each	Diameter (mm.)	Length (mm.)	Total Cross-sectional Area (cm.²)
Cartilaginous					
Trachea	0	1	18	120	2.5
Main bronchi	1	2	12	47.6	2.3
Lobar bronchi	2	4	8	19.0	2.1
Segmental bronchi	3	8	6	7.6	2.0
Subsegmental bronchi	4	16	4	12.7	2.4
Transitional					
Small bronchi	5-10	1,024°	1.3°	4.6°	13.4°
Noncartilaginous					
Bronchioles	11-13	8,192°	0.8°	2.7°	44.5°
Terminal bronchioles	14, 15	32,768°	0.7°	2.0°	113°
Respiratory bronchioles	16-18	262,144°	0.5°	1.2°	534°
Alveolar ducts	19-22	4,194,304°	0.4°	0.8°	5,880°
Alveolar sacs	23	8,388,608	0.4	0.6	11,800
Terminal alveoli	24	300,000,000	0.2		

° Numbers refer to last generation in each group.

Table 17-5. Approximate dimensions of the airways of the lung in an adult human. The lungs were inflated to about two thirds of the total lung capacity. (Modified from Comroe, J.H., Jr.: Physiology of Respiration. P. 117. Chicago, Year Book Publishers, 1965)

During forced expiration, for example, the following values are typical:

Intrathoracic pressure	+4
Pressure due to recoil	+2
Pressure within alveolus	+6
Pressure loss between alveolus and bronchiole	1
Pressure within bronchiole	+5
Intrathoracic pressure	+4
Transmural pressure across bronchiole = +5 − +4 = +1	

A loss of some of the elastic recoil of the lungs can produce a negative transmural pressure and, as a consequence, occlusion of the bronchiole during expiration. This is essentially what occurs in *emphysema*. In other words, in emphysema, flows may reach 500 liters per minute during the height of inspiration because all channels remain patent but during expiration, channels without supporting cartilage collapse when a maximum expiratory effort is attempted; the maximum flow rates may therefore remain below 25 liters per minute.

Any decrease in the diameter of the lumen of a respiratory channel produces an increase in its resistance. This may result from (1) the contraction of smooth muscle, (2) the development of

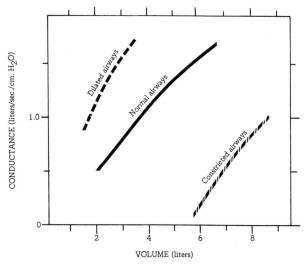

Fig. 17-25. Effect of acute changes in lung volume on conductance in dilated, normal, and constricted airways. Apparently acute increases in lung volume produce a distention of the alveoli (increased lung elastic pressure) which raises the intrapulmonary pressure and dilates the respiratory passages during expiration. (*Redrawn from* Dubois, A. B.: Resistance in breathing. *In* Fenn, W. O., and Rahn, H. (eds.): Handbook of Physiology. Sec. 3 (Respiration), vol. I, p. 459. Washington, American Physiological Society, 1964)

edema, or (3), as mentioned above, a loss of elastic pressure in the lung. An increase in resistance in one channel, even though it is parallel with a number of others, increases the total resistance. If, for example, we have four channels—each with a resistance of 4, the total resistance is:

$$\frac{1}{\Omega_T} = \frac{1}{4} + \frac{1}{4} + \frac{1}{4} + \frac{1}{4} = \frac{4}{4} = 1$$

If one of the channels collapsed or became plugged, however, the total resistance would increase 33 per cent.

$$\frac{1}{\Omega_T} = \frac{1}{4} + \frac{1}{4} + \frac{1}{4} + \frac{1}{\text{infinity}} = \frac{3}{4}$$
$$\Omega_T = 1.3$$

The resistance of the respiratory system is also dependent on the velocity of air flow. As the velocity increases, *turbulence* becomes more pronounced and resistance increases (Fig. 17-26). Increases in velocity are produced by more forceful contractions of respiratory muscles and by the narrowing of parts of the respiratory system. Increases in the density of the air, increases in the radius of the respiratory channel, branching of the respiratory channel, and decreases in the viscosity of the air can also lead to turbulence. The density of the air increases with the atmospheric pressure. The viscosity varies with the composition of the gas. If we assume air to have a viscosity of 1, then the viscosity of 100 per cent O_2 would be 1.13, and that of 80 per cent He + 20 per cent O_2 would be 1.11. Factors producing turbulence are discussed more fully in the section on murmurs in Chapter 7 ("Heart—Contractile Properties").

Air flow. In the study of respiration we are interested in the maximum volume of air that can be inspired or expired (vital capacity), and the rate at which this air can be moved. The vital capacity is dependent on (1) the strength of the respiratory muscles, (2) their state of innervation, (3) the amount of functional lung tissue, and (4) the distensibility of the lungs and chest wall. In Figure 17-28 I have indicated some of the conditions (restrictive ventilatory insufficiency) in which the vital capacity is low. Also indicated in Figure 17-28 are some conditions (obstructive ventilatory insufficiency) in which the vital capacity may be normal but the rate of air flow in milliliters per minute low. A young healthy per-

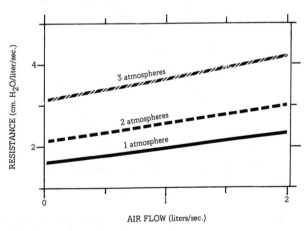

Fig. 17-26. The effect of increases in air flow on resistance (pressure/flow in cm. H₂O/liter/sec.) at 1, 2, and 3 atmospheres of pressure. Note that with increases in either air flow or atmospheric pressure the resistance to flow increases. These 2 factors increase resistance by increasing the amount of turbulent flow. (*Redrawn from* Dubois, A. B.: Resistance in breathing. *In* Fenn, W. O., and Rahn, H. (eds.): Handbook of Physiology. Sec. 3 (Respiration), vol. I, p. 457. Washington, American Physiological Society, 1964)

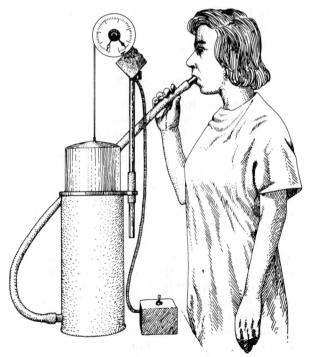

Fig. 17-27. The spirometer is used to measure the tidal volume, vital capacity (pointer 2 on the calibrated dial above spirometer) and various timed vital capacities (pointer 1) such as the 1 second forced expiratory volume. (Webb, W. R.: Management of postoperative pulmonary complications. Hosp. Med., *1*:17, 1965)

son, for example, should be able to move between 30 and 40 liters of air into and out of the lungs in 15 sec. (120 to 160 liters per minute). This procedure has been used by clinicians as a test of *pulmonary function* and is called the *maximum voluntary ventilation* test.

Maximum flow rates are also frequently determined by recording the maximum volume of air that can be inspired or expired in a period of 0.5 to 3 sec. (Figs. 17-27 and 17-28). A normal person, for example, should be able after maximum inspiration to expire 83 per cent of his vital capacity in the first second of expiration (forced expiratory volume, FEV^1), or 97 per cent during the first 3 sec. (FEV^3). In emphysema the forced inspiratory volumes for 1 and 3 sec. are frequently normal, whereas the forced expiratory volumes are reduced.

Work of breathing. The work done on the lung during a single respiratory cycle can be calculated by simultaneously studying on an x-y recorder the intrapleural pressure and the volume of air inspired and expired (Fig. 17-29). We note that during inspiration a pressure-volume curve, AIC, is inscribed. If we draw a horizontal line from point C to a point that represents atmospheric pressure (D′), and another horizontal line from point A to another point that represents atmospheric pressure (D), we form a figure (AICD′D) whose area equals the work performed during inspiration. This figure is made up of (1) a rectangle (AA′D′D) whose area equals the work performed during inspiration to overcome the elastic recoil of the lungs, (2) a triangle (ACA′), whose area equals the work done in producing distention of the lungs, and

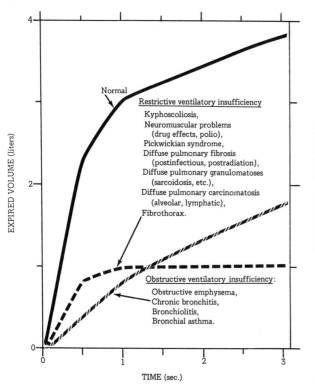

Fig. 17-28. The timed maximum expiratory volume. In these studies the subject took a maximum inspiration and then expired maximally into a recording spirometer. In restrictive ventilatory insufficiency the vital capacity is most markedly reduced and in obstructive ventilatory insufficiency the maximum flow (in ml./sec.) is most reduced. (*Modified from* Webb, W. R.: Management of postoperative pulmonary complications. Hosp. Med., *1*:18, 1965)

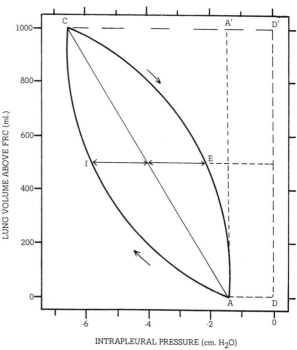

Fig. 17-29. Relationship between intrapleural pressure and change in lung volume above functional residual volume (FRC). Curve AIC is produced during inspiration and CEA during expiration in a resting individual. The area represented by AICD′D equals the work performed during inspiration. The area CEA represents energy stored during inspiration and released during expiration to overcome the inertia of the air and the resistance of the airway. (*Redrawn from* Cotes, J. E.: Lung Function. P. 92. Oxford, Blackwell Scientific Publications, 1965)

(3) a figure (AIC) whose area represents the work performed in overcoming resistance and inertia during inspiration.

$$(AICD'D) = (AA'D'D) + (ACA') + (AIC)$$

During expiration a second pressure-volume curve (CEA) is inscribed. If we draw a straight line between points A and C we form a slope which approximates the compliance of the lungs. The area represented by the figure CEA equals the work performed during expiration to overcome resistance and inertia. In quiet breathing this represents energy stored in the elastic elements of the lung during inspiration and released during expiration. The work of inspiration, on the other hand, is accomplished by muscle contraction.

Respiratory minute work. The work performed by the respiratory muscles in one minute varies with the respiratory rate and depth and the ventilation. During quiet breathing, the cost to a healthy person in terms of O_2 *consumption* is about 0.5 ml. of O_2 per liter of ventilation. In other words, about 1.5 per cent of the O_2 consumed by such a person is used by the respiratory muscles. When the ventilation is increased to 40 liters of air per minute, the O_2 cost is about 1.2 ml. O_2 per minute (3% of total O_2 consumption). Or, increases in ventilation above the resting level result in decreases in efficiency:

Efficiency (%) =
$$\frac{\text{work performed (gm.-cm./min.)}}{\text{energy required (gm.-cm./min.)}} \times 100$$

Low respiratory efficiency may also occur at a low pulmonary minute volume in patients in whom the work of breathing is markedly increased due to obstructive lung disease, pulmonary fibrosis, or obesity, and may be sufficiently great to lead to respiratory failure.

In Figure 17-30 we note that when the minute volume in a healthy person is kept constant and the respiratory rate and depth modified, for each

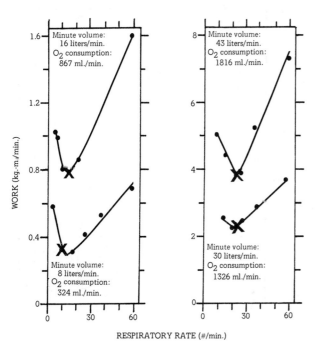

Fig. 17-30. The work of breathing at different respiratory frequencies and different fixed pulmonary minute volumes and oxygen consumptions. The X represents the frequency spontaneously chosen by the subject at each level of O_2 consumption. Note that in these studies this frequency lies between 10 and 30 per minute and represents the most efficient frequency for each work load. (*Redrawn from* Milic-Emili, G., and Petit, J. M.: Il lavoro meccanico della respirazione a varia frequenza respiratoria. Arch. Sci. Biol. *43*:328, 1956)

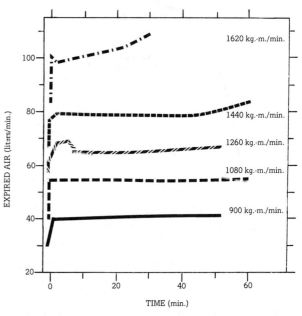

Fig. 17-31. Changes in the volume of the expired air at different work intensities of subjects on a bicycle ergometer. The response to exercise is an increase in tidal volume and respiratory rate which produces an increased pulmonary minute volume. (*Modified from* Nielsen, M.: Untersuchung über die Atemregulation beim Menschen. Skand. Arch. Physiol. Suppl. *10*: 154-157, 1936)

minute volume there is a rate and depth at which the least amount of respiratory work is performed. At 8 liters per minute the most efficient rate is about 15 per minute; at 43 liters per minute it is about 30 per minute. The fact that the optimal respiratory rate for man lies at about 15 per minute during rest is due to the relationship between respiratory rate and depth. As the rate decreases, the tidal volume must increase to maintain a constant minute volume. Increases in tidal volume require more energy per minute in order to distend the alveoli, but also require less energy per minute to perform non-elastic work (i.e., to overcome airway resistance). In other words, at a respiratory rate of 15 per minute, a fixed minute volume can be maintained at the lowest energy cost, because the sum of the energy needed to (1) distend the alveoli (produce a certain tidal volume) and (2) overcome airway resistance (produce a certain flow) is minimal. This is not to say that either (1) or (2) by itself is minimal, but that the sum of (1) and (2) is minimal at these rates.

During strenuous exercise the minute volume may go as high as 140 liters of air per minute in well-trained athletes. This is the result of an increase in respiratory rate to about 40 per minute, and in depth, to about 3.5 liters (Fig. 17-31):

Minute volume =
respiratory rate × tidal volume =
140 liters/min. =
40 breaths/min. × 3.5 liters/breath

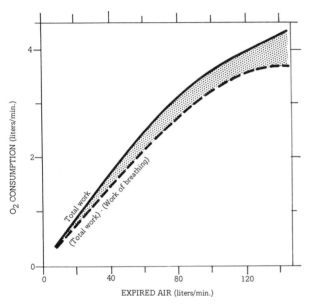

Fig. 17-32. Relation between the volume of air expired and the O_2 consumption of a subject during exercise. The upper curve represents the total oxygen consumption and the width of the shaded area the oxygen cost of breathing. Note that with an increase in O_2 consumption there is also an increase in minute volume. The increase in minute volume results in a progressive increase in the oxygen cost of breathing until a point is reached where a further increase will serve no useful purpose for the body. (*Redrawn from* Otis, A. B.: The work of breathing. *In* Fenn, W. O., and Rahn, H., (eds.): Handbook of Physiology. Sect. 3 (Respiration), vol. I, p. 474. Washington, American Physiological Society, 1964)

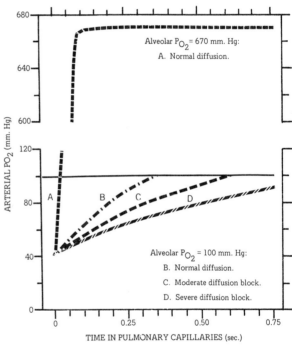

Fig. 17-33. Rate of O_2 uptake by the blood when the alveolar P_{O_2} is (A) 670 mm. Hg and when it is (B) 100 mm. Note that the blood comes into equilibrium much faster at (A), a high alveolar P_{O_2} (about 0.1 sec.), than at (B), a low alveolar P_{O_2} (about 0.3 sec.). Since blood usually remains in the pulmonary capillaries for 0.75 second, the blood under both circumstances will reach equilibrium with the alveolus before it leaves the lungs. With moderate (C) to severe (D) impairment of diffusion due to thickening of the alveolar epithelium or pulmonary edema, the rate of diffusion may decrease to a point where the blood leaving the lung has not reached equilibrium with the alveolar air. (*Redrawn from* Physiology of Respiration by J. H. Comroe, Jr. Copyright © 1965, Year Book Medical Publishers, Inc.; used by permission)

Increases in minute volume above the resting level increase the work of breathing and decrease the respiratory efficiency, but are essential for strenuous exercise. Eventually a point is reached at which the cost of breathing is so high that a further increase in minute volume serves no useful function (Fig. 17-32).

DIFFUSION

The function of ventilation is to increase the P_{O_2} and decrease the P_{CO_2} in the respiratory portions of the respiratory tree. This creates a situation favorable for the movement of O_2 into and CO_2 out of the blood of the pulmonary capillaries. This movement is called diffusion, and is a passive process resulting from the random motion of molecules in a fluid (gas or liquid). Net diffusion for a particular substance is always from the area of higher partial pressure to that of lower partial pressure. Thus O_2 moves from the alveolar lumen, where its P_{O_2} is about 100 mm. Hg, to the plasma, where its P_{O_2} may be only 40. The total number of O_2 molecules passing into the plasma each minute depends on (1) the difference between the P_{O_2} in the alveolus

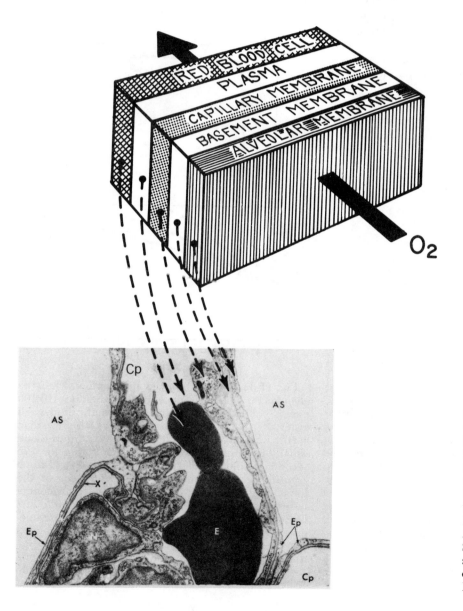

Fig. 17-34. Electron micrograph of an alveolus (AS) and its associated blood supply (Cp, capillary). The diagram shows the barriers through which the O_2 must diffuse if it is to combine with the hemoglobin of the red blood cell (E). Diffusion will decrease if this barrier is enlarged as in pulmonary edema, thickening of the alveolar epithelium (Ep), or increase in the basement membrane (X). It will also decrease if the blood supply to the pulmonary capillaries is decreased or if the surface area of the respiratory channels is decreased (obstruction or lobectomy). (Electronmicrograph from Porter, K. R., and Bonneville, M. A.: An Introduction to the Fine Structure of Cells and Tissues. 3rd ed. p. 40. Philadelphia, Lea & Febiger, 1968)

and in the plasma (Fig. 17-33), (2) the diffusion characteristics of the alveolar-capillary membrane, and (3) the ability of the erythrocytes to pick up O_2.

Normally, the erythrocyte is in the pulmonary capillary about 0.75 sec. As can be seen in Figure 17-33, this is usually more than enough time for the blood to reach equilibrium with the alveolar air. That is, 7.5 sec. is sufficient time for the plasma and erythrocytes to reach the same PO_2 as the alveolus, and for the erythrocytes to become about 98 per cent saturated with O_2. On the other hand, a decrease in (1) the number of patent capillaries, (2) the number of erythrocytes, or (3) the quantity of available hemoglobin decreases the amount of O_2 carried away from the alveoli. Distention of the capillaries too interferes with diffusion by increasing the distance O_2 must diffuse in order to enter the erythrocytes.

The amount of O_2 entering the plasma is also dependent on the diffusion characteristics of the alveolar-capillary membrane. Decreases in the surface area of the lungs available for gas diffusion, increases in the thickness of the alveolar-capillary barrier (Fig. 17-34), and changes in the character of that barrier can all interfere seriously with diffusion.

It is important to realize that the alveolar-capillary barrier restricts the movement of some gases more than others. In Table 17-6, for example, we find that CO_2 diffuses from the blood to the alveolus or in the reverse direction twenty times more rapidly than does O_2. Part of the reason for this is that CO_2 is much more soluble in aqueous solutions than is O_2. Because of these differences, although O_2 diffusion may cause serious clinical problems in pulmonary edema (Fig. 17-35) and in thickening of the alveolar epithelium, the movement of CO_2 into the alveolus is seldom a problem. When O_2 diffusion is so severely impaired that life can be maintained only by breathing pure O_2, CO_2 diffusion as well may be affected.

Tests for Pulmonary Diffusion and Blood Flow

The rate at which a gas is removed from the lungs depends on (1) how rapidly it is transferred to the blood and (2) how rapidly the blood carries it away. We can see from Table 17-6 that *acetylene*, CO_2, and *nitrous oxide* diffuse through the *alveolocapillary barrier* much more readily than O_2 and *CO*. On the other hand, O_2 and *CO* are picked up by the blood much more readily

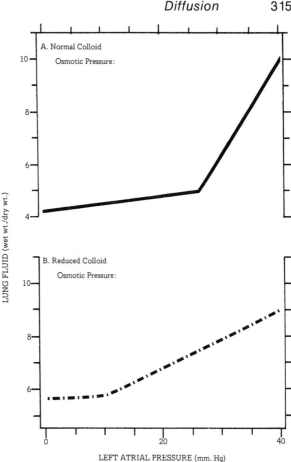

Fig. 17-35. Left atrial pressure and fluid in the lungs. At A the left atrial pressure was elevated to various levels in 19 dogs for a period of 30 minutes and the wet weight and dry weight of the lungs were determined. At B the plasma proteins were reduced to 0.47 times normal and the experiments repeated on 33 dogs. It is concluded that pulmonary edema can be produced by an increase in the blood pressure in the pulmonary capillaries and that this edema is further facilitated by a decrease in colloid osmotic pressure. Pulmonary edema can seriously interfere with the diffusion of O_2 into the blood. (Guyton, A. C., and Lindsey, A. W.: Effect of elevated left atrial pressure and decreased plasma protein concentration on the development of pulmonary edema. Circ. Res., 7:651, 1959; by permission of the American Heart Association, Inc.)

than either nitrous oxide or acetylene. In other words, pathological changes in the alveolar-capillary barrier (*alveolar-capillary block*) decrease the movement of O_2 or CO out of the alveolus, and have little or no effect on the movement of nitrous oxide out of the alveolus. Thus if the amount of O_2 or CO that moves from al-

Gas	Molecular Weight	Solubility* at 37°C.	Rate of Diffusion Relative to O_2
Oxygen	32	0.0239	1.00
Acetylene	26	0.749	34.8
Argon	40	0.0259	0.97
Carbon dioxide	44	0.567	20.3
Carbon monoxide	28	0.0184	0.83
Ethylene	28	0.0784	3.43
Helium	4	0.0085	1.01
Krypton	85	0.0449	1.15
Nitrogen	28	0.0123	0.55
Nitrous oxide	44	0.388	13.9
Xenon	133	0.085	1.75

* Solubility in milliliters of gas dissolved at standard temperature, pressure, and dryness per milliliter of solution.

Table 17-6. Rate of diffusion of respired gases relative to that of oxygen. (From Cotes, J.E.: Lung Function. Oxford, Blackwell Scientific Publications, 1965)

veolus to blood is reduced, but the removal of nitrous oxide normal, there is a defect in the alveolar-capillary barrier.

On the other hand, if the removal of nitrous oxide is impaired but that of O_2 and CO normal, the defect does not involve the barrier. The latter circumstance probably results from an inadequate pulmonary blood flow, since the blood has, in the erythrocyte, a much more effective system for carrying away O_2 and CO than for carrying away nitrous oxide. The uptake of nitrous oxide from the lungs, in other words, is limited primarily by the pulmonary blood flow, for which reason nitrous oxide is preferred in the study of pulmonary blood flow changes. In CO and O_2 we have gases whose uptakes are limited primarily by the alveolocapillary barrier, and which are used to study changes in this barrier. The movement of CO_2, on the other hand, is limited for the most part by neither the barrier nor the blood flow.

Diffusing capacity. Earlier in this section we defined conductance as follows:

$$\text{Conductance} = \frac{\text{flow}}{\text{driving pressure}}$$

In studying the diffusion characteristics of the lung we also measure a type of conductance called the diffusing capacity:

$$\text{Diffusing capacity} = \frac{\text{flow}}{\text{mean driving pressure}}$$

The diffusing capacity for O_2 (D_{O_2}) can be calculated if we know the flow of O_2 from the alveolus to the blood in milliliters per minute (I_{O_2}) and the mean alveolar P_{O_2} ($P_{A_{O_2}}$) and mean pulmonary capillary P_{O_2} ($P_{C_{O_2}}$):

$$D_{O_2} = \frac{I_{O_2}}{P_{A_{O_2}} - P_{C_{O_2}}}$$

Getting adequate data from the above calculation is difficult and involves taking blood samples. A simpler method is to determine the carbon monoxide diffusing capacity (D_{CO}). In this technique a sample of air containing 0.3 per cent CO is inspired and held for 10 sec. and the per cent of CO in the alveolus is determined at the beginning and end of the 10 sec. From these data plus the volume of the alveolar gas we can estimate the flow of CO from the alveolus to the blood (I_{CO} in ml./min.). We also calculate the mean alveolar P_{CO} ($P_{A_{CO}}$). The mean capillary $P_{C_{CO}}$ is usually assumed to be sufficiently close to zero not to warrant a blood sample:

$$D_{CO} \text{ (ml./min./mm. Hg)} = \frac{I_{CO} \text{ (ml./min.)}}{P_{A_{CO}} - 0 \text{ (mm. Hg)}}$$

D_{O_2} can be estimated from the D_{CO} by multiplying the D_{CO} by a constant:

$$D_{O_2} = (D_{CO})(1.23)$$

Unfortunately, determination of the D_{O_2} is not a specific diagnostic procedure, since the P_{O_2} is reduced in so many different types of lung disease (decreased lung volume, emphysema, chronic obstructive disease). It is of value in the rare case in which all other aspects of pulmonary function appear normal.

	Age (yrs.)			
	25	45	65	75
Vital capacity (liters)	5.25	4.28	4.05	3.20
Residual volume (liters)	1.66	1.48	1.72	1.92
Dead space (liters)	0.144		0.235	
Total capacity (liters)	6.91	5.76	5.77	5.12
Resid. vol./total cap. (%)	24	26	30	37
Respiratory rate (no./min.)	14	16	17	19
Alveolar ventilation (liters/min.)	5.3	6.9	6.7	6.9
Maximum breathing capacity in males (liters/min.)	126	109	91	54
Maximum breathing capacity in females (liters/min.)	94	89	73	
O_2 Consumption (liters/min.)	0.232	0.215	0.193	0.188

Table 17-7. Changes in the respiratory system with age. All data were obtained from standing subjects. (From Richards, D.W.: Pulmonary changes due to age. *In* Fenn, W.O., and Rahn, H. (eds.): Handbook of Physiology. Sec. 3 (Respiration), vol. II, pp. 1525-1529. Washington, American Physiological Society, 1965; and Boren, H.G., et al.: Lung volume and its subdivisions in normal men. Am. J. Med., *41*:103, 1966)

AGING

In the fetus, the lungs are collapsed and occupy a volume of 40 ml. Immediately after birth the infant takes his first breath and, in so doing, increases his lung volume fivefold and the blood flow to his lungs five-to tenfold. The increase in pulmonary blood flow is brought about by the closure of the foramen ovali between the right and left atria, and closure of the ductus arteriosus between the pulmonary artery and aorta. Large volumes of blood are thus shunted into the pulmonary system. The mechanisms initiating this are not clearly understood.

During the first 3 days of life O_2 consumption increases from 4 to 7 ml. per kilogram per minute. By the age of 8, the child has developed his maximum number of alveoli (304×10^6), but continues to increase his total lung capacity and maximum breathing capacity past puberty. After 25 years of age (Table 17-7) the muscles usually begin to weaken, O_2 consumption at rest decreases, the vital capacity decreases, and the maximum breathing capacity decreases. After the age of 30 or 40, the respiratory bronchioles and alveolar ducts enlarge and, as a result, the residual volume and dead space increase. There is also a decrease in the maximum diffusing capacity for O_2 from age 20 to age 70 of about 7.7 ml. per minute per mm. Hg of O_2 per decade, and the lung shows increased concentrations of collagen and elastin and a decrease in lipid.

Chapter 18

MEDULLARY CENTERS
PONTINE CENTERS
OTHER CENTERS
NEUROLOGICAL LESIONS IN MAN
CHEMORECEPTORS
STRETCH RECEPTORS
RECEPTORS IN THE UPPER RESPIRATORY TRACT
RESPONSE OF RECEPTORS TO INJECTED AGENTS
OTHER RECEPTORS

RESPIRATORY CONTROL

The respiratory system, like the circulatory system, involves movement of a fluid by means of the contraction of a muscular pump. In the case of the circulatory system the heart is the major pump. It contains its own pacemaker, which sends impulses to all the cardiac fibers. The role of the central nervous system in the control of the heart is to (1) modify the rate of firing of the cardiac pacemaker (chronotropic action), (2) modify the rate of conduction along the cardiac fibers (dromotropic action), and (3) modify the contractile response of these fibers (inotropic action). These messages are transmitted from the central nervous system by the cardiac sympathetic and parasympathetic neurons.

The major pump for the respiratory system is the diaphragm. It does not contain its own pacemaker, but rather is dependent upon the right and left *phrenic nerves* for the initiation of its contraction. Each muscle fiber that contracts receives its stimulation from a neuron branch — rather than from a neighboring muscle fiber. The phrenic neurons originate from the third, fourth, and fifth cervical portions of the cord. A transection of the cord below C-5 has little effect on respiration, but a transection anywhere from C-3 to the fourth ventricle of the medulla oblongata (Figs. 18-1 and 18-4) causes the cessation of breathing. If, on the other hand, the transection is made between the medulla and pons, a person continues to breathe — though it should be made clear that such a transection produces marked changes in the character of the respirations. The *medullary animal* (transection between the medulla and pons) shows an abnormal respiratory pattern characterized by spasmodic, usually maximal inspiratory efforts which terminate abruptly (gasping). The respiratory intervals also vary considerably. Wang et al. have demonstrated in three medullary cats that respiration continues even if the cord is also cut at the sixth cervical segment, and all cervical dorsal (sensory) roots and cranial nerves are severed. From data such as these it has become increasingly clear that there are respiratory centers in the medulla oblongata which have an *intrinsic rhythmic activity* serving to initiate inspiration and expiration. In short, they serve as pacemakers for the respiratory system which, like the cardiac pacemaker, are modified by activity in other parts of the body. In the case of the medullary respiratory centers, some of the more important modifying influences come from areas in the pons, midbrain, diencephalon, cerebral cortex, and sensory neurons in the ninth and tenth cranial nerves.

MEDULLARY CENTERS

A center is a portion of the brain rich in cell bodies and synapses, the functional integrity of which is essential for certain patterned responses. It is an area in which many incoming and frequently antagonistic impulses are integrated into a series of synergistic stimuli. In the medulla, two overlapping areas, the respiratory centers, exhibit these characteristics. In Figure 18-2(A), for example, we note one area in the medulla (part of the *inspiratory center*) which discharges action potentials that initiate and maintain the inspiratory act, and in Figure 18-2(B) we see a record from another area (part of the *expiratory center*) which discharges action

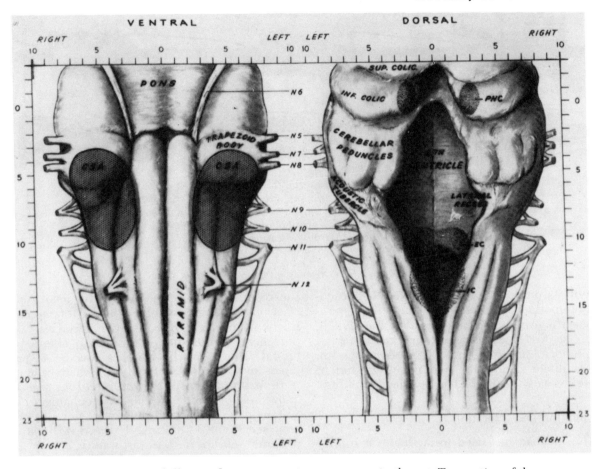

Fig. 18-1. Medullary and pontine respiratory centers in the cat. Transection of the brain stem at the lower border of the acoustic tubercle dorsad or transection at the lower border of the trapezoid body ventrad is usually followed by a continuation of rhythmic respiration. This is apparently due to the fact that these transections leave the inspiratory (IC) and expiratory (EC) centers intact and in connection with the respiratory muscles. A chemosensitive area (CSA) lies dorsal and rostral to the respiratory centers, and a pneumotaxic center (PNC) lies in the pons. These areas as well as others serve to modify the activity of the inspiratory and expiratory centers. The cranial nerves (N-6 through N-12) and sterotaxic coordinates in mm. (on borders) are also shown. (Severinghaus, J. W., and Larson, C. P., Jr.: Respiration in anesthesia. *In* Fenn, W. O., and Rahn, H. (eds.): Handbook of Physiology. Sec. 3 (Respiration), vol. II, p. 1222. Washington, American Physiological Society, 1965)

potentials that initiate and maintain the expiratory act. In Figure 18-3(E) and (F) we have evidence that stimulation of these areas produces changes in respiratory rate and depth.

Thus, on the basis of the study of (1) the effect of brain lesions, (2) the histology of the medulla, (3) neuronal activity in the medulla, and (4) the effects of local stimulation, we have evidence of an inspiratory and an expiratory center which control respiratory rate and depth. These centers are part of a richly interconnected neuron network in the reticular formation which extends from the medulla through the pons and has been called by various authors the respiratory mechanism, apparatus, substrate, network integrator, oscillator, and modulator. It is through this respiratory mechanism and impulses coming to it from sensory neurons, the midbrain, the hypothalamus, and the telencephalon that a respiratory act integrated with the rest of body

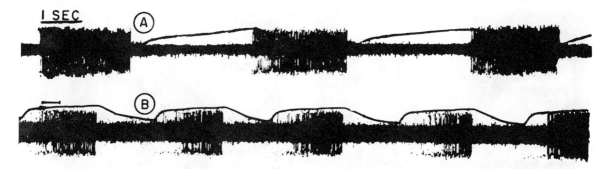

Fig. 18-2. Neuronal discharges from the medulla of the cat. At A we note a locus which discharges during inspiration (*downward deflection of the upper trace*). At B we note a different locus which fires during expiration (*upward deflection of pneumogram*). (Haber, E., et al.: Localization of spontaneous respiratory neuronal activities in the medulla oblongata of the cat: A new location of the expiratory center. Am. J. Physiol., *190*:351, 1957)

activity is produced. This integration is particularly important in respiration, since the respiratory act is used not only to ventilate the alveoli, but also to (1) eliminate an irritant from the respiratory tract (coughing), (2) produce speech, (3) facilitate deglutition and vomiting, and in some animals, (4) eliminate excess heat (panting).

Facilitation and Inhibition

The mechanisms used to facilitate or inhibit the respiratory centers are still poorly understood, but one generally accepted view is that facilitation and inhibition are accomplished by modifying the *transmembrane potentials* of the key neuron or neurons in the center. In swallowing, for example, the inspiratory center probably receives impulses which *hyperpolarize* (makes less excitable) a cell body or bodies. In hypoxia, on the other hand, the inspiratory center receives impulses which decrease the transmembrane potential (make cells more excitable). In other words, a pacemaker neuron in the inspiratory center probably has an intrinsic discharge pattern which is modified by the neurons impinging upon it. Hyperpolarization of this neuron decreases the respiratory rate by decreasing excitability and by increasing the time needed to produce a pacemaker potential. Decreasing the resting transmembrane potential increases the frequency with which it discharges.

The respiratory rate depends on the number of volleys of action potentials sent to the inspiratory muscles—not just on the number of action potentials. It should be emphasized here that the respiratory muscles, unlike the heart, contract not by single twitches, but by tetanic contractions initiated by volleys of impulses from peripheral nerves rather than by a single potential. The means by which a pacemaker produces such a volley is highly speculative, but it is probably the result of *reverberating circuits* (closed-chain circuit, feedback mechanisms, or "circus movement") in the central nervous system (see Fig. 2-14). Normally the reverberating patterns set up in the inspiratory center are terminated by impulses from other parts of the nervous system.

The depth of inspiration, or the tidal volume, is dependent on (1) the number of motor units stimulated (i.e., the degree of *recruitment*), (2) the duration of the volley of action potentials in each motor unit, and (3) the frequency of action potentials (number/sec.) in each motor neuron. Presumably all these factors are determined by the sum of the facilitatory and inhibitory impulses impinging upon the inspiratory mechanism. When, for example, the Pco_2 of the arterial blood increases, impulses impinge upon the inspiratory center and produce an increase in inspiratory rate and depth. The increase in depth is accomplished in part by an increase in the frequency of neuronal potentials, in part by an increase in the duration of the inspiratory volley of potentials, and in part by the stimulation of more motor neurons. As we progressively increase the Pco_2, the recruitment of motor neurons at first involves primarily those to the diaphragm, but eventually the spread of excitation involves neurons to the accessory muscles of

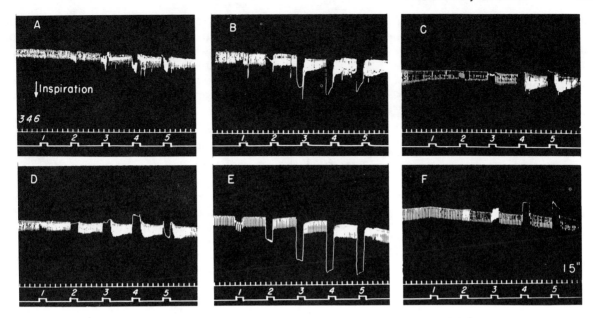

Fig. 18-3. Respiratory response of the cat to electrical stimulation of various areas in the pons and medulla. Each stimulus had a strength of 5 volts and a duration of 2 msec. The stimuli were applied at a frequency of (1) 5 cycles/sec., (2) 25 cycles/sec., (3) 50 cycles/sec., (4) 100 cycles/sec., (5) 200 cycles/sec. in each study. At A the stimuli were applied to a portion of the rostral pons and elicited respiratory acceleration, at B to the middle and caudal pons (maintained inspiration), at C to a part of the pons that produced maintained expiration, at D to the rostral medulla (maintained expiration), at E to a part of the medulla that produced a maintained inspiration, and at F to a part of the medulla caudal to the obex (acceleration and maintained active expiration). (Ngai, S. H., and Wang, S. C.: Organization of central respiratory mechanisms in the brain stem of the cat: localization by stimulation and destruction. Am. J. Physiol., *190*:345, 1957)

inspiration. In other words, we have in the inspiratory center neurons capable of activating most or perhaps all of the inspiratory muscles. Under resting conditions the inspiratory pacemaker activates only a small per cent of this neuron pool, but under conditions, for example, in which the threshold of the neurons in the pool has been lowered by facilitatory stimuli, the spread of activation may be extensive. Wang and Ngai express this as follows:

Excitation of a small part of one of the centers may lead through synaptic interconnection to activity of the whole of that center.

Reciprocal Inhibition

The relationship between the inspiratory and expiratory centers is probably one of reciprocal inhibition. When the inspiratory center fires, it transmits not only facilitatory impulses to the anterior horn cells of the phrenic nerve, but also inhibitory impulses to the expiratory center. In a normal resting expiration, on the other hand, there is little or no stimulation of expiratory muscles by the expiratory center. There is only inhibition of the inspiratory center by the expiratory center. If the P_{CO_2} in the arterial blood increases, however, the facilitatory impulses to the expiratory center increase and, as in the case of the inspiratory center, the spread of activation through the center may increase. As a result, expiration involves not only inhibition of the inspiratory center, but also the stimulation of expiratory muscles. In the case of a rising P_{CO_2}, then, we have a single stimulus capable of (1) increasing the respiratory rate by increasing the number of volleys per minute that our hypothesized inspiratory pacemaker sends out, (2) increasing the inspiratory depth by lowering the threshold of the other neurons in the inspiratory center, and (3) increasing the expiratory depth by lowering the thresholds of the neurons in the expiratory center.

Central Chemoreceptors

It is generally recognized that after the denervation of known peripheral chemoreceptors, the respiratory centers continue to increase the rate and depth of respiration in response to increases in P_{CO_2}, H^+ concentration, and HCO_3^- concentration. Some speculate that this is the result of the direct action of these substances on the inspiratory and expiratory centers, while others postulate the existence of a chemosensitive area in the medulla, lying dorsal and rostral to the expiratory and inspiratory centers (Fig. 18-1).

Spasmodic Respiratory Center

In Figure 18-4 we have a summary of some of the important respiratory centers found in the medulla and pons. Included here is the hypothesized spasmodic respiratory center. Part of the evidence for the existence of such a center is the observation that in the cat, stimulation of the dorsolateral portion of the medulla produces a spasmodic respiratory response. It has been suggested by some that this area is important in the control of respiratory rhythm, and by others that it is not a center but rather fiber tracts which initiate the respiratory acts associated with coughing, sneezing, and retching.

PONTINE CENTERS

Illustrated in Figure 18-4 are the pneumotaxic center and the apneustic center. The neurons which make up these centers are apparently diffusely organized, but most of those associated with the *pneumotaxic center* are concentrated in the rostral pons, and those of the apneustic center in the caudal pons. The pneumotaxic center is probably concerned with *facilitating the transition from inspiration to expiration*. Its stimulation results in a sustained expiration (Fig. 18-3(C)) and, when its loss (transection level 1 in Fig. 18-4) is associated with bilateral vagal section, a maintained inspiration (*apneusis*) ensues. Also consistent with the view that the pneumotaxic center facilitates the transition from inspiration to expiration is the observation that neurons in the rostral pons become excited during the firing of the inspiratory center and produce potentials which last part way through the period of expiration (Fig. 18-5(A)).

The *apneustic center* is the counterpart of the pneumotaxic center. During expiration it becomes excited and continues to fire throughout the initial period of inspiration (Fig. 18-5(B)). In other words, it is concerned with *facilitating the transition from expiration to inspiration*. It probably sends to and receives from the pneumotaxic center inhibitory impulses. The vagus

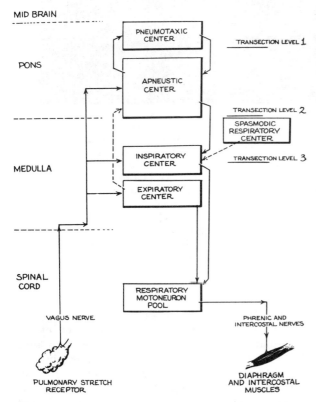

Fig. 18-4. Schematic representation of the organization of the respiratory mechanism in the brain stem of the cat. A transection in the middle of the pons (level 1) when associated with bilateral vagotomy results in a maintained inspiration. This is thought to be due to the loss of the inhibitory influences of the pneumotaxic center and the vagus on the inspiratory act. If the above procedure is followed by a transection (level 2) between the pons and medulla, respiration becomes rhythmic, but gasping. A transection at level 3 produces apnea (a cessation of breathing). (Wang, S. C., Ngai, S. H., and Frumin, M. J.: Organization of central respiratory mechanisms in the brain stem of the cat: genesis of normal respiratory rhythmicity. Am. J. Physiol., *190*:341, 1957)

Fig. 18-5. Respiratory neuronal discharges in the pons of the cat and their relationship with the phrenic action potentials. In A we see the action potentials from some *inspiratory-expiratory neurons* (*lower trace*) in the rostral pons and their relationship with potentials in the phrenic nerve (*upper trace*). In B we find the action potentials from some *expiratory-inspiratory neurons* in the caudal pons. On the basis of data such as these it is speculated that inspiration produces a buildup of potentials in the pneumotaxic center (rostral pons) which will facilitate a transition from inspiration to expiration, and that expiration produces a buildup in the apneustic center (caudal pons) which will facilitate a transition from expiration to inspiration. (Cohen, M. I., and Wang, S. C.: Respiratory neuronal activity in pons of cat. J. Neurophysiol., 22:37, 39, 1959)

nerves also send inhibitory impulses to it, initiated by the stretch of the lungs during inspiration. Stimulation in the area of the apneustic center results in a sustained inspiration (Fig. 18-3(B)), as does the removal of inhibitory influences from both the vagi and the pneumotaxic center. Transection of the brain stem below the apneustic center (transection level 2 in Fig. 18-4) eliminates the apneustic respiration caused by a middle pontine transection and an associated bilateral vagotomy.

OTHER CENTERS

Mesencephalon

Decerebration in cats by means of a transection between the superior and inferior colliculi of the midbrain results in a greater decrease in both the respiratory minute volume (hypoventilation) and the responsiveness to changes in PCO_2 than is found in response to a transection rostral to the superior colliculi. It has also been observed that electrical stimulation of parts of the midbrain results in increases in respiratory rate and depth (Fig. 18-6), while stimulation of other areas of the midbrain decreases ventilation. Some have interpreted these and other data as indicative of respiratory centers in the midbrain. Others have attributed these results to the stimulation of tracts from the cerebral cortex and diencephalon, also suggesting that they indicate the presence in the midbrain of an important part of the reticular activating system.

The *reticular activating system* lies in the reticular formation of the medulla, pons, and midbrain and consists of myriad short neurons in a complex arrangement of intertwining nets (L. *rete,* net). Collaterals funnel into it from (1) the long ascending sensory tracts, (2) the trigeminal and auditory nerves, and (3) the optic and olfactory systems. These impulses produce a nonspecific facilitation of activity in this area, which in turn results in facilitation in other parts of the central nervous system. It seems likely that the role of the midbrain in the control of the respiratory system is more concerned with the general alerting function of the reticular activating system than with the regulation of res-

Fig. 18-6. Effects of mesencephalic reticular stimulation on the phrenic discharge of the cat. In both *a* and *b* the bottom trace is the directly recorded discharge from the central end of a phrenic nerve and the upper trace is the same signal after integration in a resistance-capacitance circuit. A record prior to reticular stimulation is seen at *a* and one during stimulation at *b*. The results are interpreted as an indication that the reticular activating system of the midbrain plays an important role in the control of respiration. (Annals of The New York Academy of Sciences, Vol. 109, Art. 2, Fig. 1, p. 588, A. Hugelin, and M. I. Cohen; copyright © by The New York Academy of Sciences, 1963; reprinted by permission)

piration alone. Consistent with this view is the observation that anesthetics and other agents which depress the reticular activating system usually depress respiration as well. We also note that in deep sleep the alveolar PCO_2 rises by about 6 mm. Hg and that arousal from sleep is associated with an increase in the activity of the activating system, an increase in respiratory ventilation, and a return of the alveolar PCO_2 to 40 mm. Hg.

Cerebral Cortex

It is common knowledge that respiration can be altered voluntarily or in association with emotional states. It is not, therefore, surprising to find that there are numerous areas in the cerebral cortex that modify ventilation. For example, stimulation of a region surrounding the olfactory tubercle produces instantaneous expiratory apnea (cessation of breathing) lasting up to 30 sec. Other areas when stimulated accelerate the rate and decrease the amplitude, increase both amplitude and rate, or decrease amplitude and rate. These areas in the cortex do not usually, however, participate in the minute-to-minute control of respiration, but represent a type of control which can be superimposed on the system.

Control of Deglutition, Vomiting and the Circulation

A number of centers exist in the brain which are not primarily concerned with respiration but which, while performing their functions, send potent facilitatory or inhibitory impulses to the respiratory centers. In swallowing, for example, there is a maintained expiration. In vomiting and retching there are rhythmic spasmodic respiratory movements followed by a contraction of the diaphragm and abdominal muscles.

The cardiovascular and respiratory centers both receive impulses from the peripheral baroreceptors and chemoreceptors. The carotid and aortic *baroreceptors* act on these centers in such a way as to decrease ventilation, vasoconstriction, and heart rate, while the carotid and aortic *chemoreceptors* increase ventilation, vasoconstriction, and heart rate. The integration of respiratory and cardiovascular reactions is not, however, limited to peripheral receptors. Stimulation in the preoptic area of the *hypothalamus* brings about both vasodilation and a decrease in ventilation. Stimulation of an area in the posterior hypothalamus initiates vasoconstriction associated with increased ventilation. In the medulla, on the other hand, no area has been found that consistently elicits an integrated respiratory and circulatory change. Although both circulatory and respiratory centers exist in the reticular formation of the medulla, there is apparently little interplay at this level.

NEUROLOGICAL LESIONS IN MAN

I have in the preceding sections emphasized the importance of the brain in respiratory control. This is further underscored by a review of some of the clinical data available on the modification of brain function. In Figure 18-7, for example, we note that a number of agents which depress the brain (analgesics such as morphine, and hypnotics such as thiopental) also decrease the responsiveness of the respiratory system to increases in arterial P_{CO_2}. Overdoses of these agents may produce serious hypoventilation or complete cessation of breathing (apnea). Hypoventilation can also result from medullary tumors, bulbar poliomyelitis, certain types of meningitis, cerebellar lesions, increases in intra-cranial pressure, and thrombosis and embolism. Obstruction of the vertebral arteries is a considerably more dangerous situation for the maintenance of respiration than obstruction of the internal carotid arteries. It quickly results in a series of respiratory changes associated with a doubling of arterial pressure and bradycardia, and terminates in apnea.

Hyperventilation may result from the irritation of the facilitatory areas or destruction of inhibitory areas in the brain. It is seen in some types of meningitis and in certain pontile lesions caused by thrombosis of the midpontile portion of the basilar artery. Hyperventilation is also sometimes associated with periodic cycles of progressively increasing and decreasing tidal

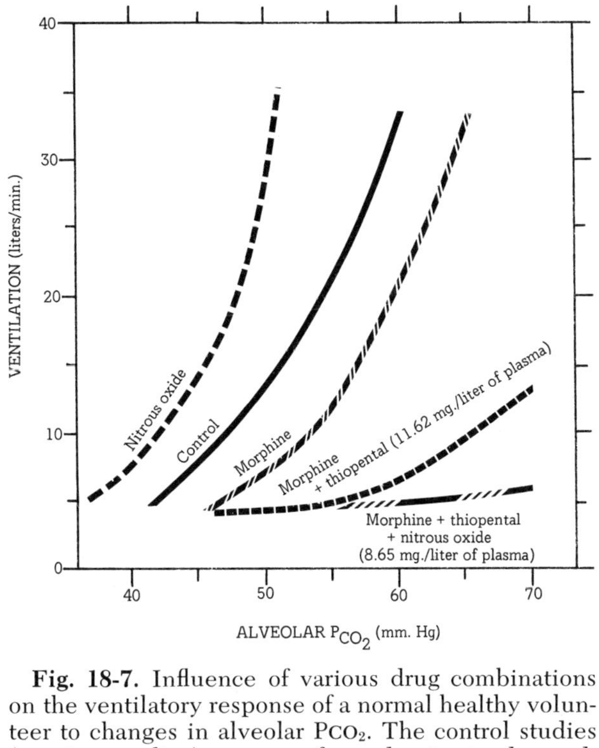

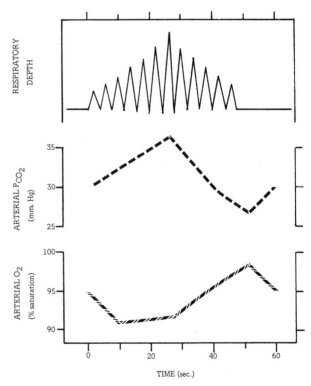

Fig. 18-7. Influence of various drug combinations on the ventilatory response of a normal healthy volunteer to changes in alveolar P_{CO_2}. The control studies (*continuous line*) were performed prior to drug administration. The numbers in parentheses represent the plasma thiopental concentration. (*Modified from* Eckenhoff, J. E., and Hebrich, M.: Effect of narcotics, thiopental and nitrous oxide upon respiration and respiratory response to hypercapnea. Anesthesiol., *19*:242, 245, 1958)

Fig. 18-8. Cheyne-Stokes respiration in man. Upper pattern shows oscillations in tidal volume and lower records demonstrate the associated variations in arterial P_{CO_2} and O_2 saturation. (*Redrawn from* Physiology of Respiration by J. H. Comroe, Jr. Copyright © 1965, Year Book Medical Publishers, Inc.; used by permission. Drawn from the data for patient A. K. Brown, H. W., and Plum, F.: Am. J. Med., *30*:8-49, 1961)

volumes (Fig. 18-8). This type of pattern is called Cheyne-Stokes breathing and has been produced experimentally by prolonging the time it takes for blood to pass from the lungs to the peripheral and central chemoreceptors. It is sometimes, but not always, noted in persons with sluggish circulation (as in congestive heart failure, for example) and is usually associated with an alteration in both arterial P_{O_2} and P_{CO_2}. It should be emphasized, however, that it usually cannot be eliminated by increasing an arterial P_{O_2} that is associated with already lowered P_{CO_2}. This, plus the observation that persons with Cheyne-Stokes breathing show increased responsiveness to CO_2, has led most investigators to conclude that the disorder is primarily a neurological one due to overcorrection in the respiratory mechanism. This overcorrection, of course, is exaggerated in persons with sluggish circulation.

CHEMORECEPTORS

Within limits, increases in the arterial P_{CO_2} and decreases in the pH and P_{O_2} can increase respiratory minute volume. If a normal person at rest keeps his alveolar P_{O_2} constant and raises his P_{CO_2} by 5 mm. Hg, he will double his minute volume. If, on the other hand, he keeps his P_{CO_2} constant it requires a decrease in P_{O_2} of 50 mm. Hg to double his minute volume. There are also times when both an increase in P_{CO_2} and a decrease in P_{O_2} are associated. This occurs in *asphyxia* and results in a greater increase in minute volume than would occur if the body were merely adding the hypoxic (decreased P_{O_2}) to the hypercapnic (increased P_{CO_2}) response. This phenomenon is called *potentiation* and is due to the fact that hypoxia not only initiates hyperventilation but also increases the sensitivity of the system to CO_2 (Fig. 18-9).

The changes in ventilation that result from the variations in P_{O_2} and P_{CO_2} mentioned above are in part or in total the result of stimulation of structures called chemoreceptors. We shall throughout this book use the term chemoreceptor to refer to specialized nerve endings that respond to natural changes in the environment. This includes sensory endings in the tongue associated with taste, but not sensory nerves which respond only to certain drugs or smooth muscle responding to hypoxia. The respiratory chemoreceptors include (1) the aortic bodies in the wall of the ascending aorta, (2) the carotid bodies just cephalad to the division of the common carotid

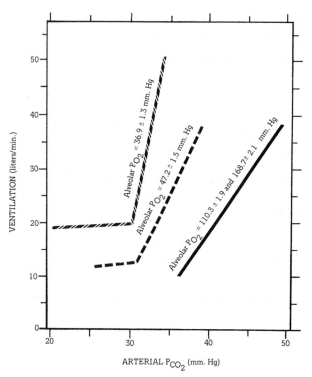

Fig. 18-9. Effect of acute hypoxia on the ventilatory response of a volunteer to CO_2 inhalation. Note that changing the arterial P_{O_2} from 110 to 169 mm. Hg had no effect on the response to a changing P_{CO_2}, but decreasing the P_{O_2} to 47 or 37 mm. Hg had a pronounced effect. (*Modified from* Nielsen, M., and Smith, H.: Studies on the regulation of respiration in acute hypoxia with an appendix on respiration control during prolonged hypoxia. Acta Physiol. Scand., 24:298, 1952)

arteries into internal and external carotid arteries (Fig. 18-10), and (3) an area in the medulla. Evidence that central chemoreceptors exist in the medulla, separate from the inspiratory and expiratory centers, has increased in the past decade. It is now recognized that the central (or medullary) chemoreceptors are more sensitive to changes in P_{CO_2} and pH than the carotid and aortic bodies, and under most conditions are responsible for the sensitivity of a person to changes in P_{CO_2} and pH. They are, however, considerably less responsive to changes in P_{O_2} than are the peripheral chemoreceptors (Fig. 18-11).

The *carotid bodies* are small nodules containing epithelial-type cells surrounded by nerve endings and associated with an abundant supply of sinusoidal capillaries. These sinusoids receive their blood from branches of the internal

Chemoreceptors

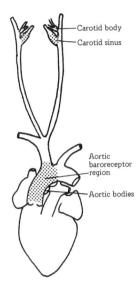

Fig. 18-10. Approximate location of the carotid and aortic chemoreceptors and baroreceptors. (*Redrawn from* Physiology of Respiration by J. H. Comroe, Jr. Copyright © 1965, Year Book Medical Publishers, Inc.; used by permission)

or external carotid artery. Their neurons join with those from the carotid sinus baroreceptors and ascend in the glossopharyngeal nerve. Both have their cell bodies in the petrosal ganglion. The aortic bodies are less accessible but seem, like the carotid bodies, to consist of chemoreceptor cells in close association with a rich supply of sinusoids. Their sensory neurons are numerous and ascend to the medulla with the sensory fibers from the baroreceptors of the aorta. Their blood supply in the adult comes from either a small branch of the coronary artery or the aortic arch.

Hypoxia

Hypoxia can be divided into four general classes: that due to (1) lowered arterial P_{O_2} (*hypoxic hypoxia*), (2) a decrease in the amount of available O_2 carried by the red blood cells (*anemic hypoxia*), (3) low blood flow (*stagnant hypoxia*), and (4) the inability of a tissue to fully utilize O_2 (*histotoxic hypoxia*). Hypoxic hypoxia may result from the failure of adequate amounts of O_2 to reach the alveolus (hypoventilation) due, for example, to breath-holding or occlusion of the respiratory tract. It may also result from failure of the blood in the lungs to reach equilibrium with the alveolar air due, for example, to pulmonary edema (Table 18-1). Anemic hypoxia results from either a reduction in available hemoglobin in the blood or from a decrease in the ability of hemoglobin to give up O_2 at a low P_{O_2}. Carbon monoxide, for example, decreases the amount of hemoglobin available to combine with O_2 without changing the total amount of hemoglobin present. In stagnant and histotoxic hypoxia we sometimes have the situation in which although the amount of O_2 in the arterial blood is normal, the amount delivered to the tissues is below normal. This may result from a reduction of the cardiac output (stagnant hypoxia) or from poisons such as *KCN* which prevent the

Type	Primary Site	Causes
Hypoxic	Inspired gas	High altitude
		Re-breathing expired air
		Hypoventilation
	Alveolus	Diffusion impairment (edema, fibrosis)
		Inadequate pulmonary blood supply
Anemic	Hemoglobin	Low hemoglobin concentration
		Excessive affinity of hemoglobin for O_2 (alkalosis, hypothermia)
		CO poisoning
		Methemoglobin formation
Stagnant	Blood flow	Hypotension
		Cardiac failure
		Vascular spasm
		Thrombosis
Histotoxic	Tissues	Diffusion impairment (edema, fibrosis, calcification)
		Poisons (KCN)

Table 18-1. Types of hypoxia.

	O_2 Tension (mm. Hg)		O_2 Content (ml. O_2/100 ml.)		Blood Flow	Respiration
	Art.	Vein	Art.	Vein		
Normal	100	88	20	19.5	normal	normal
Anemia	100	80	10.2	9.6	normal	normal
Severe anemia	100	55	4.2*	3.7*	normal	increased
Hypotension	100	60*	20	18	reduced*	increased
Hypoxic hypoxia	27*	25*	10.1	9.6	normal	markedly increased

Table 18-2. Oxygen tension and content (ml. O_2/100 ml. of blood) of carotid body blood. The carotid and aortic bodies increase ventilation when the P_{O_2} of their environment is low. This may result from a marked decrease in O_2 content (*) in the blood perfusing the bodies (severe anemia), from a reduced blood flow (*), or from a decrease in arterial P_{O_2} (*). (Data from Comroe, J.H., Jr.: Physiology of Respiration. Chicago, Year Book Publishers, 1965)

cell from utilizing O_2. The cyanides (including hydrocyanic acid) combine readily with the trivalent iron of cytochrome oxidase and in this way inhibit cellular respiration.

Not all types of hypoxia are associated with stimulation of the carotid and aortic bodies. Decreases in the arterial P_{O_2} (hypoxic hypoxia) from 100 mm. Hg to 80 mm. Hg, for example, stimulate the carotid and aortic chemoreceptors. There are cases, however, in which no change in the arterial P_{O_2} occurs, but in which a local hypoxia develops which stimulates these receptors (Table 18-2). This occurs in cyanide poisoning and severe anemia. In anemia the arterial P_{O_2} may be normal, but the O_2 content of the blood may be so reduced due to decreased amounts of hemoglobin as to result in a mild reduction of the O_2 delivery and a resulting stimulation of the chemoreceptors. We also find that a reduction of blood flow produces mild stimulation.

A note of caution is advisable, however. Blood flow to the chemoreceptors is considerably greater than is warranted by their metabolic needs. It has been observed in the cat that there is virtually no arteriovenous O_2 difference through the carotid bodies, and that reducing the flow to one fourth normal produces an A-V difference of only 2 ml. of O_2 per 100 ml. of blood. With such a relatively high resting flow the chemoreceptors are not so susceptible to decreases in either blood flow or the total O_2 content of the blood as are many other tissues. Thus when the arterial P_{O_2} is normal, only a severe reduction in the O_2 content of the blood or in blood flow stimulates these receptors.

In Figure 18-11 we see that it is primarily the carotid and aortic bodies that initiate man's response to hypoxic hypoxia. In these experiments a subject breathed a mixture of 8 per cent O_2 and 92 per cent N_2. Under these hypoxic con-

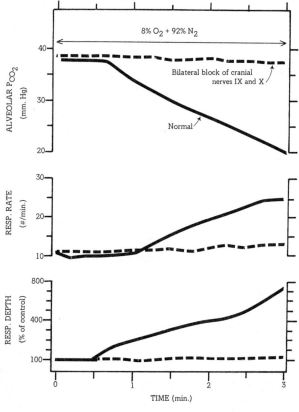

Fig. 18-11. Response of a subject to hypoxia (8% O_2) before (*continuous lines*) and after (*interrupted lines*) bilateral block of the 9th and 10th cranial nerves. (*Modified from* Guz, A., et al.: Peripheral chemoreceptor block in man. Resp. Physiol., *1*:39, 1966)

ditions his respiratory rate and depth increased and, as a result of the increase in ventilation, his alveolar PCO_2 decreased. These responses could be eliminated by blocking the sensory nerves from the carotid (glossopharyngeal nerves) and aortic (vagus nerves) bodies.

Cyanosis. Frequently but not always associated with hypoxia is a blueness (cyanosis, from Gk. *kyanós*, blue) of the skin due to abnormally large amounts of *reduced hemoglobin* in the capillaries. Cyanosis is first noted when the concentration of reduced hemoglobin in the capillaries reaches 5 gm. per 100 ml. of blood, and increases as the concentration becomes greater. It can be produced in a subject with a normal concentration of hemoglobin by lowering the O_2 saturation of the arterial blood to below 80 per cent, or by a vasoconstriction which reduces the blood flow through the capillaries. Exposing the skin to cold, for example, can produce cyanosis due to a reflex decrease in flow through the skin and, as a consequence, a greater A-V O_2 difference. On the other hand, there are a number of forms of hypoxia not associated with a sufficient increase in the amount of reduced hemoglobin to produce cyanosis. In most pulmonary diseases, for example, the O_2 saturation of the blood seldom falls below 85 per cent. In shock there is usually *pallor* of the skin rather than cyanosis. This is due to an intense vasoconstriction which produces generalized loss of blood by the capillaries of the skin. Pallor is also characteristic of severe anemia. Cyanide and CO poisoning, on the other hand, produce a hypoxia associated with neither pallor nor cyanosis.

Hypercapnea

Figure 18-12 compares the effects of increasing CO_2 (hypercapnea), increasing H^+ (acidosis), and decreasing O_2 concentrations on ventilation. Note that an increase in the per cent of CO_2 in the inspired air from 0 to 4 per cent causes ventilation to increase from 7 to 13 liters per minute. A decrease in O_2 from 20 per cent to 12 per cent, on the other hand, has by itself little effect on ventilation. It is apparent, then, that considerably greater changes in the concentration of O_2 than in that of CO_2 are required to modify ventilation.

The greater sensitivity of the body, under most circumstances, to changes in PCO_2 than to changes in PO_2 is apparently due to the action of CO_2 on the central chemoreceptors of the

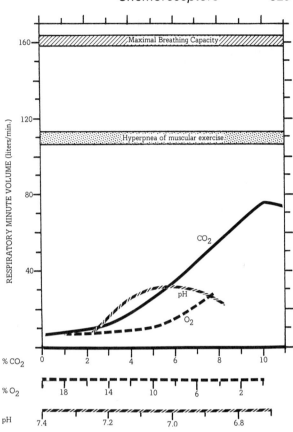

Fig. 18-12. Change in ventilation in man in response to (1) breathing an increased concentration of CO_2, (2) breathing decreased concentrations of O_2, and (3) a decreased pH. The ventilation obtained during strenuous exercise and a maximum ventilation test are included for reference. (*Redrawn from* The Lung, Second Edition, by J. H. Comroe, Jr., et al. Copyright © 1962, Year Book Medical Publishers, Inc., used by permission)

medulla. We know, for example, that increases in arterial PCO_2 that have no effect on ventilation when they involve only the peripheral receptors cause an increase in respiratory minute volume when they occur in the arterial blood supply to the brain as well. It has also been noted that denervation of the *peripheral chemoreceptors* has little effect on a person's response to hypercapnea.

In summary, the peripheral chemoreceptors are apparently most important in those few circumstances in which there is a marked decrease in O_2 or a marked increase in CO_2 in the body. Marked decreases in O_2 may occur at high altitudes. Marked asphyxia (hypoxia associated with hypercapnea), on the other hand, may result from respiratory depression due to morphine, barbiturates, or a number of general anesthetics (cyclopropane, Fluothane, etc.). There is also some evidence that these receptors play an important role during sleep, when the respiratory centers have lost most of the facilitatory impulses from the reticular activating system.

The *central chemoreceptors* probably lie on the outer surface of the medulla where they are close to or bathed by the cerebrospinal fluid (CSF). In fact, it is probably the P_{CO_2} or the H^+ concentration of the CSF, or both, that usually stimulate these receptors. This suggestion is based on the observation that although increases in P_{CO_2} and decreases in P_{O_2} that are sufficient to stimulate the peripheral receptors produce their effect on ventilation in a few seconds, hypercapnea acting on the central receptors produces its effect after a few minutes. It appears, then, that in order for mild hypercapnea to produce hyperventilation, the CO_2 must first diffuse through the blood-CSF barrier. Carbon dioxide, unlike H^+ and HCO_3^-, diffuses through the barrier readily. Once in the cerebrospinal fluid, some of the CO_2 combines with water to form H_2CO_3, HCO_3^-, and H^+:

$$CO_2 + H_2O \rightleftharpoons H_2CO_3 \rightleftharpoons H^+ + HCO_3^-$$

A number of investigators feel that it is the H^+ formed in the CSF and not CO_2 per se that is the more important stimulus to the central receptors. It should be remembered, in this regard, that CSF because of its low protein content is a much more poorly buffered solution than blood. As a consequence, small increases in P_{CO_2} lower the pH of the CSF much more than similar changes lower the pH of blood. It has been suggested, for example, that certain inflammatory diseases of the membranes of the central nervous system (meningitis and encephalitis) produce the abnormal ventilatory patterns sometimes associated with them because of an increase in the permeability of their membranes. The entry of buffers into the CSF can depress the action of CO_2 on the center, and in this way cause hypoventilation.

Let us now consider the effective range at which CO_2 affects the central chemoreceptors and hence the minute volume. In the unanesthetized person we find that progressive decreases in P_{CO_2} decrease ventilation but do not produce apnea (cessation of breathing). In a person under the influence of an anesthetic, mor-

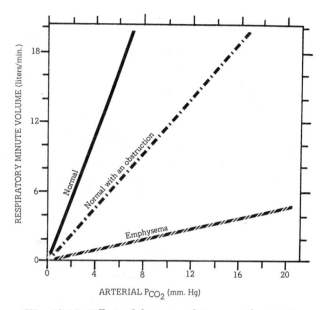

Fig. 18-13. Effect of the arterial P_{CO_2} on the respiratory minute volume of a normal subject, a subject breathing through an artificial airway obstruction, and a patient with pulmonary emphysema. Note that the author has extended his curves to a minute volume of 0 liters/min. Many physiologists now maintain that such an extrapolation is unwarranted and that in unanesthetized man, unlike the anesthetized individual, respirations will continue when the P_{CO_2} is 0. (*Redrawn from* Cherniack, R. M.: Work of breathing and the ventilatory response to CO_2. *In* Fenn, W. O., and Rahn, H. (eds.): Handbook of Physiology. Sec. 3 (Respiration), vol. II, p. 1470. Washington, American Physiological Society, 1965)

phine, a barbiturate, or some other agent that may depress the respiratory mechanism, either an increase in P_{O_2} or a decrease in P_{CO_2} can produce apnea. It is also important to realize that as we increase the P_{CO_2} we eventually reach a point at which further increases initiate a decrease in respiratory minute volume (Fig. 18-12) and generalized depression of the central nervous system. If, for example, one breathes a gas that is 5 per cent CO_2, an increase in respiratory minute volume results, but at this high concentration there is depression of the corneal reflex as well. At concentrations of CO_2 above 10 per cent there is a decrease in minute volume and the sensation becomes extremely distressing. At a concentration of 30 per cent, there is a state of depression similar to that seen under surgical anesthesia, and at concentrations of CO_2 in the inspired air which exceed 40 per cent, death generally results.

Acidosis

The acidity of the blood increases in response to an increase in P_{CO_2} or in acids absorbed from the gastrointestinal tract. Acidosis is also found in patients with such disturbances as diabetes mellitus and uremia. In severe acidosis there is deep, labored breathing (*Kussmaul respiration*) due apparently to the stimulation of chemoreceptors by the H^+. The carotid bodies are sensitive to a decrease of 0.1 pH unit, and produce a two- to threefold increase in ventilation in response to a decrease of 0.4 pH units. It is also recognized that decreases in the pH of CSF increase ventilation.

It is important to realize, however, that decreases in blood pH are not always associated with decreases in the pH of the CSF. We have already noted that there is a diffusion barrier between the blood and CSF that is far more effective in keeping H^+ from passing from the blood to the CSF than it is in preventing the passage of CO_2. The significance of this can be demonstrated by a simple experiment. Have a subject swallow NH_4Cl. You will note that the blood pH soon decreases from 7.42 to perhaps 7.34, and that a hyperventilation results which decreases the P_{CO_2} of the blood. There is then a decrease in the P_{CO_2} of the CSF and an associated increase in the pH of the CSF from 7.32 to 7.34. In other words, the NH_4Cl produces a decrease in the pH of the blood which eventually becomes associated with an increase in the pH of the CSF. The overall effect is an increase in ventilation which would be more pronounced if the pH and P_{CO_2} of the CSF had not also changed.

STRETCH RECEPTORS

Inflation and Deflation Receptors in the Lung

In 1868, Hering and Breuer reported that maintained distention of the lungs in anesthetized animals produced a decrease in inspiratory frequency, and that maintained deflation produced an increase in inspiratory frequency. Both responses disappeared when the vagal afferent fibers from the lungs were blocked or cut (Fig. 18-14). These reactions have come to be called the *Hering-Breuer inflation and deflation reflexes*. They can be eliminated by cutting the vagi or cooling the vagi to below 8°C. When these two reflexes are blocked by cooling the vagi to 5°C., some of the neurons concerned with other reflexes continue to function. The *paradoxical reflex of Head* (Fig. 18-14) is one of these. This is an inflation reflex that is apparently masked by the inflation reflex described by Hering and Breuer. It is referred to as paradoxical because it, unlike the Hering-Breuer reflex, produces facilitation rather than inhibition of inspiration in response to inflation.

At present, our understanding of the *paradoxical inflation reflex* is fragmentary. The experiments on cooling the vagi (Fig. 18-14) are good evidence that in the animals in which it has been studied, this reflex is separate from the Hering-Breuer reflexes. Its receptors are thought to be in the lungs, possibly in or near the alveoli. It has been further suggested that when a few alveoli collapse, inspiration stimulates inflation receptors, which stimulate vagal afferent fibers to prolong and intensify the firing of the inspiratory center. In 1960 Cross et al., on the basis of their studies on the newborn, suggested that this reflex might be particularly active during the first five days of life and might serve to effectively aerate collapsed alveoli in the neonatal lung. A number of other workers have suggested that in the adult too, periodic deep inspirations might be initiated by this reflex and serve as an effective way of preventing progressive collapse of the alveoli. I would emphasize, however, that most of this is only speculation based on animal experimentation.

The receptors for the Hering-Breuer inflation and deflation reflexes lie in the bronchi and bronchioles. They do not adapt readily and are capable of initiating as many as 300 impulses per second in a single afferent neuron. Sensory neurons activated by lung inflation and associated with the Hering-Breuer reflex ascend in the tractus solitarius to the apneustic center, where they inhibit its action in facilitating inspiration. The relative importance of the inflation reflex in controlling eupnic (resting) respiration has been studied in a number of species. On the basis of these studies we suggest the following descending order for the importance of the inflation reflex in controlling eupnic respiration: rabbit, rat, guinea pig, monkey, mouse, dog, cat, man. In the dog, for example, bilateral section of the vagi may triple the duration of the inspiration and produce a decrease in the respiratory rate. The respiratory minute volume remains normal. In man the inflation reflex is probably considerably less important during eupnea, but may play a

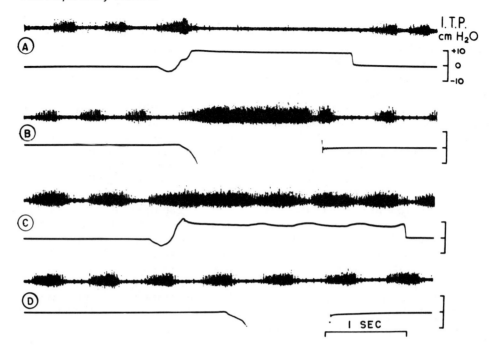

Fig. 18-14. Hering-Breuer inflation and deflation reflexes. The effects of inflation (A and C) and deflation (B and D) of the lungs of a rabbit before (A and B) and after (C and D) bilateral cooling of the vagus nerves to 5°C. In each study the upper pattern is the electromyogram from the diaphragm and the lower pattern is the intratracheal pressure (I.T.P.). Note that before block (A) an increase in intratracheal pressure depressed the diaphragm but after block (C) it increased the activity of the diaphragm. The phenomenon at C is called the paradoxical reflex of Head and is apparently due to neurons in the vagi that were not blocked at 5°C. We also note that before vagal block, deflation (B) increased the activity of the diaphragm. (Widdicombe, J. G.: Respiratory reflexes. In Fenn, W. O., and Rahn, H. (eds.): Handbook of Physiology, Sec. 3 (Respiration), vol. I, p. 591. Washington, American Physiological Society, 1964)

progressively more important role as inspiratory depth increases. When pulmonary compliance is low (pulmonary vascular congestion, pulmonary edema, atelectasis, alveolar fibrosis), the respirations are generally rapid and shallow. It has been suggested that this is due to greater participation of the inflation reflex in the total control of respiration in these cases. It has also been suggested that certain drugs (ether, chloroform, trichloroethylene) produce sensitization of the inflation receptors and, as a consequence, rapid, shallow breathing.

Prior to the discovery of the respiratory chemoreceptors it was generally agreed that the inflation reflex played a dominant role in the regulation of ventilation. After this discovery it became increasingly clear that it was little concerned with the control of the amount of air inhaled per minute, but influenced rather the efficiency of the respiratory system. We noted in the previous chapter (Fig. 17-30) that for each ventilation there is a frequency and depth at which the respiratory system does the least amount of work, and that a normal person sets his system at this level. We might also note that when the afferent neurons from the lungs are severed, this mechanism is disturbed. As emphasized in the previous chapter, the disadvantage of a high respiratory rate is a high airway resistance, and the disadvantage of a high respiratory depth is increased resistance due to the elastic recoil of the system. Thus for each respiratory minute volume the sum of the recoil resistance and the airway resistance is normally minimal.

It is interesting to note in this regard that in persons with low lung compliance (congestion,

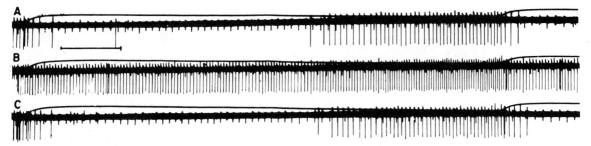

Fig. 18-15. Inflation reflex before (A), during (B), and after pulmonary congestion (C) due to the occlusion of the pulmonary vein. Record B was taken 3 minutes after the beginning of the occlusion and record C 2 minutes after the release of the occlusion. The upper line in each study is intratracheal pressure (inspiration is in a downward direction) and the lower line is action potentials from a single afferent vagal fiber. The horizontal marker represents 0.5 second. Note in B that the decrease in compliance which results from the congestion of the lungs is associated with a more marked stimulation of inflation receptors. These, in turn, will reflexly decrease inspiratory depth. (Costantin, L. L.: Effect of pulmonary congestion on vagal afferent activity. Am. J. Physiol., *196*:50, 1959)

edema, atelectasis, fibrosis) and persons with a high airway resistance (asthma), the sum of the recoil resistance and the airway resistance may also be minimal. Characteristic of those with reduced lung compliance is rapid, shallow respiration. Characteristic of the asthmatic is slow, deep respiration. Thus, with the vagi intact, the respiratory system seems to set its rate and depth at a level sufficient to maintain (1) an adequate respiratory minute volume and (2) optimal efficiency. The chemoreceptors are at least partly responsible for the former, while the stretch receptors are partly responsible for the latter. When lung compliance decreases there is probably a greater stretch of the inflation receptors (Fig. 18-15), a lowered threshold to stretch, or a combination of the two. This results in shallow respiration. The mechanisms responsible for the deep inspirations in an asthmatic are less well understood, but the result seems consistent with the general pattern discussed above.

The Hering-Breuer deflation reflex is probably not active during eupnea, but may be brought into play during marked deflation of the lungs. For example, it may be the mechanism responsible for increases in ventilation during a pneumothorax. In man we find that pneumothorax frequently initiates such a marked increase in ventilation that the alveolar and arterial Pco_2 decrease. In short, the deflation reflex, like the paradoxical inflation reflex, may serve to aerate collapsed alveoli.

Proprioceptive Impulses from Skeletal Muscle

Muscle contraction and stretch stimulate receptors in the muscle which send impulses to the central nervous system. It has been suggested by a number of workers that these impulses may in part be responsible for the hyperpnea (increased ventilation) seen in running and various other types of exercise (Fig. 18-16). The proprioceptors in skeletal muscle that seem to have the most profound influence on the respiratory system are those in the diaphragm and intercostal muscles. In some animals, unilateral section of the sensory roots to those parts of the cervical cord from which the phrenic neurons originate (C-3 through C-6) produces complete paralysis of half the diaphragm. Apparently the elements in the sensory root whose loss is responsible for the paralysis are the neurons from the muscle spindles of the diaphragm. The spindles are sensory endings joined to small muscle fibers and enclosed by a sheath, the fusiform sheath (Fig. 2-11). Stretching of the sensory endings in the spindle causes the activation of sensory neurons which impinge on the anterior horn cells of the phrenic nerve and facilitate their firing. This stretching (1) may be associated with the distention of adjacent (extrafusal) fibers or (2) may result from contraction of the intrafusal fibers. The intrafusal fibers are in series with the sensory element and are innervated by small *gamma efferent* neurons,

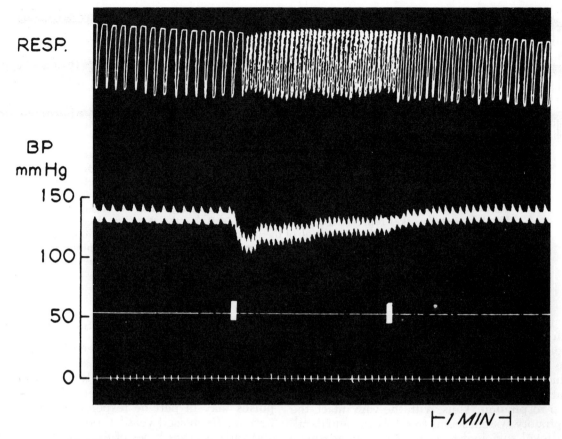

Fig. 18-16. Response of the anesthetized dog to passive movements (*between 2 white marks*) of the hind limbs. The limbs were connected to the rest of the body only by blood vessels and nerves. These movements resulted in an increase in respiratory rate, an increase in heart rate, and a decrease in arterial pressure. (Comroe, J. H., Jr., and Schmidt, C. F.: Hyperpnea of exercise and reflexes from limbs. Am. J. Physiol., *138*:543, 1943)

whereas the extrafusal (outside-the-spindle) fibers are innervated by the larger *alpha efferent* neurons (Fig. 18-17).

One of the suggestions made concerning the importance of the sensory potentials from the muscle spindle is that the spindle is constantly sending facilitatory impulses to the alpha fibers and that this facilitation is important if the impulses coming from the inspiratory center are to stimulate the alpha fibers. Another hypothesis is that inspiration is initiated by stimulation of the gamma fibers, which causes (1) contraction of the intrafusal fibers, (2) stretching of the sensory element, (3) impulses in the sensory neuron, and (4) stimulation of the alpha fiber. This is called the *gamma loop*. A more attractive

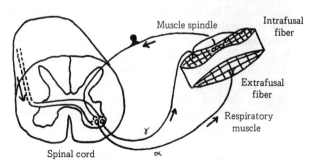

Fig. 18-17. Muscle spindles in the diaphragm and intercostal muscles. (*Redrawn from* Cotes, J. E.: Lung Function. P. 237. Oxford, Blackwell Scientific Publications, 1965)

hypothesis as far as man is concerned is that the impulses from the inspiratory center stimulate both the alpha and gamma efferent neurons. In this system, increased resistance to the contraction of the diaphragm would result in increased activation of the gamma loop and more forceful contraction. Decreased resistance to contraction would decrease the facilitatory influence of the gamma loop. Thus the gamma loop would serve to adjust the contractile response of the diaphragm to changing mechanical conditions such as changes in compliance and airway resistance.

Apparently a similar system exists in the intercostal muscles. Section of the sensory roots at T-4 through T-7 results in a decrease in intercostal muscle activity.

Arterial Baroreceptors

Baroreceptors can be demonstrated in the aortic arch and carotid sinuses. They respond to increases in arterial pressure by reflexly slowing the heart, producing arteriolar dilation, and decreasing the respiratory rate, depth, or rate and depth. There is some question as to the importance of the respiratory effect in man. Certain investigators, however, believe that in hypo-

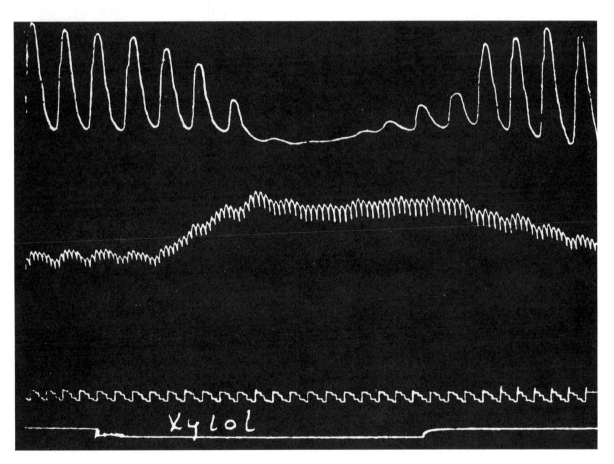

Fig. 18-18. Insufflation of air containing xylol vapor into the nose of a tracheotomized, vagotomized rabbit. Note the apnea (*upper trace*), bradycardia, and increased arterial pressure (*second trace*). Also shown is a 1-second interval indicator and a signal line (*lower trace*). (Allen, W. F.: Effect on respiration, blood pressure and carotid pulse of various inhaled and insufflated vapors when stimulating one cranial nerve and various combinations of cranial nerves. III: Olefactory trigeminals stimulated. Am. J. Physiol., 88:119, 1929)

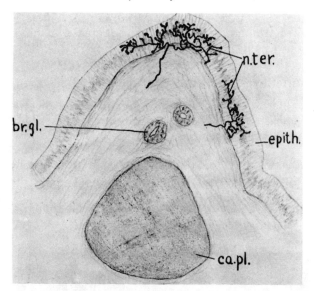

Fig. 18-19. Epithelial receptor in the bronchial wall of a child. Note the nerve terminations (n.ter.) between the columnar cells of the epithelium (epith.). Also shown is a bronchial gland (br.gl.) and a cartilaginous plate (ca.pl.). (Larsell, O., and Dow, R. S.: The innervation of the human lung. Am. J. Anat., 52:138, 1933)

tension an increase in respiratory rate and depth results from the decreased stimulation of the baroreceptors. This increased ventilation not only serves to increase the alveolar PO_2 and decrease the PCO_2, but possibly also to increase venous return.

RECEPTORS IN THE UPPER RESPIRATORY TRACT

Associated with the upper respiratory tract are a number of reflexes which serve to expel

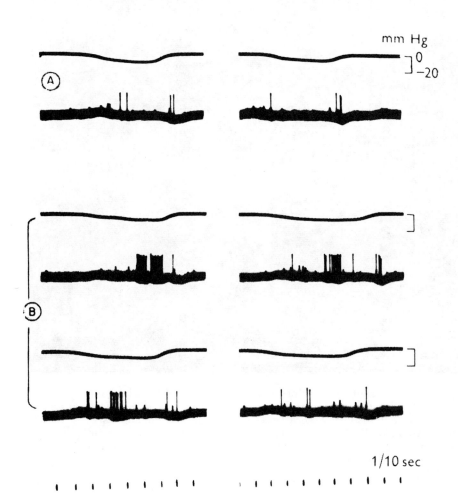

Fig. 18-20. Action potentials (*lower traces*) from a cat's tracheal receptor during deflation (*upper traces*) of the trachea before (A) and after (B) powdered talc was blown into the trachea. (Widdicombe, J. G.: Receptors in the trachea and bronchi of the cat. J. Physiol., 123:93, 1954)

irritant substances from, or prevent their entry into, the respiratory system. In diving mammals such as the seal, for example, immersion of the nose in water initiates (1) a reflex apnea which protects the lungs from water inhalation, (2) reflex bradycardia, and (3) vasoconstriction in skeletal muscle and the abdominal viscera. The bradycardia and vasoconstriction tend to decrease the metabolic requirements of the heart and to shunt most of the cardiac output to the heart and brain—the two organs least able to survive long periods of ischemia.

Mechanical and chemical irritants in the respiratory passages of man frequently have an effect similar to that discussed above for diving mammals (Fig. 18-18). Generally small degrees of irritation produce rapid, shallow respiration associated with bronchial constriction. As the degree of irritation increases, bronchial constriction also increases and eventually either apnea, a cough, or a sneeze occurs. Cigarette smoke, for example, can cause sufficiently intense bronchial constriction to result in a ten to twenty fold increase in airway resistance.

The receptors responsible for some of the actions of irritants on respiration lie in the epithelium of the trachea and bronchi (Fig. 18-19). Other receptors are found in the larynx and nasal cavity. It has also been suggested that the stretch receptors in the intrinsic muscles of the larynx and respiratory passages may be affected by irritants. Normally these stretch receptors fire during deflation of the trachea (that is, when the intratracheal pressure is lowest) and serve to

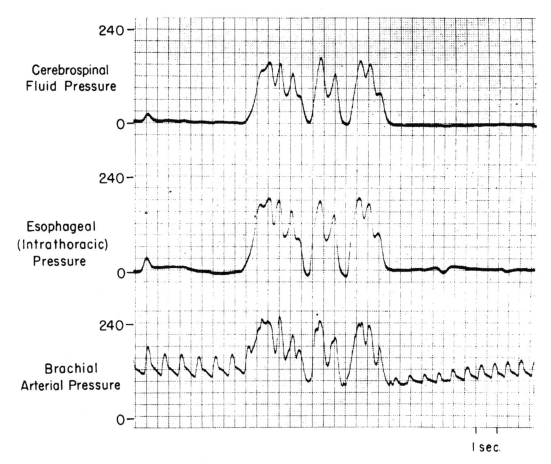

Fig. 18-21. Mechanical effects of coughing in man. Note that the cough caused an increase in cerebrospinal fluid pressure, intra-esophageal pressure, and brachial artery blood pressure. (McIntosh, H. D., Bates, E. H., and Warren, J. V.: Mechanism of cough syncope. Am. Heart J., 52:75, 1956; by permission of the American Heart Association, Inc.)

inhibit the inspiratory center. When irritants are present their output increases (Fig. 18-20). The areas most sensitive to mechanical stimulation lie in the larynx, the carina, and the branches of the first-order bronchioles. Sensitivity to chemical stimuli is greatest in the lower airways. Here, however, irritation produces hypotension rather than hypertension.

A *cough* results from irritation of the respiratory tract and is usually preceded by deep inspiration followed by a strong expiratory effort against a closed glottis. In man the glottis remains closed for about 0.2 sec. It has been suggested that during this interval stretch receptors are stimulated which feed back to the expiratory muscles and which facilitate a progressively more forceful contraction of these muscles and an even greater buildup in pressure within the respiratory tract. This pressure may exceed 300 mm. Hg, and once the glottis is opened, peak flows as high as 6 liters per second and peak velocities approaching the speed of sound may be obtained. The high pressure developed in the thoracic cavity decreases the venous return of blood to the heart, increases the cerebrospinal fluid pressure, and temporarily increases arterial pressure (Fig. 18-21).

RESPONSE OF RECEPTORS TO INJECTED AGENTS

In 1867 von Bezold demonstrated that an intravenous injection of veratrine caused apnea, hypotension, and bradycardia. It was later observed that this same triad of changes resulted from stimulation of the central end of the vagus, and evidence was presented that the stimulation of afferent neurons from the lungs also initiated the triad. Then in 1937, Jarisch showed that the reflex could arise from the heart alone. The combination of apnea, hypotension, and bradycardia initiated by the chemical stimulation of receptors in the pulmonary and coronary circulations is sometimes referred to as the (1) *von Bezold-Jarisch reflex*, (2) thoracic reflex, (3) pulmonary chemoreflex, or (4) coronary chemoreflex. It is referred to as a chemoreflex rather than a chemoreceptor reflex because it is initiated by chemical changes that do not normally occur in the body and may involve receptors which are not usually classified as chemoreceptors. In some species it can also be initiated by veratridine (Fig. 18-22), nicotine, serotonin, adenosine triphosphate,

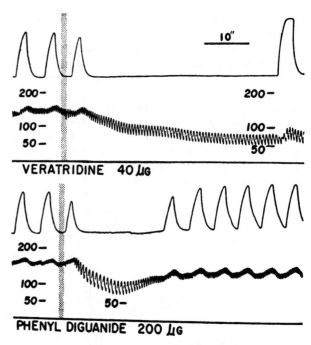

Fig. 18-22. Changes in respiration (*upper traces*) and carotid arterial pressure (*lower traces*) in response to the injection of veratridine and phenyl diguanide (*shaded area*) into the right atrium of a cat. (Dawes, G. S., and Comroe, J. H., Jr.: Chemoreflexes from the heart and lungs. Physiol. Rev., 34:168, 1954)

phenyldiguanide, and a number of antihistaminics. I would emphasize, however, that there may be differences from species to species. In the cat, serotonin produces apnea, hypotension, and bradycardia. In the dog it produces hyperpnea, hypertension, and tachycardia. There is a difference too in the primary site of action of these drugs. The action of veratridine, for example, can be blocked by cooling the vagi to 8°C. Some of the other agents continue to produce the triad of changes until the vagi have been cooled to 2°C.

The full clinical import of the above observations has not been completely investigated, but there are numerous suggestions in this regard. It is known that the rapid injection numerous drugs can initiate hypotension. In a number of such cases it has been hypothesized that this is the result of a chemoreflex. It has also been suggested that the inhalation of irritant gases, the occlusion of a coronary artery, pulmonary embolism, or the handling of nerves around the hilum of the lung during surgery might initiate the reflex. In coronary occlusion and pulmonary embolism it is further hypothesized that it is the accumulation of agents from metabolically active or degenerating cells which initiates the reflex.

OTHER RECEPTORS

In 1921 Ranson wrote: "It is clear that the respiratory center has connections with all the afferent cranial and spinal nerves." The validity of this statement can be demonstrated by the stimulation of sensory neurons in almost any nerve. If, for example, we stimulate the central end of a cut nerve from the arm or leg, we find an increase in respiratory rate and depth. This is apparently due to the stimulation of sensory fibers from the stretch receptors (muscle spindles, etc.), temperature receptors, and pain endings in the appendage. Apparently all of these fibers can facilitate ventilation. It has been shown that cold showers (Fig. 18-23) and somatic pain are both associated with increases in ventilation. Sometimes the obstetrician uses some of these reflexes to initiate breathing in the newborn. Slapping the newborn child, splashing cold water on him, or stretching his anal sphincter have all proven effective in facilitating the onset of respiration.

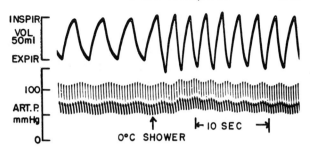

Fig. 18-23. Response of a decerebrate cat to cold water applied to the chest. (Widdicombe, J. G.: Respiratory reflexes. *In* Fenn, W. O., and Rahn, H. (eds.): Handbook of Physiology. Sec. 3 (Respiration), vol. I, p. 618. Washington, American Physiological Society, 1964)

On the other hand, visceral *pain* is associated with initial apnea followed by hyperpnea. This is seen when (1) the gallbladder biliary ducts or small intestines become distended, (2) a blow is applied to the testes or solar plexus, (3) the arterial wall is irritated, or (4) the eyeballs are pressed into their sockets.

Chapter 19

EXERCISE
BREATH-HOLDING
DIVING
DROWNING
HYPERVENTILATION
BAROMETRIC PRESSURE
OXYGEN TOXICITY

RESPIRATORY ADJUSTMENTS

In the preceding chapter we characterized the control of the respiratory system as concentrated in the medulla oblongata (Fig. 19-1). We suggested that one of the medullary centers, the inspiratory center, contains a pacemaker cell which initiates a volley of nervous impulses. These impulses are carried to the diaphragm, where they produce a tetanic contraction. Some of them are also carried to other areas in the brain stem where they either facilitate or inhibit activity. Feeding into both the inspiratory center and a second center, the expiratory center, are

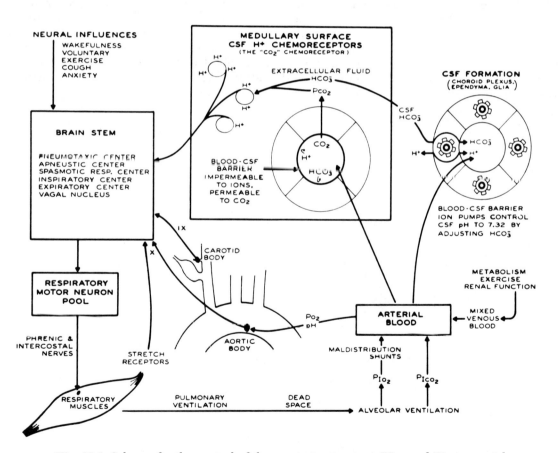

Fig. 19-1. Scheme for the control of the respiratory system. PI_{O_2} and PI_{CO_2}, partial pressure of O_2 and CO_2 in the inspired air. (Severinghaus, J. W., and Larson, C. P., Jr.: Respiration in anesthesia. *In* Fenn, W. O., and Rahn, H. (eds.): Handbook of Physiology. Sec. 3 (Respiration), vol. II, p. 1223. Washington, American Physiological Society, 1965)

numerous facilitatory and inhibitory impulses from elsewhere in the central nervous system — the effect of which is modification of respiratory rate and depth. We referred to this complex system of facilitation and inhibition as the respiratory mechanism, and noted that it can be markedly affected by anesthetics, changes in its environment (PCO_2 and pH for example), and messages entering it by way of the sensory nerves. We further recognized that practically any message to the central nervous system affects the respiratory mechanism, but emphasized the importance of messages from arterial chemoreceptors in the thorax and neck, from the lung stretch receptors, and from muscle and joint receptors.

The result of this complex interplay at rest is a respiratory rate of approximately 14 per minute, a tidal volume of 500 ml., a respiratory minute volume of 7,000 ml. per minute, an arterial PO_2 of 95 mm. Hg, and an arterial PCO_2 of 40 mm. Hg. The (1) inspiratory center fires primarily in response to facilitatory impulses from the reticular formation and the central chemoreceptors, which are in turn responding to the PCO_2, pH, and HCO_3^- of their environment. The volleys of impulses originating at the inspiratory center result (2) in volleys of impulses in internuncial neurons, phrenic neurons, and intercostal neurons which (3) produce a submaximal tetanic contraction of the diaphragm and external intercostal muscles. The (4) resulting decrease in intrapleural pressure is followed by (5) distention of the lung, which produces not only (6) inspiration but also (7) stimulation of the stretch receptors in the lung, which (8) initiate propagated impulses in afferent neurons in the tenth cranial nerve. These afferent fibers (9) cause stimulation of the expiratory center, which (10) produces inhibition of the inspiratory center and (11) stimulation of the internal intercostal muscles. The (12) relaxation of the diaphragm resulting from inhibition of the inspiratory center causes (13) the return of intrapleural pressure to its previous level, (14) an increase in intrapulmonary pressure, (15) expiration, and (16) a decrease in lung size.

We shall in this chapter be concerned with how this basic pattern of control is modified in response to the changing needs of the body resulting from such stresses as increases in metabolic activity, breath-holding, hyperventilation, changes in the barometric pressure, and hypoxia. In some cases we shall emphasize the importance of the mechanisms discussed above, but in many we shall also emphasize mechanisms not important in a resting person.

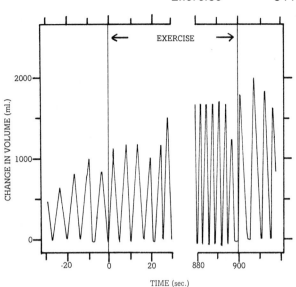

Fig. 19-2. Spirogram of a subject at rest, during the transition from rest to exercise, and during the transition from exercise to recovery. The exercise started without warning, was performed on a bicycle for a period of 15 minutes, and involved an O_2 consumption of 1.62 liters/min. Note the immediate rise in respiratory rate and depth at the beginning of exercise and the immediate decrease in rate at its end. (*Redrawn from* Dejours, P.: La régulation de la ventilation au cours de l'exercice musculaire chez l'homme. J. Physiol. (Paris), *51*:190, 1959)

EXERCISE

The most marked increase in the body's metabolic activity occurs in response to exercise. It is associated with (1) the shunting of blood to active muscles of the body, (2) an increase in cardiac output, and (3) an increase in the respiratory minute volume. All these changes begin in the first second of exercise (Fig. 19-2) or just before exercise begins. In either case the response is too rapid to be due to the accumulation of metabolites at any of the known chemoreceptor areas. In short, the initial response is undoubtedly of nervous origin.

Initial Response

The observation that increases in cardiac output and respiratory minute volume sometimes occur when a runner is told to "get on his mark" and to "get set" is similar to that of Ivan Petrovitch Pavlov (1849-1936), who in 1906 described his now-famous discovery of the conditioned reflex. He found that if he rang a bell and then presented a dog with food ("unconditioned stimulus"), the dog would salivate ("unconditioned response"). Eventually, if this procedure were repeated enough, the bell by itself ("conditioned stimulus") would initiate salivation ("*conditioned response*"). In other words, he found that one could learn to respond in a certain way to a number of "get set" stimuli.

Conditioning is but one of many factors that play a role in the initial response to exercise. It has been suggested, for example, that the same areas in the cerebrum that initiate the contractions and relaxations of skeletal muscle which we associate with exercise also send impulses to the cardiovascular and respiratory centers, which in turn dilate arterioles, increase cardiac output, and increase ventilation. In Figure 19-3 we note another important nervous mechanism. We see here that if an investigator flexes and extends the legs of a supine subject (passive movement), there is a reflex increase in ventilation and an associated decrease in alveolar P_{CO_2} due to the stimulation of proprioceptors (stretch receptors, etc.) in the parts being moved. Note that when the subject actively moved his own legs a still greater increase in ventilation occurred, associated with a less marked change in the arterial P_{CO_2}.

Secondary Responses

Following the initial, rapid increase in minute volume at the onset of exercise, relatively stable ventilation is maintained for 20 to 30 sec., usually followed by a gradual increase in the respiratory minute volume. The mechanisms responsible for this secondary elevation in the respiratory minute volume which occurs while exercise continues are not well established. One common suggestion is that it is due to accumulated metabolites. In Figure 19-4, for example, we see that in strenuous exercise (O_2 consumption greater than 2 liters/min.) the concentrations of arterial lactate and H^+ increase, but that there is an associated decrease in CO_2 and HCO_3^- due to increased ventilation. In mild exercise there is frequently a small increase in arterial HCO_3^- and CO_2, as well as in H^+. In other words, looking solely at the mean changes in arterial blood during exercise, the only fairly consistent change we find is in the pH.

The situation in the systemic veins is considerably different. Here we find increases in both the CO_2 and H^+ concentration during exercise, but no one has been able to demonstrate chemoreceptors sensitive to P_{CO_2}, pH, or P_{O_2} in the limbs, systemic veins, right heart, or pulmonary arteries. The chemoreceptors, that is, lie solely in the systemic arteries and brain. In these areas the mean changes in P_{CO_2} and P_{O_2} during exercise do not seem to be important in the normal response to exercise. Physiologists, however, have been reluctant to give up the concept that at least some of the changes in ventilation are due to changes in P_{CO_2}. For example, we know that in exercise there is a much more consistent increase in the venous P_{CO_2} than is found in the systemic arteries. This, plus the observation by Yamamoto that one could produce a fivefold increase in ventilation by increasing venous P_{CO_2} even though the average arterial P_{CO_2} remained constant, forms a strong case for P_{CO_2} as a respiratory stimulant during exercise. On the basis of these data it is tempting to hypothesize the presence of venous chemoreceptors, despite evidence to the contrary, but there is another suggestion more consistent with the evidence. It has been hypothesized that although increases

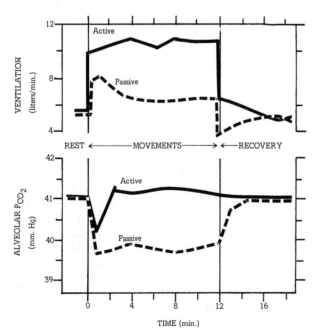

Fig. 19-3. Changes in expiratory minute volume and alveolar P_{CO_2} in response to active (*continuous line*) and passive (*discontinuous line*) flexion and extension of the lower legs at frequencies of 60/min. in a supine subject. (*Modified from* Dejours, P.: La régulation de la ventilation au cours de l'exercice musculaire chez l'homme. J. Physiol. (Paris), *51*:195, 1959)

in the venous P_{CO_2} are frequently not associated with increases in the average arterial P_{CO_2}, they are associated with greater *fluctuations of the arterial* P_{CO_2} *about the mean*. It has also been suggested that the chemoreceptors are more susceptible to the fluctuations in the arterial P_{CO_2} which occur during each respiratory cycle (Fig. 17-5) than they are to the mean P_{CO_2}. Unfortunately none of the exercise studies as yet reported involves the measurement of fluctuations in P_{CO_2} in arterial blood or cerebrospinal fluid; most have been more concerned with the mean arterial P_{CO_2} for several respiratory cycles.

Another factor emphasized in Figure 19-4 is the increase in catecholamines (epinephrine and norepinephrine) which occurs in severe exercise. These substances not only affect the cardiovascular system but also increase the respiratory minute volume. If, for example, the arterial P_{CO_2} is kept constant and 3 μg. of epinephrine plus 9 μg. of norepinephrine are injected per minute, there will be an increase in ventilation of 10 liters per minute. This is apparently not due to a direct action of the catecholamines on the respiratory centers.

Termination of Exercise

The end of exercise, like its beginning, is marked by an abrupt change in ventilation (Fig. 17-12). This decrease in the respiratory minute volume (Fig. 19-2) is followed by a gradual, further decrease until respiration returns to the resting state. The rapid decrease in minute volume is apparently due to a decrease in the number of messages from the stretch receptors in the active muscles, whereas the more gradual decrease results from the return of the blood pH, P_{CO_2}, and catecholamine concentration to normal. It has been shown, for example, that if a tourniquet is placed on an exercising appendage at the time exercise is stopped, and released after ventilation has returned to normal, there will be another increase in ventilation. This is apparently due to the release of metabolites by the appendage during its recovery phase.

BREATH-HOLDING

Breaking Point

Breath-holding is an important respiratory act used to protect the lungs from a noxious external environment. It occurs, for example, during swimming and during the use of ether for the purpose of anesthesia, and may be either a voluntary or an involuntary act. In a normal man, breath-holding is involuntarily terminated before

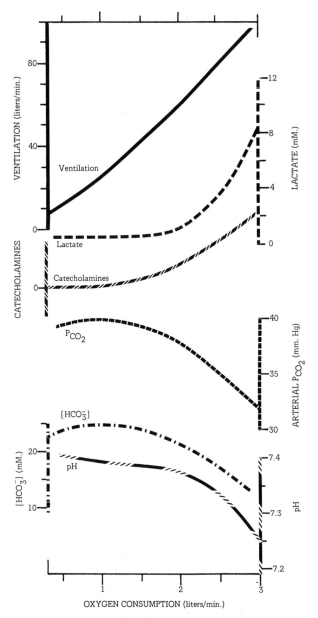

Fig. 19-4. Diagram of the effect of increases in O_2 consumption due to exercise on ventilation, and the concentration of lactate, epinephrine and norepinephrine (catecholamines), CO_2, HCO_3^-, and pH in the arterial blood. There were no important changes in arterial P_{O_2}, regardless of the intensity of the exercise. (*Modified from* Dejours, P.: La régulation de la ventilation au cours de l'exercice musculaire chez l'homme. J. Physiol. (Paris), *51*:167, 1959)

any damage to the body occurs. The time at which this happens is termed the breaking point. Rabbits, on the other hand, have been known to hold their breath in response to a noxious stimulus until they die.

The breaking point in man is determined by a number of factors. Two of these are the P_{CO_2} and P_{O_2} of arterial blood. In Figure 19-5, for example, we note that if a person takes several breaths of pure O_2 before holding his breath he can extend his breaking point. We also note that if he lowers his alveolar P_{CO_2} as well, the breaking point can be still further extended. Another important influence on the breaking point is the degree of lung distention at the beginning of breath-holding. A lung at its vital capacity does, of course, have a greater reservoir of O_2, and is able to accumulate larger amounts of CO_2 with a smaller increase in P_{CO_2}, than a lung at a smaller volume; this in itself tends to delay the breaking point. There is another factor, however. The distended lung causes an increase in the body's tolerance for hypercapnea (Fig. 19-6). This is apparently due to the fact that lung distention stimulates the Hering-Breuer stretch receptors, which elicit depression of the respiratory centers. Thus after a maximum inspiration one generally reaches his breaking point in about

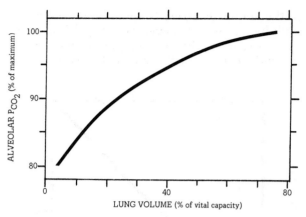

Fig. 19-6. The effect of lung volume (in terms of % of vital capacity) on the alveolar P_{CO_2} at which the breaking point for breath holding occurs. The effect of a changing P_{O_2} was minimized by having the subject inhale 100% O_2. The lung volume was measured at the breaking point. This was a point at which the lung would be smaller than at the beginning of breath holding. If the breath is held at vital capacity, the lung volume at the breaking point will, for example, be reduced by 25 per cent. (*Redrawn from* Mithoeffer, J. C.: Breath holding. *In* Fenn, W. O., and Rahn, H. (eds.): Handbook of Physiology. Sec. 3 (Respiration), vol. II, p. 1012. Washington, American Physiological Society, 1965)

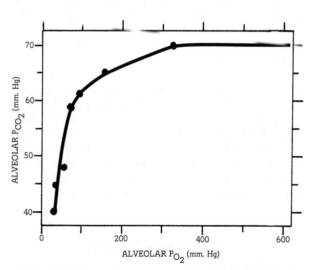

Fig. 19-5. Partial pressures of O_2 and CO_2 in the alveolar gas when the breath was held to the breaking point. The variation in P_{O_2} was produced by having the subject breathe various O_2 mixtures prior to breath holding. (*Redrawn from* Stroud, R. C.: Combined ventilatory and breath-holding evaluation of sensitivity to respiratory gases. J. Appl. Physiol., 14:354, 1959)

80 sec. If, on the other hand, he holds his breath after the usual resting inspiration, he reaches the breaking point in about 30 sec.

DIVING

With the added interest in swimming throughout the world, physiologists have recently begun to study the effects of diving as well as the associated phenomenon of breath-holding. Figure 19-7 is a summary of some of the effects of a simulated dive to 33 ft. below the surface of the water. Note in the upper segment of the figure that as one descends after taking a single deep breath, one's alveolar P_{CO_2} and P_{O_2} increase due to the doubling of pressure on one's body and in one's alveoli. As a person swims at this depth he is constantly using O_2, so that the P_{O_2} in his alveoli and arterial blood decreases. When the P_{O_2} decreases to 130 mm. Hg and the P_{CO_2} reaches 60 mm. Hg, the person feels the desire to ascend and does so. As he ascends the ambient pressure decreases, resulting in decreased P_{O_2} and P_{CO_2}; by the time he reaches the surface he has almost reached the breaking point, or the

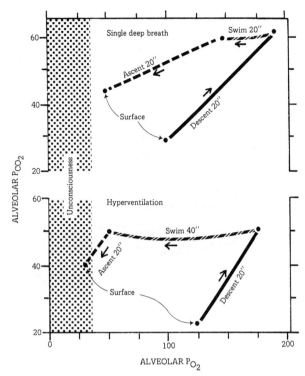

Fig. 19-7. Effect of a simulated dive to 33 feet on the alveolar P_{O_2} and P_{CO_2}. These experiments were performed in a chamber where the atmospheric pressure was slowly doubled (20 sec.) to simulate a dive of 33 feet. The period of swimming represented a period when the O_2 consumption was doubled. In the upper segment a single deep breath was taken before breath holding and in the lower one the subject hyperventilated before breath holding. Note that hyperventilation represents a potential danger in diving in that it decreases the P_{CO_2} and in so doing may result in a subject's staying submerged so long (40 sec. in this case) because of a reduced respiratory drive that when he ascends the associated decrease in P_{O_2} may cause unconsciousness. (Redrawn from Mithoeffer, J. C.: Breath holding. In Fenn, W. O., and Rahn, H. (eds.): Handbook of Physiology. Sec. 3 (Respiration), vol. II, p. 1023. Washington, American Physiological Society, 1965)

point of unconsciousness. If, on the other hand, he had hyperventilated before descending (lower segment of Fig. 19-7) he would have started his descent at a lower P_{CO_2} and a higher P_{O_2}. The lower P_{CO_2} would enable him to swim longer underwater and reach a lower P_{O_2} before feeling the need to ascend, but as he ascended, his P_{O_2} and P_{CO_2} would decrease so much that he would lose consciousness before surfacing, and as a result inhale water.

DROWNING

The lungs, unlike the skin, stomach, and large intestine, are poorly protected from hypotonic and hypertonic solutions. The unanesthetized person, so long as a certain P_{O_2} and P_{CO_2} are maintained, can usually hold his breath while submerged—or for that matter while swallowing. In those rare instances when food, water, or vomitus is aspirated, such mechanisms as the cough prevent foreign material from reaching the lungs. The anesthetized or unconscious subject, or the person who has been submerged for a long time, loses these mechanisms and inhales various liquids.

Unfortunately we know little at present about the effects of drowning on humans, so I shall rely heavily in this presentation upon studies of drowning in dogs. The effects on the dog of drowning in fresh water are summarized in Figure 19-8. The initial effects of submersion are apnea, hypoxia, hypercapnea, acidosis, bradycardia, and an increase in the arterial and venous pressures. The mechanisms involved in these changes have already been discussed. As the fresh water enters the lungs it is rapidly picked up by the blood, producing hypervolemia and hemodilution. As hemodilution progresses, an electrolyte imbalance ensues associated with decreased Na^+ and plasma protein concentrations. With the hemolysis of the red blood cells, however, the plasma protein concentration may even rise due to the increase in plasma hemoglobin. The decrease in plasma Na^+, along with the developing hypoxia, causes any of a number of cardiac arrhythmias (atrial fibrillation, ventricular tachycardia, atrioventricular block, etc.) which eventually lead to ventricular fibrillation and irreversible brain damage.

Immersing a dog in sea water (Fig. 19-9) results in a series of events leading to hemoconcentration and hypovolemia. In other words, hypertonic sea water in the lungs takes water out of the blood and tends to increase the plasma protein, Na^+, and chloride concentrations. Immersed dogs seldom develop a cardiac arrhythmia—due apparently to the fact that hyponatremia (decreased Na^+) is an important cause of fibrillation in freshwater drowning and that in saltwater drowning there is hypernatremia.

It has been suggested on the basis of postmortem examinations of human drowning victims that, although large amounts of water may enter the stomach, less liquid enters the lungs in man than does in the case of the dog. This, it is believed, is due to a laryngospasm in the drowning man. Thus the hemodilution or hemoconcentration associated with drowning is probably less in man than in dogs.

Chances for survival after drowning depend on a number of factors. The victim has a better chance if treated before the arterial pressure starts to decrease. It is also true that in pro-

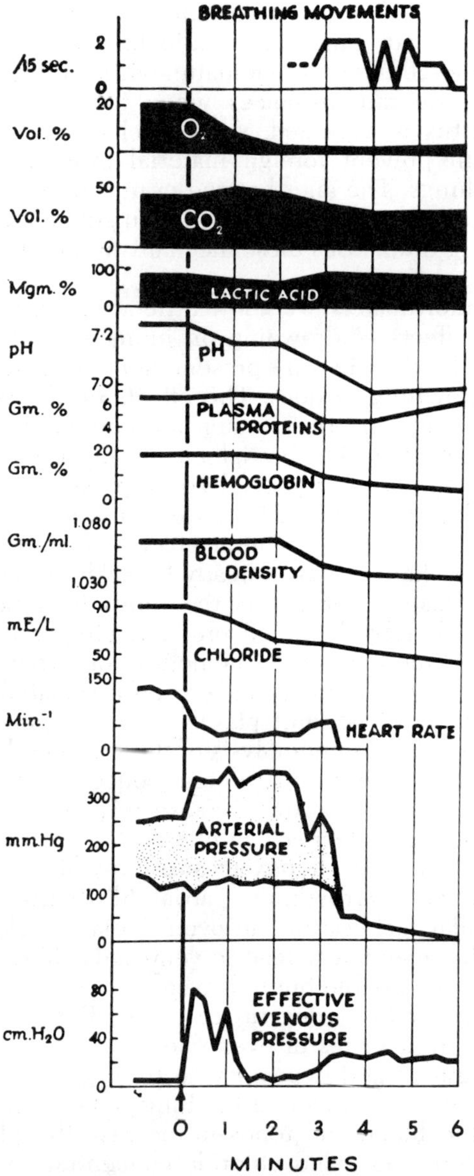

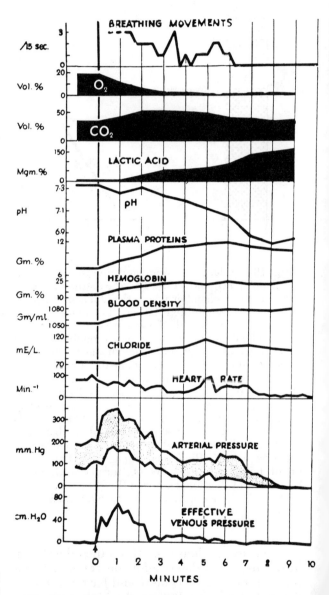

Fig. 19-8. Response of a conscious dog to drowning in fresh water. Most of the above symptoms are the result of a reflex apnea followed by hemodilution and hemolysis. Associated with the hemodilution was an atrial fibrillation, atrioventricular dissociation, and eventually ventricular fibrillation. (Swann, H. G., and Brucer, M.: The cardiorespiratory and biochemical events during rapid anoxic death. VI. Fresh water and sea water drowning. Tex. Rep. Biol. Med., 7:605, 1949)

Fig. 19-9. Response of a conscious dog to drowning in salt (hypertonic) water. Most of the above symptoms are the result of a reflex apnea followed by hemoconcentration. (Swann, H. G., and Brucer, M.: The cardiorespiratory and biochemical events during rapid anoxic death. VI. Fresh water and sea water drowning. Tex. Rep. Biol. Med., 7:606, 1949)

longed submersion a person who drowns in cold water has a better chance of surviving than one who drowns in warm water. Although after 6 min. of cardiac arrest and apnea under most conditions artificial respiration and cardiac massage are not recommended, there are reports of persons who have been submerged for 22 min. in cold water (−8°C. in one case) who have been revived and who have recovered with minimal nerve damage. In this regard I would emphasize that the first and most important treatment for drowning is reinflation of the lungs, artificial respiration, and, when necessary, external cardiac massage. In some cases exchange transfusions and intravenous $NaHCO_3$ are also recommended.

HYPERVENTILATION

Hyperventilation is a movement of air greater than is necessary to maintain the arterial Po_2 at 100 mm. Hg and the Pco_2 at 40 mm. Hg. When it occurs it results in hyperoxia, hypocapnea, or a combination of the two. It sometimes occurs during pain, fear, or anxiety, and has been noted in certain types of brain malfunction (acute hyperventilation syndrome, epidemic encephalitis, etc.), as well as in such conditions as rheumatoid spondylitis, diffuse pulmonary disease, and fever. *Fever* increases the body's utilization of O_2, but usually increases the ventilation to such an extent that there is hypocapnea associated with either a normal or an elevated arterial Po_2.

There are also numerous examples of cases in which ventilation results in hypocapnea associated with hypoxia. This is noted when the Po_2 of the inspired air is below normal, in anemia, and in histotoxic hypoxia. It is important to realize in these cases that the respiratory system does not regulate just the arterial Po_2 or just the arterial Pco_2, but rather both—by the simple mechanism of inspiration and expiration. Thus there are situations in which the body (1) in compensating for hypoxia produces dangerous hypocapnea, or (2) in compensating for hypocapnea produces dangerous hypoxia. Both hypoxia and hypocapnea can produce the deterioration of psychomotor performance and unconsciousness. When, for example, an alveolar Pco_2 of 27 mm. Hg is associated with an alveolar Po_2 of 29 mm. Hg, unconsciousness occurs (Table 19-1). In other words, the respiratory system must sometimes compromise its functions of maintaining a constant arterial Pco_2, Po_2, or both Pco_2 and Po_2 in order to maintain survival.

In Figure 19-10 we have a demonstration of a person's response to voluntary hyperventilation. Note that although the alveolar Po_2 increases, there is little change in the amount of O_2 carried by the blood. This is due, of course, to the fact that prior to hyperventilation the hemoglobin in the blood is almost completely saturated with O_2. Probably the most important changes in hyperventilation are the decrease in the arterial and alveolar Pco_2 and the associated increase in pH (alkalosis). As the arterial Pco_2 approaches 15 mm. Hg and the blood becomes more alkaline there is a generalized decrease in the *threshold* of irritable structures and a decrease in their tendency to *accommodate*. This affects first (1) the touch receptors, then (2) the proprioceptive fibers, (3) the somatic motor neurons, and (4) cold, heat, and rapid-pain neurons, and finally (4) the slow-pain fibers. This progression of changes results in a series of muscle spasms due

Type of Ventilation	Electroencephalogram			
	6 cycles/sec. (Deterioration)		3 cycles/sec. (Unconsciousness)	
	Po_2 (mm. Hg)	Pco_2 (mm. Hg)	Po_2 (mm. Hg)	Pco_2 (mm. Hg)
Spontaneous ventilation	29.8	28.1	28.7	26.6
Voluntary hyperventilation	38.0	18.0	34.5	20.0

Table 19-1. Effects of hyperventilation and hypoxia on the electroencephalogram. At 6 c.p.s. there was deterioration of psychomotor performance, and at 3 c.p.s. unconsciousness was imminent. Note that it is not just the alveolar Po_2 that determines the level of brain function, but a combination of alveolar Po_2 and Pco_2. (Luft, U.C.: Aviation physiology—The effects of altitude. *In* Fenn, W.O., and Rahn, H. (eds.): Handbook of Physiology. Sec. 3 (Respiration), vol. II, pp. 1099-1145, Washington, American Physiological Society, 1965)

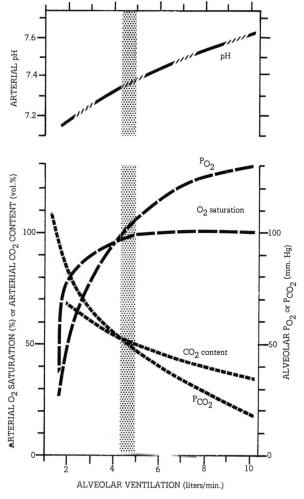

Fig. 19-10. Effect of changes in ventilation on the composition of the alveoli and arterial blood. The shaded area represents values for a normal man with an O_2 consumption of 250 ml./min. (*Redrawn from* The Lung, Second Edition, by J. H. Comroe, Jr., et al. Copyright © 1962, Year Book Medical Publishers, Inc.; used by permission)

to the firing of the proprioceptive fibers, and eventually leads to muscle fasciculation due to the spontaneous firing of somatic efferent neurons.

Hypocapnea produces hyperreflexia and increased muscle tone primarily because of the associated alkalosis and its action on the affected cells. The alkalosis results in a loss of Ca^{++} from the cell membrane and a decrease in the serum Ca^{++} level, which increases the irritability of the cells as alkalosis progresses. This is probably not, however, the main reason an increased pH produces hyperirritability, since in hypoparathyroidism considerably more marked decreases in serum Ca^{++} levels occur before similar increases in irritability are noted.

BAROMETRIC PRESSURE

At sea level the barometric pressure is 760 mm. Hg; at the top of Mount Everest (29,028 ft. above sea level) it is 235 mm. Hg; 33 ft. below the surface of the sea it is 1,520 mm. Hg (2 atmospheres); and 66 ft. below the surface of the sea it is 3 atmospheres. From these data we see that there are a great variety of pressures on earth to which man may be exposed. The use of modern aircraft and diving equipment, as well as of the hyperbaric chamber in certain clinical conditions, has broadened the range even more. The United States' X-15, for example, cruises at 50,000 ft., where the atmospheric pressure is 87 mm. Hg and the temperature −67°F., and has

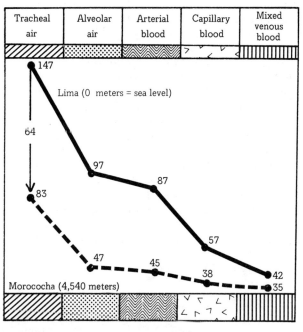

Fig. 19-11. Mean P_{O_2} of inhaled air in 80 natives of Lima (sea level) and 80 natives of Morococha (14,850 feet above sea level), Peru. Note the large differences between the P_{O_2} of tracheal air in these 2 groups as opposed to the differences in the mixed venous blood. (*Modified from* Hurtado, A.: Animals in high altitudes: resident man. *In* Dill, D. B., et al. (eds.): Handbook of Physiology. Sec. 4 (Adaptation to the Environment), p. 845. Washington, American Physiological Society, 1964)

Altitude		Pressure		P$_{A-47}$	Po$_2$
(feet)	(meters)	(p.s.i.)	(mm. Hg)	(mm. Hg)	(mm. Hg)
0	0	14.7	760	713	149
2,000	610	13.7	707	660	138
4,000	1,220	12.7	656	609	127
6,000	1,830	11.8	609	562	118
8,000	2,440	10.9	564	517	108
10,000	3,050	10.1	523	476	100
12,000	3,660	9.3	483	436	91
14,000	4,270	8.6	446	399	83
16,000	4,880	8.0	412	365	76
18,000	5,490	7.3	379	332	69
20,000	6,100	6.8	349	302	63
22,000	6,710	6.2	321	274	57
24,000	7,320	5.7	294	247	52
26,000	7,930	5.2	270	223	47
28,000	8,540	4.8	247	200	42
30,000	9,150	4.4	226	179	37
32,000	9,760	4.0	206	159	33
34,000	10,370	3.6	187	140	29
36,000	10,980	3.3	170	123	26
38,000	11,590	3.0	155	108	23
40,000	12,200	2.7	141	94	20
42,000	12,810	2.5	128	81	17
44,000	13,420	2.2	116	69	14
46,000	14,030	2.0	106	59	12
48,000	14,640	1.8	96	49	10
50,000	15,250	1.7	87	40	8
63,000	19,215	0.9	47	0	0

Table 19-2. Altitude-pressure table, showing the partial pressure exerted at different altitudes by dry gas in the alveoli (P$_{A-47}$) and the partial pressure that would be exerted by O$_2$ in air if that air were warmed to 37°C. (body temperature) and saturated with water vapor (Po$_2$). (Luft, U. C.: Aviation physiology—The effects of altitude. *In* Fenn, W. O., and Rahn, H. (eds.): Handbook of Physiology. Sec. 3 (Respiration), vol. II, pp. 1099-1145. Washington, American Physiological Society, 1965)

reached heights in excess of 314,000 ft. Mountaineers have been able, in the absence of supportive procedures, to live for several weeks at heights up to 21,000 ft., and some villages exist at heights up to 18,000 ft. For most of us, however, these heights would initially result in a series of symptoms called *mountain sickness* (breathlessness, cardiac palpitations, headache, dizziness, weakness, nausea, impairment of mental activity, and dimness of vision) and possibly death.

In Table 19-2 I have summarized the pressures for various altitudes. At 14,000 ft. (4,270 meters), for example, the atmospheric pressure is 446 mm. Hg and the O$_2$ (20.8%) exerts a partial pressure of 92.8 mm. Hg (446 × 0.208). This would not, however, be the Po$_2$ in the trachea, since the air inspired into the trachea is warmed and saturated with water vapor. In the trachea the water vapor contributes a partial pressure of approximately 47 mm. Hg. Hence a person living at this altitude inspires into his upper respiratory tract air with a Po$_2$ not of 92.8 mm. Hg, but of 83 mm. Hg [(446 − 47) × 0.208 = 399 × 0.208]. In other words, the dilution of the inspired air with water vapor is an important factor in decreasing its Po$_2$.

As the inspired air moves down the trachea and into the alveoli its O$_2$ concentration becomes further decreased due to its mixing with alveolar air. This results in an alveolar Po$_2$ of about 100 mm. Hg at atmospheric pressure and 50 mm. Hg at 14,850 ft. (Fig. 19-11). Some of the other effects of high altitude on the body are presented in Table 19-3. Note the greater ventilation of Peruvians living in Morococha (14,850 ft.) than of those living in Lima (sea level). This, in turn, results in a reduction in the alveolar and arterial

	Lima (sea level)	Morococha (14,850 ft.)
Body weight (kg.)	63 ± 1.02	52.7 ± 0.6
Respiratory system		
Ventilation (liters/min./kg., BTPS)	0.13 ± 0.003	0.19 ± 0.005
Respiratory rate (number/min.)	14.7 ± 0.52	17.3 ± 0.46
Tidal volume (liters, BTPS)	0.6 ± 0.02	0.59 ± 0.03
Alveolar P_{O_2} (mm. Hg)	104.4 ± 0.66	50.5 ± 0.74
Alveolar P_{CO_2} (mm. Hg)	38.6 ± 0.35	29.1 ± 0.52
Whole blood		
Blood volume (ml./kg.)	79.6 ± 1.49	100.5 ± 2.29
Red cell count (millions/mm.^{3})	5.11 ± 0.02	6.44 ± 0.09
Hematocrit (% red cells in blood)	46.6 ± 0.15	59.5 ± 0.68
Reticulocytes (thousands/mm.^{3})	0.4 ± 0.02	1 ± 0.07
Red cell volume (ml./kg.)	37.2 ± 0.71	61.1 ± 1.93
Hemoglobin (gm./100 ml./kg.)	12.6 ± 0.25	20.7 ± 0.62
O_2 Capacity (mM./liter)	9.3 ± 0.06	12.29 ± 0.17
O_2 Saturation of arterial blood (%)	97.9 ± 0.12	81 ± 0.49
Plasma		
Plasma volume (ml./kg.)	42 ± 0.99	39.2 ± 0.99
Bilirubin, total (mg./100 ml.)	0.76 ± 0.03	1.28 ± 0.13
Bicarbonate CO_2 (mM./liter)	25.29 ± 0.13	19.7 ± 0.22
pH of arterial plasma	7.41 ± 0.003	7.39 ± 0.005
Arteries		
Pulse rate (number/min.)	72 ± 1.52	72 ± 1.97
Arterial pressure: Systolic (mm. Hg)	116 ± 1.6	93 ± 4.03
Diastolic (mm. Hg)	79 ± 1.73	63 ± 2.6
Net efficiency during exercise (%)	19.9 ± 0.66	22.2 ± 0.66

Table 19-3. Comparison of Peruvian natives living at sea level (760 mm. Hg atmospheric pressure) and 4,540 meters (14,850 ft.) above sea level (431 mm. Hg). The respiratory volumes are corrected to body temperature, 760 mm. Hg pressure, and saturation with water at body temperature (BTPS). Because of the difference in weights of the two groups, much of the data are given in units per kilogram of body weight. All values are expressed as means ± the standard error and are based on between 10 and 280 different subjects. (Hurtado, A.: Animals in high altitudes: Resident man. In Dill, D.B., Adolf, E.F., and Wilber, C.G. (eds.): Handbook of Physiology. Sec. 4 (Adaptation to the Environment), pp. 843-860. Washington, American Physiological Society, 1964)

P_{CO_2}. Note too that the people of Morococha have an increased blood volume, hematocrit, hemoglobin, O_2-carrying capacity of the blood, and concentration of young red cells (reticulocytes). The disadvantages of a high hematocrit and the associated increase in the viscosity of the blood have been discussed in Chapter 10 (Fig. 10-16). One mechanism which initiates these changes is the action of hypoxia (arterial P_{O_2} = 50 mm. Hg) on the hemopoietic tissue (through the release of erythropoietin into the blood).

Decompression (Caisson Disease)

If one uncaps a bottle containing a carbonated beverage, gas dissolved in the beverage comes out of solution and forms bubbles. This phenomenon is due to the reduction of pressure in the beverage when it is uncapped and may be similar to what happens when the pressure exerted on a man is suddenly reduced. It has been noted, for example, that a man working 90 ft. below the surface of the sea (pressure, 3.8 atmospheres) for 100 min. cannot safely move to the surface unless he is slowly decompressed to atmospheric pressure over a period of 57 min. (Table 19-4). If he is not slowly decompressed there is expansion of the gases in the digestive tract (barometeorism), middle ear (barotitis), paranasal sinuses (barosinusitis), teeth (barodontalgia), and lungs which can be painful and dangerous. In time other symptoms (Table 19-5) develop, probably as a result of gases coming out

Depth (ft.)	Pressure (mm. Hg)	Exposure (min.)	Decompression (min.)
33	1,520	1,440	2
90	2,840	100	57
90	2,840	540	720
100	3,060	25	rapid
150	4,220	10	rapid

Table 19-4. Recommended period of decompression for divers or caisson workers breathing air, who have remained at certain depths for specific intervals. (Behnke, A.R., Jr.: B. Bends, decompression, and recompression. *In* Fenn, W.O., and Rahn, H. (eds.): Handbook of Physiology. Sec 3 (Respiration), vol. II, pp. 1161-1176. Washington, American Physiological Society, 1965).

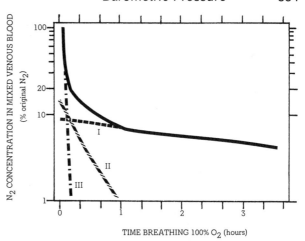

Fig. 19-12. N_2 washout from the peripheral stores of a dog after shifting from breathing air to breathing 100% O_2. The ordinate is plotted on a logarithmic scale in terms of N_2 concentration in mixed venous blood. The curve is divided into 3 components (III, initial rapid loss of N_2 from body water; II, reduced N_2 loss; I, slow, sustained loss of N_2 from fat depots. (*Redrawn from* Farhi, L. E.: Gas stores of the body. *In* Fenn, W. O., and Rahn, H. (eds.): Handbook of Physiology. Sec. 3 (Respiration), vol. I, p. 883. Washington, American Physiological Society, 1964)

of solution in a manner similar to that described for the uncapped beverage. No doubt these symptoms are in part the result of the *ischemia* produced when the bubbles interfere with blood flow, and in part due to their formation in the interstitial spaces.

Now let us take a look at the source of the bubbles. Under normal circumstances we have three major gases dissolved in the body fluids—O_2, CO_2, and N_2. At 3 atmospheres the arterial blood would have a P_{O_2}, P_{CO_2}, and P_{N_2} considerably greater than at 1 atmosphere. On rapid decompression (e.g., during ascent to sea level or the ascent of a plane) the partial pressures of these gases are reduced; as a result, the gases tend to come out of solution (form bubbles). Both O_2 and CO_2 diffuse from one area to another more rapidly than does N_2 (Table 17-6), and therefore tend to be expired more rapidly than the excess N_2. This plus the fact that there are specialized systems for removing O_2 and CO_2 from an area make the problem of N_2 bubbles the more serious one.

As we saw in Table 19-4 it is safe for a subject to ascend rapidly from 150 ft. below the sea if he has remained there for less than 10 min., but

Central Nervous System	Cardiorespiratory System	Extremities	Skin—Systemic
Unconsciousness	Substernal distress	Pain	Rash, mottling
Spastic paralysis	Paroxysmal coughing	Numbness	Pruritus
Visual field defects	Tachypnea, dyspnea	Paresthesia	Pallor, lowered temperature
Vertigo, aphasia	Asphyxia (choking)	Weakness	Fatigue, malaise
Sensory loss	Shock (circulatory obstruction)	Bone infarcts	Fever, sweating
Bladder and bowel Dysfunction	Hemoconcentration	Cartilage loss	

Table 19-5. Signs and symptoms of decompression sickness (caisson disease). (Behnke, A.R., Jr.: B. Bends, decompression, and recompression. *In* Fenn, W.O., and Rahn, H. (eds.): Handbook of Physiology. Sec 3 (Respiration), vol. II, pp. 1161-1176. Washington, American Physiological Society, 1965)

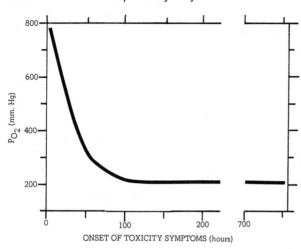

Fig. 19-13. Levels of P_{O_2} and exposure time at which O_2 toxicity develops in man. These data are a compilation of many studies which used varying indications of pulmonary irritation as the criterion for toxicity. (*Redrawn from* Lambertsen, C. J.: Effects of oxygen at high partial pressure. *In* Fenn, W. O., and Rahn, H. (eds.): Handbook of Physiology. Sec. 3 (Respiration), vol. 2, p. 1029. Washington, American Physiological Society, 1964)

not if he remains there for longer periods. Apparently it takes considerably more than 10 min. to saturate the body with N_2. It is also apparent, from Figure 19-12, that it takes longer than 10 min. to remove excess N_2 from the body. These are but a few of the indications that decompression sickness is in part concerned with diffusion. It may be that initially on decompression the body fluids become supersaturated, allowing large volumes of the rapidly diffusing gases to be exhaled before bubbles form. Another suggestion is that the body can tolerate the small degrees of embolization which occur in those

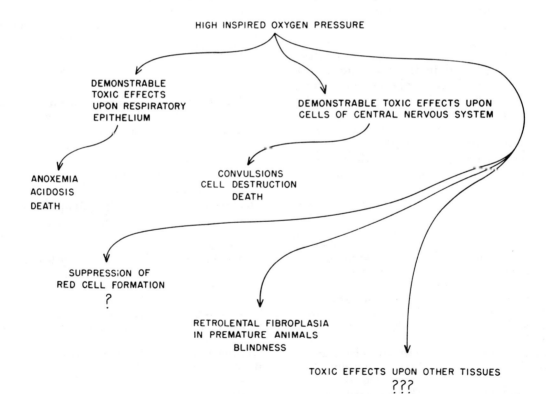

Fig. 19-14. Toxic effects of hyperoxia on the body. The most commonly recognized symptoms of O_2 toxicity are generalized convulsions, pulmonary damage, and, in the premature infant, retinal damage. (Lambertsen, C. J.: Effects of oxygen at high partial pressure. *In* Fenn, W. O., and Rahn, H. (eds.): Handbook of Physiology. Sec. 3 (Respiration), vol. 2, p. 1028. Washington, American Physiological Society, 1964)

cases of rapid decompression not associated with outward symptoms of distress.

In Figure 19-12 we have a demonstration of the speed with which N_2 passes from its reservoirs. On the basis of experiments on lean men (70-kg. man with 7 kg. of fat) it has been estimated that about 75 per cent of the body's total N_2 is eliminated in a period of 2 hr. In such a man at 1 atmosphere there are about 400 ml. of N_2 in the body liquids, 100 ml. in bone and spinal cord, and 350 ml. in adipose tissue. The adipose tissue has an N_2 elimination half-life of about 69 min., bone marrow 85 min., and low-lipid organs from less than 1 min. to 20 min. Since N_2 is about five times more soluble in fat than in the body's aqueous solutions, and since many of the high fat areas have a poor blood supply, it is easy to see why it takes an extended period at high pressures to saturate the body with N_2 and an extended period at low pressures to eliminate it.

OXYGEN TOXICITY

An additional problem for persons at high pressures is the toxic action of O_2. In Figure 19-13, for example, we note that prolonged exposure to a P_{O_2} as low as 300 mm. Hg can cause toxicity symptoms. Although normal healthy persons have been safely exposed to pure O_2 at a pressure of 187 mm. Hg for 14 to 30 days, exposure to pure O_2 at 760 mm. Hg for 24 hours has caused pulmonary and nasopharyngeal irritation. Exposure for longer periods has resulted in bronchopneumonia. In the premature infant, as opposed to in the child or adult, exposure to pure O_2 at 1 atmosphere may cause *retrolental fibroplasia* and blindness due to an extensive vascular growth and fibroblastic infiltration of the retina. Some of the other effects of O_2 toxicity are given in Figure 19-14. We also find that persons breathing pure O_2 have a tendency to become atelectatic (i.e., suffer partial collapse of the lung) in response to respiratory infection or irritation. This is due to the fact that in these conditions mucus production may be excessive and cause obstruction of a number of alveoli. When this happens the alveolus with a high percentage of O_2 collapses much more rapidly than one containing a high percentage of a poorly diffusing inert gas such as N_2 when a person breathes air, or He when he breathes an artificial mixture of gases.

Many of the symptoms of O_2 toxicity are apparently the result of the depressant action of O_2 on

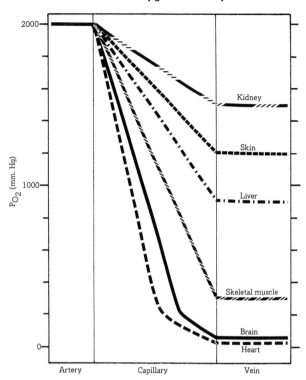

Fig. 19-15. The effect of breathing O_2 at a pressure of 3.5 atmospheres (2,600 mm. Hg) on the arterial, capillary, and venous P_{O_2}. The data for the brain are based on samples from the arterial and internal jugular vein of 16 conscious volunteers. The other data were calculated from measured arterial values and tables of O_2 consumption and blood flow for man. (*Redrawn from* Lambertsen, C. J.: Effects of oxygen at high partial pressure. *In* Fenn, W. O., and Rahn, H. (eds.): Handbook of Physiology. Sec. 3 (Respiration), vol. 2, p. 1034. Washington, American Physiological Society, 1964)

critical cellular enzyme systems. In Figure 19-15, for example, we see an estimate of the effect of breathing O_2 at a P_{O_2} of 3.5 atmospheres (2,600 mm. Hg). Note that although the arterial P_{O_2} goes to 2,000 mm. Hg in these experiments, the venous blood leaving the brain and heart has a P_{O_2} approaching normal. This is apparently the result of vasoconstriction in these areas, which in part protects them from the hyperoxia. Lambertsen (Fig. 19-16) attributes the vasoconstriction not directly to the hyperoxia, but rather to the hypercapnea which the hypoxia produces. He and others have shown that if hypercapnea is prevented, cerebral blood flow will not decrease during hyperoxia. It has also been noted that when hyperoxia is associated with hyper-

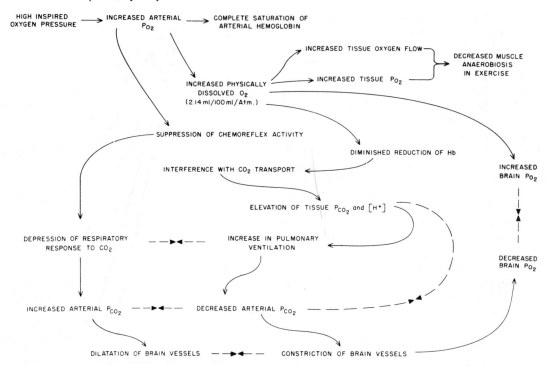

Fig. 19-16. Sequence of the acute effects of hyperoxia in man. Each of the effects shown in the diagram has been demonstrated at a P_{O_2} of one atmosphere and, with the exception of the chemoreflex suppression, at 3.0 to 3.5 atmospheres. It will be noted that 5 sets of opposing arrows exist in the diagram. This is an indication that the conflicting actions of hyperoxia lead to a new state of dynamic balance in the body. (Lambertsen, C. J.: Effects of oxygen at high partial pressure. *In* Fenn, W. O., and Rahn, H. (eds.): Handbook of Physiology. Sec. 3 (Respiration), vol. II, p. 1043. Washington, American Physiological Society, 1964)

ventilation the toxicity symptoms develop more rapidly, and when the P_{CO_2} of the inspired air is increased, the symptoms develop more slowly. All of this is consistent with the view that hypercapnea protects certain critical areas from hyperoxia by producing vasoconstriction.

Chapter 20

ANATOMY
FILTRATION
REABSORPTION
SECRETION
CLINICAL EVALUATION
OF KIDNEY FUNCTION

URINE FORMATION

ANATOMY

Urine is produced by the kidneys and carried by the ureters to the bladder, where it is stored until forced into the urethra and out of the body.

Although the two kidneys together weigh 300 gm. and constitute 0.4 per cent of the body weight, they receive from 20 to 25 per cent of the cardiac output in a resting person. In most people, almost all of this blood passes via (1)

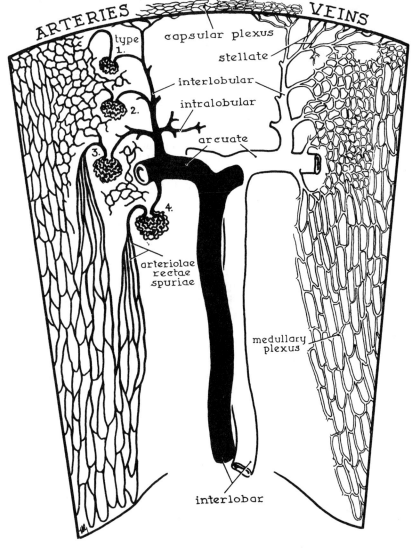

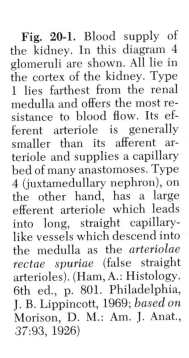

Fig. 20-1. Blood supply of the kidney. In this diagram 4 glomeruli are shown. All lie in the cortex of the kidney. Type 1 lies farthest from the renal medulla and offers the most resistance to blood flow. Its efferent arteriole is generally smaller than its afferent arteriole and supplies a capillary bed of many anastomoses. Type 4 (juxtamedullary nephron), on the other hand, has a large efferent arteriole which leads into long, straight capillary-like vessels which descend into the medulla as the *arteriolae rectae spuriae* (false straight arterioles). (Ham, A.: Histology. 6th ed., p. 801. Philadelphia, J. B. Lippincott, 1969; *based on* Morison, D. M.: Am. J. Anat., 37:93, 1926)

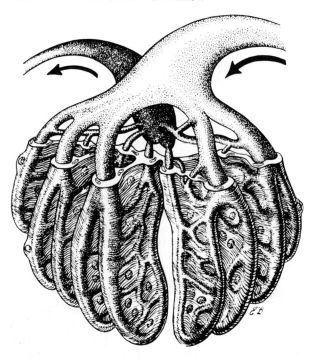

Fig. 20-2. The glomerulus. Note that the glomerulus is formed by the branches of a single efferent arteriole. Each of these branches forms a system of capillary loops, which together with its associated basement membrane forms a glomerular lobule. Also shown is the efferent arteriole (*heavily shaded vessel*). (Elias, H. A., et al.: Blood flow in the renal glomerulus. J. Urol., 83:795; copyright © 1960, The Williams and Wilkins Company, Baltimore)

the right and left renal arteries to (2) the interlobular arteries, to (3) the arcuate arteries, and eventually to (4) the *afferent arterioles* (Fig. 20-2), containing two to five capillary loops. The capillary system formed by a single renal afferent arteriole and its associated basement membrane is called a *glomerulus*.

There are in each kidney from 1 million to 1¼ million glomeruli. Each is surrounded by the expanded blind end of the uriniferous tubule (or nephron). It is at the glomeruli that approximately 180 liters of liquid are *filtered* into the nephron each 24 hr. Blood passes from the glomerulus via the efferent arteriole to a second series of capillaries (peritubular capillaries) which surround the rest of the nephron (Fig. 20-4) and the collecting duct into which the nephron leads. It is into the peritubular capillaries that most of the filtrate formed by the glomerular capillaries is ultimately *reabsorbed*. It is also here that certain substances pass from the peritubular blood into the nephron. The latter process is called *secretion*. Thus the kidneys, through the processes of (1) filtration, (2) reabsorption, and (3) secretion, regulate the composition of the urine and in so doing control that of the blood.

The water, electrolytes, acids, bases, waste products (urea, uric acid, creatinine, etc.), and foreign material (such as barbiturates) which remain in the nephron pass into (1) the collecting ducts, where the process of reabsorption continues. The collecting ducts eventually come together in the renal papilla to form (2) the papillary ducts of Bellini. There are from 8 to 10 renal papillae in each kidney (Fig. 20-3), and the apex of each papilla is pierced by 18 to 24 bearly visible openings which drain urine from the ducts of Bellini into (3) a minor calyx (Gk. *kalyx*, cup), then into (4) either the superior or inferior major calyx, and finally into (5) the renal pelvis, (6) the ureter, and (7) the urinary bladder. In the urinary bladder the urine is stored until the bladder contracts so as to propel the urine through (8) the urethra and out of the body. In this way the kidneys and associated structures eliminate certain substances and help maintain body homeostasis.

Cortical and Juxtamedullary Nephrons

The nephron and its collecting duct are the functional unit of the kidney. It is here that an ultrafiltrate of plasma is received and modified. The ultrafiltrate is formed from the glomerular blood at the end of the nephron called Bowman's capsule. Modification of this ultrafiltrate occurs throughout the nephron (Fig. 20-4) and collecting duct in conjunction with the peritubular blood vessels. About 85 per cent of the nephrons are classified as *cortical*, since they lie primarily in the cortex of the kidney (Fig. 20-5). The other type, the *juxtamedullary* (L. *juxta*, near) nephron, has its glomerulus near the medulla and sends its tubule deep into the medulla. These units are responsible for the greater tonicity of the peritubular fluid in the medulla than in the cortex (Fig. 20-18). The peritubular vessels of the cortical nephrons anastomose freely, whereas those of the juxtamedullary nephrons form loops. A second difference is that in the cortical nephrons, the efferent arteriole is usually smaller than the afferent, whereas in the juxtamedullary nephrons the reverse is true.

Fig. 20-3. Coronal section of the kidney. (Ham, A.: Histology. 6th ed., p. 775. Philadelphia, J. B. Lippincott, 1969)

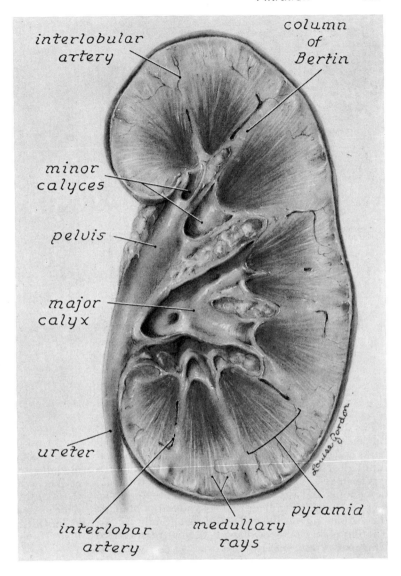

FILTRATION

Filtration is the movement of a liquid and its dissolved particles through a semipermeable membrane, any suspended particles being left behind. In the case of filtration through the systemic and glomerular capillaries not only the suspended cells and cell fragments are left behind, but also most of the colloids (albumin, globulin, etc.). In these capillaries an *ultrafiltrate* (blood minus formed elements and large dissolved substances) is formed. The character of the ultrafiltrate depends in part on the composition of the blood and the permeability of the membrane. Skeletal muscle capillaries function as though they contained pores of 60 A (0.006 μ), and the glomerular capillaries, pores 75 to 100 A in diameter. This is not to say that such *pores* actually exist, but that these capillaries can be superficially thought of in this way. As can be seen in Figure 20-6, there is no one simple membrane separating the capillary lumen (left) from the lumen of the nephron, nor is there a simple system of pores or holes leading from the capillary to the nephron. There are open fenestrae (windows) with diameters of about 400 to 900 A in the capillary endothelium, but there is also the adjacent, continuous, nonporous basement membrane (Table 12-1) which is about 0.1 μ thick. This seems to be a far more effective

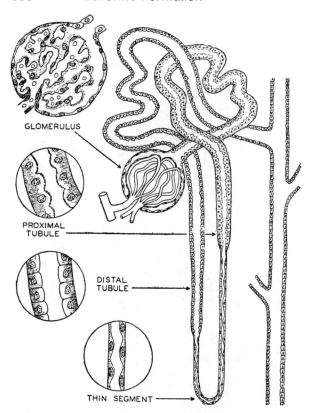

Fig. 20-4. The nephron. The nephron consists of (1) Bowman's capsule and its associated glomerular capillaries and (2) the tubule and its associated peritubular capillaries. The tubule consists of (1) a short connecting segment, (2) a proximal convoluted portion, (3) descending and ascending thick and thin limbs of the loop of Henle, and (4) a distal convoluted portion. At the extreme right is the collecting duct. (Smith, H. W.: The Physiology of the Kidney. P. 6. New York, Oxford University Press, 1937)

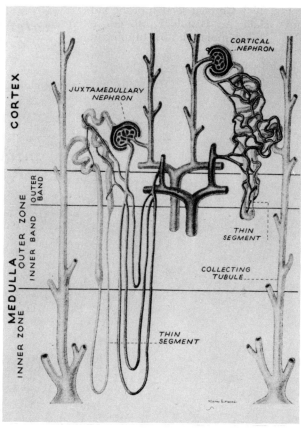

Fig. 20-5. Comparison of the cortical and juxtamedullary nephrons. Note that the peritubular vessels of the cortical nephron freely anastomose, whereas those of the juxtamedullary apparatus form long loops which extend well into the medulla. (Smith, H. W.: The Kidney Structure and Function in Health and Disease. Plate II. New York, Oxford University Press, 1951)

barrier than the capillary endothelium. The epithelial cells (podocytes), whose pedicles or foot processes also attach to the basement membrane, constitute another potential barrier, since the spaces between pedicles is about 100 A. It has been suggested by some that the disruption of the basement membrane is primarily responsible for the proteinuria in nephrotic syndrome.

The glomerular filtrate normally contains the same concentration of electrolytes, glucose, urea, uric acid, and creatinine that plasma does (Table 20-1). Substances of considerably larger molecular size, however, appear in the filtrate at concentrations less than that found in the plasma (Table 20-2). Albumin, for example, appears in the filtrate of the proximal tubule only in very minute quantities (less than 3 mg./100 ml. in the dog) and is absent in the lumen of the loop of Henle in a normal person. One of the factors responsible for this low concentration is albumin's molecular dimensions. It is only 36 A thick, so it should be able to pass through a pore 75 to 100 A in diameter. Its length, however, is 150 A, and in a system of streamlined flow molecules of this dimension generally come into contact with the capillary wall.

The phenomenon of filtration should be distinguished from that of *active transport*. In the latter, metabolic energy is expended by the cell in moving a particle or particles through a mem-

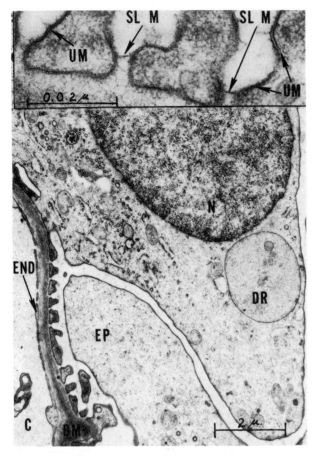

Fig. 20-6. Ultrastructure of the glomerulus of a 12-year-old boy. The lumen of the glomerular capillary (C) is separated from the lumen of the nephron by an endothelium (END) with its associated fenestrae and nuclei, a basement membrane (BM), and epithelial cells (EP). The epithelium attaches to the basement membrane by means of pedicles. Each pedicle contains a unit membrane (UM) and is separated from adjacent pedicles by a split pore membrane (SL M). Also shown is a droplet (DR) and nucleus (N) within an epithelial cell. The upper insert is × 122,000 and the lower picture is × 975. (Spargo, B. H.: Structure of the kidney. *In* Mostofi, F. K., and Smith, D. E. (eds.): The Kidney. P. 30. Washington, International Academy of Pathology, 1966)

brane or membranes. In filtration, the energy comes from the hydrostatic pressure of the blood in the glomerular capillaries. If, for example, the mean aortic pressure goes below 70 mm. Hg, the hydrostatic pressure in the glomerular capillaries becomes so low that filtration ceases and *anuria* follows. Normally in man the mean aortic pressure is about 100 mm. Hg and the glomerular

Filtration 359

Substance	Glomerular Filtrate	Urine	Urine/Filtrate
	mEq./liter	mEq./liter	
Na^+	142	128	0.9
K^+	5	60	12
Ca^{++}	4	4.8	1.2
Mg^{++}	3	15	5
Cl^-	103	134	1.3
HCO_3^-	28	14	0.5
$H_2PO_4^-$ and HPO_4^{--}	2	50	25
SO_4^{--}	0.7	33	47
Ammonia	0	27	
	mg./100 ml.	mg./100 ml.	
Glucose	100	0	0
Urea	26	1,820	70
Uric acid	3	42	14
Creatinine	1.1	196	140

Table 20-1. Comparison of the concentration of substances in the glomerular filtrate and urine. All the substances listed have the same concentration in the filtrate as in the plasma. The values vary, of course, with diet and plasma pH.

capillary pressure 70 mm. Hg. Under these conditions the following data are characteristic for a nephron:

Glomerular capillary pressure	70 mm. Hg
Intracapsular pressure	15 mm. Hg
Effective Hydrostatic Pressure	55 mm. Hg
Capillary colloid osmotic pressure	30 mm. Hg
Intracapsular colloid osmotic pressure	0 mm. Hg
Effective Colloid Osmotic Pressure	30 mm. Hg
Filtration Pressure	25 mm. Hg

In short, the high effective hydrostatic pressure pushing fluid into Bowman's capsule more than compensates for the colloid osmotic pressure in the glomerular blood. As a result of the above conditions, the average person filters into the nephrons of the two kidneys about 125 ml. of liquid per minute, or 180 liters per day. If we divide this figure by a hypothesized 2 million active nephrons in the kidneys, we find that each active nephron filters about 0.09 ml. a day. The total daily filtration of NaCl into the nephrons is about 1,100 gm.; of $NaHCO_3$, 410 gm.; glucose, 150 gm.; urea, 53 gm.; creatinine, 1.4 gm.; and uric acid, 8.5 gm.

Substance	Mol. Wt. (gm.)	Dimensions (A)		Conc. in Nephron / Conc. in Capillaries
		Radius from Diffusion Coef.	Length and Thickness from X-Ray Diffraction	
Water	18	1		1
Urea	60	1.6		1
Glucose	180	3.6		1
Sucrose	342	4.4		1
Inulin	5,500	14.8		0.98
Myoglobin	17,000	19.5	54 × 8	0.75
Egg albumin	43,500	28.5	88 × 22	0.22
Hemoglobin	68,000	32.5	54 × 32	0.03
Serum albumin	69,000	35.5	150 × 36	<0.01

Table 20-2. Effect of molecular dimensions on the filtration of various substances. Note that as molecular size increases, the ratio between the concentration of the substance in the nephron and that in the blood decreases.

Micropuncture Technique

Studies on the concentrations of substances within the nephron gave us the first definitive evidence that glomerular filtration is the initial step in urine formation. These data were collected using the micropuncture method developed by A. N. Richards and his colleagues in 1930's and revived twenty years later by Wirz and Gottschalk (Fig. 20-7). In this technique (1) the kidney is exposed and (2) punctured with a glass or quartz micropipette, (3) the tubule is blocked by the injection of oil or mercury, (4) 0.0001 μ liters to 0.5 μ liters are withdrawn, (5) the site is marked by means of a dye or latex, and (6) the locus is identified by microdissection. This method does not, however, tell us anything about the rate of filtration.

Inulin Clearance

In order to estimate the rate of glomerular filtration we usually use a marker in the plasma which moves freely through Bowmann's capsule but to which the rest of the nephron and the collecting duct are impermeable. Hence the quantity that enters the nephron is equal to the quantity found in the urine, and its concentration in the plasma equals that in Bowmann's capsule. If, for example, the marker is maintained at a concentration of 0.1 mg. per milliliter in the plasma, and at this plasma concentration 10 mg. per minute enter the urine, we conclude that 100 ml. of glomerular filtrate are being formed per minute:

$$\frac{10 \text{ mg./min.}}{0.1 \text{ mg./ml. filtrate}} = 100 \text{ ml. filtrate/min.}$$

If the same plasma concentration is maintained and 20 mg. of the marker are excreted per minute,

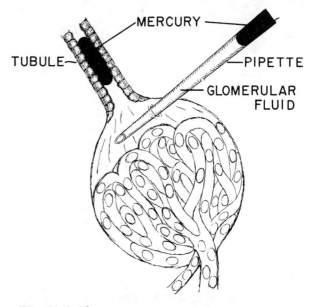

Fig. 20-7. The micropuncture technique. This is a technique that has been successfully used to study kidney function in necturus, frog, rat, guinea pig, opossum, and dog. It has not been used in the study of the human kidney. (*From* Physiology of the Kidney and Body Fluids, Second Edition, by R. F. Pitts. Copyright © 1968, Year Book Medical Publishers, Inc.; used by permission)

then 200 ml. of glomerular filtrate per minute are being formed.

The best marker currently available for such studies is inulin. It (1) is a polysaccharide with a molecular weight of 5,500, (2) has no known pharmacological action in the body, (3) is not bound to the plasma proteins and therefore is not prevented from passing through the glomerulus, (4) does not modify renal function, (5) is not reabsorbed from the tubule and back into the blood, and (6) is not secreted from the peritubular blood into the tubule. In addition, (7) when injected into the blood over 90 per cent of it can be recovered from the urine and identified. These characteristics make inulin ideal for the study of the glomerular filtration rate in all vertebrates. Creatinine has also been used in many animals for this purpose, but is inappropriate in man since his system not only filters it into the nephron at the glomerulus, but also secretes it elsewhere along the tubule.

Inulin clearance (C_{In}) is the term applied to studies in which inulin is used to estimate the rate of glomerular filtration. In an inulin clearance test the volume of plasma completely cleared of inulin in one minute is determined. First, a priming dose of inulin is injected intravenously to bring the inulin concentration to 10 to 20 mg. of inulin per 100 ml. of plasma. Intravenous infusion is then continued throughout the study to maintain a constant inulin concentration in the plasma. Urine is collected during the study, usually by means of a catheter. Let us say that in such a study a plasma inulin concentration (P_{In}) of 10 mg. per 100 ml. of plasma is maintained, the urine inulin concentration (U_{In}) is 1,250 mg. per 100 ml. of urine, and the urine production (V) is 1 ml. of urine per minute. From these data we can estimate the amount of filtrate formed per minute:

$$C_{In} = \frac{U_{In}}{P_{In}} \times V = \frac{1,250 \text{ mg.}/100 \text{ ml. urine}}{10 \text{ mg.}/100 \text{ ml. plasma}} \times$$

$$1 \text{ ml. urine/min.} = 125 \text{ ml./min.}$$

Inulin clearance can also be defined as follows:

$$C_{In} = \frac{\text{inulin excreted (mg./min.)}}{P_{In} \text{ (mg./ml. plasma)}}$$

In Figure 20-8 we note that as the plasma inulin concentration is increased, the excretion of inulin in the urine increases proportionately. This is what we would expect if the movement of inulin into the nephron were due to filtration alone. If, on the other hand, it were due in part to an active metabolic process, the slope of the line (C_{In}) would decrease after the point at which the capacity of the metabolic system responsible for active transport had been exceeded. In other words, the evidence is consistent with the view that inulin is filtered but not reabsorbed or secreted.

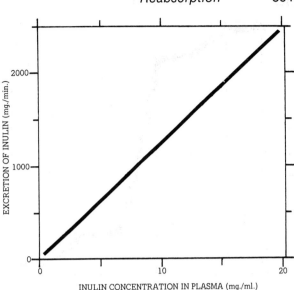

Fig. 20-8. Relationship between the rate of excretion of inulin in the urine and the plasma concentration of inulin. The slope of the line is equal to the inulin clearance. Since the slope remains constant when P_{IN} changes, we know that inulin clearance also remains constant. (*Redrawn from* Physiology of the Kidney and Body Fluids, Second Edition, by R. F. Pitts. Copyright © 1968, Year Book Medical Publishers, Inc.; used by permission)

REABSORPTION

We pointed out in the section on glomerular filtration that tremendous quantities of water and dissolved substances enter the nephron each day. It should be apparent that if man is to survive, most of this material must be conserved. We find, for example, that over 99 per cent of the Na^+, Cl^-, HCO_3^-, water, glucose, and amino acids, and over 92 per cent of the K^+ that is filtered into the nephron is reabsorbed into the blood. On the other hand, in man only from 40 to 70 per cent of the urea in the filtrate is reabsorbed. You will note in Figure 20-9 that by

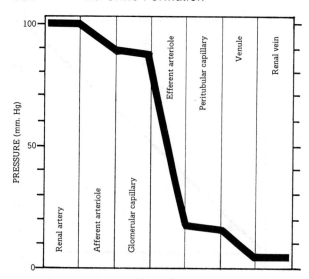

Fig. 20-9. Blood pressure in the renal circulation. (*Redrawn from* Physiology of the Kidney and Body Fluids, Second Edition, by R. F. Pitts. Copyright © 1968, Year Book Medical Publishers, Inc.; used by permission)

the time the blood has passed to the peritubular capillaries its pressure has reached a level considerably lower than that found in the glomerular capillaries. This low intracapillary pressure facilitates the reabsorption of the fluid from the peritubular space and the nephron.

Passive Reabsorption

There are two types of reabsorption going on similtaneously in the nephron—passive and active. In passive reabsorption no energy is directly expended by the kidney cells to move a particular substance via a carrier system from the nephron back to the blood. The movement of substances is determined rather by differences in hydrostatic pressure, osmotic pressure, concentration, and electrical potential, as well as by the characteristics of the substance itself. The reabsorption of water and urea, for example, is thought by most investigators to be a passive process occurring throughout the nephron. The reabsorption of Cl^- is probably passive only in the proximal tubule.

To say water, urea, and Cl^- are passively reabsorbed in the proximal tubule is not to imply that they are unaffected by active reabsorption. There is, for example, active transport of Na^+ from the nephron to the peritubular fluid at the proximal tubule. This would result in a higher peritubular osmotic pressure and a positive electrical charge if the membrane were not permeable to water and anions such as Cl^-. The transport of Na^+ to the peritubular space therefore facilitates the movement of water and Cl^- out of the nephron, since the proximal tubule is freely permeable to these two substances.

The importance of Na^+ in the movement of water out of the nephron is demonstrated in Figure 20-10. In these experiments the investigators perfused four different concentrations of NaCl into the tubule of an amphibian, *Necturus*. The solutions were all kept isotonic to *Necturus* plasma by the addition of mannitol. Note that as the NaCl concentration in the tubule increases, the percentage of water reabsorbed also increases.

Urea. We have noted in this chapter under filtration that inulin clearance in man is usually about 125 ml. per minute, i.e., that 125 ml. of plasma are completely cleared of inulin each minute. Urea clearance can be determined in a manner similar to that used for inulin.

$$C_{urea} = \frac{U_{urea}}{P_{urea}} \times V = \frac{E_{urea}}{P_{urea}}$$

Here C_{urea} is the urea clearance in milliliters of plasma per minute, U_{urea} is the urea concentration in urine in milligrams of urea per milliliter of urine, P_{urea} is the urea concentration in plasma in milligrams urea per milliliter of plasma, V is the urine production in milliliters of urine per

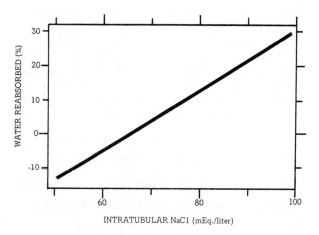

Fig. 20-10. The effect of intratubular NaCl concentration on the reabsorption of water from the nephron of necturus. The solutions studied were made isotonic to necturus plasma by the addition of mannitol. (*Redrawn from* Windhager, E. E., et al.: Single proximal tubules of the Necturus kidney. III. Dependence on H_2O movement on NaCl concentration. Am. J. Physiol., *197*:315, 1959)

minute, and E_{urea} is the amount of urea excreted in milligrams per minute. Using this technique most investigators report urea clearances ranging from 25 to 90 ml. per minute, the value most reported being 75 ml. per minute. That is to say that in man under cetain conditions 75 ml. of plasma are completely cleared of urea each minute. The reason the urea clearance is less than the inulin clearance, of course, is that inulin is filtered but not reabsorbed, whereas urea is both filtered and reabsorbed. Or, the amount of urea excreted is less than that filtered.

There are two pieces of evidence for the passive reabsorption of urea in man. One is that the urea clearance does not change with changes in plasma concentration. If we were dealing with an active transport system we would expect that as the plasma urea concentration increased, a point would eventually be reached at which the system would be saturated and, as a result, clearance would increase. In other words, the ratio of U_{urea} to P_{urea} would increase.

A second observation consistent with the theory that urea is reabsorbed only passively is that the urea clearance increases as the quantity of urine excreted does (Fig. 20-11). Urea reabsorption, in other words, is directly related to water reabsorption.

Active Reabsorption

Active reabsorption is an energy-requiring phenomenon in which a substance is moved out of the nephron against an electrochemical concentration gradient. In Figure 20-12 we have an example of how this may occur. It is hypothesized here that the active transport mechanism lies at the peritubular side of the cell. This mechanism is poorly understood, but probably involves the combination of intracellular Na^+ with a carrier molecule (CARRIER+Na^+ → CARRIER-Na^+) which diffuses near the surface of the cell, where it is broken down and releases the Na^+ (CARRIER-Na^+ → carrier + Na^+) to the peritubular fluid. In this hypothesized system I am calling the molecule with the high affinity for Na^+ *CARRIER*, and the molecule with the reduced affinity for Na^+ *carrier*. An alternate approach would be to call these two molecules carrier-I and carrier-II, or carrier-E (E, efferent)

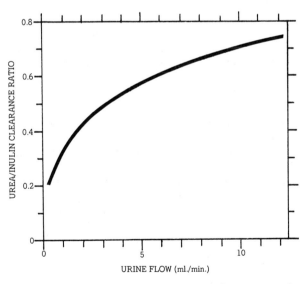

Fig. 20-11. The effect of urine production on the urea clearance to inulin clearance ratio. It is concluded that an increase in the reabsorption of water from the tubule (a decrease in urine production) increases the reabsorption of urea. (*Modified from* Chasis, H., and Smith, H. W.: The excretion of urea in man and in subjects with glomerulonephritis. J. Clin. Invest., 17:347, 1938)

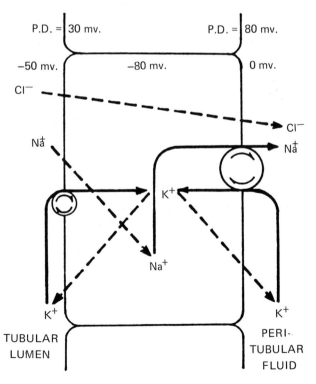

Fig. 20-12. Ion transport in a cell of the proximal tubule. Solid lines represent active transport and interrupted lines passive transport. (Pitts, R. F.: A comparison of the modes of action of certain diuretic agents. Progr. Cardiovasc. Dis., 3:540, 1961)

and carrier-A (A, afferent). The reason both molecules are designated "carrier" is that both seem to serve a transport function—the former combining with Na⁺ and certain other substances to move them toward the cell membrane, and the latter combining with K^+, H^+, NH_4^+, and so on to move them into the cell. Thus the CARRIER molecules would be continuously produced in the cytoplasm (carrier → CARRIER) and broken down at the peritubular surface (CARRIER → carrier). By having the enzymes which catalyze Na⁺ release and CARRIER breakdown (the efferent process) at the peritubular surface, we are able to move large quantities of Na⁺ through the cell without increasing the intracellular Na⁺ concentration.

It is quite apparent in Figure 20-12 that the Na⁺ extrusion mechanism affects much more than the movement of Na⁺ into the peritubular fluid. The pump, by lowering the intracellular Na⁺ concentration, facilitates the diffusion (interrupted lines) of Na⁺ into the cell from the lumen of the nephron. It also creates an electrical concentration gradient that favors the diffusion of Cl^-, and an osmotic gradient that favors the diffusion of water, through the cell and into the peritubular space. In portions of the medulla of the kidney, the tubules, ducts, and capillaries allow so little water to pass that the Na⁺ pump causes the peritubular fluid to become markedly hypertonic to the blood and urine (Fig. 20-17). In the cortex, on the other hand, the Na⁺ pump results in large quantities of water passing from the nephron into the peritubular space.

Some evidence is also accumulating that substances previously thought to have their own active transport systems are dependent on the Na⁺ system for their reabsorption. It may be, for example, that a substance we shall call X combines with the sodium carrier complex and is released from this complex when the Na⁺ is released. This piggyback arrangement can be diagramed as follows:

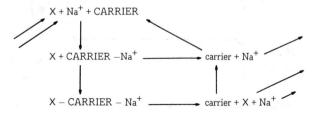

The movement of Na⁺ into the peritubular fluid requires the expenditure of energy. Possibly it is the reformation of a carrier molecule that has an affinity for Na⁺ that uses this energy:

$$\text{carrier} \xrightarrow{E} \text{CARRIER}$$

In other words, as the amount of Na⁺ moved into the peritubular fluid increases, the energy expenditure of the cell increases. This relationship is not true for our hypothesized substance X. Its presence or absence does not influence the energy expenditure of the cell. In short, the Na⁺ is actively transported, while substance X, though dependent upon an active transport system, is characterized as being passively moved into the peritubular fluid.

Also of interest in Figure 20-12 is the observation that the outward transport of Na⁺ may be *coupled with* the inward transport of some other cation. In the proximal tubule the cation is usually K^+, (carrier + K^+ → carrier-K^+), but in other areas it may be H^+ or NH_4^+.

The transport of Na⁺ out of, and K⁺ into, a cell can occur either in conjunction or as separate mechanisms. I do not mean to imply by this statement, however, that all transport systems work independently of one another. We have good evidence that (1) glucose, fructose, galactose, and xylose are all transported by a single system. We also find that (2) sulfate and thiosulfate and (3) arginine, lysine, ornithine, and cystine have a separate transport system. Part of the evidence that system (1), for example, is a single mechanism is that one can markedly decrease the reabsorption of xylose by saturating the system with glucose. One can also decrease the reabsorption of all four of these carbohydrates without decreasing the reabsorption of substances in system (2) or (3) by the intravenous injection of the blocking agent *phlorizin*. Phlorizin probably binds the membrane carrier for system (1).

Glucose and other carbohydrates. Glucose, like fructose, galactose, and xylose, is actively reabsorbed from the proximal tubule (Fig. 20-13) by a transport system which is blocked by phlorizin and whose greatest affinity is for glucose. Under most circumstances this system enables complete reabsorption of all the glucose in the glomerular filtrate, i.e., glucose clearance is usually zero. The plasma glucose concentration is usually about 100 mg. glucose per 100 ml. plasma. It may, however, rise above this level soon after a meal. If it goes above 200 mg. glucose per 100 ml. plasma and glomerular filtration remains above 100 ml. per minute, some

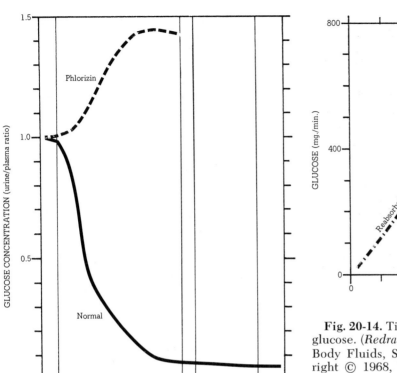

Fig. 20-13. Reabsorption of glucose in the normal and phlorizinized frog. Note that phlorizin prevents some of the reabsorption of glucose. In other words, it prevents the decrease in the ratio of $U_{glucose}$ to $P_{glucose}$. (*Modified from* Walker, A. M., and Hudson, C. L.: The reabsorption of glucose from the renal tubule in amphibia and the action of phlorizin upon it. Am. J. Physiol., *118*:133, 137, 1937)

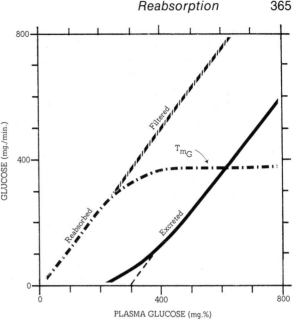

Fig. 20-14. Titration of the renal tubules of man with glucose. (*Redrawn from* Physiology of the Kidney and Body Fluids, Second Edition, by R. F. Pitts. Copyright © 1968, Year Book Medical Publishers, Inc.; used by permission)

glucose is usually found in the urine. The lowest plasma glucose concentration at which glucose spills over into the urine is called the *renal plasma threshold* for glucose. As the plasma glucose concentration increases above the renal plasma threshold, progressively more glucose is excreted in the urine and progressively more is reabsorbed (Fig. 20-14). Eventually, however, a plasma glucose concentration is reached at which the reabsorption of glucose is maximal. A constant reabsorption has been demonstrated over a range of 400 to 2,200 mg. of glucose per 100 ml. of plasma. The maximum quantity of glucose that can be reabsorbed in 1 min. is called the Tm_G, the *tubular maximum* reabsorptive capacity for glucose.

Note that in the case of glucose the renal plasma threshold (200 mg. glucose/100 ml. plasma) is lower than the minimum plasma glucose level at which there is maximum reabsorption (425 mg. glucose/100 ml. plasma). This may be due in part to differences in efficiency among the tubules, and in part to the kinetics of the carrier system itself. It is important to realize that at the renal plasma threshold for most substances, reabsorption is not yet maximum. In some cases of renal glucosuria both the Tm_G and threshold are reduced. In others the Tm_G is normal and the threshold is reduced.

The curve in Figure 20-14 is a "glucose titration of the renal tubules" and was performed as follows. The plasma glucose level was progressively increased while the plasma glucose concentration (P_G), the urine glucose concentration (U_G), the urine flow (V), and the inulin clearance (C_{In}) (glomerular filtration) were simultaneously determined.

Glucose filtered (mg./min.) =
  C_{In} (ml./min.) × P_G (mg./ml.)
Glucose excreted (mg./min.) =
  V (ml./min.) × U_G (mg./ml.)
Glucose reabsorbed =
  glucose filtered − glucose excreted

From these data the curves were calculated. The rounding of the titration curve is called its *splay*. Note that if the reabsorption curve were to break sharply, the renal plasma threshold would be 300 mg. glucose per 100 ml. plasma rather than the 200 mg. shown. Procedures which increase the difference between the level at which the T_m is reached and that at which the renal plasma threshold is reached are said to increase the splay.

Amino acids. The amino acids, like glucose and other carbohydrates, are usually reabsorbed by transport systems of limited capacity. The plasma threshold and Tm for a number of amino acids are summarized in Table 20-3. In those cases in which the symbol > is used, the level of toxicity is lower than the presumed threshold and Tm, so these values have not been determined in man. As in the case of glucose, the threshold for all amino acids is higher than is usually found in the body.

There are at least three different renal transport systems for the amino acids. These are (1) the lysine, arginine, ornithine, cystine, and possibly histidine system, (2) the glutamine and aspartic acid system, and (3) one or more systems which transport the rest of the amino acids. It has been suggested, for example, that glycine, alanine, and creatine are reabsorbed by means of a single transport system. The problem in identifying the amino acid transport systems is that they seem to interact, which often leads to confusion. The infusion of large quantities of lysine, for example, markedly depresses the reabsorption of other amino acids transported by the same mechanism, and mildly depresses the absorption of amino acids transported by other mechanisms.

A study of amino acid reabsorption is presented in Figure 20-15. You will note that in this case (lysine), the renal plasma threshold (0.09 mM./min.) occurs almost at the same plasma lysine concentration at which renal reabsorption is maximal (0.092 mM./min.). In other words, this pattern has very little splay. Quite a different situation, however, exists for some of the other amino acids. Glycine exhibits a marked splay and has a Tm fifteen times that for lysine.

Proteins. In man, approximately 0.5 per cent of the albumin is filtered into the nephron. This results in a concentration in the filtrate of 20 mg. albumin per 100 ml. filtrate. If all of this albumin were lost in the urine we would lose about 32 gm. a day. In point of fact, less than 100 mg. are lost each day, due to the presence in the proximal tubule of an albumin transport system with a Tm of 30 mg. per minute. The renal plasma threshold is 6.5 gm. albumin per 100 ml. plasma—considerably higher than the usual plasma level.

Hemoglobin is another protein small enough to be filtered (Table 20-2). Normally, however, the α_2-globulin in the plasma combines with up to 128 mg. of hemoglobin per 100 ml. of plasma. Hence it is not until hemoglobin exceeds this

Substance	Level in Plasma	Renal Plasma Threshold	Tm
Glucose	100 mg./100 ml.	200 mg./100 ml.	375 mg./min.
Lactate	10 mg./100 ml.	60 mg./100 ml.	?
Malate	0.8 mg./100 ml.		5 mg./min. (dog)
β-Hydroxybutyrate		20 mg./100 ml.	
Amino acids	3 mM./liter		
Lysine		0.09 mM./min.	0.092 mM./min.
Arginine		0.09 mM./min.	0.092 mM./min.
Histidine		>0.9 mM./min.	>0.9 mM./min.
Glycine		0.6 mM./min.	1.5 mM./min.
Tyrosine		>0.9 mg./100 ml.	>0.9 mg./100 ml.
Albumin	4 gm./100 ml.	6.5 gm./100 ml.	30 mg./min.
Hemoglobin		150 mg./100 ml.	1 mg./min.
Vitamin C	1 mg./100 ml.		1.8 mg./min.
Phosphate	1 mM./liter		0.13 mM./min.
Sulfate	1.2 mM./liter	0.09 mM./min.	0.1 mM./min.
Uric acid	4 mg./100 ml.		15 mg./min.

Table 20-3. Some of the substances actively reabsorbed from the nephron.

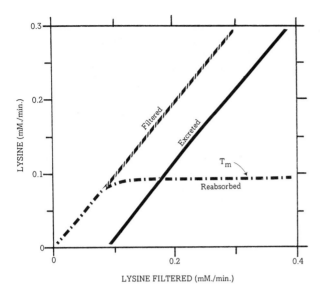

Fig. 20-15. Effect of changes in the quantity of lysine filtered on its reabsorption and excretion. (*Redrawn from* Physiology of the Kidney and Body Fluids, Second Edition, by R. F. Pitts. Copyright © 1968, Year Book Medical Publishers, Inc.; used by permission)

concentration that it is seen in the filtrate. About 1 mg. of hemoglobin per minute can be transported back to the blood.

Uric acid. In man, uric acid results from the breakdown of nucleoproteins. The uric acid in the urine contains about 5 per cent of the urinary nitrogen. In birds and certain reptiles, however, it is the major nitrogenous product found in the urine. In these animals the urine passes to a cloaca and is then moved into the intestine where additional water is reabsorbed, leaving a paste, rather than a solution, to be eliminated. In man it is important that uric acid be kept in solution while in the urine. When it precipitates in the kidney or ureter it can form urate calculi which are painful and can cause damage. Another matter of concern is the deposition of urate crystals in the joints in *arthritis*. This is usually associated with an elevation in the plasma urate concentration. Reasons for such a high plasma concentration might include (1) increased production or (2) decreased filtration in the kidney. Since uric acid is actively reabsorbed from the proximal tubule and secreted into the distal tubule, two other possibilities might be (3) increased reabsorption and (4) decreased secretion. A certain degree of success in the treatment of chronic gout has been achieved with the drug probenecid, which blocks the reabsorption of uric acid and in this way increases its excretion.

Citrate. Citrate, discussed earlier under the anticoagulants (Chap. 14), is an organic anion which is also a normal constituent of the urine. Its excretion in the urine is increased in alkalosis and decreased in acidosis. Through its property or binding Ca^{++}, it helps prevent the formation of calcium phosphate stones in the ureter.

Electrolytes. Up to this point we have characterized the reabsorption of nutrients (glucose, lactate, amino acids, and proteins) as being almost complete at plasma levels considerably higher than those usually noted in the body. It would appear that organs other than the kidneys (liver, muscle, etc.) play the predominant role in controlling the concentration of these nutrients in the blood. In the case of certain waste products (urea, uric acid, and creatinine) the situation in man is different. They are incompletely reabsorbed from the ultrafiltrate of the glomeruli; as a result, their plasma level is controlled primarily by renal function. Most of the electrolytes (chloride, bicarbonate, phosphate, sulfate, sodium, postassium, calcium, and magnesium) are also, under most conditions, only partially reabsorbed. The control of their reabsorption, however, is much more complex. The excretion of these electrolytes may be modified not only by changes in plasma concentration, but also by changes in plasma pH and the output of hormones from the endocrine glands. In man, for example, little bicarbonate is excreted when the pH of the urine is 5.2; 1.5 mEq. per liter of urine is excreted at pH 6.0; 6 mEq. per liter at 7.0; and 36 mEq. per liter at 7.4. The hormones act by changing (1) the output of certain electrolytes from cells and intercellular reservoirs, (2) the filtration rate of the kidney, (3) the Tm (tubular maximum rate of reabsorption), (4) the renal plasma threshold, or (5) a combination of these.

Phosphate. Phosphate is found in the plasma as HPO_4^{--} (approximately 80% of the plasma phosphate) and $H_2PO_4^-$ (20%), in bone as $Ca_3(PO_4)_2$, in the cell as organic phosphate, and in the urine as HPO_4^{--} and $H_2PO_4^-$. In bone it is part of an important mineral reserve, in the cell it is an essential part of energy transformations, and in the urine it is the major buffer system. Usually between 7 and 20 μM. per minute are excreted (Fig. 20-16), the daily loss equaling approximately the oral intake. This excretion is, however, quite variable. During acidosis the

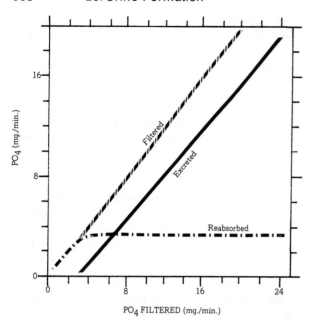

Fig. 20-16. The effect of the rate of phosphate filtration on phosphate reabsorption and excretion in the dog. (*Redrawn from* Pitts, R. F., and Alexander, R. S.: The renal reabsorptive mechanism for inorganic phosphate in normal and acidotic dogs. Am J. Physiol., *142*:651, 1944)

Tm_{PO_4} remains fairly constant but the splay of the titration curve increases, as does the phosphate mobilization from soft tissue and bone. Parathormone, a hormone from the parathyroid glands, on the other hand, increases phosphate excretion in the urine by elevating the renal filtration rate and decreasing the Tm_{PO_4}. Excess cortisone, due for example to hyperadrenalcorticism, also decreases the Tm_{PO_4}.

It is of interest that large quantities of glucose, alanine, and acetoacetate decrease phosphate reabsorption. Although the phosphate transport mechanism is for the most part separate from the transport mechanisms of these substances, it may be that one or more steps are shared with the other systems. Some have suggested that the kidney cell does not have enough energy available for the maximum reabsorption of phosphate and glucose at the same time. Although saturation of the renal transport system with glucose decreases the amount of phosphate that can be reabsorbed, it is important to realize for the sake of our thesis that phlorizin block prevents the active reabsorption of glucose but increases the capacity for phosphate reabsorption.

Sodium. Sodium is the major extracellular cation. It is through the delicate control of its reabsorption that the kidneys are able to regulate not only the quantity of Na^+ in the urine but also that of water and Cl^-. Most of the active reabsorption of substances other than Na^+ takes place in one part of the nephron. The active reabsorption of the carbohydrates, amino acids, proteins, uric acid, phosphate and sulfate occurs in the proximal tubule; that of Cl^- occurs in the distal tubule. The active reabsorption of Na^+, on the other hand, occurs in the proximal tubule, the ascending limb of Henle, the distal tubule, and the collecting duct (Fig. 20-17). In the collecting duct the reabsorption of Na^+ is frequently associated with the secretion of K^+, H^+, or NH_4^+ into the lumen of the duct. Since the proximal tubule and the descending limb of Henle are freely permeable to water, the active reabsorption of substances from the lumen of the proximal tubule to the peritubular fluid results in the greatest efflux of water found anywhere in the

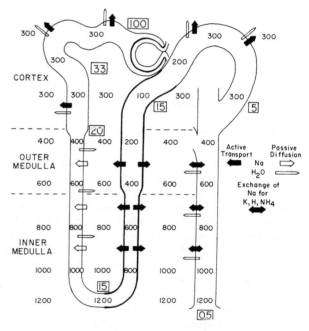

Fig. 20-17. Production of hypertonic urine by the juxtamedullary nephron. The boxed numbers represent estimates of volume of glomerular filtrate left in the nephron. Other numbers represent the concentration of substances in the peritubular fluid and tubular urine expressed in mOsm./liter. (*From* Physiology of the Kidney and Body Fluids, Second Edition, by R. F. Pitts. Copyright © 1968, Year Book Medical Publishers, Inc.; used by permission)

nephron. It is from these structures that about 80 per cent of the glomerular filtrate is reabsorbed.

The ascending limb, on the other hand, is relatively impermeable to water and, unlike the proximal tubule and descending limb, has an osmolar concentration below that of the peritubular fluid which surrounds it. In the ascending tubule the active reabsorption of Na$^+$ from the nephron results in a progressive decrease in the osmotic pressure of the urine. It is also of great importance to note that the transport of Na$^+$ into the peritubular liquid results in the development of a hypertonic environment in those parts of the nephron extending into the renal medulla (Fig. 20-18). Apparently the peritubular capillaries serving this area do not remove Na$^+$ and Cl$^-$ rapidly enough to maintain an isotonic environment. You will note in Figure 20-17 that the hypertonicity of the medullary peritubular liquid results in an increase in tonicity in the descending limb in this area. In other words, the active Na$^+$ transport system expels Na$^+$ into the peritubular space and this Na$^+$ diffuses into the descending limbs in the medulla. The further the tubule extends into the renal medulla, the greater is its tonicity at the loop of Henle. The greater the Na$^+$ concentration at the loop of Henle, the greater the amount of Na$^+$ in the ascending limb available for active transport and the potential of the active transport system to increase the osmolar concentration of the medulla. This system of producing marked increases in the tonicity of the peritubular liquid is called *countercurrent multiplication*.

In summary, countercurrent multiplication consists of the following:

1. Active transport of Na$^+$ through the relatively impermeable membrane of the ascending limb
2. Diffusion of this Na$^+$ through the peritubular liquid and back into the descending limb (*countercurrent*)
3. Flow of the now-hypernatremic (i.e., having a higher concentration of Na$^+$ than is found in plasma) liquid of the descending limb back to the ascending limb
4. Further increase in the tonicity of the peritubular fluid by the active transport of Na$^+$ from the now-hypernatremic ascending limb (*multiplication*)

As the liquid travels toward the distal convoluted tubule and away from the renal pelvis, active Na$^+$ transport continues until a hypotonic urine is produced. In the distal convoluted tubule and the collecting duct we come to an area in which the permeability to water is controlled by a hormone from the posterior pituitary, the antidiuretic hormone. You will note in Figure 20-17 that the collecting duct dips back into the renal medulla in its passage to the renal pelvis. This is the area that the long nephrons (*juxtamedullary nephrons*) have made markedly hypertonic through their countercurrent multiplication system. If in this area we increase the per-

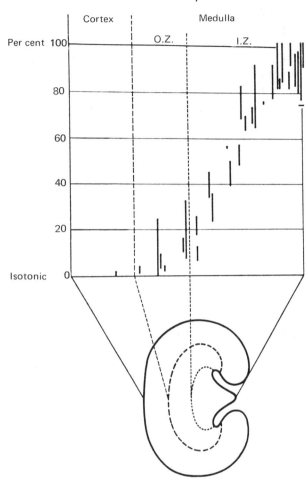

Fig. 20-18. Osmolarity of tissue slices from the kidney of the hamster. Note that slices from the cortex are isotonic with plasma (0%), those from the outer zone (O.Z.) of the medulla are slightly hypertonic, and those from the inner zone (I.Z.) are maximally hypertonic (100%). (Wirz, H., et al.: Lokalization des Konzentriertungsprozessen in der Niere durch directe Kryoskopie. Helv. Physiol. Pharmacol. Acta., 9:200, 1951)

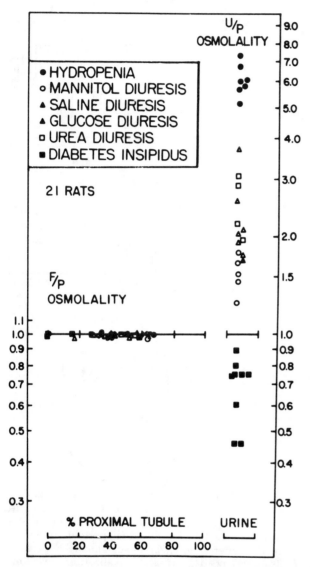

Fig. 20-19. Effect of various conditions on the relative osmolarity of the urine of the rat. The data on the ordinate is expressed in terms of a ratio between the osmolarity of urine and plasma. Note that the ratio in the proximal tubule is 1 and that this changes by the time the filtrate reaches the ureters (urine). (Gottschalk, C. W.: Micropuncture studies of tubular function in the mammalian kidney. Physiologist, 3:37, 1961)

we keep the collecting duct impermeable to water and maintain active transport of Na^+ to the peritubular space, we produce a hypotonic urine. Thus by the control of Na^+ transport and of the permeability of both the distal convoluted tubule and the collecting duct we are able to produce a urine with an osmolarity one tenth that of plasma or four times that of plasma. The former has a specific gravity of 1.001, and the latter, one of 1.040.

It is apparently the ADH of the posterior pituitary which is most concerned with the control of permeability in the distal tubule and collecting duct. Under normal circumstances its release results in the production of between 1 and 2 liters of urine a day. If no ADH is released (diabetes insipidus) there is a decrease in the permeability of these distal structures, less water is reabsorbed, and a urine production as high as 30 liters a day may result.

This is not to say, however, that all polyuria (elevated urine production) is due to a decrease in circulating ADH. Another cause of polyuria is an increase in the quantity of osmotically ac-

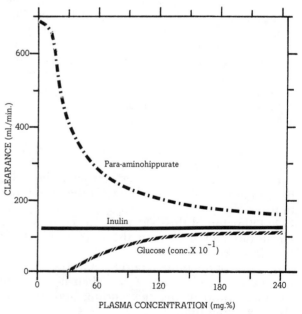

Fig. 20-20. The effect of plasma concentration on the clearance of para-aminohippurate (filtered and secreted), inulin (filtered), and glucose (filtered and reabsorbed) in man. (*Redrawn from* Physiology of the Kidney and Body Fluids, Second Edition, by R. F. Pitts. Copyright © 1968, Year Book Medical Publishers, Inc.; used by permission)

meability of the collecting ducts (see hydropenia in Fig. 20-19) by stimulating the release of the antidiuretic hormone (ADH), water diffuses into the hypertonic peritubular spaces and its capillaries and a hypertonic urine is produced. This is shown in Figure 20-17. If, on the other hand,

tive particles in the glomerular filtrate. We can, for example, increase the volume of urine produced by injecting sucrose, mannitol, sorbitol, urea, NaCl, or large quantities of glucose into a vein. When these particles appear in the glomerular filtrate they prevent the reabsorption of water from the filtrate. If they remain only in the filtrate of the proximal tubule, they result in the delivery of a greater water load to the distal tubule and the collecting duct. If they remain in the filtrate throughout the nephron and collecting duct, they cause not only the delivery of a greater water load to the distal system, but also a greater osmotic load. In either case polyuria may result.

SECRETION

Tubular secretion, like tubular reabsorption, involves the movement of a substance through or between the cells which surround the lumen of the nephron. Reabsorption is a movement from the ultrafiltrate in the lumen to the peritubular fluid. Secretion is in the opposite direction. A substance which is filtered into the nephron at the glomeruli and later reabsorbed (Fig. 20-20 and Table 20-4) has a clearance at low plasma concentrations less than that for inulin. One that is filtered and secreted but not reabsorbed has a clearance, at low plasma concentrations, greater than that for inulin. At high plasma concentrations of certain substances the Tm (maximum reabsorptive or secretory capacity of the tubules) is frequently reached and, as the concentration of the substances increases further, their filtration so exceeds their reabsorption or secretion that they begin to demonstrate a clearance similar to that of inulin. In other words, within certain limits reabsorption substracts measurable quantities of material from the glomerular filtrate, whereas secretion adds significant quantities to it.

Creatinine, Na^+, K^+, H^+, weak acids, and weak bases are filtered, reabsorbed, and secreted in man. Since the K^+ usually has a clearance less than that for inulin, we conclude that more is usually reabsorbed than secreted. We do find, however, that when the plasma concentration of K^+ increases (hyperkalemia), K^+ clearance exceeds inulin clearance. In man creatinine clearance is greater than inulin clearance. In other words, as in the case of hyperkalemia, secretion exceeds reabsorption. In dogs and many other animals, however, creatinine clearance, like inulin clearance, is used to measure the glomerular filtration rate, since in these animals there is neither reabsorption nor secretion of creatinine.

Stop-Flow Analysis

We have till now emphasized two methods for studying the action of the kidney on its

	Clearance (ml./min.)	Plasma Concentration (mg./ml.)
Filtered only		
Inulin	125	0->20
Filtered and reabsorbed		
Glucose	0	0-0.2
Urea	57-75	0.02->2
Urate°	9	0.05
Filtered, reabsorbed, and secreted		
Creatinine	175	0.01
Filtered and secreted		
Diodrast	700	<0.07
Para-aminohippurate	246	0.6
	660	0.01-0.06
Phenol red	400	<0.01

° Possibly secreted also.

Table 20-4. Clearance of various substances in men with a glomerular filtration rate of 125 ml. per minute and a surface area of 1.73 m.²

filtrate: (1) micropuncture and (2) clearance. A third method, developed by Malvin, Sullivan, and Wilde, is called stop-flow analysis. In this technique (1) a ureter of the subject (a dog in Fig. 20-21) is catheterized and (2) substances to be studied are infused in such a way as to reach and maintain the appropriate plasma concentrations. In the analysis shown in Figure 20-21 the substances infused were creatinine (Cr, filtered but not reabsorbed or secreted in the dog), para-aminohippurate (PAH, filtered and secreted), and glycine (AN, or amino-nitrogen, filtered and reabsorbed). In these particular experiments the creatinine and PAH served for reference—glycine being the substance under study.

Next, (3) hypertonic mannitol is infused into the blood to increase the urine flow to 10 ml. per minute and (4) the catheterized ureter is clamped. During the period of clamping the pressure in the tubules and ducts of the kidney rises to 90 mm. Hg and filtration into the nephron almost ceases. After 3 to 8 min. (5) the clamp is released and (6) a marker (inulin or sodium ferrocyanide) is injected into a peripheral vein. Then (7) about thirty 1-ml. samples are collected from the catheter. In the study shown in Figure 20-21, three 3-min. samples were collected before occlusion of the catheter (at the right of each graph). The first sample collected represents fluid from the renal pelvis. Note that by the time 20 per cent of the colume in the kidney had been sampled, NH_4^+ had reached maximal concentration. Later in the collection, the concentration of PAH relative to that of creatinine increased (secretion of PAH), while the relative concentration of amino-nitrogen decreased (reabsorption). The increases in the concentration of creatinine represent a loss of water from the nephron rather than a change in the quantity of creatinine present. The appearance of NH_4^+ in the early samples is taken as evidence of its secretion in the distal portions of the nephron and the collecting duct. The decrease in the concentration of amino-nitrogen in the later samples at a point at which PAH is increasing is used as evidence of the reabsorption of glycine in the proximal tubule. One should remember, however, that the tubules vary considerably in length. Thus samples collected at the 60 per cent point, for example, may represent urine from the proximal tubules of some nephrons and the distal tubules of others.

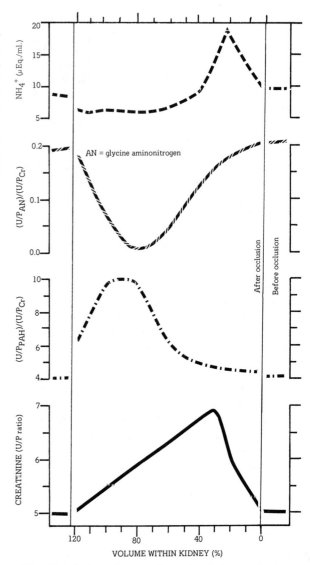

Fig. 20-21. Stop-flow analysis for the site of glycine (AN, aminonitrogen reabsorption) reabsorption in the dog. The first samples collected are on the right and the last on the left. Three samples (*right*) were collected before occlusion of the ureteral catheter and 3 samples (*left*) after normal glomerular filtration had resumed. The ratio for urinary and plasma creatinine (U/P_{Cr}), PAH (U/P_{PAH}), and aminonitrogen (U/P_{AN}) were used in the analysis. These data are consistent with the view that NH_4^+ is secreted into the distal tubule and collecting duct, PAH is secreted into the proximal tubule, and glycine is reabsorbed in the proximal tubule. The increase in U/P_{Cr} is due to a loss of water from the nephron during the stop-flow procedure. (*Modified from* Brown, J. L., et al.: Localization of amino-nitrogen reabsorption in the nephron of the dog. Am. J. Physiol., 200:371, 1961)

Passive Secretion

There are three general categories of secretion—(1) passive, (2) Tm-limited active, and (3) gradient-limited active (Table 20-5). Active secretion, like active reabsorption, is an energy-requiring movement of a substance against an electrochemical concentration gradient. Passive secretion includes the movement into the nephron or collecting duct of all substances which are not directly dependent upon energy-requiring reactions.

Weak acids and bases, for example, probably move into the urine of the collecting duct by a passive process called *diffusion trapping*. Since the plasma in the peritubular blood has a pH of about 7.4, the *weak bases* in the plasma exist in both the cationic and free base form. The free base is lipid-soluble and readily diffuses into the urine. If the urine is acid, most of the free base that diffuses combines with H^+ to form cations which are not readily diffusible. The acidity of the urine, then, tends to maintain a concentration gradient more conducive to diffusion into than out of the urine by free bases. This phenomenon, diffusion trapping, is directly related to the acidity of the urine. When the pH of the urine is 8, the clearance of weak bases is less than that for inulin, i.e., reabsorption exceeds secretion. When, on the other hand, the pH of the urine is 5, the clearance of weak bases exceeds that for inulin. Since the pH of plasma stays at about 7.4 and the pH of urine can go as low as 4.4, it is possible to have a ratio for the H^+ concentration in the urine and that in the plasma of 1,000:1 ($10^{-4.4}/10^{-7.4} = 10^3 = 1000$). This would seem to be a rather favorable condition for diffusion trapping.

Diffusion trapping of *weak acids* is similar to that of weak bases. An un-ionized weak acid diffuses into an alkaline urine and forms anions which do not readily move through the charged cell membranes and which become trapped in the urine. Since the urine does not normally reach a pH higher than 8.2, the best H^+ gradient for the secretion of weak acids is less than 10:1 ($10^{-8.4}/10^{-7.4} = 10^{-1}$). This gradient can, however, result in the extraction of substantial amounts of weak acid if the urine flow is high.

Tm-limited Active Secretion

In the section on tubular reabsorption we defined Tm as the maximum rate at which a substance can be secreted or reabsorbed by the kidneys. In the case of glucose we noted that in a healthy man with a surface area of 1.73 m.² up to 375 mg. per minute are reabsorbed (Table 20-3), or, that $Tm_{glucose} = 375$. Phenol red and PAH, on the other hand, are actively secreted. $Tm_{phenol\ red}$ is 36 mg. per minute per 1.73 m.², and Tm_{PAH} is 80 (Fig. 20-22).

At present we know of only two Tm-limited active transport mechanisms. One (see mechanism 1 in Table 20-5) transports a number of substances, many of which are organic acids, and the other transports a group of strong organic bases. All of these substances are both filtered

---

Passive Secretion (distal tubule and collecting duct)
  Weak acids: phenobarbital and salicylic acid
  Weak bases: ammonia (?), chloroquine, quinine, quinacrine, neutral red, and procaine

Active Secretion
  Tm-limited (proximal tubule)
    Mechanism 1: acetylated sulfonamides, carboxylic acid, chlorothiazide, creatinine, iodopyracet (Diadrast), various glucuronides, p-aminohippurate, penicillin, phenol red, sulfonic acids, sulfate acid esters, iodopyridone acetate, iodomethamate
    Mechanism 2: choline, guanidine, hexamethonium, histamine, piperidine, tolazoline (Priscoline), tetraethylammonium, and thiamine
  Gradient-limited (distal tubule and collecting duct)
    K^+ and H^+

Table 20-5. List of substances secreted into the nephron and their sites of secretion.

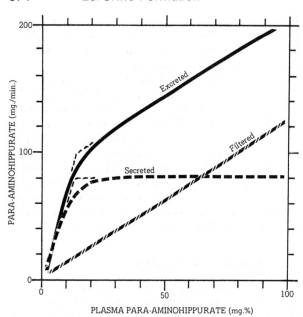

Fig. 20-22. Filtration, secretion, and excretion of para-aminohippurate in man. The abscissa is in terms of per cent of free PAH. This represents from 80 to 90 per cent of the total PAH. The rest is bound to the plasma proteins. (*Redrawn from* Physiology of the Kidney and Body Fluids, Second Edition, by R. F. Pitts. Copyright © 1968, Year Book Medical Publishers, Inc.; used by permission)

into the nephron and secreted. Both mechanisms are localized to the proximal tubule and are blocked by procedures which interfere with either oxidative processes or the coupling of oxidation with phosphorylation. Cold, an anaerobic environment, and small concentrations of arsenite, azide, cyanide, and *dinitrophenol* (DNP) all block the secretion of para-aminohippurate (PAH), phenol red, etc. The administration of small amounts of DNP to dogs, for example, decreases their capacity to secrete PAH (decreased Tm_{PAH}) by more than one third without affecting the tubular reabsorption of either amino acids or glucose. Since DNP uncouples oxidation and phosphorylation it is concluded that active secretion is much more dependent upon high concentrations of the high energy substances—ATP and creatine phosphate—than is the active reabsorption of nutrients.

Competition for tubular transport. A similarity between the active reabsorptive and secretory mechanisms is that the various substances reabsorbed or secreted by a given mechanism compete for a site on the carrier molecule. If, for example, we saturate mechanism 1 in Table 20-5 with phenol red and then infuse Diodrast or PAH, the excretion of phenol red decreases. In a similar manner, saturation of the system with PAH or Diodrast followed by the infusion of phenol red decreases the excretion of the PAH or Diodrast. The results of the first experiment are more spectacular, however, since the transport system has a greater affinity for PAH and Diodrast than for phenol red.

The concept of competitive inhibition of secretion was put to very good use by Karl Beyer and his associates toward the end of World War II. At this time penicillin was being produced by industry in quantities insufficient to meet the demand for it. Beyer noted that PAH interfered with the excretion of penicillin by the kidney and suggested that it be administered to prolong the action of penicillin in the body. Eventually, on the basis of these observations, two compounds were synthesized (carinamide and probenecid) which were given to patients receiving penicillin. After the war penicillin became readily available and inexpensive and the use of PAH, carinamide, and probenecid to block penicillin extraction was discontinued.

Extraction. In man the normal kidney is capable of extracting up to 17 per cent of the inulin, 70 per cent of the phenol red, 90 per cent of the Diodrast, and 90 per cent of the PAH that passes through it. It does this by filtration and secretion. Reabsorption, on the other hand, serves to return particles to the plasma. The capacity of filtration to remove molecules from the plasma is demonstrated by the study of a substance which is filtered but not secreted or reabsorbed (i.e., inulin). We have noted that the kidneys can extract in one minute the inulin contained in 125 ml. of plasma. They can also extract in one minute the PAH contained in 660 ml. of plasma, provided the concentration is kept below 6 mg. of PAH per 100 ml. of plasma. The reason for this higher clearance is that PAH is not only filtered but also secreted.

If we increased the blood flow through the kidneys and kept the plasma PAH concentration constant, the extraction of PAH would stay at about 90 per cent. Its clearance would increase, in other words. Apparently at low plasma concentrations the proximal tubules secrete all the PAH that comes to them. The 10 per cent which is not extracted is in the renal blood that goes to areas other than the proximal tubule.

Effective renal plasma flow (PAH clearance). Since PAH clearance at low PAH concentrations is related to blood flow to the proximal tubule, its measurement has been used as an index (effective renal plasma flow) of total renal flow. Later in the chapter we will find that it is not a good index, however, in cases in which there is tubular damage. Under these circumstances the ratio of C_{PAH} to Tm_{PAH} is a better indication of changes in flow. The technique of using C_{PAH} as an indication of flow is similar to the technique developed by *Fick* for the measurement of cardiac output. Fick noted that he could calculate the blood flow through the lungs if he knew the quantity of O_2 removed from the lungs by the blood (O_2 consumption in ml./min.) and the concentration of O_2 in the blood (ml. of O_2/ml. of blood) leading to and from the lungs. We now believe that this calculated pulmonary blood flow is equal to the cardiac output (C.O.):

$$C.O. = \frac{O_2 \text{ consumption (ml. } O_2/\text{min.)}}{\text{arteriovenous } O_2 \text{ difference (ml. } O_2/\text{ml. blood)}}$$

A similar calculation can be performed in the case of renal plasma flow (RPF):

$$RPF = \frac{\text{PAH excretion (mg. PAH/min.)}}{\text{arteriovenous PAH difference (mg. PAH/ml. plasma)}}$$

In order to determine the renal blood flow (RBF) only the renal plasma flow and the hematocrit need be known:

$$RBF = RPF \times \frac{1}{1 - \text{hematocrit}}$$

The determination of renal plasma flow is made particularly easy if we assume that the kidney removes all the free PAH passing through it. And this assumption would be valid if the plasma concentration of free PAH were kept below 6 mg. per 100 ml. of plasma and if all the blood going to the kidney passed through functional peritubular capillaries going to the proximal tubules. In the normal person, however, only about 90 per cent of the renal blood passes via this route, and in a person with chronic renal disease there is destruction of the tubules that may result in considerably less extraction than the usual 90 per cent.

When one, in using the formula for renal plasma flow given above, assumes a PAH concentration of zero in the renal vein he is said to be measuring effective renal plasma flow (ERPF). This determination requires that one obtain only urine samples (PAH excretion) and venous blood samples (arteriovenous PAH difference), since the concentration of PAH in the renal artery is the same as that in an arm, a leg, or a neck vein. In short, only the kidneys remove substantial quantities of PAH from the blood. Thus the formula for the effective renal plasma flow (ERPF) is:

$$ERPF = \frac{\text{PAH excretion (mg. PAH/min.)}}{\text{PAH conc. in cephalic vein (mg. PAH/ml. plasma)}}$$

You will note that the formula for ERPF is also the formula for PAH clearance (C_{PAH}):

$$C_{PAH} = \frac{U_{PAH}}{P_{PAH}} \times V = \frac{E_{PAH}}{P_{PAH}} = ERPF$$

Here U_{PAH} is the concentration of PAH in the urine; P_{PAH}, its concentration in the plasma; V, the volume of urine formed, in milliliters per minute; and E_{PAH}, the PAH excretion in milligrams per minute.

In other words, a normal person with a C_{PAH} of 660 ml. per minute and a hematocrit of 45 per cent would have an effective renal blood flow (ERBF) of:

$$ERBF = 660 \times \frac{1}{1 - 0.45} = 1,200 \text{ ml./min.}$$

Since this usually represents only 90 per cent of the total renal flow (RBF) we would expect the following:

$$RBF = \frac{1,200}{0.9} = 1,333 \text{ ml./min.}$$

In a person with a cardiac output of 5,600 ml. per minute a renal blood flow of 1,333 ml. per minute represents about 24 per cent of the total blood pumped from the heart.

Gradient-limited Active Secretion

In Tm-limited active secretion or reabsorption the clearance of substances such as PAH or glucose is fairly constant at low plasma concentrations, but as the plasma concentration increases, the tubular capacity for transport (Tm) is exceeded and secretion and reabsorption become incomplete (Fig. 20-20). In the case of K^+ and

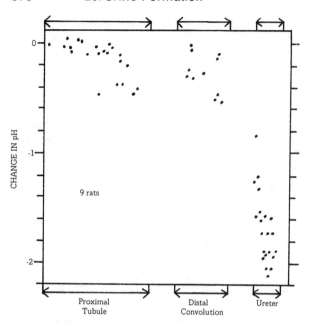

Fig. 20-23. Changes in the pH of the glomerular filtrate as it passes along the nephron and collecting duct of the rat. (*Modified from* Gottschalk, C. W., et al.: Localization of urine acidification in the mammalian kidney. Am. J. Physiol., *198*:582, 1960)

H^+ secretion and Na^+ reabsorption, on the other hand, there is no well-defined Tm. In these cases tubular transport seems to be more affected by changes in the concentrations of K^+, H^+, and Na^+ in the body than the intrinsic characteristics of the transport system. In most persons, for example, K^+ clearance is less than that for inulin. In other words K^+ is filtered and then reabsorbed. In hyperkalemia and alkalosis, however, K^+ clearance exceeds that for inulin.

Apparently in the case of K^+ and H^+ we have a potential secretory mechanism in the collecting duct and possibly the distal tubule. As the filtered K^+ and H^+ pass through the first parts of the nephron their movement is associated with passive phenomena (Fig. 20-23). In the collecting duct, however, there is a transport mechanism responsive to certain hormones concerned with the control of pH and electrolyte balance. It is speculated that this mechanism is similar to that illustrated in Figure 20-24. In this illustration it is emphasized that the cell membrane adjacent to the tubular lumen is associated with a transport system in which the pumping of Na^+ into the cell is coupled with the extrusion of H^+

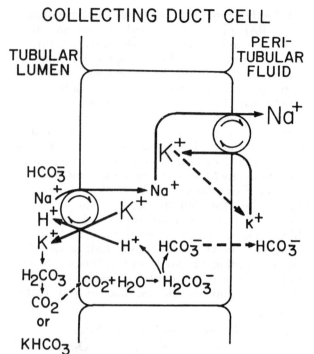

Fig. 20-24. The active transport systems of the collecting duct. Note at the tubular lumen an active transport system (gradient-limited) in which the inward movement of Na^+ is coupled with either the outward movement of H^+ or K^+. The H^+ is produced from H_2CO_3, which is produced from CO_2 and H_2O in the presence of the intracellular enzyme carbonic anhydrase. The K^+ is brought into the cell by an active transport system on the peritubular side of the cell. In this latter system, K^+ infusion is coupled with Na^+ extrusion. (*From* Physiology of the Kidney and Body Fluids by R. F. Pitts. Copyright © 1963, Year Book Medical Publishers, Inc.; used by permission)

or K^+ from it. The extrusion part of the mechanism has its greatest affinity for H^+, but excretes large quantities of K^+ in alkalosis or hyperkalemia. You will also note from the diagram that there is an Na^+ extrusion and K^+ infusion system at the peritubular end of the cell. The H^+ comes from the H_2CO_3 formed in the cell when CO_2 diffuses in. The cells of the collecting duct contain substantial quantities of the enzyme *carbonic anhydrase*. This catalyst is apparently very important in the H^+ secretory mechanism, since H^+ secretion is blocked when an agent (certain sulfonamide derivatives) which inhibits carbonic anhydrase action is administered to an animal.

CLINICAL EVALUATION OF KIDNEY FUNCTION

C_{IN}, Tm_{PAH}, and C_{PAH}

We have been concerned in this chapter with four basic aspects of kidney function: (1) filtration, (2) reabsorption, (3) secretion, and (4) renal plasma flow. Kidney disease may initially involve a disruption of any one or combination of these functions. If the disease is progressive and remains untreated it leads to generalized kidney damage, uremia, and death. Some kidney diseases are associated during the early stages with poor filtration (acute glomerulonephritis and lipoid nephrosis), some with malfunction of reabsorption and secretion (pyelonephritis and tubular necrosis), and some with a decrease in renal plasma flow (nephrosclerosis).

Fortunately, tests are now available to aid the physician in his diagnosis. The *glomerular filtration* rate can be estimated from the inulin clearance (C_{In}) test, the capacity for *reabsorption* and *secretion* by the determination of the tubular maximum for glucose reabsorption (Tm_G) or PAH secretion (Tm_{PAH}), and the effective renal plasma flow by the determination of the PAH clearance (C_{PAH}). By comparing the C_{In}, Tm_G or Tm_{PAH}, and the C_{PAH} of a patient with tables of standards (Table 20-6) one can gain insight into kidney function.

Derived Parameters

Acute glomerulonephritis is primarily a disease of the glomerulus, and pyelonephritis, of the tubules, yet a disease which disrupts one aspect of kidney function frequently leads to disruption of other functions as well. In other words, in both diseases C_{In}, Tm_{PAH}, and C_{PAH} are decreased. To distinguish diseases affecting chiefly the glomeruli from those acting mainly on the tubules, certain derived parameters can be calculated and used.

Filtration fraction (C_{IN}/C_{PAH}). In man 16 to 20 per cent of the effective renal plasma flow passes from the glomerular capillaries into the nephron ($C_{In}/C_{PAH} = 0.16 - 0.20$). In acute glo-

	Normal Males	Normal Females	Acute Glomerulo-nephritis	Pyelonephritis
C_{In} (ml./min.) (filtration)	125	110	decreased	decreased
Tm_G (mg./min.) (reabsorption)	375	303		
Tm_{PAH} (mg./min.) (secretion)	80	77	decreased	decreased
C_{PAH} (ml./min.) (effective blood flow)	660	570	decreased	decreased
Derived Parameters				
C_{In}/C_{PAH} (filtration fraction)	0.19	0.19	decreased	increased
C_{In}/Tm_G (glomerulo-tubular preponderance)	0.33	0.36		
C_{In}/Tm_{PAH} (glomerulo-tubular preponderance)	1.56	1.43	decreased	
C_{PAH}/Tm_G (renal blood flow)	1.75	1.88		
C_{PAH}/Tm_{PAH}	8.19	7.40	increased	decreased

Table 20-6. Renal function tests in human subjects with a surface area of 1.73 m.² These tests are used to assess such processes as filtration, reabsorption, and secretion, and blood flow.

merulonephritis this per cent is markedly decreased. In pyelonephritis, on the other hand, the ability of the proximal tubule to secrete PAH is far more affected than the filtration of inulin. Therefore, since C_{PAH} is reduced more than C_{In}, the filtration fraction is increased. This is an indication not of greater filtration, but that secretion is more extensively reduced than filtration.

Glomerulotubular preponderance (C_{In}/Tm_{PAH}). Another index of the relative damage to the glomeruli and proximal tubules is the ratio between the inulin clearance and the tubules' capacity to secrete PAH. In acute glomerulonephritis the ratio is decreased, indicating that the glomeruli have been more seriously affected than the tubules.

Renal blood flow (C_{PAH}/Tm_{PAH}). The clearance of PAH is directly related to renal blood flow to the proximal tubule, and to the capacity of the secretory mechanism. In acute glomerulonephritis there is an increase in renal flow associated with destruction of the tubular cells. In short, there is a greater decrease in the Tm_{PAH} than in the C_{PAH}, so that the ratio of C_{PAH} to Tm_{PAH} increases. This is used as an indication of increased flow. In pyelonephritis, on the other hand, the ratio decreases markedly as a result of ischemia.

Chapter 21

RENAL REGULATION OF WATER AND ELECTROLYTE BALANCE

RENAL BLOOD FLOW
ELECTROLYTE BALANCE
WATER BALANCE

The kidneys are concerned with (1) the maintenance of water, electrolyte, and acid-base balance, (2) the conservation of nutrients, (3) the elimination of waste products, and (4) the destruction, elimination, or destruction and elimination of a number of other substances. Some of these functions are carried out by virtue of the response of the kidney itself to the blood it receives (autoregulation). Many others occur in response to the stimulation of specialized receptors scattered throughout the body. Most of these receptors modify or initiate kidney activity by stimulating or inhibiting the release of hormones (antidiuretic hormone, aldosterone, parathormone, etc.).

The role of the nerve supply to the kidney, however, is uncertain. Part of the confusion stems from the fact that many anesthetics initiate an adrenergic sympathetic tone to the renal vessels that is not normally present in a resting person. It is believed, however, that during exercise, hypotension, or psychological stress renal sympathetic neurons produce vasoconstriction in the kidney, which serves to shunt part of the renal blood flow to other parts of the body. This vasoconstriction is usually not sufficient to interfere with urine production (Fig. 21-1), except during extreme exercise, hypotension, or stress.

It has also been shown that stimulation of the peripheral end of the renal nerve causes the increased release of renin by the kidney.

RENAL BLOOD FLOW

In a resting person, approximately 90 per cent of the blood supply to the kidneys goes to its cortex, where it (1) is filtered in the glomeruli and (2) modified in the peritubular capillaries by the processes of secretion and reabsorption, and (3) serves a nutritive function. The rest of the blood supply to the kidneys does not pass through glomerular capillaries. That to the medulla represents from 8 to 10 per cent of the total flow, and that to the renal pelvis, 1 to 2 per cent. This means that there is a cortical flow of about 4 ml. per gram per minute, a medullary flow of 0.8 ml. per gram per minute, and a pelvic flow of 0.2 ml. per gram per minute.

The total renal blood supply serves two functions. One of these is the maintenance of body homeostasis, through the processes of filtration, reabsorption, and secretion. The other is to sup-

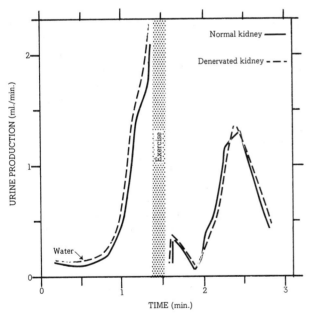

Fig. 21-1. Response of the normal and denervated kidneys of a dog to water given by mouth and later to exercise. Note that the exercise caused a temporary decrease in urine production and that the denervated and normal kidneys responded similarly to both procedures. (*Redrawn from* Physiology of the Kidney and Body Fluids, Second Edition, by R. F. Pitts. Copyright © 1968, Year Book Medical Publishers, Inc.; used by permission)

ply the substances necessary for growth, repair, and metabolic activity, and to remove waste products. You will note in Figure 11-5 (Chapter 11, "Circulatory Control") that in a resting person the blood supply to the kidney is about 730 ml. per minute per 100 gm. of tissue, whereas that to skeletal muscle is only 3 ml., and that to the heart 70. In Figure 12-10 (Chapter 12, "Capillaries and Sinusoids") we find that these flows result in an arteriovenous O_2 difference in the kidneys of 1.7 ml. of O_2 per 100 ml. of blood; in skeletal muscle, of 5.8 ml.; and in the heart, of 15.6 ml. The kidneys, in other words, receive much more O_2 than is necessary to meet their needs.

Oxygen Consumption

A 70-kg. resting man consumes approximately 250 ml. of O_2 per minute. The kidneys, though making up less than 0.5 per cent of the body weight, consume approximately 8 per cent of this, or 20 ml. of O_2 per minute. The actual level of O_2 consumption, however, varies with the circumstances, decreasing when filtration decreases and increasing when filtration increases. The kidney is one of the few organs in the body which increases its O_2 consumption in response to an increase in blood flow and decreases it in response to ischemia. This is due to the high energy cost to the body for reabsorption. It is estimated that the kidneys burn 1 μM. of O_2 for each 25 μM. of Na^+ reabsorbed from the glomerular filtrate. For example, in the dog the normal O_2 consumption of the kidneys is 5 μM. per gram per minute. Cessation of filtration, and hence of reabsorption, brings the O_2 consumption to 1.2 μM. per gram per minute, a reduction of 76 per cent. Homer Smith characterized this arrangement as follows:

There is enough waste motion here to bankrupt any economic system other than a natural one, for nature is the only artificer who does not need to count the cost by which she achieves her ends.

Extrinsic Control

The kidneys are innervated by nerves originating from the fourth thoracic through the second lumbar segments of the spinal cord, the most abundant supply coming from the tenth through twelfth thoracic segments. These nerves are composed of pain fibers from the capsule and adrenergic sympathetic fibers to the blood vessels. There is also some evidence that sympathetic fibers to the kidney facilitate its release of renin. The sympathetic fibers, during periods of hypoxia, hypotension, exercise, and forms of stress, produce a vasoconstriction which decreases the renal blood supply. Epinephrine and norepinephrine have a similar effect.

The blood flow to the kidneys is in part a result of (1) the pressure head in the aorta and renal arteries, and in part of (2) the resistance to flow provided by the renal vessels. You will remember from our discussion in Chapter 10 that:

$$\text{Resistance} = \frac{\text{change in pressure}}{\text{flow}} = \frac{P_1 - P_2}{I}$$

In Figure 20-9 we note that the greatest changes in pressure occur at the afferent arteriole, efferent arteriole, and venule. These three vascular elements in series with each other represent areas of high resistance in the renal vessels. Their constriction increases the resistance to flow, i.e., decreases renal flow, and their dilation increases renal flow.

The importance of the pressure in the renal artery to renal flow and filatration is illustrated in Figure 21-2. You will note that at arterial pressures below 80 mm. Hg both plasma flow and filtration are directly related to renal arterial pressure, but at pressures between 80 and 180 mm. Hg, flow and filtration tend to stabilize. This stabilization is apparently due to an intrinsic system which produces a progressive vasoconstriction of the kidney's resistance vessels in response to rising aortic pressures. That is to say that the kidneys are dependent upon the heart and circulatory system to provide a working pressure head of between 80 and 180 mm. Hg. Within this range, filtration, reabsorption, and secretion are controlled by intrinsic mechanisms, the composition of the blood, and the hormones released into the blood.

At rest there is apparently little or no nervous tone to the blood vessels of the kidney. During periods of hypoxia, hypotension, or exercise, however, adrenergic sympathetic fibers to the kidney are stimulated. They, in turn, produce a renal vasoconstriction which can shunt up to 1,000 ml. of blood per minute away from the kidneys and into the general circulation (Fig. 11-4). The release of epinephrine from the adrenal medulla has a similar effect. An example of the response of a person to a stressful situation is seen in Figure 21-3. Note that when the

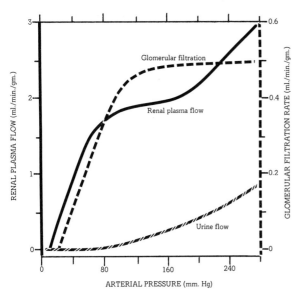

Fig. 21-2. Effect of renal arterial pressure on glomerular filtration rate and renal plasma flow in the dog. Note that between a pressure of 80 and 180 mm. Hg, the kidneys function essentially independently of arterial pressure. Apparently at these pressures intrinsic systems maintain a fairly constant flow through the kidneys. (*Modified from* Shipley, R. E., and Study, R. S.: Changes in renal blood flow, extraction of inulin, glomerular filtration rate, tissue pressure and urine flow with acute alterations of renal artery blood pressure. Am. J. Physiol., *167*:681, 682, 1951)

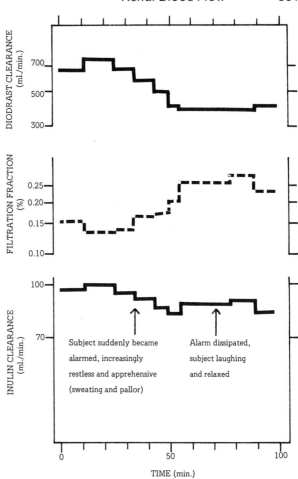

Fig. 21-3. Changes in renal function in response to a psychic shock. On becoming alarmed, the subject decreased his renal blood flow and there resulted a decrease in Diodrast clearance from 700 ml./min. to 350 ml./min., an increase in the filtration fraction from 15 per cent to 25 per cent, and a decrease in the inulin clearance from 105 ml./min. to 87 ml./min. (*Modified from* Smith, H. W.: The physiology of the renal circulation. *In* Harvey Society of New York (ed.): *The Harvey Lectures.* P. 206. Lancaster, Science Press Printing Co., 1939-40)

subject became excited both his Diodrast clearance (Diodrast is filtered and secreted) and inulin clearance (inulin is only filtered) decreased. This is apparently the result of a vasoconstriction induced in the arterioles of the kidneys by sympathetic nerves and possibly by epinephrine, which results in a decrease in renal blood flow. You will also note that there is an increase in the filtration fraction ($C_{In}/C_{Diodrast}$). This is due to the fact that the kidneys can, within certain limits, compensate for the decreased renal flow by constriction of the efferent arteriole or dilation of the afferent arteriole. Either tends to increase the filtration pressure in the glomerulus and therefore partially compensates for a decrease in glomerular flow. The Diodrast clearance, on the other hand, is more dependent on renal flow and therefore decreases more in this study than does inulin clearance.

Exercise. Moderate to heavy exercise has been demonstrated in numerous studies on human subjects to decrease renal plasma flow, filtration, and secretion. For example, subjects running for 16 min. at 3 m.p.h. on a treadmill at zero slope show C_{PAH} decreases of 6 per cent, but a relatively normal C_{In}. On the other hand, subjects running at 3 m.p.h. at a 5 per cent grade and a room temperature of 50°C. (122°F.) have a decrease in C_{PAH} of 36 per cent, and in C_{In} of 16.5 per cent. Apparently even more marked changes in kidney function occur in cardiac patients. In many of these cases the pa-

Subject's Heart Size and Condition	Cardiac Index (ml./min./m.²)	C_{In} (ml./min.)	C_{PAH} (ml./min.)	Filtra. Fraction (%)	Renal Plasma Flow (ml./min.)	Renal Flow (% C.O.)	Renal Vasc. Resistance $\left(\dfrac{\text{dyne-sec.}}{\text{cm.}^5}\right)$
512 ml./m.²							
Rest	3,590	108.7	404.8	27.8	731	12.2	105.6
Exercise	+70	−2.7	−48.5	+3.1	−90	−3.4	+22.9
796 ml./m.² (no right heart failure)							
Rest	2,110	76.3	216.4	34.9	398	10.8	183
Exercise	+860	−4	−36.8	+3.4	−71	−3.4	+24.8
807 ml./m.² (right heart failure)							
Rest	1,940	83	204.6	43.3	403.6	11.2	199.7
Exercise	+260	−8	−23.7	+2.6	−48	−1.8	+65.6

Table 21-1. Response of 28 cardiac patients to mild exercise (subject recumbent, pedaling a bicycle ergometer at a work level of 70 kg.-m./min.). Note that in all cases the exercise resulted in an increase in the cardiac index, a decrease in inulin clearance (C_{In}), a decrease in PAH clearance (C_{PAH}), an increase in the filtration fraction, a decrease in renal plasma flow, a decrease in the per cent of the cardiac output (C.O.) going to the kidneys, and an increase in renal vascular resistance. (From Werkö, L.E., Varnauskas, H.E., Ek, J., Bucht, H., Thomasson, B., and Bergström, J.: Studies on the renal circulation and renal function in mitral valvular disease. I. Effect of exercise. Circulation, 9:687-699, 1954)

tient does not have the cardiac reserve to call upon in exercise and compensates by shunting even larger quantities of blood away from the kidneys to the heart and brain. In Table 21-1 we have characterized the response of cardiac patients to a level of exercise which had little or no effect on renal function in normal subjects. Note the marked decreases the exercise produced in C_{PAH}, the filtration fraction, and renal plasma flow, and the increase in renal vascular resistance.

Ischemia. The marked decreases in renal flow that result from hypotension and, to a lesser degree, certain other conditions, logically lead to the question of the capacity of the kidneys to tolerate either a decrease in blood flow or a cessation of flow. Studies on dogs show that they can tolerate periods of complete ischemia up to an hour. After 30 min. of ischemia an additional 30 min. is required for kidney function to return to normal. Two hours of ischemia produces a tubular damage which persists for 3 months (Table 21-2). After more prolonged periods of ischemia, death due to uremia is common.

Intrinsic Control

In the absence of an extrinsic nerve supply the kidneys, at renal artery pressures ranging from 80 to 180 mm. Hg, can regulate their own blood flow. In Figure 21-4, for example, we note that an increase in pressure in the renal artery from

	Before	3-4 Hr.	24 Hr.	5-8 Days	2 Weeks	3 Months
C_{Cr} (ml./min.)	59	1.9	8.6	13.5	12.7	63
C_{PAH} (ml./min.)	182	5.2	27.9	44.7	62.5	174
Filtration fraction	32	47	40	20.9	20.3	36.6
Excretion$_{PAH}$ (%)	66	4.8	14.2	26.3	32.7	84
C_{PAH}/E_{PAH} (ml./min.)	276	108	196	170	191	208
Tm$_{PAH}$ (mg./min.)	13.8		1.4	3.17	3.23	6

Table 21-2. Effect of 2 hr. of renal ischemia on unilaterally nephrectomized, anesthetized dogs. (From Friedman, S.M., Johnson, R.L., and Friedman, C.L.: The pattern of recovery of renal function following renal artery occlusion in the dog. Circ. Res., 2:231-232, 1954)

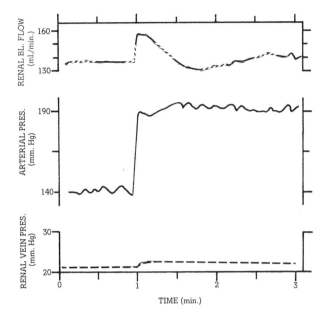

Fig. 21-4. Effect of an increased pressure in the renal artery on renal blood flow, and renal vein pressure. The initial response is due to the increase in the perfusion pressure. The secondary decrease in flow is the result of an intrinsic mechanism which produces an increased resistance to flow. (*Redrawn from* Semple, S. J. G., and de Wardener, H. E.: Effect of increased renal venous pressure on circulatory "autoregulation" of isolated dog kidneys. Circ. Res., 7:643, 1959; by permission of the American Heart Association, Inc.)

140 mm. Hg to 190 mm. Hg initially produces an increase in renal blood flow, but that within 40 sec., although the increase in pressure is maintained, the flow returns to what it was prior to the increased pressure. The initial increase in flow is what we would expect in response to an increased pressure head, but the secondary decline apparently represents an intrinsic control system that produces an increased resistance to flow. The nature of this system is not known. Some suggest it is (1) a response of the blood vessels to the change in the environment (hyperoxia, hypocapnea, increased pH, etc.) brought about by the increased flow ("*metabolic theory*"). Others point to (2) a type of *axon reflex*, (3) an increase in the tone of the smooth muscles of the vessels, in response to stretch ("*myogenic theory*"), (4) an increase in interstitial pressure which decreases the size of the lumen of the renal vessels ("*tissue pressure theory*"), and (5) changes in the *viscosity* of blood which result from the turbulence associated with increased flow velocity. The importance of preventing an excessive flow to the kidney is that if allowed to persist, it decreases the hypertonicity of the peritubular fluid of the medulla and in this way interferes with the ability of the kidneys to produce a hypertonic urine.

Renal Hypertension

High blood pressure can be produced by (1) disturbance of kidney function, (2) malfunction of the nerves controlling the heart and circulation, (3) malfunction of the cardiovascular system itself, (4) a disturbance in the production of hormones, and (5) some unknown cause (essential hypertension). We shall be concerned in this section only with hypertension produced at least in part by distrubances in kidney function. In 1898 Tigerstedt and Bergman noted that the injection of a saline extract from the kidney produced a slowly developing, sustained increase in arterial pressure. In 1934 *Goldblatt* and co-workers showed that partially constricting clamps on the main renal arteries of the kidney resulted in severe hypertension. Later Page in the United States and Braun-Mendez in Argentina found that the active principle (renin) from the kidney that produced the hypertension was a proteolytic enzyme that acted upon a plasma component to produce a pressor agent. Apparently the *renin* acts on plasma α_2-globulin to form a decapeptide (angiotensin I) which is changed into an octapeptide (angiotensin II) by a plasma enzyme (Fig. 21-9). The antiotensin II is inactivated by the enzyme *angiotensinase,* which is found in the plasma, kidney, intestine, and liver. The injection of angiotensin II results in an increase in arterial pressure lasting 5 to 10 min., whereas renin produces hypertension for about 40 to 60 min. These reactions have been discussed in Chapter 11, "Circulatory Control," and are summarized in Figure 11-13.

Juxtamedullary apparatus. It is now generally agreed that renin is produced and released by the juxtamedullary apparatus (Fig. 21-5). This includes the granular juxtaglomerular cells of the thickened wall of the afferent arteriole (polkissen or polar cushion), and a group of cells with prominent nuclei which lies at the point where the distal tubule, afferent arteriole, and glomerulus are adjacent to each other (macula densa). The granular cells are thought by many to produce, store, and release renin in response to the proper stimuli. It has further been suggested that (1) a

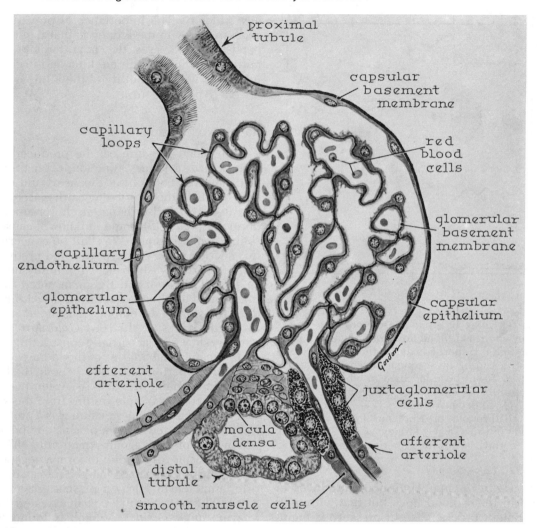

Fig. 21-5. Semidiagrammatic drawing of the juxtaglomerular apparatus and the renal corpuscle. The juxtaglomerular cells are thought to be the site for the formation, storage, and release of renin. (Ham, A.: Histology, 6th ed., p. 778. Philadelphia, J. B. Lippincott, 1969)

decrease in renal pressure, (2) renal ischemia, (3) epinephrine, (4) renal nerves, (5) an acute depletion of Na^+, and (6) ureteral obstruction can all cause the release of renin. Figure 21-6 summarizes some of the speculations of Vander on the control of renin release. He visualizes the renin system as a regulatory mechanism for the maintenance of blood volume, arterial pressure, renal flow, and Na^+ balance.

In Figure 21-7 we have some evidence of the importance of the kidneys in the control of arterial pressure. In these experiments it was demonstrated that rats with kidneys were better able to counter the hypotensive effects of hemorrhage than those which, just prior to hemorrhage, had had their kidneys removed. It will also be noted that the rats with kidneys have a higher initial arterial pressure. Some feel this is due to the release of renin by the kidneys in response to the anesthetic agent used in the experiments.

Renoprival hypertension. There is also evidence that the kidneys not only produce an agent that leads to an increase in arterial pressure, but also destroy certain poorly defined extra-renal circulating hypertensive agents. We find, for example, that if the kidneys are removed and the

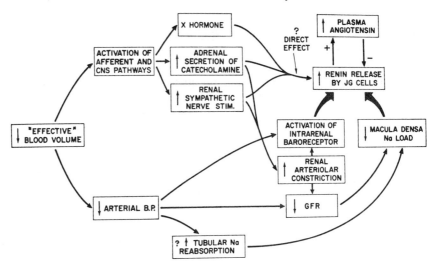

Fig. 21-6. Summary of the pathways proposed for the control of renin release. (Vander, A. J.: Control of renin release. Physiol. Rev., 47:377, 1967)

plasma dialyzed so as to maintain a fairly homeostatic environment, a sustained elevation in arterial pressure develops. It has also been reported that when a normal kidney is grafted into a hypertensive subject the blood pressure frequently decreases. In short, the kidney may play a dual role in the control of arterial pressure by (1) producing an agent that causes vasoconstriction and (2) destroying certain circulating hypertensive agents.

ELECTROLYTE BALANCE

The functions of the cells of the body are dependent upon a relatively constant intercellular and extracellular environment. Lowering the K^+ concentration of the intercellular fluid, for example, increases the resting membrane potential of the cell and therefore decreases its excitability (Fig. 3-5). Lowering the Na^+ concentration in the intercellular fluid decreases the amplitude of the action potential and therefore interferes with intercellular conduction (Fig. 3-3). Lowering the Ca^{++} in the intercellular fluid decreases the stability of the cell membrane and eventually leads to hyperreflexia and tetany. High levels of Hg^{++} depress the nervous and muscular systems. The cell's electrolyte environment is dependent, in part, on the content and quantity of the foods we eat, and the sweat, vomitus, and feces we eliminate. It is also dependent upon the cell itself and the character and quantity of the urine we produce.

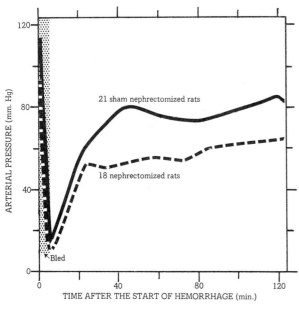

Fig. 21-7. Response of the rat to an acute blood loss. It is suggested that the difference in response between nephrectomized and non-nephrectomized rats lies in the fact that the latter can release renin in response to the blood loss. (*Modified from* Tobian, L.: Shock and Hypertension. L. C. Mills and J. H. Moyer (eds.). P. 348. New York, Grune & Stratton, 1965)

Sodium, Potassium, and Chlorine

Sodium salts account for over 90 per cent of the osmotically active particles in the plasma and intercellular fluid. They are, therefore, of prime importance in the control of (1) the extracellular fluid volume and (2) the amplitude of the action potential. Since the quantity of sodium salts entering the blood from the digestive tract may vary from day to day and from person to person, the kidneys and the systems that control the elimination of sodium salts take on great importance. We find, for example, that a normal adult with a glomerular filtration rate of 125 ml. of plasma per minute, a plasma bicarbonate concentration of 27 mEq. per liter, and a plasma Na^+ concentration of 145 mEq. per liter filters 18.125 mEq. of Na^+ ($125 \times 0.145 = 18.125$) into the nephrons of the kidneys per minute. In order for an Na^+ balance to be maintained, 96 to 99 per cent of this Na^+ must be reabsorbed from the nephrons and collecting ducts back into the blood. In the case of a person who reabsorbs 18.060 mEq. per minute, the average associated reabsorption is:

Cl^- 14.585 mEq./min.
HCO_3^- 3.375 mEq./min.

In addition, there is a movement into the nephron of K^+ and H^+. Collectively, K^+ and H^+ account for the reabsorption of 0.100 mEq. of cation per minute.

The above figures vary, of course, with the circumstances. In the case of a person on a low NaCl diet, the amount of Na^+ excreted in the urine may be as low as 1 mEq. per day. Persons with a high salt intake, on the other hand, may excrete more than 400 mEq. of Na^+ a day in their urine. The excretion of progressively larger quantities of Na^+, of course, is associated with the excretion of progressively larger quantities of anions such as Cl^- and HCO_3^-, the reabsorption of progressively larger quantities of H^+ and K^+, and the excretion of increased quantities of water. You will remember from the previous chapter (Fig. 20-17) that there is active reabsorption of Na^+ throughout much of the nephron and collecting duct, and that in the proximal tubule this is associated with the passive reabsortion of Cl^-. In the collecting duct, on the other hand, Na^+ reabsorption is also associated with the secretion of K^+, H^+, and NH_4^+, as well as with the reabsorption of Cl^-.

The relative concentrations in the urine of the ions mentioned above depend upon a complex, poorly understood interplay of control mechanisms. Increases in urinary HCO_3^-, for example, facilitate decreases in urinary Cl^-. Increases in plasma K^+ facilitate decreases in urinary H^+ and Na^+. Increases in plasma H^+ cause decreases in urinary K^+. Part of the control of the relative concentrations of these ions is exerted through the hormones released by the adrenal cortex. We find, for example, that removal of the cortex results in 2 per cent of the Na^+ that is filtered appearing in the urine:

0.02×18.1 mEq./min. = 0.362 mEq./min. = 520 mEq./day

It also results in an increase in urinary Cl^-, water diuresis, a decrease in plasma Na^+ and Cl^-, and in many cases a decrease in plasma volume and arterial blood pressure. In addition, there is a decrease in the urinary K^+, an increase in the plasma K^+, an increase in intracellular K^+, and an inability to handle a water load without developing the symptoms of water toxicity.

All the symptoms listed above, except the difficulty in handling a water load (Fig. 31-8), can be successfully treated by the use of a high Na^+ low K^+ diet or by the injection of aldosterone. Aldosterone, like a number of other hormones produced by the adrenal cortex, increases the reabsorption of Na^+ from the collecting duct, and the secretion of K^+. In other words, it serves to decrease the Na^+-to-K^+ ratio in the urine and

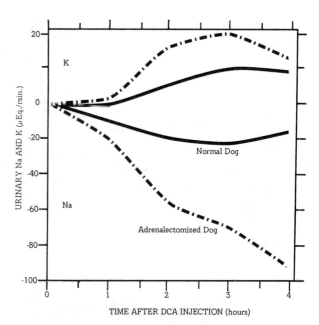

Fig. 21-8. Response of normal and adrenalectomized dogs to 800 μg. of intravenous deoxycorticosterone acetate (DCA). Note the increase in K^+ and decrease in Na^+ excretion in response to this mineralocorticoid. (*Redrawn from* Davis, J. O., et al.: Humoral factors in the regulation of renal sodium excretion. Fed. Proc., 26:61, 1967)

increase the ratio in the blood. The adrenocortical hormones that perform this function are called mineralocorticoids (Fig. 31-3 and Tables 31-1 and 31-3). Two such substances are aldosterone and desoxycorticosterone (Fig. 21-8). Aldosterone is the more potent, being thirty times more effective in inducing Na^+ reabsorption and five times more effective in inducing K^+ secretion. It should be emphasized, however, that most investigators believe that the loss of K^+ in the urine is secondary to Na^+ reabsorption in the collecting duct. Part of the evidence for this is that (1) K^+ is completely reabsorbed in the proximal tubule and (2) in order to show the effect of aldosterone on K^+ excretion one must have an adequate Na^+ load delivered to the collecting duct. Other factors capable of changing the excretion of Na^+ include changes in (1) the glomerular filtration rate, (2) the plasma pH, (3) the plasma concentration of hormones such as oxytocin and the glucocorticoids, and (4) possibly the stimulation of renal nerves.

Control of the release of the mineralocorticoids. Aldosterone, the major mineralocorticoid, is—in a resting person—continually released by the adrenal cortex and carried by the blood to the kidney. Here it facilitates the transport system of the distal tubule and collecting duct in removing Na^+ from the glomerular filtrate and adding K^+ and H^+. The amount of aldosterone released is increased when the Na^+ concentration in the plasma perfusing the adrenal cortex is decreased or when the K^+ concentration is increased. Aldosterone secretion is decreased when either the Na^+ concentration goes up or that of K^+ goes down. It is also apparent that a hormone produced by the anterior pituitary, adrenocorticotrophic hormone (ACTH), facilitates the release of mineralocorticoids. In fact this hormone facilitates to a greater degree the release of certain other adrenal cortical hormones, the glucocorticoids, but we shall leave our discussion of this relationship for a later chapter (see p. 533).

Playing an important role in the control of mineralocorticoid release is the substance renin. As we have seen, it is released in response to reduced renal blood flow, renal blood pressure, and plasma Na^+ (Fig. 21-9). I have also indicated, in Figure 21-6, that it is released in response to epinephrine (the major adrenal catecholamine), decreases in the effective blood volume, and stimulation of the sympathetic nerves to the kidney. I might add that there is some evidence that denervation of the kidney leads to a decrease in renal secretion. It is presently hypothesized that the stimulation of *volume receptors* (stretch receptors) in the left atrium, *baroreceptors* in the carotid sinus, and *osmoreceptors* in the hypothalamus of the brain all tend to decrease the secretion of renin.

The renin, after it is released, converts α_2-globulin to angiotensin II. The angiotensin stimulates the adrenal cortex to release additional aldosterone and other mineralocorticoids. In the preceding section we emphasized the vasoconstrictor capacity of angiotensin, and in this section, its action in facilitating the release of aldosterone. Aldosterone secretion results in the conservation of Na^+ and therefore an increase in plasma Na^+, which leads to an increase in blood volume and a consequent increase in arterial pressure. The vasoconstrictor action of angiotensin also results in an increase in arterial pressure, but by a different mechanism. The point I would like to stress here is that the renin-angiotensin system under some circumstances increases aldosterone secretion without producing vasoconstriction. In fact there are many who feel that in a normal person the only role of angiotensin is to control mineralocorticoid secretion and that angiotensin concentrations become high enough to produce vasoconstriction only in renal disease or adrenal cortical malfunction (e.g. tumors). There is also evidence that angiotensin at low concentrations produces vasoconstriction if a person has a positive salt balance. If he is on a low-salt diet, on the other hand, the same dose of angiotensin has no vasoconstrictor action. It is because of this, we think, that certain forms of hypertension are so effectively treated by placing the patient on a low-salt diet.

Congestive heart failure. Congestive heart failure is sometimes called heart failure, cardiac decompensation, congestive failure, myocardial failure, and dropsy. It is a condition in which the cardiac output is insufficient to prevent the accumulation of blood and elevation of pressure in either the pulmonary veins (left heart failure), systemic veins (right heart failure), or both. It may be associated with either a lower than normal cardiac output (low output failure) or an elevated cardiac output (high output failure). It may result from stenosis, valvular failure, hypertension, or myocardial malfunction (infarction, myocarditis, etc.). Evidence is accumulating that the renin-angiotensin system plays an important

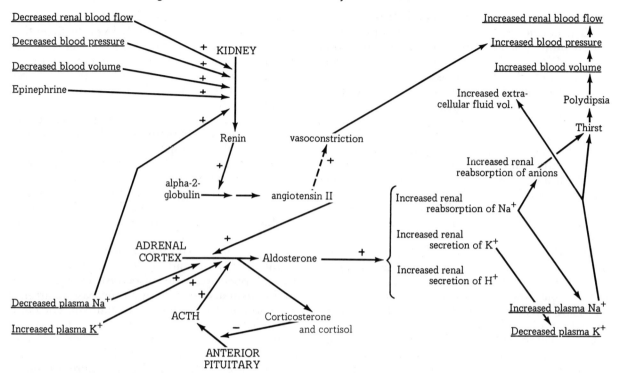

Fig. 21-9. The renin-angiotensin-aldosterone system for the control of blood flow, blood pressure, blood volume, plasma Na⁺ concentration, and plasma K⁺ concentration. The following symbols are used: ⟶, produces; $\xrightarrow{+}$, facilitates; ⟶, inhibits; $\dashrightarrow{+}$, facilitates under some conditions, but not all).

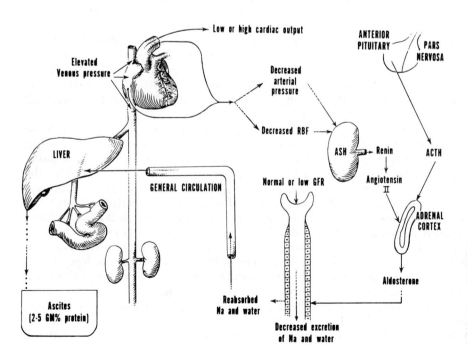

Fig. 21-10. Summary of the mechanisms which may lead to a hypersecretion of aldosterone, Na⁺ and water retention, and ascites in cardiac failure. (Davis, J. O.: The physiology of congestive heart failure. *In* Hamilton, W. F. (ed.): Handbook of Physiology. Sec. 2 (Circulation), vol. III, p. 2113. Washington, American Physiological Society, 1965)

role in producing some of the edema that results from heart failure. Cardiac failure, for example, is associated with increased levels of aldosterone in the body, as well as NaCl and water retention. Many have suggested that in cases in which the heart is unable to prevent venous congestion there is also a decrease in arterial pressure, a decrease in renal blood flow, or both which leads to the release of increased amounts of renin and aldosterone (Fig. 21-10). This, in turn, produces the retention of salt and water in the body which causes further congestion of the veins, an increase in capillary hydrostatic pressure, and generalized edema. As a result, the liver may acquire a nutmeg appearance due to its congestion, and fluid may accumulate in the abdominal cavity (ascites), the alveoli (pulmonary congestion), the body tissues, or all the three.

Calcium, Phosphate, and Magnesium

The concentration of calcium, phosphate, and magnesium in the plasma is regulated in part by a hormone released from the parathyroid glands (see Chap. 32). The amount of this hormone, parathormone (PTH), that is released is controlled by the Ca^{++} concentration in the plasma—increasing when the plasma Ca^{++} concentration decreases and decreasing when the plasma concentration increases (Fig. 21-11). Parathormone increases the plasma concentration of Ca^{++} by first increasing the reabsorption of Ca^{++} by the kidney, then increasing the absorption of Ca^{++} from the digestive tract, and finally, stimulating the osteoclasts of bone to release Ca^{++} into the plasma. All of this may result in a progressive increase in the plasma Ca^{++} level to a point at which secondary hypercalciurea develops. We find, for example, in Table 21-3, that the injection of a parathyroid extract (PTE) can result in a sufficiently marked increase in plasma Ca^{++} that there is also an increase in urinary Ca^{++}. It should be emphasized, however, that the direct effect of PTH on the kidney is to decrease urinary Ca^{++}, and that this is generally the first mechanism to start increasing plasma Ca^{++} in response to the PTH.

Parathormone also affects the renal reabsorption of phosphate. It has been shown, for example, that the injection of PTH into a renal artery increases the concentration of phosphate in the urine. Apparently this is the result of a lowered Tm for phosphate. When parathormone stimulates osteoclast activity, on the other hand, it causes a loss of both Ca^{++} and HPO_4^{--} from the bone. In other words the action of PTH on bone tends to increase plasma phosphate, and the action on the kidneys tends to decrease it. The long-term effect is usually a decrease in plasma phosphate.

Thus the injection of PTH leads to (1) hypercalcemia associated with (2) an initial decrease in urinary Ca^{++} followed by hypercalciurea and (3) hyperphosphaturia. If the plasma parathyroid hormone levels are kept high for a prolonged period, the resultant decalcification of the bones may lead to serious weakening of the skeleton (*osteitis fibrosa cystica*). The removal of the parathyroid glands, on the other hand, leads to (1) hypocalcemia associated with (2) an initial increase in urinary Ca^{++} followed by hypocalciuria, and (3) an initial hypophosphaturia followed by a return to normal urinary phosphate levels and an associated hyperphosphatemia. As the plasma Ca^{++} levels decrease there is an increase in the excitability of nerve and muscle and the person goes into tetany. This is a condition of increased muscle tone, hyperreflexia, and intermittent tonic muscular contractions.

The importance of parathormone in the con-

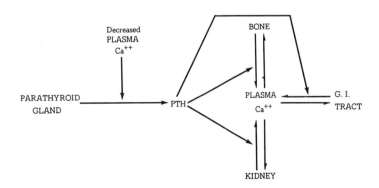

Fig. 21-11. The role of parathyroid hormone in the control of Ca^{++} balance.

	Before PTE	During PTE Treatment			After PTE
		Day 1-2	Day 3-6	Day 7-10	
Normal Subjects					
Calcium					
Serum (mg./100 ml.)	9.9	10.8	11.0	11.2	10.0
Urine (mg./day)	272	538	671	845	392
Magnesium					
Serum (mg./100 ml.)	1.90	1.94	1.91	1.90	1.99
Urine (mg./day)	88	100	103	109	57
Hypoparathyroid Subjects					
Calcium					
Serum (mg./100 ml.)	8.3	10.4	11.2	10.9	8.2
Urine (mg./day)	147	107	258	323	189
Magnesium					
Serum (mg./100 ml.)	1.85	2.06	1.99	1.97	1.76
Urine (mg./day)	79	69	85	82	65

Table 21-3. Response of two normal women (17 to 22 years old) and five women (39 to 51 years old) with surgically induced hypoparathyroidism to 600 units a day of parathyroid extract (PTE). The data labeled "After PTE" were collected 3 to 8 days after treatment had been stopped. (From Gill, J.R., Jr., Bell, N.H., and Bartter, F.C.: Effect of parathyroid extract on magnesium excretion in man. J. Appl. Physiol., 22:136-138, 1967)

trol of the plasma magnesium concentration is uncertain. Most of our information at present indicates that parathormone causes the release of magnesium from bone and other tissues and decreases its loss in the urine. In other words, PTH has an action on magnesium storage and release similar to that on calcium storage and release. You will note in Table 21-3, however, that hypoparathyroidism is associated with a more marked decrease in serum Ca^{++} than in serum Mg^{++} levels.

WATER BALANCE

Water is lost from the human body by a number of routes. These include the skin, lungs, kidneys, digestive tract, and, in the lactating female, the mammary glands (Table 21-4). The amount of water lost via the skin increases as the body temperature increases (pp. 475-477). The amount lost in the expired air, for example, increases during exercise, and the amount lost from the alimentary system rises in diarrhea and vomiting. The amount of water excreted by the kidneys is determined in part by the organism's need for water, but during water deprivation is also determined by the kidneys' capacity to produce a hypertonic urine. In other words, during water deprivation the water-conservation function of the kidney comes in conflict with its function in the excretion of waste products and the maintenance of an electrolyte balance.

The amount of time a person can survive water deprivation varies, of course, with the circumstances. Men have been known to survive without food or water for periods of 18 days when the temperature and humidity are moderate. In extremes of heat, however, survival may be decreased to 2 to 3 days. Water normally enters the body in our food and drink. An additional 200 ml. per day is formed in the body as a result of catabolic reactions. Our water intake is regulated, in turn, by various factors associated with the

	Normal Temperature (ml.)	Elevated Temperature (ml.)	Prolonged Exercise (ml.)
Insensible loss			
Skin	350	350	350
Lungs	350	250	650
Urine	1,400	1,200	500
Sweat	100	1,400	5,000
Feces	200	200	200
Total	2,400	3,400	6,700

Table 21-4. Daily water loss under varying conditions. Not included in this table are the mammary glands of the lactating female. These may result in an additional water loss of about 900 ml. per day.

sensation of thirst. A person who leads a fairly sedentary life has a daily intake of between 880 and 2,440 ml. of water (average 2,290), whereas one doing moderate work ranges between 2,225 and 4,550 ml. of water a day (average 3,700). Hunt has reported that in India a person may drink as much as 13 liters of water a day.

The control of water intake is still a controversial subject, but there does seem to be a great deal of evidence that dryness of the mouth and throat causes *thirst*. It is generally recognized, for example, that thirst can be temporarily relieved by moistening the mouth without increasing the water intake. It is also reported that a number of procedures which increase the flow of saliva relieve thirst, e.g., placing a pebble or nail in the mouth or placing acid substances in the mouth, and that factors causing decreased secretion of saliva causes thirst. Such factors include (1) fear and rage, (2) hypovolemia, and (3) hypertonicity of the blood. Apparently in each of these conditions there is a sufficient change in the autonomic nervous tone to the secretory glands of the mouth to produce dryness. Some of the causes of the hypovolemia are hemorrhages insufficient to produce a change in the arterial blood pressure, water deprivation, profuse sweating, and large meals. A meal produces hypovolemia by the withdrawal of large quantities of water from the blood for use in the intestinal tract. We find, for example, that when a meal is consumed without water, the salivary secretions decrease toward the end of the meal and the mouth becomes dry. Apparently hypovolemia is sensed by receptors in the atria of the heart, and hypertonicity by receptors in or near the hypothalamus.

It is important to remember, however, that mouth dryness is not the only factor controlling fluid intake. A thirsty dog, for example, drinks for a while after his mouth has been moistened, and the dog with an esophagotomy drinks (sham drinking) for even longer periods. Apparently it is (1) the relief of mouth dryness followed by (2) distention of the stomach by the water that elicits a sensation of satiety. This has been shown in sham-drinking experiments in which a balloon is placed in the stomach. When the balloon is inflated soon after the onset of sham drinking, the drinking stops sooner than if the balloon is not inflated. Inflation of the balloon does not by itself eliminate thirst, though infusing water directly into the stomach does eliminate the sensation after a period of 10 to 15 min.

Antidiuretic Hormone

The kidneys constitute the chief means the body has of maintaining a water balance. When there is excessive water intake the urine volume increases, and when there is water deprivation it decreases. If, for example, a normally hydrated person drinks 1,000 ml. of water, he will have eliminated approximately 92 per cent of it in the urine within 3 hr. (Fig. 21-12). If, on the other hand, he drinks 1,000 ml. of 1 per cent NaCl solution, the increase in urine production is considerably less rapid. The difference in the response of the body to these two solutions has to do with their different osmotic properties. The hypotonic solution apparently reduces the osmotic pressure of the blood sufficiently that this change is sensed by an area in or near the supraoptic nuclei of the hypothalamus. This area is referred to as having *osmoreceptors*. These so-called receptors have not been differentiated from the rest of the hypothalamus by histologic techniques, but have been localized on the basis of the injection of hypotonic (increasing urine volume) and hypertonic (decreasing urine vol-

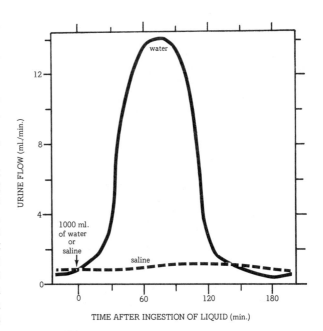

Fig. 21-12. Response of man to 1,000 ml. of water and to 1,000 ml. of 1% NaCl solution (saline). In the experiment with water, 92 per cent of the water appeared in the urine during the first 3 hours. (*Redrawn from* Smith, H. W.: The Physiology of the Kidney. P. 156. New York, Oxford University Press, 1937)

ume) solutions. We find, for example, that the injection of small volumes of hypertonic or hypotonic solutions of saline or sucrose change urine production when the injection is made into the common carotid artery or some other artery carrying blood to the hypothalamus, but that the same volumes have little or no effect when injected into vessels such as the femoral artery or jugular vein.

Apparently the osmoreceptors, when exposed to a hypotonic solution, act on the posterior pituitary gland to decrease its release of an antidiuretic hormone (ADH), and when exposed to a hypertonic solution cause an increase in the release of the hormone.

Since ADH increases the permeability of the distal portion of the tubular system (i.e., the collecting ducts and the distal convoluted tubule) to water, inhibition of its release would facilitate diuresis by preventing a loss of water from the urine to the hypertonic peritubular fluid of the renal medulla.

Water deprivation, on the other hand, increases the release of ADH. Through fluctuations in the plasma ADH one can (1) vary the specific gravity of the urine from 1.002 to 1.035, (2) vary the osmolarity of the urine from 40 to 1,400 milliosmols per liter, and (3) vary the urine volume from 0.5 to 20 liters per day. One clinical test for a normal renal function is the *water-deprivation test*. In this test the subject receives no water from 8:00 P.M. to 8:00 A.M. the next morning, at which times he voids and discards his urine sample. He continues his water deprivation until noon, collecting and storing each urine sample. Under these circumstances the kidney should be able to produce a urine of a specific gravity greater than 1.025. Serious impairment of the kidney's ability to produce a concentrated urine would result in a urine of a specific gravity near 1.010. This is the specific gravity of the glomerular filtrate. One can also test for the ability of the kidneys to produce a dilute urine by having the subject void and then collect subsequent samples after drinking a large volume of water. Under these circumstances the kidney should be able to produce a urine with a specific gravity less than 1.005.

There is also evidence that ADH affects kidney function other than by merely increasing the permeability of the collecting duct to water. One of these is a vasoconstrictor action on arterioles leading to the vessels (vasa recta) of the renal medulla. This serves to reduce the flow in this area and in this way prevents a washing out of the hypertonic fluid which makes up the medullary peritubular space. A third action of ADH is to facilitate greater activity by the Na^+ transport system. In other words, ADH may facilitate the concentration of the urine in three ways, i.e., by (1) increasing the permeability of the collecting duct to water, (2) preventing the peritubular capillaries from reducing the hypertonicity of the liquid surrounding the collecting ducts, and (3) facilitating the transport of additional Na^+ into the hypertonic medullary fluid.

Solute Load

We emphasized in the preceding section that we increase the volume of urine formed each day when we increase our water intake. This results, in part, from a hypotonicity produced in the blood which acts via osmoreceptors to decrease the secretion of ADH. It is also important to realize that marked increases in the concentration of solutes in the distal tubule reduce the osmotic gradient between the tubular urine and the hypertonic peritubular fluid in the renal medulla and therefore reduce the movement of water out of the collecting duct. We see, for example, in Figure 21-13, that if a dog swallows 60 gm. of NaCl there is an increase not only in

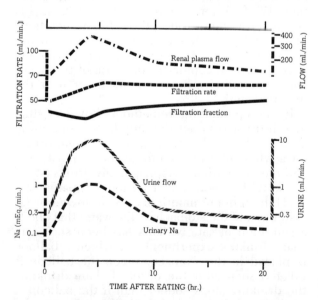

Fig. 21-13. Response of the dog to a meal containing 60 Gm. of NaCl. (*Modified from* Ladd, M., and Raisz, L. G.: Response of the normal dog to dietary NaCl. Am. J. Physiol., *159*:151, 1949)

urinary NaCl, but also in urine volume. This apparently occurs despite the fact that a hypertonicity of the blood results that would initiate the secretion of additional ADH.

Increases in urine volume also result from the injection of such organic substances as mannitol, urea, sucrose, and glucose. In diabetes mellitus, for example, inadequate amounts of insulin are released by the pancreas, causing such an increase in blood sugar that the $Tm_{glucose}$ is exceeded and large quantities of glucose appear in the urine; polyuria (increased urine volume) results. This type of diabetes was for a long time confused with diabetes insipidus, a polyuria due to the formation of inadequate amounts of ADH. In time it was noted that in some cases of polyuria the urine tasted sweet (diabetes mellitus), while in others it did not (diabetes insipidus).

Other Factors

There are a number of substances which can be given by the physician or which are taken in in our diet that increase the urine volume. Many of these act by interfering with Na^+ reabsorption. By increasing the osmotic pressure of the urine, that is, they increase urine volume. These osmotic diuretics include a number of organic mercurials, chlorothiazide (Diuril), and the aldosterone antagonist, spironolactone (Aldactone). In addition the coffee, tea, and cocoa we drink contain xanthines which may, by dilating renal afferent arterioles, increase glomerular filtration to such an extent that the Tm for Na^+ is markedly exceeded.

Some of the hormones other than ADH that affect water balance are (1) the mineralocorticoids and glucocorticoids of the adrenal cortex, (2) the thyroid hormones, and (3) estrogen. I have discussed the role of the mineralocorticoid aldosterone earlier in this chapter, and have summarized its actions in Figure 21-9. It, through its control of the electrolyte concentration in the urine and blood, can modify the osmotic properties and hence the volume of these two fluids. The blood volume and osmotic pressure, in turn, can affect volume, osmotic and pressure receptors, which reflexly influence the secretions of the mouth and hence the sensation of thirst.

Glucocorticoids also play a role in the control of water loss. You will note, for example, in Figure 31-8, that a subject suffering from adrenal cortical insufficiency needs a glucocorticoid in order to promptly increase his urine volume in response to a water load. Apparently under a variety of circumstances the glucocorticoids play a role in (1) maintaining a high glomerular filtration rate, (2) facilitating Na diuresis, and (3) preventing water reabsorption from the renal tubules.

Another hormone that can play an important role in the control of water balance is *estrogen*. We find, for example, that in a large number of women prior to menstruation there is (1) decreased urinary output of water and NaCl, (2) increased thirst, (3) an increased blood volume, (4) edema, and (5) increased body weight. It is suspected that these changes are at least partly responsible for the *premenstrual tension* about which many women complain. Their appearance and disappearance also correlate well with the estrogen levels in the blood, and can be produced by the intravenous injection of estrogen and reduced by the use of diuretics and the reduction of salt intake. The means of action of estrogen are complicated. It (1) apparently has no effect on aldosterone secretion, but does retard aldosterone destruction by increasing its plasma protein binding, and (2) decreases the secretion of the glucocorticoid cortisol. Estrogen also, however, brings about fluid and salt retention in dogs which have had their adrenal glands and gonads removed.

Other factors in the control of water balance include the liver, the microcirculatory system, the lymphatics (pp. 220-222), and the thyroid gland. Androgens too may facilitate the retention of Na, Cl, K, and water, but these actions appear to be of little importance. The thyroid hormone thyroxine, on the other hand, apparently plays an important part in the control of water balance. In hypothyroidism there is a generalized increase in the volume of water and Na in the extracellular spaces which is associated with a decrease in blood volume. Apparently these changes are due mostly to the accumulation of proteins in the extracellular space. The proteins tend to combine with Na and draw water from the blood and cells. The administration of thyroxine, on the other hand, (1) facilitates the breakdown of these proteins, (2) increases the ingestion of food and its catabolism, and (3) increases the solute load in the urine to such an extent that there is an increase in urine volume. Thyroxine given to a hypothyroid subject increases the nitrogen, Na, and water in the urine. On the other hand, thyroxine given to a euthyroid person increases urinary nitrogen, K, and water. Apparently in the first case thyroxine has its most profound effect on the intercellular protein that has been allowed to accumulate, whereas in the second case it has a more generalized catabolic action. This is not to say, however, that the only role of thyroxine is to increase the solute load. You will note in Figure 29-5, for example, that thyroxine markedly increases the oxygen consumption of the kidney. Through its control of metabolism in the kidney and throughout the circulatory system, it apparently tends also to (4) increase glomerular filtration, (5) increase renal blood flow, and (6) under some circumstances increase the urinary calcium and phospahte concentrations. In small doses thyroxine does not affect the Na-to-K ratio.

Distribution of Body Water

The quantity of water in the body can be determined by administering a known amount of a substance (A) which distributes itself evenly throughout the body's water but is neither broken down, bound, nor taken up by non-aqueous compartments. Some of the more commonly used agents are antipyrine and the heavy isotopes deuterium oxide and tritium oxide. After one of these is injected urine is collected, and when an equilibrium has been established between the aqueous compartments, the concentration of the agent in the plasma (C) and the amount eliminated by the kidneys (E) is determined. From these data the total volume of water in the body (V) can be calculated:

$$V\ (ml.) = \frac{A\ (mg.) - E\ (mg.)}{C\ (mg./ml.)}$$

Mannitol is a tracer substance which neither enters the cell nor readily penetrates connective tissue fluid, bone water, or the cerebrospinal fluid. It has been used to estimate that part of the extracellular fluid that does not include the compartments mentioned above. Evans blue (T-1824) and iodinated albumin, on the other hand, tend to remain in the plasma water and are used for the estimation of plasma volume.

On the basis of the use of such indicators and the study of the change in total body and tissue weight after dessication (prolonged heating at about 105° C.), the data shown in Figure 21-14 can be derived. Total body water generally represents 500 to 600 ml. per kilogram of body weight. Dessication studies, however, generally

	% Body Weight	% Water	Liters of Water/70 kg.
Blood	8	83	4.65
Kidneys	0.4	82.7	0.25
Heart	0.5	79.2	0.28
Lungs	0.7	79	0.39
Spleen	0.2	75.8	0.10
Muscle	41.7	75.6	22.10
Central nervous system	2	74.8	1.05
Intestine	1.8	74.5	0.94
Skin	18	72	9.07
Liver	2.3	68.3	1.03
Skeleton	15.9	22	2.45
Adipose tissue	10	10	0.70
Total or average	101.5	63	43.01

Table 21-5. Distribution of water in various tissues of man. (From Skelton, H.: The storage of water by various tissues of the body. Arch. Int. Med., 40:141, 1927)

yield a value somewhat higher (640 ml./kg.). It should be noted that lower values are obtained in an obese subject and higher ones in a lean subject. If, however, we express the total body water in milliliters per kilogram of lean body

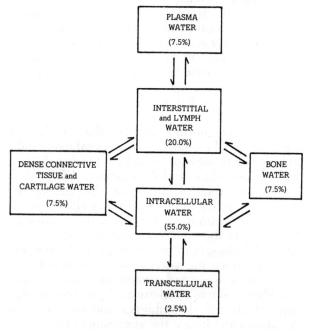

Fig. 21-14. Distribution of body water. Figures in parentheses represent per cent of total body water found in a particular segment. The transcellular compartment includes water in the cerebrospinal fluid, ocular fluid, saliva, mucous, pancreatic juice, and bile.

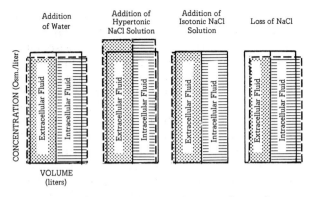

Fig. 21-15. Changes in the volume and osmolar concentration of the extracellular and intracellular fluids. Columns formed by solid lines represent the initial state. The interrupted lines represent the changes brought about by each experimental condition.

mass we obtain a value, 700 ml. per kilogram, fairly constant from person to person.

You will note in Figure 21-14 that the body water in the various compartments are in equilibrium with one another. The distribution of the water depends partly on the osmotic pressure gradient and partly on the hydrostatic pressure gradient between compartments. If a person drinks water there is an increase in the volume of water in all compartments, and a concomitant decrease in osmotic pressure (Fig. 21-15). These changes result in the stimulation of volume and osmotic receptors, which in turn causes a decreased release of ADH and a resultant increase in urine volume. You will remember from Figure 21-12, however, that it usually takes the kidneys 2 or more hours to eliminate excess water consumed. In other words, there is an interval even when the kidneys are functioning normally during which water intake increases extracellular and intracellular volume.

Infusing a hypertonic NaCl solution or taking NaCl by mouth has quite a different effect. The Na^+ and Cl^- are both extracellular ions and tend to remain in the extracellular compartments, where they increase the osmotic pressure. The increase in osmotic pressure results in the movement of water from the intracellular to the extracellular compartments. The infusion of isotonic saline, on the other hand, has little or no effect on the intracellular volume, since it remains in the extracellular fluid until eliminated by the kidneys. Sodium chloride depletion can also produce marked changes in water distribution. A man working in a hot environment, for example, can lose more than 2 liters of sweat per hour. *Sweat* is a hypotonic solution containing Na^+, K^+, Cl^-, etc. If sweating is accompanied—as it frequently is—with the intake of water but no salt, it may produce sufficient depletion of the extracellular ions Na^+ and Cl^- to create a marked decrease in extracellular fluid volume and an increase in intracellular fluid volume. Other conditions in which dehydration is associated with a marked electrolyte loss are prolonged vomiting, diarrhea, and diabetes mellitus. In each of these conditions a large decrease in the extracellular fluid volume is associated with mild changes in the intracellular fluid volume.

In cases of water and food deprivation, on the other hand, there is dehydration associated with an increase in the concentration of electrolytes. In the early phases the loss of water from the body via the lungs, skin, and feces tends to produce an increase in the osmotic pressure of the extracellular fluids that pulls water from the intracellular spaces. The kidney responds to this stress by increasing its reabsorption of water and its elimination of electrolytes. As the stress continues, however, tissue breakdown becomes an increasingly serious problem. The results of this are the release of cell water and such cellular factors as K^+ into the extracellular space. Eventually glomerular filtration is seriously depressed and the concentration of K^+ in the extracellular fluids reaches fatal levels.

Shifts in water from the plasma to the rest of the extracellular space also occur in the above example, as well as in response to malnutrition not associated with water deprivation, and liver damage. In each of these conditions the colloid osmotic pressure of the plasma decreases due to a decrease in the concentration of albumin and other plasma colloids. This results in a decrease in plasma volume due to the movement of plasma water to the other compartments. A similar situation occurs when the hydrostatic pressure in the capillaries increases. Hemorrhaging, on the other hand, may result in a decrease in hydrostatic pressure in the capillaries and an eventual increase in plasma volume.

The *transfusion* of whole blood frequently decreases the total plasma volume due to an increase in capillary hydrostatic pressure which tends to move plasma water into the other compartments. Transfusion of high-molecular-weight substances such as *dextran* (a plasma expander), on the other hand, increases the colloid osmotic pressure of plasma and can produce increases in plasma volume sufficient to cause up to a 40 per cent increase in total blood volume.

Dehydration. When water loss markedly exceeds intake the skin loses its elasticity and becomes hard and leathery and eventually the heat regulatory mechanisms of the body (insensible perspiration, sweating, shunting of large volumes of blood to the skin) may become so impaired that high fever results. Continued dehydration causes the blood to become sufficiently concentrated (anhydremia) that a general circulatory failure and anuria develop. Acidosis, cerebral disturbances (delirium and coma), and death represent the terminal events.

Water intoxication. The movement of excess quantities of water into the cells of the body can occur in reponse to a positive water balance or a negative salt balance, i.e., it may be associated with either an increase or a decrease in the total body water. In cases of excess sweating, for example, the subject may replace the water loss without replacing the salt loss. This can lead to water intoxication. Most of the more serious symptoms of this condition are associated with the movement of water into cells of the central nervous system. The resultant symptoms include salivation, nausea, vomiting, restlessness, asthenia (weakness), muscle tremors, ataxia (loss of muscle coordination), tonic and clonic convulsion, and eventually stupor and death.

Chapter 22

ACID-BASE BALANCE
FOREIGN SUBSTANCES
UREMIC SYNDROME
MICTURITION

ELIMINATION OF ACIDS, BASES, WASTE PRODUCTS, AND FOREIGN SUBSTANCES

Normal cellular activity in man is dependent upon a fairly constant extracellular environment. In Chapter 21 we considered the role of the kidney in the maintenance of electrolyte balance. In this chapter we shall be concerned primarily with its contribution to the maintenance of an acid-base balance and its role in the elimination of waste products and foreign substances.

ACID-BASE BALANCE

The H^+ concentration of the blood plasma generally remains at 0.00004 mEq. per liter. This represents a pH of 7.4. An increase above this pH is termed alkalosis and a decrease below it acidosis. The cells of the body usually function normally at plasma pH's between 7.35 and 7.45, but a pH below 7.0 (0.0001 mEq. of H^+/liter) or above 7.7 (0.00002 mEq. of H^+/liter) is not compatible with life. One of the problems in keeping the pH of the blood and other intercellular fluids constant is the metabolic activity of the cells. They, through the production of CO_2, place an H^+ load on the body of approximately 12,500 mEq. per day:

$$CO_2 + H_2O \rightleftharpoons H_2CO_3 \rightleftharpoons H^+ + HCO_3^-$$

Some cells also release large quantities of lactic acid into the blood during exercise, and, during diabetic ketosis, large amounts of acetoacetic acid and beta-hydroxybutyric acid. A second problem in maintaining a constant pH is the H^+ that enters the body in the food we eat. The catabolism of dietary protein, for example, results in the release of H_2SO_4 and H_3PO_4. This causes the addition of approximately 150 mEq. of H^+ to the blood each day. Fruits, on the other hand, are probably our main source of dietary alkali. Other sources of dietary acid are the salts, NH_4Cl and $CaCl_2$. $NaHCO_3$, on the other hand, is a salt that produces system alkalosis. A final problem in the maintenance of an acid-base balance is the conservation of acidic and basic body secretions (see Table 22-1). Vomiting, for example, can result in the loss of sufficient gastric juice to produce systemic alkalosis.

Buffers

The first line of defense against a change in pH are the buffer systems of the body. These systems minimize a change in H^+ concentration by giving up H^+ when the pH is high and taking up H^+ when the pH is low. Some buffer systems are made up of pairs of molecules, one member of which is an H^+ donor and another member of which is an H^+ receiver. Examples of this include the $NaHCO_3$-H_2CO_3 and the Na_2HPO_4-NaH_2PO_4 systems. The addition of acid to these causes some of the H^+ to combine with the $NaHCO_3$ and Na_2HPO_4 to form H_2CO_3 and NaH_2PO_4. The addition of base, on the other hand, causes the H_2CO_3 and NaH_2PO_4 to release H^+ into the solution. Another type of buffer system is represented by the proteins of the body,

	H^+ (mol/liter)	pH
Gastric secretion	1.6×10^{-1}	0.8
Urine: Acidosis	3×10^{-5}	4.5
Alkalosis	1×10^{-8}	8
Cerebrospinal fluid	5×10^{-8}	7.3
Plasma: Acidosis	1×10^{-7}	7
Normal	4×10^{-8}	7.4
Alkalosis	2×10^{-8}	7.7
Aqueous humor of eye	3×10^{-8}	7.5
Pancreatic juice	1×10^{-8}	8

Table 22-1. Hydrogen ion concentration in various body fluids under normal conditions and during extreme acidosis and alkalosis.

which are amphoteric. That is to say, each protein molecule has the capacity to both release H^+ and take it up (Fig. 16-10). We find, for example, that if one adds acid to 1 liter of blood until the pH has changed from 7.5 to 6.5, 27.5 mEq. of H^+ are taken up by the hemoglobin of the red blood cells, and 4.24 mEq. of H^+ are taken up by the plasma proteins. Other buffers include (1) the organic phosphate complexes—adenosine triphosphate, adenosine diphosphate, creatine phosphate, etc., and (2) the hydroxyapatite crystals of bone.

Respiratory Compensation

The respiratory system's role in acid-base balance is controlling the elimination of the acid-forming gas CO_2. This gas is stored in the blood as CO_2, HCO_3^-, and H_2CO_3. The equilibrium between these forms is characterized in the Henderson-Hasselbalch equation as follows:

$$pH = pK_{H_2CO_3} + \log \frac{HCO_3^-}{H_2CO_3}$$

If we let H_2CO_3 represent the extracellular fluid concentration of carbonic acid plus dissolved CO_2, in mEq. per liter, we find in a normal person at a pH of 7.4 an HCO_3^--to-H_2CO_3 ratio of 20:1 and a pK of 6.1. In other words,

$$pH = 6.1 + \log 20 = 7.4$$

The 20:1 ratio stays relatively constant as long as the body's major acid stress is the formation of CO_2. If a strong acid is added to the blood, however, the ratio decreases. This is illustrated in Figure 22-1. In the first bar-graph we have a representation of a normal person. His ratio of HCO_3^- to H_2CO_3 is 20:1 and his concentration of HCO_3^- is 28 mEq. per liter of extracellular fluid. If we now add to each liter of extracellular fluid 14 mEq. of strong acid (second bar-graph in Fig. 22-1) we find that the ratio changes from 20:1 to 0.9:1, and that the pH decreases to 6.06. If next the respiratory system eliminates the additional H_2CO_3 formed in response to the increased H^+ concentration, the pH will increase to 7.10. Generally, due to the low pH, there is sufficient hyperventilation to decrease the

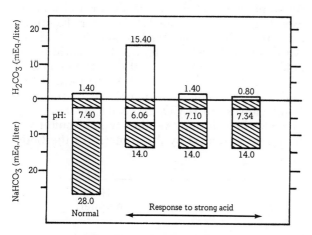

Fig. 22-1. The effect of the addition of 14 mEq. of strong acid/liter of extracellular fluid. The first graph represents the condition of the subject before the addition of acid. The second graph represents the primary response (the change in the bicarbonate buffer system). The third graph shows the results of the respiratory elimination of sufficient CO_2 to reduce the H_2CO_3 to the normal state. The fourth graph represents the usual response to acidosis: hyperventilation and a more marked reduction of H_2CO_3. (Redrawn from Physiology of the Kidney and Body Fluids, Second Edition, by R. F. Pitts. Copyright © 1968, Year Book Medical Publishers, Inc.; used by permission)

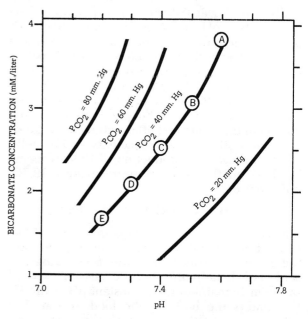

Fig. 22-2. The effect of the addition of fixed acid or base on pH and HCO_3^- concentration at different P_{CO_2}'s. The normal situation is represented at point C. The addition of fixed base shifts the curve to point B or A. The addition of fixed acid shifts the curve to point D or E. (Redrawn from Davenport, H. W.: The ABC of Acid-Base Chemistry. 4th ed., p. 40. Chicago, University of Chicago Press, 1958)

amount of H_2CO_3 to less than was found before the addition of acid. This results in a pH of about 7.34 and a decrease in the arterial P_{CO_2} from 40 mm. Hg to about 24 mm. Hg. In other words, the addition of fixed acid to the body decreases the HCO_3^- concentration, while the addition of fixed base raises the HCO_3^- concentration (Fig. 22-2).

Renal Compensation

The lungs are responsible for the net loss of about 13,000 mEq. of volatile acid a day, and the kidneys for a net loss in the urine of 40 to 80 mEq. of fixed acid a day. Their importance in the control of the acid-base balance, however, lies not merely in the quantity of acid they excrete, but also in the form in which the acid is excreted. We have already noted that one consequence of a respiratory response to acidity produced by a nonvolatile acid is a *decrease in the HCO_3-to-H_2CO_3 ratio*. The kidney, however, can restore this ratio to normal by (1) increasing the excretion of H^+ from the body ($H_2CO_3 \rightleftharpoons H^+ + HCO_3^-$) and (2) increasing the reabsorption of HCO_3^-. The respiratory response to injection of a fixed base, on the other hand, is an *increase in the HCO_3^--to-H_2CO_3 ratio*. The renal response is (1) a decrease in the excretion of H^+ and (2) a decrease in the reabsorption of HCO_3^-. The cost to the kidney of excreting HCO_3^- is that it must be paired with a cation such as Na^+ or K^+, and the cost of excreting H^+ is that it must be paired with an anion such as Cl^-, phosphate, or various organic substances. If, as is usually the case, these cations and anions are in excess and must be excreted anyway, the cost is unimportant. Through their control of the excretion of CO_2, H^+, and HCO_3^-, that is, the lungs and kidneys together are able to maintain a fairly constant plasma pH without upsetting the concentration of other substances in the plasma:

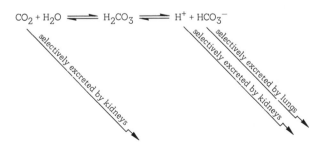

Bicarbonate reabsorption. Bicarbonate excretion generally begins at a plasma concentra-

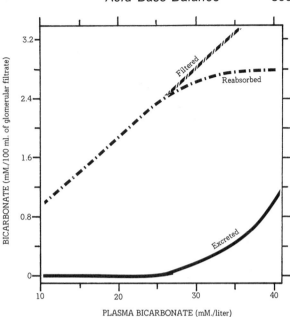

Fig. 22-3. Filtration, reabsorption, and excretion of HCO_3^- in normal men. (*Redrawn from* Pitts, R. F., et al.: Reabsorption and excretion of bicarbonate. J. Clin. Invest., 28:37, 1949)

tion of 28 mM. per liter of plasma (Fig. 22-3). Since this is the usual plasma concentration of HCO_3^-, we see that the kidneys seem to have a built-in mechanism for maintaining this concentration. If, for example, a subject were to take $NaHCO_3$ by mouth, the kidneys would increase their excretion of HCO_3^- and within a number of hours cause the plasma levels to return to normal.

It should be emphasized in reference to Figure 22-3, however, that the relationship between plasma HCO_3^- and renal reabsorption is not static. Increases in the plasma concentration of CO_2 (Fig. 22-4), decreases in the body's K^+, decreases in the plasma concentration of Cl^-, and increases in the production of the adrenocortical hormones can all elevate the quantity of HCO_3^- the kidney is capable of reabsorbing.

The mechanisms for HCO_3^- reabsorption lie in both the proximal tubule of the nephron and the collecting duct. Normally about 90 per cent of the HCO_3^- is reabsorbed in the proximal tubule, and the rest in the collecting duct. Apparently an important part of the reabsorptive mechanism is the carbonic anhydrase in the tubular and collecting duct cells. It has been demonstrated that large concentrations of car-

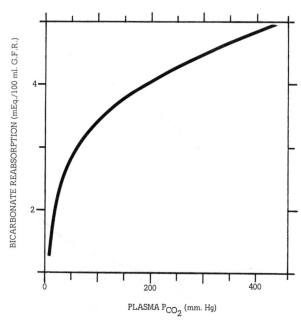

Fig. 22-4. The effect of changes in plasma CO_2 concentration on HCO_3^- reabsorption. The above relationship permits the kidney to maintain a ratio of HCO_3^-/H_2CO_3 of 20/1 during an elevated concentration of CO_2 and hence H_2CO_3. (*Redrawn from* Rector, F. C., Jr., et al.: The role of plasma CO_2 tension and carbonic anhydrase activity in the renal reabsorption of bicarbonate. J. Clin. Invest., 39:1709, 1960)

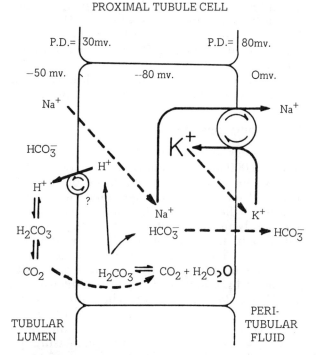

Fig. 22-5. A schematic diagram of the HCO_3^- reabsorption system for the proximal tubule. In this system carbonic anhydrase (not indicated) catalyzes the rapid formation of H_2CO_3, which serves as a ready source for H^+ and HCO_3^-. The active transport (*uninterrupted lines*) of H^+ into the tubular fluid is more characteristic of the collecting duct than the proximal tubule cell, since in the proximal tubule the active transport of Na^+ into the peritubular fluid is, by itself, probably all that is needed to facilitate the diffusion (*interrupted lines*) of H^+ and HCO_3^- out of the cell. (*Modified from* Pitts, R. F.: A comparison of the modes of action of certain diuretic agents. Progr. Cardiovasc. Dis., 3:547, 1961)

bonic anhydrase are contained in these cells and that *carbonic anhydrase inhibitors* such as acetazolamide depress HCO_3^- reabsorption, thereby increasing its excretion.

The role of carbonic anhydrase (a zinc-containing protein with a molecular weight of 30,000) in catalyzing the $H_2CO_3 \rightleftharpoons H_2O + CO_2$ reaction in the red blood cell was discussed in an earlier chapter. It apparently serves a similar function in the renal cells. The importance of this in the reabsorption of HCO_3^- in the proximal tubule is indicated in Figure 22-5. You will note that the tubule cell actively extrudes Na^+ into the peritubular fluid and that this creates a concentration gradient which facilitates the diffusion of additional Na^+ into the cell from the tubular lumen. The role of the carbonic anhydrase is to maintain a high concentration of H_2CO_3 in the cell, which serves as (1) a source of H^+ to replace the Na^+ lost from the peritubular fluid and (2) a source of HCO_3^- for the peritubular fluid. In other words, the active movement of Na^+ into the peritubular fluid facilitates the diffusion of H^+ into the tubular lumen and the diffusion of HCO_3^- into the peritubular fluid.

A few investigators have suggested that in the proximal tubule there may also be an active transport system for the movement of H^+ into the tubular lumen. Other workers, however, feel that the Na^+ pump shown in Figure 22-5 is more than adequate to facilitate the diffusion of HCO_3^- into the peritubular fluid and H^+ into the tubular lumen. In the collecting duct, on the other hand, the situation is quite different. Here the concentration gradient between tubular HCO_3^- and peritubular HCO_3^- may be considerably larger than that in the proximal tubule. In order to produce a urine with virtually no HCO_3^-, an additional pump mechanism is required.

Secretion of H^+. In the normal person most of the H^+ that enters the urine is secreted into the nephron rather than into the collecting duct. This movement, however, generally results in changes in pH of less than 0.5 pH units because of the buffer capacity of the urine in the nephron (Fig. 20-23). In the collecting duct, on the other hand, the amount of H^+ secreted is considerably less but may produce a more marked change in pH than is noted in the nephron. The small increases in urine acidity in the nephron result from the movement of large quantities of H^+ into the urine against a small concentration gradient, but in the collecting duct activate a mechanism which moves smaller quantities of H^+ against a much larger concentration gradient.

I suggested in an earlier section that the secretion of H^+ into the nephron is for the most part the result of the active transport of Na^+ out of the urine and the availability of H^+ to replace it, due to the presence of the intracellular catalyst carbonic anhydrase (Fig. 22-5). I further suggested that in the collecting duct there is, in addition to the Na^+ transport mechanism, an H^+-K^+ pump (Fig. 20-24). Apparently the K^+ and H^+ share the same mechanism, since large quantities of K^+ restrict the secretion of H^+. Other mechanisms which play a role in the acidification of the urine and the associated alkalinization of the body are the reabsorption of such basic salts as sodium dibasic phosphate (Na_2HPO_4) and sodium bicarbonate from the nephron.

Secretion of buffers. The amount of acid eliminated in the urine also depends upon the quantity and type of buffers in the urine. Usually the major buffers in the urine are the bicarbonate (pK 6.1) and dibasic phosphate (pK 6.1) systems. The approximate concentration of HCO_3^- in the glomerular filtrate is 24, and that of HPO_4^{--}, 1.5 mEq. per liter. Creatinine, β-hydroxybutyrate, and p-aminohippurate generally serve a less important buffer function due to their lower pK values (Fig. 22-6). In diabetic acidosis, on the other hand, there are greater molar quantities of β-hydroxybutyrate in the urine, which can be a more important buffer than the phosphate system.

One means of determining the amount of H^+ eliminated in the urine is to find out the amount of base which must be added to a given volume of urine to return its pH to 7.4. The value obtained is the *titratable acidity*. It should be

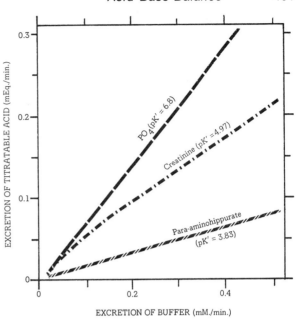

Fig. 22-6. The role of increasing quantities of buffer on the excretion of titratable acid in the urine of acidotic men. In these experiments the subjects were made acidotic by the oral administration of NH_4Cl. In one experiment neutral sodium phosphate was infused at a rate sufficient to cause equimolar increases in excretion. These experiments were repeated using creatinine and sodium p-aminohippurate. It is concluded that increasing quantities of buffer in the urine facilitates the elimination of greater quantities of titratable acid. (*Redrawn from* Schiess, W. A., et al.: The renal regulation of acid-base balance in man. II. Factors affecting the excretion of titratable acid by the normal human subject. J. Clin. Invest., 27:57-64, 1948)

emphasized, however, that this determination measures only a fraction of the acid lost in the urine, since it does not measure the H^+ combined with HCO_3^- or NH_3.

Secretion of NH_3. The movement of H^+ into or out of the urine is limited by the concentration gradient for H^+ between the urine and the cells which form the collecting duct. As a result it is impossible to produce a urine with a pH less than 4.4 or more than 8.2. This fact underscores the importance of substances in the urine which can combine with H^+ and in this way minimize decreases in pH as H^+ is moved into the urine. The bicarbonate, phosphate, and other buffer systems we have discussed do remove H^+ from solution, but unfortunately are effective only within the range of 1 pH unit on either side of their pK. It should also be remembered that the excretion of large amounts of phosphate in the form of NaH_2PO_4 can result in the depletion not only of the body's phosphate reserves, but also in that of its sodium reserves. One mechanism which the kidney uses to eliminate H^+ in the urine does not have these disadvantages. This is the production and secretion of NH_3 by the cells of the nephron, the collecting duct, or both. The site of production varies from one mammalian species to the next. The importance of NH_3 is that it combines with H^+ and forms the cation NH_4^+. Although NH_3 diffuses readily through the cell membranes of the nephron, NH_4^+ does not. In short, we have a situation which favors the movement of NH_3 into the urine as long as NH_4^+ is being produced, but which does not permit the diffusion of NH_4^+ out of the urine. The presence of large quantities of NH_4^+ in the urine also facilitates the reabsorption of other cations and is therefore an important mechanism in the conservation of body Na^+.

The ability of the kidney to form NH_3 was suggested by Nash and Benedict in 1921. They noted (1) that the concentration of ammonia in the renal venous blood is higher than that in the renal arterial blood and (2) that the arterial concentration of ammonia does not change under conditions which markedly modify the excretion of ammonia (acidosis, alkalosis, and nephrectomy). Apparently most of the NH_3 produced by the renal cells is the result of the removal of an amide group from the glutamine which the renal cells receive from the plasma (Fig. 22-7):

$$\text{glutamine} \xrightarrow{\text{(glutaminase)}} \text{glutamic acid} + NH_3$$

Other amino acids (L-asparagine, L- and D-alanine, L-histidine, L-aspartic acid, glycine, L-leucine, L-methionine, and L-cysteine—listed in order of decreasing importance as sources of NH_3) apparently also contribute to the production of NH_3 by providing amine groups to α-ketoglutarate and glutamate to form additional glutamine:

$$\alpha\text{-ketoglutarate} + \alpha\text{-amino acid} \xrightarrow{\text{transaminase}} \text{glutamate} + \alpha\text{-keto acid}$$

$$\text{glutamate} + \alpha\text{-amino acid} + ATP \xrightarrow{\text{enzymes}} \text{glutamine} + \alpha\text{-keto acid} + ADP + PO_4$$

Much of what we know concerning ammonia production is summarized in Figure 22-8. Here we see the data obtained from a young adult. Under normal conditions he was excreting in

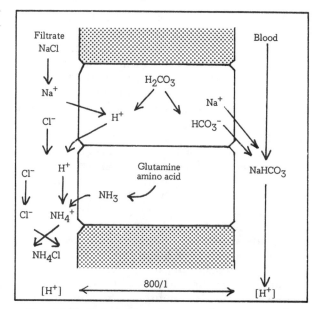

Fig. 22-7. Tubular synthesis and secretion of H^+ and NH_4^+. (*Modified from* Pitts, R. F.: Renal excretion of acid. Fed. Proc., 7:425, 1948)

his urine about 130 mEq. of chloride, 40 mEq. of ammonia, 40 mEq. of titratable acid, 125 mEq. of sodium, and 78 mEq. of potassium each day. In response to an increase in his normal fixed acid load there was an initial moderate increase in the titratable acid and ammonia in the urine and a marked increase in urinary Na^+ and Cl^-. Note the progressive decline in Na^+ excretion as the acidosis continued. This is in part due to a progressive increase in the elimination of other cations (K^+ and NH_4). Apparently as the acidosis continues there is a progressive increase in the production of ammonia.

The importance of the ammonia production in response to acidosis has also been shown in studies on subjects with chronic nephritis. In these cases ammonia production is impaired and, as a result, the subject is less able to compensate for a fixed acid load. A load that can be well tolerated by a normal person can, in chronic nephritis, result in (1) depletion of the body's intercellular and intracellular cation stores, (2) exhaustion of the bicarbonate and other buffer systems, (3) systemic acidosis, and (4) dehydration.

Metabolic Acidosis and Alkalosis

Metabolic acidosis is due to the addition of fixed acid to, or the loss of fixed base from the

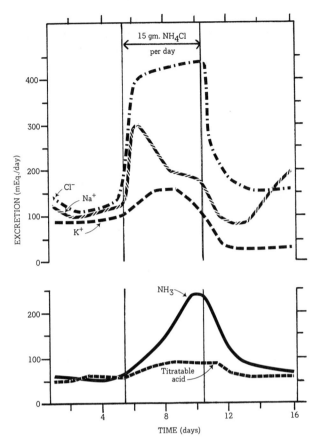

Fig. 22-8. Changes in urine composition (mEq./day) in a young healthy man in response to an acidosis produced by the ingestion of 15 Gm. of NH$_4$Cl/day (290 mEq. of a strong acid/day) for 5 days. The subject was maintained on a diet that was constant with respect to calories, proteins, and electrolytes. Note the progressive increase in the secretion of ammonia and chloride during the acidosis. Note, too, the initial increase in sodium excretion and its progressive decline as ammonia excretion increases. It is concluded that increases in the excretion of ammonia can serve not only to bring about increased excretion of acid by the formation of the ammonium cation, but also the conservation of cations such as sodium. (*Modified from Pitts, R. F.: Physiology of the Kidney and Body Fluids.* 2nd. ed., p. 208. Chicago, Year Book Medical Publishers, 1968)

body. It can be produced by the ingestion of NH$_4$Cl or by the ketosis produced in diabetes mellitus. It is associated with a decrease in the pH of the body fluids and a decrease in their HCO$_3^-$ concentration. In other words there is a shift from point C in Figure 22-2 to point D or E.

Metabolic alkalosis, on the other hand, results from the addition of fixed base to, or the loss of fixed acid from the body. It can be produced by the ingestion of NaHCO$_3$ or by prolonged vomiting. It is associated with changes in pH and HCO$_3^-$ concentration which are the opposite of those found in metabolic acidosis.

Respiratory Acidosis and Alkalosis

The excretion of CO$_2$ by the respiratory system decreases in cases of emphysema and depression of the respiratory centers. This, in turn, results in a buildup of CO$_2$, HCO$_3^-$, and H$^+$ in the body fluids. This condition is called respiratory acidosis. Like metabolic acidosis, it is associated with a decrease in pH, but unlike metabolic acidosis, is also associated with an increase in HCO$_3^-$ concentration.

Respiratory alkalosis, on the other hand, may result from living at a high altitude or from hyperventilation associated with hysteria or some other condition. In Figure 22-9(A) we have a comparison of the situations which exist in metabolic acidosis and alkalosis and in respiratory acidosis and alkalosis. Note that respiratory alkalosis is characterized in terms of an increase in pH and a decrease in HCO$_3^-$ concentration (a shift from point A to point C).

Compensated Acidosis and Alkalosis

The initial response of the body to the addition of an acid or base involves buffer systems. The addition of a fixed or volatile acid results in the hemoglobin molecule taking up H$^+$ and the HPO$_4^{--}$ forming H$_2$PO$_4^-$. The addition of a base results in the hemoglobin giving up H$^+$ and the H$_2$PO$_4^-$ forming H$^+$ and HPO$_4^{--}$. The end result is that our buffers minimize but do not prevent changes in body pH.

A further compensation for changes in pH involves the respiratory system and the kidneys. In respiratory acidosis, for example, there is an increase in HCO$_3^-$ concentration and a decrease in pH (a shift from A to B in Fig. 22-9(B)). Eventually the kidneys respond to the acidosis by reducing their HCO$_3^-$ excretion and increasing their excretion of acid in the form of NH$_4$Cl, NaH$_2$PO$_4$, and H$^+$. This results at first in a partial compensation (a shift from B to B$_1$ in Fig. 22-9(B)) or in other words, a return of the pH toward 7.4. Several days, however, may be required for complete compensation (shift from B$_1$ to B$_2$). The acidotic condition prior to the increased excre-

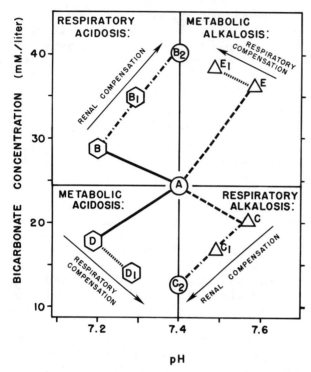

Fig. 22-9. Effect of an acid–base disturbance on HCO$_3$ concentration and pH. The capacity of the kidney to compensate for respiratory acidosis and alkalosis is represented by points B$_2$ and C$_2$ respectively. The capacity of the respiratory system to compensate for metabolic acidosis and alkalosis is indicated by points D$_1$ and E$_1$.

tion of acid by the kidney is called *uncompensated respiratory acidosis*. The condition after the kidneys have returned the pH to normal is termed *compensated respiratory acidosis*. Note here that although the pH is returned to normal, the HCO$_3^-$ concentration is abnormally high.

In compensated respiratory alkalosis the kidneys decrease their excretion of acid and increase their excretion of NaHCO$_3$ and Na$_2$HPO$_4$. This brings the pH back to normal but markedly reduces the HCO$_3^-$ concentration (shift from C to C$_1$ to C$_2$). In compensated metabolic acidosis (D$_1$), the respiratory system eliminates increased quantities of CO$_2$. This results in a decrease in the HCO$_3^-$ concentration and an increase in the pH, but is seldom associated with complete restoration of the pH to 7.4. Compensated metabolic alkalosis (E$_1$), on the other hand, is associated with decreased excretion of CO$_2$, an elevation of HCO$_3^-$, and a lowering of the pH to a value generally in excess of 7.4. In short we have found that by checking the pH and HCO$_3^-$ concentration of the blood we can characterize a pH disturbance as being uncompensated or compensated (1) metabolic acidosis, (2) metabolic alkalosis, (3) respiratory acidosis, or (4) respiratory alkalosis.

FOREIGN SUBSTANCES

A number of drugs and other foreign materials introduced into the body are eliminated by either biliary or renal excretion. The substance may be eliminated unchanged or may first undergo certain metabolic transformations. In the case of the barbiturate barbital, 65 to 90 per cent of the total administered dose appears in the urine unchanged. In this particular case the process of elimination is a very slow one, taking about 12 hr. for the excretion of the first 8 per cent and 48 hr. for the excretion of 35 to 65 per cent. Traces of barbital may be found in the urine as long as 8 to 12 days after the administration of a single dose. As much as 30 per cent of phenobarbital, diallylbarbituric acid, and aprobarbital are also excreted unchanged by the kidney.

The renal excretion of penicillin, on the other

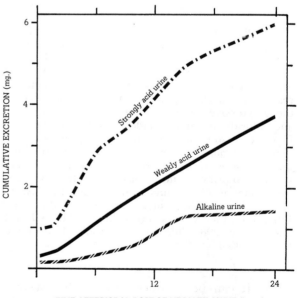

Fig. 22-10. The role of urinary pH on the excretion of a single oral dose (10 mg.) of the ganglionic blocking agent mecamylamine. (*Redrawn from* Milne, M. D.: Observations on the pharmacology of mecamylamine. Clin. Sci., *16*:604, 1957)

hand, is much more rapid than that of barbiturates. Approximately 60 to 90 per cent of an aqueous solution of intramuscular pencillin appears in the urine, most of it in the first hour after administration. Approximately 90 per cent of the urinary penicillin is secreted into the nephron, the remaining 10 per cent getting there by filtration. This secretory process is so active in 3- to 4-year-old children that detectable levels of penicillin are not found in the blood 2 to 3 hr. after injection. With increasing age, however, the secretory process becomes progressively less effective.

The excretion of drugs by the kidney sometimes varies not only with age, but also with the pH of the urine (Fig. 22-10). This is true, for example, in the case of a number of moderately strong amines (chloroquine, mecamylamine, etc.). These molecules become ionized in strongly acidic urine and this inhibits their reabsorption. The action and duration of action of many drugs, in other words, may depend to a great extent on the functioning of the kidney.

UREMIC SYNDROME

In Chapter 20 I discussed renal disease and made the point that it can lead to generalized kidney damage, uremia, and death. In this section we shall take a closer look at some of the symptoms associated with renal shutdown. These are symptoms noted in progressive renal disease, acute nephritis, tubular nephrosis, and acute urinary obstruction, and after unilateral nephrectomy.

One of the most prominent symptoms in partial or complete renal shutdown is an increase in excretory products in the blood. Piorry in 1840 invented the term "uremia" (urine in the blood) to characterize this condition. Uremia has, however, come to mean the complex of symptoms seen in kidney shutdown (headache, vomiting, dyspnea, insomnia, delirium, convulsions, coma, etc.). One of the commonest and most pronounced symptoms is an increase in blood urea. This results in an increase in the urea concentration in the various secretions of the body, which in some cases results in a urea frost on the skin from the crystalized urea in the sweat and a smell of ammonia on the breath due to the breakdown of the urea in saliva by the microorganisms of the mouth. The increases in plasma urea do not, however, seem to be the cause of the more serious symptoms in uremia.

The most serious effect of uremia is the systemic acidosis which results from partial or complete renal shutdown. In chronic renal disease, for example, there is a marked decrease in the excretion of titratable acid and an even more pronounced decrease in the excretion of ammonia. This can cause a sufficiently great systemic acidosis that the cells start to lose small quantities of intracellular K^+. Death results if the extracellular concentration reaches a value of 8 to 10 mEq. of K^+ per liter (see Fig. 6-22). Smaller increases can result in marked changes in the heart (high and peaked T waves in the electrocardiogram, broadened QRS complexes, and conduction defects such as bundle branch block) and central nervous system (psychosis, muscular twitchings, headache, neurlagic pains). In some cases, however, vomiting and diarrhea may result in such a loss of K^+ that hypokalemia results.

Usually there is also an increase in plasma phosphate and sulfate and a decrease in bicarbonate. The increase in phosphate tends to decrease the concentration of dissolved Ca^{++}, which in turn tends to produce a further increase in the excitability of muscle and nerve. If the uremic patient is not already showing muscle twitches or tetany due to increases in extracellular K^+, the decreases in Ca^{++} may now be more than enough to initiate these phenomena. Sometimes, however, there is not tetany even in the presence of hyperkalemia and hypocalcemia. In these cases the absence of tetany is due to the fact that H^+ per se decreases excitability. In these cases the use of alkali therapy may trigger a tetanic episode. In other words, the acidosis in uremia produces over a period of several days or weeks a series of changes in the electrolyte composition of the extracellular fluid which cannot be rapidly relieved by alkali therapy.

MICTURITION

The urine produced in the kidney is carried in the collecting ducts to the renal pelvis and then to the ureter, where a wave of smooth muscle contraction (peristalsis) moves it toward the urinary bladder. The right ureter from the right kidney and the left ureter from the left kidney both enter the bladder obliquely. The oblique course of the ureters, along with a flaplike fold of mucous membrane which separates the lumen of the urinary bladder from that of each ureter, serves to prevent the reflux of urine back into the ureter.

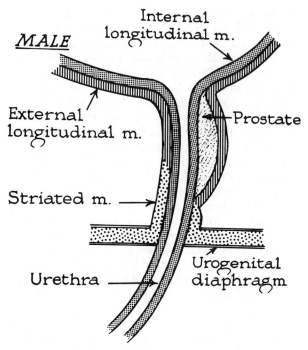

Fig. 22-11. Schematic drawing of a sagittal section of the male urethra. (Hutch, J. A., and Rambo, O. H., Jr.: New theory of anatomy of internal urinary sphincter and physiology of micturition. III. Anatomy of urethra. J. Urology, 97:701. Copyright © 1967, The Williams and Wilkins Company, Baltimore)

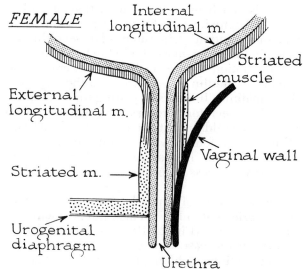

Fig. 22-12. Schematic drawing of a sagittal section of a female urethra. (Hutch, J. A., and Rambo, O. H., Jr.: New theory of anatomy of internal urinary sphincter and physiology of micturition. III. Anatomy of urethra. J. Urology, 97:702. Copyright © 1967, The Williams and Wilkins Company, Baltimore)

The urinary bladder is a musculomembranous organ which serves to store the urine it receives and, when properly stimulated, to expel it into the urethra (Figs. 22-11 and 22-12). In the initial stages of micturition parasympathetic fibers to the bladder cause its contraction. This results in (1) an increase in pressure in the bladder, (2) a widening or funneling of the first part of the urethra, and (3) an apparent shortening of the urethra (Fig. 22-13). Finally there is reflex relaxation of the striated muscle surrounding the urethra (the external urethral sphincter) and the expulsion of urine from the body.

Ureters

The peristaltic waves which carry the urine down the ureters to the bladder are a response of the smooth muscle of the ureter to distention. That is to say they are either myogenic or due to the intrinsic nervous plexi in the ureter. They generally occur 2 or 3 times a minute but their frequency may be increased to 10 a minute by the stimulation of parasympathetic fibers to the ureter, or decreased by the stimulation of sympathetic fibers. Complete denervation of the ureters, however, does not cause the peristaltic waves to disappear.

Normally the peristaltic waves produce a pressure of from 1 to 8 cm. H_2O. This is sufficient to move a bolus of urine through the ureter and to open the intravesical portion of the ureter. This portion of the ureter is normally closed at all times except when the pressure in the ureter is greater than that in the bladder. It serves to protect the upper urinary system from the pressures developed by the bladder during micturition and from infection due to *vesicoureteral reflux.* In cases of ureteral obstruction, however, the ureter may become distended with urine central to the obstruction and develop pressures as high as 70 cm. H_2O. This results in the stimulation of pain fibers in the ureter and is one of the most intense pains (*renal colic*) known to man. If not relieved promptly it may also produce renal atrophy.

Urinary Bladder and Urethra

Innervation. The smooth muscle of the urinary bladder is innervated by parasympathetic fibers traveling down the *pelvic nerves* from the sacral portion of the spinal cord (S-1 through 4) and by sympathetic fibers in the *hypogastric*

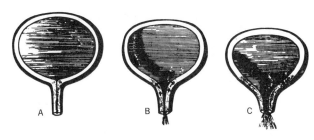

Fig. 22-13. Micturition. At A we note a distended bladder prior to contraction. At B and C we observe its changes during the contraction of the detrusor muscle. Note that during micturition the contraction of the detrusor muscle causes a widening or funneling of the vesicle outlet and an apparent shortening of the urethra. (Nesbit, R. M., and Lapides, J.: The physiology of micturition. J. Mich. Med. Soc., 58:385, 1959)

nerve from the thoracolumbar cord (T-10 through 12, L-1 and 2). The skeletal muscle which forms the external urethral sphincter is innervated by somatic efferent fibers originating from the sacral cord (S-1 through 4) and carried in the *pudendal nerve*. Sensory impulses from the bladder and urethra move up these nerves, entering the cord as high as the ninth thoracic segment. Part of the evidence for this is that patients with cord transections as high as T-10 have reported feeling urinary catheters at the neck of the bladder as well as bladder distention, and patients with transections at L-4 and 5 have reported the desire to void during bladder distention.

The specialized receptors in the urinary bladder are more similar to those found in the tendons and aponeuroses of skeletal muscle than to those in the rectum or most other organs containing smooth muscle. They have not been thoroughly studied, but are thought to include both stretch and cold receptors. The stretch receptors include both rapidly and slowly adapting *proprioceptors* which play an important role in (1) the sensation of bladder fullness, (2) the reflex depression of detrusor muscle tone, and (3) the reflex increase in detrusor muscle tone.

Reflex micturition. The capacity to expel urine from the urinary bladder appears in the fetus during the fifth intra-uterine month and usually remains an involuntary reflex act until the child reaches the age of 2½ yrs. At this time it begins to come under cortical control; by 3 yrs. of age it is usually almost completely under the control of the will. In the adult micturition usually removes all the urine from the bladder except from 0.09 to 2.34 ml., and reduces the intravesicular pressure to close to 0 cm. of water. As (1) the bladder slowly fills its pressure changes very little. This is initially the result of the high compliance of the bladder, but as the bladder continues to fill there is (2) stimulation of stretch receptors which reflexly produce an increase in the sympathetic tone to the bladder. This (3) causes a decrease in detrusor muscle tone, which also prevents marked changes in intravesicular pressure during bladder filling (Fig. 22-15). You will note, for example, in Figure 22-17(A), that in a normally innervated bladder its distention with 10 ml. of liquid produces an increase in pressure followed by a decrease. In the sympathectomized bladder, on the other hand, distention with only 4 ml. of liquid is sufficient to initiate the micturition reflex (see 22-17(B)).

As the bladder continues to fill, a point is eventually reached at which (4) the subject has his first feeling of bladder fullness. At this point the bladder contains about 50 ml. of urine and has an intravesicular pressure of less than 3 cm. of water. At a volume of approximately 200 ml. and a pressure of 3 cm. of water the subject (5) experiences his first desire to void. This is associated with a vague feeling that seems to originate from the penis or perineum. At a volume of 325 ml. the pressure in the bladder has risen to 4.5 cm. of water, and further increases in volume result in considerably more marked increases in pressure. At this volume the subject feels (6) definite discomfort. At a volume of 375 ml. the pressure rises to 8 cm. of water and the subject experiences (7) a strong desire to void and (8) a reflex stimulation of the detrusor muscle by parasympathetic neurons which he can no longer prevent.

In the case of the bladder with a small volume of urine, micturition is (1) initiated by facilitatory impulses originating from the paracentral lobule of the cerebral cortex. In the full bladder, however, micturition is initiated by a removal of cortical inhibition. In either case this permits (2) the stimulation of parasympathetics to the bladder which (3) causes a contraction of the detrusor muscle. As the muscle contracts (4) the intravesicular pressure rises to between 50 and 150 cm. H_2O (Fig. 22-14). At a pressure of between 18 and 43 cm. of water (5) the bladder neck gradually opens and (6) a bolus of urine enters the dorsal urethra (the prostatic urethra in the male). This results in (7) the stimulation of afferents in the pelvic nerve and (8) a reflex inhibition in S-2 through S-4 of somatic efferent neurons to the external sphincter. As a result, (9) the sphinc-

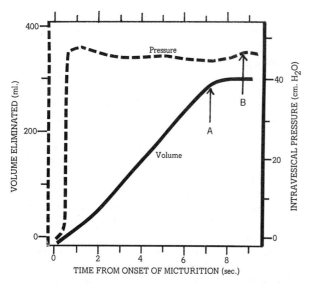

Fig. 22-14. Urine volume and pressure curves in a normal male subject during micturition. The subject eliminated 300 ml. of urine by producing a pressure of 48 cm. H_2O. At A the flow rate approached 0, but the pressure in the bladder remained high. The elevation of pressure at B ("after-contraction") is probably a reflex contraction of the striated muscle surrounding the urethra, which produces a final ejaculatory emptying of the urethra. (*Redrawn from* Hinman, F., Jr., and Cox, C. E.: Residual urine volume in normal male subjects. Trans. Am. Ass. Genito-urin. Surg., 58:84, 1966)

clude the relaxation of the levator ani and perineal muscles at the onset of the micturition reflex and a closing of the glottis followed by contraction of the diaphragm and muscles of the abdominal wall. These all serve to accelerate the ejection of urine.

Normally during micturition (11) the contraction of the detrusor muscle continues for a second or more after the expulsion of urine has ceased. This is apparently due to facilitatory impulses from the brain stem. Then there is (12) an increase in tone in the external sphincter, (13) relaxation of the detrusor muscle, and (14) gradual closing of the bladder neck. It is also possible, on the other hand, to rapidly stop micturition before the bladder has completely emptied. This is done by the cerebral cortex' causing a strong contraction of the external sphincter and perineal muscles. This results in a burning sensation which appears to originate from the membranous urethra. If this contraction is maintained there is a gradual relaxation of the detrusor muscle and a reflux of urine from the urethra back into the bladder. In cases in which the external urethral sphincter is paralyzed it takes from 10 to 15 sec. to stop micturition once it has begun.

Role of the central nervous system. Under normal conditions micturition results in the expulsion of all the urine in the bladder except for 0.09 to 2.34 ml. Transection of the cord above S 1, however, produces what is usually called a *"cord bladder,"*—i.e., a bladder and urethra whose parasympathetic and somatic efferent innervation is no longer under the influence of the brain. The transection in man initially causes the bladder to become flaccid and unresponsive. Eventually it becomes

ter relaxes and (10) urine passes through the urethra to the outside. While these reactions are occurring there may be a number of other muscular changes which are not essential for but which usually accompany micturition. These in-

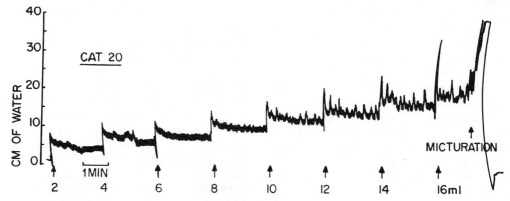

Fig. 22-15. Cystometrogram (recording of pressure in the urinary bladder) obtained by the step-wise filling of the urinary bladder of a cat under chloralose-urethane anesthesia. (Gjone, R., and Setekleiv, J.: Excitatory and inhibitory bladder responses to stimulation of the cerebral cortex in the cat. Acta Physiol. Scand., 59:340, 1963)

markedly distended and urine dribbles past the sphincters. In time, however, the micturition reflex returns. Under these conditions (1) it is no longer possible to initiate voluntary micturition by the usual techniques and (2) the *residual volume* of urine left after reflex micturition is considerably above normal. Since urine is a good medium for the growth of microorganisms, this situation markedly increases the danger of urinary infection. Some paraplegic patients, however, are able to initiate micturition by pinching their thighs or scratching the skin in the genital area.

The role played by the central nervous system in the control of micturition is a complicated one. Besides the micturition reflex centers in the sacral cord there are also, in the medulla oblongata, pons, midbrain, hypothalamus, septum pellucidum, amygdala, and cerebral cortex, areas which facilitate and inhibit micturition (Fig. 22-16). The sensation of bladder distention is carried to the brain in the dorsal funiculus of the cord, and that of active contraction travels to the brain in the lateral funiculus. Descending fibers (1) in the lateral reticulospinal tract facilitate vesicular contraction, (2) in the ventral reticulospinal tract, vesicle relaxation, and (3) in the medial reticulospinal tract, tonic contractions of the external urethral sphincter.

Automatic bladder. If the sacral reflex centers are destroyed or if the motor nerves to the bladder are cut, the bladder becomes flaccid and distended for a while but eventually becomes active and shrunken and initiates contractions which expel small quantities of urine (Fig. 22-17). These contractions are not sufficiently strong or prolonged, however, to prevent the development of a small, constantly present, and potentially dangerous residual volume of urine.

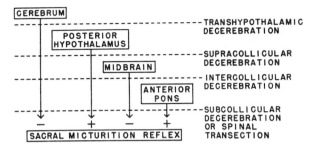

Fig. 22-16. Effects of transections at various levels of the brain on the micturition reflex. On the basis of studies such as this, it is concluded that the predominant effect of the cerebrum is depression (−) of the micturition reflex. Thus, a transhypothalamic decerebration eliminates this depression of the sacral cord and results in an initiation of the micturition reflex at a smaller bladder distention. An intercollicular decerebration also results in a micturition reflex at a lower than normal vesicle volume, but the subcollicular decerebration results in a complete loss of the reflex for several days. Thus the pons is thought to have a powerful facilitatory area. (Tang, P. C.: Levels of brain stem and diencephalon controlling micturition reflex. Neurophysiol., 18:594, 1955)

Destruction of (1) the sensory neurons from the bladder or (2) the dorsal columns of the sacral cord also prevents the micturition reflex and results in a chronically high residual volume of urine in the bladder. This condition is characteristic of the late stages of *tabes dorsalis*, a chronic progressive sclerosis of the posterior columns and spinal roots.

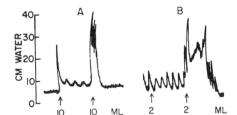 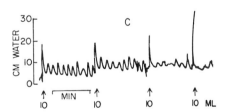

Fig. 22-17. Effects in the anesthetized cat of bladder distention on vesical pressure in the normal bladder (A), the bladder after sympathectomy (B), and the bladder after complete decentralization (C). Note that it required a bladder distention of 20 ml. to initiate micturition in the normal animal (A), but only 4 ml. to initiate micturition in the sympathectomized cat (B). The decentralized bladder (C) continues to have rhythmic contractions, but does not show the maintained contraction associated with micturition in the normally innervated animal. (Gjone, R.: Peripheral autonomic influence on the motility of the urinary bladder in the cat. Acta Physiol. Scand., 66:85, 1966)

Chapter 23

THE DIGESTIVE SYSTEM (MOTILITY)

DIGESTION
ANATOMY
CHEWING (MASTICATION)
SWALLOWING (DEGLUTITION)
THE STOMACH
THE SMALL INTESTINE
THE COLON
DEFECATION
COLECTOMY
MEGACOLON

DIGESTION

The digestive tract is a tube about $4\frac{1}{2}$ meters long which, in a sense, is surrounded by the organism and which passes from the mouth to the anus. Substances which enter it are (1) physically and chemically broken down, (2) mixed with its secretions, and (3) transported from one area of the tract to another. The breakdown of foods frequently begins outside the body when a food is cooked, ground up, or sliced. It continues in the mouth, where it is chewed, mixed with the enzyme salivary amylase, and swallowed. The food then passes into the gastrointestinal tract (G-I tract), where swallowed material is mixed with the acid secretions of the stomach and later with the more alkaline secretions of the pancreas, liver, and intestine.

The functions of the G-I tract are many, but we shall concentrate initially on its role in digestion, or the breakdown of foods. If most of the foods taken into the alimentary tract were not broken down they would pass unchanged out of the body. In other words, it is the function of the digestive system to change the foods into sufficiently small particles or molecules that they can pass or be transported across the cell membranes of the body. These membranes include those of (1) the G-I tract, (2) the capillaries, and (3) the cell that uses the food as a source of energy, protoplasm, or intercellular matrix.

Claude Bernard (1813-1878) was one of the early investigators to emphasize the role of the digestive tract in catabolism (breakdown). He showed that if a disaccharide was injected into the blood, all of it would appear in the urine. If, on the other hand, the disaccharide was first mixed with digestive juice, it would not appear in the urine. Apparently the digestive juice reduced the size of the disaccharide enough so that it could be taken into the cells of the body and stored or utilized for energy or growth.

ANATOMY

Figure 23-1 is a schematic diagram of a cross-section of the G-I tract. It shows (1) an irregular inner surface of epithelial cells, which together with (2) the underlying connective tissue layer (lamina propria) and (3) muscular layer (muscularis mucosa) constitute the mucous membrane. Within the lamina propria lie numerous blood and lymph vessels important in the absorption of substances from the lumen. Note that the epithelium invaginates to various degrees to form numerous channels of varying lengths. This results, in part, in finger-like projections called *villi*. The villi provide a large absorptive surface which is kept in motion by the contractions of the muscularis mucosa. The muscle layer is also responsible for many of the folds that appear and disappear in the mucosal lining. Note too that many of the invaginations of the epithelium form glands and associated ducts in the mucous membrane and the submucosa, and peripheral

	Mean (cm.)	Range (cm.)
Nose to pyloric sphincter	63	51-74
Nose to end of duodenum	86	64-100
Nose to ileocecal valve	341	295-411
Nose to anus	451	394-500
Duodenum	22	18-26
Jejunum and ileum	255	206-318
Colon	110	91-125

Table 23-1. Length of various segments of the digestive tract in man measured by intubation. (From Blankenhorn, D. H., et al.: Transintestinal intubation: technique for measurement of gut length and physiologic sampling at known loci. Proc. Soc. Exp. Biol. Med., 88:356-362, 1955)

Fig. 23-1. Schematic drawing of a cross-section of the gastrointestinal tract. (Ham, A. W.: Histology. 6th ed., p. 678. Philadelphia, J. B. Lippincott, 1969)

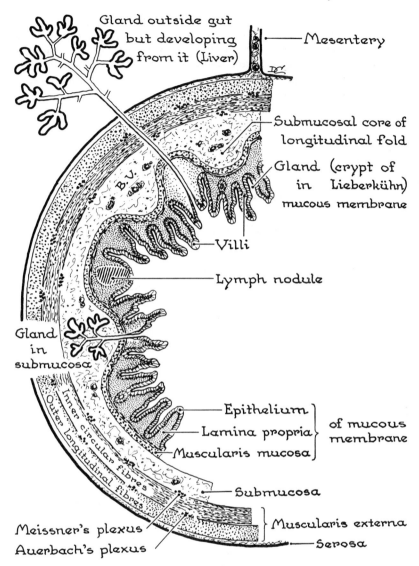

to the gastrointestinal tract. All of these channels result from the growth of the epithelial cells away from the intestinal lumen in the embryo. The salivary glands, pancreas, and liver are all, in part, the result of such a process.

The muscles of the external layer of the tract (muscularis externa) are responsible for changes in the pressure in the tract and changes in the size of the lumen. Their contractions are regulated by (1) pacemaker cells in the muscularis externa, (2) stretch and other stimuli originating from changes in the lumen, (3) hormones, and (4) autonomic neurons. You will note that numerous axons and cell bodies are concentrated in areas called Meissner's plexuses (submucosal plexuses) between the muscularis mucosa and muscularis externa and in areas called Auerbach's plexuses (myenteric plexuses) between the circular and longitudinal muscle fibers of the muscularis externa. Both of these plexuses contain unmyelinated postganglionic sympathetic neurons and areas of synapse between the pre- and postganglionic parasympathetic fibers (parasympathetic ganglia). The cell bodies of the postganglionic parasympathetic fibers, however, are in greatest concentration in the myenteric plexus. The impulses coming into and out of this plexus serve to produce coordinated waves of contraction and to integrate intestinal activity with activity elsewhere in the body. In times when skeletal muscle activity is great, for example, G-I motility and secretion are decreased.

It should be emphasized, however, that Figure 23-1 is an idealized drawing. In the mouth, esophagus, and anal canal the epithelium of the mucosa is not simple columnar epithelium, but stratified squamous epithelium. Also in the mouth, part of the esophagus, and the external anal sphincter, striated muscle rather than smooth muscle predominates. The villi shown in the diagram are found only in the small intestine.

CHEWING (MASTICATION)

The *teeth* serve to rip and grind the food and to mix it with saliva. The incisors of an adult man can exert forces on the food from 1 to 25 kg., whereas the molars can exert forces from 29 to 90 kg. The amount of chewing depends upon the size of the food and the training and nervous disposition of the individual. As long as the food can be comfortably swallowed the teeth have served their function. Apparently poorly chewed food is as fully digested in the alimentary tract as that which is excessively chewed.

Chewing is usually a voluntary act. It can be initiated by stimulation of the cerebral cortex or somatic efferent neurons, but it also occurs on an involuntary reflex basis when food is placed in the mouth of a subject. It may even occur after a transection just above the midbrain.

SWALLOWING (DEGLUTITION)

Swallowing, like chewing, is an act which can be initiated voluntarily or reflexly. For example, there are areas in the mouth and pharynx such as the anterior and posterior fauces, and the sides of the posterior wall of the hypopharynx which, when stimulated, activate nerves to the swallowing center in the medulla (Fig. 23-2). This results in a motor output from the medulla which

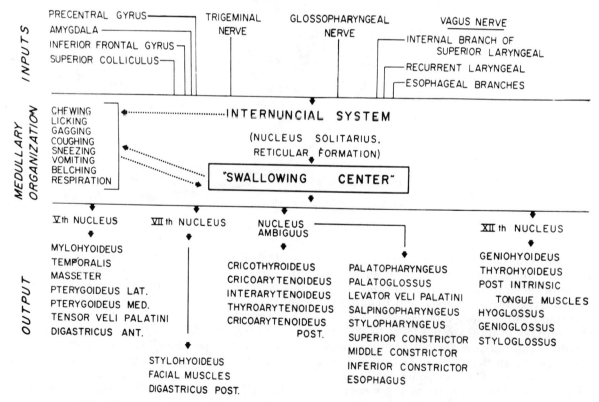

Fig. 23-2. Scheme showing the nervous inputs and outputs to and from the swallowing center. (Doty, R. W.: Neural organization of deglutition. *In* Code, C. F.: Handbook of Physiology, Sec. 6 (Alimentary Canal), vol. IV, p. 1862. Washington, American Physiological Society, 1968)

Swallowing (Deglutition) 413

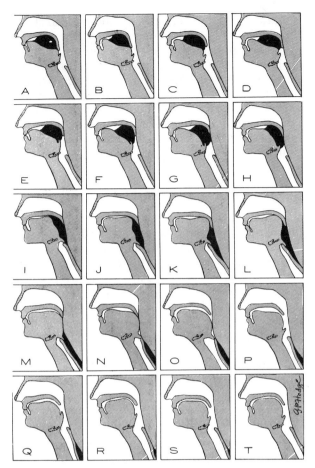

Fig. 23-3. Sequence of events during swallowing. Shown in the illustrations are the nasal cavity (*clear*), soft palate (*shaded*), the bolus of food (*black*), the tongue (*shaded*), the pharynx (*clear*), the epiglottis (*shaded*), the trachea (*clear*), and the esophagus (potential lumen posterior to trachea). Note in D the elevation of the soft palate which prevents the movement of the bolus into the nasal cavity, and in I the depression of the epiglottis over the trachea. In J and K the tongue moves backward, propelling the food into the esophagus. At K the food is delayed at the hypopharyngeal sphincter. At O and P, while the food is being carried down the esophagus by peristaltic contractions, the soft palate relaxes and the epiglottis ascends. (*From* Physiology of the Digestive Tract, Second Edition, by H. W. Davenport. Copyright © 1966, Year Book Medical Publications, Inc.; used by permission. *Adapted from* Rushmer, R. F., and Hendron, J. A., J. Appl. Physiol. 3:622-630, 1951)

causes (1) the anterior surface of the *tongue* to be brought up against the *soft palate* (see A through C in Fig. 23-3), (2) the palatopharyngeal muscles to contract, and (3) the palate to be elevated in such a way as to seal off the nasal cavity from the oral cavity (Fig. 23-3(D)). At this point (4) the tongue is more completely elevated and creates a pressure of 4 to 10 mm. Hg, which pushes the bolus into the oral pharynx (see D through I). Then there is (5) an inhibition of respiration and an elevation of the *larynx*, which produces closure of the opening into the trachea (the glottis). This persists until the bolus moves below the clavicular portion of the esophagus.

As (6) the bolus reaches the *epiglottis* (E through I) it tilts it backward until it lies over the closed glottis. This particular act, under normal conditions, is not necessary for the prevention of the passage of food and water into the trachea, since the larynx has been elevated and the glottis closed prior to the time it occurs. We find, for example, that swallowing progresses quite normally after the removal of the epiglottis.

The Esophagus

The esophagus is a tube well suited for its function in the transport of a bolus of food from the pharynx to the stomach. It has an inner covering of stratified squamous epithelium, which, in a resting person, folds on itself and completely occludes the potential esophageal lumen. In the upper third of the esophagus lie striated muscles, in the lower third smooth muscles, and in the middle third an area of transition. In the esophagus the concentration of blood vessels, lymph vessels, and glands is considerably less than that found in other parts of the alimentary tract. At the pharyngo-esophageal junction lie striated muscles (*hypopharyngeal sphincter, or superior esophageal sphincter*), which in a resting person maintain a tone sufficient to produce a pressure of about 20 mm. Hg above atmospheric pressure. In the thoracic esophagus the pressure is equal to that in the thoracic cavity (above -5 mm. Hg), but in the subdiaphragmatic esophagus the pressure is about 5 mm. Hg above the abdominal pressure (subdiaphragmatic esophageal pressure $= +10$ mm. Hg). The reason for this is the muscle tone of the smooth muscles at the *esophagogastric junction*. The pressure in the stomach is approximately that found in the abdominal cavity (5 mm. Hg). These pressures vary, of course, with respiration, the pressures in the intrathoracic structures increasing during expiration and decreasing during inspiration. The abdominal structures, on the other hand, show decreases in pressure during quiet expiration and increases during inspiration.

The pattern of control for the esophagus, then, includes a superior esophageal sphincter which prevents the esophageal contents from being regurgitated into the pharynx and an area of esophagus above the stomach which does not have the structural characteristics of a sphincter but which functions to prevent the regurgitation of the gastric contents of the stomach into the esophagus. Many physiologists refer to the esophagogastric junction as the *cardiac sphincter*, since it lies near the cardiac end of the stomach and functions like a sphincter.

In swallowing (1) the pressure in the pharynx may go up to 100 mm. Hg and (2) the hypopharyngeal sphincter relaxes for a period of 2 or more seconds. This results in (3) the bolus of food being pushed into the esophagus, and a wave of contraction developing which has a velocity of 2 to 4 cm. per second and passes from the pharyngeal to the gastric end of the esophagus in about 8 sec. This wave of contraction is sometimes called the peristaltic wave and serves to create a pressure of from 30 to 120 mm. Hg behind the bolus of food. After the food has entered the esophagus (4) the hypopharyngeal sphincter also contracts, and produces a pressure up to 100 mm. Hg behind the food. The peristaltic wave in the esophagus is sufficient to move a liquid or solid mass of 5 to 10 gm. against gravity toward the stomach. In an erect person, however, liquids reach the esophagogastric junction in about 1 sec.—well ahead of the peristaltic wave. These liquids remain there until (5) the cardiac sphincter (esophagogastric junction) relaxes. This occurs about 1.3 sec. after swallowing begins and lasts until the peristaltic wave has reached the sphincter. At this point (6) the cardiac sphincter increases its muscle tone for a short period to a point that produces a pressure of 20 to 30 mm. Hg in the sphincter area. This is well above the resting level and serves to effectively prevent regurgitation from the stomach.

When the swallowed food is neither liquid nor formed into a large, coherent mass, the peristaltic wave may leave particles of food behind in the esophagus. These particles (7) stimulate sensory nerves in the esophagus which initiate a reflex contraction of the superior esophageal sphincter and a second wave of peristalsis which descends the esophagus. This *secondary peristalsis* differs from swallowing in that it is an involuntary reflex not associated with an elevation of the tongue, soft palate, or larynx. I should also stress that in normal swallowing, as is not the case in secondary peristalsis, the sensory neurons of the esophagus play no role in the initiation of the peristaltic waves. Apparently it is impulses to the swallowing center from the mouth, pharynx, or parts of the brain, or a combination of these, which produce the machine gun-like firing of motor fibers to progressively more caudal portions of the esophagus.

Delayed Esophageal Emptying

There are a number of cases in which food taken into the esophagus remains there for extended periods (Fig. 23-4). This occurs in certain neurotic persons and in patients with an intrinsic lack of the myenteric plexus in the esophagus. It has also been produced experimentally in the dog by bilateral section of the vagus nerves. Apparently in each of these cases the cardiac sphincter fails to relax when food or water reaches it. It is sometimes treated by section of the muscle fibers which make up the cardiac sphincter.

Regurgitation into the Esophagus

Animals such as the cow and llama move food not only from the mouth to the stomach, but also from the stomach to the mouth (cud). These ruminants do this by relaxing the cardiac sphincter and inspiring against a closed glottis. This creates a sufficient negative pressure in the esophagus to draw some food into the esophagus, and then reverse peristalsis occurs. This type of activity is not characteristic of man and other nonruminants. When regurgitation into the esophagus of man occurs it may result in a type of chest pain called *heartburn,* due to exposure of the esophagus to a pH less than 4. Since it is common for the regurgitated material from the stomach to have a pH of 2.0, reflux from the stomach in man is frequently associated with heartburn.

Reflux of the contents of the stomach occurs during the last 5 months in about 50 per cent of all pregnancies. This is due, in part, to the pushing of the abdominal esophagus into the thorax. Under these circumstances, stooping and other acts that increase the intra-abdominal pressure still further increase the pressure exerted on the stomach without an associated increase in pres-

THE STOMACH

Gastric Filling

The peristaltic wave that carries the contents of the esophagus to the stomach is preceded by relaxation of the cardiac sphincter, lasting about 10 sec., and relaxation of the fundus and body of the stomach (Fig. 23-5), lasting about 20 sec., and results in a decrease in intragastric pressure. Generally before the gastric relaxation is half complete, however, the cardiac sphincter contracts enough to produce an increase in pressure at the cardiac orifice that persists for at least 30 sec. after gastric pressure has returned to normal. This prevents the regurgitation of the gastric contents.

At its smallest, the stomach in adult humans generally contains less than 50 ml. of material and has an intragastric pressure of about 5 mm. Hg. Drinking 225 ml. of water produces an increase in pressure of less than 2 mm. Hg. The addition of 600 ml. more results in the intragastric pressure going to 7 mm. Hg, but further increases in volume up to 1,600 ml. have little effect on the intragastric pressure. This means, of course, that as the stomach gets progressively more distended, progressively more tension is exerted on its walls. The relationship between

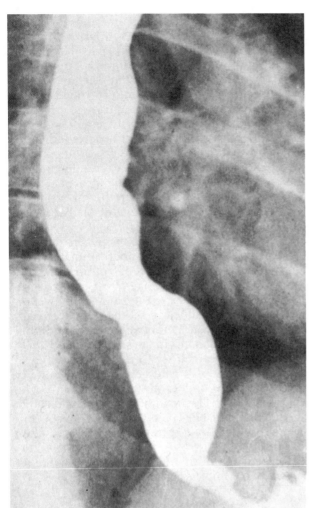

Fig. 23-4. Achalasia (failure to relax) of the esophagogastric junction in man. This is a frontal view showing the markedly dilated esophagus cephalad to the stomach. (*From* Physiology of the Digestive Tract, Second Edition, by H. W. Davenport. Copyright © 1966, Year Book Medical Publishers, Inc.; used by permission. Courtesy of F. J. Hodges and J. N. Correa)

sure on the esophagus. Thus the resistive characteristics of the cardiac sphincter are impaired.

The newborn child also has little or no abdominal esophagus. This condition is associated with a lower esophagus which has not yet developed the capacity to contract its esophagogastric junction or produce peristaltic waves. When regurgitation occurs in the infant the gastric contents frequently bubble out the mouth. In infants, however, the gastric contents are not sufficiently acid to produce heartburn.

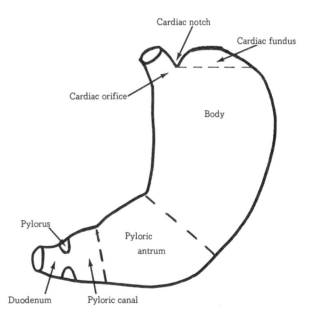

Fig. 23-5. Parts of the stomach. (*Modified from* Grant, J. C. B., and Barmajian, J. W.: A Method of Anatomy. 7th ed., p. 221. Copyright © 1965, The Williams and Wilkins Company, Baltimore)

pressure and tension was discussed in Chapter 8 under "Stroke Volume." There we noted the law of *Laplace* for a cylinder:

Tension (dynes/cm.) =
        pressure (dynes/cm.^{2}) × radius (cm.)

The contents of the stomach during a normal meal resemble a stew, containing chunks of meat and vegetables submerged in a gravy. Most of the air which is swallowed with the food does not pass the cardiac sphincter, but that which does goes into the stomach, floats up, and accumulates in the fundic portion. Particles with a density greater than the gastric contents move toward the caudal portion of the stomach. When the gastric contents are quite viscous, a solid food forms layers in the body of the stomach. Liquids, however, readily bypass these solids and move toward the pyloric antrum.

Gastric Contractions

In the muscularis externa of the stomach is a circular layer of smooth muscle which passes uninterrupted from the cephalad portion of the stomach to the duodenum. At the gastroduodenal junction this muscle is completely replaced by connective tissue which extends from the serosa to the mucosa. The longitudinal muscle also extends throughout the stomach, but is only partly interrupted at the gastroduodenal junction, a few muscle bundles extending into the duodenum.

Near the cardia in the circular muscle lies a *pacemaker* area, apparently similar in function to that found in the heart. This pacemaker fires about 3 times a minute, its impulses producing an electrical potential of about 0.9 mv. In the body of the stomach this impulse travels about 1 cm. per second, and in the antrum about 4 cm. per second. It is not, apparently, conducted into the duodenum, since we find little evidence of any coordination between the waves of contraction of the stomach and the distal duodenum. The coordination found in the proximal duodenum is thought to be due to the physical distention caused by the movement of the gastric contents into the duodenum.

The contractions of the stomach vary with the conditions. Parasympathetic fibers from the vagus tend to increase the force of contraction, and sympathetics from the celiac plexus decrease it. The autonomic nerves to the stomach, however, have little or no effect on the rate of pacemaker firing or on the conduction velocity. In addition, hormones and changes in the stomach itself affect the force of contraction. Generally the force of the gastric waves of contraction is considerably weaker in the body of the stomach than in the pyloric antrum. This is in part due to the fact that the circular muscle layer in the antrum is considerably thicker and therefore stronger than that in the body.

Contractions in the digestive tract, in general, can be placed in one of three classes (Fig. 23-6).

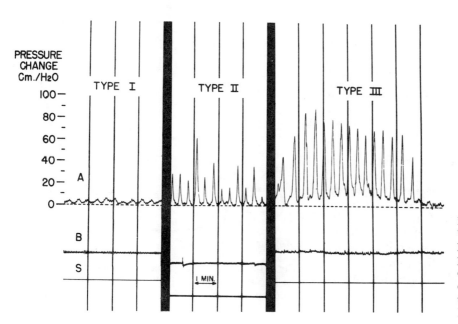

Fig. 23-6. Classification of pressure waves in the digestive tract. Trace B is a pneumograph (respiration) pattern. Vertical lines occur each minute. (Code, C. F., et al.: Motility of the alimentary canal in man. Am. J. Med., *13*:335, 1952)

Type I waves are produced by contractions which cause an increase in intraluminal pressure of 3 to 10 mm. Hg. In the stomach they begin at the cardia and move toward the pyloris, but in other parts of the G-I tract they may be nonprogressive, or stationary. For the most part they last only 5 to 10 sec. at any one spot.

Type II waves are much like type I waves except that they are associated with intraluminal pressures of 8 to 40 mm. Hg and last from 12 to 60 sec.

Type III waves are compound waves. They are associated with a rise in the base line that may last up to a few minutes, on which are superimposed waves of shorter duration. In the stomach the base line shift is less than 15 mm. Hg, but in the small intestine it may be 20 to 30 mm. Hg.

The Empty Stomach

Generally, by the time the stomach has reached a minimum volume, it shows type I waves in the antrum about 25 per cent of the time, type II about 15 per cent of the time, and type III less than 1 per cent of the time. It is quiescent, in other words, about 60 per cent of the time. As the fast continues, however, gastric activity increases. After about 3 additional hours without food, type I antral waves are converted into type II waves. These are seen for about half of the gastric cycle and are little different from those found an hour after a meal (Fig. 23-7).

It was once thought that the contractions of the empty stomach were responsible for the sensation of hunger, but it is now apparent that the sensation of hunger can occur in the absence of gastric contractions. We find, for example, that decreases in the blood glucose level in normal subjects initiate the so-called *hunger contractions* in the stomach and the sensation of hunger. In patients who have had the vagal nerves severed for the treatment of a peptic ulcer, however, the same *hypoglycemic* stimulus produces a sensation of hunger not associated with hunger contractions. On the other hand, *hyperglycemia* in the normal subject prevents hunger contractions and the sensation of hunger. On the basis of these and other observations it has been suggested that there are glucose-sensitive cells (glucostats) in or near the hypothalamus which respond to a decrease in available glucose, causing increased vagal parasympathetic tone to the stomach and the sensation of hunger. The sensation of hunger, then, is apparently not caused by increased gastric motility, but is only associated with it.

It should be emphasized that hunger contractions and the sensation of hunger can also occur in hyperglycemia, as in diabetes mellitus. Here, even though the blood sugar level is high, the problem is decreased availability of sugar to the cells of the body, i.e., the ability of cells to remove sugar from the blood and to utilize it is impaired and the body responds as it would to hypoglycemia.

Gastric Emptying

The rate at which the stomach empties depends upon (1) the fluidity and (2) chemical nature of its contents and (3) the nervous and (4) hormonal influences impinging upon it. If, for example, a person drinks 750 ml. of a liquid that is approximately isotonic, the weak type I waves of the body of the stomach are changed into type II waves in the pyloric antrum. These waves ap-

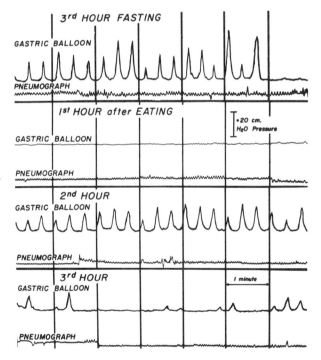

Fig. 23-7. Changes in the pressure in the pyloric antrum of the stomach of a normal human subject. (*From* Physiology of the Digestive Tract, Second Edition, by H. W. Davenport. Copyright © 1966, Year Book Medical Publishers, Inc.; used by permission. Courtesy of C. F. Code)

pear about 3 times a minute, each moving approximately 7 ml. of fluid into the duodenum. If, on the other hand, the meal consists primarily of solid food, the peristaltic waves frequently become less forceful in the antrum (Fig. 23-7, record 2) for a period of several hours. This is a period of little gastric emptying and of liquification of the stomach's contents. Eventually the peristaltic waves in the antrum become more forceful, changing from type I to type II. The type II waves produce constriction of the antrum but not complete closure. As the constriction moves toward the pyloric sphincter it produces a pressure capable of overcoming the resistance offered by the sphincter, and the more fluid contents of the gastric chyme are forced into the duodenum. Many of the solid particles are projected past the wave of constriction back into the proximal antrum. By the time the wave reaches the *pyloric sphincter* a maintained contraction of most of the antrum is apparent (systolic contraction). At this time the muscle tone of the pyloric sphincter increases sufficiently to produce complete closure, thus preventing regurgitation of the chyme that entered the duodenum. Ninety-five per cent of the time, however, the pyloric sphincter remains open. During this time, through the pressure it produces in its lumen, it impedes but does not prevent the passage of chyme through its lumen. In other words it serves to prevent flow in either direction except when the contractions of the stomach or duodenum force chyme through it.

Control of Gastric Motility

The stomach receives nerve impulses from sympathetic and parasympathetic fibers (Fig. 23-8). Contained in many of the sympathetic nerves are sensory neurons from the stomach. The parasympathetic fibers are found in nerves originating from the sacral cord and medulla (the vagus nerves). These nerves also contain sensory fibers from the stomach.

The stimulation of sympathetic fibers produces a decrease in the force of the gastric contractions, but section of these fibers apparently has little or no effect on gastric motility. Interestingly enough, bilateral preganglionic sympathectomy does abolish pain due to marked distention or irritation of the gastric wall. Stimulation of the peripheral ends of the vagi, on the other hand, produces marked increases in the force of the peristaltic wave. Bilateral section of the vagi just above or below the diaphragm, unlike sympathectomy, decreases the force of the peristaltic wave almost to zero and results in a delay in gastric emptying. These changes may persist for periods of more than 3 yrs. after vagal section.

The impulses traveling over the autonomic fibers to the stomach are markedly modified by activity elsewhere in the body. We have already noted that decreases in the arteriovenous glucose difference can act on the hypothalamus in such a way as to increase the parasympathetic impulses to the stomach. Pain, fear, rage, and so on can also affect gastric autonomic tone. We find, for example, that intense pain may almost completely suppress gastric contractions for periods up to 24 hr. We also find that the force of the gastric contractions changes as the volume of the stomach, the character of the gastric contents, and the character of the intestinal contents change.

Enterogastric reflex. The motility of the stomach is modified by stimuli in the small intestine. We find, for example, that (1) irritation of the duodenum and jejunum, (2) a duodenal pH below 3.5, (3) a hypertonic solution in the duodenum, (4) a 10 per cent ethanol solution in the duodenum, or (5) an intraduodenal pressure of 10 to 15 mm. Hg all stimulate sensory neurons in the vagi which in the medulla reflexly inhibit vagal parasympathetic fibers. These enterogastric reflexes (intestine-to-stomach messages) are unaffected by preganglionic sympathectomy, but are markedly diminished by bilateral vagotomy.

One reason the vagotomy does not completely abolish the enterogastric reflex is suggested in Figure 23-8. You will note in this illustration that a number of motor neurons in the alimentary tract are not only innervated by neurons coming from the central nervous system, but are also innervated by neurons originating in the tract itself. In other words there seems to be a primitive type of nervous control in the digestive system which can function in the absence of a central nervous system. This system has been compared to the *nerve net* found in such primitive phyla as Coelenterata (jellyfish, coral, sea anemone, and hydroid). It, after complete decentralization of the G-I tract or its removal from the body, is responsible for what peristalsis and other integrated activity remain. Its importance in the normally innervated G-I tract is not fully understood.

Enterogastrone. Enterogastrone is a hormone thought to be released by the duodenum and

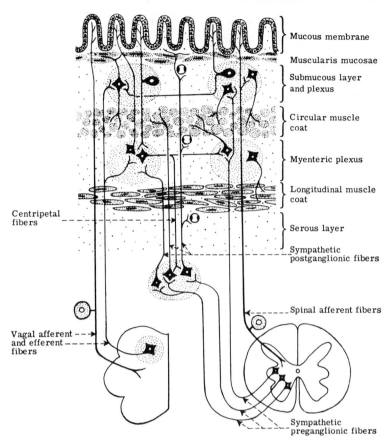

Fig. 23-8. Sensory and motor control of the gastrointestinal tract. This diagram emphasizes the following perspectives: (1) preganglionic sympathetic neurons originate from the thoracolumbar cord and innervate (a) postganglionic neurons whose cell bodies lie outside the G-I tract and (b) postganglionic neurons whose cell bodies lie in the G-I tract, (2) preganglionic parasympathetic fibers in the vagus nerve innervate postganglionic neurons whose cell bodies lie in the G-I tract, (3) sensory neurons pass in the vagus and in thoracolumbar nerves from the G-I tract to the central nervous system, (4) sensory neurons transmit impulses to ganglia inside and peripheral to the G-I tract, and (5) neurons transmit impulses from one ganglion to another. In addition, there is evidence that (1) adrenergic and cholinergic fibers impinge on the myenteric ganglia, and (2) sacral parasympathetic fibers originating from the sacral cord are of most importance in controlling the terminal portion of the large intestine. (Schofield, G. C.: Anatomy of muscular and neural tissues in the alimentary canal. *In* Code, C. F.: Handbook of Physiology. Sec. 6 (Alimentary Canal), vol. IV, p. 1611. Washington, American Physiological Society, 1968)

carried by the blood to the stomach, where it inhibits gastric motility. The "entero-" in the name signifies that it originates from the intestine (enteron), the "-gastro-" that it acts on the stomach (gastron), and the "-one" that it is inhibitory (chalone). (If it were excitatory, the suffix "-kinin" would be used.) Fatty acids, triglycerides, and phospholipids in the duodenum act on the mucosa to cause its release. It has been suggested that carbohydrates, oligosaccharides, and dextrins may also facilitate its release. Generally it takes from 3 to 5 min. after the lipid or carbohydrate is placed in the duodenum for the stomach to be affected. This hormone, as well as the enterogastric reflex, serves to delay the emptying of the stomach until alkalinization, intestinal digestion, absorption, and peristalsis have progressed to a certain point. In other words, they force the stomach to be a storage organ until the intestine is ready for its contents. They also serve to keep the gastric chyme in the stomach until it has become approximately isotonic. After surgical removal of the stomach, the quantity of food eaten at a single meal is markedly decreased and the incidence of vomiting increases.

Vomiting (Emesis). Vomiting, like swallowing, consists of a sequence of highly integrated acts which results in solids and liquids being moved up through the esophagus. In adults, vomiting is associated with a widespread discharge of the autonomic and somatic efferent neurons. It is generally preceded by an increase in the respiratory and cardiac rates, salivation, dilation of the pupils, sweating, pallor, and nausea. It begins with (1) a deep inspiration, (2) closure of the glottis, and (3) an elevation of the soft palate that closes off the nasal cavity from the pharynx. Next there is (4) a contraction of the abdominal skeletal muscles which produces an increase in the pressure in the abdomen and thorax, (5) a relaxation from the pharyngeal end of the esophagus through the esophagogastric junction and body of the stomach, and (6) the production of secretions from the stomach high in mucous and low in acid. There follow (7) a series of contractions in the antrum and duodenum toward the body of the stomach, (8) an increase in the quantity of duodenal secretion, and (9) a maintained contraction of the duodenum that may last a number of minutes and which projects the duodenal contents into the stomach.

In the meantime, (10) the pharyngo-esophageal sphincter closes, (11) the skeletal muscles of the abdomen contract, and the contents of the stomach are pushed into the esophagus. This process is termed *retching* and is followed by (12) relaxation of the abdominal muscles and a movement of the esophageal contents back into the stomach. This cycle may repeat itself numerous times and may eventually terminate in (13) a violent maintained contraction of the abdominal muscles which projects the vomitus past the pharyngo-esophageal sphincter and out the mouth. When (14) the abdominal muscles relax, (15) the vomitus remaining in the esophagus is carried back to the stomach by peristalsis, (16) the cardiac sphincter closes, and if there is still a moderate volume of material in the stomach there may be (17) a repetition of the cyclic filling and emptying of the esophagus which culminates in another episode of projectile vomiting.

Control. Vomiting of the type discussed above can be initiated by a number of stimuli from different parts of the body. It can be produced by (1) touching the back of the throat, (2) distention or irritation of the stomach or duodenum, (3) distention or irritation of the uterus, bladder, or renal pelvis, (4) an increase in intracranial pressure, (5) stimulation of the labyrinthine receptors in the inner ear (motion sickness), (6) pain in the testis and certain other types of intense pain, (7) certain substances (emetics) in the blood or cerebrospinal fluid, and even (8) certain unpleasant thoughts. All of these stimuli act indirectly on an area in the dorsolateral border of the lateral reticular formation of the medulla, the *vomiting center*. Certain drugs act indirectly on the vomiting center by stimulating a part of the area postrema in the floor of the fourth ventricle of the medulla. This area, the *chemoreceptor trigger zone*, in turn sends facilitatory impulses to the vomiting center. The destruction of the trigger zone not only eliminates the effect of a number of emetic drugs, but also prevents the vomiting associated with uremia, radiation sickness, and motion sickness.

The vomiting center itself lies near the respiratory and cardiovascular centers of the medulla, on which it has a profound influence. The vomiting center is apparently responsible for (1) stimulation of the vagal parasympathetic fibers that produce the contractions of the duodenum and gastric antrum, and (2) stimulation of the skeletal muscles of the abdomen. The relative importance of parasympathetic and somatic efferent fibers in initiating vomiting can be shown either by selective decentralization or nerve block. If, for example, we block the parasympathetic fibers to the stomach and duodenum with atropine, but leave the somatic efferent neurons intact, we find that vomiting is no longer associated with a maintained contraction of the duodenum or with strong antiperistalsis (waves of contraction traveling opposite the usual direction) in the gastric antrum, though it can still occur. Blocking the somatic efferent neurons, on the other hand, by the administration of curare, prevents vomiting even when there is a normally functioning autonomic nervous system.

Regurgitation

Vomiting is associated with a type of abdominal skeletal muscle contraction so intense that it is usually painful and leaves the muscles sore. Regurgitation, on the other hand, does not involve skeletal muscle to any important degree other than in the inhibition of the diaphragm, which prevents aspiration of the regurgitant. In regurgitation there are contractions of the stomach which propel the chyme toward the mouth. This is common in infants, in whom the esophagogastric junction is not fully developed and cannot, as a result, close completely; it is therefore not particularly effective in preventing the movement of chyme into the esophagus.

THE SMALL INTESTINE

The small intestine consists of a length of digestive tract about 3 meters long (Table 23-1), which is connected to the stomach at the pyloric sphincter and to the large intestine at the ileocecal sphincter. Its inner lining of mucosa forms many folds, each containing numerous villi. This results in a mucosal surface about 9 times greater than the serosal surface. This large mucosal surface is an important factor in the digestion and absorption of the nutrients in the chyme.

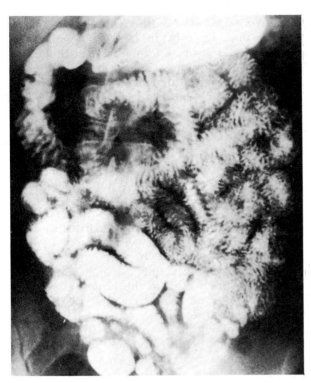

Fig. 23-9. X-ray of the stomach and small intestine of a normal man after a barium meal. Note the characteristic patterns of the mucosa. (*From* Physiology of the Digestive Tract, Second Edition, by H. W. Davenport. Copyright © 1966, Year Book Medical Publishers, Inc.; used by permission. Courtesy of F. J. Hodges and J. N. Correa)

The Duodenum

The first 22 cm. of the small intestine is called the duodenum. That part of the duodenum between the pyloric sphincter and the entrance of the common bile duct is called the *duodenal cap*. It is sometimes characterized as a receptor area, since the presence of hypo- and hypertonic solutions and acid, and various other characteristics of the chyme in this part of the intestine markedly modify the function of the rest of the gastrointestinal tract. The duodenal cap generally exhibits greater motility than the rest of the duodenum and is poorly coordinated with both the stomach and the lower duodenum. It is not until the gastric chyme passes into the lower duodenum that a propulsive wave of contraction develops in the duodenum. This wave moves toward the jejunal part of the small intestine. Electrical stimulation of the duodenum just below the duodenal cap also results in such a wave. The resultant contractions in man apparently never move toward the stomach.

The Jejunum and Ileum

The duodenum and jejunum are the parts of the alimentary system in which the major digestion and absorption occur. Once the chyme passes the duodenal cap, *weak waves* traveling at a rate of about 2 cm. per second carry it a short distance toward the terminal ileum. At the same time, however, *transverse creases* appear and disappear in the mucosa. These are the result of a synchronous stimulation of points localized within a centimeter of a line passing through the circumference of the muscularis externa. This nonpropagated contraction is generally associated with a similar contraction nearby which together with the first one serves to segment a part of the intestinal contents. The phenomenon is called *segmentation*, and chops up the contents of the intestine and mixes them with the digestive juices. The pressures which these strong segmenting contractions produce result in a mixing outside the segment as well. It has been noted, for example, that tracer substances placed in the lower ileum frequently appear in the duodenum. It was once thought that this *retrograde movement* was the result of a peristaltic wave which moved from the anal portion of the G-I tract toward the oral portion (called *antiperistalsis*), but it is now generally agreed that antiperistalsis seldom if ever exists in man. Apparently it is the pressures produced by the segmentation and occasional intestinal spasms which are primarily responsible for this retrograde movement.

The amount of time the chyme remains in the duodenum and jejunum varies from 2 hr. to considerably longer, depending on the type of food it contains and on the impulses arriving via the autonomic nerves. As the chyme passes into the terminal ileum most of the digestion and absorption of nutrients is complete, but the peristaltic and segmental contractions continue and serve to move the food toward the usually closed *ileocecal sphincter*. For the most part, chyme remains in the ileum somewhat over an hour. The opening of the sphincter is inhibited by mechanical stimulation of the mucosa of the colon or the distention of the colon. Its opening is facilitated by (1) increased gastric activity which (2) facilitates increased motor activity in the terminal ileum, which in turn (3) causes the sphincter to dilate. This is called the *gastroileal reflex*. The sphincter's opening may also result from increased motor activity in the ileum alone. It is of interest, however, that surgical removal of the sphincter does not seem to impair digestion or result in a significant amount of regurgitation into the small intestine. Some animals (the bear, racoon, and mink) are born without the ileocecal sphincter and seem to have no resultant digestive problems.

The muscularis mucosa. The motor activity we have discussed till now is primarily the result of contractions of the muscularis externa. The muscularis mucosa, on the other hand, also produces folds in the mucosa and movement of the villi. These contractions occur in response to local stimuli at the mucosal surface and seem to be independent of the contractions of the more peripheral muscularis externa. The contractions of the muscularis mucosa seem to (1) further mix the chyme with the digestive juices, (2) change the absorptive surface and move the chyme to which it is exposed, and (3) milk lymph out of and into the central lacteal of each villus.

Control

In Figure 23-8 Schofield has summarized the nervous relationships that exist in the G-I tract. We have noted in the case of the stomach, and we note again in the case of the small intestine, that sensory endings in the alimentary tract transmit impulses to the central nervous system, and that autonomic fibers transmit impulses from the central nervous system back to the intestine, where they modify function. We also note, on the other hand, that some fibers carrying messages from the digestive tract synapse with neurons in the myenteric plexus of the tract, and others synapse at the level of the sympathetic ganglion without first going to the central nervous system.

In Figure 23-10 we see some of the responses of a jejunum which has had all its extrinsic nerves severed—i.e., a jejunum without any central nervous system control. In part A the mucosa was stroked, causing an increase in the amplitude of the pressure waves in the lumen above the area stimulated and a decrease below it. At B, C, and D similar responses were obtained by the application of acid to the mucosa (B) and by the application of 5-hydroxytryptamine to the mucosa (C) and serosa (D). Apparently in the decentralized intestine, these stimuli, as well as distention, are capable of producing an increase in the force of contraction which is propagated for a short distance toward the anus as a wave of contraction. This wave begins in the longitudinal muscle layer of the muscularis externa and is followed by a wave of contraction in the circular layer. The response of the *circular muscles* to

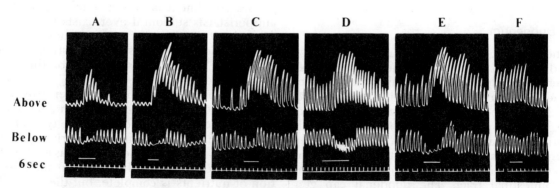

Fig. 23-10. Response of a decentralized jejunum in an anesthetized dog. Each upper trace is a record of the pressure in a balloon below the point of stimulation and each middle trace from above the point of stimulation. The lower trace is from a 6-second time marker. In A the mucosa was stimulated by stroking, in B by applying 0.05 N HCl to the mucosa, in C by applying a solution of 5-hydroxytryptamine to the mucosa, and in D by applying the same solution to the serosa. In E and F the villi were scraped and 5-hydroxytryptamine found to still produce a response (E), but HCl was now found to be ineffective (F). (Hukuhara, T., et al.: The effects of 5-hydroxytryptamine upon the intestinal motility. Japan. J. Physiol., *10*:421, 1960)

distention can be prevented by a ganglionic blocking agent, but not that of the *longitudinal muscles.*

The type of activity seen in Figure 23-10(A) and (B) is also associated with a release of *5-hydroxytryptamine* (5-HT, or *serotonin*). Since this release cannot be prevented by a ganglionic blocking agent it is hypothesized that it is the pressure-sensing system in the intestine which releases it.

Some of the other responses to stimuli that occur in the absence of extrinsic innervation include (1) increases in the motility of the ileum in response to gastric distention (gastroileal reflex), (2) closure of the ileocecal sphincter in response to distention of the cecum, (3) responses of the intestine to hypotonic and hypertonic solutions, and (4) the contraction of the muscularis mucosa in response to solids in the intestine. On the other hand, the parasympathetics also play a role in modifying the force of contraction of the muscles of the muscularis externa. The injection of antiacetylcholine esterases (which destroy the destroyer of acetylcholine) tends to increase the force of contraction of the muscularis mucosa, as does stimulation of the vagus. Apparently in fasting there is a decrease in the force of these contractions, and after a meal an increase, both of which are partly related to changes in parasympathetic tone.

In contrast to the effects of parasympathetic fibers on the muscularis externa, we find that parasympathetic stimulation produces little or no effect on the contractions of the muscularis mucosa. Sympathetic stimulation or the infusion of epinephrine or acetylcholine, on the other hand, increases the contractions of the muscularis mucosa.

THE COLON

The colon receives about 300 to 500 ml. of chyme a day, the first material entering about 4 hr. after being swallowed and the last possibly as long as 9 hr. after being swallowed. This chyme usually consists of material the body is not able to digest, as well as water and minerals. It enters the cecum and associated appendix

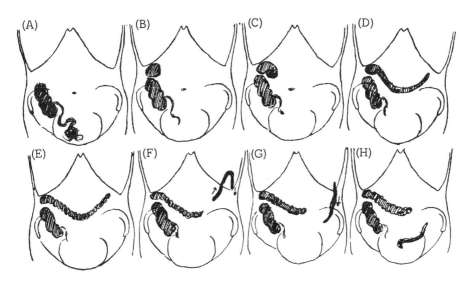

Fig. 23-11. Movements in the colon of a normal man. A breakfast containing barium sulfate was eaten at 7 A.M. Five hours later the barium was seen in the ileum, sigmoid colon, and part of the ascending colon (A). At this time the subject had a lunch of meat, vegetables, and pudding. During the lunch the ileum rapidly emptied into the colon (B), and at the end of lunch some of the barium was seen at the hepatic flexure (C). Next the diameter of the hepatic flexure became reduced and the barium passed through the transverse colon almost to the splenic flexure (D). Haustral segmentation developed (E), and 5 min. later the barium moved through the splenic flexure (F), down the descending colon (G), into the sigmoid colon (H). (Hertz, A. F., and Newton, A.: Movements of colon. J. Physiol., 47:61, 1913)

through the ileocecal valve and passes through the ascending colon to the hepatic flexure (Fig. 23-11(A) and (B)). It then moves through the transverse colon to the splenic flexure (Fig. 23-11(F)) and down the descending colon (G) to the sigmoid colon and rectum (H). The amount of time the chyme remains in the large intestine depends upon a number of factors. On an average about 70 per cent of the indigestible material we eat is eliminated from the intestines in the feces within 72 hr. after being swallowed. Total loss of swallowed material from the body, however, may require more than a week.

Every 24 hr. approximately 400 ml. of a fluid isotonic suspension of chyme enters the large intestine and is changed into a semisolid mass of 150 ml. The chyme, while in the colon, is exposed to pressures as high as 100 mm. Hg, which not only serve to move it toward the rectum, but also facilitate the absorption of water and minerals. The movement and mixing of the chyme in the colon is due to periodic changes in muscle tone, segmentation contractions similar to those seen in the small intestine, and waves of contraction that move toward the rectum at a speed of about 2 cm. per second.

Mass Movements

Three or four times during the day there are also very strong contractions which move large quantities of the colonic contents toward the rectum. These contractions are called *mass movements* and take about 30 sec. to reach their peak, during which there may be a rise in intracolic pressure up to 100 mm. Hg. Within about 45 sec. after maximum pressure has been obtained, the pressure has declined by 80 per cent, and 2 to 3 min. later the pressure is back to the base line value. The movement of the barium meal seen in Figure 23-11 is probably due largely to such a mass movement.

Control

All the contractions of the colon can occur in the absence of any connection with the central nervous system. Like the contractions of the small intestine, they are associated with the release of *5-hydroxytryptamine* by the mucosa. This is not to say, however, that the colon is not affected by the autonomic nervous system. Parasympathetic fibers in the vagus nerve innervate the proximal colon (possibly as far as the first third of the transverse colon), and parasympathetic fibers in the pelvic nerve innervate the rest of it. Sympathetic fibers to the colon are in the inferior mesenteric and hypogastric nerves. Parasympathetic stimulation results in a inhibition of the internal anal sphincter and an increase in the rate and force of the contractions throughout the colon. Sympathetic stimulation, on the other hand, inhibits motility. We find that when food or drink enters the stomach, there is facilitation of colonic motility (*gastrocolic reflex*), and during sleep there is depression of colonic motility. Apparently in the first case there is an increase in parasympathetic tone to the colon, and in the second a decrease. A person with a hyperexcitable colon or an excess of parasympathetic tone may show more than the normal number of mass movements a day, and as a result, diarrhea.

DEFECATION

In Figure 23-12 we note the response of a man whose sacral cord has been destroyed and hence also the parasympathetic outflow to the rectum and internal anal sphincter. In this man a certain degree of distention of the rectum produced a contractile response of the rectum and a decrease in tone at the internal anal sphincter. This is the type of response that produces defecation or the expulsion of feces from the body. Under normal circumstances, however, it is (1) the mass movements of the colon which deliver the feces to the rectum and hence are responsible for (2) the distension of the rectum. It has been suggested that there are (3) *stretch receptors* in series with the smooth muscle of the rectum and that (4) these receptors stimulate sensory neurons which carry signals to the spinal cord. In the cord (5) some of these afferent impulses are relayed to areas of consciousness in the cerebral cortex and to the defecation center in the medulla. Some of the afferent impulses (6) facilitate the stimulation of parasympathetic fibers which go back to the rectum and (7) help bring about more forceful contraction of the rectum and (8) more complete relaxation of the internal anal sphincter. The afferent neurons also (9) inhibit somatic motor tone to the external anal sphincter.

Voluntary Control

The response to rectal distention discussed above is, of course, the type of response one sees

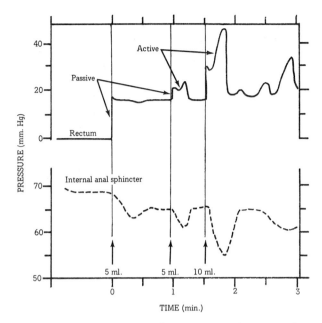

Fig. 23-12. Response of a 46-year-old man with a neurofibroma at the cauda equina to distention of the rectum. Even though his tumor had destroyed the parasympathetic innervation to the rectum and anal sphincter, note that when 5 ml. of air was injected into a balloon in the rectum (*upper trace*) there resulted a decrease in the pressure around a balloon in the internal anal sphincter (*lower trace*). An additional 5-ml. injection into the rectal balloon also produced contraction of the rectum (active increase in pressure). An addition of another 10 ml. to the rectum produced a more powerful contraction of the rectum and a more marked dilation of the internal anal sphincter. Note, too, the 3 waves of contraction of the rectum which followed the last injection. (*Redrawn from Physiology of the Digestive Tract, Second Edition, by H. W. Davenport. Copyright © 1966, Year Book Medical Publishers, Inc.; used by permission. Adapted from* Denny-Brown, D., and Robertson, E. G.: Brain, 58:256-310, 1935)

in small children. In time, however, the child learns that he can prevent defecation by voluntarily contracting the skeletal muscle of his pelvic diaphragm and *external anal sphincter*. After approximately a minute of increased tone in the external anal sphincter the rectal contraction decreases, the afferent fibers from the rectal stretch receptors stop firing, and the urge to defecate disappears until additional feces are added to the rectum.

Volitional defecation can be initiated by a series of voluntary acts characterized as "straining." There is (1) a strong, maintained contraction of the diaphragm, (2) closure of the glottis, (3) contraction of the chest muscle, and (4) contraction of the abdominal muscles, all of which increase the intra-abdominal and intrathoracic pressures to from 100 to 200 mm. Hg. This results in (1) a movement of additional feces into the rectum, (2) stimulation of the rectal stretch receptors, and (3) contraction of the longitudinal muscles of the rectum to such an extent that the angle between the distal colon and the rectum disappears and additional feces enter the rectum, (4) the internal and external anal sphincters relax, and (5) the longitudinal muscles of the rectum which extend to the perineal region pull the external anal sphincter toward the rectum, thus shortening the anal canal. A full bladder and tactile stimuli to the bladder or anus tend to facilitate this reflex. Anger, resentment, hostility and other emotional disturbances also can lead to increased motility. This relationship has for a long time been recognized in our common speech. It is also rather common among students to find bouts of diarrhea during examination periods. On the other hand, pain, fear, and anxiety frequently produce pallor, decreased secretion of mucous, and decreased motility in the colon. When such responses are extreme there may be constipation, diarrhea, spasms, abdominal pain, or some combination of these.

Straining during defecation, micturition, or lifting a weight, as previously mentioned, is potentially dangerous for a person suffering from cardiac decompensation or from any of a number of other pathological conditions. It can produce (1) an initial increase in arterial pressure, (2) a decrease in venous return and a resultant decrease in cardiac stroke volume and arterial pressure, (3) an increase in cerebrospinal fluid pressure, (4) dislodging of blood clots, (5) coronary occlusion, and (6) ventricular fibrillation.

COLECTOMY

In some persons it is necessary to remove the entire colon and bring the end of the ileum out the abdominal wall (ileostomy). The chyme expelled by the ileum is then collected in a plastic bag fastened around the opening. The patient can survive the loss of his colon if the vitamins synthesized in the large intestine are administered and if fluid and electrolyte balance is maintained. The quantity of fluid collected in the external bag can be reduced by a well-controlled diet, but the procedure of caring for such a patient is at best time-consuming and difficult.

MEGACOLON

Megacolon is a condition in which there is extreme dilation of the colon by its contents. It is similar to the condition of achalasia discussed earlier and may be due either to an area of spasticity in the colon which prevents the passage of chyme or feces through it, or to an area of colon in which there is little or no motility. It is frequently caused by a congenital absence of ganglion cells in the myenteric and submucous plexi (Hirschsprung's disease), but may be acquired in later life, possibly due to ischemia or damage to a part of the colon. Approximately 40 per cent of patients with megacolon also have dilated urinary bladders. Some of the symptoms of megacolon are abdominal distention, anorexia, lassitude, infrequent defecation, and a hypersensitivity to acetylcholine. Some patients may go longer than 3 wks. without defecation. It is frequently treated by removal of the affected part of the colon.

Chapter 24

CARBOHYDRATES
PROTEINS
FAT
WATER AND ELECTROLYTES
VITAMINS
PROTECTION OF THE
DIGESTIVE TRACT
FECES

DIGESTION AND ABSORPTION

The substances we take into our digestive tracts are (1) absorbed from the tract, (2) acted upon by the secretions and microorganisms of the tract, and/or (3) eliminated from the body in the feces. Generally over 60 per cent of the digestion and absorption of nutrients is complete by the time the chyme reaches the ileum, and almost all of it is complete when the chyme reaches the large intestine. The small amount of digestion that occurs in the large intestine is the result of the microorganisms found there. The absorption of water, minerals, and vitamins occurs throughout the G-I tract (Table 24-1).

Many of the concepts of absorption and secretion which we have developed in the chapters on the kidney are also relevant to an understanding of the G-I tract. Here, as in the case of the kidney, absorption and secretion result from diffusion or active transport. Characteristically both processes are involved with the movement of substances that are relatively small, fat-soluble, or both. We note, for example, that undigested polysaccharides stay in the tract, whereas monosaccharides are readily absorbed from it.

In most newborn animals, however, large molecules (gamma globulin, for example) have been observed to pass quite readily from the digestive tract into the circulation (Fig. 24-1). It has been suggested that this is due to greater permeability of the tract to molecules, or to transport by pinocytosis. It is important in that it represents a means of transferring the antibodies found in the mother's colostrum (the first secretion of the mammary glands after the birth of a child) and milk to the infant, which is characteristically ill-equipped to produce its own antibodies. It is also important in that it explains the greater frequency of allergic reactions in children than in adults.

In the case of human infants, however, evidence of such transfer is rather meager. Like most newborn animals, the human infant has a stomach which (1) is more permeable than that of the adult and which (2) has not developed enough to denature the proteins that pass through it. We find that anti-polio factors in the mother's milk pass through the entire digestive tract of the newborn child and into the feces unchanged. This is in part due to defective digestion in the infant, and to the presence of trypsin inhibitors in the milk itself. To date, however, although there are many examples of an infant

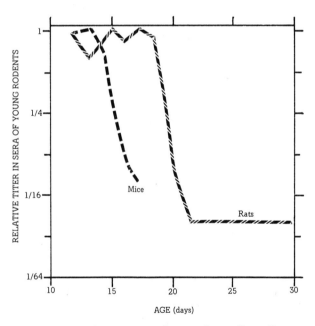

Fig. 24-1. Relative titers of anti-Salmonella pullorum agglutinins in the sera of young rats and mice fed a standard dose of homologous immune serum 24 hours previously. Note that with age the amount of agglutinins which cross the alimentary barrier decreases. (*Redrawn from* Morris, I. G.: Gamma globulin absorption in the newborn. *In* Code, C. F. (ed.): Handbook of Physiology. Sec. 6 (Alimentary Canal), vol. III, p. 1496. Washington, American Physiological Society, 1968)

Substance	Location of Absorptive Capacity			
	Small Intestine			Colon
	Upper	Mid	Lower	
Absorption (active transport)				
Sugars (glucose, galactose, etc.)	++	+++	++	0
Neutral amino acids	++	+++	++	0
Basic amino acids	++	++	++	?
Betaine, dimethylglycine, sarcosine	++	++	++	?
Gamma globulin (newborn animals)	+	++	+++	?
Pyrimidines (thymine and uracil)	+	+	?	?
Triglycerides	++	++	+	?
Fatty acid absorption and conversion to triglycerides	+++	+	+	0
Bile salts	0	+	+++	
Vitamin B_{12}	0	+	+++	0
Na^+	+++	+	+++	+++
H^+ (and/or HCO_3^- secretion)	0	+	++	++
Ca^{++}	+++	++	+	?
Fe^{++}	+++	++	+	?
Cl^-	+++	++	+	0
SO_4^{--}	++	+	0	?
Absorption (passive diffusion)				
Nucleic acid derivatives (except uracil and thymine)				
Vitamins (except vitamin B_{12})				
Water				
Secretion (active transport)				
K^+	0	0	+	++
H^+ (and/or HCO_3^- absorption)	++	+	0	0
Sr^{++}	0	0	+	?
Cl^- (under special conditions)	+	?	?	?
I^-	0	+	0	0
Secretion (passive diffusion)				
Na^+				
Cl^-				
Water				

Table 24-1. Transport of substances by the intestine. The capacity of each segment to absorb or secrete is graded from 0 to +++. The upper intestine refers primarily to the jejunum. The duodenum functions much as the jejunum does, except that it shows little absorption of NaCl. (Modified from Wilson, T. H.: Intestinal Absorption. Philadelphia, W. B. Saunders, 1962)

developing antibodies to proteins in his food, there are no well documented cases of an antibody passing unchanged from a human infant's G-I tract to his blood. I have listed in Tables 24-1 and 24-2 some of the particles which characteristically pass into and out of the G-I tract of an adult.

CARBOHYDRATES

Carbohydrates are not essential for life, but generally constitute well over 50 per cent of our diet and hence over half our caloric intake. Most of the carbohydrate we eat is in the form of starch, a polysaccharide yielding glucose upon digestion in the small intestine. We also eat carbohydrates in the form of cellulose, glycogen, sucrose, maltose, lactose, glucose, pentose, etc. The body is able to digest, absorb, or digest and absorb all the carbohydrates mentioned above except cellulose (Table 24-2). This polysaccharide passes into the colon unchanged and is there either catabolized (broken down) by the resident microorganism or eliminated from the body in the feces.

Enzymes which catalyze the breakdown of starch are released by the salivary glands and the pancreas. Those that catabolyze disaccha-

Substrate	Enzyme	Source of Enzyme	Products	Optimal pH
Carbohydrates (CHO)				
Starch	Ptyalin (SA)	Salivary glands	CHO polymers	6.7
Starch	Pancreatic amylase	Pancreas (exocrine cells)	Disaccharides	6.7-7
Maltose	Maltase	Intestinal glands	Glucose	5-7
Lactose	Lactase	″	Glucose and galactose	5.8-6.2
Sucrose	Sucrase	″	Glucose and fructose	5-7
Proteins and their products				
Proteins	Pepsin	Gastric chief cells	Polypeptides	1.6-2.4
Proteins and polypeptides	Trypsin (trypsinogen)	Pancreas (exocrine cells)	Small polypeptides	8
Chymotrypsinogen	Trypsin	″	Chymotrypsin	8
Proteins and polypeptides	Chymotrypsins (chymotrypsinogens)	″	Small polypeptides, dipeptides	8
Polypeptides	Carboxypeptidases (procarboxypeptidase)	″	C-terminal amino acids and peptides	
Nucleic acids	Nucleases	″	Nucleotides	
Nucleic acids	Ribonuclease	″	Polynucleotides	
Nucleic acids	Deoxyribonuclease	″	Polynucleotides	
Polypeptides	Aminopeptidases	Intestine	C-terminal amino acids and peptide	8
Dipeptides	Dipeptidases	″	Amino acids	
Nucleic acids	Nuclease	″	Pentoses, purine and pyrimidine bases	
Lipids				
Fats	Gastric lipase	Stomach	Glycerides, fatty acids (minor role)	
Fats	Pancreatic lipase	Pancreas (exocrine cells)	Glycerides, fatty acids	8
Fats	Intestinal lipase	Intestine	Glycerides, fatty acids	8

Table 24-2. Some of the enzymes secreted into the digestive tract and the substrates on which they act. Enzyme precursors are indicated in parentheses. SA, salivary amylase.

rides in the epithelial cells lining parts of the succus entericus). Some of the disaccharides are absorbed and broken down into monosaccharides in the epithelial cells lining parts of the digestive tract. In any case the major carbohydrates which normally enter the blood from the G-I tract are the hexoses and pentoses.

The absorption of glucose and galactose occurs throughout the small intestine. This absorption from the G-I tract, like that from the nephron of the kidney, (1) can occur against a concentration gradient, (2) is associated with increased oxygen consumption, and (3) is inhibited by the presence of certain other nutrients (competition for a shared transport system) and by the drug phlorizin. The maximum rate of D-glucose absorption in the small intestine is about 120 gm. per hour. The maximum rate of absorption for D-galactose is slightly less. That for fructose and the pentoses, L-xylose and D-ribose, is between one half and one tenth as rapid. In other words, the rate of transport of hexoses and pentoses is related not solely to molecular size, but also to the characteristics of the transport systems available to them. Apparently most of the hexoses and pentoses other than glucose and galactose pass through the alimentary barrier primarily through diffusion. There is also some evidence that some of the fructose absorbed by the epithelial cells is changed in these cells to glucose.

PROTEINS

Proteins are complex nitrogen-containing molecules used by the body as a source of energy and essential for growth, repair, and most of the body's metabolic activity. The protein

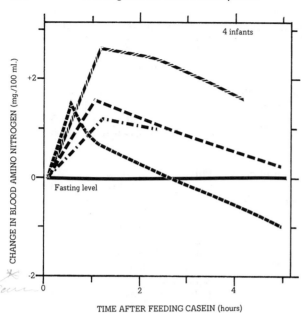

Fig. 24-2. Changes in blood amino-nitrogen in 4 normal infants after the digestion of the protein, casein. (West, C. D., et al.: Changes in blood amino nitrogen levels following injection of proteins and of a protein hydrolysate in infants with normal and with deficient pancreatic function. Am. J. Dis. Child., 72:259, 1946)

tents to such a point that any pepsin present is inactivated. In other words pepsin is at best responsible for only a small amount of protein digestion in the stomach. In cases in which the stomach produces too little HCl to produce a pH compatible with pepsin activity (hypochlorhydria), or when the stomach has been removed, protein digestion progresses normally in the small intestine and the fecal content of protein and other nitrogenous molecules remains normal.

It is the enzymes of the pancreas which are most important in protein catabolism. You will note in Figure 24-3, for example, that albumin taken in a test meal is approximately 50 per cent digested and absorbed by the time it reaches the middle of the jejunum (at approximately 140 cm.), and about 70 per cent absorbed by the time it reaches the ileum (about 180 cm.). Approximately 10 per cent escapes digestion and absorption in the small intestine. If, on the other hand, the pancreatic ducts are ligated or for some reason the pancreas fails to secrete its enzymes into the duodenum, the amount of pro-

in our diet is broken down (catabolized) into amino acids in the digestive tract. These amino acids are then transported to the cells of the body, where they are used as a source of energy or re-formed into proteins (Fig. 24-2). It is important that each person take in at least enough amino acid nitrogen to replace that which is lost. The importance of the amino acid nitrogen is that it, unlike the N_2 dissolved in the body fluids, can be utilized to produce protein. In order to maintain a nitrogen balance an adult requires approximately 0.5 to 0.7 gm. of protein every 24 hr. A growing child between 1 and 3 yrs. of age requires about 4 gm. every 24 hr. to meet his needs.

The breakdown of proteins in the digestive tract begins in the stomach with the release of pepsinogen, which is activated to pepsin by the HCl in the stomach. The pepsin at a highly acid pH catalyzes the breakdown of protein. At best, less than 15 per cent of the protein is broken down into amino acids in the stomach. Some protein remains unchanged and some is converted to polypeptides of different sizes. The secretions into the duodenum increase the pH of its con-

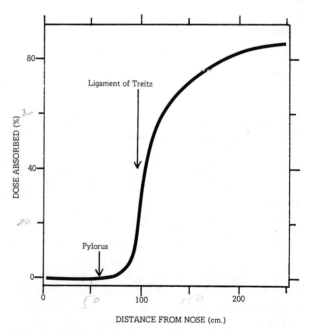

Fig. 24-3. Digestion of ^{131}I-labeled human serum albumin and absorption of its breakdown products in a normal human subject. The test meal also contained corn oil, glucose, lactose, and polyethyleneglycol. (*Modified* from Borgstrom, B., et al.: Studies of intestinal digestion and absorption in the human. J. Clin. Invest., 36:1525, 1957)

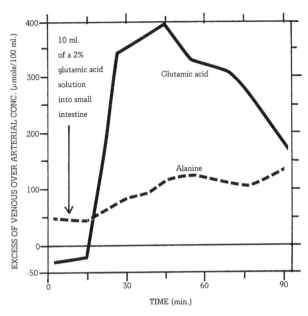

Fig. 24-4. Absorption of glutamic acid in the anesthetized dog. At the arrow, 10 ml. of a 2% glutamic acid solution was injected into the lumen of a portion of the small intestine. The rapid rise of glutamic acid which occurred in the venous blood is due to the rapid absorption of the injected amino acid. The rise in alanine concentration is due to the transamination in the cells of the intestine of some of the absorbed glutamic acid. (*Redrawn from* Neame, K. D., and Wiseman, G.: Transamination of glutamic and aspartic acids during absorption by the small intestine of the dog *in vivo*. J. Physiol. (London), *135*:447, 1957)

tein lost in the feces markedly increases. For example, in the case of a human subject whose pancreas had been removed for cancer, an average of 45 per cent of a 75-gm. protein meal appeared in the feces. This can be compensated for by increasing the protein in the diet or by feeding the subject pancreatic enzymes.

The absorption of amino acids from the G-I tract is a rapid process involving active transport (Fig. 24-4). Three transport systems for the amino acids have been described. One system transports neutral amino acids (L-methionine, D-methionine, etc.) rapidly into the mucosal cells, a second, basic amino acids (L-arginine, DL-orinthine, L-cystine), and a third transports L-proline, hydroxyproline, sarcosine, dimethylglycine, and betaine. Hydroxyproline and L-proline can be transported by either the first or third system, but seem to have a greater affinity for the latter.

FAT

The amount of fat in the diets of various people ranges from 25 gm. per 24 hr. in rice-eating Japanese workers to 150 gm. per 24 hr. in persons on a high meat diet. Their intake of unsaturated fatty acids, however, is essentially the same. Fat is normally completely digested and absorbed before it reaches the colon (Fig. 24-5). Though digestion begins in the stomach with the release of gastric lipase, the stomach is relatively unimportant as far as fat digestion is concerned, its main role being storage. You will remember from the preceding chapter that fat in the duodenum facilitates the release of a hormone, enterogastrone, which inhibits gastric motility and hence gastric emptying. A meal containing 50 gm. of fat so retards gastric emptying that some of that fat is still in the stomach 5 hr. after it was first swallowed. The control mechanisms of the duodenum and stomach, in other words, are such that in response to a meal containing 50 gm. of fat only about 10 gm. of the fat are delivered to the duodenum per hour.

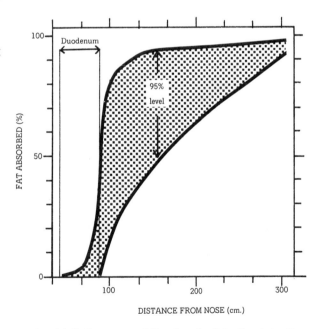

Fig. 24-5. Per cent of fat absorbed in the intestine after a meal of 30 Gm. of fat. The shaded area includes 95 per cent of the values obtained in 90 samples. (*Redrawn from* Physiology of the Digestive Tract, Second Edition, by H. W. Davenport. Copyright © 1966, Year Book Medical Publishers, Inc.; used by permission)

Digestion

As you will note in Figure 24-5, it is in the first part of the intestine that the major digestion and absorption of fat and its products occur. Approximately 10 to 20 min. after a meal is started, pancreatic secretion into the duodenum begins, and within an hour the concentration of pancreatic lipase in the intestine has reached a maximum. Thirty minutes after the beginning of the meal the gallbladder contracts and adds its bile to the duodenum. The addition of pancreatic juice and bile to the intestinal contents continues until the fat is completely digested and absorbed. In the case of the bile this may result in the addition of 4 to 5 gm. of bile salts to the duodenum. Since there is generally less than 4 gm. of bile salt in the entire body at any one time, apparently during the digestion of fat the bile salt is absorbed from the lower part of the small intestine (Table 24-1), transported back to the liver, and re-released into the bile ducts.

Role of the pancreas. The pancreas can secrete up to 4.7 ml. of pancreatic juice per minute into the duodenum, as well as up to 36 mEq. of HCO_3^- per hour per 100 gm. of tissue. Pancreatic juice has a pH between 7.6 and 8.2 and contains approximately the same concentration of Na^+ and K^+ as the blood plasma, but in addition may contain between 25 and 170 mEq. per liter of HCO_3^-, plus albumin, globulin, and carbohydrate-, protein-, and fat-splitting enzymes.

If the pancreas is removed from the body, *steatorrhea* (fatty stools) develops. In this condition the amount of fat in the stool is in direct proportion to the amount in the diet. Normally all the fat in our food is digested and absorbed before it reaches the colon, but in the absence of the pancreas there is a marked increased in *long-chain fatty acids* in the feces. Apparently after pancreatectomy there is still (1) extensive hydrolysis of triglycerides to diglycerides, monoglycerides, and fatty acids, as well as (2) absorption of the short-chain fatty acids, glycerol, and monoglycerides. It has been suggested that some of the digestion of fat after pancreatectomy is the result of the intestinal lipase, which is released from the mucosal cells of the intestine during their rapid production and desquamation.

We should not, however, place complete emphasis on the loss of pancreatic enzymes as the cause of steatorrhea. We find, for example, in pancreatectomized dogs that the addition of 160 mM. of $NaHCO_3$ to the duodenum returns to normal their digestion and absorption of oleic acid. We also find that the failure of the body to release bile into the duodenum produces steatorrhea even in the presence of a normally functioning pancreas.

Fortunately for man he has a large *pancreatic reserve*. On an average only about 15 per cent of patients with pancreatic-secretion disturbances show steatorrhea. Apparently these are all cases of very advanced pancreatic damage, since up to 90 per cent of the pancreas can be removed without malabsorption of fat.

Role of bile. Bile is produced in the liver and passed via the hepatic duct to either the gallbladder, where it is stored and concentrated, or the duodenum. The liver forms approximately 500 ml. of bile a day. The bile contains bile salts, cholesterol, lecithin, waste products resulting from the breakdown of hemoglobin (the *bile pigments*), many of the adrenocortical and other steroid hormones, and frequently drugs (Table 24-3). The bile, in other words, (1) represents a medium for elimination, but (2) is also important in digestion.

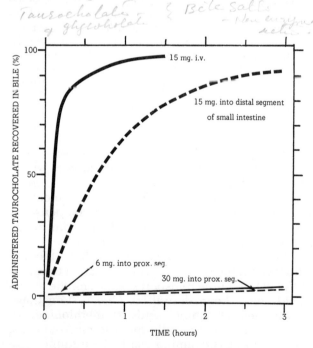

Fig. 24-6. Recovery of the bile salt, taurocholate, in the bile after injection into (1) a vein (I.V.), (2) the distal small intestine, and (3) the proximal small intestine of anesthetized guinea pigs. (*Modified from* Lack, L., and Weiner, I. M.: J. Pharmacol. Exp. Therapeutics. *139*:250; copyright © 1963, The Williams and Wilkins Company, Baltimore)

pH	7.5
Water	97%
Inorganic salts	0.7%
Bile salts (congregates of cholic and chenodeoxycholic salts with glycine and taurine)	0.7%
Bile pigments (biliverdin and bilirubin glucuronides)	0.2%
Fatty acids (palmitic, oleic, linoleic, etc.)	0.15%
Lecithin	0.1%
Fat	0.1%
Cholesterol	0.06%
Protein (mucoprotein and plasma protein)	<0.01%
Alkaline phosphatase	<0.01%

Table 24-3. Composition of human bile obtained from the hepatic duct.

That part of the bile which seems to be most important in the digestion of fat is the bile salt. It is synthesized in the liver from cholesterol, secreted into the duodenum with the rest of the bile, actively reabsorbed in the ileum, and carried by the hepatoportal circulation back to the liver, where it is stored and resecreted (Fig. 24-6). In other words, the bile salts are released into a part of the intestine in which they are poorly absorbed (the duodenum), and are absorbed in a part of the intestine in which the chyme contains little fat (the ileum).

The bile salts in the duodenum and jejunum serve several functions. They (1) lower surface tension (act as a surfactant), (2) facilitate the formation of micelles, (3) form water-soluble complexes with various lipids, (4) protect cholesterol esterase from proteolytic attack, and (5) may facilitate the synthesis of triglycerides in the mucosal cell. We will be concerned primarily with the first two functions.

The fat that enters the duodenum does not dissolve to any degree in the aqueous fluid which constitutes most of the chyme, but rather remains suspended in this solution as oil droplets. These droplets present a relatively small surface area to the lipolytic enzymes and also trap the various fat-soluble *vitamins* (vitamins A, D, E, and K) in them. Such droplets if not broken up would pass through the intestinal tract, and out in the feces in large quantities. The bile salts, along with the fatty acids, glycerol, lecithin, and monoglycerides, serve to prevent this by lowering the surface tension throughout the chyme and therefore decreasing the size of the fat droplets.

Pancreatic lipase is the lipolytic enzyme we know most about. It, and probably the other lipases as well, tends to break down triglycerides into (1) 2-monoglycerides (monoglycerides with a fatty acid in the 2 position), (2) fatty acids, and (3) glycerol. This can be diagrammed as follows:

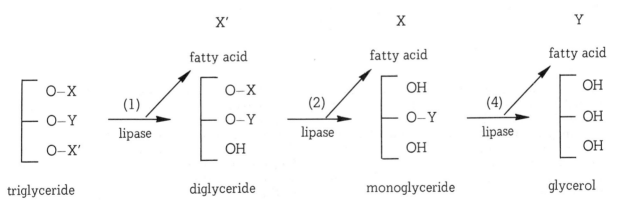

Most of the fatty acids and monoglycerides formed in this process, along with some bile salts, then, combine to form a negatively charged polymolecular aggregate called a *micelle*. Approximately 10^6 micelles are produced from one fat globule. They have a diameter of between 20 and 50 Å, and contain vitamins A, D, and K and small quantities of diglyceride, triglyceride, *cholesterol*, and *phospholipid*. The micelles apparently diffuse to the membrane of the epithelial cells. Whether some, many, a few, or none penetrate the membrane is not known. They are, however, responsible for the movement of their contents, with the possible exception of the bile

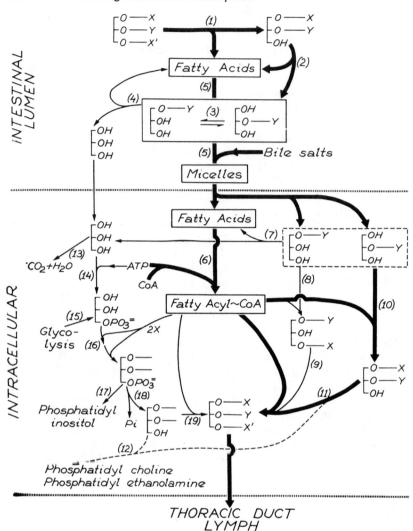

Fig. 24-7. A proposed mechanism for the digestion, absorption, and biosynthesis of fats. In this scheme triglycerides are broken down into fatty acids and monoglycerides in the intestine (reactions 1 and 2). The fatty acids and monoglycerides then combine with the bile salts to form micelles (reaction 5), which facilitate their deposition in the mucosal cell. The monoglycerides and fatty acids are then recombined to form triglycerides (reactions 6, 10, and 11) which together with the phospholipids (reactions, 14, 16, 17, 18, and 12), cholesterol (*not shown*), and protein (*not shown*) form a chylomicron which is carried by the lymphatic system to a vein. (Brown, J. L.: The role of monoglycerides in the biosynthesis of triglycerides in the intestine. Ph.D. Dissertation. Dallas, University of Texas, 1964)

salts, into the endoplasmic reticulum of the epithelial cells. Some of these concepts are summarized in Figure 24-7.

Transport of lipids to the blood. The lipids in the intestinal tract must pass a barrier of mucosal cells before entering the blood. Some of these lipids are chemically changed in transit, while others are not. After passing the barrier, some enter the blood capillaries. Others enter the lymph capillary and are carried to the blood via the thoracic duct (Fig. 24-8). At present, the details of the absorption process are not clear. Some investigators maintain, for example, that practically all triglycerides in our diet are (1) completely hydrolyzed to glycerol and fatty acids and then (2) resynthesized during absorption. Others maintain that up to 50 per cent of the triglycerides in our diet may enter the blood unchanged.

The role of bile in the movement of lipid into the blood is in part due to the micelles which it helps form. In the absence of bile negligible amounts of tripalmitin are absorbed from the intestinal tract of the lamb, and in man, inadequate amounts of *vitamins A, D, and K* are absorbed. In fact, the absence of bile is much more likely to produce a deficiency of the fat-soluble vitamins than is an absence of pancreatic juice.

Most of the monoglycerides and fatty acids deposited in the endoplasmic reticulum of the epithelial cells by the micelles are apparently resynthesized into triglycerides. If the amount

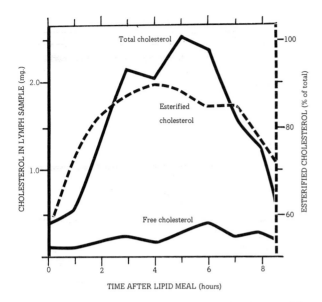

Fig. 24-8. Changes in the cholesterol content of the lymph in the rat after a meal of taurocholate, oleic acid, and 50 mg. of cholesterol. (*Redrawn from* Treadwell, C. R., et al.: Observations on the mechanism of cholesterol absorption. J. Am. Oil. Chem. Soc., 36:108, 1959)

of fatty acids absorbed exceeds that of available reaction sites on monoglyceride molecules, reactions 14, 16, 18, and 19 in Figure 24-7 are increased. In these reactions glycerol is changed into alpha glycerophosphate, which is converted into a triglyceride or a phospholipid. If there is an excess of monoglycerides, much of the glycerol that enters the cells is oxidized and an intracellular lipase catalyzes the formation of glycerol and fatty acids from monoglycerides (reaction 7). This supplies the additional fatty acids necessary for the formation of triglycerides from the remaining monoglycerides.

Parallel to triglyceride synthesis there is also a synthesis of phospholipids (lecithin, etc.), cholesterol, and protein. These substances apparently form a specific beta lipoprotein which combines with the triglycerides, and certain trace substances which are released from the cell. The resultant polymolecular aggregate, the *chylomicron*, facilitates the transport of its water-insoluble contents in the lymph and blood.

The chylomicron is between 0.1 and 0.5 microns in diameter and contains (1) the triglycerides (10 to 93%) and phospholipids (5 to 9%), synthesized in the mucosal cell, (2) free fatty acids (1 to 7%), (3) cholesterol (0.7 to 1.5%), and (4) protein (0.5%). It is probably too big to pass through the continuous basement membrane and the endothelial cells that surround the lumen of the blood capillaries. The lymph capillaries, however, do not have a continuous basement membrane and apparently receive all the chylomicrons. Whether they diffuse into the lacteal, pass in by pinocytosis, or perhaps both is still undecided.

The chylomicrons, once in the lymph ducts, may accumulate up to a concentration of 6 per cent. Some have been known to stay there for up to 13 hr., but all eventually pass from the lymph ducts into the blood, where they are transported to the cells of the body. On the other hand, glycerol is water-soluble and the short- and medium-chain fatty acids (acetic through hexanoic acid) are partially water-soluble. These substances apparently pass directly into the blood capillaries.

WATER AND ELECTROLYTES

Water moves freely between the lumen of the small and large intestine and its blood vessels. This results in the aqueous contents of the small and large intestine having the same tonicity as their associated intercellular fluids. The stomach, on the other hand, is considerably less permeable to water and its contents may remain either hypertonic or hypotonic for a number of hours (see Fig. 24-9). The significance of this is that the stomach can receive large quantities of hypertonic or hypotonic solutions without producing either a hemoconcentration or a hemodilution. The stomach then slowly releases these potentially dangerous solutions to the intestine over a period of a number of hours. The importance of the stomach in this regard can be demonstrated by placing these solutions directly into the jejunum or duodenum. When large volumes of hypertonic solutions are placed in the small intestine the result is often a series of symptoms called the *dumping syndrome*. These include nausea, a sense of epigastric fullness, pain, pallor, sweating, dizziness, and fainting.

The effect of numerous test meals on alimentary functions has been studied in human volunteers. One of these meals consists of steak, margarine, salad, and tea with sugar. When thoroughly mixed and broken up in a blender it has an osmolality of 232 milliosmoles per kilogram. This is hypotonic to plasma by about 60 milliosmoles per kilogram. Another test meal

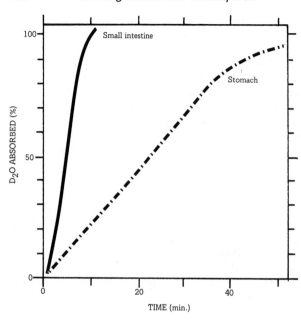

Fig. 24-9. Rate of movement of labeled water (deuterium) out of the stomach and small intestine of a healthy subject. (*Redrawn from* Scholer, J. F., and Code, C. F.: Rate of absorption of water from stomach and small bowel of human beings. Gastroenterol., *27*:572; copyright © 1954, The Williams and Wilkins Company, Baltimore)

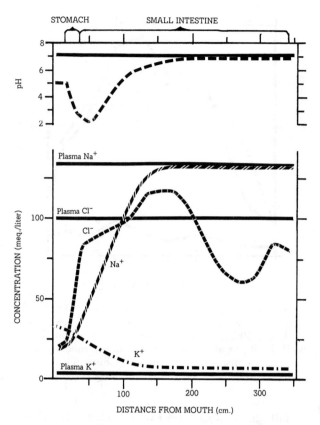

Fig. 24-10. Concentrations of electrolytes and changes in pH in a steak meal as it passes through the gastrointestinal tract of a normal man. This meal remained hypotonic until it reached the 90-cm. point in the graph. The heavy line in each graph represents the normal plasma concentration of each electrolyte or the normal plasma pH. (*Modified from* Fordtran, J. S. and Locklear, T. W.: Ionic constituents and osmolarity of gastric and small-intestinal fluids after eating. Am. J. Digest. Dis., *11*:507, 508, 509, and 510, 1966)

consists of milk and doughnuts and when homogenized in a blender has an osmolality of 630 milliosmoles per kilogram. It is markedly hypertonic to plasma. When either the hypotonic or the hypertonic meal is taken into the body it is mixed with saliva and moved into the stomach. It remains hypotonic in the case of the steak meal and hypertonic in the case of the doughnuts while in the stomach for periods in excess of 1.5 hr.

In Figure 24-10 I have summarized some of the changes in the concentration of electrolytes in the hypotonic steak meal as it passes through the stomach and small intestine. You will note that once the meal reaches the small intestine it very rapidly acquires and maintains a concentration of Na^+ and K^+ which is approximately that found in plasma. This is apparently due to free diffusion of these two ions and of water as well. In the case of the hypotonic meal there would be a net diffusion of Na^+ into and a diffusion of K^+ and water out of the upper part of the small intestine.

You will also note in Figure 24-10 that the pH in the stomach is quite low due to the high H^+ concentration of gastric secretion (Fig. 24-11). This acidity, as I mentioned earlier, is countered by the alkaline secretions of the pancreas and intestine. These secretions owe their alkalinity to their high HCO_3^- concentration. You will note in Figure 24-12 that the high HCO_3^- concentrations in the secretions of the pancreas and ileum are associated with reduced concentrations of Cl^-. The lower concentrations of Cl^- in pancreatic juice and the succus entericus from the ileum are probably the reason the Cl^- concentrations in the duodenum and ileum are lower than those found in the plasma.

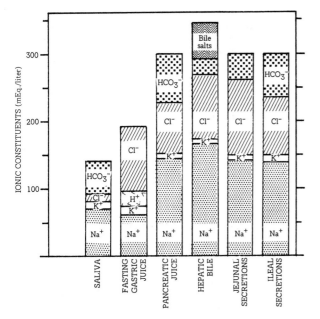

Fig. 24-11. Ionic constituents of the alimentary secretions. The pH of these secretions is directly related to the HCO$_3^-$ concentration. For example, the secretions of the pancreas and ileum have the highest HCO$_3^-$ concentration and the highest pH. Note too that both the salivary and gastric secretions are hypotonic (*Redrawn from* Fordtran, J. S., and Ingelfinger, F. J.: Absorption of water, electrolytes, and sugar from the human gut. *In* Code, C. F. (ed.): Handbook of Physiology. Sec. 6 (Alimentary Canal), vol. III, p. 1466. Washington, American Physiological Society, 1968)

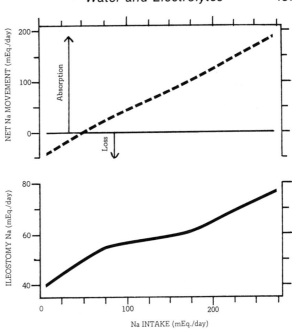

Fig. 24-12. The effect of Na intake on its content in ileostomy fluid (*lower graph*) and on the net sodium gain to (absorption) or loss from the body of ileostomy patients. (Fordtran, J. S., and Ingelfinger, F. G.: Absorption of water, electrolytes, and sugar from the human gut. *In* Code, C. F. (ed.): Handbook of Physiology. Sec. 6 (Alimentary Canal), vol. III, p. 1471. Washington, American Physiological Society, 1968)

Ileostomy

In some patients it is necessary to establish a fistula between the terminal ileum and the skin. These ileostomy patients have been studied extensively, and from such studies we have learned much about the physiology of the small intestine. Apparently the small intestine has a large capacity to absorb not only nutrients but minerals as well. We find that a subject with an Na intake of 261 mEq. per day still delivers to the colon a chyme which is isotonic with plasma and contains less than 70 mEq. of Na a day. On the other hand, a dietary intake of only 9 mEq. of Na per day results in a delivery to the colon of 38 mEq. per day. The small intestine appears not to be an important factor in maintaining constant Na concentrations in the body. Each day it delivers between 38 and 70 mEq. of Na to the colon despite the intake or needs of the body for Na. In the patient with an ileostomy the 38 mEq. represents an obligatory Na loss. Should the dietary Na be less than this there would be an Na depletion from the body. When it is more, the excess Na is excreted by the kidneys (Fig. 24-12).

Colon

In the colon there is an active absorption of Na$^+$ which results in an efflux of about 400 ml. of water from the colonic contents (Table 24-4). Associated with this movement is an increase in the K$^+$ concentration of the feces. In the ileum the ratio of Na$^+$ to K$^+$ concentration is between 12:1 and 20:1. In the feces, however, the ratio is on the order of 1:3. This means that there are approximately 5 mEq. of Na and 7 to 15 mEq. of K lost in the feces each day. As in the case of the small intestine this loss continues even during salt deprivation. During prolonged salt deprivation, for example, the loss of Na$^+$ in the fecal

	Plasma	Ileostomy Fluid	Fecal Fluid
Water loss (ml./day)		500-600	100-150
Na^+ Concentration (mEq./liter)	140	40-50	25-49
K^+ Concentration (mEq./liter)	4	3-6	80-132
Cl^- Concentration (mEq./liter)	100	20-40	15
HCO_3^- Concentration (mEq./liter)	29	30-35	32
Organic anions (mEq./liter)	6		179
Osmolality (mosm./kg.)	290	290	376

Table 24-4. Comparison of the composition of plasma, fluid collected from ileostomy patients, and fecal water. Apparently as the chyme passes through the large intestine there is an active reabsorption of Na^+ and an associated secretion of K^+ which has a net effect of facilitating the absorption of water from the colon. The high osmolality of the fecal fluid is due to its high concentration of organic anions. These have been added by the microorganisms of the colon. (Data in part from Fordtran, J.S., and Ingelfinger, F.J.: Absorption of water, electrolytes, and sugar from the human gut. *In* Code, C.F. (ed.): Handbook of Physiology. Sec. 6 (Alimentary Canal), vol. III, pp. 1457-1490. Washington, American Physiological Society, 1968)

water (about 5 mEq./day) exceeds that in the urine (between 0.4 and 3 mEq./day).

In diarrhea, on the other hand, the contents of the small intestine pass very rapidly through the large intestine, permitting an inadequate period for the absorption of Na and water from the lumen of the large intestine. This results in a markedly increased loss of body Na. You will note in Figure 24-13 that as the volume of fluid lost in diarrhea increases, the quantity of Na lost also increases. When the diarrhea becomes so severe that more than 3 liters of fluid are lost per day we find that the concentration of both Na and K in the stool water is approximately that found in the small intestine and plasma. You will also note in Figure 24-13 that the increase in K^+ loss as the diarrhea becomes more severe is not so great as that found for Na^+. This is consistent with the view that the colon removes Na^+ from and adds K^+ to the contents of the large intestine, and that in diarrhea the chyme remains in the colon for so short a time that it is little different from the chyme in the ileum.

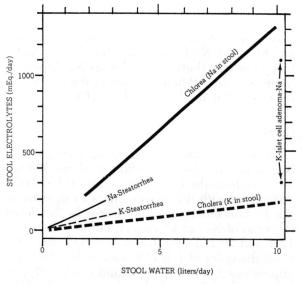

Fig. 24-13. Relation of stool water to Na^+ and K^+ loss in diarrhea in human subjects. (*Modified from* Fordtran, J. S., and Dietschy, J. M.: Water and electrolyte movement. Gastroenterol., 50:281, 1966)

Calcium, Magnesium, Phosphate and Organic Acids

Calcium absorption, unlike the absorption of chloride, magnesium, and phosphate, seems to be largely dependent upon the body's need for calcium. Apparently in hypocalcemia a hormone is released, *parathormone*, which increases the calcium absorption from the intestine. This absorption occurs throughout the small intestine but is most rapid in the duodenum. It is the result of an active transport system which is dependent upon the presence of vitamin D as well as ionized Ca^{++}. Calcium in the form of insoluble salts (as a carbonate, phosphate, or phytate or combined with fatty acids) is poorly absorbed.

Magnesium and phosphate, on the other hand, pass from the digestive tract by passive transport throughout the small intestine. The concentration of magnesium, phosphate, and calcium in a dialysis bag which was allowed to pass through the alimentary tract and out in the feces is shown in Figure 24-14. The concentrations in this bag are only in part a result of the subject's diet and the transport systems of the digestive tract. They are also to a great extent the result of the sloughing of mucosal cells into the digestive lumen and the metabolic activity of the microorganisms in the lower small intestine and colon. We find, for example, that the use of a broad spectrum antibiotic decreases or eliminates the intestinal flora and in so doing converts the fecal water from a hypertonic solution with a high concentration of organic anions to an isotonic solution with a low concentration or organic anions.

Iron

There are approximately 4 gm. of iron in the body. About 2 to 2.5 gm. are invested in hemoglobin, 0.5 to 1 gm. in storage (primarily as ferritin), 150 mg. in myoglobin, and 5 mg. in plasma transferrin; the rest is scattered throughout the body in respiratory pigments such as cytochrome, peroxidase, and so on. The normal dietary intake of iron is from 15 to 20 mg. a day. The average absorption is between 0.5 and 1.5 mg. The amount absorbed depends upon the amount in the diet and the needs of the body (Fig. 24-15). In iron deficiency, after hemor-

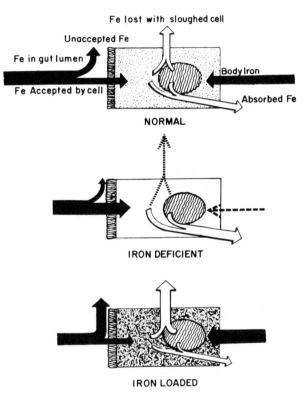

Fig. 24-15. Control of iron absorption and elimination by the intestinal mucosa. In this concept the mucosal cell is central. It regulates the amount of iron absorbed from the alimentary canal (*black arrow at left*) and the amount absorbed from the plasma (*black arrow at right*). It controls the amount of iron released to the plasma (*clear arrow at right*). These cells also store iron and are periodically sloughed into the digestive lumen. Thus, they regulate the amount of iron lost from the body. (Conrad, M. E., and Crosby, W. H.: Intestinal mucosa mechanisms controlling iron absorption. Blood, 22:411, 1963; used by permission of Grune and Stratton, Inc., New York)

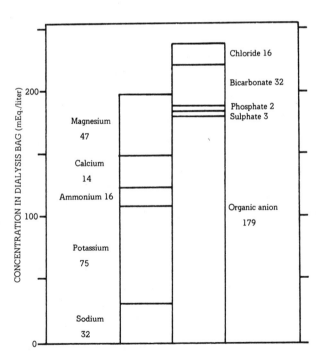

Fig. 24-14. Concentrations of ions in a bag (made of a dialysis membrane), which has passed through the alimentary tract of a man and out with his feces. (*Redrawn from* Wrong, O. M., et al.: In vivo dialysis of faeces as a method of stool analysis. I. Clin. Sci., 28:372, 1965)

rhage, and in pregnancy the absorption may increase as much as fivefold. In a person with an excess of body iron, iron absorption decreases but never ceases. This failure of iron absorption to cease when there is already an excess of iron in the body can eventually lead to a marked accumulation of two iron-containing protein complexes (ferritin and hemosiderin) in the body and a series of symptoms characterized as *hemochromatosis*. These symptoms include pigmentation of the skin, pancreatic damage with diabetes, cirrhosis of the liver, and gonadal atrophy.

The duodenum and first part of the jejunum are the areas of the digestive tract in which most of the iron absorption occurs. Most of the iron absorbed is in the ferrous state, since the columnar cells that line the intestine absorb ferrous iron 2 to 15 times more readily than ferric iron. The iron, once in the cytoplasm of these cells, may remain there as such or may combine with apoferritin to form *ferritin*. Some believe that ferritin iron remains bound in the molecule until the cell is sloughed into the intestinal lumen—i.e., that it is unavailable for iron metabolism. Others believe it is a storage form that can be called upon. In either case the mucosal cell acts as a reservoir of available iron, giving off large quantities of ferrous iron to the plasma in iron deficiency and taking on large quantities of iron from the plasma when there is an excess. The iron transported in the plasma is carried combined in a globulin called *transferrin* (siderophilin).

The body's chief means of eliminating excess iron is by the sloughing of mucosal cells into the intestinal lumen. In cases of iron overload these mucosal cells have large quantities of ferritin and, as a result, large amounts of iron are lost from the body in the feces. In cases of iron deficiency the cells have considerably less ferritin and iron loss from the body is therefore reduced.

VITAMINS

In Table 24-1 we noted that most of the vitamins pass into the blood by passive transport. The exception to this is the cobalt-containing compound vitamin B_{12}. It has a molecular weight of 1,357 and is probably the largest water-soluble essential nutrient absorbed intact from the intestinal tract. Man requires about 1 μg. of this vitamin a day. In its absence a condition called *macrocytic anemia* develops. This disease may be due to too little B_{12} in the diet or, what is more likely, to malfunction of the stomach. Apparently the normally functioning oxyntic cells of the stomach secrete a heat-labile mucoprotein with a molecular weight of 53,000 which forms a complex with B_{12} or one of its coenzyme forms. The mucoprotein is called the *intrinsic factor* of the stomach and apparently greatly facilitates the movement of the B_{12} and its coenzymes into the mucosal cells. Once in the mucosal cells the complex dissociates and the vitamin is released to the blood. Generally the B_{12} begins to appear in the blood about 3 hr. after its ingestion and reaches peak concentration about 5 hr. later.

The mechanism whereby the B_{12}-intrinsic factor complex passes into the mucosal cell is not known. One suggestion is that it is the result of pinocytosis. The absorption of B_{12} alone, however, closely parallels the absorption by the membrane of other large molecules. We find, for example, that its absorption is greatest in the newborn and smallest in the adult (Fig. 24-1). The adult rat absorbs the B_{12} complex about 10 times more rapidly than it does the free vitamin. In a patient with pernicious anemia it may take as much as 10,000 times the daily requirement of B_{12} to treat the disease if the B_{12} is given orally. If the B_{12} is injected, the dose can be substantially reduced.

PROTECTION OF THE DIGESTIVE TRACT

The function of the digestive tract is to catabolize carbohydrates, lipids, and proteins. It is, of course, important while the digestive tract is performing this function that it not also destroy the carbohydrates, lipids, and proteins which make up the tract itself. A second problem is protecting the tract from the microorganisms which invade it in the food. Protection of the tract is accomplished in part by the *mucus* produced throughout the alimentary system. This mucus is a thick solution that coats the membranes and serves both as a barrier to the enzymes and as a means of keeping these membranes at a fairly constant pH. It, on the other hand, is freely permeable to water and electrolytes.

Another important characteristic of the tract is its capacity for the rapid repair of injury. In response to a restricted *desquamation* of the stomach, for example, substantial repair occurs in the first few hours, and apparently complete recovery, within 24 hr. More extensive damage, on the other hand, may take many months to counter. Part of this capacity for repair may have to do with the high proliferative rate of the cells lining the alimentary tract. In the epithelium of the mouth and esophagus there is a complete replacement of cells every 4 to 15 days; in the pyloric gland area of the stomach, every 2 days; in the duodenum, every 5 to 6 days; in the jejunum, every 5 days; and in the ileum, every 3 days. The turnover time increases in response to trauma and decreases during starvation.

The Mouth

The secretions of the mouth include viscid fluids with a high concentration of mucin, produced by the mucous gland cells, and watery secretions without mucin but containing the enzyme ptyalin, produced by the serous gland cells. These secretions are well buffered (HCO_3^- concentration 28 mEq./liter) at a pH of 7.0 and therefore tend to maintain this pH in the mouth, keeping the membranes moist and lubricated. At a pH of 7.0 the saliva is saturated with calcium and hence does not tend to remove calcium from the teeth. The cells of the mouth release a total of about 1.5 liters of liquid per day. This rather large volume serves to wash food particles in the mouth out to the esophagus, thereby preventing bacterial growth. Patients with deficient salivation (*xerostomia*) have a higher than normal incidence of dental caries.

The Stomach and Intestine

The mucus in the stomach forms a lining 1 to 1.5 mm. thick, which maintains the pH of the mucosal cells near 7.0 despite the fact that the gastric secretions may have a pH of less than 2.0. Apparently the defense systems of the duodenum against gastric juice are also rather effective. We find that gastric juice which produces no damage to the duodenal mucosa produces damage in the jejunum if injected or allowed to get there through a gastrojejunostomy (a surgical anastomosis between the stomach and jejunum).

FECES

Feces is the semisolid mass of material expelled past the anal sphincter in the process of defecation. Its content varies with the diet of the person, the motility of the digestive tract, the secretory activity of the digestive system, and the medications a person might be taking. Normally the feces is about 75 per cent water and 25 per cent solids. The longer the chyme remains in the large intestine the lower the water content of the feces. The concentration of Na in the feces can be decreased fourfold, and that of K fivefold, by increases in the output of *aldosterone* from the adrenal cortex. Such increases occur in response to Na depletion and congestive failure. Most of the solids in the feces are organic material. This includes (1) mucus, (2) desquamated cells and their products, (3) digestive system enzymes and their products, (4) bacteria, yeasts, and fungi and their products, and (5) a number of polysaccharides that the alimentary tract cannot digest (cellulose, hemicellulose, etc.).

Microorganisms

The concentration of microorganisms in the stool varies from person to person. The newborn child has none. In the stool of the average adult they constitute about 10 per cent of the dry weight. Microorganisms are also found in the stomach, small intestine, and colon. The acidity of the gastric secretions, however, results in a very small concentration in the stomach; it is not until the chyme reaches the ileum and colon that it acquires substantial concentrations of microorganisms. Here the microoganisms break down cellulose and hemicellulose and produce such potentially useful products as riboflavin, nicotinic acid, biotin, folic acid, and vitamins K and C. The importance of the vitamins produced in the intestine can be shown by comparing the dietary vitamin requirements of a normal person with those of one who has received extensive antibiotic therapy and who has, as a result, lost his intestinal flora. The need for vitamins in the diet is less in the former.

Flatus

A second important product of the intestinal flora is the gases it produces. Normally in the alimentary tract there are about 150 ml. of gas. Fifty milliliters are in the stomach and consist of approximately 78 per cent N_2, 15 per cent O_2, and 7 per cent CO_2, and 100 ml. are in the colon and consist of N_2 (about 50%), CO_2 (about 40%), O_2, and products of fermentation—methane, hydrogen, and hydrogen sulfide (as much as 10%). There are also varying quantities of gas in the small intestine, but this is not usually substantial. The gas in the stomach gets there in swallowed food and saliva. Most of it is either passed on into the intestine or back up the esophagus by means of a belch. In a horizontal person considerably more gas goes into the intestine from the stomach than in a vertical subject. When large quantities of gas do get into the small intestine they may produce gushing sounds as they move to the colon.

Obstruction of the small intestine or colon, or stasis due to decreased motor activity, may result in an increase in the concentration of microorganisms above the obstruction or in the area of stasis. This, in turn, results in an increase in the amount of gas produced and accumulated in these areas. In one patient, for example, 3,500 ml. of gas at standard temperature, pressure, and saturation were withdrawn in a period of 24 hr. from a part of the small intestine above an obstruction.

The normal person releases between 380 and 650 ml. of flatus at standard temperature, pressure, and saturation. Approximately half of this is from swallowed air and the other half from microorganisms in the intestine. The quantity varies tremendously with the frequency of defecation and the diet. For example, a group of volunteers were placed on a diet in which 25 to 50 per cent of the caloric intake came from lima beans plus pork and beans. Before going on the diet they produced about 17 ml. of flatus per hour. After 2 wks. they were producing an average of 203 ml. per hour.

Dietary Bulk

Some foods contain large quantities of cellulose, etc., that we are not able, in the absence of microorganisms, to digest. In the case of herbivores such as the cow, the breakdown of cellulose by microorganisms is essential for survival. In man, however, the breakdown of cellulose is not an important source of calories. The reasons for this are that (1) relatively small amounts of cellulose and other roughage are broken down in the G-I tract and (2) the colon, where most of the breakdown occurs, has a poor absorptive capacity. There are, however, some exceptions to this. We find that only about 10 per cent of a 50-gm. meal of pectin (a hemicellulose found in fruits) ap-

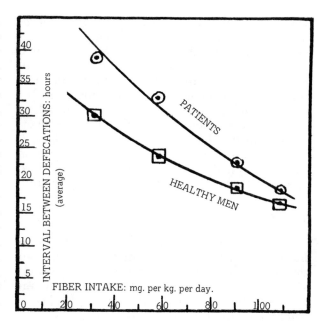

Fig. 24-16. The effect of vegetable fiber intake on the interval between bowel movements in normal and constipated men. (Cowgill, G. R., and Sullivan, A. J.: Further studies on use of wheat bran as laxative: Observations on patients. JAMA, *100*:800, 1933)

pears in the feces. With cellulose, however, most of it appears in the feces unchanged, except in cases in which the motility of the intestine has been profoundly decreased. Normally, then, cellulose and certain related substances serve primarily to increase the fecal bulk. This, in turn, results in more frequent defecation (Fig. 24-16).

Many people are much more concerned with the frequency of defecation than the facts justify. The feeling that there should be at least one bowel movement a day is widespread. All of our evidence, however, indicates that the main function of defecation is to eliminate feces (1) before distention of the intestine becomes excessive and (2) before the feces become so hard that defecation is unpleasant. Many who defecate just once a week show no bad side effects and there are cases of persons who have not defecated for a year; some people defecate after each meal, while in some cases on record persons have carried in their colon for years fecaliths weighing between 60 and 100 lbs. There is no evidence, in other words, to justify the old idea that feces kept for lengths of time in the colon release toxins into the blood.

Chapter 25

CONTROL OF THE SALIVARY GLANDS
CONTROL OF GASTRIC SECRETION
CONTROL OF PANCREATIC SECRETION
CONTROL OF BILE SECRETION
THE GALLBLADDER
THE INTESTINE

SECRETION AND ITS CONTROL

In this section we will be concerned primarily with the control of secretion. Secretion, like motility, is regulated by (1) sympathetic and parasympathetic neurons, (2) intrinsic plexuses which continue to function even when separated from the central nervous system for extended periods of time, (3) hormones produced in the digestive tract and in distant organs, and (4) the response of the cell itself to irritation or to agents to which it is directly exposed.

Many of the secretory cells are organized around a duct which leads into the digestive tract. The secretory cells of the liver are organized around a duct which carries secretions both (1) past a sphincter which leads to the duodenum and (2) into a storage organ (the gallbladder) which concentrates the secretion and pumps it into the intestine when stimulated to contract. The simplest secretory cell lies in the mucosa and releases its secretion directly into the digestive tract.

Secretion in the case of many cells involves the movement of zymogen granules from the cytoplasm of the cell into a duct or directly into the digestive tract. Here the granules dissolve, releasing enzymes or enzyme precursors. Generally the precursors are activated to their respective enzymes in the digestive tract. Digestive enzymes may also result from the disintegration of the cell and the release of its cytoplasm, enzymes, and enzyme precursors directly into the digestive tract. We have already indicated, for example, that intestinal lipase comes from the mucous cells which are sloughed from the small intestine.

Secretion is also associated with the release of a large volume of fluid. We find that the daily volume of salivary juice is about 1.5 liters, of gastric juice about 1.5 liters, of pancreatic juice about 0.5 liters, of bile about 0.8 liters, of intestinal juice about 4 liters, and of colonic juice about 0.2 liters. Some of these secretions are hypotonic and others hypertonic. We have noted in Figure 24-12 that they all differ in ionic content. For the most part, in other words, they do not seem to be simple filtrates of plasma to which zymogen granules have been added. In the case of the salivary glands, there seems to be a system of filtration, reabsorption, and secretion similar to that seen in the nephron of the kidney and, as in the nephron, producing a liquid which varies with the composition of the blood and the stimuli impinging upon it (Fig. 25-1). This system, like other secretory systems, is dependent upon its blood supply. Sometimes, however, through its active transport processes, it produces pressures in the salivary ducts greater than systolic arterial blood pressure.

CONTROL OF THE SALIVARY GLANDS

Saliva is produced by the serous cells of the parotid glands, the mucous and serous cells of the submaxillary glands and sublingual gland, and numerous mucous cells in the labial, buccal, and palatal glands. Some of the saliva produced is the result of spontaneous secretion. That is to say it is not the result of extrinsic stimuli and does not disappear on decentralization, but rather appears because of the intrinsic properties of a cell or gland.

The Parasympathetic Neurons

Main control of the salivary glands is exerted through the parasympathetic nerves which innervate them. In man, maximum secretion of saliva occurs when the parasympathetic fibers in the chorda tympani are stimulated at a rate of 10 stimuli per second. Increases or decreases in this rate decrease the release of saliva. The saliva produced by parasympathetic stimulation

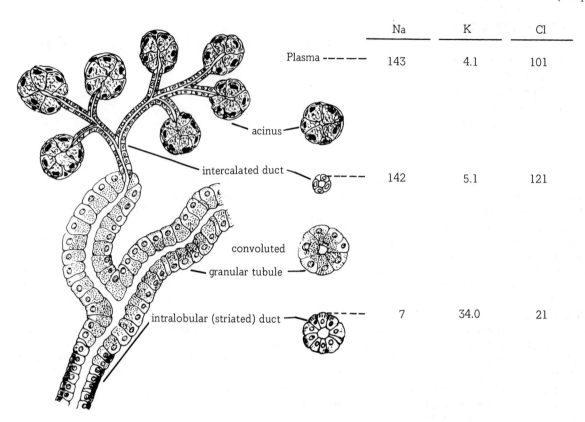

Fig. 25-1. Organization of the submaxillary gland of the adult male rat. This organization is thought to be similar to that found in the parotid and submaxillary glands of man. The data on the concentration of electrolytes in the ducts were obtained from the secreting gland by the micropuncture technique. (*Diagram from* Leeson, C. R.: Structure of salivary glands; *electrolyte concentrations from* Burgen, A. S. V.: Secretory processes in salivary glands. *In* Code, C. F. (ed.): Handbook of Physiology. Sec. 6 (Alimentary Canal), vol. II, pp. 469 and 564. Washington, American Physiological Society, 1967)

is characteristically watery (serous) with a high concentration of ptyalin and a relatively low concentration of mucin. Associated with the secretion is increased oxygen consumption and some nonpropagated changes in the transmembrane potential of the affected salivary gland cells.

Also associated with the stimulation of the chorda tympani is an increase in blood flow to the glands. Certain investigators have suggested that this is in response to the metabolites produced by the gland during secretion, but it is possible to block the secretory effect of parasympathetic stimulation by administering atropine without blocking the increase in blood flow to the gland (Fig. 25-2). Some have suggested that the increase in blood flow is due to the release of an agent, kallikrein, into the interstitial fluid. *Kallikrein*, it is believed, is a bradykinin-releasing enzyme. *Bradykinin* by itself produces vasodilation. In support of this concept is the observation that saliva obtained by parasympathetic stimulation, when injected intra-arterially, produces vasodilation. On the other hand, salivary glands made insensitive to kallikrein have been shown to increase their blood flow in response to parasympathetic stimulation. Possibly, then, there are several mechanisms for producing vasodilation by the stimulation of parasympathetic fibers.

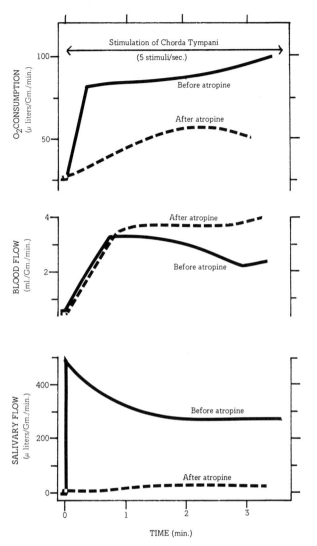

Fig. 25-2. Response of the submaxillary gland of the dog to stimulation of the chorda tympani before and after parasympathetic blockade with atropine. The horizontal arrow represents the period of parasympathetic stimulation (5 shocks/sec.). The ordinate is in terms of (*top*) oxygen consumption of the gland in microliters/Gm./min., (*center*) arterial blood flow to the gland in ml./Gm./min., and (*bottom*) flow of saliva in microliters/Gm. of gland/min. (*Modified from* Terroux, K. G.: Oxygen consumption and blood flow in the submaxillary gland of the dog. Canad. J. Biochem. Physiol., 37:13, 1959)

The Sympathetic Neurons

The most consistently demonstrated function of the sympathetic neurons to the salivary glands is the maintenance of vasomotor tone. Decreases in this tone result in vasodilation and an increase in blood flow to the salivary glands; increases in tone have the opposite effect. Some investigators have suggested the presence of cholinergic sympathetics to the salivary glands as well, and have indicated that their stimulation produces vasodilation and an increase in blood flow.

The role of the sympathetics in the control of salivary secretion, however, is less well understood. Much of the confusion is probably due to the fact that there is tremendous variability in the role of sympathetics in different glands and in different species. At one end of the spectrum is the dog, whose sympathetic innervation seems to be the least important in the control of secretion, and at the other is the cat, in which threefold increases in salivary secretion can be demonstrated in response to sympathetic stimulation, and in which extirpation of the superior cervical ganglion (a sympathetic ganglion) results in hypersensitivity to circulating catecholamines (epinephrine, norepinephrine, etc.) and a resultant increase in secretion. Most observers feel, however, that in man and most animals the sympathetics play a minor role at best in the control of salivary secretion.

In some animals sympathetic stimulation produces contraction of the myoepithelial cells of the gland and a momentary increase in the rate of salivation. Repeated stimulation, however, has progressively less effect. In these animals prior stimulation of the parasympathetics increases the response to sympathetic stimulation. In other cases, sympathetic stimulation apparently has a direct effect on secretion, being able to produce long-term increases in the flow of saliva. It has also been reported that the simultaneous stimulation of sympathetics and parasympathetics produces a more profuse flow of saliva than the sum of what the stimulation of each alone produces. This relationship—*synergism*, or potentiation—is considerably different from the *antagonistic* relationship between sympathetics and parasympathetics found in other organs.

Hormones

The salivary glands release saliva in response to food in the mouth, mouth dryness, and various other stimuli. These responses are due almost completely to nervous reflexes. The role of the hormones of the body at least as far as the salivary glands are concerned is to regulate metabolic pathways, electrolyte balance, and water balance. Alpha-stimulating catecholamines such as *epinephrine* and *norepinephrine* in high concentrations in the blood have about the same action on the salivary glands as sympathetic stimulation, but probably play no role in salivary control in man under most conditions.

The anterior pituitary gland releases hormones which (1) facilitate protein anabolism (growth hormone, or STH), (2) facilitate the release of mineralocorticoids from the adrenal cortex (adrenocorticotrophic hormone, or ACTH), and (3) facilitate the release of thyroxin from the thyroid gland (thyroid stimulating hormone, or TSH). The importance of these hormones to the body can be seen when the pituitary is removed (*hypophysectomy*). In the rat, for example, hypophysectomy results in (1) increases in Na$^+$ in the saliva, feces, urine, and sweat, (2) atrophy of many of the salivary glands, the gastric mucosa, the pancreas, and the intestinal epithelium, (3) a decrease in zymogen granules throughout the body, and (4) a decrease in amylase release by the salivary glands. In other words, many of the hormones of the body control the basal level of activity and the degree of responsiveness of the salivary glands. The saliva which results from the stimulation of motor nerves depends, then, upon (1) the character of the nervous stimulation, (2) the plasma concentration of hormones influencing the glands, and (3) other properties of the plasma. For example, when the plasma concentration of urea, ketone bodies, or even poisons such as mercury is high, their concentration in the saliva is also high. An example of the effect of the restriction of dietary Na on salivary Na concentration is presented in Figure 25-3.

Central Nervous System

Sensory neurons in the mouth, pharynx, and olfactory epithelium lead into an area in the medulla, the salivation center. These sensory neurons are stimulated by pressure during chewing, by touch receptors in the mouth, and by the taste and smell of food. They serve to stimulate salivary secretion by their action on the salivation center. Stimulation of these neurons on one side of the mouth facilitates secretion on both sides, but has its most profound effect on the ipsilateral side. Salivation centers in the cerebral cortex have also been postulated for man and dog, but apparently in man play little or no role in the control of salivation. This is quite surprising

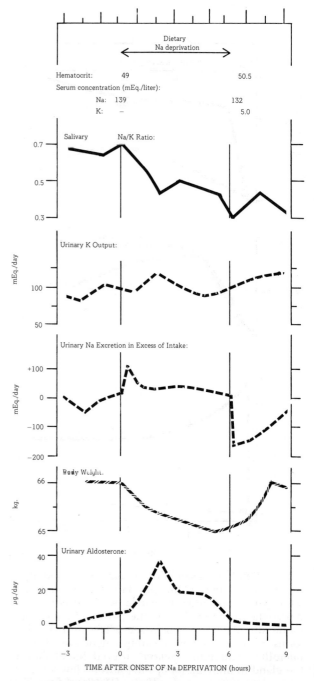

Fig. 25-3. Response of a normal man to an abrupt reduction in dietary Na from 210 to 3 mEq./day. Note that there resulted (1) a negative Na balance (*third line graph*), (2) a decrease in serum Na from 139 to 132 mEq./liter, (3) a reduction of the Na/K ratio in the saliva (*first line graph*), and (4) an increase in mineralocorticoid (aldosterone, etc.) output (*see lower graph*). Changes in hematocrit, serum K, urinary K, and body weight are also shown. (*Modified from* Crabbe, J., et al.: The significance of the secretion of aldosterone during dietary sodium deprivation in normal subjects. J. Endocr. Metab., *18*:1162, 1958)

when one remembers the classic experiments on salivation in the dog performed by Pavlov. He showed that he could train a dog to salivate in response to some stimulus, such as a bell, which the dog had learned to associate with food. In man, however, dinner bells and words like "sizzling sirloin steak" have not been shown to elicit a measurable increase in salivary secretion.

CONTROL OF GASTRIC SECRETION

The stomach, unlike the salivary glands, is controlled by many different influences. These include those arising from the stomach (gastric factors), the intestine (intestinal factors), and the head (cephalic factors). The cephalic factors seem to modify gastric secretion primarily through the parasympathetic and sympathetic nerves. Hostility, depression, and the smell and taste of food all seem to affect these nerves. Food in the stomach, on the other hand, acts (1) directly on secretory cells, (2) on the intrinsic nerve plexuses, (3) on sensory neurons to the brain, and (4) on gastric cells which release hormones that facilitate gastric secretion. Food in the intestine can cause the release of (1) a hormone that facilitates gastric secretion or (2) one that inhibits it. Some of these factors are summarized in Figure 25-4.

Cephalic Phase

The stimulation of areas in the hypothalamus or frontal cortex elicits action potentials in the parasympathetic neurons in the vagus nerve, which, acting through the myenteric plexus of the stomach, can elicit a mucous secretion 9 times the resting level, plus a considerably greater increase in the secretion of acid by the oxyntic cells (parietal cells) and pepsinogen from the chief cells. The acid activates the pepsinogen to pepsin. Some of the stimuli which act through the hypothalamus to increase the secretion of mucus, acid, and enzymes include the taste, smell, or sight of an appetizing meal, the sound of a steak frying, or any one of a number of stimuli which one might associate with an appetizing meal. Interestingly enough, however, an unappealing meal may elicit little or no gastric secretion. Note, for example, the experiments in Figure 25-5. Here a woman with complete closure (stenosis) of the esophagus and a gastrostomy was studied. Her usual manner of eating was to taste, swallow, and regurgitate her food. The regurgitant was then placed in the stomach through a gastrectomy opening. When an appealing meal was chewed but not placed in the stomach, gastric secretion as measured by acid production was markedly increased. When an unappealing meal was eaten in the same manner, there was little or no change in gastric secretion.

Another factor in the control of gastric secretion is the emotional state of the individual. Single incidents eliciting hostility and aggression have been found to increase the gastric secretion of acid for as much as 2 wks. Incidents which cause depression or the turning of the subject's hostility in on himself, on the other hand, seem to result in prolonged periods of

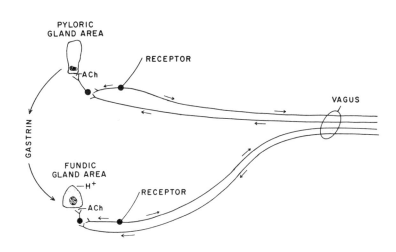

Fig. 25-4. A schematic diagram showing some of the complex relationships that modify the secretion of acid by the stomach. Note that (1) impulses originating from the central nervous system can (a) act directly on the fundic gland area to increase acid secretion or (b) act on the pyloric gland area to cause the release of gastrin which will act to increase acid secretion. Also note that (2) impulses originating in the stomach can (a), through intrinsic nerve fibers, stimulate acid secretion or (b), through sensory neurons in the stomach, stimulate neurons in the central nervous system to stimulate acid secretion. (Grossman, M. I.: Integration of neural and hormonal control of gastric secretion. Physiologist, 6:255, 1963)

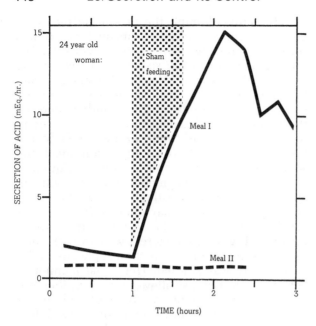

Fig. 25-5. Sham feeding and its effect on gastric secretion. In these studies a 24-year-old woman with an occluded esophagus and a gastric fistula ate a meal (8 oz. of cereal gruel) which she disliked (meal I) and a more appealing meal (meal II) of fresh vegetables, salad with dressing, bread and butter, potatoes, meat (fried chicken, lamb chops, fried ham steak, or 2 fried eggs), and ice cream and cake. The data presented are based on 4 experiments with meal I and 12 with meal II. Note that the more appealing meal produced the more pronounced gastric response even though neither meal reached the stomach. (*Modified from* Janowitz, H. D., et al.: A quantitative study of the gastric secretory response to sham feeding in a human subject. Gastroenterology, *16*:106; copyright © 1950, the Williams and Wilkins Company, Baltimore)

inhibited gastric secretion. Chronic stimulation of, or certain lesions in, the hypothalamus have been shown to produce such profound increases in gastric secretion as to cause marked ulceration of the stomach and intestine. Experiments have also been performed on monkeys in which stress alone has been shown to result in ulceration. In these experiments two monkeys were paired. Both sat in a chair in front of a control panel. The first, the executive monkey, could prevent himself and his partner from being electrically shocked if he pushed the proper buttons in response to a certain stimulus. The second, the control monkey, had the same buttons and stimuli, but could not prevent the shock. The four executive monkeys studied died within 9 to 48 days of the beginning of the experiments. They had extensive ulcerations of the stomach and intestine and associated hemorrhages. The control monkeys showed no ill effects.

Since all the stimuli associated with the cephalic facilitation of gastric secretion are ineffective after bilateral vagotomy or atropinization, they are generally regarded as acting by means of increasing parasympathetic tone to the stomach. The sympathetic fibers to the stomach, on the other hand, seem to be more important in the control of its blood supply. Increases in sympathetic tone can decrease the blood supply to the stomach to such a degree that the volume of the gastric juice is markedly decreased. In rage, pain, or strenuous exercise, for example, there is a pronounced increase in sympathetic tone and a decrease in gastric secretion which may outlast the stimulus. Decreases in sympathetic tone below the resting level result in a greater than resting blood supply to the stomach but do not increase gastric secretion over the resting level.

Gastric Phase

Food placed directly into the stomach produces increases in secretion which last as long as 4 hr. and which cause the secretion of as much as 600 ml. of gastric juice. This secretion is not dependent upon stimuli originating from the head, but may be facilitated or inhibited by these stimuli. The food in the stomach facilitates secretion by (1) mechanical stimulation (gastric distension and physical contact), (2) stimulation provided by chemicals in the food (secretogogues), and (3) stimulation provided by the products of protein digestion. If the secretion produces a pH in the gastric antrum of less than 2.0, there will also be an inhibition of further secretion.

Extrinsic neurons. Carbohydrates and fats, while in the stomach, facilitate gastric secretion primarily or perhaps solely by means of distension and mechanical contact. Proteins and their products, caffeine, and dilute alcohol (10% ethanol has the maximum effect), on the other hand, facilitate secretion both by virtue of their volume and their chemical characteristics. As can be seen in Figure 25-4 these foods act by means of a number of mechanisms. The distension they produce stimulates sensory neurons which travel up the vagus nerves and reflexly increase parasympathetic tone to the stomach. In this reflex preganglionic parasympathetic neurons stimulate postganglionic parasympathetic neurons in the myenteric plexus.

Intrinsic neurons. If we section the vagi which contain both the sensory and motor elements of the reflex discussed above, we find that the secretory response to distension is decreased but that the person still responds to distension with secretion. The mechanism here is apparently direct stimulation of the neurons of the intrinsic myenteric plexus. Part of the evidence for this can be seen in Figure 25-6. Here all the nerves going to and coming from the stomach of a dog were severed. The resultant decrease in secretory activity was then compensated for by subcutaneous injections of histamine every 15 min. Under these circumstances distension resulted in an increase in secretion that could be prevented by the injection of atropine, a postganglionic cholinergic blocking agent.

Gastrin. Distension, meat extracts, and 10 per cent ethanol all act on the pyloric antrum to cause the release of decapeptides into the circulation. These decapeptides have been isolated and synthesized and are referred to as gastrin I and gastrin II, or collectively as gastrin. Injections of as little as 50 μg. of gastrin increase acid secretion by the oxyntic cells, and somewhat larger doses also increase the secretion of pepsinogen by the stomach, facilitate a more forceful gastric contraction, and increase the production of pancreatic juice and enzymes.

Vagal stimulation, as well as the injection of acetylcholine, also facilitates the release of gastrin. We find, for example, that if we inject enough insulin to decrease the blood sugar level to below 45 mg. per 100 ml. of plasma, gastrin is released and a secretion of acid and pepsinogen that lasts about 90 min., as well as an increase in gastric motility, occurs. This response is not seen if all the vagal fibers to the stomach have been severed; an absence of gastric secretion after an *insulin-induced hypoglycemia* is therefore frequently used to test for complete vagal decentralization.

In Figure 25-7 we have a study of the effect of meat extract in a decentralized antral pouch on the gastric secretion of acid. You will note that even though the pouch has no extrinsic innervation, it still produces acid in response to the appropriate stimulus. This response is apparently due to the release of gastrin, but is depressed by atropine, by painting the mucosa with 2 per cent cocaine, and by procedures which destroy the intrinsic nerve plexuses. From this we conclude that stimuli which elicit the release of gastrin are most effective when acting on the intrinsic neurons. It should also be emphasized in this regard that the secretogogues in meat extract facilitate the release of gastrin only if they act on the gastric mucosa. They do not elicit the release of gastrin when injected intravenously. If the pH in the pyloric antrum goes below 2.0, gastrin release is inhibited.

Histamine. For a long time many believed that histamine and gastrin were the same. This was based on the fact that (1) histamine produces increases in gastric secretion similar to those produced by gastrin, (2) the antral mucosa contains high concentrations of the mast cells which

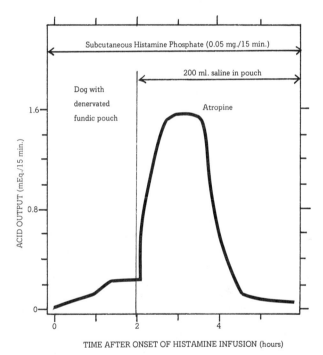

Fig. 25-6. Response of the decentralized oxyntic gland area of the stomach of the dog to distention before and after the injection of atropine. Secretory activity was facilitated by the subcutaneous injection of histamine every 15 min. Distention was produced by the injection of 200 ml. of isotonic saline into a pouch formed by the oxyntic gland area. On the basis of these experiments it is concluded that distention directly stimulates intrinsic cholinergic fibers in the stomach, which, through the release of acetylcholine, cause the secretion of hydrochloric acid. (*Redrawn from* Grossman, M. I.: Stimulation of acid secretion. Gastroenterology, *41*:389; copyright © 1961, The Williams and Wilkins Company, Baltimore)

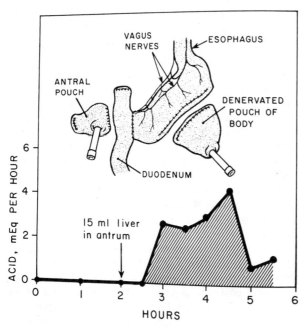

Fig. 25-7. The effect of injecting 15 ml. of homogenized beef liver into a decentralized antral pouch of the dog's stomach. Note that the meat produced an increase in acid secretion in the antrum. Apparently this is the result of the release by the antral mucosa of a hormone, gastrin, which is released into the blood, circulated, then returned to the antrum to facilitate acid secretion. This response is decreased by the injection of atropine. (*From* Physiology of the Digestive Tract, Second Edition, by H. W. Davenport. Copyright © 1966, Year Book Medical Publishers, Inc.; used by permission. *Adapted from* Oberhelman, H. A., Jr., et al.: Am. J. Physiol., *169*:738, 718, 1952)

produce histamine and heparin (10,000 to 20,000 mast cells/mm.³), (3) the gastric mucosa contains large quantities of the enzyme histidine decarboxylase, which is responsible for histamine synthesis but contains little or none of the enzyme responsible for histamine destruction, diamine oxidase, and (4) the concentrations of histamine in the stomach correlate well in many cases with the amount of acid present. With the isolation and synthesis of gastrin I and II, however, it was shown that gastrin is a separate entity and, unlike histamine, does not decrease arterial pressure.

It was later suggested that histamine is the agent which acts directly on the gastrin-secreting cell. It was noted, for example, that atropine does not prevent the gastrin-like action of histamine. On the other hand, antihistaminics in dosages which do not destroy the gastric mucosa do not seem to inhibit gastrin release. We also find that hypoglycemia increases the release of gastrin but not histamine. Why then do we see such a good correlation in most cases between gastrin and histamine concentrations in the stomach? Possibly the most logical suggestion is that microorganisms in the small intestine use the meats we eat for the synthesis of histamine. In other words, with increased meat there is increased gastrin release and increased histamine release by microorganisms, not mast cells. In support of this view is the observation that if we separate the antrum from the small intestine, we find that meat placed in the antrum causes acid secretion but not histamine release. The same sort of results are obtained using antibiotics which destroy the intestinal flora.

Intestinal Phase

The character of the chyme in the small intestine and particularly in the duodenum also plays a role in modifying the secretions of the stomach. In Figure 25-8, for example, we note that chyme placed in the jejunum of a dog increases the quantity of gastric juice and its concentration of acid in a decentralized pouch formed from the stomach (Heidenhain pouch). Some of the stimuli to which the intestine or duodenum respond by increasing gastric secretion include protein and its products, dilute alcohol, acid, distension, irritation, and a high osmotic pressure. Some have suggested that the intestine, in response to these stimuli, releases a hormone similar to gastrin, sometimes called "intestinal gastrin." The release of this hormone is prevented by the application of a local anesthetic or procaine to the intestinal mucosa exposed to the stimuli mentioned above.

Chyme in the small intestine can also inhibit gastric secretion. The degree of this inhibition depends upon the amount of vagal tone to the stomach and the concentration of circulating gastrin. Acid in the duodenum, for example, frequently increases secretion in a decentralized stomach but decreases it in a stomach with a moderate to high degree of vagal tone. It has been suggested that at least part of this response is due to the release of an inhibitory hormone from the intestinal mucosa. The hormone secretin has a profound effect on the formation and release of bile and pancreatic juice, and can inhibit the action of the gastrin on the stomach. Possibly secretin is important in the inhibition

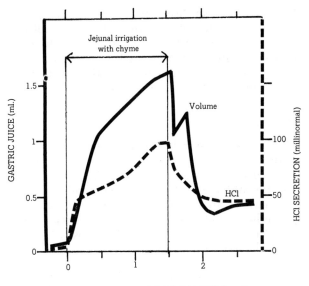

Fig. 25-8. Secretion of gastric juice in a decentralized gastric pouch in response to chyme placed in a jejunal loop of a dog. It has been suggested that the intestine releases an "intestinal gastrin" into the blood and that it is this hormone which facilitates gastric secretion. (*Modified from* Grossman, M. I.: Neural and hormonal stimulation of gastric secretion of acid. *In* Code, C. F. (ed.): Handbook of Physiology. Sec. 6 (Alimentary Canal), vol. II, p. 846. Washington, American Physiological Society, 1967)

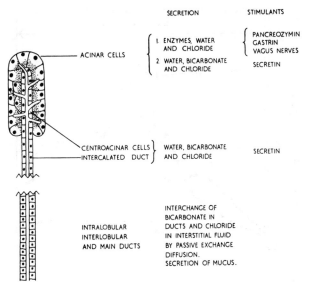

Fig. 25-9. Sites of action of pancreozymin, gastrin, secretin and the vagus nerves in the pancreas. (Harper, A. A.: Hormonal control of pancreatic secretion. *In* Code, C. F. (ed.): Handbook of Physiology. Sec. 6 (Alimentary Canal), vol. II, p. 980. Washington, American Physiological Society, 1967)

of the stomach by the intestine. There also has been an agent, gastrone, isolated from the gastric mucosa which inhibits gastric secretion. It has been suggested that acid in the intestine might cause its release.

CONTROL OF PANCREATIC SECRETION

The pancreas produces an alkaline fluid containing enzymes important for the digestion of carbohydrates, fats, and proteins. The quantity of pancreatic juice produced per day is between 200 and 800 ml. Both the volume of pancreatic juice and its contents are controlled by (1) the hormones pancreozymin, gastrin, and secretin, and (2) the extrinsic nerves that go to the pancreas, as well as the intrinsic plexuses associated with them (Fig. 25-9). There is evidence too that diet may have an effect on the exocrine function of the pancreas. We find that the pancreas of rats kept on a diet containing 18 per cent casein for several months contains twice the proteolytic enzymes as that of rats fed 6 per cent casein for the same period. High starch diets can result in the secretion of 7 times as much amylase as a low carbohydrate diet, but the quantity of fat in the diet apparently has little or no effect on the quantity of pancreatic lipase.

Autonomic Nervous System

The pancreas receives both sympathetic and parasympathetic innervation. The adrenergic sympathetic neurons control the degree of tone of the arterioles. Marked increases in sympathetic tone can so decrease pancreatic flow that secretion too decreases. It has been suggested that there are cholinergic sympathetics to the pancreas, but the evidence is not well documented.

The major direct efferent nervous control of the pancreas is apparently through the parasympathetic fibers in the vagi. In Figure 25-10, for example, we note the effect of sham feeding on pancreatic secretion. Apparently there is a cephalic phase of pancreatic control, just as there is a cephalic phase in gastric control. In these experiments, placing food in the mouth produced an increase in both the volume of pancreatic juice

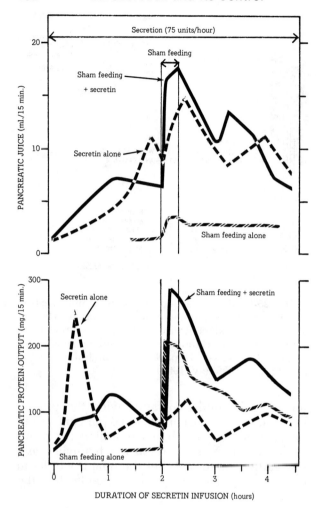

Fig. 25-10. Pancreatic flow and protein output in 3 dogs with esophagostomy and gastric and pancreatic fistulas. In these investigations (1) sham feeding alone (6 experiments), (2) secretin alone (3 experiments), and (3) secretin plus sham feeding (3 experiments) were studied. (*Modified from* Preshaw, R. M.: Intergration of nervous and hormonal mechanisms for external pancreatic secretion. *In* Code, C. F. (ed.): Handbook of Physiology. Sec. 6 (Alimentary Canal), vol. II., p. 998. Washington, American Physiological Society, 1967)

released and its protein content. We also find that insulin-induced hypoglycemia increases vagal tone not only to the stomach but to the pancreas as well. In other words, many of the stimuli that increase vagal tone to the stomach, and through this mechanism increase the volume and acidity of the gastric juice, also increase vagal tone to the pancreas, thereby increasing the volume, alkalinity, and enzyme content of the pancreatic juice. These responses all occur even when the pyloric sphincter of the stomach is tied off—so are not the the result of the response of the duodenum to the secretions of the stomach. They are prevented by either the injection of atropine or bilateral section of the vagal nerves.

Hormones

The three major hormones affecting the exocrine function of the pancreas are *secretin*, *pancreozymin*, and *gastrin*. Their effect on the volume and content of the pancreatic juice is summarized in Figure 25-11. Secretin is found in highest concentration in the parts of the mucosa of the small intestine closest to the stomach. As we pass from the duodenum to the terminal end of the jejunum we find that its concentration progressively declines. Secretin was first reported by Bayliss and Starling in 1902 and has the distinction of being the first hormone discovered. It is produced in response to a duodenal or jejunal pH of less than 4.5 and is released into the blood, where it has a half-life of 18 min. It serves to (1) depress gastric secretion, (2) facilitate a pancreatic secretion high in bicarbonate and low in enzyme concentration, and (3) facilitate the secretion of bile.

Pancreozymin is also produced by the intestinal mucosa, but facilitates the release of a secretion higher in enzyme and lower in bicarbonate concentration than secretin. Its release is probably stimulated by peptones, amino acids, oleate, water, starch, and dextrose in the small intestine. These same substances may also stimulate the release of gastrin from the antral mucosa of the stomach. You will note that gastrin produces a pancreatic juice higher in enzyme and lower in bicarbonate than secretin.

CONTROL OF BILE SECRETION

Man produces between 250 and 1,100 ml. of bile a day. The bile has a pH above 7.0 and contains Na^+, K^+, Cl^-, HCO_3^-, bile salts, bile pigments, lipids, and various synthetic compounds. The bile salts are important in the digestion of fat (see the preceding chapter), the HCO_3^- helps neutralize acid in the duodenum, and the bile pigments are excretory products resulting from the breakdown of hemoglobin. Various synthetic compounds taken in medications are also removed from the blood by the liver and eliminated from the body in the bile.

Control of Bile Secretion

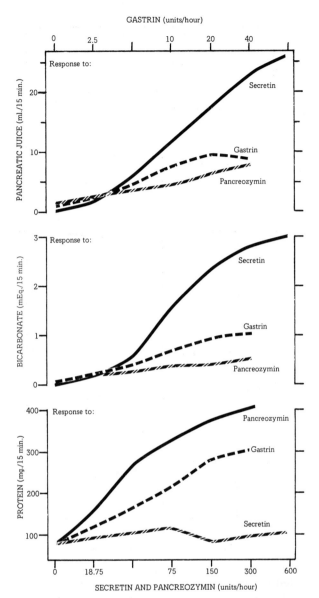

Fig. 25-11. Response of 4 dogs with chronic pancreatic fistulas to continuous intravenous infusions of gastrin, secretin, and pancreozymin. (Modified from Harper, A. A.: Hormonal control of pancreatic secretion. In Code, C. F. (ed.): Handbook of Physiology. Sec. 6 (Alimentary Canal), vol. II, p. 977. Washington, American Physiological Society, 1967)

Sulfobromophthalein

One of the synthetic compounds which the liver excretes in the bile is sulfobromophthalein (bromsulphalein, or BSP). Its removal from the plasma and secretion into the bile is used as a test of liver function. At BSP concentrations greater than 3 mg. per 100 ml. plasma, the liver of a normal man should be able to secrete about 9.5 mg. per minute into the bile. In women the secretion is about 7.1 mg. per minute. Each of these values is called the normal transport maximum (T_m) for BSP. The concept of a T_m was discussed in the chapters on the kidney. Since patients with severe liver disease may have a T_m for BSP near zero, one can test for liver function by injecting 5 mg. of BSP per kilogram of body weight and sampling the plasma 45 min. later. A normal person should have less than 5 per cent of the BSP left in his plasma.

Liver blood flow may also be approximated by the intravenous infusion of BSP. In this test the plasma level of BSP is kept constant by continuous infusion. The effective hepatic plasma flow (EHPF) is then calculated as follows:

$$\text{EHPF} = \frac{\text{BSP infused (mg./min.)}}{\text{arteriovenous difference (mg./liter of plasma)}}$$

The arterial concentration of BSP in the plasma is obtained by analyzing a blood sample from an artery or arm vein. It is assumed that this concentration is the same as that going to the liver in the hepatic artery and portal vein. The venous sample is obtained from a catheter in the hepatic vein. This technique is similar to the Fick principle used for the determination of cardiac output and is discussed more fully in Chapter 8 (p. 131). The EHPF can be changed to the effective hepatic blood flow (EHBF) if one knows the subject's hematocrit (% of red blood cells in the blood):

$$\text{EHBF} = \text{EHPF} \times \frac{100}{100 - \text{hematocrit}}$$

Normally the EHBF is about 1,400 ml. of blood per minute per 1.73 m.² of body surface. This is about 25 per cent of the cardiac output.

Bile Secretion

The amount and character of the bile formed by the liver is dependent upon (1) the hormones in the blood, (2) the concentration of bile salts in the blood, (3) the state of the hepatic circulation, and (4) the pressure in the bile cannuliculi. Bile is formed by an active process which can produce pressures in the bile cannuliculi 4 to 5 times greater than those found in the hepatic sinusoids (Fig. 25-12). The stimulation of sympathetic fibers

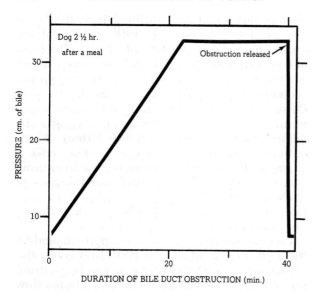

Fig. 25-12. Secretory pressure in the bile duct of an unanesthetized dog with the duct to the gallbladder (cystic duct) and the duct to the duodenum obstructed. Apparently the rate of bile production remains constant until the secretory pressure of the liver is reached. (*Redrawn from* McMaster, P. D., and Elman, R.: On the expulsion of bile by the gallbladder. J. Exp. Med., *44*:176, 1926)

to the liver, on the other hand, can, through a vasoconstriction which decreases hepatic blood flow, markedly decrease bile production. Epinephrine from the adrenal medulla can have a similar effect.

The vagus. The vagus controls the secretion of bile in much the same way it controls gastric and pancreatic secretion. During a meal and during insulin hypoglycemia there is an increase in hepatic secretion that is blocked by atropine or bilateral vagotomy and is the result of the stimulation of vagal efferent neurons to the liver. Stimulation of the peripheral end of the vagus of dogs and monkeys usually results in a two- to fourfold increase in bile secretion.

Secretin. The response of a dog to the hormone secretin is illustrated in Figure 25-13. Apparently in the dog, as well as man, secretin produces greater increases in bile flow and HCO_3^- production than in the release of bile salts. Note in the illustration that secretin produced a decrease in the concentration of bile salts in the bile. The HCO_3^- concentration in the bile can be decreased by the injection of a *carbonic anhydrase* inhibitor such as acetazolamine. An injection of 65 mg. of this drug per kilogram of body

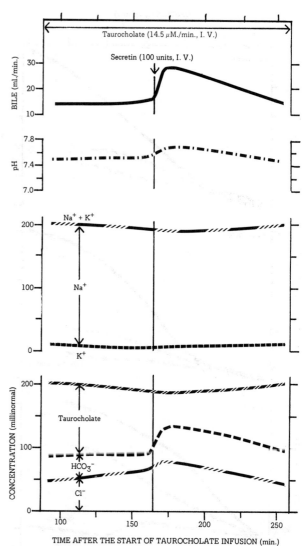

Fig. 25-13. Composition and volume of the bile secreted by the liver of a dog infused with a bile salt (taurocholate). Note that the injection of secretin increased bile secretion and the concentration (in millinormal) of HCO_3^- and Cl^-, but decreased the concentration of the bile salt, taurocholate. (*Modified from* Wheeler, H. O., and Ramos, O. L.: Determinants of the flow and composition of bile in the unanesthetized dog during constant infusions of sodium taurocholate. J. Clin. Invest., *39*:165, 1960)

weight markedly reduces the bicarbonate in the secretions of both the pancreas and the liver.

Bile salts. Bile salts are (1) synthesized in the liver from cholesterol, (2) released into the bile, where their concentration ranges from 10 to 20 mM. and where they form negatively charged ions and micelles, (3) released into the duodenum with the bile, (4) absorbed into the blood of the enterohepatic circulation, and (5) reabsorbed by the liver, where they are stored and released again. In addition, approximately 25 per cent of the body's bile salts are lost in the feces each day.

In response to a moderately fatty meal, a person will secrete from the liver somewhat more than twice the quantity of bile salts contained in the body. This is the result of the bile salts being released with the bile into the duodenum, absorbed from the intestine into the blood, reabsorbed by the liver, and resecreted into the liver cannuliculi. If the bile produced in response to a meal is collected, and the bile salts are not permitted to be absorbed into the blood, bile salt production will increase fourfold but bile salt release and the volume of bile produced will markedly diminish. Apparently bile salts in the plasma inhibit the production of more bile salts and facilitate the secretion of bile. Any agent which increases the the volume of bile secreted by the liver is called a *choleretic*. Vagal stimulation, secretin, and bile salts, therefore, are all choleretics.

THE GALLBLADDER

Man, dogs, cats, ducks, and chickens contain an organ, the gallbladder, which stores and concentrates the bile (Fig. 25-14). Animals such as the horse, rat, and pigeon have no gallbladder, whereas the pig, goat, sheep, cow, bush rat, rabbit, and guinea pig have gallbladders with a poor concentrating capacity. In man the gallbladder has an average capacity of about 35 ml. It receives some but not all of the bile produced by the liver. Bile that does not go to the gallbladder is stored in the hepatic ducts or the bile duct and is released periodically past sphincters of the bile duct to the duodenum. This release occurs during starvation as well as in response to a meal.

The bile which enters the gallbladder has Na^+, Cl^-, HCO_3^-, and water removed and mucin added. As a result, gallbladder bile differs from liver bile in that it has a lower concentration of Cl^- and HCO_3^- and a lower pH. In fact, its pH has been reported to go as low as 5.6, but this is not typical. Gallbladder bile, on the other hand, has a higher concentration of bile salts, bile pigments, and cholesterol (each may become 5 to 10 times more concentrated in the gallbladder). These functions of the bladder are important but not essential for fat digestion. *Cholecystectomized* patients generally show good health and adequate fat digestion. They digest fried foods quite well, but do have to avoid meals with a particularly high fat content. Part of their adaptation to cholecystectomy is a progressive dilation of the bile duct that increases the amount of bile they can store and release in response to a fatty meal.

Gallstones

The high concentration of certain dissolved particles in the bile is generally not a problem, but in cases of infection or biliary stasis, the cholesterol, bile pigments, and calcium may precipitate out in the gallbladder, cystic duct, hepatic duct, or bile duct. The preciptiant is usually called a gallstone. Most gallstones contain about 94 per cent cholesterol, 1 per cent calcium, and 4 per cent bile pigment. A few may be primarily calcium carbonate. If a gallstone brings about obstruction of the bile duct, the liver continues to form bile, a high biliary pressure develops, and the blood starts to absorb the bile from the bile duct. The result is jaundice. Its effect on digestion and absorption is also quite marked. There is less complete absorption of fat and the fat-soluble vitamins. The decrease in the absorption of vitamin K may lead to bleeding disorders, and the decrease in the

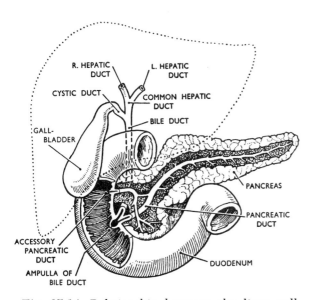

Fig. 25-14. Relationship between the liver, gallbladder, pancreas, and duodenum. (Bell, G. H., et al.: Textbook of Physiology and Biochemistry. 7th ed., p. 320. Edinburgh, E & S Livingstone, 1968)

bile pigments in the feces results in a clay-colored stool.

Control

The gallbladder and the sphincters which constrict the bile duct (sphincter of Oddi and Boyden) are controlled by parasympathetic nerves in the vagus and by hormones. The vagus causes the bladder to contract in response to a meal and apparently also in response to the thought of food. Its role in the control of the sphincters, however, is not certain. Chyme in the duodenum which contains fat, egg yolk, or meat also causes contraction of the bladder. This response, unlike the vagal response, is not blocked by atropine or bilateral vagotomy and is probably due to the release of a hormone from the small intestine. Some refer to this hormone as *cholecystokinin*. Others feel it is pancreozymin.

THE INTESTINE

The small intestine is the structure in which most of the digestion and absorption of nutrients occurs. It is protected from the enzymes contained in its lumen by the mucus produced by the goblet cells. The secretions released by pouches formed from (1) the duodenum have a pH of about 6.8, (2) the jejunum, 6.8, (3) the ileum, 7.6, and (4) the colon, 8.0. It is not known whether the intestine secretes any enzymes into its lumen, and it has been suggested that possibly all the enzymes added to the intestine by the mucosa are the result of desquamation. That is to say that they are intracellular enzymes released when the cell membrane of the mucosal cell is broken down.

The control of secretion in the small intestine seems to be similar to that discussed for the liver and pancreas. Vagal stimulation has been shown to increase the secretion of duodenal pouches from a resting level of 3 ml. per hour to one of 6 ml. per hour, but hypoglycemia apparently has no effect on secretion. Irritation of the mucosa, on the other hand, has been shown to increase secretion in the decentralized intestine. Since atropine blocks this response it is thought to be due to local nerve nets. Neither atropine nor decentralization, however, blocks increases in secretion in response to food in the small intestine. Some attribute this response to the release of a hormone, *enterocrinin*, secreted into the blood by the mucosa. Secretin has also been shown to increase duodenal secretion up to threefold, but unlike enterocrinin, has no effect on secretion by the jejunum.

In the large intestine no enzymes are secreted into the lumen. Here the major secretion is mucus and the major control of it is through direct contact with the mucosa. Extrinsic nerves, intrinsic nerve nets, and hormones are apparently of minor importance in the control of secretion.

Chapter 26

CATABOLISM
ANABOLISM
CALORIMETRY
CALORIC INTAKE

ENERGY BALANCE

Man, unlike chlorophyll-containing plants, is totally dependent upon the foods he eats for energy and nutriment. These foods are essential for the maintenance of such body processes as growth, repair, excitability, contraction, and secretion. They act as a source of (1) chemical energy which the body can convert into mechanical, electrical, or thermal energy and (2) matter which the body uses to produce protoplasm and extracellular material. In this chapter we will be concerned with the relation between nutritive intake and utilization.

CATABOLISM

The sum of the chemical reactions occurring in the organism is called metabolism and can be divided into reactions associated with (1) the breakdown of a molecule (catabolism) and (2) the combination of molecules to form larger molecules (anabolism). During catabolism, molecules are changed into other, smaller molecules capable of passing readily through cell membranes and other barriers. Catabolism is also the process by which we destroy microorganisms and foreign molecules, and by which we obtain the energy needed to maintain the cell and perform work.

Carbohydrates

Carbohydrates constitute a major source of energy for the cell. They enter the blood from the digestive tract primarily in the form of hexoses and pentoses and are then picked up by various cells, which either (1) break them down to pyruvates, lactates, CO_2 and H_2O, and energy, (2) incorporate them into a structural molecule, or (3) form them into a storage molecule such as glycogen or fat (Fig. 26-1). The major carbohydrates in the plasma are glucose and lactic acid (Table 15-1). Under resting conditions the concentration of glucose is 85 mg. per 100 ml. of plasma, and of lactic acid, 10 mg. per 100 ml. of plasma. During strenuous exercise, however, active muscles catabolize tremendous quantities of glycogen and glucose to lactic acid plus en-

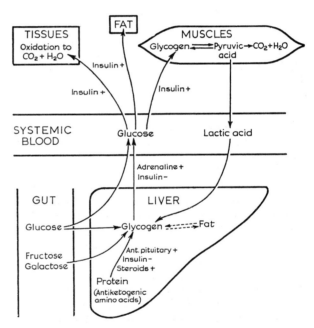

Fig. 26-1. Role of the liver in carbohydrate metabolism in man. Monosaccharides pass from (1) the gut to the portal blood to (2) the sinusoids of the liver and then into (3) the systemic circulation. The liver stores CHO in the form of glycogen. The glycogen concentration increases when the concentration of CHO in the blood is increased, and decreases when the CHO in the blood is decreased. Certain hormones can also facilitate (+) and inhibit (−) the catabolism and formation of glycogen in the liver and other organs. Note that glycogen and glucose can be formed from fructose and galactose, fat, and protein (gluconeogenesis), and that fat can be formed from CHO. (Bell, G. H., et al.: Textbook of Physiology and Biochemistry. 7th ed., p. 357. Edinburgh, E & S Livingstone, 1968)

ergy; the lactate concentration of the arterial and venous blood may increase more than eightfold (Fig. 5-14). Part of this lactate is removed from the blood by the heart, which under such conditions catabolizes more blood lactate than blood glucose (Table 7-1). The rest is picked up by the liver and other organs, which use it for nutrition or change it into a storage form such as glycogen or fat.

The organ most concerned with maintaining constant blood sugar and blood lactate concentrations is apparently the liver. It can (1) convert the carbohydrate (CHO) absorbed from the digestive tract to glucose, (2) convert lactic acid from the blood to glucose, (3) produce glucose from glycerol, dicarboxylic acids, amino acids, and odd-chain fatty acids, (4) produce glucose from its own glycogen stores, (5) produce glycogen from glucose, (6) produce fat, glycerol, fatty acids, and ketone bodies from glucose, and (7) release glucose, ketone bodies, triglycerides, amino acids, and certain proteins into the blood. The balance existing between these reactions depends in part on the nutritive composition of the plasma and the hormones released into the plasma.

During acidosis and in prolonged starvation, however, the kidneys may also serve as an important source of blood sugar. Under normal conditions they will use up more glucose than they produce, but Owen et al. in 1969 were able to show in obese humans starved for 35 to 40 days that their kidneys produced from circulating lactate, pyruvate, glycerol, and amino acid carbons approximately 45 per cent of the glucose found in the blood.

The dependence of organs upon a fairly constant blood sugar level varies with (1) their glycogen stores and (2) their dependence upon carbohydrate as a nutritive source. The liver in an average individual contains about 100 gm. of glycogen. This represents the highest concentration found anywhere in the body (Table 17-2). The heart, on the other hand, contains about 10 per cent of the concentration found in the liver; skeletal muscle, 7 per cent; and the brain, about 0.9 per cent.

Under most conditions the brain and the red blood cell appear to be dependent upon blood glucose as their sole fuel. A rapid fall in the blood sugar level, like a fall in the PO_2, may so affect the central nervous system as to produce confusion, lethargy, coma, or irreversible brain damage, the latter occuring at plasma concentrations of sugar below 40 mg./100 ml. On the other hand, fasts lasting 4 to 6 weeks result in some basic changes in brain metabolism which may provide some protection against hypoglycemia. After such a fast, men are able to utilize ketone bodies, acetoacetate, and beta hydroxybutyrate as fuels to supplement the low concentrations of glucose available to the brain under starvation conditions. The heart, skeletal muscle, and most other organs, on the other hand, are must less dependent than the brain and red blood cells on blood sugar because their concentrations of glycogen are higher and their catabolism of fatty acids is a more important source of energy. You will note in Table 17-1, for example, that the respiratory quotient for the brain is 1, whereas that for the heart, skeletal muscle, and skin is approximately 0.85.

The healthy heart differs from skeletal muscle in its greater dependence on aerobic catabolism. During exercise, for example, when skeletal muscle is releasing large quantities of lactate into the blood, the heart is producing little or no lactate of its own and is, in addition, (1) removing lactate from the blood, (2) converting the lactate to pyruvate, and (3) converting the pyruvate to CO_2 and water. In cardiac ischemia, hypoxia, and infarction, on the other hand, lactate is produced by the myocardium. This change in metabolism is studied clinically by comparing arterial blood samples (blood going to the coronary arteries) with coronary sinus blood samples (blood coming from the coronary veins). Either (1) a decrease in the cardiac withdrawal of lactate from the blood or (2) an elevation of the lactate to pyruvate ratio in the coronary sinus blood is an index of cardiac pathology. Other changes associated with cardiac ischemia include (3) a loss of intracellular K^+ (see Fig. 26-2), (4) a decrease in intracellular glycogen, ATP, and creatine phosphate, and (5) an increased reliance on the hexosemonophosphate shunt (HMP, phosphogluconate shunt, pentose-phosphate pathway). The HMP shunt is a pathway for the catabolism of glucose that represents an alternative to the classical Embden-Meyerhof scheme of glycolysis. Both pathways result in the production of about 35 moles of ATP from ADP per mole of glycogen catabolized to CO_2 and water, but the HMP shunt is also important in the production of such compounds as ribose, deoxyribose, galactose, glucosamine, etc. Normally,

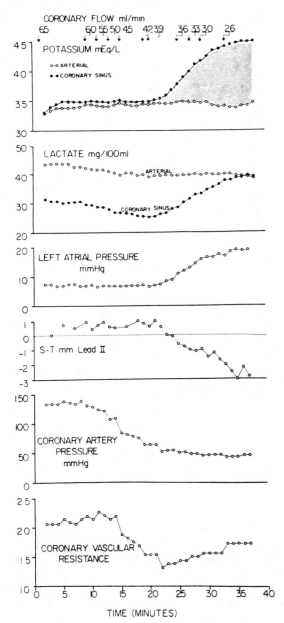

Figure 26-2. The dog's response to cardiac ischemia. Apparently a step-wise reduction of blood flow to the perfused left main coronary artery of a dog causes (1) a loss of potassium from the heart, (2) an elevation of the coronary sinus lactate concentration, (3) an elevation in the left atrial pressure, (4) a depression of the S-T segment of electrocardiogram lead II, (5) a depression in coronary artery pressure, and (6) a decrease in coronary vascular resistance. (Case, R. B., et al.: Biochemical aspects of early myocardial ischemia. Am. J. Cardiol., 24:768, 1969)

less than 10 per cent of the glucose catabolized will follow the HMP shunt.

Cardiac ischemia lasting longer than 5 minutes will also produce degenerative changes in the nuclei and after 20 minutes of ischemia there will be a mitochondrial swelling that can lead to a rupture of the mitochondrial membrane. After several hours of ischemia, increases in the concentration of various enzymes in the blood will begin. This includes increases in the concentration of (1) serum creatine phosphokinase (CPK) and the mitochondrial enzymes, (2) serum alpha-hydroxybutyrate dehydrogenase (SHBD), (3) serum glutamic oxalacetic transaminase (SGOT), and (4) lactic dehydrogenase (LDH). The CPK is elevated within 6 hours after an infarct and stays at a peak concentration for 18 to 24 hours. It is also elevated during cerebral and skeletal muscle necrosis. The SHBD is elevated within about 12 hours and obtains levels 4 to 5 times normal values after 48 to 72 hours of infarction. It will remain at a high level for 1 to 3 weeks. The SGOT reaches a peak concentration after about 24 hours but returns to normal levels in 2 to 7 days. It also increases in liver and skeletal muscle disease as do many of the other enzymes. The LDH will increase to a peak value and remain there for 1 to 3 weeks. Because of this and the relative simplicity with which its concentration is determined, LDH has been widely used as an indicator of cardiac necrosis. LDH can be further differentiated on the basis of chromatography into 5 components or isoenzymes. LDH4 and LDH5 are both present in heart muscle. They can be differentiated from the other components in that they are heat stable (are not changed when kept at a temperature of 65° C. for 30 min.) and migrate most rapidly during electrophoresis. An increase in their plasma concentration is characteristic of cardiac necrosis.

Lipids

Lipids, like carbohydrates, are an important source of nutriment for the cells of the body. They are carried from the intestinal tract in the blood and lymph and are picked up by the liver, adipose tissue, muscle, and other structures, in which they are (1) catabolized, (2) stored, or (3) used to produce various molecules (Fig. 26-3). In these processes they may be catabolized to CO_2, H_2O, and energy, or converted to cholesterol, fatty acids, glycerol, acetyl derivatives, or

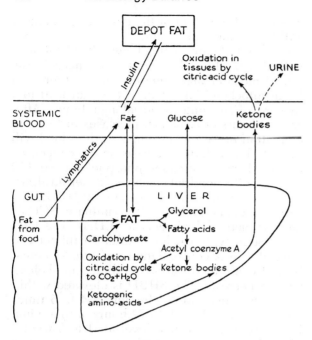

Fig. 26-3. Role of the liver in fat metabolism. (Bell, G. H., et al.: Textbook of Physiology and Biochemistry. 7th ed., p. 381. Edinburgh, E & S Livingstone, 1968)

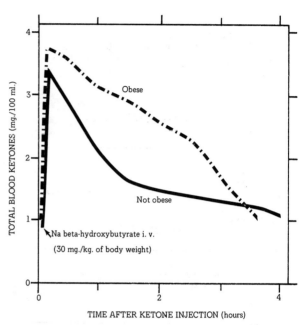

Fig. 26-4. Utilization of ketone bodies. In these experiments 30 mg. of sodium beta-hydroxybutyrate (a ketone body) per kg. of body weight were injected. This produced a rise in the blood ketone level which progressively declined over a period of 4 hours. Since the urine output of ketone bodies over the observed period was less than 1 per cent of the quantity of ketones injected, the author concluded that the disappearance of the ketone bodies was due to their uptake by the tissues. (Redrawn from Kekwick, A.: Adiposity. In Renold, A. E., and Cahill, G. F., Jr. (eds.): Handbook of Physiology. Sec. 5 (Adipose Tissue), p. 623. Washington, American Physiological Society, 1965)

ketone bodies (acetoacetic acid, acetone, and beta-hydroxybutyric acid).

Ketone bodies. Ketogenesis (the formation of ketone bodies) occurs primarily in the liver and is the result of the catabolism of fatty acids and some proteins (those containing ketogenic amino acids). The ketone bodies are released by the liver into the plasma and carried to all the tissues of the body, where they are catabolized to CO_2, H_2O, and energy (Fig. 26-4). The plasma ketone level in man is usually about 1 mg./100 ml., but this may rise markedly (1) after exercise, (2) during exposure to the cold, (3) during pregnancy, (4) in response to trauma, (5) during starvation, (6) in hypoglycemia, (7) in diabetes mellitus, (8) in hyperthyroidism, or (9) in response to the injection of epinephrine or growth hormone. In starvation the glucose concentration of the blood is low and in diabetes mellitus the cells' ability to utilize glucose is decreased. As a result, the body's energy requirements are met by increased fat catabolism. This is associated with such an increase in the liver's output of ketone bodies that their concentration in the blood rises (*ketonemia*), the smell of acetone is detected on the subject's breath and in his urine, and the beta-hydroxybutyric acid in the urine may change from a normal value of 5 to 10 mg. per day to a value of more than 200 mg. per day.

Plasma lipids. In addition to glucose, lactic acid, and ketone bodies, a healthy person also has lipids in his blood. These come primarily from the liver and adipose tissue and represent a total plasma lipid concentration of from 400 to 1,240 mg./100 ml. They include free fatty acids (12 mg./100 ml.), triglycerides (140 mg./100 ml.), cholesterol (200 mg./100 ml.), and the phospholipids (110 mg. of lecithins, plus 80 mg. of cephalins, plus 25 mg. of sphingomyelins in each 100 ml. of plasma). Most of the free fatty acids, cholesterol, and phospholipids are bound to albumin, alpha globulin, and beta globulin while in the blood. Adipose tissue is primarily responsible for maintaining a constant concentration of fatty acids in the blood and releasing them when their concentration in the plasma falls. The liver, on the other hand, is primarily responsible for the release of triglycerides and cholesterol into the blood when their concentration falls.

Exercise and starvation. Examples of conditions in which the fatty acid content of the plasma rises are exercise and starvation. In exhaustive exercise of short duration in dogs, plasma free fatty acids supply between 20 and 30 per cent of the energy expended, and plasma glucose about 10 to 15 per cent. There are in addition marked increases in plasma lactate. In prolonged exercise, however, the plasma free fatty acids supply 70 to 90 per cent of the energy. The response of man to prolonged exercise is illustrated in Figure 26-5, showing changes in the concentration of free fatty acids and glycerol in arterial blood before, during, and after exercise. The response has been divided into three phases. In the first, the circulatory phase, the free fatty acid concentration of the blood decreases, apparently because the amount of fatty acids used by muscle exceeds that released by adipose tissue. This is followed by the second, or metabolic, phase, in which there is a gradual increase in the concentration of both fatty acids and glycerol. In the third phase, the recovery phase, the exercise

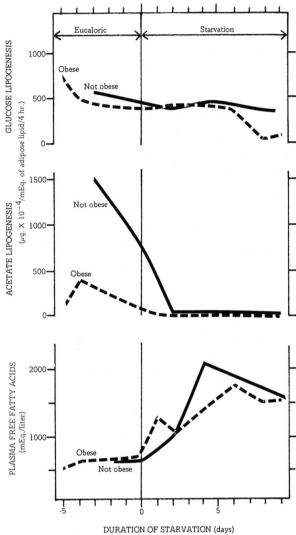

Fig. 26-6. Effects of total caloric deprivation in man on (1) glucose incorporation into adipose glycerides, (2) acetate incorporation into adipose glycerides, and (3) the levels of plasma free fatty acids. Note that the production of lipid from glucose was not abolished by starvation but that its production from acetate was abolished. Note, too, the increased output of fatty acids during starvation was sufficient to maintain a high plasma level. (*Modified from* Hirsch, J., and Goldrick, B.: Metabolism of human adipose tissue in vitro. *In* Renold, A. C., and Cahill, G. F., Jr. (eds.): Handbook of Physiology. Sec. 5 (Adipose Tissue), p. 468. Washington, American Physiological Society, 1965)

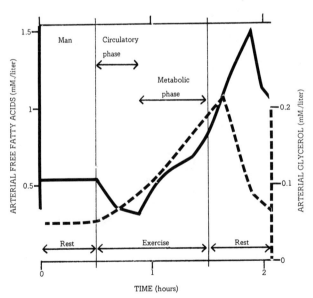

Fig. 26-5. Changes in the arterial concentration of free fatty acids and glycerol in men in the fasting state during and after prolonged exercise. (*Modified from* Carlson, L. A., Boberg, J., and Hogstedt, B.: Some physiological and clinical implications of lipid mobilization from adipose tissue. *In* Renold, A. E., and Cahill, G. F., Jr. (eds.): Handbook of Physiology. Sec. 5 (Adipose Tissue), p. 631. Washington, American Physiological Society, 1965)

ceases but the concentration of free fatty acids and glycerol in the blood continues to rise for a period of time. They eventually return to normal levels.

In starvation or caloric deprivation there is also a mobilization of fatty acids from adipose tissue. This is associated with (1) less glucose being utilized in the production of lipids (a decrease in glucose lipogenesis) and (2) less acetate being utilized in the production of lipids (Fig. 26-6).

Nitrogen-containing Nutrients

The nitrogen-containing compounds in our diet include (1) proteins, (2) amino acids, (3) imino acids, (4) nucleic acids, and (5) creatine. Most of the proteins are catabolized in the alimentary tract to their constituent amino acids, which are then absorbed into the blood. From here they move into the various cells of the body, where they are used to produce structural elements. Those not so used either remain in the blood or are removed by the liver and deaminated (Fig. 26-7). This is followed by the incorporation of the amino acid nitrogen into a waste product called *urea* ($CO(NH_2)_2$) and the utilization of the rest of the molecule for energy or as a constituent of fat and carbohydrate.

Under basal conditions the plasma amino acids are at a concentration of about 4 mg./100 ml. of plasma, but after a meal this concentration rises and then returns to a basal level (Fig. 24-3). The organ most responsible for the maintenance of this relatively constant plasma amino acid concentration and the formation of urea from amino acids is the liver. You will note, for example, in Figure 26-8, that the removal of the liver in a dog results in an increase in plasma amino acids and a decrease in plasma urea.

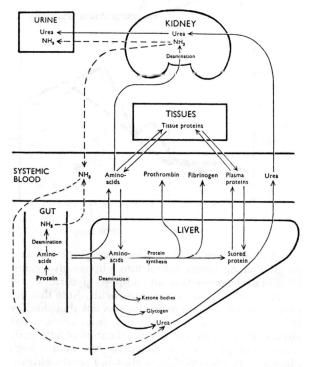

Fig. 26-7. A summary of some of the metabolic pathways utilized in protein metabolism. (Bell, G. H., et al.: Textbook of Physiology and Biochemistry. 7th ed., p. 427. Edinburgh, E & S Livingstone, 1968)

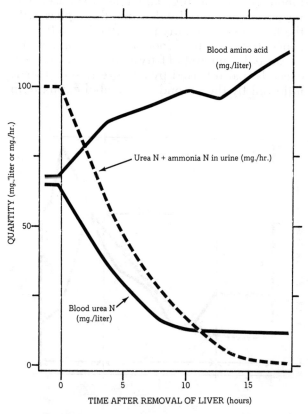

Fig. 26-8. Effect of the removal of the liver on blood urea concentration, blood amino acid concentration, and urea nitrogen plus ammonia nitrogen concentration in the urine. In these experiments the liver of a dog was removed at 8 A.M. and the per cent change in the above parameters was followed for a period of 20 hours. (*Redrawn from* Mann, F. C.: Modified physiologic processes following total removal of the liver. JAMA, p. 1473, 1925)

Nucleic acids, on the other hand, are broken down in the digestive tract to purine bases, nucleosides, pentoses, and phosphoric acid. Some of these substances remain in the intestine and are eliminated in the feces. Some are absorbed. The nucleosides that are absorbed into the blood are catabolized to purine and pyrimidine bases. The pyrimidines are then further broken down in the liver, forming the waste product urea. The purines adenine and guanine, on the other hand, are catabolized throughout the body and form the nitrogen-containing waste product uric acid:

The amount of urea and uric acid formed per day depends in part on the quantity of protein and purine in the diet. A person on a purine-free diet, for example, will excrete in his urine as little as 0.1 gm. of uric acid per day, whereas one on a high purine diet may excrete as much as 2 gm. of uric acid per day. Normally the uric acid concentration in the plasma ranges between 2 and 4 mg./100 ml., but when there is (1) excessive destruction of cells, as in leukemia, starvation, and polycythemia, (2) chronic renal disease, or (3) gout, the plasma values increase. In *gout*, for example, the plasma uric acid may rise to values as high as 15 mg./100 ml. At this concentration, uric acid crystals will be deposited in tissues throughout the body.

Table 26-1 is a record of the effect of a high protein diet on the composition of the urine. In this particular case an increase in the protein in the diet resulted in (1) an increase in the urea excreted each day, (2) an increase in the sulfate excreted, and (3) an increase in the uric acid excreted. The increased urea nitrogen is due to the increased catabolism of amino acids, and the increased sulfate is due to the increased catabolism of sulfur-containing amino acids. It is interesting to note, however, that there was little increase in the excretion of ammonia or creatinine. This is because the production of NH_3 by the kidneys is primarily a response to the H^+ concentration of the blood (Fig. 22-8) rather than to increased nitrogen excretion, and most of the creatinine in the urine results from the destruction of the body's own native creatine and creatine phosphate. Changes in the concentration of uric acid, on the other hand, depend in part on the amount of purine or potential purine in the diet. A high protein diet is not necessarily a high purine diet.

	High-protein Diet	Low-protein Diet
Urine volume	1,170 ml.	385 ml.
Total nitrogen	16.8 gm.	3.6 gm.
Urea nitrogen	14.7 gm.	2.2 gm.
Uric acid nitrogen	0.18 gm.	0.09 gm.
Ammonia nitrogen	0.49 gm.	0.42 gm.
Creatinine nitrogen	0.58 gm.	0.6 gm.
Inorganic sulfate	3.27 gm.	0.46 gm.
Ethereal sulfate	0.19 gm.	0.10 gm.
Neutral sulfur	0.18 gm.	0.20 gm.

Table 26-1. Effect of the protein content of the diet on the composition of the urine. (Data from Keele, C.A., and Neil, E.: Samson Wright's Applied Physiology. New York, Oxford University Press, 1965)

ANABOLISM

Anabolic activity is constantly taking place. It is associated with (1) the formation of such nutrient stores as glycogen and fat, (2) the formation of hormones, enzymes, and the plasma proteins, and (3) growth, hyperplasia, hypertrophy, and repair. The amount of nutrient stored in the body depends upon the balance between energy input and output. An average 70-kg., 25-year-old male has about 9.9 kg. (14%) of adipose tissue. This increases to 17.5 kg. (25%) when he reaches 55 years of age (Table 26-2). In the average female the fat content is somewhat higher and the distribution somewhat different. In both male and female, however, adipose tissue is found in high concentration in the retroperitoneal and subcutaneous areas, where it serves as a cushion and as insulation for the body core from the external environment. It is estimated that adipose tissue is responsible in an average adult for con-

	25 Years Old (kg.)	40 Years Old (kg.)	55 Years Old (kg.)
Males			
Weight	70	70	70
Adipose tissue	9.9 (14%)	15.4 (22%)	17.5 (25%)
Females			
Weight	60	60	60
Adipose tissue	15.6 (26%)	19.2 (32%)	23 (38%)

Table 26-2. Approximate relationship between body mass and adipose tissue mass in man. (From Kekwick, A.: Adiposity. *In* Renold, A.E., and Cahill, G.F. (eds.): Handbook of Physiology. Sec. 5 (Adipose Tissue), p. 618. Washington, American Physiological Society, 1965)

verting approximately 30 per cent of all dietary carbohydrate into triglycerides.

Another type of anabolic activity occurring in the body is the production of molecules not primarily a part of the nutritive reserve. The liver, for example, must produce plasma proteins to keep their concentration in the blood constant. The lymph nodes respond to certain foreign molecules by producing antibodies. Many cells of the body (the cells of the skin, alimentary tract, germinal epithelium, and blood, for example) are constantly being destroyed and replaced. Muscle, in response to certain stresses, forms additional cytoplasmic elements (hypertrophy). Blood vessels, in response to hypoxia and distension, may form additional cells and cytoplasmic elements. Hormones and other secretory products are constantly being released and destroyed. This means that even to maintain the status quo a considerable amount of anabolic activity is essential.

In man anabolism requires (1) an energy source and (2) a source of protoplasm. Proteins serve as a general source of nitrogen-containing molecules, whereas carbohydrates and lipids serve as sources of a number of other types of molecules. Apparently there are certain limitations, however, on the types of amino acids and lipids that can be formed in the body from the general pool of carbohydrates, lipids, and amino acids. For example, man cannot form sufficient methionine from a pool of carbohydrates, lipids, and glycine to meet his needs. On the basis of numerous dietary studies it has been suggested that there are eight amino acids which man cannot synthesize in his body in sufficient quantities. These are called the *essential amino acids* and include (1) phenylalanine, (2) valine, (3) tryptophan, (4) threonine, (5) lysine, (6) leucine, (7) isoleucine, and (8) methionine. A deficiency of any of these may result in poor growth and development or in some form of malfunction (Fig. 26-9).

It has also been shown that animals fed a fat-free diet develop skin and kidney lesions, become infertile, and stop growing. These symptoms can be prevented by the inclusion of linolenic, linoleic, and arachidonic acids in the diet. Although similar lesions have not been un-

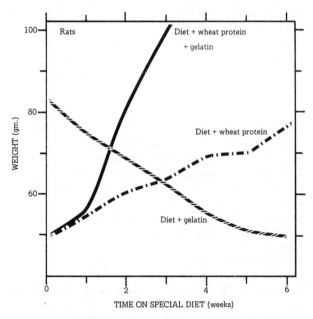

Fig. 26-9. Influence of various proteins on growth in the normal rat. The rats were fed nonprotein basal diets supplemented with (I) wheat protein (deficient in lysine), (II) gelatin (high in lysine but deficient in many other amino acids), and (III) wheat protein plus gelatin. (*Redrawn from* Keele, C. A., and Neil, E.: Samson Wright's Applied Physiology. 10th ed., p. 425. London, Oxford University Press, 1961)

equivocally demonstrated in man, there are data indicating that certain unsaturated fatty acids are required in the diets of children and possibly adults as well.

CALORIMETRY

In the preceding sections we have discussed the utilization of carbohydrates, lipids, amino acids, proteins, nucleotides, and nucleic acids by the body. We have noted that most of our energy is stored in the form of glycogen and fat, but that small quantities of glucose, amino acids, proteins, nucleotides, and nucleic acids can be utilized as energy sources without seriously interfering with body function. In this section we shall concentrate on the utilization of these nutrients by the body.

The first law of thermodynamics states that energy can be neither created nor destoyed. In terms of the human body this means that the amount of energy taken in minus the amount of energy lost equals the amount of energy stored. In a growing child the energy stored is incorporated into new cytoplasm and interstitial material. In a sedentary adult it is found mostly in the adipose tissue.

The unit used in biology for the measurement of energy is the kilocalorie (kcal.). It is the amount of heat required to raise 1 kg. of water from 15°C. to 16°C., or 1 centigrade degree. The number of calories contained in a particular food can be determined by placing that food in a bomb calorimeter and measuring the amount of heat released when the food is burned to CO_2 and H_2O. You will notice in Table 26-3 that 1 gm. of starch gives off 4.2 kcal.; 1 gm. of fat, 9.5 kcal.; and 1 gm. of protein, 4.4 kcal. when burned in a bomb calorimeter. You will also note that 1 gm. of starch requires 829 ml. of O_2; 1 gm. of fat, 2,013 ml.; and 1 gm. of protein, 957 ml. These differences are in part due to the relative concentrations of hydrogen, oxygen, and carbon in the molecules being catabolized, as well as to the presence of nitrogen, sulfur, and phosphorus atoms, for example.

In the case of carbohydrate the ratio of hydrogen to oxygen is 2:1. To burn a carbohydrate to CO_2 and water, in other words, all one needs is one molecule of O_2 for each atom of carbon in the molecule. Hence for a carbohydrate the ratio of CO_2 produced (in ml.) to O_2 utilized (in ml.) is 1. This ratio is called the *respiratory quotient* (RQ). Note in Table 26-3 that the respiratory quotient for fat is 0.71, and for protein 0.81.

Human Subject

One can measure the amount of heat produced by a human by placing him in a modified bomb calorimeter (Fig. 26-10). Another technique, indirect calorimetry, does not involve the direct measurement of heat production, but rather the measurement of the RQ of the subject and his O_2 consumption. If one then assumes a certain caloric value for each liter of O_2 consumed by the subject, one can calculate his heat production.

If a reclining subject breathed from a spirometer such as that shown in Figure 26-11 we

	Heat per gm. (kcal.)	CO_2 Produced per gm. (ml.)	O_2 Required per gm. (ml.)	RQ CO_2/O_2	Value of 1 liter	
					O_2 (kcal.)	CO_2 (kcal.)
Starch	4.20	829.3	829.3	1.00	5.06	5.06
Cane sugar	3.96	785.5	785.5	1.00	5.04	5.04
Dextrose	3.74	746.6	746.2	1.00	5.01	5.01
Lactic acid	3.62	745.9	745.9	1.00	4.85	4.85
Animal fat	9.50	1,431.1	2,013.2	0.71	4.72	6.64
Protein	4.40	773.8	956.9	0.81	4.60	5.69
Acetone	7.43	1,157.2	1,542.9	0.75	4.82	6.42
Ethyl alcohol	7.08	972.9	1,459.5	0.67	4.85	7.28

Tables 26-3. Heat production, O_2 utilization, and CO_2 production resulting from the burning of different nutrients to CO_2 and water. (Modified from Carpenter, T. M.: Tables, Factors, and Formulas for Computing Respiratory Exchange and Biological Transformations of Energy. Washington, Carnegie Institute of Washington, 1939)

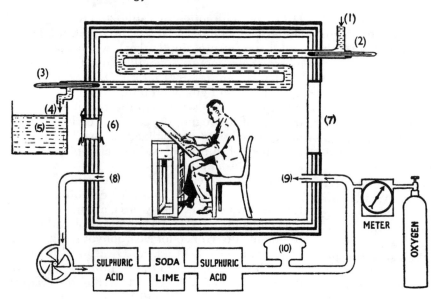

Fig. 26-10. The Atwater-Benedict respiration calorimeter. In this equipment the amount of heat produced is measured by determining the volume and change in temperature of water passing in at 1 and out at 4. There is a porthole at 6 for bringing in food or taking out excreta, and a window at 7. The air is removed from the room at 8 and is passed through sulfuric acid and soda lime to remove water and CO_2. Known quantities of O_2 are then added and the air returned to the room at point 9. (Bell, G. H., et al.: Textbook of Physiology and Biochemistry. 7th ed., p. 210. Edinburgh, E & S Livingstone, 1968)

could measure his O_2 consumption. Let us say he consumes 250 ml. of O_2 per minute at a temperature of 23°C. and at an atmospheric pressure of 744 mm. Hg. To calculate the number of calories this represents we would first have to correct the volume of O_2 consumed for standard temperature, pressure, and dryness (STPD). At 25°K. the pressure of the water vapor in this spirometer would be 23.8 mm. Hg (Table 26-4); the pressure exerted by the O_2 would therefore be:

744 − 23.8 = 720.2 mm. Hg corrected for dryness

Next we would correct the volume for standard pressure (760 mm. Hg) and temperature (0°C. = 273°K.).

$$250 \text{ ml.} \times \frac{720.2 \text{ mm. Hg}}{760 \text{ mm. Hg}} \times \frac{273°K.}{273 + 25°K.} = 217 \text{ ml.}$$

If the subject's RQ was 0.82, then each liter of O_2 burned would represent 4.825 kcal. (Table 26-5) and in 1 min. he would produce 1.047 kcal.

217 ml./min. × 0.004825 kcal./ml. = 1.047 kcal./min.

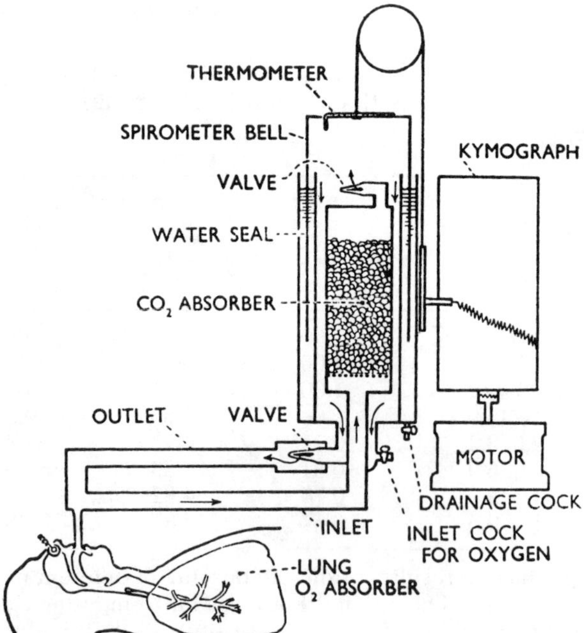

Fig. 26-11. A recording spirometer. A subject's oxygen consumption can be measured by placing a clip on his nose and having him inhale and exhale into a spirometer filled with O_2 or sometimes air. The CO_2 that the subject expires is absorbed by soda lime, so the amount of O_2 used in a period of time can be determined by noting how much the spirometer bell falls or by how much the associated kymograph pen rises. The amount of CO_2 produced can be determined by removing the soda lime (volume change with soda lime minus volume change without soda lime). The volume changes obtained are then expressed at standard temperature (0°C.), pressure (760 mm. Hg), and dryness (0% humidity). (Weir, J. B. de V. *In* Bell, G. H., et al.: Textbook of Physiology and Biochemistry. 7th ed., p. 212. Edinburgh, E & S Livingstone, 1968)

Temperature of gas (°C.)	15	17	19	20	21	22	23	24	25	37
Pressure exerted by water vapor (mm. Hg)	12.79	14.54	16.49	17.55	18.67	19.84	21.09	22.40	23.78	47

Table 26-4. Pressure exerted at different temperatures in air and O_2 by water vapor over water.

RQ	0.707	0.75	0.80	0.85	0.90	0.95	1
kcal.	4.686	4.739	4.801	4.862	4.924	4.985	5.047

Table 26-5. Caloric value of 1 liter of O_2 at various respiratory quotients.

This would represent:

$$1.047 \times 60 = 62.82 \text{ kcal./hr.}$$

Basal metabolic rate. If the above data were recorded for a resting person 12 hr. or more after the last meal, they would represent the subject's basal heat production. Most physiologists and physicians, however, prefer to express basal heat production in terms of kilocalories per square meter of body surface per hour. One can estimate the body surface area of a subject by obtaining his height and weight and using a standard chart such as that in Figure 26-12. If, for example, the subject weighs 60 kg. and is 160 cm. tall, he has a body surface area of about 1.62 $m.^2$.

$$\frac{62.82 \text{ kcal./hr.}}{1.62 \text{ m.}^2} = 38.8 \text{ kcal./m.}^2\text{/hr.}$$

If the subject were a 35-year-old man he would

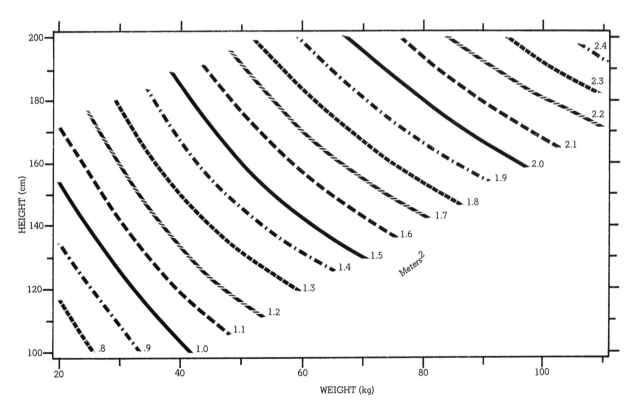

Fig. 26-12. Chart for determining the surface area of an individual from his body weight and height. The chart is derived from the following formula: Surface area (cm.^{2}) = (Wt.$^{0.425}$)·(Ht.$^{0.725}$)·(7,184). (*Redrawn from* DuBois, D., and DuBois, E. F.: Clinical calorimetry. X. A formula to estimate the approximate surface area if height and weight be known. Arch. Intern. Med., *17*:865, 1916)

Age (yrs.)	BMR (kcal./m.²/hr.)		BMR (kcal./24 hr.)		TMR* (kcal./24 hr.)	
	Men	Women	Men	Women	Men	Women
Birth	30	30	288	288	440	440
1	55	52	660	624	1,000	1,000
2	57	53	780	725	1,200	1,200
5	53	52	915	886	1,600	1,600
8	51.8	47	1,143	993	2,000	2,000
11	47.2	45.2	1,268	1,193	2,500	2,500
14	46.4	41.5	1,537	1,391	3,200	2,800
16	46.7	38.8	1,764	1,434	3,500	2,600
18	43.2	37.5	1,783	1,440	3,800	2,500
20	41.6	36.3	1,756	1,437	3,500	2,400
25	40.3	36	1,760	1,442	3,000	2,400
40	38	35	1,641	1,344	3,000	2,400
55	37.5	35				
65	36.5	34				
75	35.5	33				

* TMR, total metabolic rate.

Table 26-6. Basal metabolic rate and total metabolic rate at different ages. (From Cruickshank, National Research Council, K. W. Cross, and Aub and DuBois)

be well within the normal range (Table 26-6). It is also common practice to express this value (basal metabolic rate) in terms of per cent deviation from normal. An average 40-year-old man would have a basal metabolic rate (BMR) of 38.0 kcal. per square meter per hour according to the table. If his recorded BMR were 38.8, his rate would be 2 per cent higher than this standard:

$$\frac{38.8 - 38.0}{38.0} \times 100 = +2\%$$

The usually acceptable range of normal is ±20 per cent (Fig. 26-13). A BMR of +2 per cent is well within this range.

Modification of basal metabolic rate. You will note in Table 26-6 that the BMR is considerably different from the total energy expenditure for the individual. This is to be expected since the energy expenditure during sleep (about 1.17 kcal./min.) is less than the BMR (1.26 kcal./min.), and the energy expenditure during shivering and exercise (over 20 kcal./min.), as well as after eating, is more than the BMR. We also find that a person with an O_2 consumption of 240 ml. per minute in a room at 30°C. will have an O_2 consumption of 330 ml. in a room at 0°C., and one of 260 ml. at 40°C. The increase at 0°C. is apparently due to the mobilization of some of the

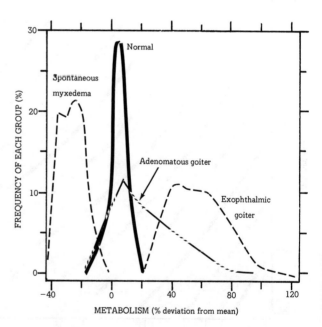

Fig. 26-13. Basal metabolism in the normal subject and in the patient with (1) myxedema (deficiency in thyroid hormone), (2) goiter with and without hyperthyroidism, and (3) exophthalmic goiter (increased thyroid hormone). (*Redrawn from* Boothby, W. M.: The variability of basal metabolism, p. 148. *In* Transactions of the American Association for the Study of Goiter, 1937)

body's mechanisms for maintaining a constant body temperature (shivering, for example). The increase at 40°C. is probably due also, at least in part, to the mobilization of homeostatic mechanisms (sweating, for example), as well as to the increased chemical activity occurring at higher body temperatures. It has been estimated that for each increase in body temperature above the normal resting value there is a 13 per cent increase in the metabolic rate.

Metabolic rate and exercise. Exercise is probably the most important factor in increasing the metabolic rate above basal levels. A man with a basal requirement of 2,000 kcal. of energy per day, for example, needs an additional 500 kcal. when he gets out of bed and participates in a sedentary occupation. If he is a lumberjack, on the other hand, he will use up about 5,000 kcal. per 24 hr. (basal requirement of 2,000, plus activity requirement of 3,000).

Metabolic rate and food. One's metabolic rate also depends upon the food one eats. We find, for example, that a person who has not eaten for 12 hr. expends 375 kcal. over a period of 5 hr. If we now feed this person 375 kcal. of protein, for the next 5 hr. he will expend 450 kcal. In other words the protein meal increased his caloric output by 75 kcal., or 20 per cent of the kilocalories he took in. Apparently most of this increased metabolism is due to the oxidative deamination of the additional amino acids by the liver. The effect of the foods we eat on the caloric output of the body is called the *specific dynamic action* of the food. The specific dynamic action of protein is an average increase in the caloric output by 20 per cent of the caloric input of protein. The specific dynamic action of carbohydrate is about 12 per cent, and of fat, 9 per cent.

CALORIC INTAKE

The regulation of caloric intake is still a poorly understood phenomenon, though we have evidence that it is regulated at least in part by caloric output and by the quantity of stored nutrient. There is no well-established mechanism that we can point to, however, as controlling caloric intake in the sense, for example, that the baroreceptor reflex controls arterial pressure.

Part of the evidence for a caloric control system is that when the exercise level of subjects increases there is also an increase in caloric intake. In Figure 26-14 we note that when military

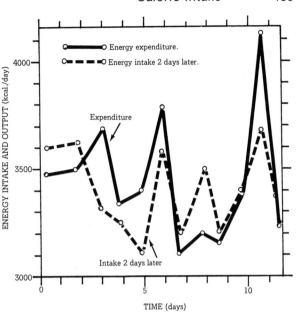

Fig. 26-14. A comparison of caloric output and input in military cadets. It is concluded that the daily intake of food is influenced by the activity of 2 days previous. (*Redrawn from* Edholm, O. G., et al.: The energy expenditure and food intake of individual men. Brit. J. Nutr., 9:298, 1955)

cadets exercise to varying degrees there is a direct correlation between energy expenditure and caloric intake 2 days later. There is also evidence that dogs over a 3-hr. period have the same caloric intake on a high-calorie diet as on a low-calorie one (that is, a diet high in water, cellulose, and clay, with or without mineral oil). Also indicating the existense of a caloric control system are observations on force-feeding. In experiments on rats, Cohn and Joseph (1962) showed that they could force-feed rats until they were 200 gm. heavier than their litter mates. When taken off this regime and allowed to eat ad libitum, the obese rats ate considerably less than their litter mates and lost weight, while the litter mates eating ad libitum continued to gain weight. After 40 days of ad libitum feeding the two groups weighed almost the same.

The Hypothalamus

There has been considerable interest in the role of the hypothalamus in regulating caloric intake since Hetherington and Ranson in 1940 reported that lesions in the hypothalamus that do not injure the pituitary gland produce obesity.

We find that 200- to 300-gm. rats with such lesions initially gain from 5 to 10 gm. per day if allowed to eat ad libitum, while the control rats gain less than 1 gm. per day. On the other hand it has been shown that stimulation of the ventromedial region of either the right or left hypothalamus causes an animal to lose interest in food. The region which when stimulated produces aphagia (cessation of food intake) and when destroyed produces hyperphagia has been called by some the *satiety center*. Also in the hypothalamus is a region which when stimulated produces hyperphagia, and when destroyed, aphagia. This has been called the *feeding center*.

The evidence cited above is hardly sufficient to establish the existence of centers in the hypothalamus which control caloric intake. The same results might be obtained were we dealing merely with fiber tracts concerned with the control of food intake. It is interesting to note that if animals with destroyed "feeding centers" are force-fed for a period of from a few days to few weeks, they regain their interest in food and once again have a normal intake. If the "satiety center" is destroyed, the animals gain weight until they have reached a certain level of obesity and then tend to maintain this new weight. If food is restricted they lose weight, but when allowed to feed freely again they reach their previous level of obesity. Loss of the "feeding center," in other words, seems to produce a resetting of the mechanisms responsible for the control of caloric intake.

Sensory Control

Our nutritive intake is controlled in part by two sensations: hunger and appetite. Hunger is a rather diffuse sensation usually associated with an unpleasant feeling of emptiness in the abdomen. It is apparently experienced from birth and is triggered by either a caloric deficit or deficiency of a particular nutrient. Appetite, on the other hand, is a desire for a particular food or drink and is apparently learned from previous pleasant gustatory, olefactory, and visual experiences. Examples of conditions in which specific appetites have survival value are numerous. We find that normal rats, given a choice of water or of a 3 per cent NaCl solution, choose the water, while rats losing excessive quantities of NaCl in their urine (adrenalectomized rats) choose the 3 per cent NaCl. Parathyroidectomized rats (losing excessive quantities of calcium from the body) drink sufficient volumes of a calcium lactate solution to compensate for the calcium loss, and a hypoglycemic rat will choose a 30 per cent glucose solution over a 10 per cent glucose solution. Animals have also been shown to perfer a meal

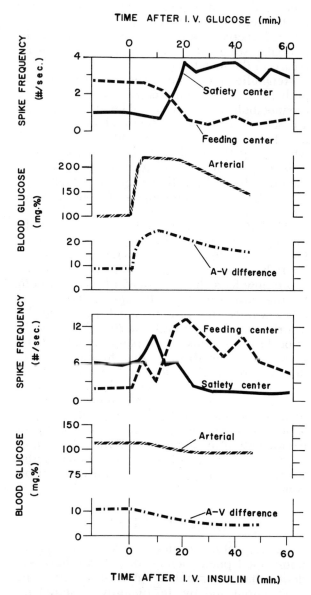

Fig. 26-15. A study of the effects of changes in the arterial blood glucose and the arteriovenous difference in glucose concentration on the spike frequency from 2 different areas in the hypothalamus of 4 animals. Increases in blood sugar were produced by the intravenous infusion of glucose, and decreases were produced by the injection of insulin. (Modified from Anand, B. K., et al.: Activity of single neurons in the hypothalamic feeding centers: Effect of glucose. Am. J. Physiol., 207:1149, 1151, 1964)

with an adequate amino acid content over one deficient in a particular amino acid.

The sensation of hunger is apparently less well directed. There is evidence that it can be initiated by (1) a decrease in the amount of available glucose (Fig. 26-15), (2) a decrease in the amino acid content of the plasma, and (3) a cooling of the hypothalamus. Some have suggested that it can be initiated by impulses from the stomach, but it should be borne in mind in this regard that the sensation of hunger continues to occur in man after the stomach has been removed.

The role of changes in the temperature of the hypothalamus in controlling food intake has proven to be of great interest. We find that the temperature of the hypothalamus increases after a meal—more dramatically after a meal high in protein than after one high in carbohydrate. On the basis of this it has been suggested that the specific dynamic action of foods serves to stimulate the satiety center and to depress the feeding center and that changes in hypothalamic temperature are important in the regulation of caloric intake.

It has also shown that soldiers stationed where the temperature is 95°F. have an average daily caloric intake of 3,100 kcal. At environmental temperatures of 70°F. their intake is 3,500 kcal.; at 25°F., 4,200 kcal.; and at −20°F., 4,950 kcal. For what reason the temperature of the hypothalamus in these examples affects caloric intake we cannot at present state with any degree of certainty.

The preceding discussion has emphasized that elevations in the level of nutrients in the plasma and the specific dynamic action of foods may stimulate the satiety center. Neither of these factors, however, satisfactorily explains why a hungry man or animal stops eating at some point during his meal. In many cases the feeling of fullness after eating and the loss of desire for additional food apparently occurs before the level of nutrients in the blood has increased or the specific dynamic action of the food has been sufficiently expressed to affect hypothalamic function. Some have suggested the presence of specialized chemoreceptors and stretch receptors in the stomach and abdomen. In the case of the dog we find that the volume of food consumed is an important factor. If fed a low-calorie meal with a high-bulk content one day and a high-calorie meal with a low-bulk content another day but under similar conditions, a dog tends to consume the same volume of food each time. A rat, under similar conditions, tends to consume the same number of calories each time. The rat, that is, seems to eat for calories rather than volume, and the dog for volume rather than calories. On the other hand, if a dog is placed on a high-bulk diet and allowed to eat ad libitum, he tends to gain weight at the same rate as a dog on a low-bulk diet. In the dog the mechanisms responsible for limiting the amount eaten (1) at a single meal and (2) over a period of 24 hr. may be considerably different.

Chapter 27

REGULATION OF BODY TEMPERATURE

HEAT LOSS
THE CORE TEMPERATURE
THE CONTROL OF BODY TEMPERATURE

One's body temperature is a result of the balance between heat produced in the body through the catabolism of nutrients and heat lost from or gained by the body through the processes of convection, radiation, vaporization, and conduction (Fig. 27-1). Much of the heat added to the body results from physiological processes not directly concerned with the regulation of body temperature. Such processes include the digestion and assimilation of food (specific dynamic action), catabolic reactions associated with maintaining the basal state (the sodium pump, contractions of the heart, contractions of the diaphragm), and contractions of skeletal muscle associated with exercise. Heat production may also result from a type of skeletal muscle contraction called shivering. Heat loss is associated with the relationship existing between the body and its external environment and, like shivering, is controlled by a body temperature-regulating mechanism.

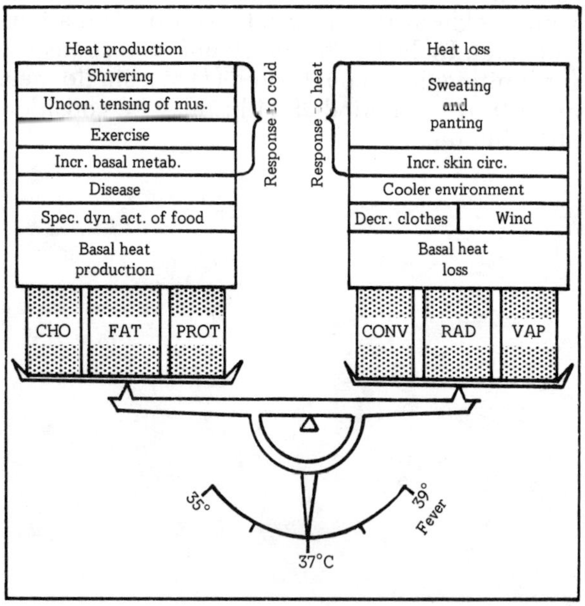

HEAT LOSS

Heat loss is in part controlled by regulating the relationship between man and his environment or by controlling the environment itself. Essentially, control of the environment means regulating environmental temperature, humidity, and air movement. Control of the relationship between man and his environment involves modifying the barrier between man or man's internal organs (his core) and his external environment.

The critical feature of a good temperature barrier is low conductivity. In conduction, heat is transferred from one molecule to the next. This requires contact, and the speed with which it occurs depends on (1) the difference in temperature between two molecules and (2) the conductivity of the molecules. The conductivity of silver at 18°C. is 1.006; that of water at 17°C. is 1.31×10^{-3}; that of air at 0°C., 5.68×10^{-5}; and that of wool, 5.4×10^{-5}. In other words, one could better cool off rapidly by jumping nude into water at 80°F. than by walking around nude in a room with the air at 80°F.

For the purpose of our discussion we will divide the body into two parts: (1) a *shell* and (2) a *core*. The shell for the most part comprises the skin, subcutaneous tissue, hands, feet, and most of the arm and leg. Its temperature and conductivity vary considerably more than these same parameters in the body core. In cold weather, for example, the hands may without damage or pain go to temperatures below 22°C. (71.6°F.).

Fig. 27-1. Diagram of the balance which exists in the body between heat production and heat loss. Heat production is attributed primarily to carbohydrate, fat, and protein catabolism, and heat loss to convection, radiation, and vaporization. (*Modified from* DuBois, E. F.: Harvey Lecture Bull. N. Y. Acad. Med., 15:145, 1939)

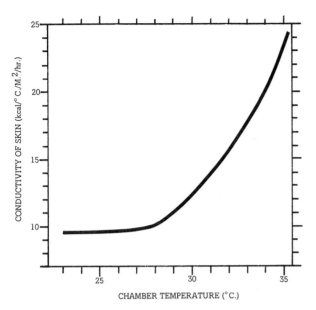

Fig. 27-2. Changes in skin conductivity with changes in environmental temperature. These data were obtained from 2 nude men who lay motionless and whose heat production remained constant during the study. (*Modified from* Hardy, J. D., and Soderstrom, G. F.: Heat loss and peripheral blood flow. J. Nutr., *16*:495, 1938)

During exercise skin temperature may go to 39°C. The conductivity of the skin varies with the amount of blood it contains. Cutaneous vasoconstriction is generally characteristic of cool environments, and vasodilation and an associated increase in skin conductivity, of warm environments (Fig. 27-2).

Conservation of Heat

The body shell serves as one of the major heat regulating structures in the body. During exposure to cold there is vasoconstriction throughout the skin, which by reducing the amount of blood in, and the blood flow through the skin reduces the movement of heat from the core (body cavities) to the shell and hence to the external environment. In animals such as the dog and cat, pilo-erection also occurs—increasing the thermal barrier even more. In man, all that remains of this reflex is the production of "gooseflesh."

In the appendages there is a heat conservation mechanism similar to the countercurrent mechanism in the juxtamedullary nephron (Fig. 20-17). We might find that in a hand and arm exposed to the cold, for example, the blood in the artery leading to the arm has a temperature of 37°C. The blood in the hand will have a temperature of 22°C., but the blood in the vein leaving the arm will be 36°C. Apparently the warming of the venous blood coming from the hand results in part from a transfer of heat from the artery to the vein. This mechanism (the countercurrent heat mechanism) results in a conservation of the body's heat.

Besides setting up a thermal barrier, the body may react to cooling by increasing its metabolic activity. Skeletal muscle is particularly important in this regard. During mild cooling the electromyograph shows increases in skeletal muscle activity which increase as cooling gets more severe, until a type of skeletal muscle contraction called *shivering* occurs. In very severe episodes of shivering the caloric output of the body can increase fivefold.

A number of the endocrine glands also react to cooling. The adrenal medulla, for example, may release epinephrine, which increases the caloric output of the body. The thyroid too may increase its output of hormone, as well as its size, in response to low temperatures. It has been suggested that when a nude man is exposed to temperatures below 33°C., the heat regulatory center in the hypothalamus (1) causes the release of thyrotropin releasing factor, which (2) causes the release of thyroid-stimulating hormone, which (3) causes the release of thyroxine (pp. 497-498). Approximately 4 hr. later (the period of latency for thyroxine) (4) the metabolic rate (heat production) of many of the body's tissues begins to increase.

Many of these mechanisms have been well documented in the rat, in which the release of thyroxine has been proven important not only maintaining a constant body temperature in response to chronic exposure to cold, but also in the very survival of the rat during such exposure. Studies of fully clothed man are not so straightforward. One cannot assume, for example, that an Eskimo living in Alaska is exposed to the cold in the same sense that a rat placed in a refrigerated cage is. In fact, some have suggested that the Eskimo, because of his warm clothing and warm home, is more like a person in a tropical environment. On the other hand, it has been shown in some cultures that the decreases in environmental temperature occurring in winter are associated with increases in metabolic rate (increased heat production). An example of this is seen in Figure 27-9.

Elimination of Heat

The body eliminates excess heat in one of four ways. Least used among these is conduction. This is important when the nude body comes in contact with a considerably better conductor of heat than air, such as water or metal. In Figure 27-3 we see the relative importance of the three other means of heat dissipation in a room at 75°F. Note that in the resting condition before exercise about 66 per cent of the heat is being dissipated by radiation, 15 per cent by convection, and 19 per cent by vaporization. With the beginning of exercise there is so great an increase in heat production that an increase in core temperature occurs (rectal temperature) even though heat dissipation by the skin is increased almost sixfold. During strenous exercise the major means of dissipating heat is vaporization, which results from the increased production of sweat during exercise; it is so effective that there is actual cooling of the skin despite an increased core temperature.

Radiation. In radiation, heat travels from one object to another in the form of electromagnetic waves. Heat waves are quite similar in this respect to light waves. Like light, they travel 186,000 miles per second and can be reflected, refracted, and polarized. Visible light has a wavelength between 4,000 and 7,500 A, whereas heat waves have a wavelength between 7,500 A and 0.04 cm. The net amount of heat radiated from one object to another depends upon (1) the duration of exposure, (2) the difference in temperature between the two objects, (3) the amount of surface exposed, and (4) the absorption characteristics of the object. If a body absorbs all the radiant energy that strikes it, it is called a perfect black body. Air is a poor absorber of heat, so heat is freely radiated through it. We find, for example, on a sunny day that the temperature of a sidewalk may be considerably higher than that of the air.

The skin of man, as far as heat is concerned, is a 07 per cent black body. This is true of both the white and black man. Outside the visible range, in other words, the absorption characteristics of both white and black skin are the same. Du Bois has stated: "Man does all his radiating at wave lengths beyond the visible, and therefore, there is certainly no difference in color between a white man and a black man in a dark room in the middle of a moonless night."

Loomis (1967) has stated that one physiological difference between black and white skin is that black skin prevents some of the ultraviolet light in a wavelength between 290 and 320 mμ from penetrating to that part of the skin containing the provitamin 7-dehydrocholesterol. In other words, black skin produces less *vitamin D* from its provitamin for any given quantity of ultraviolet light striking its surface than does white skin. Loomis further suggests that this characteristic, like suntanning, has a survival value in tropical climates where one is exposed to much sunlight in that it prevents the production of excess quantities of vitamin D and therefore the toxicity symptoms of hypervitaminosis D.

Some of the ways in which we increase heat

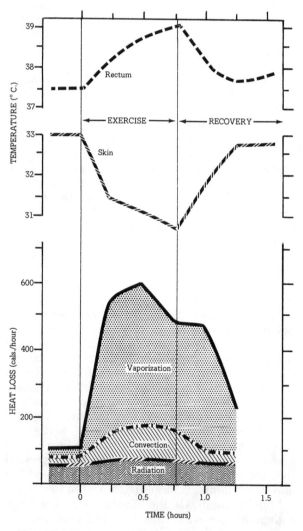

Fig. 27-3. Response to strenuous exercise. In these studies a subject was followed during three 12-minute games of squash racquets. (*Modified from* Hardy, J. D., and DuBois, E. F.: The technique of measuring radiation and convection. J. Nutr., *15*:476, 1938)

loss due to radiation are (1) by exposing larger areas of skin and (2) by increasing the temperature difference between our skin and the objects to which we are transferring heat by radiation. In this regard, we saw in Figure 11-4 that during all but the most strenuous exercise there is a marked increase in the amount of blood flowing to the skin, and hence that there is an increase in the amount of heat moved from the core to the shell. In Figure 11-4 we note an increase in this blood flow from 500 ml. per minute to 1,900 ml. per minute. There have been cases in which a twentyfold increase in flow to the hand was noted. Apparently these increases in flow are brought about by two means: (1) an arteriolar vasodilation and (2) an opening up of arteriovenous shunts in the skin. These responses are characteristic not only when increased metabolic activity produces an increase in body temperature, but also when there is an increase in the temperature of the external environment.

Convection. Convection is the process by which one fluid changes places with another. The layer of air or liquid immediately adjacent to the skin is approximately the same temperature as the skin. If that fluid can be replaced by a cooler one, the result will be cooling of the skin. This process is accelerated when we swim, walk, or turn on a fan.

Vaporization. In vaporization, liquid water is converted into gaseous water. The heat required to produce this change at 1 atmosphere of pressure is 539 cal. per gram of water (539 kcal./kg. of water) and is called the *heat of vaporization*. In the human body, however, one usually figures a heat loss of 580 cal. for each gram of water lost by evaporation. This represents the heat loss due to the cooling action of evaporation, plus that due to the heat carried from the body in the vapor itself.

One aspect of the importance of vaporization in the cooling of the skin is that it is not totally dependent upon the temperature of the external environment. In order for conduction, radiation, and convection to be effective in heat loss, the temperature of the external environment must be cooler than that of the skin. In order for vaporization to be effective, the relative humidity must be less than 100 per cent. You will note in Table 27-1, for example, that as the temperature of the external environment increases, the amount of heat lost by evaporation eventually becomes 100 per cent of the total heat lost by the body. The effectiveness of sweating in maintaining the body temperature at a survival level was demonstrated in 1775 by Blagden, who showed that men could survive at a temperature of 250°F. in a dry atmosphere for 15 min. During this same period and at this same temperature it is possible to cook a beefsteak.

Insensible perspiration. Under normal resting conditions at a room temperature of 75°F. (24°C.) man secretes little or no sweat. Any that is secreted comes from glands in the feet, hands, and axillae. Despite this, an average man vaporizes about 20 gm. of H_2O per hour (30 gm. × 0.58 kcal./gm. = 17.4 kcal.). This is true of normal persons as well as of those with *ichthyosis*—a condition in which there is decreased ability to produce sweat and sebum. Sanctorius, in 1614, called this water loss "insensible perspiration." We now know that approximately one third of it is due to evaporation from the respiratory mucosa, and the rest from the interstitial fluid in the dermis. The amount of heat loss due to respiration is directly proportional to respiratory ventilation (liters of air expired/min.). The amount of water lost from the skin depends upon the circulation in it.

The sweat glands. In man there are approximately 2,500,000 sweat glands, each innervated by cholinergic sympathetic neurons. These

Temperature (°C.)		Heat Loss (% of total)		
Air	Wall	Evaporation	Radiation	Convection
17.1	19	10	40	50
16	49.1	21	0	79
22.8	22.8	17	13	70
29.4	52.4	78	0	22
35.4	36.6	100	0	0

Table 27-1. Character of the body's heat loss under different environmental conditions. (From Winslow, C. E. A., and Herrington, L. P.: Temperature and Human Life. Princeton, Princeton University Press, 1949)

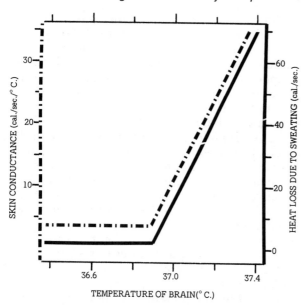

Fig. 27-4. Changes in the rate of sweating and skin conductance in man in response to changes in the temperature of the interior of the head. It has been suggested that much of the increased skin conductance associated with sweating is due to the release of a vasodilator, bradykinin, with the sweat. (*Modified from* Benzinger, T. H.: Receptor organs and quantitative mechanisms of human temperature control in a warm environment. Fed. Proc., *19*:40, 1960)

sweat glands usually represent an unimportant source of skin moisture at hypothalamic temperatures below 37°C., but at higher core temperatures can represent the major cooling mechanism in the body (Fig. 27-4). It is interesting to note that dogs and cats have sweat glands only in their footpads, and that when they become overheated they must depend almost totally on panting for evaporation cooling.

The amount of sweat lost by a person depends upon the conditions. Pugh et al., for example, studied athletes competing in a marathon race (distance, 42 km.; average speed, 16 km./hr.; temperature, 23°C.; humidity, 58%) and found that the sweat rate of the first four to finish the race varied from 0.9 to 1.8 liters per hour, the winner producing a total of approximately 4.8 liters of sweat during his 158-min. run. This volume of sweat represented 6.9 per cent of his body weight.

In producing sweat we are losing substances from the body other than water. The importance of this observation is that when we replace the water we have lost through sweating, it is sometimes also important to replace the NaCl lost with the sweat. When the rate of sweat formation is low, the concentration of Na^+ ranges from 9 to 53 mEq. of Na^+ per liter of sweat, but when the rate is high, it ranges from 52 to 118 mEq. of Na^+ per liter of sweat. In other words, as the rate of sweating increases, the concentration of Na^+ in the sweat approaches that in the interstitial fluids and plasma.

THE CORE TEMPERATURE

The functioning of man's internal organs is dependent upon the maintenance of a fairly constant core temperature. This temperature is frequently estimated by placing a recording device under the tongue, in the rectum, or in the esophagus. The oral temperature is generally about 0.65°C. below rectal temperature and varies in normal healthy relaxed persons from 97° to 99°F. (36° to 37.4°C.). The usual range of normal for rectal temperature in the relaxed person is between 97° and 100°F. (36° to 37.8°C.). Esophageal temperature is usually about 0.2°F. lower than rectal temperature.

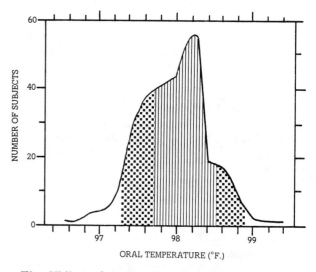

Fig. 27-5. Oral temperatures found in 276 medical students seated in a class between 8 and 9 A.M. These data have a mean of 98.1°F. (36.7°C.) and a standard deviation of 0.4°F. Sixty-eight per cent of the population lies within the middle shaded area and 95% of the population within the combined shaded areas. (*Modified from* Ivy, A. C. What is normal or normality? Quart. Bull. Northwestern Univ. Med. School, *18*:25, 1944)

Elevations of Core Temperature

One's core temperature increases in a warm environment, after a meal, during exercise, during emotional crises, in hyperthyroidism, and during infection. Of the physiological factors mentioned above, exercise is probably most frequently responsible for marked increases in core temperature. During prolonged strenuous exercise, the rectal temperature has been found to go as high as 40.5°C.

Similar elevations in temperature also result from toxins called *pyrogens*. These may come from invading microorganisms or may be in solutions injected into the blood. They apparently act on granulocytes in the blood, causing them to release an endogenous pyrogen that affects the thermoregulatory centers in the hypothalamus in such a way as to increase the core temperature. Prior to the advent of antibiotics, infections were sometimes treated by raising the core temperature to 41°C. The apparent effectiveness of this procedure is an indication that fever is part of a defense response to infection, rather than an unwanted symptom. With the exception of heat stroke and brain lesions, neither in fever nor during exercise does rectal temperature exceed 40.5°C. This is a fortunate situation, since at rectal temperatures maintained for several hours at 41°C. (106°F.), permanent brain damage results. At temperatures above 43°C. death is common. These observations are summarized in Figure 27-6.

Lowering Core Temperature

Rectal temperature not only fluctuates with body activity, but also shows a diurnal rhythm, being in most people who work during the day and sleep at night 0.5° to 0.6°C. lower at 6 A.M. than at 12 P.M. or 12 A.M. It also decreases in response to numerous anesthetics and hypnotics if the environmental temperature is not kept high. Body temperature is sometimes intentionally reduced during surgery on the brain or heart to reduce the risk of hemorrhage and to decrease the O_2 consumption of the tissues. This enables the tissues to survive longer periods of ischemia than they could at a normal core temperature.

Human subjects tolerate well rectal temperatures of between 21° and 24°C. How much lower the body temperature can safely be taken in humans is not known, but experiments on dogs

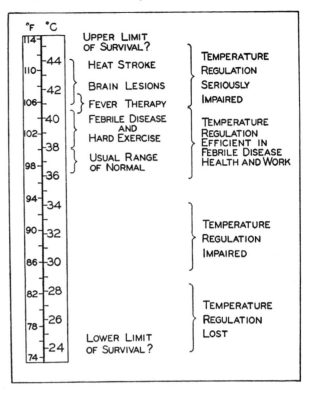

Fig. 27-6. The extremes of rectal temperature. (DuBois, E. F.: Fever and the regulation of body temperature. J. Neurophysiol., *12*:9, 1948)

and other warm-blooded animals have shown that if care is taken to prevent the formation of ice crystals in the tissues, the rectal temperature may be taken below 0°C. These animals, when returned to a normal temperature, generally show no ill effects from their cooling experience. In both humans and other animals, however, rectal temperatures below 28°C. destroy a subject's ability to spontaneously restore his normal core temperature.

THE CONTROL OF BODY TEMPERATURE

All the data we now have on mammals and man are consistent with the view that peripheral and central thermoreceptors transmit impulses to centers in the hypothalamus which, through their control of the sympathetic nervous system and the thyroid gland, control (1) body metabolism, (2) shivering, (3) cutaneous blood flow, (4) sweating, (5) panting, and (6) the release of catecholamines from the adrenal medulla. We find, for example, that transection of the thoracic cord in human subjects results in a loss of the ability to shiver in those parts of the body innervated by anterior-horn cells below the transection. If the leg of a subject with such a transection is cooled, shivering occurs in neither the cooled nor the contralateral limb—since the transection has eliminated the influence of the hypothalamus on these areas, but does occur in muscles innervated by somatic efferent neurons above the transection. If the transection is made in the cervical region, upper pons, or midbrain, there is also loss of the ability to sweat or produce cutaneous vasodilation in response to an increase in the environmental or body temperature. Discrete lesions in the caudal portion of the hypothalamus produce similar results. All available evidence, in other words, is consistent with the view that the maintenance of body temperature is dependent upon an area above the midbrain, the hypothalamus.

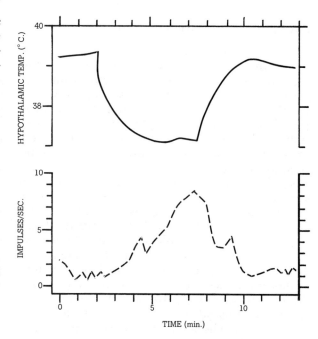

Fig. 27-8. A "cold cell" in the rostral hypothalamus of a lightly anesthetized dog. Note that decreases in hypothalamic temperature increase the number of impulses per second in this neuron. (*Redrawn from* Hardy, J. D., Hellon, R. F., and Sutherland, K.: Temperature-sensitive neurones in the dog's hypothalamus. J. Physiol., *175*:246, 1964)

The Heat Loss Center

There is in the rostral portion of the hypothalamus an area that responds to warming by increasing the activity of some of its neurons (Fig. 27-7), and to cooling by increasing the activity of other neurons (Fig. 27-8). The increased activity of this area, "the heat loss center," in response to warming results in (1) cutaneous vasodilation, (2) panting, in the dog and cat, (3) sweating, in man, and (4) inhibition of shivering. The sweating is in response to stimulation of cholinergic sympathetic neurons to the sweat glands. The vasodilation is in response to inhibition of adrenergic sympathetic tone and possibly the stimulation of cholinergic sympathetic fibers to the skin. It may also occur in response to a vasodilator, *bradykinin*, released by the sweat glands.

The heat loss center responds to cooling by producing (1) an increase in adrenergic sympathetic tone to the cutaneous blood vessels, (2) shivering, (3) an increase in plasma epinephrine, and (4) an increase in plasma protein-bound

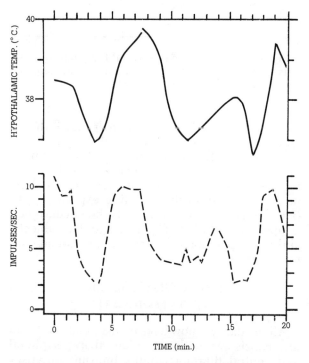

Fig. 27-7. A "warm cell" in the rostral hypothalamus of a lightly anesthetized dog. Note that increases in hypothalamic temperature increase the number of impulses per second in this neuron. (*Redrawn from* Hardy, J. D., Hellon, R. F., and Sutherland, K.: Temperature-sensitive neurones in the dog's hypothalamus. J. Physiol., *175*:245, 1964)

iodine (PBI). The epinephrine is secreted from the adrenal medulla, and the rise in PBI is the result of the secretion of thyroxine by the thyroid gland. Both hormones increase heat production by the body—i.e., they are both calorigenic. Their difference lies in the speed with which they produce a response: epinephrine can be released within minutes of hypothalamic cooling and acts on the tissues almost as soon as it reaches them, while protein-bound iodine (an index of thyroxine concentration) does not start to rise in the blood until about 30 min. after the onset of hypothalamic cooling. After about 3 hr. it has reached a concentration 55 to 125 per cent higher than before cooling, and after 4 to 6 hr. it reaches its maximum plasma concentration. Even then it has little or no calorigenic effect on the tissues however; this is delayed still longer. It has been suggested that the higher calorie production during cold weather is due at least in part to the increased production and release of thyroid hormone (Fig. 27-9).

The Heat Maintenance Center

In the caudal hypothalamus is an area which is sensitive to neither direct cooling nor warming but which, when destroyed bilaterally in monkeys and cats, results in poikilothermia (Gk. *poikilis*, varied; plus *thermē*, heat), or loss of temperature regulation. It is apparently an area which receives (1) impulses from the peripheral thermoreceptors and (2) inhibitory impulses from the more rostral heat loss center. Its primary function is thought to be the prevention of hypothermia, while that of the more rostral center is the prevention of hyperthermia. A lesion in the rostral center produces a loss of one's defenses against hyperthermia and a moderate decrease in one's defenses against hypothermia. The poikilothermia resulting from lesions in the caudal center is probably due in part to the destruction of fiber tracts going to the so-called heat loss center, and in part to the loss of reflexes preventing hypothermia.

Integration of Peripheral and Central Influences

The thermoregulatory mechanism is apparently not set to maintain a particular core temperature under all conditions. It has been suggested that both pyrogens and stimuli associated with exercise act on the thermoregulatory mechanism to reset it at a level that maintains a tempo-

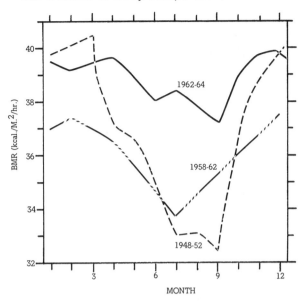

Fig. 27-9. Changes in basal metabolism during the year. Note that the basal metabolic rate is highest when the weather is cold (December through February) and lowest when the weather is warm (June through August). (*Redrawn from* Sasaki, T.: Relation of basal metabolism to changes in food composition and body composition. Fed. Proc., 25:1166, 1966)

rarily higher core temperature. It has been speculated that the elevation in core temperature that pyrogens produce is useful in fighting invading microorganisms. In the case of exercise it is hypothesized that the higher temperature results in a muscle capable of more efficient contraction. In fact, in some athletic competitions the temperature of active muscles has been noted to be as much as 1.5°C. higher than the already elevated rectal temperature. DuBois commented on this as follows: "Is it not significant that athletic records are established and football games are won by young men tolerating easily temperatures that we clinicians would designate as high fever."

Another aspect of the role of the thermoregulatory centers is their integration of sensory information from peripheral warm and cold receptors. The data in Table 27-2 are consistent with the hypothesis that the thermoregulatory centers also respond to sensory input from cold and warm receptors outside the hypothalamus. In these studies an attempt was made to keep the temperature of the hypothalamus of a dog constant while varying the room temperature.

Temperature (°C.)				
Air	Hypothalamus	Rectum	Skin	Effect on Skin
27	38.2	38.2	37	Control
12	38.2	37.8	15	Vasoconstriction
40	38.4	38.6	40	Vasodilation

Table 27-2. The role of changes in the ambient temperature on the thermoregulatory mechanism in the resting, fasting dog. (From Hammel, H. T., Jackson, D. C., Stolwijk, J. A., Hardy, J. D., and Strxmme, S. B.: Temperature regulation by hypothalamic proportional control with an adjustable set point. J. Appl. Physiol. 18:1146-1154, 1963)

You will note that a decrease in the temperature of the room air without an associated change in hypothalamic temperature did produce a decrease in skin temperature, in part due to cutaneous vasoconstriction. An increase in air temperature, on the other hand, produced cutaneous vasodilation. Studies on human subjects using the temperature at the tympanic membrane as an index of hypothalamic temperature tend to confirm the observation that the regulation of body temperature is affected both by changes in hypothalamic temperature changes outside the central nervous system.

REFERENCES

Adler, R. H.: Spontaneous pneumothorax. Hosp. Med., 1:2-6, 1965.

Adolph, E. F.: Physiology of Man in the Desert. New York, Interscience Publishers, 1947.

Anand, B. K.: Central chemosensitive mechanisms related to feeding. In Code, C. F. (ed.): Handbook of Physiology. Sec. 6 (Alimentary Canal), vol. I, pp. 249-263. Washington, American Physiological Society, 1967.

Anand, B. K., and Pillai, R. V.: Activity of single neurones in the hypothalamic feeding centres: Effect of gastric distension. J. Physiol., 192:63-77, 1967.

Asmussen, E.: Muscular exercise. In Fenn, W. O., and Rahn, H. (eds.): Handbook of Physiology. Sec. 3 (Respiration), vol. II, pp. 939-978. Washington, American Physiological Society, 1965.

Barach, A. L.: Pulmonary emphysema. Hosp. Med., 2:12-16, 1965.

Behnke, A. R., Jr.: Bends, decompression, and recompression. In Fenn, W. O., and Rahn, H. (eds.): Handbook of Physiology. Sec. 3 (Respiration), vol. II, pp. 631-648. Washington, American Physiological Society, 1964.

Bell, G. H., Davidson, J. N., and Scarborough, H.: Textbook of Physiology and Biochemistry. 7th ed. Baltimore, Williams & Wilkins, 1968.

Benzinger, T. H.: Receptor organs and quantitative mechanisms of human temperature control in a warm environment. Fed. Proc., 19:32-41, 1960.

Bergman, H. (ed.): The Ureter. New York, Hoeber Medical Division (Harper & Row), 1967.

Berliner, R. W.: Some recent developments in the physiology of the kidney. In Mostifi, F. K., and Smith, D. E. (eds.): The Kidney. Pp. 60-68. Baltimore, Williams & Wilkins, 1966.

Bernstein, L.: Respiration. Ann. Rev. Physiol., 29: 113-140, 1967.

Blair-West, J. R., Coghlan, J. P., Denton, D. A., and Denton, R. D.: Effect of endocrines on salivary glands. In Code, C. F. (ed.): Handbook of Physiology. Sec. 6 (Alimentary Canal), vol. II, pp. 633-664. Washington, American Physiological Society, 1967.

Blankenhorn, D. H., Hirsch, J., and Ahrens, E. H., Jr.: Transintestinal intubation: technique for measurement of gut length and physiologic sampling at known loci. Proc. Soc. Exp. Biol. Med., 88:356-362, 1955.

Boyarsky, S., Labay, P., Kirshner, N., and Gerber, C.: Does the ureter have nervous control? J. Urol., 97: 627-632, 1967.

Brodie, B., Maickel, R. P., and Stern, D. N.: Autonomic nervous system and adipose tissue. In Renold, A. E., and Cahill, G. F., Jr. (eds.): Handbook of Physiology. Sec. 5 (Adipose Tissue), pp. 583-600. Washington, American Physiological Society, 1965.

Brown, G., and Dobson, R. L.: Sweat sodium excretion in normal women. J. Appl. Physiol., 23:97-99, 1967.

Brown, J. L., Samiy, A. H., and Pitts, R. F.: Localization of amino-nitrogen reabsorption in the nephron of the dog. Am. J. Physiol., 200:270-372, 1961.

Bunag, R. D., Page, I. H., and McCubbin, J. W.: Neural stimulation of release of renin. Circ. Res., 19:851-858, 1966.

———: Influence of dietary sodium on stimuli causing renin release. Am. J. Physiol., 211:1383-1386, 1966.

Burgen, A. S. V.: Secretory processes in salivary glands. In Code, C. F. (ed.): Handbook of Physiology. Sec. 6 (Alimentary Canal), vol. II, pp. 561-579. Washington, American Physiological Society, 1967.

Burgen, A. S. V., and Seeman, P.: The role of the salivary duct system in the formation of the saliva. Canad. J. Biochem., 36:119-143, 1958.

Callender, S. T.: The intestinal mucosa and iron absorption. Brit. Med. Bull., 23:263-265, 1967.

Campbell, M. F.: Urology. Vol. I. Philadelphia, W. B. Saunders, 1963.

Carlson, L. A., Boberg, J., and Högstedt, B.: Some physiological and clinical implications of lipid mobilization from adipose tissue. *In* Renold, A. E., and Cahill, G. F., Jr. (eds.): Handbook of Physiology. Sec. 5 (Adipose Tissue), pp. 625-644. Washington, American Physiological Society, 1965.

Case, R. B., Nasser, M. G., and Crampton, R. S.: Biochemical aspects of early myocardial ischemia. Am. J. Cardiol., 24:766-775, 1969.

Chasis, H., and Smith, H. W.: The excretion of urea in normal man and in subjects with glomerulonephritis. J. Clin. Invest., 17:347-358, 1938.

Cherniack, R. M.: Work of breathing and the ventilatory response to CO_2. *In* Fenn, W. O., and Rahn, H. (eds.): Handbook of Physiology. Sec. 3 (Respiration), vol. II, pp. 1469-1474. Washington, American Physiological Society, 1965.

Chernick, V., Hodson, W. A., and Greenfield, L. J.: Effect of chronic pulmonary artery ligation on pulmonary mechanics and surfactant. J. Appl. Physiol., 21:1315-1320, 1966.

Clements, J. A., and Tierney, D. F.: Alveolar instability associated with altered surface tension. *In* Fenn, W. O., and Rahn, H. (eds.): Handbook of Physiology. Sec. 3 (Respiration), vol. II, pp. 1565-1583. Washington, American Physiological Society, 1965.

Code, C. F., Hightower, N. C., Jr., and Morlock, C. G.: Motility of the alimentary canal in man. Am. J. Med., 13:328-351, 1952.

Comroe, J. H., Jr.: Physiology of Respiration. Chicago, Year Book Publishers, 1965.

Comroe, J. H., Jr., Forster, R. E., Dubois, A. B., Briscoe, W. A., and Carlsen, E.: The Lung. Chicago, Year Book Publishers, 1956.

Comroe, J. H., Jr., and Schmidt, C. F.: Reflexes from limbs as factor in hyperpnea of muscular exercise. Am J. Physiol., 138:536-547, 1943.

Costantin, L. L.: Effect of pulmonary congestion on vagal afferent activity. Am J. Physiol., 196:49-53, 1959.

Cotes, J. E.: Lung Function. Oxford, Blackwell Scientific Publications, 1965.

Cowgill, G. R., and Sullivan, A. J.: Further studies on use of wheat bran as laxative: Observations on patients. JAMA, 100:795-802, 1933.

Crabbe, J., Ross, E. J., and Thorn, G. W.: The significance of the secretion of aldosterone during dietary sodium deprivation in normal subjects. J. Endocr. Metab., 18:1159-1177, 1958.

Crane, R. K.: Absorption of sugars. *In* Code, C. F. (ed.): Handbook of Physiology. Sec. 6 (Alimentary Canal), vol. III, pp. 1323-1351. Washington, American Physiological Society, 1968.

Creamer, B.: The turnover of the epithelium of the small intestine. Brit. Med. Bull., 23:226-230, 1967.

Crosby, W. H.: Iron absorption. *In* Code, C. F. (ed.): Handbook of Physiology. Sec. 6 (Alimentary Canal), vol. III, pp. 1553-1570. Washington, American Physiological Society, 1968.

Cross, K. W.: Respiration and oxygen supplies in the newborn. *In* Fenn, W. O., and Rahn, H. (eds.): Handbook of Physiology. Sec. 3 (Respiration), vol. II, pp. 1329-1343. Washington, American Physiological Society, 1965.

Daniel, E. E., and Wiebe, G. E.: Transmission of reflexes arising on both sides of the gastroduodenal junction. Am. J. Physiol., 211:634-642, 1966.

Davenport, H. W.: The ABC of Acid-Base Chemistry. 4th ed. Chicago, University of Chicago, 1958.

———: Physiology of the Digestive Tract. 2nd ed. Chicago, Year Book Publishers, 1966.

Davis, J. O.: The physiology of congestive heart failure. *In* Hamilton, W. F. (ed.): Handbook of Physiology. Sec. 2 (Circulation), vol. III, pp. 2071-2122. Washington, American Physiological Society, 1965.

Davis, J. O., Binnion, P. F., Brown, T. C., and Johnston, C. I.: Mechanisms involved in the hypersecretion of aldosterone during sodium depletion. Circ. Res. (Suppl. I), 18 and 19:143-157, 1966.

Davis, J. O., Johnston, C. I., Howards, S. S., and Wright, F. S.: Humoral factors in the regulation of renal sodium excretion. Fed. Proc., 26:60-69, 1967.

Dawson, A. M.: Absorption of fats. Brit. Med. Bull., 23:247-251, 1967.

Dejours, P.: Control of respiration in muscular exercise. *In* Fenn, W. O., and Rahn, H. (eds.): Handbook of Physiology. Sec. 3 (Respiration), vol. I, pp. 631-648. Washington, American Physiological Society, 1964.

Doty, R. W.: Neural organization of deglutition. *In* Code, C. F. (ed.): Handbook of Physiology. Sec. 6 (Alimentary Canal), vol. IV, pp. 1861-1902. Washington, American Physiological Society, 1968.

DuBois, A. B.: Resistance to breathing. *In* Fenn, W. O., and Rahn, H. (eds.): Handbook of Physiology. Sec. 3 (Respiration), vol. I, pp. 451-462. Washington, American Physiological Society. 1964.

DuBois, D., and DuBois, E. F.: Clinical calorimetry. X. A formula to estimate the approximate surface area if height and weight be known. Arch. Intern. Med., 17:863-871, 1916.

DuBois, E. F.: Fever and the Regulation of Body Temperature. Springfield, Ill., Charles C Thomas, 1948.

Dutton, R. E., Hodson, W. A., Davies, D. G., and Fenner, A.: Effect of the rate of rise of carotid body P_{CO_2} on the time course of ventilation. Resp. Physiol., 3:367-379, 1967.

Edholm, O. G., Fletcher, J. G., Widdowson, E. M., and McCance, R. A.: The energy expenditure and food intake of individual men. Brit. J. Nutr., 9:286-300, 1955.

Edvardsen, P.: Nervous control of urinary bladder in cats. Acta Physiol. Scand. 72:157-193, 234-247, 1967.

Elam, J. O.: Respiratory and circulatory resuscitation. *In* Fenn, W. O., and Rahn, H. (eds.): Handbook of Physiology. Sec. 3 (Respiration), vol. I, pp. 1265-1312. Washington, American Physiological Society, 1965.

Elias, H., Hassmann, A., Barth, I. B., and Solmar, A.: Blood flow in the renal glomerulus. J. Urol., 83: 790-799, 1960.

Emmelin, N.: Nervous control of salivary glands. In Code, C. F. (ed.): Handbook of Physiology. Sec. 6 (Alimentary Canal), vol. II, pp. 595-632. Washington, American Physiological Society, 1967.

Farhi, L. E.: Gas stores of the body. In Fenn, W. O., and Rahn, H. (eds.): Handbook of Physiology. Sec. 3 (Respiration), vol. I, pp. 873-885. Washington, American Physiological Society, 1964.

Faridy, E. E., Permutt, S., and Riley, R. L.: Effect of ventilation on surface forces in excised dogs' lungs. J. Appl. Physiol., 21:1453-1462, 1966.

Fordtran, J. S., and Ingelfinger, F. J.: Absorption of water, electrolytes, and sugar from the human gut. In Code, C. F. (ed.): Handbook of Physiology. Sec. 6 (Alimentary Canal), vol. III, pp. 1457-1490. Washington, American Physiological Society, 1968.

Fordtran, J. S., and Saltin, B.: Gastric emptying and intestinal absorption during prolonged severe exercise. J. Appl. Physiol., 23:331-335, 1967.

Fourman, J., and Kennedy, G. C.: An effect of antidiuretic hormone on the flow of blood through the vasa recta of the rat kidney. J. Endocr., 35:173-177, 1966.

Fox, I. J., et al.: Effects of the Valsalva maneuver on blood flow in the thoracic aorta in man. J. Appl. Physiol., 21:1553-1560, 1966.

Friedberg, C. K.: Coronary heart disease. In Beeson, P. B., and McDermott, W. (eds.): Texbook of Medicine. 12th ed., pp. 634-657. Philadelphia, W. B. Saunders, 1967.

Friedman, S. M., Johnson, R. L., and Friedman, C. L.: The pattern of recovery of renal function following renal artery occlusion in the dog. Circ. Res., 2:231-235, 1954.

Ganong, W. F.: Review of Medical Physiology. 2nd ed. Los Altos, Lange Medical Publications, 1965.

Ganong, W. F., and Boryczka, A. T.: Effect of a low sodium diet on aldosterone-stimulating activity of angiotensin II in dogs. Proc. Soc. Exp. Biol. Med., 124:1230-1231, 1967.

Giebisch, G.: Measurements of electrical potential differences on single nephrons of the perfused *Necturus* kidney. J. Gen. Physiol., 44:659-678, 1961.

Gill, J. R., Jr., Bell, N. H., and Bartter, F. C.: Effect of parathyroid extract on magnesium excretion in man. J. Appl. Physiol., 22:136-138, 1967.

Gjone, R.: Peripheral autonomic influence on the motility of the urinary bladder in the cat. Acta Physiol. Scand., 66:64-90, 1966.

Gjone, R., and Setekleiv, J.: Excitatory and inhibitory bladder responses to stimulation of the cerebral cortex in the cat. Acta Physiol. Scand., 59:337-348, 1963.

Glynn, I. M.: Sodium and potassium movements in human red cells. J. Physiol., 134:278-310, 1956.

Gottschalk, C. W.: Micropuncture studies of tubular function in the mammalian kidney. Physiologist, 3: 35-55, 1961.

Gottschalk, C. W., Lassiter, W. E., and Mylle, M.: Localization of urine acidification in the mammalian kidney. Am. J. Physiol., 198:581-585, 1960.

Greene, D. G.: Drowning. In Fenn, W. O., and Rahn, H. (eds.): Handbook of Physiology. Sec. 3 (Respiration), vol. II, pp. 1195-1204. Washington, American Physiological Society, 1965.

———: Pulmonary edema. In Fenn, W. O., and Rahn, H. (eds.): Handbook of Physiology. Sec. 3 (Respiration), vol. II, pp. 1585-1600. Washington, American Physiological Society, 1965.

Grossman, M. I.: Stimulation of acid secretion. Gastroenterology, 41:385-390, 1961.

———: Neural and hormonal stimulation of gastric secretion of acid. In Code, C. F. (ed.): Handbook of Physiology. Sec. 6 (Alimentary Canal), vol. II, pp. 835-863. Washington, American Physiological Society, 1967.

Guz, A., Noble, N. I. M., Widdicombe, J. G., Trenchard, D., and Mushin, W. W.: Peripheral chemoreceptor block in man. Resp. Physiol., 1:38-40, 1966.

Hallenbeck, G. A.: Biliary and pancreatic intraductal pressures. In Code, C. F. (ed.): Handbook of Physiology. Sec. 6 (Alimentary Canal), vol. II, pp. 1007-1025. Washington, American Physiological Society, 1967.

Hamilton, C. L.: Food and temperature. In Code, C. F. (ed.): Handbook of Physiology. Sec. 6 (Alimentary Canal), vol. I, pp. 303-317. Washington, American Physiological Society, 1967.

Hammel, H. T., Jackson, D. C., Stolwijk, J. A., Hardy, J. D., and Stromme, S. B.: Temperature regulation by hypothalamic proportional control with an adjustable set point. J. Appl. Physiol., 18:1146-1154 1963.

Hardy, J. D., and Du Bois, E. F.: The technique of measuring radiation and convection. J. Nutr., 15: 461-475, 1938.

Hardy, J. D., Hellon, R. F., and Sutherland, K.: Temperature-sensitive neurones in the dog's hypothalamus. J. Physiol., 175:242-253, 1964.

Hardy, J. D., and Soderstrom, G. F.: Heat loss from the nude body and peripheral blood flow at temperatures of 22°C. to 35°C. J. Nutr., 16:493-510, 1938.

Harper, A. A.: Hormonal control of pancreatic secretion. In Code, C. F. (ed.): Handbook of Physiology. Sec. 6 (Alimentary Canal), vol. II, pp. 969-995. Washington, American Physiological Society, 1967.

Hashim, S. A., Felch, W. C., and van Itallie, T. B.: Lipid metabolism in relation to physiology and pathology of atherosclerosis. In Hamilton, W. F. (ed.): Handbook of Physiology. Sec. 2 (Circulation), vol. II, pp. 1167-1195. Washington, American Physiological Society, 1963.

Hertz, A. F., and Newton, A.: Movements of colon. J. Physiol., 47:57-65, 1913.

Hinman, F., Jr., and Cox, C. E.: Residual urine volume

in normal male subjects. Trans. Am. Ass. Genitourin. Surg., 58:82-86, 1966.

Hirsch, J., and Goldrick, B.: Metabolism of human adipose tissue in vitro. *In* Renold, A. E., and Cahill, G. F., Jr. (eds.): Handbook of Physiology. Sec. 5 (Adipose Tissue), pp. 455-470. Washington, American Physiological Society, 1965.

Hirschowitz, B. I.: The control of pepsinogen secretion. Ann. N. Y. Acad. Sci., 140:709-723, 1967.

Hokin, L. E., and Hokin, M. R.: Studies on the carrier function of phosphatidic acid in sodium transport. J. Gen. Physiol., 44:61-85, 1960.

Hugelin, A., and Cohen, M. I.: The reticular activating system and respiratory regulations in the cat. Ann. N. Y. Acad. Sci., 109:586-603, 1963.

Hukuhara, T., Nakayama, S., and Nanba, R.: The effects of 5-hydroxytryptamine upon the intestinal motility. Jap. J. Physiol., 10:420-426, 1960.

Hurtado, A.: Animals in high altitudes: Resident man. *In* Dill, D. B., Adolf, E. F., and Wilber, C. G. (eds.): Handbook of Physiology. Sec. 4 (Adaptation to the Environment), pp. 843-860. Washington, American Physiological Society, 1964.

Hutch, J. A.: New theory of anatomy of internal urinary sphincter and physiology of micturition. IV. Urinary sphincteric mechanism. J. Urol., 97:705-712, 1967.

Hutch, J. A., and Rambo, O. N., Jr.: New theory of anatomy of internal urinary sphincter and physiology of micturition. III. Anatomy of urethra. J. Urol., 97:696-704, 1967.

Ivy, A. C.: What is normal or normality? Quart. Bull. Northwestern Univ. Med. School, 18:22-23, 1944.

Jacobson, E. D.: Secretion and blood flow in the gastrointestinal tract. *In* Code, C. F. (ed.): Handbook of Physiology. Sec. 6 (Alimentary Canal), vol. II, pp. 1043-1062. Washington, American Physiological Society, 1967.

Janowitz, H. D.: Pancreatic secretion of fluid and electrolytes. *In* Code, C. F. (ed.): Handbook of Physiology. Sec. 6 (Alimentary Canal), vol. II, pp. 925-933, Washington, American Physiological Society, 1967.

Janowitz, H. D., Hollander, F., Orringer, D., and Margolin, S. G.: A quantitative study of the gastric secretory reponse to sham feeding in a human subject. Gastroenterology, 16:104-116, 1950.

Johnston, J. M.: Mechanism of fat absorption. *In* Code, C. F. (ed.): Handbook of Physiology. Sec. 6 (Alimentary Canal), vol, III, pp. 1353-1375. Washington, American Physiological Society, 1968.

Keele, C. A., and Neil, E.: Samson Wright's Applied Physiology. New York, Oxford University Press, 1965.

Kekwick, A.: Adiposity. *In* Renold, A. E., and Cahill, G. F., Jr. (eds.): Handbook of Physiology. Sec. 5 (Adipose Tissue), pp. 617-624. Washington, American Physiological Society, 1965.

Kelley, M. L., Jr., Gordon, E. A., and Deweese, J. A.: Pressure responses of canine ileocolonic junctional zone to intestinal distention. Am. J. Physiol., 211:614-618, 1966.

Kellogg, R. H.: Central chemical regulation of respiration. *In* Fenn, W. O., and Rahn, H. (eds.): Handbook of Physiology. Sec. 3 (Respiration), vol. I, pp. 507-534. Washington, American Physiological Society, 1964.

Kirby, W. M. M. (ed.): Modern management of respiratory diseases. Med. Clin. N. Am., 51:267-578, 1967.

Krahl, V.: Anatomy of the mammalian lung. *In* Fenn, W. O., and Rahn, H. (eds.): Handbook of Physiology. Sec. 3 (Respiration), vol. I, pp. 213-284. Washington, American Physiological Society, 1964.

Kügler-Podelleck, I., Rodewald, G., Horatz, K., Kügler, S., and Müller-Brunotte, P.: Resuscitation after drowning in freezing water. German Med. Monthly, 11:232-236, 1966.

Kuru, M.: Nervous control of micturition. Physiol. Rev., 45:425-494, 1965.

Lambertsen, C. J.: Effects of oxygen at high partial pressure. *In* Fenn, W. O., and Rahn, H. (eds.): Handbook of Physiology. Sec. 3 (Respiration), vol. II, pp. 1027-1046. Washington, American Physiological Society, 1964.

Laragh, J. H.: Renin, angiotensin, aldosterone and hormonal regulation of arterial pressure and salt balance, Fed. Proc., 26:39-41, 1967.

Leeson, C. R.: Structure of salivary glands. *In* Code, C. F. (ed.): Handbook of Physiology. Sec. 6 (Alimentary Canal), vol. II, pp. 463-495. Washington, American Physiological Society, 1967.

Levein, R., and Haft, D. E.: Carbohydrate homeostasis. New Eng. J. Med., 283:175-183, 237-246, 1970.

Levy, M. N., DeGeest, H., and Zieske, H.: Effects of respiratory center activity on the heart. Circ. Res., 18:67-78, 1966.

Loomis, W. F.: Skin-pigment regulation of vitamin D biosynthesis in man. Science, 157:501-506, 1967.

Lotspeich, W. D.: Renal tubular reabsorption of inorganic sulfate in the normal dog. Am. J. Physiol., 151:311-318, 1947.

———: Metabolic aspects of acid-base change. Science, 155:1066-1075, 1967.

Luft, U. C.: Aviation physiology—The effects of altitude. *In* Fenn, W. O., and Rahn, H. (eds.): Handbook of Physiology. Sec. 3 (Respiration), vol. II, pp. 1099-1145. Washington, American Physiological Society, 1965.

Mason, G. R., and Nelsen, T. S.: Hypothalamic stimulation and gastric secretion in the dog. Am. J. Physiol., 213:21-24, 1967.

Milne, M. D.: Observations on the pharmacology of mecamylamine. Clin. Sci., 16:599-614, 1957.

Mithoeffer, J. C.: Breath holding. *In* Fenn, W. O., and Rahn, H. (eds.): Handbook of Physiology. Sec. 3 (Respiration), vol. II, pp. 1011-1025. Washington, American Physiological Society, 1965.

Morris, I. G.: Gamma globulin absorption in the newborn. *In* Code, C. F. (ed.): Handbook of Physiology. Sec. 6 (Alimentary Canal), vol. III, pp. 1491-1512. Washington, American Physiological Society, 1968.

Mountcastle, V. B.: Medical Physiology. Vol. I. St. Louis, C. V. Mosby, 1968.

Nesbit, R. M., and Lapides, J.: The physiology of micturition. J. Mich. Med. Soc., 58:384-388, 1959.

Ochwadt, B.: Relation of renal blood supply to diuresis. Progr. Cardiovasc. Dis., 3:501-510, 1961.

Olson, R. E.: Diet and coronary artery disease. *In* Blumgart, H. L. (ed.): Symposium on Coronary Heart Disease. 2nd ed. New York, American Heart Association, 1968.

Ono, H., Inigaki, K., and Hashimoto, K.: A pharmacological approach to the nature of the autoregulation of the renal blood flow. Jap. J. Physiol., 16:625-634, 1967.

Otis, A. B.: The work of breathing. *In* Fenn, W. O., and Rahn, H. (eds.): Handbook of Physiology. Sec. 3 (Respiration), vol. I, pp. 463-476. Washington, American Physiological Society, 1964.

⸺: Quantitative relationships in steady-state gas exchange. *In* Fenn, W. O., and Rahn, H. (eds.): Handbook of Physiology. Sec. 3 (Respiration), vol. I, pp. 681-698. Washington, American Physiological Society, 1964.

Owen, O. E., et al.: Liver and kidney metabolism during prolonged starvation. J. Clin. Invest. 48:574-583, 1969.

Page, I. H., and McCubbin, J. W.: The physiology of arterial hypertension. *In* Hamilton, W. F. (ed.): Handbook of Physiology. Sec. 2 (Circulation), vol. III, pp. 2163-2208. Washington, American Physiological Society, 1965.

Parsons, D. S.: Salt and water absorption by the intestinal tract. Brit. Med. Bull., 23:252-257, 1967.

Paul, P., and Issekutz, B., Jr.: Role of extramuscular energy sources in the metabolism of the exercising dog. J. Appl. Physiol., 22:615-622, 1967.

Pitts, R. F.: The function of components of the respiratory complex. J. Neurophysiol., 5:403-413, 1942.

⸺: A renal reabsorptive mechanism in the dog common to glycine and creatinine. Am. J. Physiol., 140:156-167, 1943.

⸺: Renal excretion of acid. Fed. Proc., 7:418-426, 1948.

⸺: A comparison of the modes of action of certain diuretic agents. Progr. Cardiovasc. Dis., 3:537-562, 1961.

⸺: Physiology of the Kidney and Body Fluids. 2nd ed. Chicago, Year Book Medical Publishers, 1968.

Pitts, R. F., and Alexander, R. S.: The renal reabsorptive mechanism for inorganic phosphate in normal and acidotic dogs. Am. J. Physiol., 142:648-662, 1944.

Pitts, R. F., Ayer, J. L., and Schiess, W. A.: Reabsorption and excretion of bicarbonate. J. Clin. Invest., 28:35-44, 1949.

Porter, K. R., and Bonneville, M. A.: An Introduction to the Fine Structure of Cells and Tissues. 3rd ed. Philadelphia, Lea & Febiger, 1968.

Preshaw, R. M.: Integration of nervous and hormonal mechanisms for external pancreatic secretion. *In* Code, C. F. (ed.): Handbook of Physiology. Sec. 6 (Alimentary Canal), vol. II, pp. 997-1005. Washington, American Physiological Society, 1967.

Proctor, D. F.: Physiology of the upper airway. *In* Fenn, W. O., and Rahn, H. (eds.): Handbook of Physiology. Sec. 3 (Respiration), vol. I, pp. 309-345. Washington, American Physiological Society, 1964.

Pugh, L.G.C.E., Corbett, J. L., and Johnson, R. H.: Rectal temperatures, weight losses and sweat rates in marathon running. J. Appl. Physiol., 23:347-352, 1967.

Radford, E. P.: Static mechanical properties of mammalian lungs. *In* Remington, J. W. (ed.): Tissue Elasticity. P. 185. Washington, American Physiological Society, 1968.

Rahn, H.: The sampling of alveolar gas. *In* Handbook of Respiratory Physiology. Randolph Field, Texas, Air University School of Aviation Medicine, 1954.

Rector, F. C., Jr., Seldin, D. W., Roberts, A. D. Jr., and Smith, J. S.: The role of plasma CO_2 tension and carbonic anhydrase activity in the renal reabsorption of bicarbonate. J. Clin. Invest., 39:1706-1721, 1960.

Redgate, E. S.: Hypothalamic influence on respiration. Ann. N. Y. Acad. Sci., 109:606-618, 1963.

Richards, D. W.: Pulmonary changes due to age. *In* Fenn, W. O., and Rahn, H. (eds.): Handbook of Physiology. Sec. 3 (Respiration), vol. II, pp. 1525-1529. Washington, American Physiological Society, 1965.

Ruch, T. C.: Central control of the bladder. *In* Field, J., Magoun, H. W., and Hall, V. F. (eds.): Handbook of Physiology. Sec. 1 (Neurophysiology), vol. II, pp. 1207-1223. Washington, American Physiological Society, 1960.

Rushmer, R. F.: Cardiovascular Dynamics. Philadelphia, W. B. Saunders, 1970.

Rushmer, R. F., and Henredon, J. A.: The act of deglutition: A cinefluorographic study. J. Apply. Physiol., 3:622-630, 1951.

Rutishauser, W. J., Banchero, N., Tsakiris, A. G., Edmundowicz, A. C., and Wood, E. H.: Pleural pressures at dorsal and ventral sites in supine and prone body positions. J. Appl. Physiol., 21:1500-1510, 1966.

Saltin, B., and Hermansen, L.: Esophageal, rectal, and muscle temperature during exercise. J. Appl. Physiol., 21:1757-1762, 1966.

Sartorius, O. W., Roemmelt, J. C., and Pitts, R. F.: The renal regulation of acid-base balance in man. IV. The nature of the renal compensations in ammonium chloride acidosis. J. Clin. Invest., 28:423-439, 1949.

Sasaki, T.: Relation of basal metabolism to changes in food composition and body composition. Fed. Proc., 25:1165-1168, 1966.

Scheuer, J., and Leonard, J. J.: Pathophysiology of

ischemic heart disease. *In* Gordon, B. L., Carleton, R. A., and Faber, L. P. (ed.): Clinical Cardiopulmonary Physiology. 3rd ed., pp. 229-244. New York, Grune & Stratton, 1969.

Schiess, W. A., Ayer, J. L., Lotspeich, W. D., and Pitts, R. F.: The renal regulation of acid-base balance in man. II. Factors affecting the excretion of titratable acid by the normal human subject. J. Clin. Invest., 27:57-64, 1948.

Selkurt, E. E.: The renal circulation. *In* Hamilton, W. F. (ed.): Handbook of Physiology. Sec. 2 (Circulation), vol. II, pp. 1457-1516. Washington, American Physiological Society, 1963.

———: Physiology. Boston, Little, Brown & Co., 1966.

Severinghaus, J. W., and Larson, C. P., Jr.: Respiration in anesthesia. *In* Fenn, W. O., and Rahn, H. (eds.): Handbook of Physiology. Sec. 3 (Respiration), vol. II, pp. 1219-1264. Washington, American Physiological Society, 1965.

Sharpey-Schafer, E. P.: Effect of respiratory acts on the circulation. *In* Hamilton, W. F. (ed.): Handbook of Physiology. Sec. 2 (Circulation), vol. III, pp. 1875-1886. Washington, American Physiological Society, 1965.

Shipley, R. E., and Study, R. S.: Changes in renal blood flow. Am. J. Physiol., 167:676-688, 1951.

Sleight, P., and Widdicombe, J. G.: Action potentials in fibres from receptors in the epicardium and myocardium of the dog's left ventricle. J. Physiol., 181:235-258, 1965.

Smith, H. W.: The Physiology of the Kidney. New York, Oxford University Press, 1937.

———: The kidney structure and function in health and disease. New York, Oxford University Press, 1951.

Smyth, D. H., and Whittam, R.: Membrane transport in relation to intestinal absorption. Brit. Med. Bull., 23:231-235, 1967.

Sparge, B. H.: Structure of the kidney. *In* Mostofi, F. K., and Smith, D. E. (eds.): The Kidney. Pp. 17-59, Baltimore, Williams & Wilkins, 1966.

Tabaqchali, S., and Booth, C. C.: Relationship of the intestinal bacterial flora to absorption. Brit. Med. Bull., 23:285-289, 1967.

Tenney, S. M., and Lamb, T. W.: Physiological consequences of hypoventilation and hyperventilation. *In* Fenn, W. O., and Rahn, H. (eds.): Handbook of Physiology. Sec. 3 (Respiration), vol. II, pp. 979-1010. Washington, American Physiological Society, 1965.

Thomas, J. E.: Neural regulation of pancreatic secretion. *In* Code, C. F. (ed.): Handbook of Physiology. Sec. 6 (Alimentary Canal), vol. II, pp. 955-968. Washington, American Physiological Society, 1968.

Tobian, L.: Renin release and its role in renal function and control of salt balance and arterial pressure. Fed. Proc. 26:48-54, 1967.

Treadwell, C. R., and Vahouny, G. V.: Cholesterol absorption. *In* Code, C. F. (ed.): Handbook of Physiology. Sec. 6 (Alimentary Canal), vol. III, pp. 1407-1438. Washington, American Physiological Society, 1968.

Turaids, T., Nobrega, F. T., and Gallagher, T. J.: Absorptional atelectasis breathing oxygen at simulated altitude: Prevention using inert gas. Aerospace Med., 38:189-191, 1967.

Ullrich, K., Kamer, K. J., and Boylan, J. W.: Present knowledge of the counter-current system in the mammalian kidney. Progr. Cardiovasc. Dis., 3:395-430, 1961.

Vander, A. J.: Control of renin release. Physiol. Rev., 47:359-382, 1967.

Vander, A. J., and Luciano, J. R.: Neural and humoral control of renin release in salt depletion. Circ. Res. (Suppl. II), 20 and 21:69-77, 1967.

Wang, S. C., and Ngai, S. H.: General organization of central respiratory mechanisms. *In* Fenn, W. O., and Rahn, H. (eds.): Handbook of Physiology. Sec. 3 (Respiration), vol. I, pp. 507-534. Washington, American Physiological Society, 1964.

Webb, W. R.: Management of postoperative pulmonary complications. Hosp. Med., 1:16-22, 1965.

Weibel, E. R.: Morphometrics of the lung. *In* Fenn, W. O., and Rahn, H. (eds.): Handbook of Physiology, Sec. 3 (Respiration), vol. I, pp. 285-307. Washington, American Physiological Society, 1964.

Weiner, I. M., and Lack, L.: Bile salt absorption. Enterohepatic circulation. *In* Code, C. F. (ed.): Handbook of Physiology. Sec. 6 (Alimentary Canal), vol. III, pp. 1439-1455. Washington, American Physiological Society, 1968.

Werkö, L. E., Varnauskas, E., Ek, J., Bucht, H., Thomasson, B., and Bergström, J.: Studies on the renal circulation and renal function in mitral valvular disease. I. Effect of exercise. Circulation, 9:687-699, 1954.

West, C. D., Wilson, J. L., and Eylis, R.: Changes in blood amino nitrogen levels following injection of proteins and of a protein hydrolysate in infants with normal and with deficient pancreatic function. Am. J. Dis. Child., 72:251-273, 1946.

Wheeler, H. O., and Ramos, O. L.: Determinants of the flow and composition of bile in the unanesthetized dog during constant infusions of sodium taurocholate. J. Clin. Invest., 39:161-170, 1960.

Widdicombe, J. G.: Respiratory reflexes. *In* Fenn, W. O., and Rahn, H. (eds.): Handbook of Physiology. Sec. 3 (Respiration), vol. I, pp. 585-630. Washington, American Physiological Society, 1964.

Wilson, T. H.: Intestinal Absorption. Philadelphia, W. B. Saunders, 1962.

Windhager, E. E., Guillermo, W., Oken, D. E., Schatzmann, H. J., and Solomon, A. K.: Single proximal tubules of the *Necturus* kidney. III. Dependence of H_2O movement on NaCl concentration. Am. J. Physiol., 197:313-318, 1959.

Winslow, C. E. A., and Herrington, L. P.: Temperature and Human Life. Princeton, Princeton University Press, 1949.

Wiseman, G.: Absorption of amino acids. *In* Code,

C. F. (ed.): Handbook of Physiology. Sec. 6 (Alimentary Canal), vol. III, pp. 1277-1307. Washington, American Physiological Society, 1968.

Wright, L. D., Russo, H. F., Skeggs, H. R., Patch, E. A., and Beyer, K. H.: The renal clearance of essential amino acids: Arginine, histidine, lysine, and methionine. Am. J. Physiol., *149*:130-134, 1947.

Wyatt, A. P.: The relationship of the sphincter of Oddi to the stomach, duodenum and gall bladder. J. Physiol., *193*:225-244, 1967.

PART VI

CONTROL—THE ENDOCRINE SYSTEM

The famous French physiologist Claude Bernard (1813-1878) added a new perspective to physiology when he stated:

Animals have really two environments: a *milieu extérieur* in which the organism is situated, and a *milieu intérieur* in which the tissue elements live. The living organism does not really exist in the *milieu extérieur* (the atmosphere if it breathes, salt or fresh water if that is its element) but in the liquid *milieu intérieur* formed by the circulating organic liquid which surrounds and bathes all the tissue elements.

The two systems most involved in control of the internal environment are the autonomic nervous system and the endocrine system. In the former are a group of motor neurons whose control is centered in the hypothalamus and other areas throughout the central nervous system. The sensory information that goes into these centers either (1) modifies their intrinsic activity, (2) facilitates their activity, or (3) inhibits their stimulation. The output of the centers is in the form of action potentials traveling over neurons. Each of these action potentials triggers the release of a neurohumor (acetylcholine, norepinephrine, and perhaps others) at the neuron ending. These neurohumors act on the membrane primarily of smooth muscle, cardiac muscle, and gland cells. In some cases the neurohumors hyperpolarize the membrane, and in others they depolarize it. In some cases they change the rate of conduction along the membrane, change the permeability of the membrane to Na^+, K^+, Ca^{++}, and other substances, or both. In producing these changes the neurohumors sometimes initiate a series of events that leads to contraction or secretion, and sometimes merely modify excitability, conductivity, and the response of the cell to extrinsic and intrinsic stimuli.

The endocrine system functions in a manner similar to that of the autonomic nervous system (ANS). Like the ANS, the endocrine system has its highest integrative center in the hypothalamus and is concerned with the control of smooth muscle, cardiac muscle, and glands by the release of a chemical substance. Its mode of action is not completely understood, but in many cases the endocrine humors, like the neurohumors, produce their effect by acting on the cell membrane. It has also been suggested that some act on enzyme systems in the cell and on the rate of RNA and protein synthesis at the level of the gene.

The endocrine humors are called hormones. The hormones are steroids, amines, and polypeptides produced in various parts of the body and released into the blood or lymph, which carries them to their target cells; here the hormones in small concentrations (10^{-7}-10^{-12} molar) modify activity. (The hormones and their sources are listed in Table 28-1). Some of them have a relatively small number of target cells. Oxytocin, for example, apparently acts solely or primarily on the uterine muscle and the myoepithelial cells of the lactating gland. Growth hormone and thyroxine, on the other hand, affect activity in most of the cells of the body.

Chapter 28

THE PITUITARY GLAND

CONTROL
THE POSTERIOR PITUITARY
THE ANTERIOR PITUITARY

The pituitary gland or hypophysis of an adult man weighs approximately 66 gm. (slightly more in women) and lies in a concavity—the sella turcica—of the sphenoid bone. It is encapsulated by the dura mater and lies immediatly behind the optic chiasma and below the hypothalamus, to which it is attached by the infundibular stalk. It can be divided into three parts: (1) an anterior lobe (pars distalis plus pars tuberalis), (2) a posterior lobe (pars intermedia plus processus infundibuli), and (3) the infundibulum (pediculus infundibularis plus bulbus infundibularis plus labrum infundibularis). Collectively the anterior lobe plus the pars intermedia are called the adenohypophysis, while the processus infundibuli plus the infundibulum are called the neurohypophysis. In the infant the pars intermedia is clearly demonstrable between the pars distalis and pars nervosa, but in the adult it fuses with the pars nervosa and becomes obsure.

CONTROL

Hormones are released in response to stimuli from the internal or external environment or on the basis of an intrinsic pattern of production and release. We find, for example, that the beta cells of the islets of Langerhans of the pancreas release insulin in response to an increase in blood sugar. Insulin, then, acting on the liver and on other cells of the body, brings about a removal of sugar from the blood. This response can be characterized as follows:

Stimulus (blood sugar)
Sense organ (pancreas)

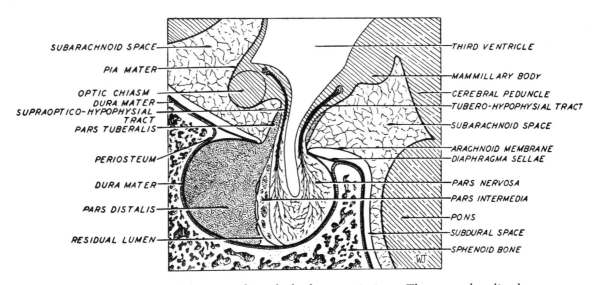

Fig. 28-1. A sagittal section through the human pituitary. The pars tuberalis plus distalis on the left form the anterior lobe and the pars nervosa (processus infundibuli), plus pars intermedia on the right constitute the posterior lobe. (Turner, C. D.: General Endocrinology. 4th ed., p. 110. Philadelphia, W. B. Saunders, 1966)

Endocrine Gland, Hormone, and Chemical Nature of Hormone	Discovered	Purified	Structure Learned
Stomach (polypeptide)			
Gastrin	1902	1964	1964
Intestine (polypeptides)			
Secretin	1902	1961	1964
Cholecystokinin	1928	1962	—
Liver (polypeptide)			
Angiotensin	1939	1955	1956
Kidney			
Prostaglandin E$_2$ (acidic lipid)	—	—	—
Erythropoietin (polypeptide)	1906	1959	—
Renin (polypeptide)	1960	—	—
Pancreas (polypeptides)			
Insulin (beta cells)	1889	1926	1953
Glucagon (alpha cells)	1930	1953	1957
Parathyroid (polypeptide)			
Parathormone	1925	1959	—
Anterior Pituitary (peptides, polypeptides, proteins)			
Follicle-stimulating hormone (glycoprotein)	1926	1931	—
Luteinizing hormone (glycoprotein)	1926	1940	—
Luteotropic hormone or prolactin (peptide)	1929	1933	—
Growth hormone	1921	1948	1966
Thyroid-stimulating hormone	1922	1948	—
Adrenocorticotropic hormone	1924	1948	1956
Intermediate Lobe of Pituitary (polypeptide)			
Melanocyte-stimulating hormone	1922	1950	1956
Posterior Pituitary or Hypothalamus			
Antidiuretic hormone (polypeptide)	1901	1954	1954
Oxytocin (polypeptide)	1901	1954	1954
Melatonin (amine)	1954	1958	1958
Corticotropin releasing factor (polypeptide?)	1955	1955	1962
Hypothalamus			
Luteinizing hormone releasing factor (polypeptide?)	1960	1961	—
Growth hormone releasing factor (polypeptide?)	1959	1965	—
Thyrotropin releasing factor (nonpolypeptide)	1962	1962	—
Follicle-stimulating hormone releasing factor(?)			
Prolactin inhibiting factor(?)			
Intermediate lobe inhibiting factor(?)			

    Endocrine gland (pancreas)
        Hormone (insulin)
            Effector organ (liver, etc.)

Other endocrine responses may also include the central nervous system and associated motor nerves:

    Stimulus (hypotension)
        Sense organ (peripheral baroreceptors) and sensory neurons
            Central nervous system (hypothalamus, etc.)
                Effector neuron (preganglionic sympathetic)

        Endocrine gland (adrenal medulla)
           Hormone (epinephrine)
               Effector organ (arterioles and heart)

Still another type of endocrine response involves the release of a *neurosecretory hormone* (a blood-borne agent released by a neuron in response to stimulation). Neurosecretory responses are divided into first-, second-, and third-order responses. In the first-order response the neurohormone is released into the blood and carried to distant target organs:

    Stimulus (hyperosmotic blood)

Endocrine Gland, Hormone, and Chemical Nature of Hormone	Discovered	Purified	Structure Learned
Thyroid			
Thyrocalcitonin (polypeptide)	1964	1965	—
Thyroxine (amino acid)	1895	1915	1926
Triiodothyronine (amino acid)	1951	1952	1953
Adrenal Medulla (amines)			
Norepinephrine	1948	1904	1904
Epinephrine	1895	1897	1901
Adrenal Cortex (steroids)			
Glucocorticoids	1935	1938	1940
Mineralocorticoids	1927	1953	1954
Androgens	1935	1939	1942
Corpus Luteum of Ovary			
Relaxin (polypeptide)	1929	—	—
Progesterone (steroid)	1925	1934	1934
Ovary (steroid)			
Estrogen	1925	1929	1931
Testis (steroid)			
Testosterone	1911	1931	1935
Prostate and many other organs (acidic lipid)			
Prostaglandins	1930	1956	1966

Table 28-1. The mammalian hormones. Although the effects of castration in man and lower animals were observed and studied for hundreds of years prior to the first century (Aristotle in 300 B.C., for example), it was not until 1891 (Brown-Séquard and d'Arsonval) that it was clearly stated that some organs regularly release into the blood substances which are necessary for development and function. In 1902 Bayliss and Starling reported on such a substance, secretin, and in 1923 Starling suggested that hormonal mechanisms regularly participate in the reactive systems of the body. These suggestions were preceded by some of the following important observations.

1849 — Berthold showed that atrophy of the capon's comb could be prevented by grafting testicular tissue onto the bird's body.
1889 — Von Mering and Minkowski produced diabetes in the dog by pancreatectomy.
1894 — Oliver and Schaffer demonstrated the pressor effects of pituitary extracts.
1895 — Syzmonowicz and Cybulski and Oliver and Schaffer showed the pressor effects of adrenal gland extracts.
1895 — Magnus-Levy reported an increased BMR in hyperthyroidism.
1901 — Magnus and Schaffer discovered the antidiuretic action of a posterior lobe extract.

  Sense organ (osmoreceptors)
    Central nervous system (hypothalamus)
      Neurosecretory hormone (antidiuretic hormone)
        Storage gland (posterior pituitary)
          Effector organ (kidney, etc.)

In a second-order response the neurosecretory hormone travels a short distance and stimulates the release of a second hormone:
  Stimulus (intrinsic activity)
    Central nervous system (hypothalamus)
      Neurosecretory hormone (growth hormone releasing factor)
        Endocrine gland (anterior pituitary)
          Hormone (growth hormone)
            Effector organ (epiphyseal plate, etc.)

In a third-order response the neurosecretory hormone travels a short distance and stimulates the release of a hormone which stimulates the release of another hormone:
  Stimulus (estrogen)
    Central nervous system (hypothalamus)
      Neurosecretory hormone (luteinizing hormone releasing factor)
        Endocrine gland (anterior pituitary)
          Hormone (luteinizing hormone)
            Endocrine gland (ovary)
              Hormone (estrogen)
                Target organ (uterus, etc.)

The Hypothalamus

In four of the five examples cited above we see that the hypothalamus plays an important part in the response of the endocrine system to stimuli. In one case the hypothalamus responds to incoming stimuli by producing a hormone which is picked up by the posterior pituitary, stored, and released into the blood. In other cases it responds to stimuli by releasing various neurosecretory hormones which facilitate (we shall talk about one that inhibits later on) the release of hormones from the anterior pituitary. Some of the hormones from the anterior pituitary, in turn, regulate the activity of other endocrine glands. In other words, the hypothalamus and pituitary are an important link between (1) nervous impulses originating from peripheral and central receptors and various parts of the central nervous system and (2) many of the other endocrine glands of the body:

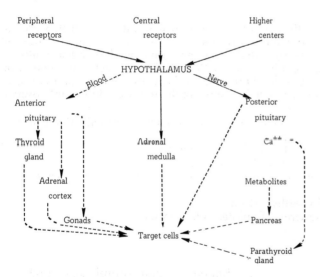

The preceding diagram is an oversimplification but does emphasize the nervous impulses (solid arrow) entering and leaving the hypothalamus and the hormones (dotted arrow) leaving the hypothalamus and the glands it effects.

THE POSTERIOR PITUITARY

The posterior pituitary releases two hormones: (1) antidiuretic hormone (vasopressin, or Pitressin) and (2) oxytocin (Pitocin). These hormones are synthesized in the supraoptic and paraventricular nuclei of the hypothalamus. The supraoptic nucleus contains more antidiuretic hormone (ADH) than oxytocin; in the case of the paraventricular nucleus the reverse is true. Apparently once the hormones are synthesized they are attached to or enclosed by a protein carrier and migrate down the axons to the posterior pituitary, where they are stored or released (Fig. 28-2). Transection or compression of the pituitary stalk in man and other vertebrates, however, prevents this migration and storage and results in an accumulation of granules containing ADH and oxytocin at the hypothalamic side of the lesion.

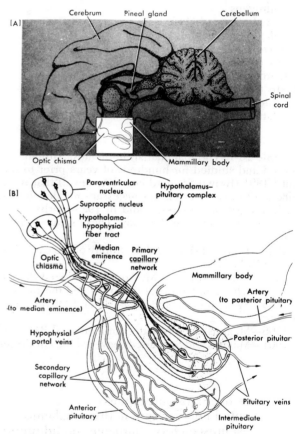

Fig. 28-2. The relationship between the hypothalamus and the pituitary. At A we see a section of the brain and at B an enlargement of the pituitary gland and hypothalamus. Note that the neurons originating in the paraventricular and supraoptic nuclei terminate (1) in the posterior pituitary and (2) near blood vessels in the median eminence of the infundibulum. The blood vessels in the median eminence constitute a part of a hypophyseal (pituitary) portal system, which contains capillaries in the infundibulum which pick up neurosecretions and carry them through veins to a second capillary network in the anterior pituitary. (Frye, B. E.: Hormonal Control in Vertebrates. p. 42. New York, Macmillan, 1967)

Electrical stimulation of these hypothalamic nuclei or the axons leading from them to the posterior pituitary leads to depolarization of the axons and a release of the hormones into the vessels of the posterior lobe. Apparently these nuclei are also closely associated with the thirst area of the hypothalamus since the stimulation of its anterior part results in (1) decreased production of urine (ADH), (2) ejection of milk (oxytocin), and (3) polydipsia (frequent drinking due to great thirst).

The Antidiuretic Hormone

In 1894 Oliver and Schaffer noted that extracts of the pituitary increased arterial blood pressure, and in 1897 Howell localized this effect to the posterior pituitary. The hormone producing these effects was called vasopressin and in 1901 was reported by Magnus and Schaffer to have an antidiuretic effect as well. By 1913 it was being used to treat diabetes insipidus and by 1928 it was shown to be a substance different from oxytocin, the other posterior pituitary hormone.

At present, ADH (vasopressin) is used in man both to treat diabetes insipidus and to elevate arterial pressure. Under physiological conditions, the amount of ADH released by the pituitary plays little or no part in the control of arterial pressure but has an essential role in the control of urine volume. This role was more fully discussed in Chapter 21. Remember simply that the quantity of ADH released into the blood depends upon the impulses impinging upon the hypothalamus. Osmoreceptors, volume receptors, and pain, all through their effect on the hypothalamus, modify the amount of ADH released. The half-life of this ADH in the blood of man is between 10 and 20 min., the kidneys and liver being the most important organs in its destruction and extraction.

Oxytocin

Oxytocin, like AHD, is produced in the hypothalamus and released by the posterior pituitary. In man it has one hundredth the vasopressor activity of ADH and is one hundred times more effective in facilitating milk ejection. Both oxytocin and ADH also facilitate the contractions of the uterus in the final stages of pregnancy and diminish urine production (ml./min.)—oxytocin being many times more potent than ADH in the case of uterine contractions, and ADH, considerably more potent in its antidiuretic effect than oxytocin. On the other hand, ADH has more effect on uterine contractions in the nonpregnant uterus than does oxytocin (see p. 155). Gaiton, Cobo, and others (1965, 1964), working with women during the last weeks of their pregnancy or soon after delivery, have provided evidence of the independent release and functioning of these two hormones. They noted that suckling induced pronounced oxytocin-like and milk-ejecting activities but negligible antidiuresis, and that hypertonic saline produced marked antidiuresis but negligible oxytocin-like or milk-ejecting activities.

The role played by oxytocin in man is not clear. As far as we know, oxytocin has no function in the male or in the female who has not been pregnant. Nor has there yet been a demonstration that it is essential in either labor or lactation, though it is sometimes injected to initiate labor or facilitate difficult lactation. Patients suffering from diabetes insipidus (impaired release of ADH), for example, generally have a normal labor and period of lactation. Although the evidence is far from definitive, oxytocin is generally regarded as playing a role in labor and lactation. One suggestion is that labor begins in response to a rapid decrease in the output of estrogen and progesterone from the placenta. The resulting uterine contractions stimulate sensory neurons to the hypothalamus, which in turn facilitate the release of oxytocin, which futher facilitates the contractions of the myometrium. Stimulation of the nipple of the breast by suckling may also signal oxytocin release, by stimulating sensory fibers to the hypothalamus. This occurs after the birth of a child and therefore at a time when the uterus is insensitive to oxytocin. Under these circumstances the effect of the oxytocin is solely on the myoepithelial cells of the mammary gland, in which it facilitates the flow of milk.

THE ANTERIOR PITUITARY

The anterior pituitary releases six hormones, possibly more. These include (1) growth hormone (GH, somatotropic hormone, or STH), (2) thyroid-stimulating hormone (TSH, or thyrotropin), (3) adrenocorticotropic hormone (ACTH, or corticotropin), (4) follicle-stimulating hormone (FSH), (5) luteinizing hormone (LH, interstitial cell-stimulating hormone, or ICSH), and (6) luteotropic hormone (LTH, prolactin, or lactogenic hormone). On the basis of numerous animal experiments it is generally agreed that each of these hormones is produced and secreted by a different cell type in the anterior pituitary. The association of acidophilic tumors of the human pituitary with abnormal growth (acromegaly and giantism) has clearly linked acidophilic cells with the production of growth hormone. Follicle-stimulating hormone (FSH) and luteinizing hormone (LH), for example, are secreted by basophils, the FSH releasing cells being large and rounded with spherical secretory granules and the LH-releasing cells being small with scanty cytoplasm.

Control of the release of these hormones is exerted primarily by the hypothalamus. If, for example, the pituitary is removed from its normal location and transplanted to another portion of the body, the proudction of some of its hormones drops to 10 per cent of normal and its capacity to produce hormones in response to environmental changes is lost. Lesions in, or stimulation of various parts of the hypothalamus also markedly change anterior pituitary function (Fig. 28-3). We find, for example, that lesions in the hypothalamus can block the release of GH, TSH, ACTH, FSH, and LH, as well as increase the release of prolactin.

Apparently the control of the anterior pituitary is exerted at least in part by the release of neurosecretions called releasing factors from axons in the median eminence of the infundibulum. These releasing factors (1) are picked up by a primary capillary network in the median eminence of the infundibulum, (2) are carried by hypophyseal portal veins to the anterior pituitary, and (3) diffuse into the interstitial spaces by means of a secondary capillary network (Fig. 28-2). In other words, this hypophyseal portal system allows the neurosecretions of the median eminence to arrive at the anterior pituitary at a fairly high concentration before being diluted in the blood of the great veins and heart.

To date five releasing and two release-inhibiting hormones (neurosecretions) have been tentatively isolated. These include growth hormone releasing factor (GRF, or SRF), thyrotropin releasing factor (TRF), corticotropin releasing factor (CRF), follicle-stimulating hormone releasing factor (FRF), luteinizing hormone releasing factor (LRF), prolactin inhibiting factor (PIF), and melanocyte-stimulating hormone inhibiting factor (MSH-IF).

Growth Hormone

Growth hormone is a polypeptide (Fig. 28-4) that facilitates protein anabolism, fat catabolism, and an increase in the length of one's long bones

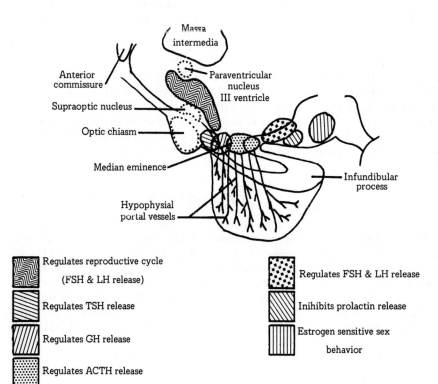

Fig. 28-3. Generalized diagram of certain areas in the hypothalamus which have been implicated as important in the control of the anterior pituitary. The control of sex behavior and the release of FSH and LH show considerable interspecies variation. (Redrawn from Metabolic and Endocrine Physiology, Second Edition, by J. Tepperman. Copyright © 1968, Year Book Medical Publishers, Inc. Used by permission. Major sources: Sawyer, C. H., and Kawakami, M., in Villee, C. A. (ed.): Control of Ovulation. New York, Pergamon Press, 1961; Ganong, W. F., in Gorbman, A. (ed.): Comparative Endocrinology. New York, John Wiley & Sons, 1959; Harris, G. W., ibid; and personal communication from C. H. Sawyer, and Seymour Reichlin for suggesting revisions of the diagram based, in part, on his recent work)

prior to puberty. It reaches its highest concentration in the plasma of the neonatal child and progressively decreases in concentration until about the second year of life, at which time its resting level stabilizes between 0 and 5 mμg. per milliliter of plasma. Hypoglycemia, however, results in an increased concentration of GH (Fig. 28-5).

The process of growth is a complex phenomenon involving the enlargement of some cells (hypertrophy) and reproduction of others (hyperplasia). It includes enlargement of the bones, a laying down of protein, and a change in body proportions. In a newborn child, for example, the head represents 25 per cent of the body length, whereas in the adult the head is only 13 per cent of the body length. In females, growth eventually involves the development of breasts, widening of the hips, and a characteristic distribution of fat, whereas in the male there is broadening of the shoulders and an increase in muscle mass. These changes are only in part the result of growth hormone. Thyroxine from the thyroid gland, insulin from the pancreas, and androgen from the testes, ovaries, and adrenal cortex also play essential roles. You will note in Figure 28-6, for example, two periods of rapid growth: the first at a time when the growth hormone is at its highest concentration in the blood, and the second when the sex hormones are being released in increasing concentrations at the onset of puberty.

Nutritive intake and a person's health also affect growth. We find that illness or injury results in a decrease in protein anabolism, but that following an illness children sometimes show a growth spurt during which the growth rate may increase by as much as 400 per cent.

Hypopituitarism. Hypopituitarism may result from accidents, infarction (destruction of tissue due to the occlusion of an artery), infections, and noninfectious granulomas. Destruction of the pituitary is also sometimes performed in the treatment of disseminated carcinoma or diabetes with degenerative vascular complications. If hypopituitarism is severe and remains untreated it can result in death of the person. It is associated with slowing of growth, a decreased capacity of organs to hypertrophy in response to stress, malfunction of the temperature regulating mechanism, increased susceptibility to infection, adrenal cortical atrophy and insufficiency, thyroid atrophy and hypothyroidism, and gonadal atrophy and sterility.

When pituitary malfunction is associated primarily or solely with a decrease in the release

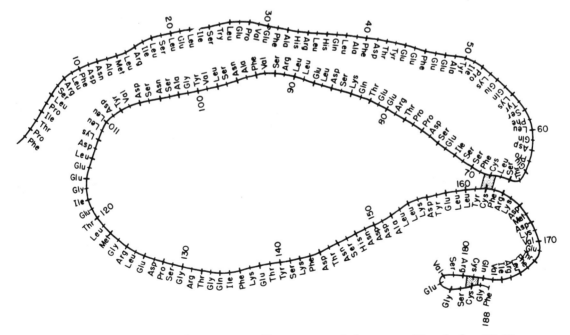

Fig. 28-4. Amino acid sequence of human growth hormone. (Daughaday, W. H.: The Adenohypophysis. *In* Williams, R. H. (ed.): Textbook of Endocrinology. 4th ed., p. 46. Philadelphia, W. B. Saunders, 1968; *after* Li, C. H., et al.: J. Am. Chem. Soc., 88:2050, 1966)

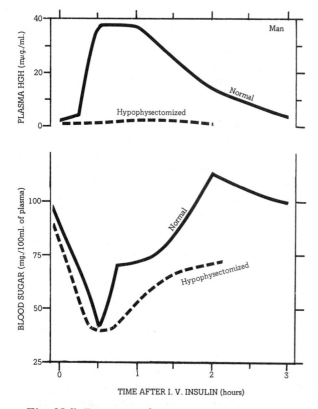

Fig. 28-5. Response of a normal and a hypophysectomized man to insulin (0.1 units/kg. of body weight). (*Modified from* Roth, J. S., et al.: Hypoglycemia, a potent stimulus to secretion of growth hormone. Science, *140*:987; copyright © May, 1963 by the American Association for the Advancement of Science.)

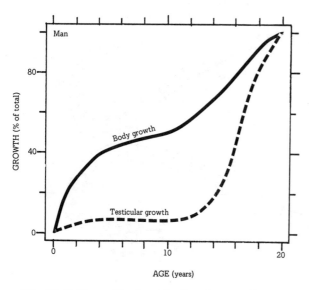

Fig. 28-6. Normal body and testicular growth in man. (*Modified from* Harris, J. A., et al.: The Measurement of Man. p. 193. Minneapolis, University of Minnesota Press, 1930)

of growth hormone (acidophilic tumors of the pituitary, for example), the symptoms are less severe. Fetal growth is normal, but from the age of one year physical growth may be less than one half the normal rate (Fig. 28-7). The child exhibits good health but may have large amounts of fat over the iliac crest and lower abdomen. About 10 per cent have hypoglycemia, and most show low fasting blood sugar levels. They exhibit normal intelligence but seem rather smart

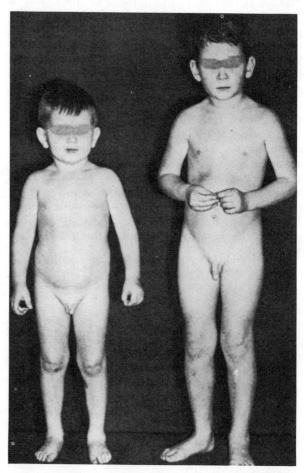

Fig. 28-7. A comparison of a boy with panhypopituitarism (*left*, 5 years old) and a normal boy (*right*, 4.3 years old). The boy with panhypopituitarism had organic brain disease associated with small genitalia, a low PBI (protein-bound iodine level of 2.6 μg./100 ml. of plasma), adrenal insufficiency (low 17-ketosteroid excretion and a low serum Na^+), a low fasting blood sugar (20 mg./100 ml.), and a bone age of 20 months. The body was treated with cortisone and desiccated thyroid. (Daughaday, W. H.: The Adenohypophysis. *In* Williams, R. H. (ed.): Textbook of Endocrinology. 4th ed., p. 64. Philadelphia, W. B. Saunders, 1968)

for their size. They go through a fairly normal puberty and become fertile, but continue growing until they reach 30 or 40 years of age, at which time they obtain a height of 4 to 5 ft. Injections of human GH as small as 5 mg. per day have been quite successful in treating this condition, producing increases in height of 3 to 5 in. during the first year of treatment.

Human GH is presently extracted from cadaver pituitaries, and its use is restricted to clinical investigation only. In some patients a high antibody titer develops toward the hormone; its use at that time must be discontinued for several months or even indefinitely. Androgens (testosterone, for example) have also been used, but less successfully, to treat growth failure due to too little GH. In low concentrations in patients over 15 yrs. of age they produce a growth spurt that need not be associated with abnormal virilization. Androgens do not increase the height a person will eventually obtain, but merely help him obtain it sooner.

Hypersecretion of growth hormone. If a hypersecretion of growth hormone develops before closure of the epiphyseal (cartilaginous) plates of the long bones, it produces marked increases in proliferative activity at the plate (Fig. 28-8) and, as a consequence, an increase in height. This is associated with increased protein content in the body and a series of symptoms characterized as giantism (Fig. 28-9). If, however, the hypersecretion develops after most growth is complete, a disproportionate growth occurs (acromegaly). This includes excessive acral (Gk. *akra*, extremity) and soft tissue growth, as well as enlargement of the sella turcica (93% of cases), headache (87%), visual impairment (62%), increased metabolism (70%), an increase in body water (60%), an increase in body hair (53%), and enlargement and protrusion of the lower jaw. Weight gain, however, is noted in only about 39 per cent of such cases.

Thyroid-Stimulating Hormone

Thyroid-stimulating hormone is a polypeptide produced and released by the anterior pituitary gland. It acts on the thyroid gland to facilitate

Fig. 28-8. Effect of growth hormone on the proximal epiphyseal plate of the tibia of a hypophysectomized rat. At the left (A) we note the thickness of the epiphyseal plate (*arrow*) in the untreated rat and at the right (B) that of the rat treated with growth hormone. The long bones increase in length in the child by a proliferation of the plate which is facilitated by GH. An excessive widening of this plate results in a weakening of the bone that may result in a slipped epiphysis (a breaking away of the epiphysis from the shaft of the bone). (Evans, H. M., et al.: Bioassay of the pituitary growth hormone. Width of the proximal epiphyseal cartilage of the tibia in hypophysectomized rats. Endocrinology, 32:14, 1943)

its growth and its production and release of thyroxine (T_4) and triiodothyronine (T_3). These two hormones (T_4 and T_3) are in turn essential for normal growth, development, and metabolic activity. They also act directly on the anterior pituitary to decrease its production and release of TSH. This relationship between the anterior pituitary and thyroid is called negative feedback and can be diagrammed as follows:

ANTERIOR PITUITARY
↓
TSH
↓ +
Thyroid growth
↓ +
Production of T_4 and T_3
↓ +
Release of T_4 and T_3

In other words, an excessive output of TSH can result in hyperthyroid (increased output of T_4 and T_3) goiter (enlarged thyroid gland). A decrease in the output of T_4 and T_3 can result in the release of increased quantities of TSH and as a consequence either a hypothyroid goiter or a euthyroid (release of normal concentrations of T_4 and T_3) goiter. In the case of the euthyroid goiter the enlargement of the thyroid compensates for the initial deficiency of the gland. Normally TSH exists in the plasma of man at a concentration of less than 4 mμg. per milliliter, but in primary hypothyroidism its concentration may range from 5 to 110 mμg. per milliliter. In other words, when the thyroid releases subnormal quantities of its hormones, the plasma concentration of TSH increases.

Another factor important in the control of the release of TSH is the hypothalamus. Here lie the heat control centers which apparently act on that part of the hypothalamus concerned with the control of the release of the thyrotropin releasing factor (TRF). Apparently in response to prolonged exposure to cold there is an increase in the release of TRF that facilitates an increase in the release of TSH. If the anterior pituitary is removed from the sella turcica and transplanted

Fig. 28-9. A pituitary giant. Robert Wadlow at birth weighed 9 pounds, at 6 months 30 pounds, and at 1 year 62 pounds. After 2 years of age, growth continued at constant rate until he died at 22 years of age from cellulitis of the feet. At this time he was 8 feet, 11 inches tall and weighed 475 pounds. (Williams, R. H. (ed.): Textbook of Endocrinology. 4th ed., p. 75. Philadelphia, W. B. Saunders, 1968; courtesy of Drs. C. M. Charles and C. M. MacBryde)

elsewhere in the body, T_4 and T_3 continue to depress the release of TSH but the quantity of TSH released per day decreases, as well as the basal metabolic rate, and the person no longer responds to a cold environment with an increase in the release of TSH. In other words, the hypothalamus — through the release of TRF — exerts both a tonic control over the release of TSH and a reflex control in response to its sensory input (Fig. 28-11).

Adrenocorticotropic Hormone

Adrenocorticotropic hormone (ACTH) is a polypeptide produced and released by the anterior pituitary. It facilitates the growth of the adrenal cortex and its production and release of hormones. These hormones include a number of steroids which (1) facilitate protein catabolism and glucose and glycogen synthesis (glucocorticoids), (2) facilitate Na^+ retention in the body

Fig. 28-10. Progressive changes in acromegaly. At A the subject is 9 years old and shows no symptoms of pathology. At B she is 16 years old and may have started to show some early coarsening of her features due to the disease. At C she is 33 and at D 52 years of age and shows that gross disfigurement associated with well established acromegaly. Note the enlargement of the hands and lower jaw and the bags under the eye (one sign of edema). (Mendeloff, A. I., and Smith, D. E. (eds.): Acromegaly, diabetes, hypermetabolism, proteinuria and heart failure. Clinicopathological conference. Am. J. Med., 20:135, 1956)

(mineralocorticoids), and (3) are associated with the development of the secondary sexual characteristics (estrogens and androgens). The plasma concentration of ACTH shows a diurnal rhythm, its lowest concentration during the day being well below 0.1 milliunit/100 ml. of plasma, and its highest resting concentration being about 0.4 milliunit/100 ml. During stress (exercise, surgery, an accident), however, it may reach a level of 2 milliunits/100 ml., and during Cushing's disease it has been reported to go as high as 400 milliunits/100 ml.

Hypophysectomy results in a marked decrease in the resting plasma levels of the glucocorticoids but little or no change in the level of the mineralocorticoids. Increases in the release of mineralocorticoids (aldosterone, for example) continue to occur in response to decreased concentrations of plasma Na^+, increases in K^+, and the production of angiotensin (Fig. 21-9). The individual, however, no longer responds to stress with an increase in the output of glucocorticoids or mineralocorticoids. Apparently the function and control of the release of ACTH are similar to those for TSH. The ACTH is responsible for maintaining a resting level of glucocorticoid in the body, just as TSH maintains a plasma concentration of T_4 and T_3, but at resting con-

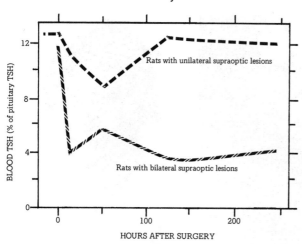

Fig. 28-11. Response of the rat to unilateral and bilateral lesions of the supraoptic nucleus. Determinations were made 8, 48, 120, and 240 hours after surgery. (*Redrawn from* Panda, J. N., and Turner, C. W.: Hypothalamic control of thyrotrophin secretion. J. Physiol., *192*:4, 1967)

centrations it has little or no effect on the release of mineralocorticoids (Fig. 28-12). The glucocorticoids, in turn, through a negative feedback, inhibit the further release of ACTH by the pituitary. In response to trauma, highly emotional experiences, or other forms of stress, however, there is a nervous input into the hypothalamus that results in the release of corticotropin releasing factor (CRF) and a sufficient increase in the level of ACTH in the blood to increase the plasma concentration of both glucocorticoids and mineralocorticoids.

In other words, under resting conditions ACTH probably functions as follows:

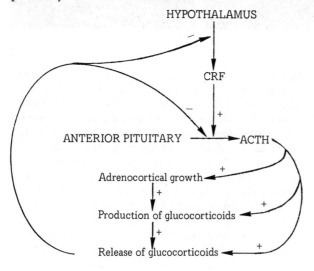

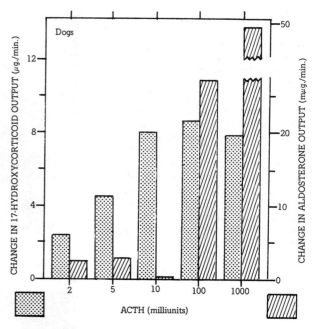

Fig. 28-12. Change in the adrenal output of 17-hydroxycorticoid (the glucocorticoid, cortisol) and aldosterone (mineralocorticoid) in response to ACTH in the nephrectomized hypophysectomized dog. Note that 5 milliunits of ACTH result in the release of 4.7 micrograms of 17-hydroxycorticoid and only 2.5 millimicrograms of aldosterone.

Nervous impulses from the limbic system, under conditions of emotional stress, and from the reticular activating system, in response to trauma, act on the hypothalamus to increase its output of corticotropin releasing factor.

The Gonadotropins

Men and women release two polypeptides from the anterior pituitary that affect solely the gonads (ovaries and testes). These hormones are follicle-stimulating hormone (FSH) and luteinizing hormone (LH). The release of FSH and LH is at least in part facilitated by two neurosecretions from the hypothalamus—follicle-stimulating hormone releasing factor (FRF) in the case of FSH, and luteinizing hormone releasing factor (LRF) in the case of LH. Secretions of the gonads (estrogens, androgens, and possibly others) also act on the hypothalamic-pituitary axis to modify the secretion of the pituitary gonadotropins.

Neither FSH nor LH can be demonstrated in the pituitary, plasma, or urine of infants, and

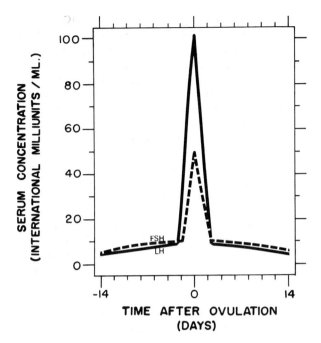

Fig. 28-13. Changes in the concentration of serum FSH and LH during the menstrual cycle of a normal woman. (*Redrawn from* Lloyd, C. W.: The ovaries. *In* Williams, R. H. (ed.): Textbook of Endocrinology. 4th ed., p. 473. Philadelphia, W. B. Saunders, 1968)

mones estrogen and progesterone. This can be summarized as follows:

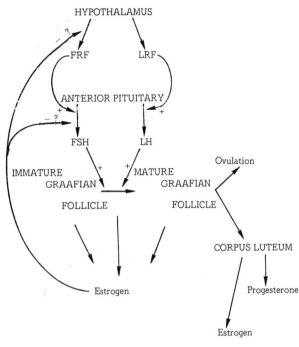

In the male, FSH acts on the germinal epithelium of the seminiferous tubules of the testis to bring about its development and to stimulate its production of spermatozoa. Selective loss of the function of the seminiferous tubules or castration results in an increase in the FSH concentration in the plasma. It has been speculated that the seminiferous tubules release some chemical which inhibits the further release of FSH by the pituitary, but to date the evidence for this is poor.

LH in the male acts on the Leydig cells of the testis to increase their output of androgens and estrogen. The androgens released by the testis are testosterone, delta-4-androstenedione, and dehydroepiandrosterone. Testosterone is the most potent of these steroids. It is responsible for the secondary sexual characteristics of the male (accelerated protein anabolism, enlargement of the larynx, bone structure, hair distribution, etc.) and apparently inhibits the release of LH from the pituitary. Over 95 per cent of the testosterone in the body of a normal male is produced in the testes, the rest coming from the adrenal cortex. Apparently LH has little or no effect on the amount of testosterone or other

their concentration in the prepuberty child is little or none. Their concentration at and after puberty reaches a high level, however. In the adult man they maintain a fairly constant level throughout his life, but in the adult female their concentration varies with her period in the menstrual cycle (Fig. 28-13). At the end of her reproductive life (postmenopause), her plasma FSH changes from about 20 international milliunits per milliliter to 200 milliunits per milliliter. Her plasma concentration of LH changes from 15 international milliunits per milliliter to 60 milliunits.

In the female, FSH acts on the ovary to bring about the maturation of one or more of its graafian follicles. In this process the graafian follicle releases estrogen, which inhibits the further release of FSH and facilitates the release of LH. The LH brings about further maturation and results in the rupture of the graafian follicle and an associated extrusion of a part of the follicle containing the ovum (ovulation). The part of the follicle that remains reorganizes to form a corpus luteum (luteinization), which releases the hor-

androgens produced by the adrenal cortex. These facts can be summarized as follows:

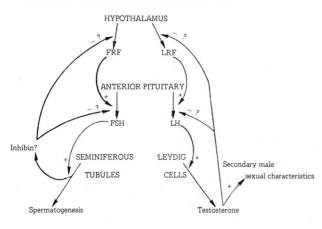

Prolactin

Prolactin is a hormone released by the anterior pituitary of many animals and probably also of man. Unfortunately, extracts from the human pituitary with prolactin activity have up to now been contaminated with growth hormone. Thus much of what we know about the physiology of prolactin in man is based on conclusions drawn from animal experiments, the injection of non-human prolactin into man, and the use of impure human prolactin. In some animals (the rat, for example) prolactin is important for the maintenance of the secretory activity of the corpus luteum and for this reason is frequently called the luteotropic hormone (LTH), but in ungulates, rabbits, guinea pigs, and probably man it is probably not important for corpus luteum function.

In man and other mammals prolactin acts on the mammary glands of the female to initiate and sustain milk production in the mother after the birth of her child. In the male it may facilitate growth of the prostate and facilitate the action of testosterone. Before prolactin can be effective in facilitating milk production, however, the mammary glands must go through a series of changes. Estrogen and progesterone released from the ovary during and after puberty facilitate proliferation of the glands and their alveoli. Growth hormone, thyroid hormone, and corticosteroids from the adrenal cortex further facilitate these changes, but apparently in man the GH is not essential for mammary development. The final step in the maturation of the mammary glands is brought about by hormones secreted by the placenta during pregnancy. These include estrogens, progesterones, and gonadotropins similar to those released by the anterior pituitary. The importance of the placenta in this regard is probably not the uniqueness of its hormonal output, but rather its quantity.

The mammary glands, having been exposed to these hormones, are now ready to produce milk. The final stimulus in this process is suckling. Tactile stimulation of the nipples of the breast stimulates sensory nerves which, acting on the hypothalamus, inhibit the release of prolactin inhibiting factor (PIF). This inhibition results in the release of prolactin, which facilitates the secretion of milk. Apparently in man this is all that is necessary though in some animals the release of oxytocin from the posterior pituitary is also important. In these animals as well as in man, suckling, again, probably elicits the release of oxytocin. The oxytocin causes a contraction of the myoepithelial cells of the mammary gland which forces the milk into the collecting ducts and cisternae of the breasts.

The preceding observations can be summarized as follows:

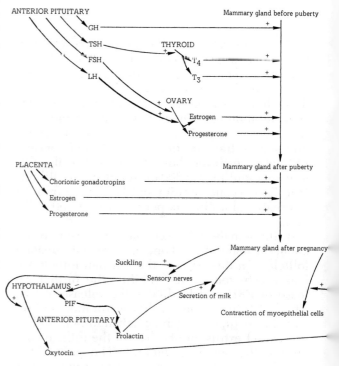

Chapter 29

THE THYROID GLAND

FORMATION OF
TRIIODOTHYRONINE
AND THYROXINE
SECRETION
PLASMA TRANSPORT
FUNCTIONS OF THE
THYROID HORMONES
CLINICAL CORRELATES

The thyroid gland, like the anterior pituitary, adrenal cortex, pancreas, and gonads, releases hormones that are important in regulating body metabolism, growth, and development. The important hormones released by the thyroid include (1) thyroxine, (2) triiodothyronine, and possibly (3) thyrocalcitonin. The thyrocalcitonin is thought to be released by the thyroid gland in response to an increase in the plasma Ca^{++} concentration and apparently acts on bone to facilitate its uptake of Ca^{++} from the plasma. We shall discuss thyrocalcitonin more fully when we consider the parathyroid gland and its role in Ca^{++} metabolism. We will be concerned primarily in this chapter with thyroxine and triiodothyronine.

FORMATION OF TRIIODOTHYRONINE AND THYROXINE

The Iodide Pump

Triiodothyronine (T_3) and thyroxine (T_4) are both iodinated hormones produced in the thyroid gland. One of the steps in their formation is the movement of I^- from the blood plasma into the thyroid cell (Fig. 29-1). This movement is against both a concentration gradient of 25 to 1 and an electrical gradient of -50 mv. (i.e., the cytoplasm is negative with respect to the intercellular fluid). The biochemical reactions responsible for the movement of I^- against this electrochemical gradient are called the iodide pump. Removal of the anterior pituitary reduces by about 5 times the capacity of the pump to maintain a concentration gradient. The subsequent injection of thyroid stimulating hormone (TSH), on the other hand, can return the concentrating capacity of the pump to normal or even twice normal. In other words, the pump continues to function in the absence of TSH but is dependent upon TSH release for normal activity.

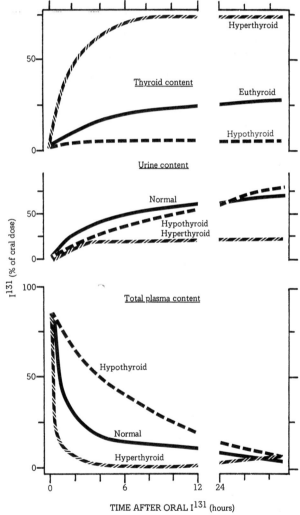

Fig. 29-1. The fate of an oral dose of I^{131} in man. Note that in the hyperthyroid the uptake of I^{131} by the thyroid gland is more marked than in the euthyroid individual. As a result the I^{131} content of the urine and plasma of the hyperthyroid rapidly becomes lower than that found in the normal subject. (*Modified from* Williams, R. H. (ed.): Textbook of Endocrinology. 4th ed., p. 144. Philadelphia, W. B. Saunders, 1968)

Iodination of the Tyrosyl Radical of Thyroglobulin

The thyroid gland contains numerous cells organized around a fluid-filled lumen. The vesicular structure that results from this organization is called a thyroid follicle and has an average diameter of 200 μ. The clear mucoid fluid of varying viscosity in the lumen of the follicle, known as thyroid colloid, has as its major constituent the glycoprotein thyroglobulin. For the most part the thyroglobulin is produced in the cells surrounding the colloid. Here also I^- is converted to I_2, and I_2 combines with the tyrosyl radicals of thyroglobulin to form monoiodotyrosine (MIT) and diiodotyrosine (DIT) radicals. Some of these radicals are then changed to triiodothyronine (T_3) and tetraiodothyronine (T_4, or thyroxine) radicals:

[Chemical reaction diagram showing conversion of Tyrosine → 3-monoiodotyrosine (MIT) → 3,5-diiodotyrosine (DIT), with coupling to form 3,5,3'-triiodothyronine (T_3) and Thyroxine (T_4)]

Approximately 90 per cent of the organic iodine in the human thyroid is in combination with thyroglobulin (molecular weight of about 680,000). Normally 17 to 28 per cent of this is in the form of MIT; 24 to 42 per cent, DIT; 5 to 8 per cent, T_3; and 35 per cent, T_4. The rate of organification of I^- is under the control of TSH.

We find, for example, that TSH increases both the T_4-to-T_3 and the DIT-to-MIT ratios.

SECRETION

The thyroid differs from other endocrine glands in that it stores a great deal of readily available hormone. Apparently in the other glands, secretion of a hormone in response to an appropriate stimulus must be associated with either a prior or a concomitant synthesis of that hormone. In the case of the thyroid gland, proteolytic enzymes stored in the gland are released in response to TSH. As a result, thyroglobulin releases MIT, DIT, T_3, and T_4 into the follicular fluid. The MIT and DIT do not normally move into the blood, but are acted upon by dehalogenases in the gland and therefore serve as a readily available source of iodine and tyrosine in the formation of additional T_3 and T_4. The T_3 and T_4 molecules, on the other hand, are resistant to the dehalogenases and diffuse through the follicular cells and into the blood. Since this process causes the T_4 molecule to be approximately 100 times more concentrated in the thyroid follicle than in the plasma, its efflux is extensive.

PLASMA TRANSPORT

Once T_4 gets into the plasma it is rapidly bound to the plasma proteins: (1) thyroxine-binding globulin (TBG), (2) thyroxine-binding prealbumin (TBPA), and (3) albumin. Approximately 99.96 per cent of the thyroxine in the plasma is bound to these proteins and 0.04 per cent (6×10^{-11} mols of free thyroxine/liter of plasma) is free in the plasma. That which is bound is not metabolically active. apparently does not affect the secretion of TSH, and for the most part stays in the bloodstream. In other words, the bound thyroxine serves as a reservoir to replace any free thyroxine that is removed from the blood. The binding also serves to decrease the destruction of thyroxine by the tissues and its elimination in the urine.

Approximately 50 per cent of plasma T_4 is bound to TBG (200 μg. of thyroxine/liter of plasma), 40 per cent of TBPA, and 10 per cent to albumin. On the other hand, 99.6 per cent of the blood T_3 is bound to TBG and 0.4 per cent is found free in the plasma. The relative concentrations of free T_3 and T_4 in the plasma are not known, but the evidence is consistent with the

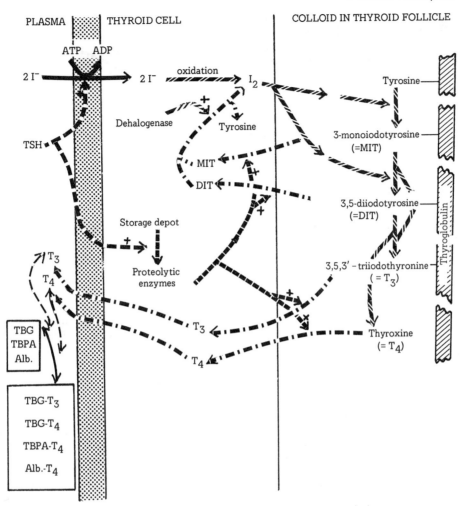

Fig. 29-2. Formation and secretion of iodinated compounds in the thyroid.

view that both the bound and the free T_3 are at a lower blood concentration than is free T_4. These relationships are shown in Figure 29-2.

Protein-Bound Iodine

Since the thyroid gland is the major secretor of iodinated compounds in the body, and since most of these compounds are bound to plasma proteins, it has become common practice for workers studying thyroid function to determine the protein-bound iodine in the plasma (PBI). In the normal subject this is about 5.8 μg./100 ml. of plasma (range, 3.5-8.0). In hypothyroidism it is lower (2.0-3.6), and in hyperthyroidism it is elevated (9.4-18.9). Thyroid function may also be evaluated by determining the butanol-extractable iodine (BEI) in the plasma. This gives the iodine in plasma thyroxine but not in MIT, DIT, and thyroglobulin. Its value in a normal person is 4.0 μg./100 ml. of plasma (range, 2.2-5.6).

It should be emphasized, however, that these tests are not always a reliable indication of thyroid function. We find, for example, that (1) T_3, (2) androgens, (3) diphenylhydantoin (Dilantin), (4) salicylates, and (5) dinitrophenol can produce a decrease in the PBI not associated with any other signs of hypothyroidism. We have noted that a much greater percentage of plasma T_3 is in the free form than is thyroxine. By inhibiting the release of TSH from the anterior pituitary, in other words, concentrations of T_3 sufficient to maintain normal body metabolism tend to bring down the release of T_4 and T_3, and as a consequence the

PBI. Androgens act in a different way. They decrease the PBI by decreasing the concentration of thyroxine-binding globulin (TBG). Dilantin, on the other hand, competes with thyroxine for sites on the TBG molecule and in this way lowers the PBI without producing any other symptoms of hypothyroidism. Binding of iodide by TBPA is depressed by salicylates and dinitrophenol. Gold therapy and the use of mercurials, on the other hand, produce chemical interference with the PBI test.

Factors other than hyperthyroidism that increase the PBI are (1) pregnancy, (2) estrogenic hormones, including contraceptive agents, (3) acute hepatocellular disease, (4) acute intermittent porphyria, and (5) idiopathic increases in TBG. In each of these cases there is usually an elevation in the TBG that increases the reservoir of plasma thyroxine without markedly affecting the amount of free thyroxine.

FUNCTIONS OF THE THYROID HORMONES

The thyroid hormones increase O_2 consumption in a number of the body's cells and facilitate growth and development. The mechanisms by which they do this, however, are not well understood. We do know that there is a latent period of several hours from the time the thyroxine is injected until the O_2 consumption begins to rise, and that this rise lasts for over 6 days (Fig. 29-5). We also know that thyroxine prior to producing an increase in O_2 consumption produces (1) increased turnover of nuclear and cytoplasmic ribonucleic acid (RNA), (2) increased incorporation of amino acids into protein by mitochondria (Fig. 29-3), and (3) increased incorporation of amino acids and RNA into ribosomes. Inhibitors of protein synthesis, on the other hand, prevent the rise in O_2 consumption in response to T_4. In other words, the mode of action of T_4 may be through its facilitation of protein (enzyme) anabolism, possibly by an action directly on the genes. Other suggestions concerning its mode of action are that thyroxine (1) influences enzyme activity by coupling with certain divalent ions (Cu^{++}, Mg^{++}, and Mn^{++}), (2) uncouples oxidative phosphorylation and in this way decreases the number of high-energy phosphate compounds formed, while simultaneously increasing the oxidation of foods, and (3) acts directly on the oxidative enzymes and their formation.

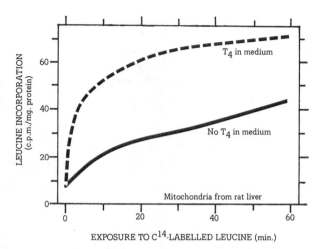

Fig. 29-3. Rate of mitochondrial uptake of C^{14}-labeled leucine. The mitochondria were prepared from rat liver and were free of soluble RNA and ribosomes. Note that the mitochondria exposed to 4.5×10^{-5} M. L-thyroxine incorporated the amino acid, leucine, much more rapidly than the mitochondria not exposed to thyroxine. (*Modified from* Buchanan, J., and Tapley, D. F.: Stimulation by thyroxine of amino acid incorporation into mitochondria. Endocrinology, 79:83, 1966)

Calorigenic Action

Both thyroxine (T_4) and triiodothyronine (T_3) produce an increase in O_2 consumption in the body and an associated increase in body temperature. You will note in Figure 29-4 that the injection of 10 μg. of T_3 per kilogram of body weight per day produces an increase in O_2 consumption of 38 ml./100 gm. per day in thyroidectomized rats, whereas the same dose of thyroxine produces an increase of only 10 ml./100 gm. per day. In the case of athyroid humans, the dose of T_3 required to maintain a normal metabolism is 0.05-0.1 mg. per day, and the dose of T_4, 0.1-0.4 mg. per day. The higher potency of T_3 is apparently due to (1) the greater sensitivity of cells to T_3 and (2) the fact that a greater percentage of the T_3 remains unbound than is the case with T_4.

You will note in Figure 29-5 that all the tissues of the body do not respond to T_4 to the same degree. Some organs markedly increase their O_2 consumption in response to T_4 (heart, liver, and kidney), whereas others show little or no change in O_2 consumption. In structures with increased O_2 consumption there are, in addition, some changes secondary to the increased metabolic rate. Thyroxine has been used, for example, to

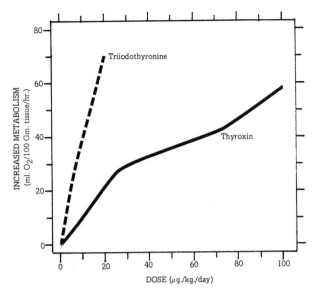

Fig. 29-4. Response of thyroidectomized rats to subcutaneous injections of thyroxine and triiodothyronine. (*Redrawn from* Barker, S. B.: Peripheral actions of thyroid hormones. Fed. Proc., *21*:637, 1962)

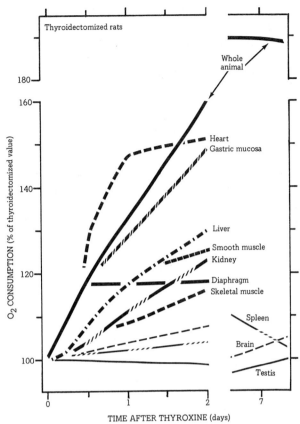

Fig. 29-5. Response of the thyroidectomized rat to a single dose of thyroxine. The various tissues were studied prior to thyroxine and 0.25, 0.5, 1, 4, 6, and 8 days after the injection. Note that most but not all of the tissues increased their O_2 consumption in response to T_4. (*Data from* Barker, S. B., and Klitgaard, H. M.: Metabolism of tissues excised from thyroxine-injected rats. Am. J. Physiol., *170*:84, 85, 1952)

increase milk production in the cow. Its general effect on the body is to increase the body's need for vitamins and other nutrients. If this need is not met by an appropriate increase in nutrient intake, there is frequently a vitamin deficiency and a general depletion of the body's fat and protein. The depletion of the protein in muscle is associated with creatinuria and muscle weakness. Its depletion in bone results in a demineralization of the bone (osteoporosis) and associated hypercalcemia and hypercalciuria.

Carbohydrate Metabolism

One of the other actions of T_4 on the body is an increase in the rate of absorption of carbohydrate from the intestinal tract. In a hyperthyroid person this absorption may be rapid enough to raise the glucose level in the blood above the renal threshold. This does not, however, result in an increase in liver glycogen, since there is also an increase in the epinephrine concentration of the blood, which, along with the increased metabolism, keeps the liver glycogen at a low level.

Cholesterol Metabolism

Thyroxine also stimulates cholesterol synthesis and the liver's uptake of cholesterol. The overall effect is to decrease plasma cholesterol. Since the plasma cholesterol level begins to drop before an increase in O_2 consumption is seen, it is believed that this action too is independent of the increase in O_2 consumption produced by T_4.

The Nervous System

In the hypothyroid adult mental activity is sluggish, the reflexes are depressed, and the reaction time prolonged. The administration of T_4 alleviates these symptoms and can, if the dose is large enough, produce hyperreflexia and nervousness. The mechanism of these actions is poorly understood, since the blood-brain barrier is well developed in the adult and T_4 apparently has little or no effect on the O_2 consumption of the brain. They may be in part the result of increased responsiveness to the catecholamines epinephrine and norepinephrine, which T_4 produces, but this is apparently only part of the answer.

In infants the blood-brain barrier is only poorly developed; T_4 therefore plays an even more important role than in the adult. Here it facilitates the myelination of neurons and the development of the mental capacities of the infant. If thyroid therapy for the hypothyroid infant is not initiated before the blood-brain barrier develops, the mental deterioration produced by hypothyroidism becomes permanent.

Growth and Development

Thyroxine is also essential for normal physical growth and sexual development. It is essential for the maturation of certain tissues in utero during the latter stages of pregnancy and its presence is especially critical for development in the early postnatal days. In hypothyroidism, bone growth is depressed, epiphyseal closure in the long bones is delayed, and the development of sexual maturity is prevented. These symptoms can be partially remedied by the injection of T_4 if pituitary and pancreatic function are otherwise normal. Apparently T_4 influences growth and development only when it acts in conjunction with growth hormone and insulin. It has been shown to potentiate the actions of growth hormone and to facilitate its release by the anterior pituitary.

CLINICAL CORRELATES

Hypothyroidism

Two types of hypothyroidism are recognized: (1) that due to the hyposecretion of TSH (pituitary hypothyroidism) and (2) that due to the hyposecretion of T_3 and T_4 in the absence of any pituitary or hypothalamic malfunction. The first type responds well to the injection of TSH. The second type does not.

Hypothyroidism is seldom apparent at birth, probably because sufficient quantities of hormones pass the placental barrier to prevent marked symptoms. If the infant has marked impairment of the thyroid function, however, symptoms begin to develop after the maternal hormones have disappeared. If this condition remains untreated, the child becomes a *cretin* (a dwarfed idiot). If the hypothyroidism is less severe or if it occurs later in childhood, the dwarfing and mental retardation are less marked and the condition is termed "juvenile hypothy-

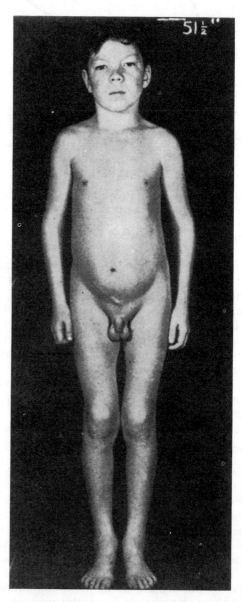

Fig. 29-6. Juvenile hypothyroidism in a 17-year-old boy. Dwarfism and delayed sexual development (absence of pubic hair, narrow shoulders, etc.) are some of the characteristics of this condition. (Ingbar, S. H., and Woeber, K. A.: The thyroid gland. *In* Williams, R. H. (ed.): Textbook of Endocrinology. 4th ed., p. 242. Philadelphia, W. B. Saunders, 1968)

roidism" (Fig. 29-6). In either case if permanent mental retardation is to be prevented, the condition must be recognized before the symptoms become marked. Some of the early signs of hypothyroidism include (1) persistence of the physiological jaundice of the newborn, (2) feeding problems, (3) constipation, (4) a hoarse cry, (5) excessive drowsiness (somnolence), (6) protuberance of the abdomen, (7) dry skin, (8) delayed development of the deciduous teeth, and (9) a delay in mental and physical development.

When hypothyroidism develops in the adult it is called myxedema (Gk. *myxa* mucus; plus *oidema*, swelling). This term is used to refer to the whole complex of symptoms in adult hypothyroidism, as well as to the edema of the face (incidence, 79%), eyelids (90%), and periphery (55%). Some of the other symptoms include muscle weakness (99%), dry, coarse skin (97%), lethargy (91%), slow speech (91%), a sensation of cold associated with a deterioration of the thermoregulatory responses to cold (89%), decreased sweating (89%), a thick tongue, coarseness of hair (76%), skin pallor (67%), impairment of memory (66%), loss of hair (57%), and nervousness (35%). If the loss of thyroid function is complete, the basal metabolic rate will be -40. Less severe hypothyroidism is associated with a less marked decrease in the BMR. The changes in the skin mentioned above are, for the most part, due to the accumulation of a variety of protein-carbohydrate complexes (hyaluronic acid and chondroitin sulfuric acid) in the skin that facilitate the retention of water there and produce the skin puffiness so characteristic of the disease. The administration of thyroxine brings about the catabolism of fat and protein throughout the body and a disappearance of the edema.

Hyperthyroidism

Hyperthyroidism, or thyrotoxicosis, is a condition in which the concentration of thyroxine and therefore the PBI, BEI, and uptake of iodine by the thyroid are elevated. Some of its symptoms include nervousness (99% of cases), goiter (more than 95%), increased sweating (91%), hypersensitivity to heat (89%), fatigue (88%), weight loss (85%), dyspnea (75%), weakness (70%), increased appetite (65%), and eye complaints (54%).

In addition, the metabolic rate is elevated from $+10$ per cent to $+100$ per cent. This results in an increased load on the heart because of the hypermetabolism and a need to dissipate the excess heat produced. The peripheral vascular beds dilate and the heart rate, stroke volume, cardiac output, and pulse pressure increase. There is also a release of catecholamines (epinephrine and norepinephrine) into the blood which tends to exaggerate the above changes and to increase the excitability of the heart. The efficiency of the heart decreases and its need for increased coronary flow is markedly elevated.

The arteriovenous O_2 difference is usually normal in hyperthyroidism, but this should not be taken as an indication of an adequate blood flow to all the tissues, since the opening up of arteriovenous shunts in the skin in response to the elevated body temperature would tend to obscure an elevated A-V difference in tissues other than the skin. Atrial fibrillation is found in approximately 10 per cent of all thyrotoxic patients.

One type of hyperthyroidism is *exophthalmic goiter* (*Graves' disease*). This is found most often

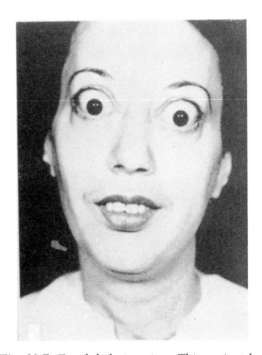

Fig. 29-7. Exophthalmic goiter. This patient had a mild hyperthyroidism associated with a slight diffuse enlargement of the thyroid. (Ingbar, S. H., and Woeber, K. A.: The thyroid gland. *In* Williams, R. H. (ed.): Textbook of Endocrinology. 4th ed., p. 205. Philadelphia, W. B. Saunders, 1968)

in patients between the ages of 30 and 50 and is characterized by all the symptoms of hyperthyroidism plus protrusion of the eyeballs (exophthalmos). The exophthalmos is apparently not directly due to T_4, T_3, or TSH, but rather to a gamma globulin produced in hyperthyroidism. It has been suggested that it is an antibody produced by the patient against some component in his own thyroid gland. For example, autoantibodies have been demonstrated in hyperthyroid patients against (1) thyroglobulin, (2) the microsomal component of thyroid cells, and (3) the nuclear component of the thyroid cells. We also find that 98 per cent of patients with lymphadenoid goiter, 82 per cent of patients with myxedema, and 63 per cent of patients with thyrotoxicosis have a weak to strong positive precipitin test against autoantibodies.

Chapter 30

FORMATION OF INSULIN
SECRETION OF INSULIN
ACTIONS OF INSULIN
DESTRUCTION OF INSULIN
DIABETES MELLITUS
GLUCAGON

ENDOCRINE FUNCTIONS OF THE PANCREAS

The pancreas is both an endocrine and an exocrine gland. In Chapter 24 we noted its function in the digestion of carbohydrates, lipids, and proteins. We will be concerned in this chapter primarily with its role in regulating carbohydrate, lipid, and protein metabolism in the cells of the body.

The pancreas produces three polypeptides with endocrine functions: (1) insulin (Fig. 30-1), (2) glucagon (Fig. 30-2), and (3) gastrin. These hormones are produced by cells in the islets of Langerhans. There are approximately one to two million of these ovoid islets in the pancreas, each approximately 75 by 175 μ and each with a copious blood supply. They constitute about 1 per cent of the weight of the pancreas and contain three cell types: alpha, beta, and D cells. The D cells contain granules which are less dense and more homogeneous than those of the alpha and beta cells, constitute 1 to 8 per cent of the total cell number, and secrete gastrin. The beta cells stain bluish-purple with modified Mallory aniline blue stain, secrete insulin, and constitute 60 to 90 per cent of the cells. The alpha cells stain red with Mallory stain, secrete glucagon, and constitute the rest of the cells in the islets.

FORMATION OF INSULIN

Insulin production begins in the fetus after the third month of pregnancy. In the adult male it may reach a level of about 50 units per day, and its concentration in the pancreas is about 4 units per gram (200 units per pancreas). Considerable evidence indicates that insulin synthesis takes place primarily in the ribosomes. It probably, like the synthesis of many other polypeptides, involves (1) the activation of amino acids in the cytoplasm by various enzymes and ATP, (2) the combination of these activated amino acids with soluble RNA (ribonucleic acid), and (3) the transfer of the amino acids to ribosomal RNA. Here, through the influence of messenger RNA and certain enzymes, they are (4) aligned in proper sequence and position. The polypepetide then (5) becomes the center of an endoplasmic vesicle with a covering of ribonucleoprotein granules, (6) becomes dense, and (7) loses it outer covering—becoming a beta granule. In response to the proper stimulus, this granule will (8) be evacuated through the cell membrane by emiocytosis, a process whereby granules are ejected into the extracellular fluid. These processes are summarized in Figure 30-3.

SECRETION OF INSULIN

Insulin secretion, of course, depends partly upon insulin production, but it is also dependent upon other factors. For example, insulin secretion can occur after insulin production has been blocked. The pancreas in a normal person releases insulin in response to small increases in blood glucose without a decrease in granulation, i.e., insulin production apparently keeps pace with insulin release. When high blood sugar levels are maintained, however, insulin release is associated with degranulation.

The stimuli that elicit insulin secretion are many, including monosaccharides in the intestine and plasma and agents such as pancreozymin that increase the plasma glucose level. In Figure 30-4, for example, we note the responses of a 23-year-old man to the infusion of 60 gm. of glucose given intravenously and given into the jejunum. In these experiments it was found that glucose administered intravenously produced and increase in plasma insulin within the first minute. The insulin than acted on the cells of the body to facilitate their withdrawal of glucose from the plasma; the plasma glucose level decreased as soon as glucose infusion stopped. You will note that with the decline in blood

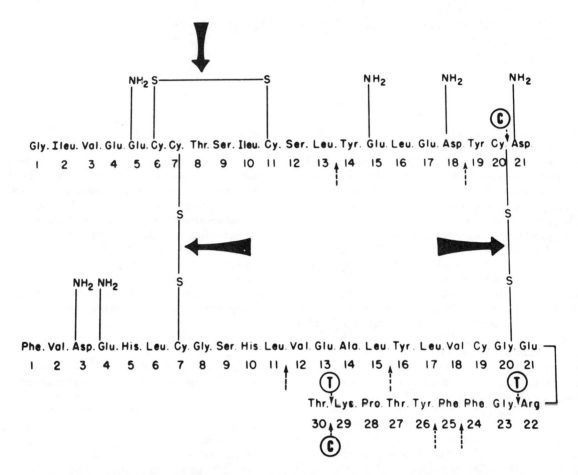

Fig. 30-1. Structure of human insulin (molecular weight, 6000). Destruction of insulin in the body is catalyzed by liver glutathione-insulin-transhydrogenase, which acts to cleave the disulfide bonds (*large arrows*). Adipose tissue, on the other hand, contains enzymes which cleave the peptide linkages at the 13-14, 18-19, 11'-12', 15'-16', 24'-25', and 25'-26' positions (*interrupted arrows*). Trypsin (T) and carboxypeptidase (C) are also found in numerous tissues and serve to break down various peptide bonds. The structure of pig, dog, rabbit, beef, goat, and sheep insulin differs from that of human insulin only with the amino acid found at position 8 (alanine for beef and goat), 9 (glycine for sheep), 10 (valine for beef, goat, and sheep), and 30 (alanine for pig, dog, beef, and goat, and serine for rabbit). The difference between the structure of human and beef insulin, for example, means that a human receiving beef insulin will usually develop antibodies to the injected insulin within a period of 2 months. Fortunately the antibody titer usually remains low and presents no apparent clinical problem. (Williams, R. H.: The pancreas. *In* Williams, R. H. (ed.): Textbook of Endocrinology. 4th ed., p. 669. Philadelphia, W. B. Saunders, 1968)

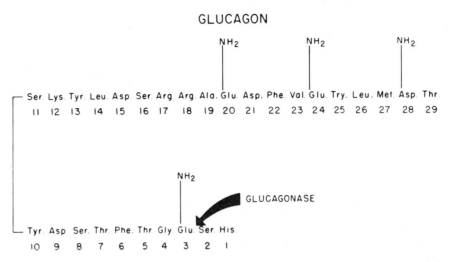

Fig. 30-2. Amino acid sequence of glucagon. Also shown is the site of cleavage by glucagonase. (Williams, R. H.: The pancreas. *In* Williams, R. H. (ed.): Textbook of Endocrinology. 4th ed., p. 630. Philadelphia, W. B. Saunders, 1968)

glucose there was also a decline in blood insulin. The reduction in insulin, however, lagged somewhat behind that of glucose, and by the time the blood glucose reached a normal level (100 mg./100 ml. of plasma) the insulin level was still high—producing a hypoglycemia that persisted for about an hour. This lag between the decrease in glucose in the blood and the decrease in insulin is partly due to the 30-min. half-life of insulin. These relationships can be summarized as in the diagram which follows on page 514.

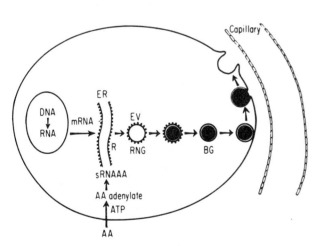

Fig. 30-3. Diagram of some of the changes which occur during insulin synthesis and secretion. DNA, desoxyribonucleic acid; RNA, ribonucleic acid; mRNA, messenger RNA; ER, endoplasmic reticulum; R, ribosomes; AA, amino acids; sRNAAA, soluble ribonucleic acid-amino acid complex; EV, endoplasmic vesicle; RNG, ribonucleoprotein granules; BG, beta granule. (Williams, R. H.: The pancreas. *In* Williams, R. H. (ed.): Textbook of Endocrinology. 4th ed., p. 622. Philadelphia, W. B. Saunders, 1968)

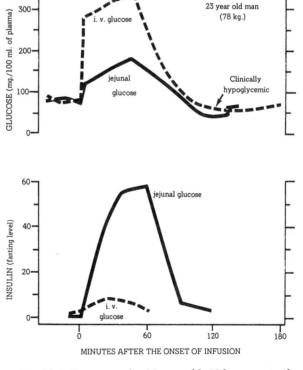

Fig. 30-4. Response of a 23-year-old, 78-kg. man to the infusion of 60 Gm. of glucose into a vein and its infusion into his jejunum. (*Modified from* McIntyre, N., et al.: Intestinal factors in the control of insulin secretion. J. Clin. Endocrinol. Metab., 25:1321, 1965)

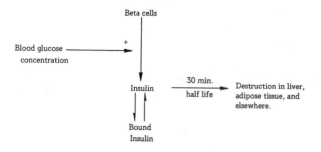

The response to glucose in the jejunum is somewhat different. You will note that although the same amount of glucose was infused into the jejunum as into the blood, the plasma glucose level increased less and the insulin level more than was the case in the intravenous infusion. It has been speculated that the intestine releases an agent in response to glucose that acts on the beta cells to augment their release of insulin. In support of this view is the observation that a substance can be extracted from the duodenal mucosa which upon injection increases plasma insulin-like activity.

There are, besides increases in plasma and intestinal glucose, other conditions that elevate insulin secretion. Some of these (the release of pancreozymin) are associated with an increase in plasma glucose and may act on insulin secretion by this route. Others (injection of secretin, amino acids, and glucagon), however, can be shown to increase insulin output even when there is no change in blood sugar. Table 30-1 lists some of the many factors that both facilitate and inhibit insulin secretion in vivo and in vitro. Epinephrine, for example, decreases insulin secretion while acting elsewhere to increase blood sugar—i.e., it facilitates an increase in blood sugar by (1) acting to facilitate the addition of sugar to the blood by the liver and (2) inhibiting the output of insulin from the pancreas.

ACTIONS OF INSULIN

Although there are many hormones that increase the blood sugar level, insulin is the only one known to decrease it. It apparently (1) acts on the cell membrane to facilitate glucose, amino acid, and K^+ transport into the cell, (2) facilitates the activity of various enzymes important in the formation of glucose-6-phosphate and its conversion into glycogen, pyruvic acid, and fatty acids, (3) inhibits enzymes that facilitate the catabolism of protein and fats, (4) inhibits the release of glucose from the liver, possibly by the inhibition of glucose-6-phosphatase (Fig. 26-1), and (5) facilitates DNA and RNA synthesis in the nucleus (Fig. 30-5). Its action on glucose trans-

	Increase Insulin Secretion						Decrease Insulin Secretion		
	Vivo	Vitro		Vivo	Vitro			Vivo	Vitro
Glucose	+	+	Estrogens	+			Epinephrine	+	+
Fructose	+	+	Isopropylnorepinephrine	+			Norepinephrine	+	
Mannose	+	+					Insulin	+	
Ribose	+	±	Phentolamine	+			Starvation	+	+
Amino acids	+	+	Insulin antibodies	+	+		Hypoxia		+
Ketones	+	±	Sulfonylureas (acutely)	+	+		2-Deoxyglucose	+	+
Glucagon	+	+					Glucosamine		+
Somatotropin	+	±	Calcium		+		D-Mannoheptulose	+	+
Placental			Magnesium		+		Phenethyl-		
lactogen	+		Potassium	+	+		biguanide	+	
ACTH	+	+	Adenosine triphosphate		+		Diazoxide	+	+
Glucosteroids	+	±					Vagotomy	+	
Thyroxine	+	±	Cyclic $3^1, 5^1$-AMP		+				
			Vagus stimulation	+					
			Secretin	+					
			Pancreozymin	+					
			Gastrin	+					

Table 30-1. Some of the factors modifying insulin secretion. (From Williams, R.H.: The pancreas. *In* Williams, R.H. (ed.): Textbook of Endocrinology. 4th ed. Philadelphia, W.B. Saunders, 1968)

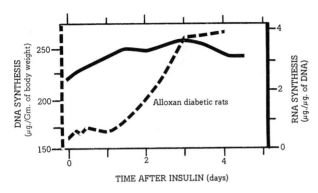

Fig. 30-5. The effect of insulin on rats made diabetic (hypoinsulinism) by the injection of alloxan on DNA and RNA synthesis. (*Modified from* Steiner, D. F.: Insulin and the regulation of hepatic biosynthetic activity. Vitamins Hormones, 24:1–61, 1966)

port has been demonstrated in skeletal muscle, the heart, adipose tissue, and fibroblasts. On the other hand it plays no role in the transport of glucose into the red blood cells, liver, brain, kidney, or intestinal cells.

Tissues shown to be sensitive to one or more of these insulin actions include (1) skeletal and cardiac muscle, (2) adipose tissue, (3) liver, (4) leukocytes, (5) mammary glands, (6) seminal vesicles, (7) cartilage and bone, (8) skin, (9) the lens of the eye, (10) the pituitary, (11) the peripheral nerve, and (12) the aorta. In adipose tissue and muscle it has been suggested that insulin facilitates the formation of (1) glucose-6-phosphate from glucose, (2) pyruvic acid, (3) glycogen, (4) proteins, and (5) fatty acids. It depresses the formation of (1) acetyl-Co A from fatty acids (Fig. 26-3), (2) the catabolism of triglycerides, and (3) the catabolism of proteins. In adipose tissue and muscle, that is, insulin facilitates the catabolism of glucose for energy while facilitating the formation of glycogen and depressing the catabolism of other nutrients (proteins and fats).

In the liver the effect of insulin is similar but not identical. In both the liver and skeletal muscle, for example, insulin initiates changes that tend to decrease the blood sugar. In skeletal muscle insulin acts both on the cell membrane to produce a rapid increase in the movement of sugar into the cell and on the intracellular enzymes. In the liver it has no such action on the cell membrane but does decrease the intracellular breakdown of glycogen and the associated release of glucose into the blood. Since this effect apparently requires the insulin to get into the cell, it takes a considerably longer time to develop than the initial action of insulin on muscle.

DESTRUCTION OF INSULIN

Figure 30-1 lists some of the points on the insulin molecule at which the various enzymes of the body act to destroy insulin. *Glutathione-insulin-transhydrogenase* cleaves the insulin molecule into an A and a B chain. This disulfide-disrupting system in the *liver* is apparently also important in the *cleavage* of the disulfide bonds in growth hormone, prolactin, antidiuretic hormone, and oxytocin. There are a number of other enzymes in the body that destroy peptide bonds as well. These systems are effective in the inactivation of glucagon and ACTH.

The *half-life* of insulin (30 min.) is considerably shorter than that of thyroxine (6.7 days) or triiodothyronine (1 day), being comparable to that of growth hormone (20-30 min.) and TSH (30 min.), and longer than that of glucagon (10-15 min.), ACTH (10 min.), and ADH (6 min.). This is probably the result of its binding characteristics (much of it is probably bound to beta globulin) as well as the effectiveness of the various catabolic systems in the body. It is interesting to note, however, that although the half-life of insulin is 30 min., it produces a lowered blood sugar for a period in excess of 1 hour. (Fig. 28-5). You will note in Figure 30-5 that it also takes insulin over 3 hr. to produce a maximum elevation in DNA synthesis in the rat. In other words, it seems likely that an important part of the hypoglycemia produced by insulin is due to the synthesis of enzymes within the cell.

DIABETES MELLITUS

Early Greek and Roman physicians used the term "diabetes" to refer to conditions in which there was an increase in urine production. Later it was noted that in some cases of polyuria the urine tasted sweet, while in others it had little or no taste. The latter condition was called diabetes insipidus and was found to result from a deficiency in the release of antidiuretic hormone; the former was called diabetes mellitus or simply diabetes.

The causes of diabetes mellitus are not fully understood. In some cases it is due to a deficiency in the production of insulin. This deficient production may or may not be preceded by a period of insulin output in which the beta cells become exhausted and as a consequence undergo degenerative changes. In some cases the exhaustion may be the result of an excessive food intake. It has also been suggested that diabetes results from (1) the production of defective insulin molecules, (2) excess of an *insulin antagonist* (synalbumin, growth hormone, glucocorticoids, aldosterone, catecholamines, glucagon, thyroxine, and triiodothyronine), or (3) tissue refractoriness to insulin.

Whatever the cause of diabetes mellitus, the basic symptoms are those that result from the destruction of the beta cells of the pancreas by a poison such as *alloxan,* or from removal of the pancreas. These symptoms are summarized in Figure 30-6. Insulin has the general effect of promoting glucose utilization (glucose catabolism and the formation of liver glycogen) while facilitating protein and lipid synthesis and depressing protein and lipid catabolism. Insulin insufficiency, then, would be expected to (1) decrease the use of glucose by the tissues, (2) increase the breakdown of glycogen by the liver and increase the movement of glucose from the liver to the blood, (3) increase gluconeogenesis (production of glucose from noncarbohydrates), and (4) decrease protein synthesis. All these changes either decrease the withdrawal of glucose from the blood by the cells or increase the addition of glucose to the blood. In other words there is a hyperglycemia.

You will also note in Figure 30-6 that the increased catabolism of lipids in diabetes results in an increase in the concentration of cholesterol, triglycerides, free fatty acids, and ketone bodies in the blood, as well as increased blood acidity. It has been suggested that one of the problems resulting from this increase in blood lipid is an arteriosclerosis that can lead to important decreases in blood flow and subsequent tissue damage. If the arteriosclerosis occurs in the kidney there may result an even more pronounced upset in acid-base and electrolyte balance than is already present. In the eye it can result in blindness, in the coronary vessels a heart attack, in cerebral vessels a stroke, and in the appendages gangrene (Fig. 30-7).

The acidosis, the ketonemia, or both can lead

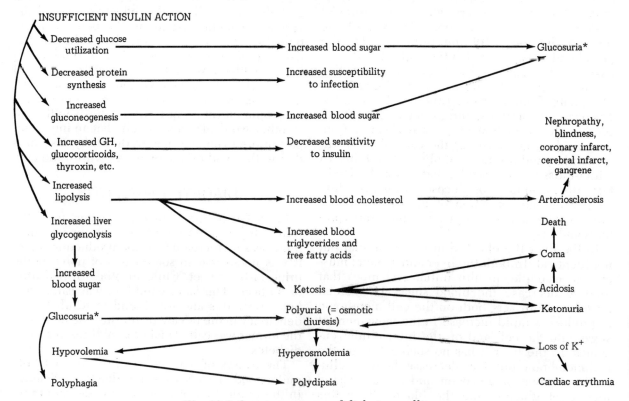

Fig. 30-6. Some symptoms of diabetes mellitus.

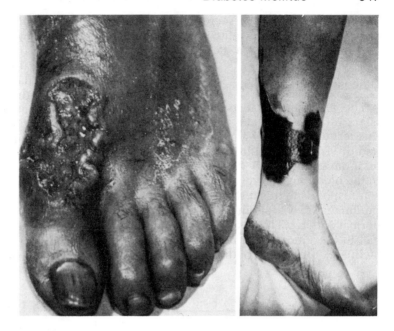

Fig. 30-7. Gangrene in the diabetic patient. At the left is a gangrene due to a destruction of the arterioles, capillaries, and venules, and at the right a gangrene due to arteriosclerosis of the large arteries of the leg. (Williams, R. H.: The pancreas. *In* Williams, R. H. (ed.): Textbook of Endocrinology. 4th ed., p. 761. Philadelphia, W. B. Saunders, 1968)

to coma and death. They, along with the glycosemia, also lead to a marked increase in the number of dissolved particles in the urine (NaH_2PO_4, KH_2PO_4, NH_4Cl, ketone bodies, glucose, etc.). These particles markedly increase the osmotic pressure of the urine and as a consequence increase its water content. This in turn results in dehydration of the body and an increase in the osmotic pressure of the blood, both of which act on the hypothalamus to increase thirst and water intake. As the condition continues, the loss of K^+ and Na^+ in the urine becomes marked. In fact, probably the most common causes of death in diabetes mellitus are the ketosis and acidosis discussed above and the disrupted K^+ balance. In diabetes there is both a loss of K^+ from the cells and an excessive loss of K^+ in the urine. In cases of kidney damage, on the other hand, there is hyperkalemia and a resultant tall, peaked T wave in the electrocardiogram. When there is hypokalemia due to the loss of more K^+ in the urine than is added to the blood, there is a prolonged Q-T interval. Either hypokalemia or hyperkalemia can lead to cardiac irregularities and death.

Another complication of diabetes is the interrelationships existing between the various endocrine glands. Apparently insulin depresses the activity of the adrenal, pituitary, and thyroid glands and desensitizes the cells of the body to some of their secretions. In the absence of insulin this depression is lost and the action of the insulin antagonists becomes exaggerated. These antagonists also tend to decrease the sensitivity of the cells to insulin.

This antagonism was first demonstrated by Houssay in 1923 when he showed in the toad that the severity of the symptoms of pancreatic diabetes could be decreased by removal of the pituitary. The animal with both the pituitary and the pancreas removed has come to be called the *Houssay animal*. In 1936 *Long* and *Lukens* showed in the cat that the symptoms of pancreatic diabetes could also be alleviated by adrenalectomy. In other words the following relationship exists among these three glands:

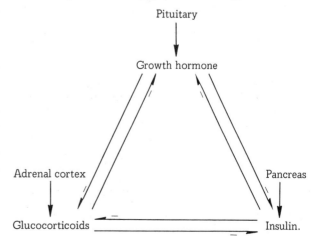

On the other hand, the secretion or injection of excess quantities of the glucocorticoids or growth hormone can produce a diabetes-like condition.

Glucose Tolerance Test

The glucose tolerance test helps a physician characterize a patient's pancreatic function. In this test a quantity of glucose is administered by mouth (1.75 gm. of glucose/kg. of body weight) or intravenously (0.5 gm./kg.) and the changes in the blood sugar level are noted. A diabetic person responds to this test with a greater and more prolonged elevation in his blood sugar than a normal subject. The response of the normal subject to the glucose is a release of insulin which facilitates the uptake of the glucose by a number of cells, and an increase in the formation of glycogen by liver and muscle.

The importance of the *liver* in this response can be demonstrated by studying the reaction of a nondiabetic person with liver disease to the same glucose stress. You will note in Figure 30-8 that his blood sugar levels change during this test in a manner similar to that of the diabetic. On the other hand, the person with liver disease exhibits decreases in serum inorganic phosphorus greater than that of the diabetic and similar to that of a healthy person. In other words, the liver disease has not impaired the action of insulin on the active transport of glucose into susceptible nonhepatic cells. Thus the diabetic shows an impaired glucose uptake by both hepatic and nonhepatic cells, while in liver damage it is only the liver cells that are effected.

Insulin Treatment

The treatment for severe diabetes is the administration of insulin. Initially this consists of the injection of progressively greater quantities of insulin until urinary sugar has almost disappeared. Later the diabetic changes to a regime of one injection of insulin just before breakfast or possibly several injections throughout the day with a daily check on the urinary sugar level. Increases in urinary glucose necessitate increases in insulin dosage. A patient who has had his pancreas removed generally requires less than 60 units of insulin per day, but patients with severe ketosis or those who have developed antibodies to insulin may require over 10,000 units per day.

One of the dangers in the initial administration of insulin is that it may so facilitate the movement of K^+ from the blood into the cell that severe hypokalemia will develop. When this happens it is sometimes necessary to administer intravenous K^+ to save the patient. Another complication of the dependence of the patient on an external supply of insulin is that this supply does not meet the stresses to which he is exposed throughout the day. Thus in severe cases the subject must be careful of his diet and his exercise. Generally, the diabetic is advised to carry some sugar with him, which he can take should he become hypoglycemic. The American Diabetes Association also provides the diabetic with a card that says, "I am a diabetic! If found in a dazed condition, administer a sweet drink, orange juice or sugar if possible, and call a physician immediately."

If the hypoglycemia becomes severe it pro-

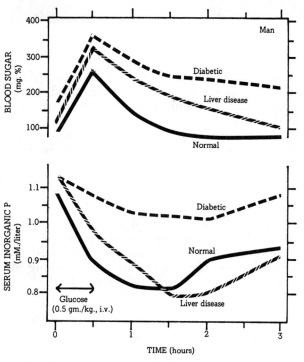

Fig. 30-8. Changes in blood sugar and serum inorganic phosphorus in man during an intravenous glucose tolerance test. The fall of serum phosphorus is related to the active entry of glucose into the glycolytic cycle. Note that liver damage impairs the removal of glucose from the blood, but not the decrease in serum phosphorus, whereas in diabetes mellitus both changes are impaired. In clinical practice such striking changes as seen above are not always encountered. (*Redrawn from* Williams, R. H.: The pancreas. *In* Williams, R. H. (ed.): Textbook of Endocrinology. 4th ed., p. 785. Philadelphia, W. B. Saunders, 1968)

duces mental confusion, weakness, dizziness, convulsions, coma, permanent brain damage, and death due to depression of the respiratory centers. A temporary rapid decline of the *blood sugar* to 20 mg./100 ml. of plasma can be tolerated until the cerebral reserves of carbohydrate have been depleted, but sustained blood glucose levels below 50 mg./100 ml. produce some of the symptoms discussed above.

GLUCAGON

Glucagon, like insulin, is produced in the pancreas and released into the blood, where it is carried first in the hepatic portal circulation to the liver and then to the blood of the general circulation (Fig. 30-9). Because of this arrangement, both insulin and glucagon normally reach the liver at a considerably higher concentration than that at which they reach the other organs in the body. Insulin facilitates hepatic glycogen synthesis and inhibits glycogenolysis, while glucagon inhibits hepatic glycogen synthesis and facilitates glycogenolysis. These relationships can be characterized as follows:

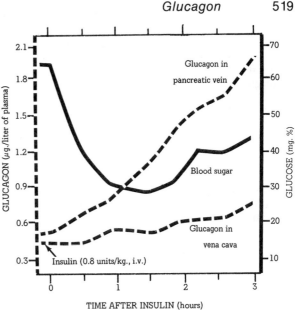

Fig. 30-9. Average response of 10 dogs to the intravenous injection of 0.8 units of glucagon-free insulin. Apparently the insulin in decreasing the blood sugar increased the output of glucagon into the hepatic portal circulation. (*Modified from* Unger, R. H. *In* Williams, R. H. (ed.): Textbook of Endocrinology. 4th ed., p. 630. Philadelphia, W. B. Saunders, 1968)

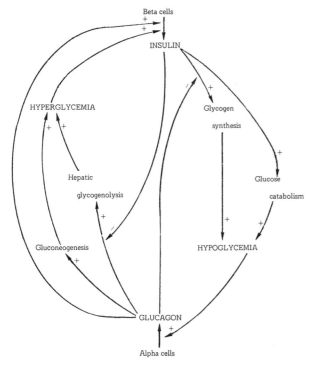

You will note in this diagram that glucagon tends to produce hyperglycemia and insulin tends to produce hypoglycemia, and that each, in performing its role, has opposite effects from the other. A more complete comparison of the roles of insulin and glucagon is presented in Table 30-2. You will note in this table and in the preceding diagram that glucagon, unlike insulin, can facilitate gluconeogenesis. It can increase the blood sugar level not only by facilitating the breakdown of liver glycogen but also by increasing the catabolism of protein and decreasing its anabolism. It can, in the absence of insulin, also increase the free fatty acid concentration of the blood. The importance of these factors (that is, the extrahepatic effects) when glucagon is added to the blood by the pancreas rather than by injection is still questionable, however. Unfortunately, only recently have our assay procedures for glucagon become sufficiently sensitive to accurately determine the concentration of glucagon in the blood outside the hepatic portal circulation.

Physiological Role

Although the extrahepatic role of glucagon is still being debated, a number of hypotheses have evolved that merit serious attention. The rate of

520   30/Endocrine Functions of the Pancreas

	Glucagon	Insulin
Blood glucose	S	I
Liver glycogen	I	S
Hepatic glucose production	S	I
Glucose utilization	I	S
Gluconeogenesis and urea production	S	I
Nitrogen balance	I	S
Fatty acid synthesis	I	S
Net triglyceride breakdown	S	I
Ketone body production	S	I
Hepatic K+ release	S	I
Gastrointestinal activity	I	S
Hunger and food intake	I	S
Cyclic AMP formation	S	I
Phosphorylase activity	S	I
Glucokinase activity	I	S
Glycogen synthetase activity	I	S
Phosphoenolpyruvate carboxykinase activity	S	I
Glucose-6-phosphatase activity	S	I

Table 30-2. Comparison of the effects of glucagon and insulin. S, stimulation or increase; I, inhibition or decrease. (From Foa, P.P.: Glucagon. Reviews of Physiology, Biochemistry, and Experimental Pharmacology. New York, Springer Verlag, 1968)

secretion of glucagon by the pancreas is approximately 50 to 200 µg. per hour, and its concentration in the plasma 1 to 8 µg. per liter. It has been estimated that the pancreas contains a total of 23 µg. and that smaller quantities may also lie in the stomach and intestine. It is apparently one of a number of hormones (epinephrine, growth hormone, and the glucocorticoids) having the capacity to increase blood sugar by decreasing glucose utilization, increasing lipolysis, and increasing glycogenolysis. These hormones are probably released between meals and during more severe fasts and serve to guarantee to the brain a fairly constant concentration of glucose. Their secretion is decreased following a meal. Although each of these hormones increases the blood sugar, their actions are not identical. Glucagon, for example, is a more effective glycogenolytic agent than the other hormones, being 100 times more effective than epinephrine in this regard. On the other hand, it is only one seventh as effective as epinephrine as a lipolytic agent. Epinephrine also produces some changes in the body that glucagon does not (Fig. 30-10). These include sweating, tachycardia, and an increase in blood lactate. Glucagon is also the only one of the above hormones that acts directly on the pancreas to produce an increase in insulin secretion.

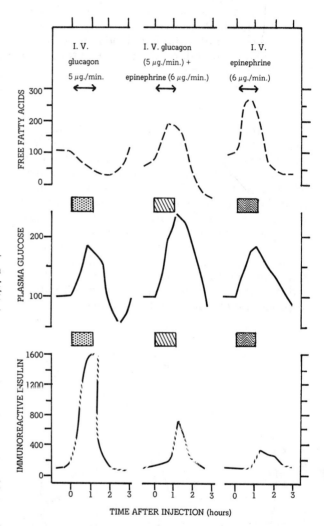

Fig. 30-10. Response (in % of control) of man to glucagon and epinephrine. Note that glucagon increases the plasma glucose level and produces an increase in plasma insulin which results in a decrease in the plasma free fatty acids. Epinephrine, on the other hand, does not produce an increase in plasma insulin. Epinephrine and glucagon together produce an increase in plasma lipids and glucose but little or no change in plasma insulin. Note also that when the infusion of epinephrine stops, the plasma concentration of insulin increases. Due to the short half life of epinephrine the cessation of its infusion permits the pancreas to now respond to the hyperglycemia by increasing its output of insulin. (Redrawn from Porte, D., Jr., et al.: The effect of epinephrine on immunoreactive insulin levels in man. J. Clin. Invest., 45:230, 231, 1966)

Chapter 31

DEVELOPMENT OF THE
ADRENAL CORTEX
DEVELOPMENT OF THE
ADRENAL MEDULLA
ACCESSORY
ADRENOCORTICAL AND
ADRENAL MEDULLARY
TISSUE
THE ADRENOCORTICAL
STEROIDS
THE ADRENAL MEDULLA

THE ADRENAL GLANDS

In adult man there exists above each kidney a gland—the adrenal gland—weighing about one twenty-eighth as much as the kidney. Each gland consists of a capsule, a cortex, and a medulla (Fig. 31-1) and receives its blood supply from numerous small arteries branching from the phrenic and renal arteries, the aorta, and occasionally the spermatic, ovarian, and intercostal arteries. This blood passes into (1) sinusoids devoid of phagocytic function in the cortex, (2) lacunae in the medulla, and finally (3) the adrenal vein. The right adrenal vein drains directly into the vena cava and the left often joins the inferior phrenic and left renal vein.

DEVELOPMENT OF THE ADRENAL CORTEX

In the fourth to sixth week of pregnancy acidophilic cells from the coelomic mesoderm are found near the genital ridge. These cells incorporate cells from this ridge to form the fetal cortex. By the tenth week smaller basophilic cells surround the cells of the fetal cortex to form the permanent cortex. By the third to fourth month the adrenals exceed the kidneys in weight, but at birth they are one third the weight of the kidneys and in the adult they are one twenty-eighth the weight of the kidneys. In the adult about 80 per cent of the adrenals is cortex.

DEVELOPMENT OF THE ADRENAL MEDULLA

The adrenal medulla, unlike the cortex, develops from ectoderm. At about the seventh week of pregnancy the fetal cortex is invaded by a group of neurogenic cells that migrate from the neural crest. In the adrenals these cells form chromaffin cells. In other areas of the body they

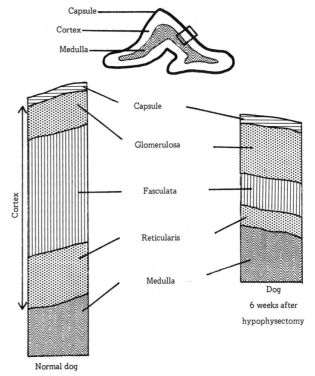

Fig. 31-1. The adrenal gland. In the upper figure we note a normal adrenal gland and its division into a capsule, cortex, and medulla. At the lower left we see a magnification of a part of the gland and the division of the cortex into 3 zones. Each of these zones can produce glucocorticoids and sex hormones, but only the zona glomerulosa produces the mineralocorticoid, aldosterone. The medulla, on the other hand, produces the catecholamines, epinephrine and norepinephrine. Note also on the right the effects of 6 weeks of hypophysectomy. Apparently the resultant loss of ACTH in the blood produces an atrophy of both the zona fasciculata and reticularis. Later there will be an atrophy of the zona glomerulosa as well. (*Modified from* Ganong, W. F.: Review of Medical Physiology. 4th ed., p. 293. Los Altos, Lange Medical Publications, 1969)

form ganglion cells (postganglionic neurons). The chromaffin cells of the adrenal medulla are responsible for its production of the catecholamines norepinephrine and epinephrine, whereas the cells of the adrenal cortex do not produce amines but produce and release steroids.

ACCESSORY ADRENOCORTICAL AND ADRENAL MEDULLARY TISSUE

Prior to birth paraganglionic masses near the abdominal aorta release more catecholamines into the blood than does the adrenal medulla. After birth, however, there is an involution of extramedullary chromaffin tissue; the adrenal medulla then becomes the source of most of the epinephrine, and the body's postganglionic adrenergic nerve endings supply most of the norepinephrine found in the blood. In some persons, extramedullary tissue is found in the bladder and near the sympathetic nerve tissue in the abdominal aorta.

As many as 20 per cent of otherwise normal adults may have accessory adrenocortical tissue. When present, this tissue is usually found in the testes or ovaries. True accessory adrenal glands are very rare, but when found usually lie in the celiac plexus, renal cortex, or testes.

THE ADRENOCORTICAL STEROIDS

Approximately fifty steroids have been isolated from the adrenal cortex, but under normal conditions fewer than eleven are found in the adrenal veins of man (Table 31-1). These hormones include (1) the glucocorticoids, which facilitate the catabolism of proteins and the synthesis of glucose and glycogen, (2) the mineralocorticoids, which facilitate the reabsorption of Na^+ from the glomerular filtrate of the kidney, and (3) the androgens, progestins, and estrogens, whose role will be discussed more fully later.

Synthesis

All the steroids produced by the adrenal cortex are derived from cholesterol, which in turn is derived from our food or is produced in the body from acetate. Cholesterol contains the cyclopentanoperhydrophenanthrene ring:

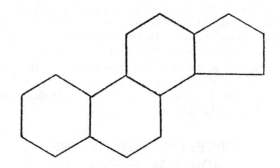

The adrenocorticosteroids produced from this ring can be divided into those with 21 carbon atoms (C 21 steroids) and those with 19 carbon atoms (C 19 steroids). The C 21 steroids have a two-carbon side chain attached to carbon 17 of the D ring of the cyclopentanoperhydrophenanthrene ring, plus some other groups (some of

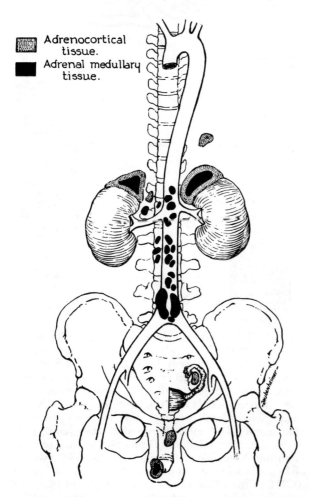

Fig. 31-2. Some areas in the body where adrenocortical and adrenal medullary hormones may be produced. (Forsham, P. H.: The adrenal cortex. *In* Williams, R. H. (ed.): Textbook of Endocrinology. 4th ed., p. 288. Philadelphia, W. B. Saunders, 1968)

which have been omitted from the following diagram for the sake of simplicity).

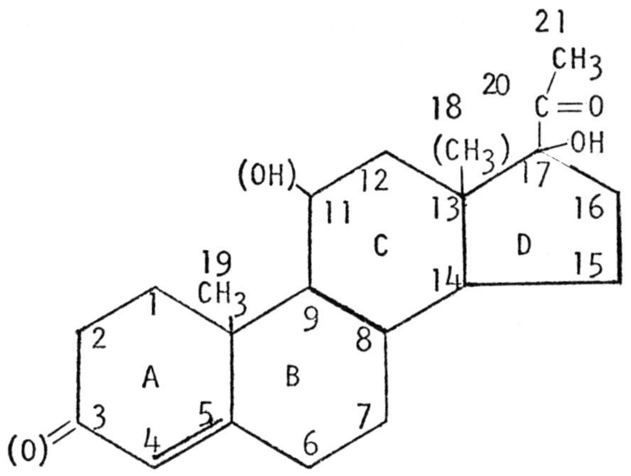

the major pathway for the formation of the 17-ketosteroids.

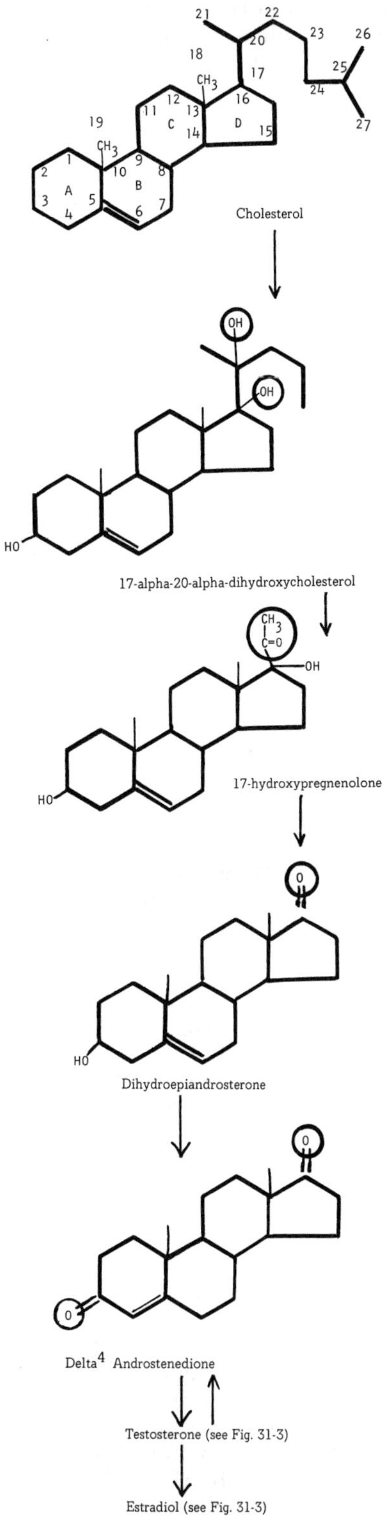

Each of the groups in parentheses represents a potential combination for the C 21 steroids. Those with an OH group in the C 17 position are sometimes called 17-hydroxycorticoids. The C 21 steroids normally secreted by the adrenal cortex include (1) the mineralocorticoids aldosterone and 11-desoxycorticosterone, (2) all the glucocorticoids, and (3) the progestins.

The C 19 steroids produced from this basic ring have either an oxygen or a hydroxyl group at the C 17 position. Those with an oxygen group are sometimes called 17-ketosteroids:

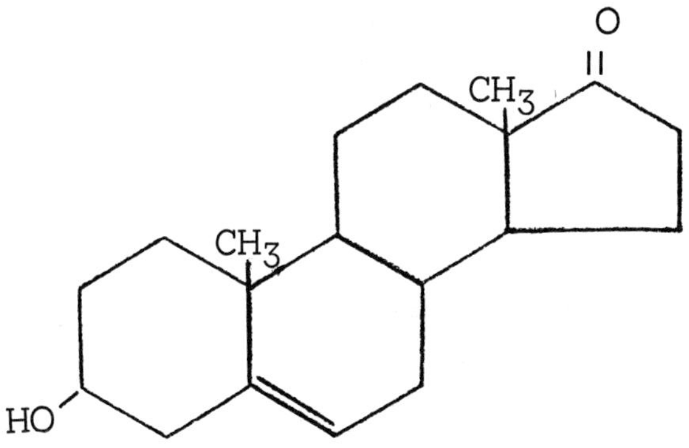

The C 19 steroids normally secreted by the cortex include estradiol and a number of androgens.

It now appears that there are two major pathways for the synthesis of the various adrenal steroids from cholesterol. One of these is shown in Figure 31-3 and proceeds through the formation of 20-alpha-hydroxycholesterol. The other is the formation of 17-alpha-20-alpha-dihydroxycholesterol. These chemical changes represent

524   31/The Adrenal Glands

In both the 20-alpha-hydroxycholesterol and the 17-alpha-20-alpha-dihydroxycholesterol pathways there is a series of cleavages of the side chain of the cholesterol molecule. The enzyme systems responsible for these cleavages, as well as the system responsible for the beta hydroxylation at C 11, lie in the mitochondria. On the other hand, the system responsible for the beta dehydrogenation at C 11 is in the microsome, and the systems responsible for hydroxylation at C 17

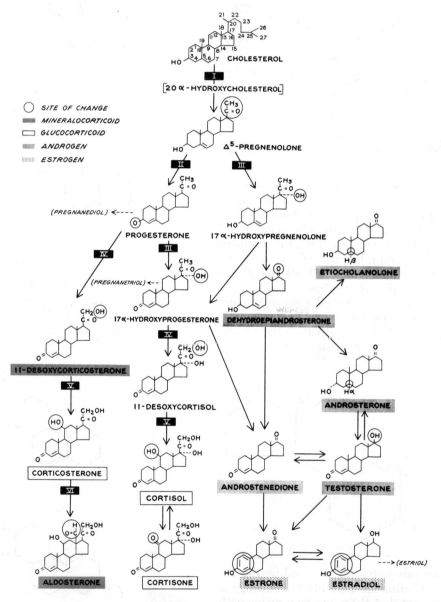

Fig. 31-3. The 20 alpha-hydroxycholesterol pathway for the formation of adrenocortical hormones. The Roman numerals represent different points in the synthesis where various hormonal blocks act. Most of the 17-ketosteroids (testosterone, estradiol, and delta[4] androstenedione, for example), on the other hand, are produced by the 17-alpha-dihydroxycholesterol pathway. (Forsham, P. H.: The adrenal cortex. *In* Williams, R. H. (ed.): Textbook of Endocrinology. 4th ed., p. 298. Philadelphia, W. B. Saunders, 1968)

and 21 are in the soluble fraction of the cytoplasm.

Secretion and Transport

The adrenal cortex releases up to five types of hormones (Table 31-1). These include (1) the glucocorticoids, which act primarily to raise the blood sugar level, (2) the mineralocorticoids, which elevate the plasma Na+ level, (3) the androgens, (4) the progestins, and (5) estradiol. Under most conditions the last three groups apparently play little or no role in the body. The glucocorticoids and mineralocorticoids, on the other hand, are released in quantities sufficient to modify body function markedly.

The glucocorticoids. Approximately 15 to 30 mg. of cortisol are released into the blood every day. The cortisol concentration in plasma is lowest at about 8 P.M. and highest at 6 A.M., at which time it has doubled (Fig. 31-4). This diurnal variation is apparently the result of a similar pattern of release for ACTH, and in the absence of stress is quite constant from day to day. We find, for example, that persons who change from day work to night work, who become blind, or who move from south to north do not change their rhythm, although persons who move from one latitude to another show gradual changes.

Most of the cortisol released into the blood is bound to an alpha globulin called *transcortin* or corticosteroid-binding globulin (*CBG*). A small amount is bound to albumin. The total amount of cortisol bound to these proteins is about 12 μg./100 ml., and that found free in the plasma is about 0.5 μg./100 ml. Apparently the free cortisol serves to inhibit the release of ACTH from the anterior pituitary, and to increase the blood glucose level, whereas the bound cortisol serves only as a reservoir:

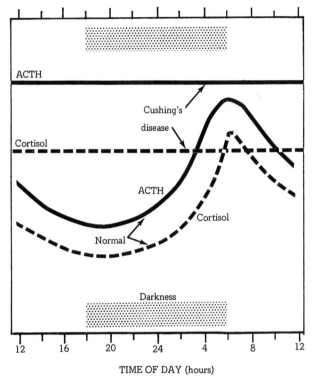

Fig. 31-4. The diurnal variation of ACTH and cortisol. ACTH varies from a low in the evening of less than 0.3 milliunits/100 ml. of plasma to a high in the morning of more than 0.5 milliunits/100 ml. of plasma. Cortisol will vary from a low of about 4 μg./100 ml. of plasma to a high of 16 μg./100 ml. of plasma. In Cushing's disease, on the other hand, the levels of ACTH and cortisol remain stable throughout the day. In Cushing's disease the glucocorticoid level is always elevated but the plasma concentration of ACTH may either be elevated or lowered. If the disease is due to an excessive output of ACTH by the pituitary, the plasma ACTH will be between 0.6 and 3 milliunits/100 ml., but if due to an excessive putput of ACTH by some other gland it may reach levels of from 29 to 100 milliunits/100 ml. If, on the other hand, the hypersecretion of glucocorticoids is not secondary to an increased output of ACTH, the plasma ACTH levels may drop below 0.15 milliunits/100 ml.

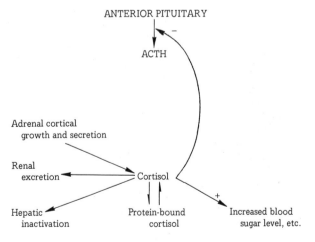

At normal glucocorticoid concentrations only about one third of the binding sites on CBG are occupied. Corticosterone is similarly bound to plasma protein but probably to a lesser degree. It has a biological half-life of 70 min. as opposed to one of 90 min. for cortisol (Table 31-2).

Group	Compound	24-Hour Adult Secretion	Blood Plasma Concentration (per 100 ml. of plasma)
Steroids			
Glucocorticoids	Cortisol	20 mg.*	12 μg. (in morning)
	Corticosterone	2–5 mg.*	0.6 μg.
Mineralocorticoids	Aldosterone	75–150 μg.*	7 mμg.
	11-desoxycorticosterone		Trace
Androgens	Dehydroepiandrosterone	20 mg.*	63 μg.
	Delta⁴, androstenedione	0–10 mg.*	
	11-beta hydroxyandrostenedione	0–10 mg.*	
Progestins	Progesterone in the male	0.75 mg.*	0.03 μg.
	Progesterone in the female	0.4–0.8 mg.*	0.1–1.5 μg.
		3–40 mg.	≤ 14 μg. (pregnancy)
	Pregnenolone	0.5–0.8 mg.*	
	17-hydroxypregnenolone	0.2–0.4 mg.*	
Estrogens	Estradiol in the male	40 μg.	2 mμg.
	Estradiol in the female	Trace*	
		90–250 μg.	6–30 mμg.
Testosterone	Testosterone in the male	4–9 mg.	0.6 μg.
	Testosterone in the female	0.2–0.5 mg.	0.05 μg.
Peptides, polypeptides, Proteins			
Pituitary Hormones	Follicle stimulating hormone in the male		20 mU.
	FSH in the premenopausal female		15 mU.
	FSH in the postmenopausal female		200 mU.
	Luteinizing hormone in the male	30 μg.	0.3 μg.
	LH in the premenopausal female	> 30 μg.	0.3–2 μg.
	LH in the postmenopausal female		1–2 μg.
	Lutetropic hormone in sheep		10 μg.
	Growth hormone	0.4–1 mg.	0–0.3 μg.
	Thyroid stimulating hormone	110 μg.	< 0.2 μg.
	Adrenocorticotropic hormone	< 10 μg.	5 μg.
	Melanocyte stimulating hormone		< 0.9 μg.
	Antidiuretic hormone	0.36 μg.	0.3 mμg.
	ADH after severe hemorrhage		135 mμg.
	Oxytocin		0.1–0.5 mU.
	Oxytocin midway through labor		20 mU.
Thyroid Hormone	Thyrocalcitonin (rabbit)		10 mμg.
Renal Hormone	Renin		7 mμg.
Pancreatic Hormones	Insulin	2 mg.	1 mU. (?)
	Glucagon		5–800 mμg. (?)
Parathyroid Hormone	Parathormone	1 mg. (?)	0.03 μg. (?)
Lipids, Amino Acids, Amines			
Unclassified Hormone	Prostaglandin F₁	1 mg. (?)	0.03 μg. (?)
Thyroid Hormones	Thyroxine	60–150 μg. (?)	9 μg.
	Triiodothyronine	25–50 μg. (?)	0.3 μg.
Adrenal Medullary Hormones	Norepinephrine	3.2 mg. (?)	15 mμg.
	Epinephrine	0.8 mg. (?)	3 mμg.

*Estimate of adrenal cortical secretion.

Table 31-1. Plasma concentration and rate of secretion of various hormones and related agents. Abbreviations include: μg., micrograms (10^{-6} Gm.); mμg., millimicrograms (10^{-9} Gm.); mU, milliunits. (In part from the data of Williams, R. H. (ed.): Textbook of Endocrinology. 4th ed. Philadelphia, W. B. Saunders, 1968; and Sawin, C. T.: The Hormones: Endocrine Physiology. Boston, Little, Brown & Co., 1969)

Hormone	Half-Life
Epinephrine	Short
Norepinephrine	Short
Angiotensin II	< 2 min.
Oxytocin	< 4 min.
Progesterone	< 10 min.
Parathormone	10 min. (?)
Thyrocalcitonin	10 min. (?)
Antidiuretic Hormone (ADH)	10 min.
Adrenocorticotrophic Hormone (ACTH)	10 min.
Glucagon	< 12 min.
Growth Hormone (GH)	25 min.
Insulin	30 min.
Aldosterone	35 min.
Thyroid stimulating hormone	< 54 min.
Leutinizing hormone	60 min.
Renin	80 min.
Testosterone	< 90 min.
Cortisol	90 min.
Melanophore stimulating hormone	< 2 hr.
Triiodothyronine (T_3)	2 days
Thyroxine (T_4)	6.7 days

Table 31-2. Half-life of different hormones.

the glucocorticoids listed. It also has a shorter half-life (20 min.) and is bound to a lesser degree.

Metabolism and Excretion

Cortisol is excreted in the urine as (1) free cortisol (0.03 mg./day), (2) tetrahydrocortisol glucuronide (5 mg./day), (3) tetrahydrocortisone glucuronide (3 mg./day), (4) 20-hydroxy derivatives of tetrahydroglucuronides (3 mg./day), and (5) 17-ketosteroids derived from cortisol and cortisone (1 mg./day, mostly as sulfates). All these derivatives are produced from cortisol in the liver (Fig. 31-5) and are released into the blood, where they are soluble and for the most part remain unbound. Only minute quantities end up in the stool. Other sources of urinary 17-ketosteroids include the adrenal cortical androgens and testosterone.

The metabolism of corticosterone is similar to that of cortisol except that it does not form a 17-ketosteroid. The concentration of free aldosterone in the urine, like that of the free glucocorticoids, is low (less than 1% of the secreted

The usual method for determining the level of unconjugated cortisol and corticosterone in the plasma measures both bound and unbound molecules. Therefore high values may be found not only when there is hyperadrenalcorticalism but also when the liver increases its output of CBG. Apparently estrogen acts on the liver to produce increases in plasma CBG, and therefore in pregnancy when the plasma estrogen is high the CBG will be high. Elevated CBG concentration produces increased binding of the free cortisol and corticosterone and therefore increased ACTH output. The ACTH, in turn, eventually brings the free cortisol and corticosterone in the plasma to a normal level but results in a marked increase in the total glucocorticoids in the plasma—i.e., the free glucocorticoids are at a normal concentration but the bound ones are at a high concentration. Low plasma glucocorticoid levels, on the other hand, may be due either to hypoadrenalcorticalism or to a low concentration of the binding proteins. The latter may result from a liver malfunction that decreases its production of plasma proteins, or to a nephrosis that increases the loss of plasma protein into the urine.

The mineralocorticoids. The major mineralocorticoid secreted by the adrenal cortex is aldosterone. You will note in Table 31-1 that both the quantity secreted and its plasma concentration are considerably less than that of either of

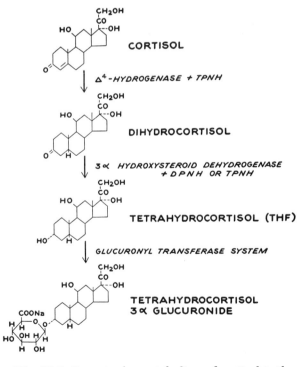

Fig. 31-5. Steps in the metabolism of cortisol in the liver to a glucuronide. (Forsham, P. H.: The adrenal cortex. *In* Williams, R. H. (ed.): Textbook of Endocrinology. 4th ed., p. 302. Philadelphia, W. B. Saunders, 1968)

aldosterone). Much of the plasma aldosterone is changed in the liver to a glucuronide or to "acid-labile conjugate" and is eliminated by the kidney in these forms.

Physiological Roles

The two hormones normally released by the adrenal cortex in sufficient quantities to play major roles in the control of body activity are cortisol (15-30 mg./day) and aldosterone (50-150 µg./day). You will note in Table 31-3 that aldosterone has a much more potent effect on Na^+-K^+ balance, and a less potent effect on glycogen synthesis and the involution of the thymus than does cortisol. As a result, aldosterone is classified primarily as a mineralocorticoid, and cortisol, primarily as a glucocorticoid. Other differences between the two hormones include the stimuli that elicit their secretion and their action on the release of ACTH from the anterior pituitary. Figure 28-12, for example, shows that ACTH is a far more potent stimulus for the release of cortisol than for aldosterone.

The roles played by the other steroids released by the adrenal cortex are less pronounced and generally less important for normal functioning. Corticosterone, for example, is released in smaller concentrations and is less potent than cortisol. On that basis one would suspect that at best it plays a minor role in regulating the blood sugar level. The androgens released by the adrenal cortex, on the other hand, have many of the properties of testosterone but are considerably less potent. They facilitate protein anabolism, muscular development, and the development of the secondary sexual characteristics of the male. It has been suggested that they also have a role in the growth and development of both boys and girls prior to puberty and that they are responsible for the virilization seen in some women after menopause. From a pathological point of view, the androgens and estrogens produced by the adrenal cortex are very important. Their excessive output in male infants can result in a precocious puberty not associated with enlargement of the testes, and in female infants can produce pseudohermaphroditism (Fig. 31-6). The excessive output of estrogens by the adrenal cortex can produce premature puberty in the female and feminization of the male infant (Fig. 31-7). The progestins are of no known significance other than as precursors of other corticoids.

Physiological Roles of the Glucocorticoids

In the absence of the adrenal glands, water, carbohydrate, protein, and fat metabolism are abnormal and death may result from minor noxious stimuli. Small amounts of cortisol or other glucocorticoids correct such abnormalities. Apparently cortisol exerts its influence in part by increasing the intracellular formation of certain enzymes. It is through this and possibly other mechanisms that cortisol (1) prevents hypoglycemia in response to fasting, (2) prevents

	Maintenance of Adrenalectomized Dogs	Change in Urinary Na-K Ratio	Liver Glycogen Synthesis	Thymus Involution
Hormones				
Cortisol (17-hydroxycorticosterone, or Compound F)	4	7	100	100
Corticosterone (Compound B)	15	14	35	28
Aldosterone (18-aldocorticosterone)	3,000	12,000	25	slight
Related compounds				
Cortisone (11-dehydro-17-hydroxy-corticosterone, or Compound E)	4	6	65	65
11-desoxycorticosterone (DOC)	100	100	<1	0
9-α-fluorocortisol	900	—	1,200	900
Prednisolone	1	—	350	400

Table 31-3. Relative potencies of some natural and synthetic adrenocortical steroids. (In part from Russell, J. A.: The adrenals. *In* Ruch, T. C., and Patton, H. D.: Medical Physiology and Biophysics 19th ed. Philadelphia, W. B. Saunders, 1965)

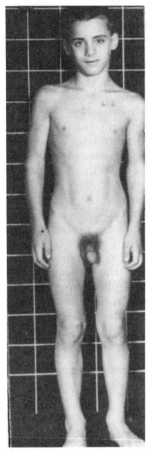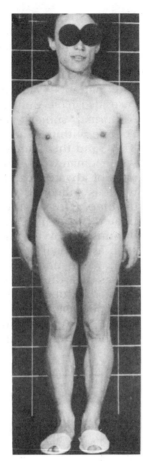

Fig. 31-6. Hypersecretion of adrenal androgens in the male (*left*) and female (*right*). The male is 5 years, 9 months old, 142 cm. tall, and produces 15 to 25 mg. of 17-ketosteroids per day. His bone structure and height are those of a 13-year-old. The prostate, penis, and pubic hair are characteristic of an adult, but the testes are small and immature. The femal patient is 23 years old, has a deep voice, shaves daily, wears a toupee, has broad shoulders and narrow hips, and a short stature (145 cm.) due to a premature epiphyseal fusion. Her output of 17-ketosteroids is 35 to 50 mg. per day. (Wilkins, L.: The Diagnosis and Treatment of Endocrine Disorders in Childhood and Adolescence. 3rd ed., pp. 431 and 433; courtesy of Charles C Thomas, Publisher, Springfield, Illinois, 1965)

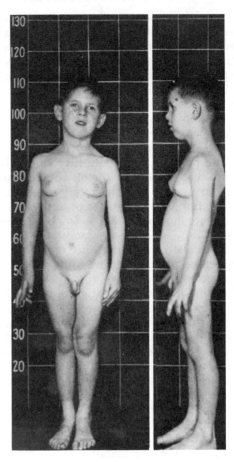

Fig. 31-7. Hypersecretion of adrenal estrogens in the male. This patient was 5½ years old, showed no acceleration of somatic growth, and, with the exception of an enlarged prostate, had normal genitalia for a child his age. Removal of the adrenal tumor resulted in a gradual decrease in the size of the breasts. (Wilkins, L.: The Diagnosis and Treatment of Endocrine Disorders in Childhood and Adolescence. 3rd ed., p. 442; courtesy of Charles C Thomas, Publishers, Springfield, Illinois, 1965)

skeletal-muscle weakness, and (3) facilitates gastric acid and pepsin secretion by the stomach. Cortisol also: (4) significantly affects brain function, (5) increases the sequestration of circulating eosinophils in the spleen and lungs, (6) decreases the number of circulating basophils, (7) increases the number of circulating erythrocytes, neutrophils, and platelets, (8) decreases the size of the lymph nodes and thymus, as well as the number of circulating lymphocytes, (9) facilitates diuresis in response to a water load, (10) inhibits the action of insulin and growth hormone, (11) facilitates the action of glucagon, norepinephrine, and epinephrine, and (12) increases resistance to stress.

Gluconeogenesis and fat metabolism. The precise action of cortisol and other glucocorticoids on metabolism in man is not well defined. In animals whose adrenal glands have been removed and in patients suffering from Addison's disease there is hypoglycemia and a reduction of liver glycogen if food intake is restricted. In addition there is loss of appetite, reduced intestinal absorption, loss of weight, and growth retardation. The administration of glucocorticoids to hypoglycemic patients produces an increase in blood sugar; in a fasting person it also produces an increase in liver glycogen associated with increased urinary nitrogen.

On the basis of these data it is generally recognized that although glucocorticoids are essential for normal growth, a very important part of their action is to facilitate a *protein catabolism* that leads to *gluconeogenesis*. In performing this function they apparently facilitate (1) extrahepatic catabolism of proteins to amino acids, (2) increased amino acid trapping by the liver, (3) increased formation in the liver of glutamic acid, purine (produced from glutamic acid), glucose, glycogen, and urea, and (4) increased release by the liver of glucose into the blood. In fact, excessive quantities of glucocorticoids may so accelerate protein catabolism that there will be (1) muscular atrophy, (2) thinning of the epidermis, (3) loss of weight, and (4) in older people a sufficient loss of protein from the matrix of bone to produce osteoporosis. The latter effect can be prevented by the administration of hormones (estrogens and androgens, for example) that facilitate protein anabolism.

The glucocorticoids also affect metabolism by modifying the response of various cells to other metabolic hormones. For example, they facilitate the glycogenolytic action of epinephrine and glucagon on the liver, as well as the release of free fatty acids from adipose tissue in response to epinephrine. They are capable of depressing the responsiveness of cells to the action of insulin and growth hormone; the injection of large amounts of cortisol in man produces a diabetes much more resistant to the action of insulin than that resulting from failure of the pancreas to release insulin. None of this is to say that the glucocorticoids and insulin are not in some respects synergistic. This is apparently true in the case of fat deposition; we find that the increase in blood sugar produced by the injection of glucocorticoids usually results in increased secretion of insulin by the pancreas. Together the circulating glucocorticoids (facilitating protein catabolism) and insulin (facilitating CHO catabolism) facilitate an increased deposition of fat in the adipose tissue of the face, suprascapular area, and trunk. Under extreme conditions this can produce the appearance of the moonface, the buffalo hump, and the pendulous abdomen seen in Cushing's disease (Fig. 31-10).

Water balance. Adrenalectomized animals and patients suffering from Addison's disease also have difficulty handling a water load. A normal subject starts increasing his urine volume within an hour after drinking water (Fig. 31-8), and this increase continues for 4 to 6 hr., while a man with Addison's disease may not dispose of a water load for over 12 hr. If the water input is sufficient, water intoxication may even develop. This problem cannot be successfully treated by the injection of aldosterone but can be diminished by the injection of a glucocorticoid such as cortisol or cortisone. Their means of action, while not fully understood, probably involves the direct action of the glucocorticoid upon the tubules of the nephron to decrease their permeability to water. In this way they decrease the reabsorption of water from the urine. Another factor that may be important is the ability

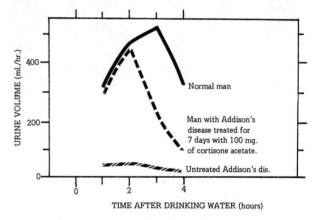

Fig. 31-8. Response to an oral dose of 20 ml. of water per kg. of body weight. Note that the normal subject produced a prompt increase in urine production in response to a water load, but that the subject with Addison's disease (adrenocortical hypofunction) did not. Pretreatment of the Addisonian patient with cortisone (100 mg. of cortisone acetate per day) for 7 days, however, resulted in a more normal response. Cortisone is produced in the liver from cortisol and has a number of properties similar to cortisol. Although the cortisone produced by the liver normally plays no physiological role in the body, its injection is a frequently used clinical procedure. (*Data from* Forsham, P. H.: The adrenal cortex. *In* Williams, R. H. (ed.): Textbook of Endocrinology. 4th ed., p. 328. Philadelphia, W. B. Saunders, 1968)

of the glucocorticoids to (1) increase the glomerular filtration rate and renal plasma flow in the patient with Addison's disease, (2) facilitate the destruction of the antidiuretic hormone by the liver, and (3) possibly decrease the reabsorption of Na^+ from the urine by competing with aldosterone for receptor sites. In the latter case it is hypothesized that although cortisol, for example, only slightly facilitates the reabsorption of Na^+ from the nephron, it might by combining with the aldosterone receptor sites decrease their stimulation by aldosterone and thus decrease Na^+ reabsorption and the associated reabsorption of H_2O.

Water metabolism. Cortisol enhances water diuresis in response to a water load; in its absence a water load may not be disposed of for 12 or more hours (Fig. 31-8). If the water load is sufficiently great in adrenal cortical malfunction, it can lead to water intoxication. The mechanism whereby cortisol facilitates diuresis is not clear. There is some evidence, however, that it is a direct antagonist of antidiuretic hormone at the level of the renal tubule and that it may also enhance its destruction by the liver.

Central nervous system. Although cortisol does not penetrate the blood-brain carrier, there are a number of nervous symptoms associated with adrenal cortical insufficiency that are successfully treated by cortisol and other glucocorticoids. These include nervousness, apathy, negativism, inability to concentrate, periods of drowsiness alternating with restlessness and insomnia, and alterations in the electroencephalogram.

Stress. When a person is exposed to any of a number of stresses, his output of ACTH from the anterior pituitary increases and as a result the cortisol concentration in the plasma also rises. Some of the acute medical stresses acting to increase the plasma concentration of cortisol are listed in Table 31-4. Other stresses producing similar elevations in plasma cortisol include (1) trauma and burns, (2) surgical operations, (3) electroshock therapy, (4) exercise, (5) hypoglycemia, (6) pyrogens, and (7) emotional stress. A study of surgical shock is presented in Figure 31-9. Characteristically after surgery there is a prompt elevation in the plasma and urinary 17-hydroxycorticoids (cortisol, cortisone, and others).

	Age (yrs.)	First Day Cortisol Secretion (mg./24 hr.)	Plasma Cortisol (μg./100 ml.)	Urinary 17-OHCS (mg./24 hr.)	Seventh Day Cortisol Secretion (mg./24 hr.)
Normal (8 A.M.)	20	21	12	9	21
Pneumonia	18	38	26	12	22
Normal	30	24	12	8.5	24
Severe diabetic ketosis	36	225	65	60	73
Myocardial infarct (minor)	39	50	19	28	25
Myocardial infarct (cardiac arrest)	46	136	65	17	38
Subarachnoid hemorrhage	49	43	24	27	22
Pneumonia, alcoholism	52	143	28	15	23
Myocardial infarct (severe)	56	72	57	24	30
Normal	60	18	10	6.1	18
Pneumonia	60	157	37	40	72
Pneumonia	72	69	31	12	21

Table 31-4. Adrenocortical and metabolic response to acute medical stress. 17-OHCS, 17-hydroxycorticosteroids. (Data in part from Wolstenholme, G.E.W., and Porter, R.: The Human Adrenal Cortex. Boston, Little, Brown & Co., 1967)

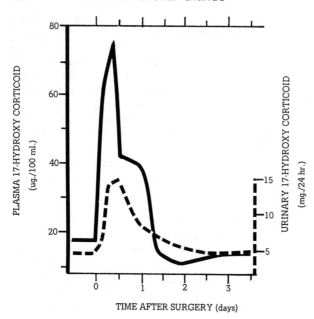

Fig. 31-9. Response of a 42-year-old man to surgery (resection of the sigmoid colon, ether anesthesia). (*Modified from* Le Femine, A. A.: The adrenal cortical response in surgical patients. Ann. Surg., *146*:29, 1957)

neoplasm that results in a large output of either ACTH or glucocorticoids. These neoplasms have been identified not only in the pituitary and adrenal cortex but also in the spleen, testes, and ovaries. All of these conditions are now called Cushing's syndrome.

Apparently many of the symptoms in Cushing's syndrome are due to the facilitation of protein catabolism by cortisol. This increased breakdown of protein produces hyperglycemia and excessive deposition of fat, as well as a decrease in the thickness of the skin and subcutaneous tissues. As the fat depots increase in size, the skin is stretched and the subdermal tissues rupture to form red and purple striae. The muscles atrophy, wounds heal poorly, bruises become common, and the hair becomes thin and straggly. The loss of protein from bone results in its demineralization. This is further accelerated by the anti-vitamin D action of cortisol and the increased glomerular filtration which the cortisol produces.

Some of the other symptoms include mental aberrations ranging from insomnia and euphoria to marked psychoses. At the high concentrations of cortisol found in Cushing's disease one also finds this hormone exerting a mineralocorticoid action (Table 31-3). About 85 per cent of such

A maximum level is generally reached in 4 to 6 hr. and there is a return to normal levels in 36 to 48 hr.

For a patient to survive a major operation it is essential that there be a functioning adrenal cortex or that glucocorticoid injections be administered. In the absence of plasma glucocorticoids the patient generally suffers a circulatory collapse. The mechanism by which the glucocorticoids prevent this collapse is not clear. The elevation of glucocorticoids in response to stress is apparently the result of nerve impulses going to the hypothalamus; we find that we can prevent the elevation of glucocorticoids in response to burning an appendage if we first denervate the appendage. The role of the glucocorticoids themselves is poorly defined, though they are known to be essential for the vasoconstrictive response of blood vessels to norepinephrine and epinephrine; possibly it is through this mechanism that they prevent circulatory collapse during stress.

Cushing's syndrome. Harvey Cushing in 1932 described a patient with "pituitary basophilism" associated with (1) abdominal abesity, (2) thin extremities, and (3) cutaneous striae (Fig. 31-10). It was later noted that these symptoms are also produced by excess stress or by any

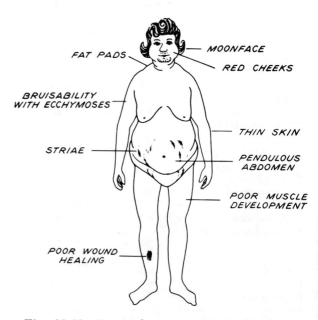

Fig. 31-10. Outstanding signs in Cushing's syndrome. (Forsham, P. H.: The adrenal cortex. *In* Williams, R. H. (ed.): Textbook of Endocrinology. 4th ed., p. 341. Philadelphia, W. B. Saunders, 1968)

patients are hypertensive, partly because of Na⁺ and water retention and partly because of the facilitation by cortisol of the action of catecholamines on the blood vessels. In some cases there is also an increase in the aldosterone and androgen output of the cortex; under these circumstances the Na⁺ retention and K⁺ loss may be exaggerated and a tendency toward virilization occurs in the female patient.

Adrenocortical hypofunction. Adrenal cortical insufficiency may be primary, as in the case of *Addison's disease,* or secondary, as in pituitary ACTH hyposecretion. It can be precipitated by (1) highly emotional experiences, (2) morphine sedation, (3) tranquilizers that impair the release of ACTH, (4) infection, (5) hemorrhage, or (6) surgery. In Addison's disease there is excessive pigmentation, while in hypopituitarism there is generally little or no increase in pigmentation. In both conditions, however, there is hypoglycemia, inability to withstand stress, hypotension, and if of long standing, a decrease in heart size. Weakness, anemia, dizziness, nervousness, and other mental symptoms are also usually present. Many of these symptoms are apparently due to (1) a decrease in the blood sugar and (2) increased sensitivity to hypoglycemia.

Physiological Roles of the Mineralocorticoids

The roles of the mineralocorticoids were discussed in Chapter 21, and their actions summarized in Figures 21-8 and 21-9. You will remember from Tables 31-1 and 31-3 that the mineralocorticoid released in greatest quantities is aldosterone, and that of the mineralocorticoids produced in the body it is the most effective in facilitating Na⁺ reabsorption from the urine, sweat, saliva, and gastric juice. Desoxycorticosterone is also released by the adrenal cortex, but is normally released only in trace amounts and is considerably less potent in its effect on Na⁺-K⁺ balance. Cortisol and corticosterone have weak mineralocorticoid action but are released by the cortex in sufficiently large quantities to have a mild Na⁺ conservation effect in addition to their glucocorticoid action. In this section, however, we shall concentrate on the most important mineralocorticoid, *aldosterone*.

Mechanism of action. Aldosterone apparently facilitates the formation of messenger RNA from DNA in the nucleus. The messenger RNA facilitates the formation of certain enzymes in the ribosomes that are important in accelerating Na⁺ transport in sensitive cells. Consistent with this view is the observation that aldosterone has a period of latency of more than 10 min. It is hypothesized that the reason there is an interval of 10 min. from the time aldosterone reaches the renal artery until Na⁺ reabsorption from the urine increases is that it takes that amount of time for the nucleus to significantly increase its production of messenger RNA and for that RNA to reach the ribosomes and affect their production of enzymes.

Na⁺-K⁺ balance. The net result of the action of aldosterone is a decrease in the loss of Na⁺ and an increase in the loss of K⁺ and H⁺ from the body. Aldosterone does these things by facilitating the movement of Na⁺ from the cells of the nephron, sweat glands, salivary glands, and stomach into the interstitial fluid surrounding the capillaries of the body. Associated with this movement of Na⁺ out of the urine, sweat, and so on is an influx of H⁺ and K⁺ into these fluids. Thus the urine in the presence of aldosterone becomes more acid, and its K⁺ concentration increases while its Na⁺ concentration decreases.

The importance of these phenomena to the body lies not only in the conservation of Na⁺ and the excretion of H⁺ and K⁺, but also in the maintenance of blood volume and therefore blood pressure. Increases in the concentration of plasma Na⁺ increase blood volume by drawing water from the intracellular compartment and act through the central nervous system to initiate thirst and an increase in water intake. On the other hand, decreases in plasma Na⁺, effective blood volume, blood pressure, and renal blood flow facilitate the release of aldosterone. This regulatory arrangement was discussed in detail in Chapter 21 (Fig. 21-9) and can be summarized as shown in the diagram at the top of page 534.

Regulation of aldosterone secretion. The release of aldosterone by the adrenal cortex is regulated primarily through the secretion of renin by the kidneys, but also by (1) the secretion of ACTH by the pituitary and (2) the direct effect of plasma Na⁺ and K⁺ levels on the adrenal cortex. Motor nerves to the adrenals apparently play little or no role in controlling their secretions. Throughout the day ACTH exerts an important influence on the release of glucocorticoids, but not usually on that of mineralocorticoids. We find that hypophysectomy or destruction of the hypothalamus has no immediate effect on the

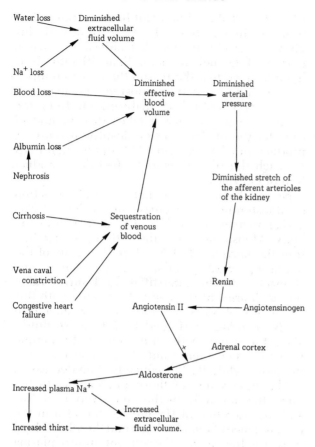

secretion of aldosterone but profoundly decreases the secretion of cortisol. After a number of weeks, the absence of ACTH from the blood results in such marked adrenal cortical atrophy that even the secretion of aldosterone decreases (Fig. 31-1). On the other hand, in cases of stress the ACTH level in the blood increases to such an extent that there are marked increases in the release not only of glucocorticoids but of mineralocorticoids and sex hormones as well.

The importance of the kidney in controlling aldosterone secretion lies in its capacity to secrete renin in response to diminished stretching of its afferent arteriole. This stretch is reduced by decreased renal arterial pressure. The receptors lie either in the afferent arteriole or in the adjacent macula densa and cause the juxtaglomerular apparatus to release renin, which eventually enters the bloodstream and catabolizes the conversion of a polypeptide produced by the liver (angiotensinogen, or α-globulin) to angiotensin I. The angiotensin I (a 10 amino acid polypeptide) is then catabolized to angiotensin II (an 8 amino acid molecule), which acts on the adrenal cortex to increase its output of aldosterone. It also weakly facilitates the secretion of corticosterone, desoxycorticosterone, and cortisol.

Hyperaldosteronism. One of the more common causes of abnormally high concentrations of aldosterone in the blood is an adrenocortical tumor. In this condition there is an increase in arterial pressure due to hypervolemia (incidence 100%), a decrease in plasma K^+ (100%), and systemic alkalosis (95%). The decrease in K^+ results in muscle weakness (58%), and if of long duration also causes severe damage to the distal tubule and collecting duct cells of the kidney. This causes the impairment of water reabsorption from the urine and a refractoriness to antidiuretic hormone. The result is polyuria (92%) and an increase in water intake (84%). Renal damage also leads to albuminuria. The decreased plasma K^+ concentration and alkalosis bring about marked changes in excitability and a type of tetany associated with cramps or positive Chvostek and Trousseau signs.

THE ADRENAL MEDULLA

The adrenal medulla is a gland containing about 0.1 mg. of norepinephrine and 0.6 mg. of epinephrine per gram of tissue. It is innervated by preganglionic sympathetic neurons, whose function is controlled largely by impulses arising in the hypothalamus. In response to the stimulation of these neurons, the adrenal medulla releases catecholamines, about 80 per cent of which are epinephrine and 20 per cent norepinephrine, the latter being a precursor of epinephrine. Both of these catecholamines produce hyperglycemia, an increase in the release of fatty acids into the blood, increased heat production, vasoconstriction of the visceral and cutaneous blood vessels, and stimulation of the central nervous system. Norepinephrine in addition produces vasoconstriction in skeletal muscle and a general increase in peripheral vascular resistance. Epinephrine, on the other hand, produces vasodilation in skeletal muscle, a decrease in peripheral resistance, an increase in heart rate, an increase in cardiac contractility, a positive dromotropic action on the heart, an increase in cardiac output, and relaxation of the bronchial musculature.

The concentration of epinephrine in the plasma is usually approximately 0.1 μg. per liter, and the concentration of norepinephrine, about 0.8 µg. per liter. When the adrenals are removed, the concentration of epinephrine falls but that of norepinephrine stays fairly constant. This is apparently due to the fact that most of the epinephrine found in the blood is released by the adrenal medulla, whereas most of the norepinephrine comes from postganglionic adrenergic sympathetic neurons. The epinephrine found in tissues other than the adrenal medulla and brain is apparently produced for the most part in the adrenal gland, released into the blood, picked up by the tissue, and stored. It has been suggested that the stimulation of cardiac sympathetics results in the release of epinephrine picked up by cardiac tissue from the blood at an earlier time, and of epinephrine produced by the postganglionic sympathetics.

The excretion of catecholamines was discussed in Chapter 4. There we noted that catecholamines are rapidly oxidized and methylated in the liver and other tissues, and are eliminated in the urine. Most of them are lost in the urine in the form of vanillymandelic acid (1.8-9.0 mg./day) — only small quantities of unchanged norepinephrine (40 µg/day) and epinephrine (0-20 µg./day) normally being lost by this route.

Pheochromocytoma

Apparently the adrenal medulla is normally a minor source of catecholamines for the body. The catecholamine content of the blood remains at a near normal level when the adrenal medulla is removed, and the individual seems to function as well without it. Tumors of the adrenal medulla, on the other hand, can be a serious problem. These tumors (pheochromocytomas) may result in sustained elevation of the level of catecholamines in the blood or may produce periodic elevations. The symptoms of pheochromocytoma are what one might expect from an excessive output of catecholamines — i.e., hypertension, tachycardia, an increased fasting blood sugar, and decreased gastrointestinal motility. Headache, loss of appetite, and psychic changes are also common.

Chapter 32

CALCIUM METABOLISM, THE PARATHYROID GLANDS, AND OTHER ENDOCRINE GLANDS

CALCIUM METABOLISM
THE PARATHYROID GLAND
THYROCALCITONIN
OTHER ENDOCRINE GLANDS

In the preceding chapters we have discussed (1) the releasing factors and prolactin-inhibiting factor, (2) oxytocin and antidiuretic hormone, (3) the six hormones released by the anterior pituitary, (4) thyroxine and triiodothyronine, (5) insulin and glucagon, and (6) the hormones produced by the adrenal cortex and medulla. Some of these hormones control the growth of glands and other organs, the production and secretion of hormones, or both. Others control metabolic activity, and still others are concerned with the contraction of smooth, cardiac, and skeletal muscles and the function of the kidney. This section will deal primarily with the control of calcium metabolism, though some of the endocrine glands that have proven important in animals other than man will also be discussed. The next chapter will have to do with hormones directly involved in the control of the reproductive system and the development of the secondary sexual characteristics.

CALCIUM METABOLISM

There are in the body approximately 1,000 gm. of calcium, distributed in the following manner:

In bone in a stable form	1,000,000 mg.
In bone in a readily exchangeable form	4,000 mg.
In the extracellular fluid	900 mg.
In the intracellular fluid	11,000 mg.

Most of the calcium in bone exists as mineral crystals of *hydroxyapatite* ($Ca_{10}(PO_4)_6(OH)_2$). These crystals measure 200 A by 30 to 70 A and are laid down on networks of collagen fiber in the bone, where they add strength and rigidity. The calcium salts serve not only as an important structural unit but also as a reservoir of calcium and phosphate for the body. Bone too is a reservoir for magnesium, sodium, and carbonate. Both the stable and exchangeable forms of the calcium-containing molecules of bone serve as dynamic reservoirs, acquiring calcium form the plasma when there is an excess and supplying calcium when there is a deficit.

The distribution of calcium in the plasma is shown in Table 32-1. Note that almost half of it exists as Ca^{++} and the rest in various bound forms. The Ca^{++} is particularly important in the blood in that it is essential for blood coagulation and moves freely into the intercellular spaces, where it modifies the excitability of the cell membrane, is used to form intercellular cement, or diffuses into the cell.

Inside the cell, Ca^{++} is incorporated into the cytoplasm and some is stored in a labile form, i.e., is sequestered. This sequestered Ca^{++} in

	Millimols/liter	% of Total
Calcium		
Free ions (Ca^{++})	1.16	47.5
Protein bound	1.14	46
$CaHPO_4$	0.04	1.6
Ca citrate	0.04	1.7
Unidentified complexes	0.07	3.2
Total	2.45	100
Phosphate		
Free HPO_4^{--}	0.50	43
Free $H_2PO_4^{-}$	0.11	10
Protein bound	0.14	12
$NaHPO_4^{-}$	0.33	29
$CaHPO_4$	0.04	3
$MgHPO_4$	0.03	3
Total	1.15	100

Table 32-1. Forms of calcium and phosphate in normal plasma. (From Walser, M.: Ion association. IV. Interactions between calcium, magnesium, inorganic phosphate, citrate and protein in normal human plasma. J. Clin. Invest., *40*:728, 1961)

nerve and gland cells is released intracellularly in response to an action potential and brings about the secretion of acetylcholine, an enzyme, a hormone, or some other compound. In skeletal, cardiac, and smooth muscles, intracellular Ca^{++} is also released intracellularly and initiates in these cases the contractile process. It is believed that within limits, the quantity of material secreted by a cell or the force of contraction exerted by a cell is dependent upon the amount of intracellular Ca^{++} released. In other words, Ca^{++} is characterized as the activation-secretion and the activation-contraction coupling principle. Hence a depletion of intracellular Ca^{++} can lead to a decrease in secretory or contractile responses to stimulation.

Calcium is also a structural part of the cell membrane and as such tends to stabilize it. Increases in extracellular Ca^{++} therefore make the plasma membrane less excitable, and decreases in Ca^{++} make it less stable and more excitable. Calcium ions, however, are not the only ones that tend to increase excitability. We find that increases in K^+, HCO_3^-, or HPO_4^-, and decreases in Ca^{++} Mg^{++}, and H^+ all tend to produce hyperreflexia and an increase in muscle tone generally called tetany:

$$\text{Tendency toward tetany} \cong \frac{(K^+)(HCO_3^-)(HPO_4^{--})}{(Ca^{++})(Mg^{++})(H^+)}$$

In hypercalcemia the resultant hypoexcitability is reflected in decreased muscle tone, dryness of the nose, difficulty in swallowing, constipation, cardiac irregularities, polyuria, decreased responsiveness to hormones and drugs, shortening of the Q-T interval of the electrocardiogram, and the formation of calcium-containing kidney stones.

Hypocalcemia, on the other hand, leads to a generalized increase in neuromuscular excitability. There may be a number of psychic changes ranging anywhere from increased irritability to delirium and delusions. Also common are hyperreflexia, increases in skeletal-muscle tone, cramps of the extremities, and carpopedal spasms (Fig. 32-1), which may in a few cases be followed by generalized convulsions and a laryngeal spasm leading to asphyxiation and death. Some of the signs of hyperreflexia frequently looked for are Chvostek's sign and Trousseau's sign. In the former the physician gives a sharp tap over the facial nerve in front of the patient's ear. A positive Chvostek sign is a twitching of the facial muscles, particularly the upper lip. In the second

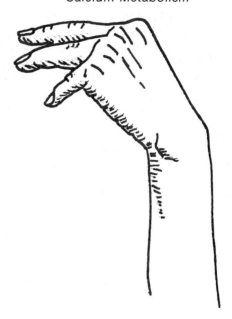

Fig. 32-1. A positive Trousseau sign. This position of the hand is found in carpopedal spasm due to hypocalcemia. (*Redrawn from* Aub, J. C.: *In* Adams, F. D.: Physical Diagnosis. 14th ed., p. 167; copyright © 1958, The Williams and Wilkins Co., Baltimore)

test, a blood pressure cuff is applied to the arm for 3 min. to completely cut off flow. If this results in a carpopedal spasm (Fig. 32-1), the patient is said to show a positive Trousseau sign. These tests are rarely positive in normal adults.

Control

The amount of calcium in the plasma and hence in the extracellular fluid is dependent upon (1) the amount of calcium in the diet, (2) the amount of vitamin D in the body, (3) kidney function, and (4) bone function. We discussed the general problem of dietary calcium in Chapter 24, and noted that Ca^{++} absorption from the intestinal tract depends upon the concentration of parathormone in the blood and the presence of vitamin D. Apparently parathormone facilitates the absorption of Ca^{++} into the blood, but under normal conditions 90 per cent of the dietary Ca^{++} remains in the GI tract and appears in the feces. Most of the excess Ca^{++} in the bloodstream, however, is either picked up by the bone or eliminated in the urine. On the average about 100 mg. of Ca^{++} are lost every 24 hr. in the urine. This is, of course, quite small when compared to the 10,000 mg. of Ca^{++} that normally appear in the glomerular filtrate each day. The reabsorption of Ca^{++} from this filtrate back into the blood was discussed in Chapter 21. There it was noted that the reabsorption of Ca^{++} from the glomerular filtrate is facilitated by parathryoid hormone. We also noted that parathyroid hormone facilitates the release of Ca^{++} from bone into the plasma.

Vitamin D. The D vitamins are a group of sterols which have a marked effect on mineral metabolism. They are regarded by many as essential dietary constituents, but it is known that a provitamin D can be changed in the skin to vitamin D_3 in the presence of sunlight. Vitamin D_3 seems to have most of the physiological properties of dietary vitamin D and can, since it is produced in the body, be considered a hormone.

There are normally between 25 and 40 mμg. of vitamin D per liter in the plasma. Most of this is bound, but some is apparently in an active form. Vitamin D, like parathormone, increases the absorption of calcium from the GI tract and increases its concentration in the plasma. It also increases the phosphate concentration in the plasma and has a growth-promoting effect. Unlike parathormone, it has little or no direct effect on the kidney, and therefore when used to treat hypoparathyroidism produces a higher concentration of urinary calcium and a lower concentration of urinary phosphate than would the injection of parathormone. You will remember that parathormone acts on the kidney to inhibit calcium excretion and to facilitate phosphate excretion.

THE PARATHYROID GLAND

There are usually two pairs of parathyroid glands in man, each situated close to the posterior surface of the thyroid glands. In some persons, however, aberrant glands may be found in the mediastinum, within the thyroid, and in some rare cases dorsal to the esophagus. The average size of a single human gland is approximately 5 mm. by 5 mm. by 3 mm., and the combined weight of the four glands is about 120 mg. They were first described by Sandstrom in 1880, but not until 1891 was it realized that parathyroidectomy results in tetany (increased muscle tone and hyperreflexia). Prior to that time it was believed that removal of the thyroid was responsible for tetany. If the thyroid gland is removed but the associated parathyroid glands are left intact, tetany does not occur.

Function

Apparently the sole function of the parathyroid glands is to produce, store, and secrete parathormone. The amount of parathormone secreted is inversely related to the plasma concentration of Ca^{++} (Fig. 32-2). In other words, within limits, plasma Ca^{++} depresses the output of parathormone into plasma. The parathormone, on the other hand, acts to increase the plasma Ca^{++} concentration. These relationships were discussed in Chapter 21.

Parathormone increases plasma Ca^{++} by (1) acting on the kidney to decrease its elimination of Ca^{++} in the urine, (2) acting on the GI tract to increase its absorption of Ca^{++} from the chyme,

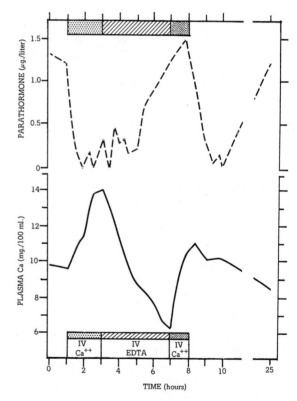

Fig. 32-2. Changes in the concentration of parathormone in the cow in response to (1) intravenous Ca^{++} and (2) the removal of Ca^{++} from the plasma by the infusion of EDTA, a chelating agent. (*Redrawn from* Care, A. D., et al.: Evaluation by radioimmunoassay of factors controlling the secretion of parathyroid hormone. Nature, *209*:52-55, 1966)

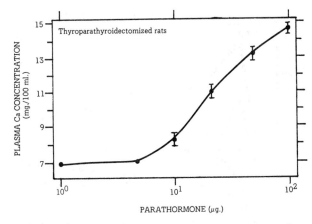

Fig. 32-3. Response of the thyroparathyroidectomized rat to parathormone. Each point represents a mean value, ± its standard error for 6 rats. (*Redrawn from* Raisz, L. D.: Bone resorption in tissue culture. J. Clin. Invest., 44:130, 1965)

and (3) acting on bone and possibly other organs to increase their contribution of Ca^{++} to the plasma (Fig. 32-3). Parathormone also affects phosphate and K^+ metabolism and excretion. It increases the movement of phosphate from bone and other tissues to plasma and increases the renal excretion of both phosphate and K^+.

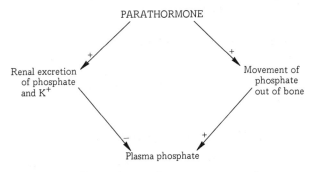

The net effect of parathormone is to decrease plasma phosphate.

Hypoparathyroidism. Removal of the parathyroid glands in man eventually changes the plasma calcium concentration from a normal value of 10 mg./100 ml. to one of 7 mg./100 ml. In most subjects it takes 2 to 3 days after the parathyroidectomy for the more serious symptoms to develop (including increases in muscle tone and hyperreflexia due to the hypocalcemia). In some cases, however, these symptoms may take more than 2 wks. to appear. There is also an initial increase in urinary Ca^{++} and a decrease in urinary phosphate and K^+, though in time these levels return to normal. Since the major problem in hypoparathyroidism is hypocalcemia, it can be quite successfully treated with parathormone, vitamin D, or Ca^{++}. Parathormone, however, has a relatively short half-life and takes only minutes to act, whereas vitamin D takes several days to produce its effect.

Hyperparathyroidism. Hyperparathyroidism is considerably more common than hypoparathyroidism. It has been estimated to be present in 1 out of every 800 patients seeking medical attention. In 83 per cent of these cases it is the result of a single adenoma. The symptoms include hypercalcemia, a decrease in plasma phosphate, increased absorption of calcium and phosphate from the intestine, inhibition of bone formation, and generally a sufficient increase in demineralization to produce osteoporosis or osteitis fibrosa cystica. The demineralization may be associated with vague skeletal pains (20% of cases) and a history of frequent fractures (3% of cases). The formation of renal calculi is also common and is frequently associated with renal colic, hematuria, "the passing of sand," dull back pain, and urinary tract infection.

THYROCALCITONIN

A thyroidectomy in which the parathyroid glands are left intact produces a more prolonged elevation of plasma Ca^{++} in response to either the intravenous injection of Ca^{++} or the injection of parathormone than is normally the case. Apparently the thyroid glands release a polypeptide, thyrocalcitonin, in response to a high plasma Ca^{++}. This substance tends to bring down the plasma Ca^{++} concentration (Figs. 32-4 and 32-5). In other words, the thyroid glands release a hormone that in many ways functions in a manner opposite to that of parathormone:

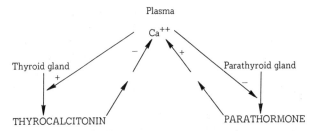

Unlike parathormone it has not been shown to have a direct action on either the kidney or gastrointestinal tract, but it does have an action on

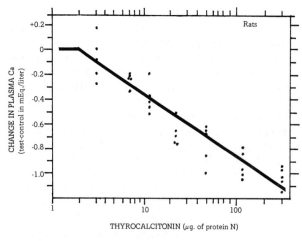

Fig. 32-4. The dose-response curve obtained by injecting thyrocalcitonin intravenously into rats and measuring the change in plasma calcium concentration 1 hour later. (*Redrawn from* Kumar, M. A., et al.: A biological assay for calcitonin. J. Endocrinol., 33:472, 1965)

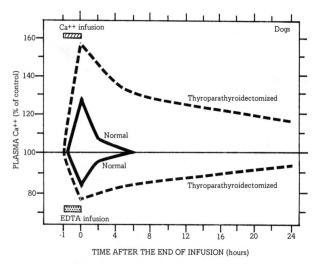

Fig. 32-5. Role of the thyroid and parathyroid glands in the control of the plasma Ca^{++} concentration. In these experiments dogs were studied before (normal) and several months after bilateral thyroparathyroidectomy. Note that the normal dog restores his plasma Ca^{++} to control levels after an increase in plasma Ca^{++} and a decrease in plasma Ca^{++} (infusion of a Ca^{++} chelating agent, EDTA) more rapidly than a dog which has had his thyroid and parathyroid glands removed. (*Data from* Sanderson, P. H.: Calcium and phosphorus homeostasis in the parathyroidectomized dog; evaluation by means of ethylenediamine tetraacetate and calcium tolerance tests. J. Clin Invest., 39:665, 667, 1960)

bone which is at least in part antagonistic to that of parathormone. It apparently acts on bone to inhibit its release of Ca^{++} and phosphate into the plasma; it is through this mechanism that it brings about decreases in both the plasma Ca^{++} and phosphate. When administered over a prolonged period it may even produce marked increases in the skeletal mass throughout the body. It is hoped that when thyrocalcitonin becomes clinically available it may be useful in preventing osteoporosis in certain diseases. Some preliminary experiments in which porcine thyrocalcitonin has been injected into human subjects have been disappointing in this regard, but human and porcine thyrocalcitonin are immunologically different compounds.

Hypersecretion and Hyposecretion

Interest in thyrocalcitonin developed several years after the pioneer experiments of Copp et al. in 1962. It is not, therefore, surprising that no specific disorders of thyrocalcitonin release have yet been established. There have been scattered reports of persons exhibiting a tetany associated with osteopetrosis (increased mineralization of the bones), and the suggestion has been made that this may be the result of hypersecretion of thyrocalcitonin. Hypercalcemia, on the other hand, is considerably more common than hypocalcemia and in past decades has frequently been attributed to hypersecretion of parathormone. As techniques develop, it will be interesting to determine if impaired secretion of thyrocalcitonin is the cause of at least some of these hypercalcemias.

OTHER ENDOCRINE GLANDS

In this chapter and in Chapters 28 through 31 we have been concerned primarily with the hormones released by the hypothalamus, the anterior pituitary, the thyroid gland, the pancreas, the adrenal gland, and the parathyroid gland. In earlier chapters we also discussed two hormones produced by the kidney. These are erythropoietin, which stimulates the production of red blood cells by the bone marrow, and renin, which helps catalyze the formation of angiotensin II from α_2-globulin. The α_2-globulin is produced in the liver and released into the blood. We have also discussed a number of hormones produced in the intestinal tract (Chap. 25) that modify gastrointestinal motility and secretion, as well as the exocrine secretions of the pancreas and liver. In the next chapter we shall discuss the secretion of three important reproductive organs: the testis, the ovary, and the placenta. Many of these glands and some of the hormones they secrete are listed in Table 28-1.

In addition to the glands discussed above there are some other structures in the body that may produce hormones. For the most part, however, they have yet to be proven of importance in normal human physiology.

The Intermediate Lobe of the Pituitary

The intermediate lobe of the pituitary produces a polypeptide, *melanocyte-stimulating hormone (MSH)*, which acts on cells (melanophores) in the skin of fish, reptiles, and amphibians to cause their dense melanin granules to move away from the nucleus of the cell and out into the cytoplasm. This results in a darkening of the cell and therefore of the skin. The secretion of this hormone is probably facilitated by an MSH-releasing factor (MIF), both of which are produced in the hypothalamus.

In man and other mammals there are melanin-containing cells called melanocytes, but apparently under physiological conditions insufficient MSH is released to affect the function. On the other hand, injecting Negroes with MSH over a long period has produced darkening of their skin. Certain tumors resulting in an excessive output of either ACTH or MSH have been reported to produce hyperpigmentation in both black and white patients.

The Pineal Gland

The pineal gland (epiphysis) arises from the roof of the third ventricle of the brain and lies under the posterior end of the corpus callosum. Because of its central position in the brain, Descartes (1596-1650) believed it to be the seat of the soul. More recently, its high content of serotonin, norepinephrine, and melatonin has been emphasized. In lower vertebrates it has been shown to contain both nerve cells and light receptors, and apparently in the brontosaurs and other ancient forms constituted a third eye. In man and other high vertebrates the pineal gland contains no light-sensitive elements. In birds and frogs, however, there is some evidence that light acts on the animal to inhibit the secretion of melatonin from the pineal gland. *Melatonin* in turn (1) produces blanching of the skin (melanophore contracting effect) and (2) inhibition of the ovary. Thus it is hypothesized that in spring when the days grow longer and the light more intense, a decrease in the secretion of melatonin in birds results in (1) darkening of the skin, (2) an increase in the release of hormones by the ovary, and (3) a migration north.

In man, on the other hand, there is considerably less evidence of a physiological role for the pineal gland, though pineal tumors in the human can produce either precocious puberty or a delay in the onset of puberty.

The Thymus

The thymus is a lymphoid organ located in the superior mediastinum and lower part of the neck. It reaches its maximum size at about the second year of life and thereafter undergoes retrograde changes. It is absent in the adult. There is, however, some evidence that in infants it produces a hormone that stimulates the development of potential antibody-forming cells in the spleen, bone marrow, and lymph nodes (see p. 268). We find that mice thymectomized in early infancy do not reject tissue transplants from other animals. Evidence has also been presented that some of the auto-immune diseases are due to a malfunction of the thymus.

Chapter 33

**BONE
DEVELOPMENT OF THE
REPRODUCTIVE SYSTEM
THE ADULT MALE**

GROWTH, DEVELOPMENT, AND REPRODUCTION

The phenomenon of growth begins with fertilization of the ovum. At first there are a series of cell divisions and a laying down of cytoplasm, and eventually segregation of the cells into a developing ectoderm, mesoderm, and endoderm. The egg 2 months after fertilization is called a fetus. At this time it is about 8 cm. long and has the appearance of a human being. After 7 months of pregnancy the fetus is 38 cm. long and weighs 0.9 kg. (2 lb.). At birth it is 48 to 54 cm. long and weighs 3 to 4 kg. (6-8 lb.). In the uterus its systems have developed, its heart has begun contracting, blood cells have been formed (Fig. 15-15), and certain cells have specialized to the extent that they have lost the capacity to divide (Fig. 1-2). Many of these cells, however, retain the capacity to lay down additional cytoplasm—i.e., to hypertrophy.

After birth, the child continues to grow and differentiate and some of its cells continue to undergo hyperplasia and hypertrophy (Fig. 33-1). This growth is in part facilitated by growth hormone, thyroxine, triiodothyronine, androgens, and estrogens, and is in part a result of the intrinsic properties of each cell in the body. Growth hormone, for example, probably reaches its highest resting plasma level soon after birth and then decreases to the adult resting level (0.5 µg./liter) in the 2-year-old. It apparently stays relatively close to this value from then on, except when there are changes in the glucose concentration of the blood. As mentioned in Chapter 28, the pituitary releases additional growth hormone in response to a decrease in the concentration of plasma glucose. Androgens and estrogens, on the other hand, remain at fairly low plasma levels until the onset of sexual puberty (Fig. 33-2).

The growth occurring in the fertilized egg, fetus, and child is disproportionate. In the erect individual at birth the distance from the top of the pubic symphysis to the top of the head (upper segment of body) is 1.7 times the distance from the top of the pubic symphysis to the floor (lower segment). At the age of 10, on the other hand, the child's upper segment is equal in length to his lower segment (Figs. 33-3 and 33-4). There is also, starting at about 6 months and lasting until 24 months of age, the appearance of 10 pairs of deciduous teeth. Starting at 6 yrs. and lasting until the age of 21, the replacement of the de-

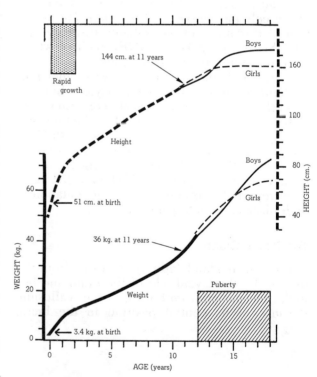

Fig. 33-1. Heights and weights of children at different ages. Note that there are 2 periods of rapid growth (*shaded area*), the first just preceding and following birth and the second at the onset of puberty. Note too that the second period of rapid growth for girls precedes that found in boys. This is due to the earlier onset of puberty in girls.

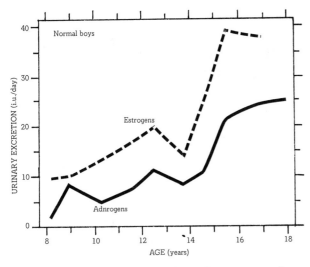

Fig. 33-2. The excretion of sex hormones in boys (i.u., international units). (*Data from* Greulich, W. W., et al.: Somatic and endocrine studies of puperal and adolescent boys. Monogr. Soc. Res. Child Develop., 7:26, 27, 1942)

	Appearance of Primary Teeth (months)	Appearance of Permanent Teeth (years)
Upper teeth		
Central incisors	6-7½	7-8
Lateral incisors	7-9	8-9
Canines	16-18	11-12
First bicuspids		10-11
Second bicuspids		10-12
First molars	12-14	6-7
Second molars	18-24	12-13
Third molars		17-21
Lower teeth		
Central incisors	6	6-7
Lateral incisors	6-7	7-8
Canines	14-16	9-10
First bicuspids		10-12
Second bicuspids		11-12
First molars	10-12	6-7
Second molars	16-20	12-13
Third molars		17-21

Table 33-1. The time at which the various teeth appear.

ciduous teeth with 16 pairs of permanent teeth occurs (Table 33-1). During adolescence there is also a general lengthening of the nose and jaw. In the hypothyroid child the ratio between the lengths of the upper and lower segments remains near the juvenile level (1.7) and the eruption of the teeth is delayed. In the eunuch, the ratio is less than that for a normal person.

BONE

There are two types of bone in the body: membranous and cartilaginous. *Membranous bone* begins as soft connective tissue membrane that becomes infiltrated with osteoblasts. These osteoblasts lay down a matrix of calcium and phosphate that changes the membrane into a hard, strong, nondistensible covering. The sites where clusters of osteoblasts first appear are called centers of ossification. In the fetus the cranium is formed by such a process. At birth, however, the process of ossification is not complete. As a result there are areas of the head in a newborn child where the brain has not yet been enclosed by a hard, calcified cranium. This occurs where two or more bones come together. Two of the largest soft spots in the cranium of the newborn are the anterior *fontanelle*, where the parietal and frontal bones meet, and the posterior fontanelle, where the parietal and occipital bones meet. In these areas the brain is left relatively unprotected. Fortunately, the former becomes calcified by 18 months of age and the latter by 2 months.

The long bones and vertebrae, on the other hand, are cartilaginous bone. In the embryo they are formed of cartilage, i.e., they are composed of chondrocytes, collagen, and amorphous intercellular substance. As the fetus develops, however, much of its cartilage begins to accumulate calcium and phosphate, and the chondrocytes become walled off from their source of nutrient and die. They are replaced by osteoblasts, which accelerate the laying down of complex calcium phosphate salts and which form an intricate pattern of haversian canals and connecting canaliculi providing links between the osteoblast and its blood supply. This transition from cartilage to bone results in an organ which, though more brittle than its predecessor, is also considerably stronger.

In the small child *cartilaginous plates* may be found at either end of the long bones separating the epiphysis of the bone from its metaphysis (Fig. 33-5). At the epiphyseal end of the plate lie rapidly proliferating cartilage cells. At the metaphyseal end lies the area of provisional cal-

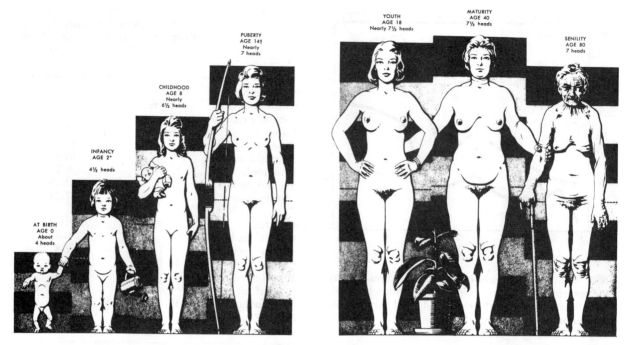

Fig. 33-3. Changes in the female with age. (*From An Atlas of Human Anatomy for the Artist* by Stephen Rogers Peck. Copyright 1951 by Oxford University Press, Inc. Reprinted by permission)

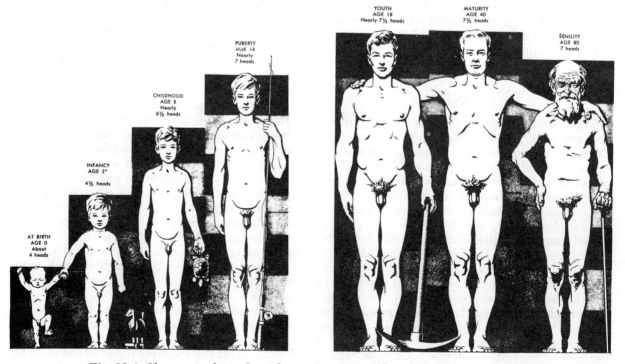

Fig. 33-4. Changes in the male with age. (*From An Atlas of Human Anatomy for the Artist* by Stephen Rogers Peck. Copyright 1951 by Oxford University Press, Inc. Reprinted by permission)

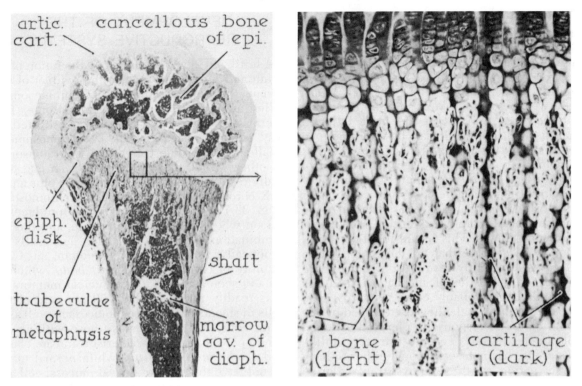

Fig. 33-5. Longitudinal section through a long bone from a growing rat. At the left there is a low-power photomicrograph showing from above downward (1) the articular cartilage of the epiphysis, (2) the cancellous bone of the epiphysis, (3) the cartilaginous plate (epiph. disk), (4) the metaphysis, and (5) the marrow cavity of the diaphysis. On the right is a high-power photomicrograph showing from above downward (1) the small rapidly proliferating chondrocytes, (2) the hypertrophied cartilage cells, and (3) an area of provisional calcification. (Ham, A. W.: Histology. 6th ed., p. 419. Philadelphia, J. B. Lippincott, 1969)

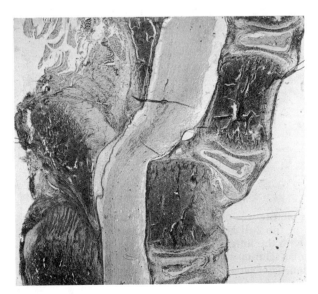

cification. Here the cartilage cells become enlarged, die, and are replaced by osteoblasts. In the shaft of the bone, we now find *osteoclasts* (cells responsible for bone resorption) destroying the calcium matrix and forming a marrow cavity.

The rapidly proliferating cartilage cells push the epiphysis further from the center of the shaft

Fig. 33-6. A slippage at the upper epiphyseal plate of the first lumbar vertebra of the rat. The dorsal part of the vertebral column is seen at the left and the spinal cord in the center. Note also the intervertebral disk at the right of the cord between each vertebra. These changes were produced by feeding young rats a diet of 50 per cent sweet pea seed and frequently produced paralysis of the lower extremities. (Ponseti, I. V., and Shepard, R. S.: Lesions of the skeleton and of mesodermal tissues in rats fed sweet pea seeds. J. Bone Joint Surg., 36A:1040, 1954)

of the bone, causing the long bones to grow in length. *Growth hormone* facilitates this process and in so doing may cause an increase not only in the length of the bone but also in the width of the cartilaginous (epiphyseal) plate. Excessive increases in the width of the plate may cause such a weakness in the bone that the epiphysis is pulled from the rest of the bone (slipped epiphysis) (Fig. 33-6). It can be a problem in teen-agers undergoing their final growth spurt during puberty. Apparently during the end of puberty they release sufficient quantities of androgens from the adrenal cortex or testis or both to bring about progressive ossification of the epiphyseal plate (epiphyseal closure) and cessation of bone growth. This occurs about two years earlier in the female than in the male, but in both cases apparently results from the release of increased quantities of androgens. In males these androgens probably come primarily from the testis. In the female they are released by the adrenal cortex in response to elevated estrogen levels in the plasma. To date estrogen, unlike testosterone, has not been shown either to have a direct action on the epiphyseal plate or to facilitate protein anabolism.

DEVELOPMENT OF THE REPRODUCTIVE SYSTEM

A fertilized egg is produced by the union of an ovum and a spermatozoan in the oviduct of the female. The ovum contains 22 chromosomes, called autosomes, plus an X chromosome. The spermatozoan may be similarly constituted or may contain instead of the X chromosome a smaller chromosome called the Y chromosome. In other words, the fertilization of an egg normally results in 22 pairs of autosomes plus a pair of X chromosomes or an X and a Y chromosome (Fig. 33-7). The XX combination produces a person who is genetically a female, and the XY combination produces a male. The cells of genetically female persons contain an extra quantity of chromatin, the *Barr body*, which is usually seen attached to the nuclear membrane. It is readily noted in about 50 per cent of the cells in smears of the buccal mucosa as well as in other cells. Normal females (those with an XX genotype) show this characteristic and are classified as *chromatin-positive*, while normal males do not have this type of buccal mucosal cell and are classified *chromatin-negative*. This test can be

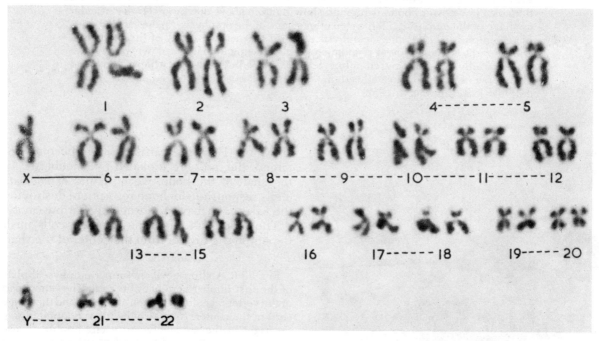

Fig. 33-7. Karyotype of a normal male. This picture was obtained by cutting out the individual chromosomes from a photograph of the cell and rearranging them. The chromosomes are arranged in pairs, starting with the longest (about 7 μ) and ending with the shortest (1.4 μ). (Ford, C. E.: Human chromosomes. *In* Hamerton, J. L. (ed.): Chromosomes in Medicine. P. 54. London, The National Spastics Society, Medical Education Information Unit, 1965)

Development of the Reproductive System 547

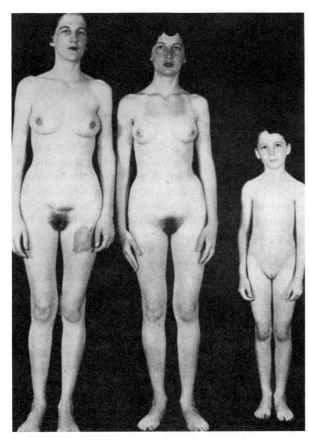

Fig. 33-8. Male pseudohermaphrodites with feminizing tumors of the testes. These 3 "sisters" had the external genitalia of a female, but had no uteri, Fallopian tubes, or ovaries. They were chromatin-negative and had testes in the inguinal area. (Wilkins, L.: The diagnosis and Treatment of Endocrine Disorders in Childhood and Adolescence. 3rd ed., p. 322. Courtesy of Charles C Thomas, Publishers, Springfield, Ill., 1965)

important when the structural characteristics of a newborn child or of an adult (his phenotype) are not characteristically male or female or when the sexual physiology seems abnormal (Fig. 33-8).

Abnormal Chromosome Combinations

In a few cases the transfer of chromosomes from one cell to another may be abnormal. If, for example, the fertilized egg contains only one X chromosome and no Y chromosome, all the cells of the body will have a genotype of XO. This results in a condition called gonadal dysgenesis. The patient is chromatin-negative and has gonads which are either rudimentary or absent. The external characteristics are those of an immature female. The incidence of the XO abnormality is estimated to be 0.03 per cent. A somewhat more common problem (incidence 0.265%) is a genotype of XXY. The resulting condition is called Klinefelter's syndrome or seminiferous tubule dysgenesis. Here the patient has external male genitalia, abnormal seminiferous tubules, and a high incidence of mental retardation, and is chromatin-positive. Some other abnormal XY genotypes include XXX (incidence 0.08%), XO/XX (containing an oviduct), and XO/XY (containing both male and female external characteristics). The latter two cases are examples of *mosaicism* (a combination of cells of unlike genetic composition). In each of these cases there was apparently an abnormal transfer (nondisjunction) of chromatin material in the embryo during cell division. Extra autosomes have also been noted, and when present usually have a deleterious effect on development. We find, for example, that an extra autosome number 21 or 22 is a common cause of *mongolism* (a type of mental retardation) and other abnormalities.

Fetal Development

In early fetal development a genital ridge appears on both sides of the body. By the sixth week a primitive gonad containing a cortex, a medulla, and two associated genital ducts (wolffian duct and müllerian duct) have developed from each genital ridge. During the seventh and eighth weeks in the male embryo the medulla of each gonad develops into a testis and releases an inductor substance that will facilitate unilateral development of the internal male genitalis (epididymis, vas deferens, seminal vesicle, prostate). While the male genitalia are developing, the cortex of the primitive gonad and its associated genital tract (müllerian duct) are regressing. The testes begin to secrete androgens from their developing Leydig cells which in turn facilitate the disappearance of the urogenital slit and the development of the typical male external genitalia. In other words, in the developing male embryo and fetus the testes release both an inductor, which has a local unilateral effect, and a hormone, which has a general effect.

In the seventh week of development in the female embryo, the cortex of the primitive gonad and its associated duct develop and the medulla and wolffian duct regress. This development of the cortex is not associated with the release of any known hormone from the gonad; as a result, the genital slit remains open and the typical female genitalia develop. In man, though not in certain other animals, the presence of a male twin in the uterus or the injection of testosterone into the mother does not affect the development of the female embryo.

The true hermaphrodite. The true hermaphrodite is a person with both an ovary and a testis or with an ovotestis (Fig. 33-9). Jones and Scott have reported on 64 true hermaphrodites, and found that in 51 of these there was an ambiguous hypospadiac penis, in 10 a normal penis, and in 3 the external genitalia of the female.

Thirty-five developed breasts and 22 menstruated. In one third there was an ovary on one side and a testis on the other. Two thirds were chromatin-positive and one third was chromatin-negative. Most had a genotype of XX, but a few were XX/XY or XX/X.

Puberty in Boys

In both boys and girls changes in the reproductive system after birth are minimal or absent until 9 or 10 yrs. of age (Fig. 33-10). In the male fetus the Leydig cells (interstitial cells of the testis) release androgens that are important in the development of the external genitalia, but these cells disappear during the first 6 months after birth; as a result, the testes and penis change very little in size for the next 10 yrs. At 1 yr., for example, the penis is usually about 3 cm. long and a single testis occupies a volume of 0.7 ml. At 8 yrs. the testis and penis are about 0.8 ml. in volume and 3 cm. long, respectively. At about age 10, on the other hand, the testis, prostate, seminal vesicles, and penis begin to enlarge (Fig. 33-11). In the 19-year-old, a single testis occupies a volume of about 16.5 ml., and the relaxed penis is from 6 to 10 cm. in length. The penis will also have acquired the capacity to become erect and to obtain a length during erection of about 13 cm.

It is important to realize, in considering the

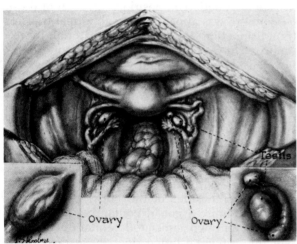

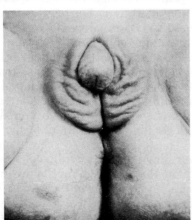

Fig. 33-9. The external and internal structures of a true hermaphrodite. Note that there is an ovary on the subject's right and both an ovary and a testis on the left. The child was chromatin-positive and subsequently raised as a female. (Wilkins, L.: The diagnosis and treatment of endocrine disorders in childhood and adolescence. 3rd ed., p. 318. Courtesy of Charles C Thomas, Publishers, Springfield, Ill., 1965)

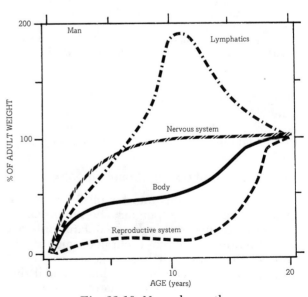

Fig. 33-10. Normal growth.

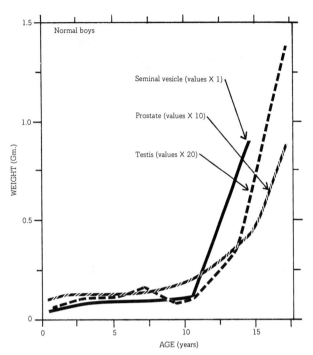

Fig. 33-11. Changes in weight of certain male reproductive organs with age. Note that an average 15-year-old boy has a seminal vesicle weighing 0.9 Gm., a prostate weighing 4 Gm., and a testis weighing 14 Gm. (*Redrawn from* Donovan, B. T., and Werff Ten Bosch, J. J. van der: Physiology of Puberty. P. 5; copyright © 1965, The Williams and Wilkins Co., Baltimore)

above averages and those to follow, that there is a wide range of ages at which these developmental changes can occur in healthy persons. This depends in part on environmental conditions and racial background. The average age at which boys acquire the ability to produce and ejaculate spermatozoa and serum—i.e., the age of puberty, is usually 15 yrs. It is not necessarily abnormal for boys to reach puberty at anywhere from 9 to 17 yrs., however, though most show a remarkably consistent pattern of change. There is usually axillary sweating and an increase in the production of odor-producing secretions from both the apocrine and sweat glands by age 11; the appearance of pubic hair by age 13; and enlargement of the larynx and the appearance of both axillary hair and hair on the upper lip at 14. The growth of the larynx may result in either an abrupt or gradual deepening of the voice. By age 16 there is a distribution of hair more characteristic of the male. This includes a bilateral indentation of the frontal hairline and the appearance of hair on the chest. The amount of chest hair depends on a person's race and does not usually become maximal until after 21 yrs. It is also in the 16-year-old that acne (a skin eruption due to the accumulation of secretion in the sebaceous glands) is frequently seen, and that there is hyperplasia of the mammary duct beneath the areolae. By age 18 there is broadening of the shoulders and a generalized muscle hypertrophy, and by age 21 practically all skeltal growth has stopped.

Puberty in Girls

A girl is said to have reached puberty after her first menstrual period. This usually occurs at about 13½ yrs., or 1½ yrs. earlier than puberty in the male. There is a wide range of normal values, however, normal menstrual periods having been reported as early as 9 yrs. and as late as 17 yrs. There is one case of a child who started menstruating at 3, became pregnant at 4 yrs. 10 months, and was delivered by cesarean section at 5 yrs. 7 months (Fig. 33-12).

Prior to puberty there is enlargement of the boney pelvis and budding of the nipples in the 10-year-old, budding of the breasts and the appearance of pubic hair in the 11-year-old, growth of the external and internal genitalia (labia minora, vagina, uterus, and oviduct) in the 12-year-old, and a filling in of the mammary glands and pigmentation of the nipples in the 13-year-old (Fig. 33-13). The onset of menstruation (menarche) generally occurs at about 13½ yrs. and is not usually associated for the first 2 yrs. with the release of an egg. By 15½ yrs., however, ovulation begins and pregnancies are possible in many cases. At 16 yrs. a deepening of the voice occurs and acne is common. By 17, skeletal growth has usually stopped.

Initiation and Control of Sexual Development

The changes in the body that initiate puberty are complicated and poorly understood. Apparently before puberty can occur there must be extensive changes in the nervous system and in the rest of the body. Two of the hormones important in facilitating these changes are growth hormone and thyroxine. They, like the androgens released by the adrenal cortex and the testis, facilitate protein anabolism, bone growth, and therefore the body's retention of nitrogen, po-

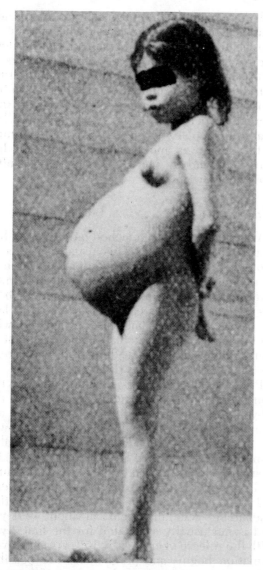

Fig. 33-12. A pregnant child at 5 years, 6 months of age. At the time of the picture the girl was 45.5 inches tall and weighed 63 pounds. The baby was delivered by Caesarean section and was a 6½-pound boy. (Escomel, E.: La plus jeune mère du monde. Presse Med., 47:875, 1939)

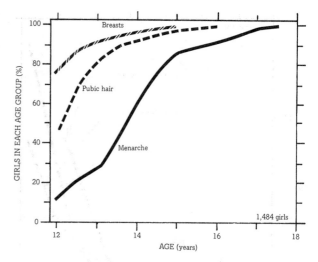

Fig. 33-13. The incidence of different developmental changes in girls of ages 12 to 18 years. (Redrawn from Land, G. M. van't, and Haas, J. H. de: Menarcheleeftijd in Nederland. Ned. Tijdschr. Geneesk., 101:1429 1957)

tassium, phosphorus, and calcium. Eventually the development of the nervous system in general and the hypothalamus and anterior pituitary in particular reaches a point at which the pituitary starts releasing follicle-stimulating hormone, luteinizing hormone, and prolactin. The importance of the hypothalamus in this process has been shown by (1) the use of experimental lesions in immature animals, (2) the destruction of areas in the anterior hypothalamus resulting in sexual infantilism, and (3) lesions in the posterior hypothalamus producing sexual precocity.

The importance of the hypothalamus and the anterior pituitary hormones in controlling (1) the hormonal output of the ovary, (2) the hormonal output of the testis, and (3) the development of the mammary glands was summarized on pages 500 to 502. Remember that follicle stimulating hormone (FSH) facilitates the production of spermatozoa and the maturation of a graafian follicle in the ovary. Associated with this FSH-induced maturation is the release of *estrogens* (estradiol, estrone, and estriol) by the follicle. These estrogens, in turn, get into the blood and facilitate the development of the breasts, nipples, labia minora, vagina, uterus, and oviduct.

Luteinizing hormone (LH) is another secretion from the anterior pituitary that acts on the ovary of the female and the testis of the male. In females it is responsible for the further development of the graafian follicle and its continued release of estrogens. It also facilitates the conversion of the follicle into a corpus luteum, a yellow body in the ovary secreting both estrogens and progesterone. The *progesterone* facilitates changes in the endometrium of the uterus and the development of the acinar lobules of the breast.

In males, LH acts on the Leydig cells of the testis to facilitate their release of testosterone and other androgens. These androgens, like growth hormone and thyroxine, facilitate the retention of nitrogen, potassium, phosphorus, and calcium by the body and an associated increase in muscle mass and bone length. Androgens, unlike growth hormone, stimulate bone calcification to a greater extent. They stimulate proliferation at the cartilaginous plate of the long bones, and as a result are responsible for a growth spurt that ends when the calcification of the bone has resulted in a closure of all the epiphyseal plates of the long bones. The androgens are also responsible for the development of pubic, axillary, and facial hair, the growth of the penis and other male genitalia, the enlargement of the larynx, the secretion of the sebaceous glands, and acne.

In the female, androgens are also released during adolescence. They are produced by the adrenal cortex, however, and do not include the most potent of the androgens, testosterone. Apparently they are released during the development of the female and are responsible both for the growth spurt characteristic of the adolescent and the calcification of the cartilaginous plates. They are also responsible for the enlargement of the larynx, the development of pubic and axillary hair, and acne.

THE ADULT MALE

The Scrotum

The testes of the fetus develop in the abdominal cavity and, as this development proceeds, descend into a sac outside the abdominal cavity between the rectum and urethra (Fig. 33-14). In the adult the temperature in the scrotal sac is about 3°C. (5°F.) lower than that in the abdomen. This lower temperature is essential for the production of the male germinal cell, the spermatozoan. Testes that remain in the abdominal cavity (cryptorchism) for prolonged periods after puberty or are kept at high temperatures (fever, for example) stop producing spermatozoa, and eventually their seminiferous tubules degenerate. When the temperature at the surface of the scrotal sac is warm, the sac is thin and characteristically hangs loosely in multiple folds or creases. When exposed to a cold environment, however, the smooth muscle in the tunica dartos under the skin of the scrotum contracts; the sac becomes smaller and draws its contents closer to the abdominal cavity. Thus a fairly constant testicular temperature is maintained.

The Testis

The testis contains (1) the Leydig cells, which secrete testosterone and estrogen and (2) the seminiferous tubules. Prior to puberty the seminiferous tubules contain only primordial germ cells. After puberty, however, these cells divide and give rise to spermatogonia, which lie next to the basement membrane. The spermatogonia contain the diploid number of chromosomes (46) and divide by mitosis. They eventually give rise to the primary spermatocyte, which by meiosis (reduction division) produces a cell with 22 autosomes plus an X chromosome or a Y chromosome. The secondary spermatocyte then divides mitotically to form a spermatid, which elongates and becomes a spermatozoan. In man it takes approximately 74 days for a spermatogonium to develop into spermatozoa. During this time the developing cell receives much of its nutrient from the Sertoli cells of the tubule.

The above process is called *spermatogenesis* and normally in the young adult results in the production of more than 10^9 (1 billion) spermatozoa per month. While in the seminiferous tubules, these spermatozoa are nonmotile. They are stored in the tubules and pass by some poorly understood mechanism into the rete testis, epididymis, and vas deferens. They are said to ripen in the epididymis and may remain there in a viable state for periods up to 60 days, but those that reach the vas deferens survive for a considerably shorter period.

The Seminal Fluid

The seminal fluid is the ejaculum minus the suspended spermatozoa. Most of it is secreted immediately before and during ejaculation (the ejection of semen through the urethral meatus) and comes from the prostate (about 60% of the seminal fluid), seminal vesicles (30%), epididymis (less than 5%), and Cowper's glands (less than 5%). These structures are illustrated in Figure 33-14. In a normal, healthy young adult who has not ejaculated for 9 days the volume of a single ejaculum will be between 2.5 and 3.5 ml. The semen (seminal fluid plus spermatozoa) is a white, viscous, opalescent, well buffered liquid with a pH of from 7.35 to 7.50. Its specific gravity is directly related to its concentration of spermatozoa. Its average sperm concentration is 100 million per milliliter, and its average specific gravity, 1.028.

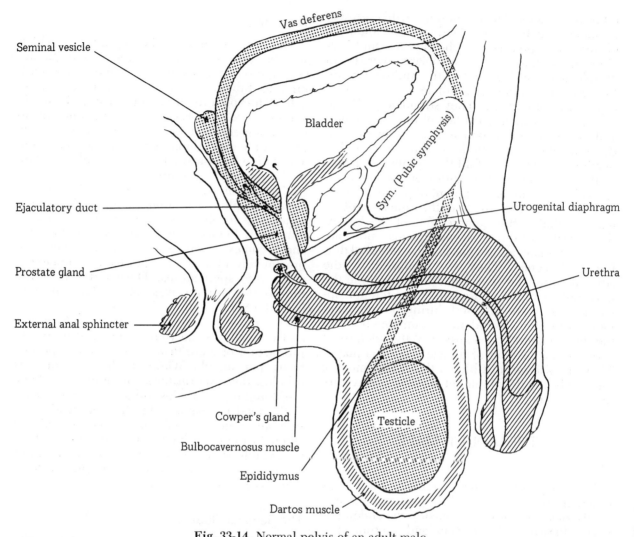

Fig. 33-14. Normal pelvis of an adult male.

The seminal fluid serves at least four important functions. It (1) acts as a suspending medium, (2) provides nutrient in the form of fructose to the sperm, (3) contains buffers that help protect the sperm from the highly acid environment of the female vagina, and (4) contains substances that change the sperm from the immotile cells contained in the epididymis to the freely swimming forms found in the ejaculum.

The prostate. The size and secretory capacity of the prostate is dependent upon the level of androgens to which it is exposed. Its secretions have a pH of about 6.4 and contain acid phosphatase, citric acid (1,400 to 6,370 μg./ml.), choline and spermine (nitrogen-containing compounds thought to facilitate sperm motility), cholesterol, cephalin, proteins (350 to 550 mg./ml.), fibrinogen, fibrinolysin, fibrinogenase, thromboplastin, and various electrolytes. In studying cancer of the prostate the physician will frequently determine the concentration of acid phosphatase in the semen as an index of the severity of the condition. Castration is sometimes an effective treatment of prostatic cancer in that it decreases the quantity of circulating androgens and in this way decreases the activity of the prostate. Estrogens also decrease the quantity of circulating androgens and thus decrease prostatic activity and the concentration of acid phosphatase in the semen.

Tests for spermine and choline constitute an important aspect of *forensic medicine.* Picric acid is used to test for spermine (the Barberio test). A positive reaction between a semen sample and picric acid indicates that the semen is from a human, whereas a negative reaction indicates a nonhuman source. Semen can be distinguished from other secretions by testing for the presence of choline. In this test (Florence reaction) iodine added to a semen stain results in the formation of brown crystals.

The seminal vesicles. The seminal vesicles are second only to the prostate in the volume of fluid they add to the ejaculum. This fluid, unlike that from the prostate, is usually alkaline. It contains fructose (an important source of nutriment for the sperm), ascorbic acid, citric acid, inorganic phosphorus, acid-soluble phosphorus, bicarbonate, potassium, traces of ergothionine (possibly facilitating sperm motility), and flavins (responsible for the yellowish quality of semen).

Spermatozoa

Spermatozoa are stored in the seminiferous tubules, rete testis, epididymis, and vas deferens in a nonmotile state. When activated by the seminal plasma, they swim in the semen. Normally, however, the semen clots—forms a gel, on leaving the male urethra. It loses its gelatin-like character 10 to 15 min. later and becomes more liquid. If the semen is deposited in the vagina of the female, the sperm swim out of the semen and into the mucus of the vagina (Fig. 33-15). Each sperm carries a negative charge, and it is speculated that this is in part responsible for its movement into the vaginal mucus.

What happens to the semen once it has entered the vagina depends upon the character of the ejaculum and the character of the secretions it is exposed to in the female reproductive system. In cases of male sterility two common causes are (1) a high concentration (over 20%) of abnormally shaped spermatozoa and (2) a high concentration of sperm with short-lived motility. The first problem may be seen by simply observing the sperm under the microscope. Tests for the second problem involve collecting fresh ejaculum and placing part of it in the center of a petrolatum ring on a slide. A cover slip is then placed over the ring and the slide is stored for 24 hr. at 60° to 70°F. At the end of this period 20 to 30 per cent of the sperm in a normal sample should be motile. In a sample

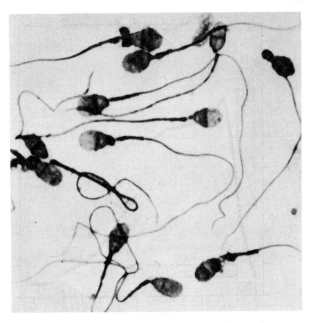

Fig. 33-15. Normal spermatozoa. Note the uniformity of size and contour. In some cases cytoplasmic masses are still attached to the neck region of a sperm. (Williams, W. W.: Sterility: The Diagnostic Survey of the Infertile Couple. 3rd ed., p. 17. Springfield, Mass., William W. Williams, Publisher, 1964)

from a sterile male, frequently none of the sperm is motile after this period.

A third problem sometimes resulting in sterility is the barrier to sperm motility set up by the secretions of the female reproductive tract. The character of the secretions of the vagina and the cervix varies with the phase of the menstrual cycle the female is in and, within limits, the duration of sexual excitation in the female that precedes the deposition of the ejaculum in the vagina. Cervical mucus, for example, increases in quantity and is least viscous and least resistant to penetration by spermatozoa 3 to 4 days before and 1 day after ovulation. Also, the pH of the vagina of the sexually unexcited female is between 3.5 and 4.0, whereas after 30 min. of petting (sexual arousal without copulation), the secretions of the vagina increase in quantity and reach a pH between 4.25 and 4.5.

Since sperm are normally immotile at a pH greater than 12 or below 3, changes in the pH of the vaginal secretions may be important, at least in some cases. In Figure 33-16 we see a study of these factors. Note that in this study the pH of the vaginal fluid was 3.7 prior to sexual

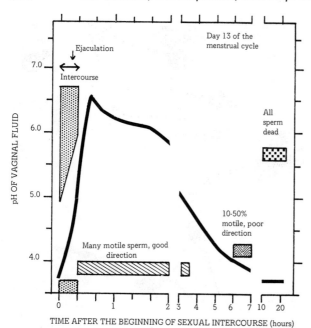

Fig. 33-16. Sperm motility and pH in the vagina after sexual intercourse. This chart is based on observation on a normal, fertile family unit. The study was preceded by 3 days of sexual continence and was performed on the 13th day of the female's menstrual cycle. There were 17 minutes of petting prior to ejaculation. The female reached orgasm 3 minutes prior to ejaculation. (*Modified from* Masters, H. W., and Johnson, V. E.: Human sexual response. P. 93. Boston, Little, Brown & Co., 1966)

relations and about 6.7 10 min. after ejaculation into the vagina. The petting prior to ejaculation raised the pH of the vaginal fluids somewhat, but the male ejaculate apparently played the major role. Note too that the pH in the vagina stayed high for several hours and that many of the sperm remained motile in the vagina for this same period.

Most of the spermatozoa deposited in the vagina stay there and are eventually destroyed. A small percentage, however, migrate into the alkaline medium of the cervix and through the uterine cavity into the oviduct (fallopian tube). The spermatozoa in the uterus increase their speed several fold, but migration from the cervix into the oviduct probably takes over an hour. Fertilization, if it occurs, takes place in the oviduct. Here the sperm must penetrate a barrier of cells surrounding the ovum—probably accomplished by release of the mucolytic enzyme hyaluronidase from the head of one or more sperm.

From the above discussion one might think that in artificial insemination in man, placing the donor's semen directly into the uterus would be advantageous. In fact, the frequency of successful fertilization is little changed by such a procedure. In addition, there is the danger that if more than 0.2 ml. of the donor semen is injected into the uterus, it will be forced up the oviduct and into the abdominal cavity where it may produce peritonitis, or that the uterus will respond to it by contracting and forcing it into the vagina.

Erection

In erection, a spongelike system of venous sinuses in a structure becomes engorged with blood, making the structure larger and less flaccid. Erectile tissue is found in the penis, nipples (only about 60% of men show nipple erection, however), and the clitoris. Normally, the relaxed penis of an adult is from 6 to 10 cm. long. It is less after severe exercise, when the penis is exposed to cold (swimming in cold water), in the aged, and in castrated patients. During erection the penis increases its length to about 13 cm. while increasing its circumference. In a young adult the penis can reach full erection from a flaccid state in 3 to 5 sec., but the reaction time is doubled or tripled in persons over 50. (Fig. 33-17).

Erection has been reported in males ranging in age from 1 day to 87 yrs. In the very young it is usually the result of irritation in the genitourinary system. In the adult male, it is usually elicited by tactile stimulation of the penis, scrotum, or rectum, or by stress applied to the perineal muscles as in straining to lift a heavy object or to defecate. Impulses originating from the brain associated with dreams or erotic thoughts are another common cause. Erection is particularly common during sleep and on awakening in the morning. Males ranging from 14 to 65 yrs. report, for example, an average of 1 to 2 erections per week when first awakening.

The sensory stimuli that produce erection apparently do so by acting on an "erection center" in the sacral cord. This results in stimulation of the parasympathetic neurons in the splanchnic nerves, which causes vasodilation of the arteries to the penis. The vasodilation produces a marked

The Adult Male 555

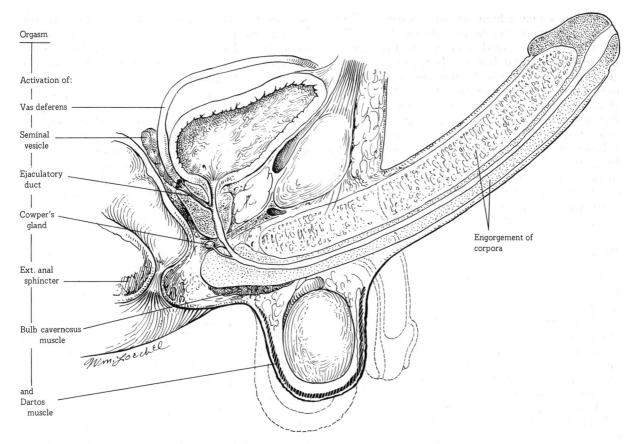

Fig. 33-17. The penis before and during erection. Also shown are some of the changes that occur immediately prior to and during orgasm.

increase in blood flow through the three cylindrical masses of erectile tissue in the penis, i.e., the two corpora cavernosa on the dorsal aspect of the penis and the corpus spongiosum surrounding the urethra and extending distally, where it enlarges to form the glans penis (Fig. 33-18).

The corpora cavernosa and corpus spongiosum have spongelike structures with irregular vascular spaces that are collapsed when the penis is flaccid. With the increase in arterial blood flow induced by parasympathetic stimulation, however, the flow of blood into the erectile tissue temporarily exceeds the flow out, and erection results. This is initially due to the high resistance to outflow offered by the valves in the veins of the penis, but as the erectile tissue becomes distended, it compresses the veins and further increases the resistance to outflow.

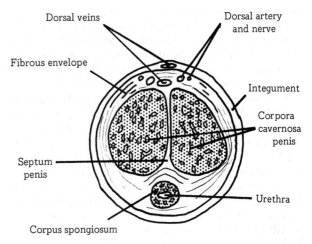

Fig. 33-18. Cross-section through the penis showing the erectile tissue (corpora cavernosa and corpus spongiosum).

Detumescence (decrease in swelling) of the penis usually occurs in two stages. In the first stage there is a rapid decrease in the size of the penis to about halfway between its flaccid and fully erect states. In the second stage the penis returns to its resting state. The latter is characteristically slower, but this depends upon the sexual stimuli still present. It can be accelerated by urination or by concentration on a subject unrelated to sex. Detumescence results from the removal of parasympathetic stimuli from, and the increase in sympathetic stimuli to the arteries of the penis, or in other words from vasoconstriction of the arteries carrying blood into the penis.

Orgasm

There is a vast array of sensory stimuli, characterized as being erotic, which produce in the adult male (1) vasocongestion of the penis, scrotum, and testes, (2) elevation of the testes due to shortening of the spermatic cords and contraction of the smooth muscle of the scrotum, (3) an increase in psychological and muscle tension, and (4) hyperventilation, tachycardia, and hypertension. Eventually these events may culminate in (1) emission followed by (2) ejaculation.

Emission. In emission there is an expulsion of seminal fluid from the prostate, the seminal vesicles, the ejaculatory duct, and other structures along the reproductive tract. This is associated with a contraction of the cremaster muscle which presses the testes against the perineum and a contraction of the smooth muscles throughout the epididymis. There results a distention of the urethra with semen and a sensation in the male of "feeling the ejaculation coming." At this point the male has no further voluntary control over the events that follow.

Ejaculation. The next series of changes is associated with ejaculation (Fig. 33-17). There is (1) an expulsive contraction of the urethra, (2) relaxation of the urogenital diaphragm (Fig. 22-11), and (3) contractions of the perineal, bulbospongiosus, and ischiocavernosus muscles. These contractions last about 0.8 sec. and are followed by relaxation and recurring periods of spasm and relaxation. The first spasm results in a sufficient pressure in the urethra of the adult below 30 yrs. of age to propel the semen 12 to 24 in. The spasms that follow become progressively less forceful and less frequent. After two or three spasms of the above-mentioned muscles, the urethra becomes insensitive. As a result, further distention of the urethra with semen and its continued movement from the urethra during the last stages of ejaculation usually goes unnoticed by the male.

Prior to ejaculation there has been a progressive, generalized increase in skeletal-muscle tone, psychological stress, and cardiorespiratory activity. Added to this during ejaculation are the spasms of the skeletal and smooth muscles associated with the male reproductive system. At the end of ejaculation there is a sudden cessation of the muscle spasms and a sudden release of the rest of the skeletal-muscle tone and psychological stress. This sudden release preceded by increases in tone is called orgasm and is associated with a pleasant feeling of fatigue. It has been compared by some of our older investigators with a sneeze. Here too there is a slow buildup in tension followed by its rapid release.

Refractory period. After the last spastic contractions of the urethra, the male goes through a refractory period. This is a time when he is insensitive to sexual stimuli. In the young adult this may be rather short-lived, whereas in the older male it is usually quite prolonged. Some young males studied had up to three orgasms within a period of 10 min. The female does not go through this refractory period after orgasm and can at most ages experience multiple orgasms much more readily and frequently than the male.

Nervous mechanisms. Orgasm involves most of the structures of the body. In fact we know very little about its central areas of control, although some evidence indicates that there are important genital centers in both the lumbar and sacral cords. The phenomenon of emission (contractions of the epididymis, vas deferens, ejaculatory duct, prostate, and seminal vesicles) is for the most part facilitated by impulses originating in the lumbar portion of the cord and passing down the hypogastric nerve through the hypogastric plexus in sympathetic neurons. The stimulation of these and other motor neurons is usually in part a response to impulses from the stimulated sensory neurons of the glans penis which travel up the internal pudendal nerve. Other important sensory impulses impinging upon the lumbosacral genital centers include those from the brain and other levels of the central nervous system. These impulses in the case of nocturnal emissions produce orgasm in the absence of an extrinsic stimulus acting on the

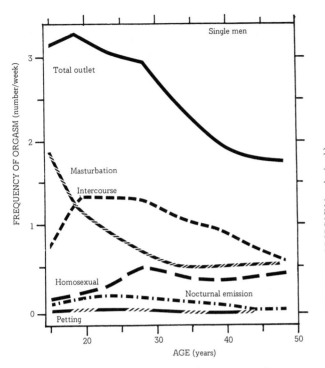

Fig. 33-19. Effect of age on the frequency of orgasm in the single male. (*Data from* Kinsey. A. C., et al.: Sexual Behavior in the Human Male. Philadelphia, W. B. Saunders, 1956)

glans. The phenomenon of ejaculation, on the other hand, involves the stimulation of somatic efferent neurons in S-1 through S-3, which stimulate the recurrent contractions of such skeletal muscles as the bulbospongiosus. There is also probably recurring depression of the somatic motor tone to the urogenital diaphragm.

There must also be marked changes in the motor output to other skeletal muscles in the body, and to the cardiovascular system. In the case of the skeletal muscles there is a generalized increase in nervous tone which depends in part on the position of the male. Respirations may increase to 40 per minute, and the heart rate has been reported to reach between 110 and 180 per minute. The systolic arterial pressure has been found to increase by 40 to 180 mm. Hg, and the diastolic pressure by 20 to 50 mm. Hg. In some persons there is also an increase in the redness and temperature of the skin (sex flush), which starts in the epigastric region and spreads to the chest, neck, and face.

Sexual practices in the male. The onset of regular sexual activity generally begins in the teen-age boy. There are cases, however, of chil-

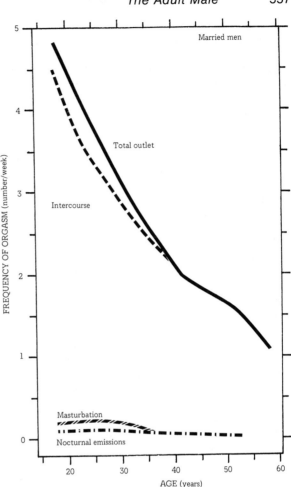

Fig. 33-20. Effect of age on the frequency of orgasm in the married male. The frequency of orgasm due to homosexual contact is not shown, but is 0.11/week at 18 years of age and less than 0.04/week at all other ages. (*Data from* Kinsey, A. C., et al.: Sexual Behavior in the Human Male. Philadelphia, W. B. Saunders, 1956)

dren experiencing orgasm without ejaculation as early as 5 months, and orgasm with ejaculation as early as 8 yrs. Over 95 per cent of adolescent males are regularly active by 15 yrs. of age, and over 99 per cent are active throughout the period from 16 to 45 yrs. of age. The frequency of orgasm in single men is illustrated in Figure 33-19, and in married men in Figure 33-20. Note that in both groups maximum frequency occurs in the teens and decreases with age. In the unmarried teen there is an average of 2.9 orgasms a week, most of these resulting from (1) self-induced stimulation of the penis, called *mas-*

turbation (responsible for an average of 2 orgasms/wk.), (2) intercourse (0.3 to 0.7 orgasms/wk.), and (3) nocturnal emissions (0.2 to 0.3 orgasms/wk.). It is estimated that 66 per cent of all males experience their first orgasm by means of masturbation. By 16 yrs. of age 87 per cent, and by 22 years of age 93 per cent of all males have masturbated at least once. The number of times a person masturbates per week decreases progressively with age and abruptly after marriage.

In nocturnal emission, the male has an erection and a subsequent ejaculation while sleeping. Twenty-two per cent of all males experience their first orgasm in this way. By age 16 about 54 per cent, and by age 22 about 80 per cent of all males have had at least one nocturnal emission. Its frequency is maximal in single men between the ages of 16 and 30, but is common in married men as well.

Another cause of orgasm is heterosexual petting to climax. This is experienced by about 13 per cent of all single males by age 16 and by 24 per cent of all single males by age 22. Homosexually induced orgasm is experienced by 32 per cent of all males by age 16 and 37 per cent by age 22. Heterosexual relations with animals are found primarily in rural areas. In this population about 12 per cent of the males experience orgasm through this route by age 16, and 15 per cent by age 22.

There is, as one might expect, wide variation in sexual performance from one week to the next in a single individual, as well as wide variation from one individual to the next. Kinsey et al.

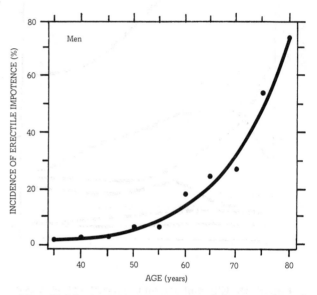

Fig. 33-21. An estimate in men of the per cent of the total population which is impotent at different ages. (*Data from* Kinsey, A. C., et al.: Sexual Behavior in the Human Male. Philadelphia, W. B. Saunders, 1956)

Age	Number of Cases	Total	Masturbation	Nocturnal Emission	Premarital Intercourse	Marital Intercourse*	Extramarital Intercourse*	Intercourse with Prostitutes	Petting to Climax	Homosexual Intercourse	Animal Intercourse
Puberty -15	3,012	29	23	12	25		7.5	2	3.5	7	8
16-20	2,868	28	15	6.5	25	25	18	4	4.5	10	4
21-25	1,535	29	12	6.5	25	29	6	7	7	11	1
26-30	550	29	9	4	16	25	4	4	4	15	0.1
31-35	195	29	7	3	13	20	4	3	1	4.5	
36-40	97	22	7	2	8.5	20	2	2.5	0.5	4	
41-45	56	15	7	1	6.5	14	2.5	1.5	0.5	5	
46-50	39	14	6	1	3.5	14	2	2.5	0.1	5	
51-55	173	7	1.5	1		6	2	1		0.1	
56-60	106	4.5	0.5	0.5		3	2			0.1	
61-65	58	4				5					

Table 33-2. Maximum frequency of orgasm per week. All data were obtained from unmarried men except those indicated by an asterisk (*). None of these data includes the extreme case in each group. (From Kinsey, A. C., et al.: Sexual Behavior in the Human Male. P. 234. Philadelphia, W. B. Saunders, 1948)

reported that in their 21- to 25-year-old age group 0.6 per cent of the population averaged over 20 orgasms per week and 1.8 per cent averaged less than 1 orgasm in 2 wks. They also noted that the maximum frequency of orgasm decreased with age (Table 33-2).

Impotence

The proportion of adult males who are incapable of having an erection (erectile impotence) or incapable of experiencing orgasm increases with age, and the frequency of erection and orgasm decreases with age. The incidence of erectile impotence in the 20-year-old is about 0.1 per cent, whereas that in the 55-year-old is 6.7 per cent. The data on persons past 55 are based on small samples, but the trend does seem to continue (Fig. 33-21). The average incidence of orgasm in the 65-year-old man still capable of this act is about 1 per week. Past this age even this small incidence tends to move rapidly toward zero. Notable exceptions to this have been reported, however. Morning erections are still common in some men past 75, and masturbation and nocturnal emission have been reported for a few men in their 80's. Probably the oldest reported to be still having sexual intercourse was an 88-year-old Negro who had intercourse with his 90-year-old wife on an average of from once a month to once a week.

Chapter 34

THE OVARY
THE UTERUS
THE CERVIX OF THE UTERUS
THE VAGINA
THE OVIDUCT
THE CLIMACTERIC AND
MENOPAUSE
SEXUAL RESPONSE

THE ADULT FEMALE

The reproductive system of the female includes the ovaries, the oviducts, the uterus, and the vagina. In the average adult female between 16 and 40 yrs. of age these structures undergo a series of changes that repeat themselves approximately every 28 days. The first day on which there is a vaginal discharge of blood is arbitrarily designated the first day of this 28-day

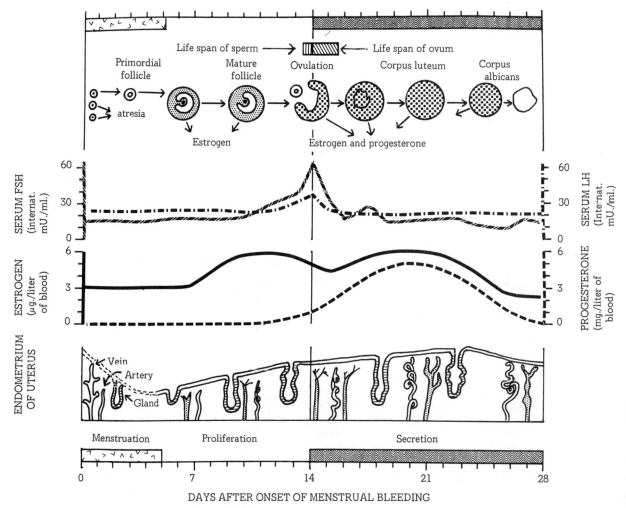

Fig. 34-1. Summary of some changes which occur during a single menstrual cycle.

560

cycle, the menstrual cycle. The blood and other materials released with it result from a sloughing of part of the lining of the uterine endometrium. This continues for about 5 days (the menstrual period) and is followed by a second phase, the proliferative period (about 9 days).

During the proliferative period there is a rapid increase in the thickness of the endometrium of the uterus and the associated development of an ovum-containing structure, the graafian follicle. The proliferative period ends when the follicle ruptures and releases the ovum and some associated cells into the oviduct. The next phase, the secretory period, is when the glands of the endometrium begin to secrete. At the same time, the ovum migrates down the oviduct. What remains of the graafian follicle in the ovary has reorganized into a hormone-producing structure, the *corpus luteum*. The hormones produced by the corpus luteum facilitate the further development of the endometrium of the uterus and prevent its necrosis and subsequent sloughing. If by the twenty-fourth day of the cycle the ovum has not been fertilized by a sperm and implanted in the endometrium of the uterus, the corpus luteum degenerates and forms in the ovary a structure incapable of producing hormones, the *corpus albicans*. The luteal hormones in the blood then decrease, the endometrium responds with a decrease in its blood supply and necrosis, the period of menstruation begins again. These events are summarized in Figure 34-1.

THE OVARY

The ovary at birth contains as many as 400,000 *primary follicles*—structures in which an ovum is surrounded by a single layer of cells. Throughout childhood some of these enlarge and develop a liquid-containing cavity, the follicular cavity. Some reach a size of 1 cm. in diameter, but none prior to puberty matures to such an extent that it releases its ovum into the oviduct. A number of these follicles prior to puberty do, however, undergo atresia (degeneration). As puberty approaches the pituitary releases increasing quantities of follicle-stimulating hormone (FSH) and luteinizing hormone (LH), and under the influence of these hormones one of the follicles matures enough to enlarge, release estrogen, and migrate toward the surface of the ovary. During this process (about the sixth day of the menstrual cycle) the ovum goes through meiotic division.

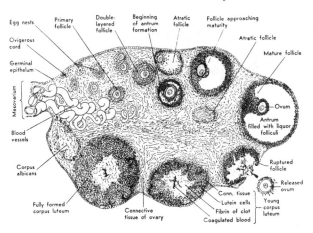

Fig. 34-2. Schematic diagram of an ovary showing the progressive development of a mature graafian follicle, a corpus luteum, and a corpus albicans. (Patten, B. M.: Human Embryology. 3rd ed., p. 18. Copyright 1958 by McGraw-Hill Book Company, New York; used by permission)

In other words, it reduces its chromosome number from 46 to 23 and in the process forms a polar body.

Under the influence of FSH and LH the developing follicle produces increasing quantities of estrogen (predominantly estradiol, but also estrone and estriol), and by the twelfth day of the cycle has apparently started to produce a second type of hormone, progesterone. By the fourteenth day of the cycle the follicle ruptures and extrudes the ovum and a few of the cells surrounding it (the cumulus cophorus). The fimbriated ends of the adjoining oviduct draw this material in and the ovum spends the next 7 days migrating down the oviduct toward the uterus. That part of the follicle remaining in the ovary fills with blood and forms a *corpus hemorrhagicum*. In some women a small amount of blood may at this time pass into the abdominal cavity, causing peritoneal irritation and temporary abdominal pain called *mittelschmerz*. Lipid-rich luteal cells replace the blood in the corpus hemorrhagicum, and along with the now rapidly proliferating theca and granulosa cells left behind in the ovary during the rupture of the follicle, form a new structure, the corpus luteum. The corpus luteum differs from the mature graafian follicle in that it produces larger quantities of progesterone than does the follicle. Like the follicle, it also produces estrogen.

In the absence of fertilization of the ovum the corpus luteum begins to degenerate between the eighth and tenth days after ovulation (twenty-third day of the menstrual cycle). In this process the corpus luteum undergoes hyalinization and becomes a considerably smaller structure, a corpus albicans secretes neither estrogen nor progesterone, the concentration of these two hormones in the blood markedly decreases. By the twenty-eight day of the cycle the progesterone concentration in the blood is close to zero and that of estrogen is at its lowest level. The blood estrogen concentration, unlike that of progesterone, never reaches zero, since by the time the corpus albicans has formed another graafian follicle has matured to the extent that it is releasing estrogen. Normally, at the beginning of each menstrual cycle a number of follicles begin to enlarge, but on an average only one matures enough to release its ovum (ovulation) and form a corpus luteum (luteinization). The others undergo atresia. Or in other words, many are called but few are chosen.

Pregnancy

If the ovum is to be fertilized, this must occur during its first 16 hr. in the oviduct. After this time the ovum is no longer viable. When fertilization does occur, cell division begins in the oviduct on about the eighteenth day of the cycle, and by the twenty-first day the egg has become a multicellular structure (containing more than 16 cells) in the uterine cavity. With further cell division it develops a large central cavity (blastocoel), and by the end of the twenty-fourth day has burrowed into the dorsal wall of the endometrium of the uterus (implantation) and started to form an endocrine-producing structure. This new structure produced by the embryo in the endometrium is responsible for the formation and release of a glucoprotein with a molecular weight of about 30,000. It is called chorionic gonadotropin and acts on the corpus luteum to (1) prevent its degeneration and (2) facilitate its growth and production of estrogen and progesterone. For most of the first trimester of pregnancy the hormones produced by the corpus luteum are essential for the development of the embryo and the prevention of endometrial sloughing and abortion. After this time a placenta (Fig. 13-13) develops which produces sufficient estrogen and progesterone by itself to meet the needs of the body. In other words, the removal

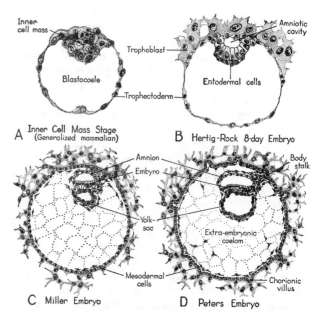

Fig. 34-3. Early stages in the development of the human embryo. At A we have a picture of the embryo at the stage when it implants in the uterus. At B (8 days after fertilization), C (12 days), and D (13 to 14 days) we see the embryo at various stages after implantation. (Patten, M. B.: Foundations of Embryology. P. 298. Copyright 1958 by McGraw-Hill Book Company, New York; used by permission)

of the ovary during the first trimester of pregnancy causes abortion, but its removal after that time is less serious.

Control

Control of the secretion of the ovarian hormones was summarized on page 500. Here we noted the importance of the hypothalamus, its releasing factors, and FSH and LH in ovarian control. We also noted that the control system was affected by the level of estrogen in the blood and by impulses originating from various parts of the central nervous system. The role of the central nervous system in the control of the hypothalamus in man has not been clearly defined, but it is apparent that nervous control of ovulation in man is not so pronounced as it is in the rabbit, for example, in which the female ovulates in response to the sight of a male rabbit or the stimulation of her external genitalia. On the other hand, marked disturbances in the menstrual cycle during stressful circumstances are not uncommon. Girls who have left home for the first time may miss several periods until they become accustomed to their new environment.

The oral contraceptive. In 1960 the oral contraceptive was first licensed for general use in the United States. By 1965, 30.5 per cent (1,241,000) of married women between 20 and 24 yrs. of age were using it regularly, and at present all estimates indicate that this figure has increased. Two basic oral contraceptive regimes are presently in use. They are (1) the daily administration for 23 days of a pill containing a progestin (a progesterone-like molecule) and an estrogen (9.85 mg. of norethynodrel plus 0.15 mg. of Mestranol, for example) followed by the withdrawal of medication for 5 days, and (2) the daily administration of an estrogen (0.1 mg. of ethinyl estradiol, for example) for 21 days followed by the daily administration of the same quantity of the estrogen plus a progestin (25 mg. of dimethisterone, for example) for 5 days.

In the combination approach (the first regime), the orally administered steroids apparently act synergistically to depress the release of luteinizing hormone (LH) by the pituitary to such an extent that the graafian follicle does not release its ovum. The withdrawal of the oral steroids, like the decrease in estrogens and progesterone occurring when the corpus luteum degenerates, results in endometrial sloughing (menstruation).

In the second type of contraceptive therapy (sequential therapy) it is apparently the inhibition of the release of follicle-stimulating hormone (FSH) that interferes with ovulation. As the contraceptive method is continued through numerous menstrual cycles, however, there may be occasional breakthrough ovulations. When this happens the corpus luteum is defective and as a result cannot prevent menstruation when estrogen and progestin are discontinued on the twenty-seventh day. Hence the full development of a placenta is prevented.

The daily use of the progestin-estrogen pill in males has similar effects on the release of FSH and LH. Here, however, there is a decrease in the size of the testis, a decrease in the production of spermatozoa and testosterone, and a decrease in libido. All these changes are reversible.

THE UTERUS

The endometrium of the uterus is a mucous membrane consisting of an epithelial lining and layers of connective tissue (endometrial stroma) which are continuous with the underlying myometrium and contain simple tubular glands opening into the lumen of the uterus. The endometrium can also be divided into (1) a basilar layer adjacent to the myometrium and (2) a functional layer adjacent to the lumen of the uterus. The basilar layer is little affected by the hormones released by the ovary, but the functional layer goes through cyclic changes in response to the concentration of estrogens and progesterone in the blood.

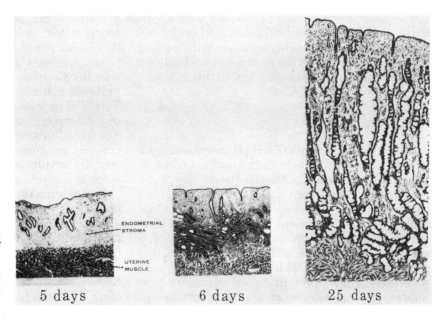

Fig. 34-4. A low-power view of the human endometrium on the 5th, 6th, and 25th days of the menstrual cycle. (Hamilton, W. J., et al.: Human Embryology. Cambridge, W. Heffer & Sons, pp. 31, 33, 1962)

Starting on about the fifth day of the menstrual cycle the functional layer responds to the estrogens in the blood by a rather rapid increase in its thickness and an associated elongation of its glands and blood vessels (Fig. 34-4). This continues until the ovary starts to release progesterone, at which time the glands begin to secrete, coil, and fold, and the endometrium becomes edematous (Fig. 34-1). In the absence of pregnancy by the twenty-seventh day of the cycle the corpus luteum will have markedly decreased its production of estrogens and progesterone, to which the endometrium responds with a series of changes in the arteris and glands. The glands regress, become narrower, and begin to buckle. There is also a buckling and coiling of the spiral arteries in the functional layer and constriction of the basal segment of the spiral artery, which together result in (1) a decrease in capillary blood flow and pressure, (2) a reduction in the extravascular endometrial fluid volume, (3) thinning of the endothelial lining, (4) ischemia, and (5) eventual necrosis. After about 2 days, bleeding and endometrial sloughing begin. The resultant fluid leaves the body through the vagina and is greatest in volume during the first 3 days. By the fifth day menstruation ceases and the functional layer of the endometrium begins to proliferate again.

The volume of menstrual fluid lost each month is on an average about 70 ml., half of this being liquid blood. When menstrual bleeding is excessive, there may also be blood clots, but the presence of clots in menstrual fluid is more characteristic of the bleeding associated with a miscarriage than with a normal menstrual cycle. Apparently when the quantity of menstrual bleeding is normal, enzymes in the endometrium facilitate the destruction of fibrinogen and in this way prevent the formation of fibrin.

Pregnancy

If the ovum is fertilized it (1) migrates down the oviduct, (2) implants in the endometrium of the uterus on about the twenty-fourth day of the menstrual cycle (tenth day of pregnancy), and (3) starts to form a *trophoblast* (Fig. 34-3). The group of cells constituting the trophoblast eventually participate in the formation of a placenta, but they are initially important in that some of them, within 24 hr. of implantation, start to release a glycoprotein called *chorionic gonadotropin*. The concentration of this hormone in the blood continues to rise after implantation, reaching a peak about 33 days after ovulation.

Pregnancy tests. Chorionic gonadotropin can be detected in the urine 12 days after ovulation and reaches a sufficient concentration by the twenty-fourth day of pregnancy (10 days after the first missed menstrual period) to be detected by any of a number of biological and chemical tests for pregnancy. One of the newer tests for pregnancy is the human chorionic gonadotropic antiserum test. In this test a drop of urine is added to human chorionic gonadotropic (HCG) antiserum on a slide and some latex particles coated with CG are added to the mixture. If the urine is devoid of HCG the mixture will acquire clumps of latex particles within 2 min. (a negative test). If it is urine from a patient in her first trimester of pregnancy, there will be no agglutination (positive test). This is due to the neutralization of the antiserum by the HCG in the drop of urine.

Another test for pregnancy is the Aschheim-Zondek test. Here 0.5 ml. of morning urine from the patient is injected twice daily for 3 days into three immature mice. On the fourth day the ovaries in the experimental mice are compared with the ovaries of a control mouse. The presence of hemorrhagic graafian follicles in the experimental ovaries indicates that the urine sample contained chorionic gonadotropin (CG urine). Other pregnancy tests include the Friedman test, in which CG urine injected into rabbits produces hemorrhagic follicles; the Kupperman hyperemia test, in which CG urine injected into prepuberal rats produces ovarian hyperemia; the Hogben test, in which CG urine injected into the South African clawed toad produces ovulation; and the Galli Mainini amphibian test, in which CG urine injected into certain male amphibians causes the release of sperm within 1 to 3 hr. These tests are generally negative in cases of miscarriage and ectopic pregnancy, and frequently after the first trimester of pregnancy. In ectopic pregnancies implantation occurs at a spot incapable of facilitating the production of large quantities of chorionic gonadotropin.

The ovarian phase of pregnancy. In the initial phase of pregnancy, the hormones from the corpus luteum of the ovary are essential for the growth and development of the endometrium and the prevention of endometrial sloughing. You will remember from Figure 34-1 that by the twenty-fourth day of the menstrual cycle the corpus luteum begins to decrease its production

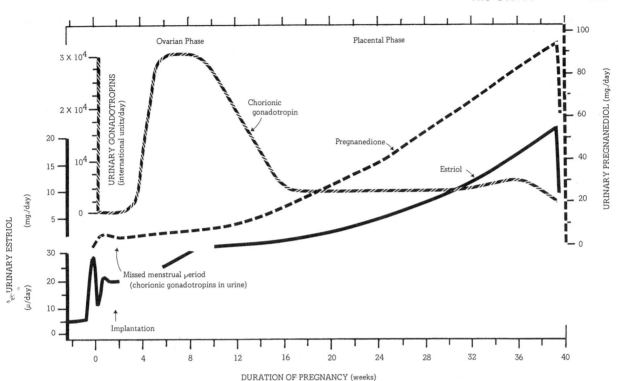

Fig. 34-5. Changes during pregnancy. In this presentation the changes in the concentration of chorionic gonadotropin, the estrogen, estriol, and the breakdown product of progesterone (pregnanedione) in the urine are shown prior to and during pregnancy. Note that 1 week before ovulation there is an elevation of estrogen production which is followed by a second elevation and an associated increase in progesterone. After implantation of the fertilized egg there is an elevation of chorionic gonadotropin which stimulates the further production of estrogens and progesterone by the corpus luteum of the ovary (ovarian phase). In time, however, the placenta begins to produce estrogens and progesterone (placental phase) and the concentration of chorionic gonadotropin decreases while the concentration of the estrogens and progesterone in the blood rises rapidly. One international unit of chorionic gonadotropin is the activity contained in 0.1 mg. of a standard preparation.

of estrogen and progesterone, and that by the twenty-eighth day it degenerates into a corpus albicans. When the ovum is fertilized, on the other hand, implantation occurs on the twenty-fourth day of the cycle and by the twenty-fifth day the trophoblastic cells of the embryo are producing chorionic gonadotropin (Fig. 34-5). Apparently the prime role of this hormone is to facilitate the functioning of the corpus luteum—i.e., to stimulate its growth and its production of estrogens (estradiol, estrone, and estriol) and progesterone. You will note in Figure 34-5, for example, that as the concentration of chorionic gonadotropin increases, so does the urinary content of estriol and pregnanediol. You will remember from Figure 31-3 that estriol is produced from estrone and estradiol, and that pregnanediol is a breakdown product of progesterone. In other words, chorionic gonadotropin increases the plasma concentration of estrogen and progesterone and in this way prevents menstruation and abortion.

The placental phase of pregnancy. During the initial phases of implantation the embryo obtains its nutrient through diffusion. By the twentieth day of pregnancy, however, the embryo has begun to develop villi (Fig. 34-3(D)) and a simple circulatory system. Each villus develops an artery, a capillary, and a vein. By the end of the third month of pregnancy the fetus begins to lose it villi at the decidua capsularis (Fig. 13-12 and 13-13) and produces an enlargement and branching of the villi at the decidua basalis that will become the placenta.

During this development there are not only marked changes in the blood supply to the endometrium (Fig. 13-15), but also a change in the character of the hormones produced there. You will note in Figure 34-5 that toward the end of the first trimester of pregnancy there is a reduction in the urinary chorionic gonadotropin but a continued increase in the excretion of the breakdown products of estrogen and progesterone. Apparently the developing placental tissue has by this time acquired the ability to produce its own estrogens and progesterone, while at the same time decreasing its production of the luteotropic hormone chorionic gonadotropin. In other words, part of the endometrium has been converted into a producer of steroid hormones, and the entire endometrium has lost its dependence upon the graafian follicle of the ovary. As a result of these changes, degenerative changes occur in the corpus luteum by the fifth month of pregnancy, though that body remains in the ovary throughout the entire pregnancy.

As pregnancy continues, other changes in the hormones occur besides those in chorionic gonadotropin, the estrogens, and progesterone. There are, for example, increases in the concentration of protein-bound iodine in the blood of the mother. This is probably due to an increase in the concentration of plasma proteins capable of binding T_3 and T_4, which is induced by the increased estrogen in the blood rather than by a major increase in biologically active T_3 or T_4. It should be remembered, however, that by the twelfth week of pregnancy the fetal *thyroid* has acquired the ability to concentrate iodine and to produce a hormone which can pass the placental barrier. In fact, the fetal thyroid can enlarge and compensate for its mother's hypothyroidism.

Other hormones that increase in concentration in the mother's plasma are *cortisone, hydrocortisone,* and *aldosterone*. These hormones, like the androgens and estrogens, are produced in part by the fetal adrenal cortex. Apparently increases in the concentration of cortisol in the mother's blood toward the end of pregnancy are compensated for by increases in the concentration of binding proteins. Increases toward the end of pregnancy in the concentration of aldosterone, on the other hand, may be responsible, along with progesterone, for the increases in salt and water retention found at this time.

Other hormones released by the placenta into the mother's blood are various androgens (dehydroepiandrosterone, dehydroepiandrosterone sulfate, androstenedione, and testosterone) and a hormone that has been termed human placental lactogen (chorionic growth hormone-prolactin). It has been suggested that this hormone acts on the mammary glands to prepare them for milk production. As for the androgens, they are found in increasing concentrations in the mother's blood toward the end of pregnancy. Testosterone exists at a concentration in the nonpregnant female of about 49 mμg./100 ml. of plasma, whereas between the fifth and ninth months of pregnancy it reaches a concentration of 114 mμg./100 ml. of plasma. Toward the very end of pregnancy the concentration of androgens in the maternal blood goes even higher. This is when the concentrations of estrogens and progesterone in the blood are decreasing and is probably due to the decreased capacity of the placenta to convert the androgens produced by the adrenal cortex of either the male or female fetus into estrone or estradiol (Fig. 31-3).

Parturition. Parturition is the process of giving birth to a child. It generally involves (1) the dilation of the cervix, (2) contractions of the smooth muscle of the uterus and the skeletal muscles of the abdomen, (3) the expulsion of the baby, and (4) the expulsion of the afterbirth. The afterbirth includes the umbilical cord and the placenta. The placenta is about one sixth the weight of the child and is disk-shaped, about

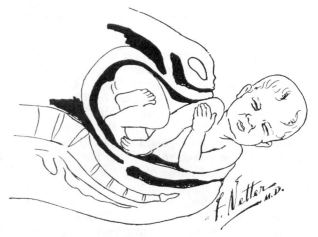

Fig. 34-6. Parturition. (Bookmiller, M. M., et al.: Textbook of Obstetrics and Obstetric Nursing. 5th ed., pp. 176, 177. Philadelphia, W. B. Saunders, 1967)

1 in. in thickness, and 7 in. in diameter. Some of the factors which probably have a role in initiating labor include (1) degenerative changes in the placenta resulting in a decrease in its production of estrogens and progesterone, (2) increasing concentrations of the posterior pituitary hormone oxytocin, and the increasing sensitivity of the uterine myometrium to this hormone, and possibly (3) the release of a hormone, relaxin, from the placenta that produces dilation of the cervix of the uterus, softening of the pelvic ligaments, and an increased sensitivity of the myometrium to oxytocin.

To emphasize any one of the above phenomena in the initiation of labor is probably a mistake. Labor is generally preceded by a decrease in the level of estrogens and progesterone in the blood, but there have been a few cases in which changes in the concentrations of these substances occurred only after labor. Another factor undoubtedly affecting the initiation of labor is oxytocin from the posterior pituitary. It has been used clinically to induce labor in women in their ninth month of pregnancy. On the other hand, it has been shown in mammals that labor can occur in its absence.

You will remember from our discussion on page 155 and from Figure 9-10 that estrogens and progesterone during pregnancy eventually cause an increase in the sensitivity of the myometrium of the uterus to stretch and oxytocin. By the thirteenth week of pregnancy there begins a progessive increase in the frequency and force of the myometrial contractions which continues up to the birth of the child and serves to "ripen" the cervix of the uterus. Labor is said to begin when the contractions occur every 15 min. This is when sensory neurons in the uterus are being stimulated and are producing an increase in the release of oxytocin by the pituitary. At first each contraction lasts only 10 to 15 sec., but as labor progresses they eventually occur at 2- to 4-min. intervals and last from 45 to 60 sec. each. It is usually during the first stage of labor that the chorion and amnion surrounding the fetus rupture and the amnionic fluid pours from the vagina. The membranes may, however, break days or weeks before the onset of labor, or they may not break until the child is born. In the latter case the baby is said to be born with a veil. At one time these veils were dried and stored for good luck or sold to sailors who thought they brought good luck at sea.

The puerperium. Throughout the period of pregnancy marked changes in the circulatory, endocrine, and metabolic characteristics of the mother occur. There are increases in cardiac output, metabolic rate, and the weight of the mother. The increase in the size of the uterus in the abdominal cavity may lead to partial occulsion of the inferior vena cava and as a result an increase in venous pressure in the legs, thighs, and hemorrhoidal veins of the rectum, as well as a diversion of vena caval flow to the azygous and epigastric veins. By the end of pregnancy the mother has usually gained in excess of 21 lb. On an average, 7 of these pounds are due to the weight of the baby, 4 to the weight of the amnionic fluid and placenta, 2 to the increase in the weight of the uterus, 3 to increased protein retention by the mother, at least 2 to increased fat deposition, and 3 to increased water retention.

Following labor the cardiac output returns to normal, as does the distribution of blood and the mother's weight. After delivery the weight of the uterus is about 1 kg. Four to five weeks later it has involuted and returned to a weight of 50 gm. In addition, there is a decrease in the body water and a slow return of the endocrine concentrations to the levels found prior to pregnancy. One of the first hormones to disappear from the blood after completion of labor is relaxin. It reaches a maximum concentration during labor, and within 24 hr. after delivery has disappeared from the blood. In mothers who do not nurse their babies menstruation begins about 2 months after delivery. In mothers who nurse for several months after delivery, however, there is a longer delay in the appearance of menstruation. It is speculated that suckling induces the production of prolactin by the pituitary at the expense of the production of the gonadotropins. The phenomenon of lactation is discussed more fully on page 502.

Some of the other postpartum changes that may occur are temporary hypothyroidism and hirsutism. The latter condition is probably due to the concentration of androgens in the blood prior to and after labor.

THE CERVIX OF THE UTERUS

That part of the uterus leading into the vagina is called the cervix and differs from the rest of the uterus in that it does not show a cyclic buildup of its lining and a subsequent desquamation. It does respond, however, to the elevation of estrogens prior to ovulation by producing a more alkaline and less viscous secretion. These changes are such that by the time of ovulation the mucus is normally optimal for the transport of spermatozoa into the uterus. After ovulation progesterone reaches a plasma concentration that causes the mucus to become more viscous and to decrease in quantity. In other words, progesterone facilitates the creation of a barrier to sperm penetration.

THE VAGINA

The buildup of estrogen occurring prior to ovulation causes proliferation of the epithelial cells of the vagina and subsequent stratification and separation of the more superficial cells from their nutrient blood vessels. As a result these cells undergo cornification. These changes protect the vagina against infection and are sometimes induced by the injection of estrogen after menopause to facilitate vaginal wound healing. The estrogen released during the follicular phase of the menstrual cycle (from the end of menstruation until ovulation) also facilitates the deposition of glycogen in the epithelial cells and subsequent increase in the production and release of lactic acid in the secretions of the vagina. This results in a decrease in the vaginal pH to a level which is less favorable for both the growth of microorganisms and the prolonged survival of spermatozoa. The progesterone released during the luteal phase of the menstrual cycle facilitates a further proliferation of the vaginal cells, but unlike the estrogens produces a decrease in the number of cornified cells. It also results in an increase in mucus secretion.

THE OVIDUCT

Estrogens released during the follicular phase of the cycle increase the frequency of contractions in the oviduct. It has been suggested that this may facilitate sperm transport. The decrease in contractions during the luteal phase of the cycle, on the other hand, is apparently due to progesterone.

THE CLIMACTERIC AND MENOPAUSE

Prior to 50 yrs. of age the female usually starts to go through a series of changes which begins with anovular menstrual cycles and which ends in a complete cessation of menstruation (menopause). One of the more important symptoms during the female climacteric is a decrease in the plasma concentration of estrogens. Since estrogens normally inhibit the release of follicle-stimulating hormone (FSH), this change is associated with a progressive increase in plasma FSH. Some of the other symptoms include the atrophy of genital tissue, an increase in arterial pressure, an increased tendency to develop diabetes mellitus, a decrease in the caloric expenditure frequently leading to obesity if the caloric intake is not decreased, and a decrease in protein anabolism possibly leading to an osteoporosis so severe that compression fractures become common.

Other symptoms include both vasomotor and emotional instability. Probably one of the most common signs of vasomotor instability is the "hot flash." During this there is a feeling of warmth beginning in the trunk associated with a redness of the skin and profuse sweating that gradually spreads to the face. All the above symptoms can usually be prevented or decreased in magnitude by the injection of estrogens. Treatment generally consists of the oral administration of an estrogen (0.5 mg. of diethylstilbestrol, for example) daily for 20 days with a repetition of the regime in a week if necessary. Larger doses can be given, but there is danger of withdrawal bleeding in these cases. When this occurs, the physician should check by a Papanicolaou smear to be sure the bleeding is not due to a malignancy.

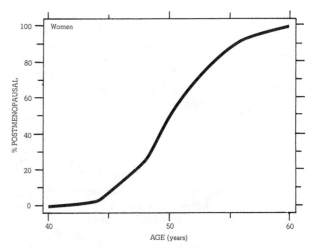

Fig. 34-7. The accumulative incidence of menopause in women. (*Data from* Kinsey, A. C., et al.: Sexual Behavior in the Human Female. Philadelphia, W. B. Saunders, 1953)

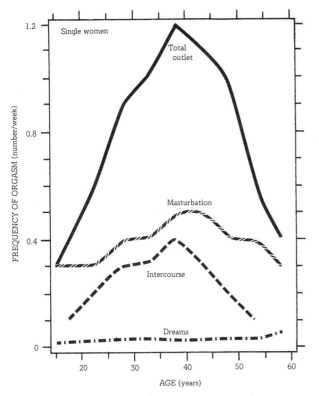

Fig. 34-8. Frequency of orgasm in the single woman. (*Data from* Kinsey, A. C., et al.: Sexual Behavior in the Human Female. Philadelphia, W. B. Saunders, 1953)

Menopause, like the removal of the ovaries in an adult, usually has little effect on the sexual activity of the female. In both the female and the male there is a decrease in sexual activity with age (Figs. 34-8 and 34-9), but there is certainly no abrupt decrease at the time of the female climacteric or menopause. Kinsey et al. found that in 123 ovariectomized women 54 per cent felt the operation had no effect on their sexual responses, 19 per cent believed their responses had increased, and 27 per cent believed they had decreased. They also found that 1 to 2 yrs. before menopause their frequency of orgams was 1.5 per week, during menopause it was 1 per week, and 1 to 2 yrs. after menopause it was 0.7 per week.

SEXUAL RESPONSE

Approximately 40 per cent of the almost 8,000 females studied by Kinsey et al. experienced their first orgams in response to self-stimulation (masturbation), 24 per cent in response to petting,

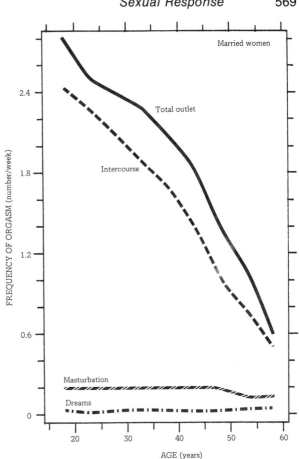

Fig. 34-9. Frequency of orgasm in the married woman. (*Data from* Kinsey, A. C., et al.: Sexual Behavior in the Human Female. Philadelphia, W. B. Saunders, 1953)

17 per cent in response to coitus after marriage, 10 per cent in response to premarital coitus, 5 per cent in response to dreams, and 3 per cent in response to homosexual contacts. Kinsey et al. also noted that only 64 per cent of the married females they interviewed had experienced orgasm prior to marriage and that the age at which the frequency of orgasm in the single female is greatest is 38. This is in marked contrast to the data obtained from the males they interviewed. Here they found that 100 per cent of the married males had experienced orgasm prior to marriage and that the age at which the greatest frequency of orgasm occurred was 18. They also noted that during sexual intercourse after marriage, although all the males experienced orgasm, the females experienced orgasm in only about 75 per cent of their coital contacts. During the first

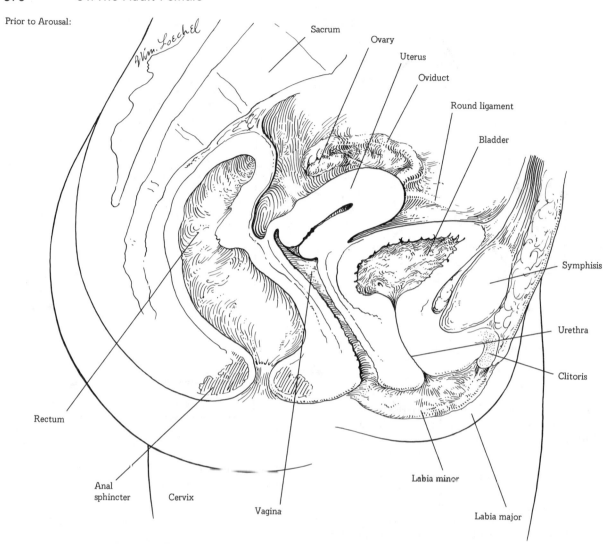

Fig. 34-10. A comparison between the normal female pelvis prior to sexual arousal and during its initial response to arousal (excitement phase).

year of marriage only 63 per cent of the females' coital experiences resulted in orgasm, whereas during the twentieth year 85 per cent of the contacts resulted in orgasm.

Although the incidence of orgasm varies between the male and the female, the phenomenon itself is quite similar in the two sexes. In both sexes the stimuli that elicit arousal and orgasm may derive from dreams, from manipulation of the penis or clitoris, from other tactile stimulation, or from visual stimulation. In both sexes the response to these stimuli include vasocongestion, secretion, an increase in both psychic and muscle tension, and increases in ventilation, heart rate, and arterial pressure.

The Excitement Phase

In the female one of the first responses during arousal is an increase in secretion in the vagina. This generally begins 10 to 30 sec. after the initial stimulus and is followed by dilation and lengthening of the *vagina* and an associated vasocongestion (Fig. 34-10). There is also a movement of the labia majora and minora away from the vaginal outlet in such a way as to make it more easily penetrated by the penis. These changes are usually also associated with vasocongestive increases in the size of the labia majora of women who have previously borne children, as well as vasocongestive increases in the size of the labia

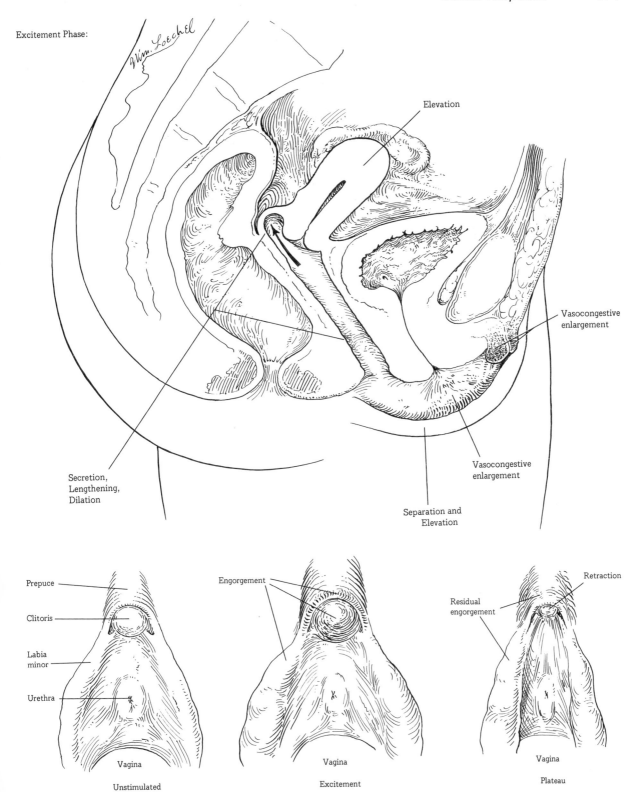

Fig. 34-11. Changes in the clitoris, prepuce, and labia minor during sexual arousal. The orgasmic and resolution phases are not shown.

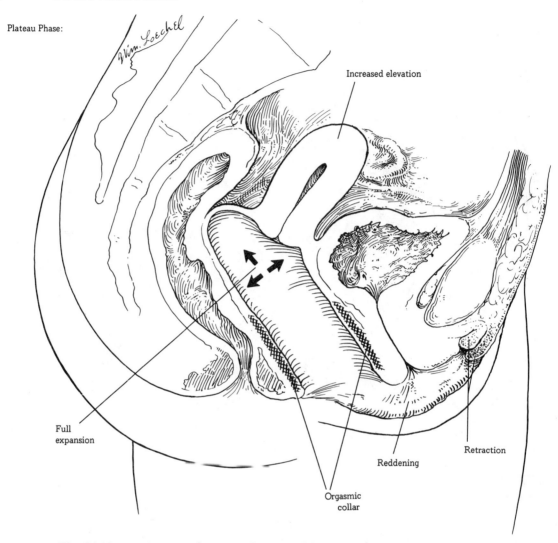

Fig. 34-12. A comparison between the normal female pelvis during the plateau and orgasmic phase of sexual arousal.

minora and clitoris of both the nulliparous and multiparous female (Fig. 34-11). There are in many females an erection of the nipples, a vasocongestive increase in breast size, an elevation of the uterus, increased muscle tone in the abdomen, chest, and appendages, and a sex flush.

The sex flush, if it occurs, starts late in the excitement phase or in a later phase. It involves first a vasocongestion of the skin in the epigastric areas that causes a feeling of warmth and the appearance of a rash. This rapidly spreads to the breasts in the next phase, and just prior to orgasm reaches it maximum distribution. This may include the thighs, buttocks, entire back, and the neck and face. Like many of the other responses to sexual stimulation, it is not found in every female or during every orgasmic encounter in those who do experience it. It has been estimated to occur in less than 75 per cent of females.

The Plateau Phase

As sexual arousal continues there is a retraction of the clitoral shaft and glans against the anterior body of the pubic symphysis, further vasocongestion of the labia majora and minora, an engorgement of the outer third of the vagina with blood to form an "orgasmic collar," a

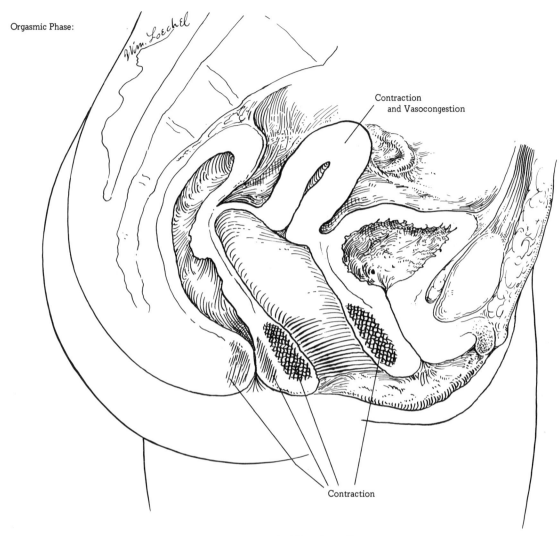

Fig. 34-12. (*Continued*)

further increase in the width and length of the vagina, a further increase in the size of the breasts, a further increase in muscle tone associated with semispastic contraction of the facial, abdominal, and intercostal muscles, an apparently inconsequential mucus secretion from some glands near the vaginal orifice (Bartholin's glands), and a further elevation of the uterus usually associated with uterine vasocongestion (Fig. 34-12). The vasocongestion of the labia minora (sometimes called the sex skin) during the plateau phase is associated with a vivid change in its color from colorless to pink or red. This change is apparently always followed by orgasm.

The Orgasmic Phase

Following the plateau phase of the sexual response there are involuntary spasms of a number of skeletal muscles, contractions of the rectal sphincter, contractions of the vaginal platform, and contractions of the uterus progessing toward the lower uterine segment and associated with a 50 per cent increase in uterine size due to vasocongestion.

The Resolution Phase

After the orgasm there is a rapid disappearance of the sex flush, the orgasmic collar, and

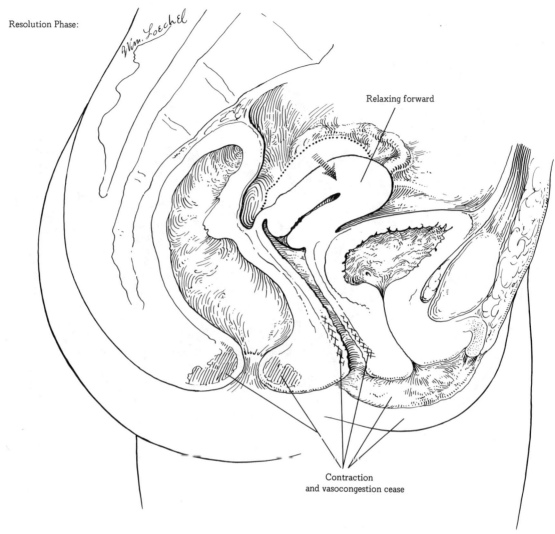

Fig. 34-13. The female pelvis during the resolution phase of sexual arousal.

the coloration of the labia minora. The vagina stops contracting and the uterus returns to the position it had prior to sexual arousal. A widespread film of perspiration appears over the skin and the increased muscle tone reached during orgasm diminishes over a period of 5 min. There is, in addition, a slow decrease in the size of the breasts, labia majora and minora, and uterus, a return of the labia majora and minora to their original postion blocking the entrance to the vagina, and disappearance of the vaginal lumen. The external cervical os remains open for the first 20 to 30 min. during the resolution phase. It is also during the resolution phase that the heart rate, arterial pressure, and respiration return to normal.

Restriction of Coitus

Attitudes among physicians vary as to when it is safe to have intercourse. Most, however, feel that intercourse during menstruation, during the first 8 months of pregnancy, and after the first postpartum month does not represent undue hazard to the woman or, if she is pregnant, to the continuance of her pregnancy. The dangers to the cardiac patient of sexual arousal and subsequent orgasm are aspects of the problem that have not been extensively investigated. In the case of pregnancy, most women who do not suffer from nausea during the first trimester report little or no change in their sexual drive or performance. On the other hand, during the second trimester of pregnancy most women report an increased interest in sex and a markedly more effective performance than they can remember from before pregnancy or during the first trimester. During the latter part of the third trimester most women lose interest in sexual stimulation. This is also a period when most physicians, fearing a premature delivery, prescribe continence.

In the postpartum period most women who do not nurse their children apparently have a reduced interest in sexual relations and a reduced performance for up to 3 months. Mothers who nurse their children, on the other hand, generally have a higher interest in sex relations and some resume sexual intercourse within 3 wks. after delivery even though most medical authorities suggest continence for at least 6 wks. A matter of concern and embarrassment to some nursing mothers is the sexual arousal they sometimes experience during nursing. Masters and Johnson have reported on thirty-four nursing mothers and noted that they frequently reported sexual stimulation to the plateau level in response to nursing. Three of their subjects even reported reaching orgasm during nursing.

REFERENCES

Adams, J. H., Daniel, P. M., and Prichard, M. L.: Observations on the portal circulation of the pituitary gland. Neuroendocrinology, *1*:193-214, 1965.

Barker, S. B.: Peripheral actions of thyroid hormones. Fed. Proc., *21*:635-641, 1962.

Barker, S. B., and Klitgaard, H. M.: Metabolism of tissues excised from thyroxine-injected rats. Am. J. Physiol., *170*:81-86, 1952.

Bookmiller, M. M., Bowen, G. L., and Carpenter, D.: Textbook of Obstetrics and Obstetric Nursing. 5th ed. Philadelphia, W. B. Saunders, 1967.

Buchanan, J., and Tapley, D. F.: Stimulation by thyroxine of amino acid incorporation into mitochondria. Endocrinology, 79:81-89, 1966.

Care, A. D., et al.: Evaluation by radioimmunoassay of factors controlling the secretion of parathyroid hormone. Nature, *209*:52-55, 1966.

Cobo, E., Gaitan, E., Mizrachi, M., and Strada, G.: Neurohypophyseal hormone release in the human. I. Experimental study during pregnancy. Am. J. Obstet. Gynec., *91*:905-914, 1965.

Escomel, E.: La plus jeune mère du monde. Presse Med., *47*:875, 1939.

Evans, H. M., Simpson, M. E., Marx, W., and Kibrick, E.: Bioassay of the pituitary growth hormone. Width of the proximal epiphyseal cartilage of the tibia in hypophysectomized rats. Endocrinology, *32*:13-16, 1943.

Foa, P. P.: Glucagon. Reviews of Physiology, Biochemistry, and Experimental Pharmacology. New York, Springer-Verlag, 1968.

Ford, C. E.: Human chromosomes. *In* Hamerton, J. L. (ed.): Chromosomes in Medicine. London, The National Spastics Society Medical Education and Information Unit, 1962.

Frye, B. E.: Hormonal Control in Vertebrates. New York, Macmillan, 1967.

Gaitan, E., Cobo, E., and Mizrachi, M.: Evidence for the differential secretion of oxytocin and vasopressin in man. J. Clin. Invest., *43*:2310-2322, 1964.

Ganong, W. F.: The central nervous system and the synthesis and release of adrenocorticotropic hormone. *In* Nalbandov, A. V. (ed.): Advances in Neuroendocrinology. Pp. 92-157. Urbana, University of Illinois Press, 1963.

⎯⎯⎯: Review of Medical Physiology. Los Altos, Lange Medical Publications, 1967.

Greulich, W. W., Dorfman, R. I., Catchpole, H. R., Solomon, C. I., and Culotta, C. S.: Somatic and endocrine studies of puberal and adolescent boys. Monogr. Soc. Res. Child Develop. 7:1-85, 1942.

Ham, A. W.: Histology, Philadelphia, J. B. Lippincott, 1969.

Hamilton, W. J., Boyd, J. D., and Mossman, H. W.: Human Embryology. Baltimore, Williams & Wilkins, 1962.

Harris, J. A., Jackson, C. M., Paterson, D. G., and Scammon, R. E.: The Measurement of Man. Minneapolis, University of Minnesota Press, 1930.

Jones, H. W., Jr., and Scott, W. W. (eds.): Hermaphroditism—Genital Anomalies and Related Endocrine Disorders. Baltimore, Williams & Wilkins, 1958.

Kinsey, A. C., Pomeroy, W. B., and Martin, C. E.: Sexual Behavior In The Human Male. Philadelphia, W. B. Saunders, 1948.

Kinsey, A. C., Pomeroy, W. B., Martin, C. E., and Gebhard, P. H.: Sexual Behavior In The Human Female. Philadelphia, W. B. Saunders, 1953.

Kumar, M. A., Slack, E., Edwards, A., Soliman, H. A., Baghdiantz, A., Foster, G. V., and MacIntyre, I.: A biological assay for calcitonin. J. Endocr., *33*: 469-475, 1965.

Land, G., van't, M., and deHaas, J. H.: Menarcheleeftijd in Nederland. Nederl. T. Geneesk., *101*: 1425-1431, 1957.

Lauritzen, C.: On endocrine effects of oral contraceptives. Acta Endocr., *124*:87-100, 1967.

Le Femine, A. A., Marks, L. J., Teter, J. G., Leftin,

J. H., Leonard, M. P., and Baker, D V.: The adrenal cortical response in surgical patients. Ann. Surg., 146:26-39, 1957.

Li, C. H., Liu, W. K., and Dixon, J. S.: Human pituitary growth hormone. XII. The amino acid sequence of the hormone. J. Am. Chem. Soc., 88:2050-2051, 1966.

Masters, W. H., and Johnson, V. E.: Human Sexual Response. Boston, Little, Brown & Co., 1966.

McIntyre, N., Haldsworth, C. D., and Turner, D. S.: Intestinal factors in the control of insulin secretion. J. Clin. Endocr., 25:1317-1324, 1965.

Nalbandov, A. V.: Advances in Neuroendocrinology. Urbana, University of Illinois Press., 1963.

Panda, J. N., and Turner, C. W.: Hypothalamic control of thyrotrophin secretion. J. Physiol., 192:1-12, 1967.

Patten M. B.: Human Embryology. 3rd ed. New York, McGraw-Hill, 1968.

Peck, S. R.: Atlas of Human Anatomy for the Artist. New York, Oxford University Press, 1951.

Ponseti, I. V., and Shepard, R. S.: Lesions of the skeleton and of other mesodermal tissues in rats fed sweet-pea seeds. J. Bone Joint Surg., 36-A:1031–1058, 1954.

Porte, D., Jr.: The effect of epinephrine on immunoreactive insulin levels in man. J. Clin. Invest., 45:228-236, 1966.

Rivarola, M. A., Forest, M. G., and Migeon, C. J.: Testosterone, androstenedione and dehydroepiandrosterone in plasma during pregnancy and at delivery: Concentration and protein binding. J. Clin. Endocr. 28:34-40, 1968.

Roth, J., Glick, S. M., Yalow, R. S., and Berson, S. A.: Hypoglycemia: A potent stimulus to secretion of growth hormone. Science, 140:987-988, 1963.

Ryder, N. B., and Westoff, C. F.: Use of oral contraception in the United States, 1965. Science, 153: 1199-1206, 1966.

Sanderson, P. H., Marshall, F., II, and Wilson, R. E.: Calcium and phosphorus homeostasis in the parathyroidectomized dog—Evaluation by means of ethylenediamine tetraacetate and calcium tolerance tests. J. Clin. Invest., 39:662-670, 1960.

Sawin, C. T.: The Hormones: Endocrine Physiology. Boston, Little, Brown & Co., 1969.

Scammon, R. E.: The prenatal growth and natal involution of the human uterus. Proc. Soc. Exp. Biol., 23:687-690, 1926.

Schmidt-Matthiesen, H. (ed.): The Normal Human Endometrium. New York, Blakiston Division (McGraw-Hill), 1963.

Sherwood, L. M., Potts, J. T., Jr., Care, A. D., Mayer, G. P. and Aurbach, G. D.: Evaluation by radioimmunoassay of factors controlling the secretion of parathyroid hormone. Nature, 209:52-55, 1966.

Sonenberg, M. (ed.): Growth hormone. Ann. N. Y. Acad. Sci., 148:291-571, 1968.

Steiner, D. F.: Insulin and the regulation of hepatic biosynthetic activity. Vitamins Hormones, 24:1-61, 1966.

Tepperman, J.: Metabolic and Endocrine Physiology. Chicago, Year Book Publishers, 1968.

Turner, C. D.: General Endocrinology. Philadelphia, Saunders, 1966.

Wells, L. J.: Descent of testis—Anatomic and hormonal considerations. Surgery, 14:436-472, 1943.

Wilkins, L.: The Diagnosis and Treatment of Endocrine Disorders in Childhood and Adolescence. 3rd ed. Springfield, Ill., Charles C Thomas, 1965.

Williams, R. H. (ed.): Textbook of Endocrinology. 4th ed. Philadelphia, W. B. Saunders, 1968.

Williams, W. W.: Sterility: The Diagnostic Survey of the Infertile Couple. 3rd ed. Springfield, Mass., Walter W. Williams, Publishers, 1964.

Wolstenholme, G. E. W., and Porter, R.: The Human Adrenal Cortex. Boston, Little, Brown & Co., 1967.

Young, C. C., Jr., Mammen, E. F., and Spain, W. T.: The effect of sequential hormone therapy on the reproductive cycle. Pacif. Med. Surg., 73:35-40, 1965.

PART VII

SENSATION AND INTEGRATION

Information concerning the internal and external environment is transmitted to the central nervous system by sensory neurons and by the blood. The sensory neurons are stimulated by either (1) a stimulus which acts directly on a group of neurons to cause a pattern of propagated action potentials, or (2) receptor cells or sense organs containing receptor cells. The receptor cell, in turn, may respond either directly to a stimulus or to some change in its environment produced by the stimulus. In the eye, for example, light first breaks down a visual pigment, and this reaction sets up a generator potential that initiates propagated impulses in the optic nerve.

Our sensory system, in other words, acts as a transducer which converts one form of energy into propagated action potentials to the central nervous system. This was discussed in more detail in Chapter 3, page 51, under receptor mechanisms. In Chapter 2 on page 23 we discussed the types of sensory transducer systems found in the body, and in Table 2-1 we divided these systems into four categories. These include (1) a group of receptors associated with macroscopic end-organs found in the head and throat—the special senses, (2) the cutaneous senses, (3) muscle senses, and (4) the visceral senses. Another classification divides the sensory transducer systems into those which have their lowest threshold to (1) mechanical changes (the pressure and touch receptors, for example), (2) thermal changes, (3) changes in light intensity, and (4) chemical changes (PO_2, odor, taste, etc.).

Once afferent neurons have been stimulated by the peripheral transducer system, they carry their impulses to numerous internuncial or motor neurons, or both, in the central nervous system. Characteristic of this sensory input is its divergence. Impulses in a single afferent neuron go to many areas of the central nervous system. One neuron often synapses with neurons carrying impulses to (1) the cerebellum, (2) the reticular activating system, (3) the cerebral cortex, and (4) numerous other areas of the brain. In addition there is (1) facilitation of ipsilateral synergistic motor neurons, (2) inhibition of ipsilateral antagonistic motor fibers, and (3) contralateral inhibition and facilitation. Some of these complex relationships were discussed in Chapter 2, and are summarized in Figure 2-15.

The pattern of motor control, on the other hand, is one of convergence. Converging on a single motor neuron are facilitatory impulses from the cerebral cortex, the basal ganglia and cerebellum, and various sensory neurons. In the chapters that follow we will be concerned with the special senses and the integration of their sensory messages in the central nervous system.

Chapter 35

LIGHT
THE REFRACTIVE
ELEMENTS OF THE EYE
THE RETINA
BINOCULAR VISION

VISION

Vision is a phenomenon requiring a peripheral receptor organ (the eye), sensory nerves from that organ, and the cerebral cortex. The eye (1) acts as an optical instrument that gathers in light waves and brings them to focus on the retina and (2) produces a pattern of action potentials that pass from the retina into the optic nerve. We shall consider first how the eye performs its optical function, and then its sensory mechanisms.

LIGHT

Light is one of a number of electromagnetic waves. These waves in order of their increasing wavelengths include (1) cosmic rays, (2) gamma rays, (3) x-rays, (4) ultraviolet waves, (5) visible light rays, (6) infrared light waves, (7) radio waves, and (8) electric waves. Electromagnetic waves with wavelengths shorter than those of ultraviolet light are biologically important because of their capacity to pass through the skin and produce tissue damage. Ultraviolet light, on the other hand, is used by the skin in producing vitamin D from its precursors, and causes tanning and burning of the skin. Infrared rays warm the objects they come in contact with.

The spectrum of visible light includes wavelengths of from 3,970 to 7,230 A (397 to 723 mμ), the shortest waves being violet. Progressively increasing, the wave lengths go from violet to blue to green to yellow to red. These colors can be produced from white light by shining the white light through a prism. When white light passes from air through the prism, the velocity of transmission for each wavelength making up the white light is changed and each wavelength is bent. The shortest waves are bent most and the longest ones least, the result being that white light entering the prism is separated according to its varying wavelengths.

Refraction

Refraction is used in optics not only to produce rainbows but also to focus a series of light rays on a single point. With this as our objective we do not use a prism to separate the various wavelengths, but we do have the light pass from air through a medium with different conductive properties than air. The amount of bending (refraction) of the light depends upon (1) the angle at which the light strikes the surface of the second transparent medium and (2) the difference between the refractive properties of the two media.

In Figure 35-1 light is characterized as a series of waves of equal duration (monochromatic light) moving from left to right. In part A, the direction of movement of the wave front is perpendicular to a second transparent medium. Under these conditions there is no refraction when the light passes through the second medium. In part B, however, the movement of the wave front is not perpendicular to the surface of the second transparent medium and the light waves are therefore refracted. The degree of refraction depends upon the conducting properties of the two media—i.e., their refractive indices. If, for example, air has a refractive index of 1, then water will have a refractive index of 1.33, since it transmits light only 75 per cent as fast as air (refractive index = 1/0.75 = 1.33). A vacuum, on the other hand, transmits light 1.00029 times faster than does air. Another method of determining the refractive index of a material is to estimate the degree to which it bends incident rays of light passing through air. In Figure 35-1 (B) we can measure

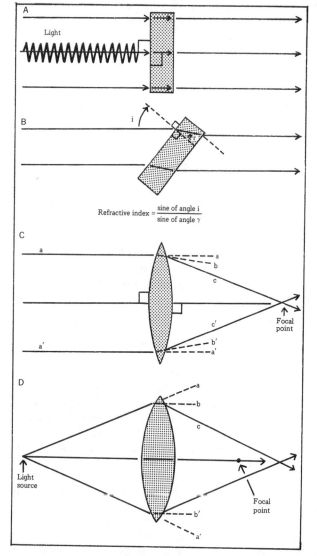

Fig. 35-1. Some basic concepts in optics. In parts A and B we see that when light passes from air through a material with different optical characteristics, the degree of refraction will depend in part upon the angle of incidence, i. In A the angle of incidence is 0 and therefore the angle of refraction, γ, is also 0. In B the angle of incidence is greater than the angle of refraction. The ratio, sine of angle i/sine of angle γ, is a constant for any homogeneous material (refractive index) except when angle i equals 0. Therefore the refractive index of the block is greater than that of air. In C and D we note how 2 different convex lenses refract light waves so as to bring them into focus at a single point. At C parallel rays are brought into focus. In ophthalmology rays of light from a point source more than 20 feet away are considered to strike the eye parallel to one another. When the object is brought closer than 20 feet to the eye (see D) the lens must be made more convex (its focal point for parallel rays must be brought closer) in order to have the rays converge at the same point.

the angle of incidence, i, and the angle of refraction, r, and determine the refractive index of the rectangle shown here:

$$\text{Refractive index} = \frac{\text{velocity of light in air}}{\text{velocity of light in block}} = \frac{\text{sine of angle i}}{\text{sine of angle r}}$$

The refractive index for aqueous humor and vitreous humor is like that of water, about 1.33, and that for the lens of the eye is about 1.4, whereas the refractive index for a diamond is 2.42.

Also of interest in Figure 35-1 is the observation that when light passes into a medium that has a higher index of refraction, it is bent toward an imaginary line at the point of contact with and perpendicular to the surface of the medium, whereas when it passes into a medium with a lower index of refraction it is bent away from the perpendicular. These relationships are taken advantage of in the lens of the eye and in the construction of lenses in general. In Figure 35-1 (C) we see a biconvex lens. Note that when parallel rays of light strike its surface, they tend to converge; when they leave the lens, the tendency to converge is even greater. In the lens pictured, all the parallel rays of light come to focus on one point, the focal point. This is because the central ray in this figure has an angle of incidence of zero both entering and leaving the lens; from the central ray to the peripheral rays the angle of incidence, and therefore the angle of refraction, increase.

It is also possible in optics to bring light rays that are not parallel to one another into focus on a single point. In Figure 35-1(D) this is done by making the lens more convex than was the case in part C. In other words, it is possible to increase the angle of incidence and therefore the amount of refraction by increasing the convexity of the lens.

The more a lens bends the light passing through it, the stronger it is considered to be. For example, a lens that brings parallel rays of light to focus 0.5 m. from its center is said to have a *focal length* of 0.5 m. and a strength of 2 *diopters*:

$$\text{Strength in diopters} = \frac{1}{\text{focal length in meters}}$$

$$2 \text{ diopters} = \frac{1}{0.5 \text{ m.}}$$

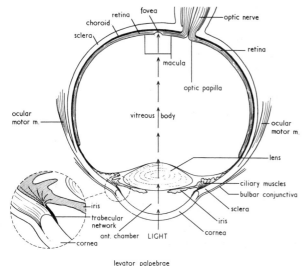

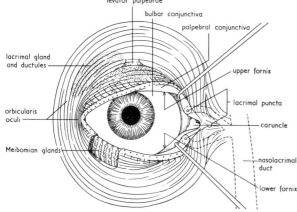

Fig. 35-2. Structure of the eye. (Heaton, J. M.: The Eye. Plate 7. Philadelphia, J. B. Lippincott, 1968)

THE REFRACTIVE ELEMENTS OF THE EYE

The light entering the eye must pass through the cornea, anterior chamber, lens, and vitreous body before it comes in contact with the retina (Fig. 35-2). Each of these substances contributes to the refraction of the light. You will note in Table 35-1 that the cornea has the greatest refractive power. The importance of the lens, on the other hand, is that its power to refract is variable. Thus the total refractive power of the eye may vary from 67 diopters when viewing a distant object to 79 diopters when viewing a near object. An example of how the eye handles light rays from an object is seen in Figure 35-3. You will note that the refraction of the light results

	Refractive Index	Distance from Cornea (mm.)	Refractive Power (diopters)
Air	1	0	
Cornea	1.376	0	+48.2
Aqueous humor	1.336	0.5	−5.9
Lens	1.386–1.406	3.6 (3)	−5
Vitreous humor	1.336	7.2	+8.3
Retina		24	

Table 35-1. Optical constants for an emmetropic eye. The refractive indices listed for the lens represent the range from the surface (first number) to the center (second number). The distances from the cornea were obtained by measuring from the anterior surface of the cornea to the anterior surface of each structure listed. The value in parentheses represents the indicated distance during accommodation. The refractive power shown for the lens is for the anterior surface only. For this reason and because the separation of the surfaces affects the cumulative refractory effect, the total of refractory powers shown above is less than the total refractory power of the non-accommodating eye (total refractory power varies from 67 to 79 diopters).

in an inversion of the image. Since the bending occurs in two dimensions, there is a reversal of the image as well:

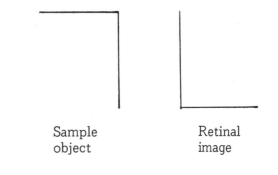

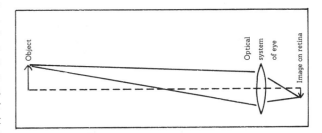

Fig. 35-3. Production of a retinal image.

Accommodation

In the optically normal eye, parallel rays of light striking the cornea are focussed on a single point on the retina. This can occur in the absence of the contraction of any of the muscles of the eye. It can occur after the infusion into the eye of a cycloplegic (a drug that paralyzes the ciliary muscle) such as homatropine. A person who in the absence of a contraction of the ciliary muscles of the eye focusses parallel rays on a single point on the retina is said to be *emmetropic*. A person who under the same circumstances focusses the light behind the retina is termed hypermetropic (mistakenly called farsighted), and a person who focusses the light in front of the retina is said to be myopic (mistakenly called nearsighted).

In the case of the emmetrope an object 20 or more feet away will be focussed clearly on the retina. As the object is moved closer to the eye, the point of focus moves to behind the retina. In other words, the rays of light from a point source now hit the retina at different points and the image is blurred. The person responds to this blurred image with a reflex stimulation of parasympathetic neurons to the ciliary muscle. This results in a contraction of the ciliary muscles that decreases the tension on the suspensory ligaments and allows the lens to become more convex (Fig. 35-4). At the same time there is a constriction of the pupil caused by decreased tension on the iris. This series of responses is called accommodation and occurs when a normal subject changes his focus from a distant object to a near object. When the subject returns his attention to an object 20 or more feet away, the ciliary muscle relaxes, the lens becomes less convex, and the pupil enlarges. In other words one cannot at the same time see distinctly an object that is 20 ft. away and one that is 10 ft. away. One must focus on one or the other.

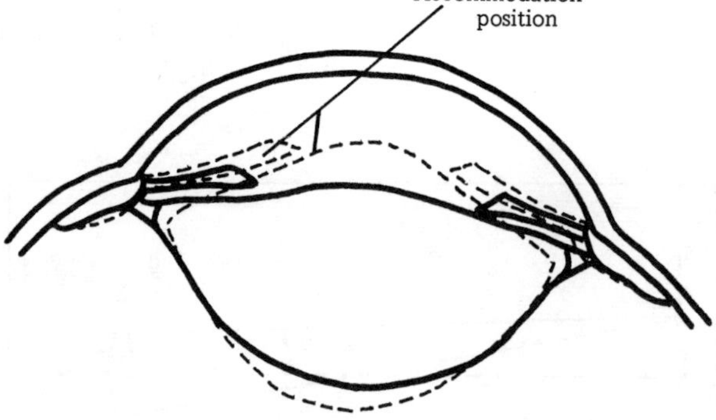

Presbyopia. The maximal optical strength of the eye can be determined by measuring the shortest distance from the optical center of the eye at which a near object can be seen distinctly. This distance is called the *near point* of vision and is usually about 10 cm. for a young adult. The optical center of the eye lies about 5 mm. from the anterior corneal surface and about 15 mm. from the fovea centralis of the retina. If we substitute in the following formula we obtain an estimate for the maximal strength of the optical system of the eye:

$$\frac{1}{\text{Object distance}} + \frac{1}{\text{image distance}} =$$

Fig. 35-4. Changes in the eye during accommodation. Note that there is an anterior bulging of the lens and an associated decrease in pupil size.

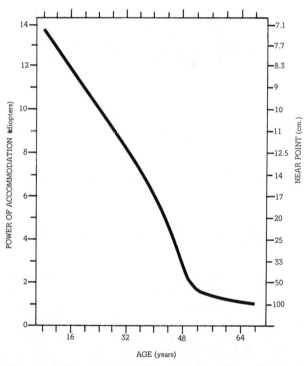

Fig. 35-5. Change in accommodation with age. Note that at 8 years of age the power of the lens can be increased by an average of 14 diopters during accommodation, whereas at 70 years of age the power can be increased by only 1 diopter. (*Modified from* Duane, A.: Studies in monocular and binocular accommodation with their clinical applications. Am. J. Ophthal., 5:870, 1922)

The Refractive Elements of the Eye

$$\frac{1}{\text{focal length}} = \text{power}$$

$$\frac{1}{0.10 \text{ m.}} + \frac{1}{0.015 \text{ m.}} = \frac{1}{0.013 \text{ m.}} = 77 \text{ diopters}$$

As we get older there is a generalized decrease in elasticity throughout the body. This loss involves the skin, blood vessels, lungs, and lens. When the ciliary muscle contracts fully, there will be a greater increase in the convexity of the lens at age 10 than at age 70 (Fig. 35-5). In other words, at age 10 the average near point is about 8 cm., while at age 70 it recedes to about 100 cm. This loss of the capacity to accommodate is called presbyopia (Gk. *presbys*, old; plus *ops*, eye) or literally "old eyes," Generally the symptoms of presbyopia become disturbing between the ages of 38 and 48. These symptoms may include the inability to see near objects distinctly or headaches and eyestrain after doing close work. At this time the physician generally prescribes bifocals. These glasses contain lenses of two different focal lengths for each eye.

Visual Acuity

The normal eye of a young adult, in the absence of accommodation, can see distinctly an object 20 or more feet away and has a near point during accommodation of less than 15 cm. Its cornea and lens transmit almost all the visual light striking them, and it can distinguish two points transmitting light to the eye that subtend an angle of at least 1 min.

The ability of the eye to detect a separation between two points is called its visual acuity. It is frequently measured by having a subject read a chart similar to that in Figure 35-6. Usually the subject starts 20 ft. away from the chart and

Fig. 35-6. A chart used for testing visual acuity. (Lebensohn, J. E.: Visual Charts. *In* Gettes, D. C. (ed.): Refraction. P. 18. Boston, Little, Brown & Co., 1965)

reads each line until he is unable to read any further. If he cannot see the chart at 20 ft., he moves up to 10 ft. away from the chart. The line marked 20/20 contains letters whose strokes subtend an angle of 1 min. when light rays from them strike the eye of a subject 20 ft. away. The line marked 20/400 contains letters whose strokes subtend an angle of 1 min. when light rays from them strike the eye of a subject standing 400 ft. away. If, for example, a subject standing 20 ft. from the chart can read no further than the line marked 20/50, his visual acuity is said to be 20/50. That is to say that he can be no further than 20 ft. from the chart to read what a normal young adult could read at 50 ft.:

$$\text{Visual acuity} = \frac{\text{distance of object in feet}}{\text{rated distance for object}}$$

In other words, a person with 20/20 vision (angle = 1 min.) has better visual acuity than a person with 20/40 vision (angle = 2 min.), but poorer acuity than a person with 20/16 vision. Defective visual acuity may be due to an inability to focus distinct images on the proper part of the retina, poor illumination, the diffraction pattern of the image on the retina, or a low density of retinal receptors.

Myopia

Myopia is a condition in which a distant object (20 or more feet away) in the absence of accommodation is brought to focus in front of the retina (Fig. 35-7). Such objects cannot be seen distinctly without corrective lenses. Myopia is usually due to the eyeball's having an excessive lens-to-retina distance. In some conditions, however, it results from excessive curvature of the cornea or lens. In either case vision can be corrected by wearing concave (causing divergence of the light rays) lenses for distance vision. In the absence of other problems, no glasses are necessary for near vision. For this reason myopia is sometimes called nearsightedness. This is not to imply that the "nearsighted" person can see objects 1 to 2 ft. from the eye any more distinctly than can the emmetropic person. One advantage he has over the emmetrope, however, is that he must accommodate less to see these objects distinctly.

Hyperopia

In hyperopia (hypermetropia) a person must accommodate to see both near objects and dis-

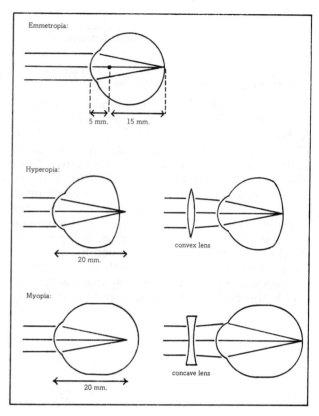

Fig. 35-7. A comparison of an emmetropic (normal eye of a young adult), hyperopic, and myopic eye. Note that the defect in hyperopia is that in the absence of accommodation parallel rays of light come to focus behind the retina. This problem, within limits, can be corrected by increasing the convexity of the lens or by wearing glasses with a convex lens. In myopia the fundamental defect is that parallel rays of light come into focus in front of the retina. This can be corrected by the use of a concave lens.

tant objects distinctly. In other words in the absence of accommodation, distant objects are focussed behind the retina. This condition is usually due to the eyeball's being too short, but in some cases is the result of an insufficient refractive system. In severe conditions, the subject may not be able to accommodate enough to see near objects distinctly. This characteristically poor near vision in hyperopia has led to the condition's being called farsightedness—a particularly misleading term since the hyperopic person cannot see distant objects any more clearly than the emmetropic one, and in fact must accommodate to see distant objects. As a result a hyperopic person who uses his eyes all day for distant vision, such as a bus driver, may at the end of

The Refractive Elements of the Eye

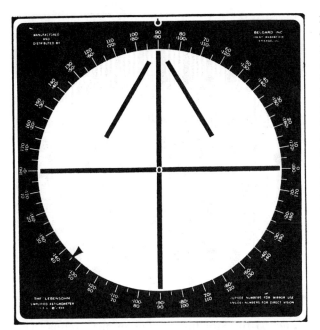

Fig. 35-8. A simplified astigmometer (actual size is 13½ × 14 inches). In using this chart the physician determines whether the patient sees all of the lines as being equally clear. If he does he is not suffering from astigmatism. If, for example, the vertical line is clearer, the arrowhead is pointed to 90° and moved so that its 2 wings are equally clear. The arrow will then point to the axis which needs correction. Different cylindrical lenses are then placed at right angles to the arrow-shaft until all of the lines on the chart are equally clear. (Courtesy of the Uhlemann Optical Co., Chicago, Ill.)

rays in another plane are focussed either in front of or behind it. The person with regular astigmatism, for example, sees one of the lines in Figure 35-8 more distinctly than the other. In order to correct for this one must refract the rays in one plane while not refracting those in a plane perpendicular to the first. This is done by the use of a cylindrical lens. If, for example, a patient saw the horizontal line in Figure 35-8 as blurred and the vertical line as distinct, he would need a corrective lens for the horizontal plane (180°) but not for the vertical plane (90°). This would be accomplished by orienting a cylindrical lens in the manner shown in Figure 35-9. In the case of irregular astigmatism cylindrical lenses cannot be used for correction purposes. Instead this condition is treated using contact lenses. The spherical surface of the contact lens tends to mold an irregular cornea into the shape of the artificial lens.

Lens Prescriptions

In prescribing glasses the ophthalmologist is concerned with correcting the optical problems of each eye. For presbyopia he may prescribe bifocals, for myopia concave lenses, and for hyperopia convex lenses. In order to determine

the day experience eyestrain or a headache, whereas the emmetrope who need not contract his ciliary muscles for distant vision will have no such symptoms. The symptoms of hyperopia can be prevented by using convex lenses. These may be used solely for close work or for both near and distant vision.

Astigmatism

Astigmatism is an optical defect usually due to abnormal curvature of the cornea. It may also result from abnormal curvature of the lens. Normally the cornea and lens have a spherical surface, but in astigmatism there is an irregularity in the surface. This may be an increase in the convexity of the eye in one particular meridian (regular astigmatism) or a variation in the convexity at a single locus (irregular astigmatism). Regular astigmatism results in parallel light rays in one plane coming to focus on the retina while

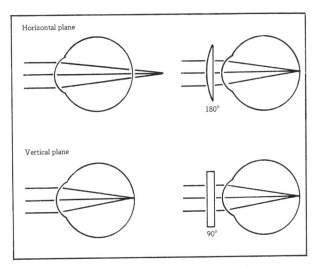

Fig. 35-9. Corrective lenses for astigmatism. The patient with astigmatism has a defective refraction in either the vertical or horizontal plane but not in the perpendicular plane. A cylindrical lens is used to make a correction for this problem. In the case above, the cylindrical lens was oriented in the horizontal plane. This resulted in a diffraction of light rays in the horizontal plane but not of those in the vertical plane.

the degree of correction necessary he may administer a cycloplegic to prevent accommodation and then use lenses of different strengths until a distant object can be seen clearly. Thus he might prescribe a +3 diopter lens (a convex lens) or a −4.5 diopter lens (a concave lens). In cases of astigmatism he must state not only the power of the lens (+2 or −1 diopters) but also its axis (90°, 95°, etc.). Thus he might prescribe a +1.5 diopter cylindrical lens at an axis of 180° (+1.5 axis 180°). When astigmatism is combined with myopia or hyperopia, he first determines the lens necessary to correct for the myopia or hyperopia and then the lens necessary to correct for the astigmatism. Thus a person with myopia and astigmatism might get the following prescription: −2 +3 axis 180°. The −2 would correct for the myopia and the +3 axis 180° would correct for the astigmatism.

THE RETINA

The retina has the thickness of a piece of paper, lies between the vitreous body and the sclera, and covers approximately 180° of the posterior pole of each eye (Fig. 35-2). The part of the retina immediately adjacent to the vitreous body is pigmented epithelium (layer 1 in Fig. 35-10). This layer is transparent, separating the vitreous body from the layers of rods and cones (layers 2 through 5b). Between the layers of rods and cones and the choroid lie a layer of ganglion cells (layer 8), a layer of optic nerve fibers (layer 9), and an internal limiting membrane.

The Rods and Cones

The rods and cones are elongate light-sensitive structures in the retina below the layer of pigmented epithelium. The cones are responsible for our vision during medium to intense illumination, and the rods for our vision during dim illumination. The cones are found in highest concentration in a yellow spot (the macula lutea, or the macula retinae), which lies in the usual visual axis. At the center of the macula is the fovea centralis, an area about 0.3 mm. in diameter (a visual angle of 60 min.) which is devoid of rods and contains the highest concentration of cones found in the retina. You will note in Figure 35-11 that in one area of the fovea the cone concentration is as high as 147,000 per square millimeter.

In the area of the fovea with the greatest visual

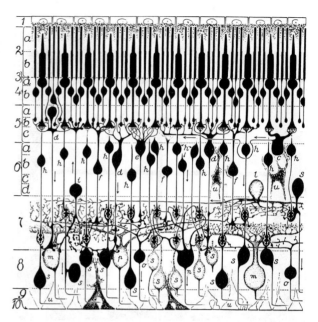

Fig. 35-10. Structure of the primate retina. The layers, starting at the layer adjacent to the vitreous body, are: (1) pigment epithelium, (2a) outer segment of rods and cones, (2b) inner segment of rods and cones, (3) outer limiting membrane, (4) outer nuclear layer containing cell bodies of (a) cones and (b) rods, (5) outer plexiform layer, (6) inner nuclear layer, (7) inner plexiform layer, (8) ganglion cell bodies, (9) optic nerve fibers, and (10) inner limiting membrane. The cell types are as follows: (c) horizontal cells, (d, e, f, h) various types of bipolar cells, (i, l) amacrine cells, (m, n, o, p, s) ganglion cells, and (u) radial glial cells of Muller. (Polyak, S. L.: The Retina. Figure 96. Chicago, University of Chicago Press, 1941)

acuity each cone is served by a single sensory neuron and lies between 0.0020 and 0.0025 mm. from an adjacent cone. As one passes peripherally, however, the ratio of cones to sensory neurons increases. The total number of cones in the retina is approximately 6.5 million, and the total number of rods is between 110 and 125 million, whereas the total number of neurons in the optic nerve is only 1 million. In other words the average ratio of receptors (cones and rods) to sensory neurons is about 120:1. This ratio is characteristically greater peripheral to the fovea centralis.

Scotopic vision. Those of us who have lived most of our lives in the city have acquired the habit of using the fovea centralis and its associated cones for our vision. Most of our visual world is in color. In dim illumination, however, the cones are not sensitive enough to perceive objects and we must therefore rely on our more

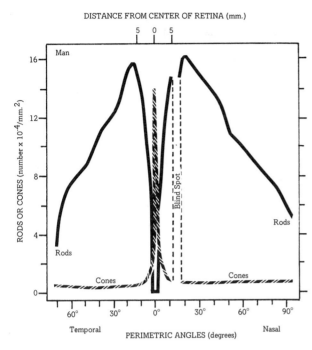

Fig. 35-11. Distribution of the rods and cones on the human retina. Note that the highest concentration of cones (147,000 per mm.²) is found in the center of the retina in the fovea centralis, and the highest concentration of rods (160,000 per mm.²) is found 5 to 6 mm. from the center of the retina. The blind spot represents the point where there are neither rods nor cones and where the optic nerve leaves the retina. (*Modified from* Osterberg, G.: Topography of the layer of rods and cones in the human retina. Acta Ophthal., Supp., 6:fig. 5, 1935)

sensitive receptors, the rods. Scotopic vision is vision that involves the rods. Objects seen with rods are characteristically poorly defined and vary in character from black to various shades of gray. If we look directly at an object in dim light, it frequently disappears because the rods lie peripheral to the fovea. In other words, in light too dim for cone function, we must focus the light to one side or another of the fovea to see the object.

In Figure 35-12 we have a comparison of the sensitivity of the rods and cones to light. Note that on exposure to darkness (removal of white light) the cones increase their sensitivity to light by twenty to fifty fold in the first 5 min. The rods, on the other hand, prior to their exposure to darkness, had been bleached by the light and were not functioning in vision. After 5 to 10 min. of darkness they start to function again in vision,

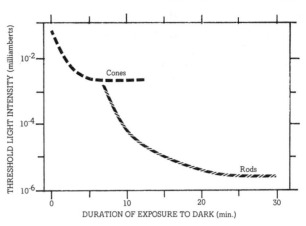

Fig. 35-12. Adaptation of the rods and cones to darkness. (*Redrawn from* Willmer, E. N.: *In* Best, C. H., and Taylor, N. B. (eds.): The Physiological Basis of Medical Practice. 8th ed., p. 297; copyright © 1966, The Williams and Wilkins Co., Baltimore)

and after 25 to 30 min. of darkness have almost reached maximum sensitivity.

The sensitivity of the rods to light is due to the reddish-purple pigment they contain. This pigment (*rhodopsin*, or *visual purple*) is most sensitive to light having a wavelength of 505 mμ (green), but also responds to wavelengths ranging from 400 (violet) to 560 mμ (yellow). You will note in Figure 35-13 that it is relatively in-

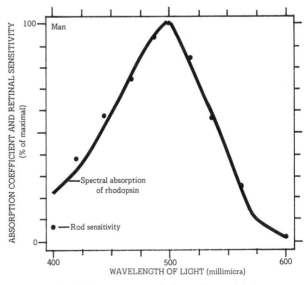

Fig. 35-13. Comparison of the rod sensitivity curve with the spectral absorption curve of rhodopsin. (*Redrawn from* Crescitelli, F., and Dartnall, H. J. A.: Human visual purple. Nature, *172*:196, 1953)

sensitive to light with a wavelength greater than 600 mμ (red, for example). For this reason sailors getting *dark-adapted* in readiness for a night watch, or x-ray technicians getting ready for fluoroscopy, frequently wear red glasses for periods of about 30 min. prior to exposure to the dark. They also put on red glasses when moving from a dark to a light environment in order not to lose their dark adaptation.

The chemical reactions leading to the stimulation of sensory neurons by the dark-adapted rods when they are exposed to light are as follows:

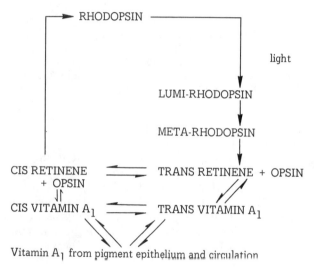

In other words light causes the breakdown of rhodopsin to a protein, opsin, and a chromophore, *trans*-retinene. The retinene is then reduced to vitamin A_1, much of which is stored in the pigment epithelium. The conversion of the *trans*-retinine to *cis*-retinine requires an energy source, but the formation of rhodopsin from *cis*-retinene and opsin occurs spontaneously.

Photopic vision. Photopic vision is vision due to stimulation of the cones. It is centered around the fovea and differs from scotopic vision in that it is sensitive to a broader spectrum of wavelengths (from 400 to 700+ mμ for cone vision v. from 400 to 625 mμ for rod vision) and has its maximum sensitivity at a higher wavelength (555 mμ for cone vision v. 507 mμ for rod vision). This change in maximum sensitivity as one moves from rod vision to cone vision was noted by Johannes Purkinje in 1825 and is called the *Purkinje shift*.

We have noted in the previous discussion that rod vision is due to chemical changes in a single pigment, rhodopsin, in the rods, The chemical changes responsible for color vision, however, are less well understood. The most generally accepted theory of color vision is the *Young-Helmholtz trichromatic theory*. In the current statement of the theory it is proposed that there are three different types of cones: those containing (1) a pigment which is most sensitive to blue light (wavelength, 445 mμ), (2) a pigment which is most sensitive to green light (wavelength, 535 mμ), and (3) a pigment whose sensitivity extends into the red wavelengths. These cones have been identified in goldfish, carp, the monkey, and man.

Part of the evidence favoring the trichromatic theory is presented in Figure 35-14. Note that

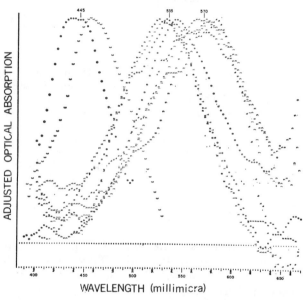

Fig. 35-14. Difference spectra of visual pigments of monkey cones (the point of maximum absorption for each type of cone is shown) and human cones. In these experiments the absorption spectra were determined for each cone before and after the cone was bleached with a strong light. The values for each wave length shown above represent the difference obtained for each cone (difference value equals value before bleaching minus value after bleaching). The points represented by numbers are from monkey cones (Macaca nemistrina and M. mulata) and those represented by open parentheses are from human cones. (Marks, W. B., Dobelle, W. H., and MacNichol, E. F., Jr.: Visual pigments of single primate cones. Science, *143*:1182; copyright March, 1964 by the American Association for the Advancement of Science)

the spectra presented for each type of cone overlap the spectra for other cone types. Thus in some cases a single wavelength may stimulate two or even three types of cone. In other words each wavelength will elicit a unique ratio of responses among the three fundamental receptor classes, and our distinction of the various hues (color sensations) is based on the recognition of the different ratios. An even stimulation of the blue, green, and red cones results in the sensation of white or various shades of gray, depending upon the intensity of the light. If we shine a combination of red and green light into the fovea, we experience a sensation of yellow. In other words, it is proposed that the three primary colors are red, green, and blue and that all the hues are the result of the stimulation of the red, green, and blue receptors in the cones.

Color blindness. In the preceding section we have been discussing normal color vision. There are, however, a small number of people who have abnormal color vision. This includes less than 0.5 per cent of all women and about 8 per cent of men. Most of these persons are classified as *anomalous trichromats*. Their three color pigments are functional but they show a weakness for one of the primary colors. The person who is red-deficient is classified as protanomalous; whereas the normal trichromat requires certain combinations of red and green light to experience the sensation of yellow, the protanomalous person requires greater intensities of red to experience the same color. The deuteranomalous person is a trichromat with a weakness for the color green. He requires a greater than normal intensity of green to experience the sensation of yellow.

Most of the rest of the so-called color-blind persons (2.6% of all men) are dichromats. They can produce any of the colors they experience by a mixture of two primary lights. They are classified as protanopic if they lack the pigment for red (pigment 1), deuteranopic if they lack the pigment for green (pigment 2), and tritanopic if they lack the pigment for blue (pigment 3). For the most part these deficiencies result from a sex-linked recessive gene. Since both the red and the green pigments show a considerable overlap in the frequencies that stimulate them (Fig. 35-14), both the protanope and the deuteranope are classified as being red-green blind. Tritanopia is considerably more rare than the other forms of dichromia. Finally, there is a small number of persons (about 0.003% of all men) who are classified as monochromats. They either lack cones and are therefore blind in the area of the fovea centralis or have only one type of color receptor. These are persons who can match light with any other light by merely adjusting the light intensity. The rod monochromat (cone deficient) differs from the cone monochromat in that the visual acuity of the former is poorer.

Critical fusion frequency. We have defined visual acuity as the ability of a person to distinguish two different points in space and have noted that the normal visual acuity of a young adult is such that he can distinguish two separate points if they subtend an angle of at least 1 min. In other words, visual acuity is the ability to resolve two different stimuli in space. The critical

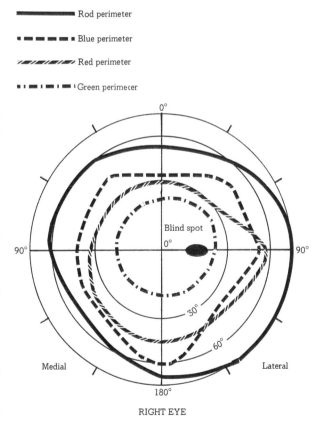

Fig. 35-15. Perimeter chart showing the normal visual fields for the right eye. Note that the green cones extend only to approximately the 20° perimeter, whereas the red and blue cones in some areas extend out past the 50° perimeter. Note that in the case of lateral vision the red perimeter extends out to 90°. (*Modified from* Hartridge, H.: Visual acuity and the resolving power of the eye. J. Physiol., 57:52, 1923)

fusion frequency has to do with a subject's ability to resolve two different objects in time. In the case of stimuli forming on the fovea the critical fusion frequency is approximately 50 stimuli per second. If 60 flashes of light strike the same point on the fovea per second they will be perceived as a steady stream of light. In the case of rod vision, just as acuity is less, so too is critical fusion frequency. These observations provide the physiological basis for the production of motion pictures.

BINOCULAR VISION

Up to this point in our discussion we have been concerned with vision in one eye. Normally, however, man uses two eyes for vision. Some of the advantages binocular vision offers are (1) an increase in the visual field, (2) a lower threshold for light, and (3) improved depth perception. The visual field of each eye (Fig. 35-15) is that part of the external world which can be seen from that eye while it is focussed on a single distant object. The two eyes working together give us a visual field on either side of approximately 180°. The cephalocaudal field, on the other hand, is only 120°.

The importance of binocular vision in depth perception relates to the fact that a single object forms slightly different images on the two retinas. Since the closer an object is to the retina the greater are the differences in the two images, the observer learns to associate these differences with variations in distance between himself and the object. An understanding of these principles is the basis for the construction of the old-time stereopticon and for the production of three-dimensional films. In both cases the producer uses a technique that exposes each eye to slightly different objects, which produces an exaggerated three-dimensional effect.

Corresponding Points

If a single point of light is focussed on the retina of one eye, and at the same time another single point of light is focussed on the retina of the other eye, two types of sensory experience are possible: (1) one may see a single point, or (2) one may see two points or a blurred image. The points on the retina of each eye which when stimulated simultaneously produce the sensation of a single stimulus are called corresponding

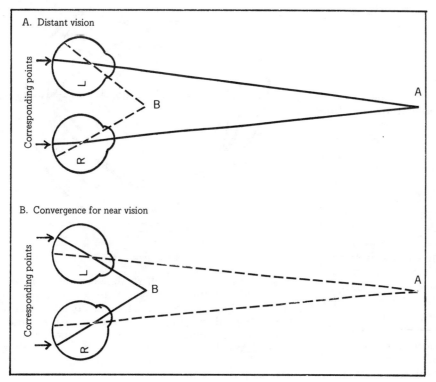

Fig. 35-16. Convergence. Note that when the eye is focused on a distant object (A), light rays from a single point source on that object are focused on corresponding points on the 2 retinas. On the other hand, when the eye is focused on a near object (B), the 2 eyes must converge in order to have the light from a single point source focused on corresponding points of the 2 retinas. Corresponding points are 2 areas which when stimulated simultaneously result in a single sensation rather than 2 sensations. When the eye is focused on a distant object, light from a near object is not focused on corresponding points and therefore the near object is seen as blurred or double and as indistinct.

points. The neuron from one of these points sends impulses up the optic nerve that cross over and stimulate a point on the contralateral occipital cortex. The neuron from the corresponding point in the other eye also sends its impulses up the optic nerve, but this message does not cross over. Rather, it is transmitted to a point on the cortex adjacent to that stimulated by the impulse from the contralateral eye.

Convergence

In Figure 35-16 we see one of the problems resulting from the use of two eyes simultaneously. If the eyeballs are oriented in such a way that light from a distant point is focussed on corresponding points on the retinas of the two eyes, it is impossible for light from a near point to focus on corresponding points of the two retinas. In other words each of us when we change our focus from a distant object to a near object must not only make the lens of each eye more convex but also move each eyeball in its socket so that the pupil of each eye is directed more medially. The increase in convexity of the eye is brought about by the contraction of a smooth muscle called the ciliary muscle. The convergence of the eye is brought about by the contraction and relaxation of the extrinsic muscles of the eye. These are skeletal muscles and include (1) the lateral and medial rectus muscles, (2) the superior and inferior rectus muscles, and (3) the superior and inferior oblique muscles (Fig. 35-17).

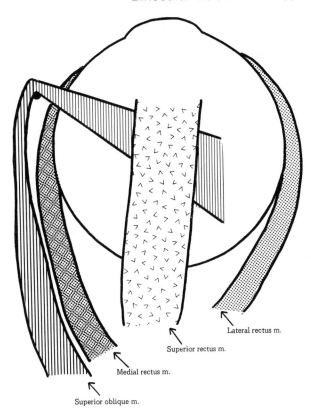

Fig. 35-17. Extrinsic muscles of the right eyeball viewed from above. Not shown in this illustration are the inferior oblique muscle (antagonistic to superior oblique) and the inferior rectus muscle (antagonistic to superior rectus). By moving the eyeball these muscles serve to direct the pupil toward the object which the individual wishes to view.

Chapter 36

VISUAL FIELDS AND THE
OPTIC PATHWAYS
THE STRIATE AREA OF THE
OCCIPITAL CORTEX
OTHER VISUAL CENTERS
VISUAL REFLEXES
LIGHT DEPRIVATION

NEURAL MECHANISMS IN VISION

Light enters the eye and is normally focussed on photoreceptors, resulting in a series of chemical reactions which probably leads to the production of a generator potential. It is speculated that the generator potential facilitates the stimulation of bipolar cells, which eventually stimulate or inhibit a group of ganglion cells. The general organization of the photoreceptors, bipolar cells, and ganglion cells is seen in Figure 35-10. Apparently this organization involves extensive convergence and divergence, as well as facilitation and inhibition of impulses before they even leave the retina. We find, for example, that if light is focussed on one spot on the retina, there follows facilitation of the discharge of a particular retinal ganglion cell. If we focus the light on an adjacent spot, however, the responsiveness of this same ganglion cell is depressed.

Nerve impulses are conducted out of the retina in the axons of the ganglion cells (Fig. 36-1). These axons are in the right and left optic nerves and all meet at the optic chiasma. It is at this point that fibers from the medial portion (representing the temporal field of vision) of each eye cross over to the contralateral side, and the fibers from the lateral portion (representing the nasal field of vision) of each eye continue along their ipsilateral tract. This arrangement is such that an optic tract is formed containing fibers from the contralateral nasal field and the ipsilateral temporal field. In other words, each optic tract contains a large number of pairs of neurons from corresponding points. The optic tract conducts its impulses toward the lateral geniculate body in the thalamus, but as it passes the midbrain, gives off collateral fibers that synapse in the superior colliculus of the midbrain. The fibers that remain synapse in the lateral geniculate body in the thalamus. From the lateral geniculate body the impulses are transmitted in the geniculocalcarine tract to the occipital lobe of the cerebral cortex, the primary visual receiving area being principally on the sides of the calcarine fissure in Brodmann's area 17.

VISUAL FIELDS AND THE OPTIC PATHWAYS

We have noted that light striking the retina initiates the conduction of a series of impulses to the occipital cortex and numerous other areas of the central nervous system. These impulses begin in the retina and are transmitted from its bipolar cells to its ganglion cells. As result of this transmission there is a combination of inhibition, facilitation, divergence, and convergence before the impulses leave the retina. When the impulses do leave, they are carried in the axons of the ganglion cells in the optic nerve. These axons are arranged such that impulses from the upper and lower temporal quadrants of the retina (representing the nasal field of vision) are carried in the upper lateral and lower lateral quadrants of the optic nerve, respectively, and impulses from the upper and lower nasal quadrants of the retina are carried in the upper and lower and medial quadrants of the nerve. The impulses in the medial quadrant of each optic nerve cross over at the optic chiasma, while those in the temporal quadrants enter the ipsilateral optic tract and join with the nasal quadrants of the contralateral side.

This arrangement of fibers means that a lesion of the optic nerve (Fig. 36-1, lesion A) produces a visual defect only in the ipsilateral eye, whereas a lesion at the chiasma or of the optic tract has bilateral consequences. Complete destruction of the optic nerve results in blindness in the ipsilateral eye. A discrete section of the optic chiasma, such as lesion B in Figure 36-1, on the other hand, produces loss of temporal vision in both eyes. A person suffering from such a defect

Visual Fields and the Optic Pathways

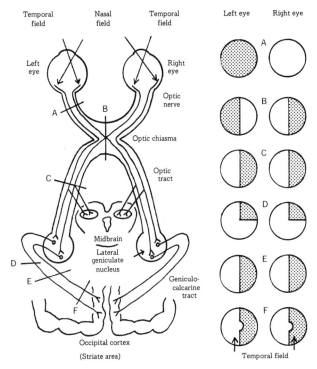

Fig. 36-1. Lesions in the optic pathways. Note that sensory messages going to the cerebral cortex from the retina synapse in the lateral geniculate nucleus and give off a few collaterals to the superior colliculus of the midbrain. The superior colliculus is an important reflex center for the light reflex, controlling the degree of pupillary constriction and dilation. The geniculate nucleus relays impulses to the striate cortex, which is an area of conscious sensation and which relays impulses to other areas of the cortex. These other cortical areas are important in integrating responses to visual stimuli with total body activity. Note that the lesions illustrated above produce rather well-defined effects. This is due to a well-defined pattern of localization within the fiber tracts. Lesion A results in blindness in one eye, lesion B in heteronymous hemianopsia, lesions C and E in homonymous hemianopsia, lesion D in homonymous superior quadrantic hemianopsia, and lesion F in homonymous hemianopsia with macular sparing. (*Modified from* Ganong, W. F.: Review of Medical Physiology. 4th ed., p. 95. Los Altos, Lange Medical Publications, 1969)

is said to have heteronymous *hemianopsia*. The noun "hemianopsia" means that half the visual field in one or both eyes has been lost, and the adjective "heteronymous" (different function), that the loss involves nonoverlapping functions in the two eyes. In other words, in this example there is a narrowing of the visual fields due to bilateral loss of function in the nasal quadrants of the retina.

If, on the other hand, the lesion is in the optic tract (lesion C) a homonymous (same function) hemianopsia results. There is an ipsilateral loss of the nasal field and a contralateral loss of the temporal field. The major defect is therefore a decrease in the temporal field of vision on the side of the body opposite the lesion. Since neither lesion B nor C produces complete blindness in either eye, it may go unrecognized by the patient. But since these lesions do decrease peripheral vision, they can lead to the increased frequency of accidents, as in an automobile driver for example.

In the optic tract, fibers from the upper quadrants of the retina tend to lie ventrolaterally, and those from the lower quadrants, ventromedially, while the foveal fibers lie dorsolaterally. Most of these fibers go to the lateral geniculate nucleus in the thalamus, but a few of the fibers from parts of the retina other than the fovea also send collaterals to the superior colliculus in the midbrain. It is in the lateral geniculate nucleus that the internuncial neurons of the optic tract synapse with the neurons of the geniculocalcarine tract. In man the ratio of optic fibers to geniculate cells is approximately 1:1. The nucleus is arranged in six orderly layers of cell bodies separated by interlaminar bundles of optic tract fibers. The most ventral layer, layer 1, and layers 4 and 6 carry impulses from the contralateral hemiretina to the cortex, while layers 2, 3, and 5 carry impulses from the ipsilateral hemiretina.

Hemianopsia

Up to this point we have discussed only two forms of hemianopsia: heteronymous and homonymous. There are also various forms of quadrantic hemianopsia (lesion D illustrates superior homonymous quadrantic hemianopsia), incongruous quadrantic hemianopsia, and hemianopsia with various types of sparing. In incongruous hemianopsia the visual field loss is different in the two eyes. In one eye the loss may represent a quadrant, while in the second eye it may represent 1½ quadrants. One of the more common forms of hemianopsia with sparing results from lesion F in Figure 36-1. This is an example of homonymous hemianopsia with macular sparing, which occurs in both heteronymous and homonymous hemianopsia but is most frequently associated with lesions near the cerebral cortex.

It has been suggested that in some cases it is due to (1) parts of the fovea having a bilateral representation on the cerebral cortices and (2) fibers carrying foveal impulses being diffusely arranged in the various tracts from the retina to the striate cortex. Actually there is very little evidence to favor either speculation.

THE STRIATE AREA OF THE OCCIPITAL CORTEX

The axons of the cells of the lateral geniculate nucleus form the geniculocalcarine tract and terminate wholly within the cortical region of the ipsilateral striate area (Brodmann's area 17). Axons from the large intermediate segment of the geniculate nucleus pass around the lateral side of the ventricle in their passage to the striate cortex. Axons from the medial portion of the nucleus lie above these and axons from the lateral portion of the nucleus lie below them. The latter fibers loop down into the temporal lobe and join the rest of the geniculocalcarine tract near the striate cortex. The organization within the striate cortex (Fig. 36-2) can be characterized as one in which the foveal field of the contralateral retina is represented by a large area of the posterior portion of the striate cortex, and the peripheral area of the retina is represented by a smaller, more anterior area. Corresponding points for the retina of the other eye are represented on the cortex at adjacent points.

The striate cortex is morphologically distinct from adjacent parts of the cortex and is apparently responsible for the recognition of color, form, position, and brightness. Its bilateral destruction in man leads to complete and permanent blindness and degeneration of the geniculate neurons. In the monkey, however, the same lesions leave a primitive awareness to changes in light intensity not found in man. This difference between the importance of the striate cortex of man and monkey is apparently due to the more highly developed retinomesencephalic projections in the monkey. In man the retinomesencephalic projection is apparently primarily concerned with the reflex control of the size of the pupil and blinking in response to changes in light intensity.

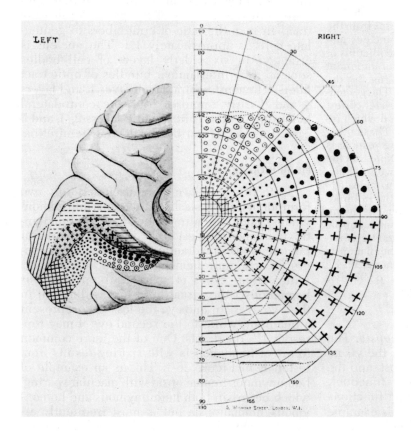

Fig. 36-2. Cortical representation of the contralateral visual field. Shown are the striate area of the left hemisphere and the visual field of the right eye. The macular region of the retina is represented in the posterior portion of the brain and is relatively large, whereas the peripheral retina is represented more anterior and is relatively small. (Duke-Elder, S.: A Textbook of Ophthalmology. Vol. I, p. 270. St. Louis, C. V. Mosby, 1952)

OTHER VISUAL CENTERS

The striate cortex has a receptive function. It is the area of the brain responsible for the conscious awareness of visual stimuli. It is not, however, an area where transmission ceases. Neurons originating from the striate cortex pass over the corpus callosum to the contralateral cortex, some going to various integrative centers in the ipsilateral cortex. Going to the striate cortex, on the other hand, are neurons that facilitate and inhibit activity there. The reticular formation of the brain stem, for example, serves during our waking hours to provide varying degrees of facilitation to the cells of the striate cortex.

Brodmann's areas 18 and 19 are two cortical areas that tend to integrate visual functions. Area 18 adjoins the striate cortex (area 17) in the occipital lobe. Area 19 is located rostrally and extends into the parietal and temporal lobes. In primates, pathways have been demonstrated from area 17 to 18, from 18 to 19, and from 18 on one side through the corpus callosum to the other side. The stimulation of areas 18 and 19 results in the conjugate deviation of the contralateral eye apparently through their action on the oculomotor centers of the brain stem. The bilateral removal of 18 and 19, on the other hand, has no effect on the recognition of objects, but does result in a disturbance of spatial judgment and a confusion when viewing moving objects.

Another area important in visual performance lies in the inferior convexity of the temporal lobe. There is some evidence that this area receives impulses from either the striate cortex or the prestriate cortex (areas 18 and 19) as well as from the contralateral visual centers. When it is removed bilaterally in monkeys, visual learning functions are disturbed but there seems to be no disturbance in either tactile or auditory learning functions. The monkey has difficulty not only in learning new visual tasks but also in performing ones learned before the operation.

VISUAL REFLEXES

The motor nerves to the eye involve the control of (1) its intrinsic smooth muscle (ciliary muscle, pupillary sphincter, and pupillary dilator), (2) extrinsic striated muscle attached to the sclera (Fig. 35-17), (3) the levator palpebrae superioris muscle (elevates eyelid), and (4) the lacrimal glands. These structures are concerned with focussing light on corresponding points on the two retinas and protecting the eye.

Protective Mechanisms

The *lacrimal glands* of the two eyes secrete about 1 ml. of liquid (tears) each day (pH 7.4). This liquid contains lysozyme, a mucolytic enzyme with bactericidal action, and serves to (1) protect the eye from infection, (2) lubricate the surface of the eye, (3) provide the eye with a good optical surface, and (4) act as an emergency wash in cases of irritation of the cornea. Its secretion is stimulated by parasympathetic fibers in the fifth cranial nerve. These fibers constitute the efferent limb of a *protective reflex*. The reflex is initiated by (1) receptors in the lids and conjunctiva, (2) irritation of the cornea, (3) bending the cilia of the eyelid, or (4) bright lights. Stimulation of the parasympathetics to the lacrimal glands may also result from certain psychic experiences (the death of a friend, etc.).

The *lids* of the eye also serve a protective function and are a means of decreasing the amount of light reaching the retina. During sleep there is a tonic contraction of the striated orbicularis oculi muscles (Fig. 35-2) which prevents about 99 per cent of the incident light from reaching the retina. The eyeballs also move up due to a change in the tone of the skeletal muscles attached to the sclera. During the waking period, the orbicularis muscles contract every few seconds for about one quarter of a second, at which time the eyeballs move up as they do in sleep. This constant blinking is bilateral and serves to keep the corneal surface free of mucus and distribute the lacrimal secretion evenly over the cornea. Blinking also occurs in response to (1) irritation of the cornea, (2) sudden bright light (dazzle reflex), or (3) rapidly approaching objects (menace reflex). The levator palpebrae muscle is responsible for the elevation of the eyelid. Its paralysis can be produced by destruction of the cervical sympathetic fibers passing up to the eye and results in ptosis (drooping of the upper eyelid).

Control of Pupil Size

The diameter of the pupil varies from 2 mm. in intense light to 8 mm. in total darkness. Its constriction (miosis) may be initiated by either an increase in parasympathetic tone to the pupillary sphincter of the iris or the stimulation of parasympathetic fibers to the ciliary muscle. Pupillary dilation (mydriasis) is initiated by a decrease in parasympathetic tone to the eye or stimulation of the sympathetic fibers to the pupillary dilator fibers of the iris. Agents that produce mydriasis are called mydriatics and include atropine (blocks parasympathetic tone to the iris) and epinephrine (stimulates the dilator fibers). Agents that produce miosis are called miotics and include eserine (an anticholinesterase), pilocarpine (a cholinergic drug), and morphine (pinpoint pupils are pathognomonic for morphine poisoning). Miosis also occurs during sleep and the intermediate states of anesthesia. Mydriasis, on the other hand, is characteristic of cerebral asphyxia and is noted in patients who have experienced several minutes of ventricular fibrillation or apnea. When a physician administers cardiorespiratory resuscitation to a patient, the disappearance of this mydriasis is taken as a sign of success.

Two reflexes that produce constriction of the pupil are (1) the near reflex and (2) the light reflex. In the near reflex a subject looks at a near object and in so doing produces stimulation of parasympathetic neurons to both the constrictor muscle of the iris and the ciliary muscle. The sensory components of the reflex include impulses traveling from the retina to the striate cortex. The motor components probably include impulses from the striate cortex to the superior colliculus and thence to the parasympathetic fibers of the third cranial nerve. The effect of the reflex is to decrease the amount of light entering the retina and in so doing decrease peripheral vision.

The light reflex also produces pupillary constriction, but does not necessarily involve the striate cortex. Here the sensory pathways are from the retina to the superior colliculus of the midbrain (Fig. 36-1) and then to the Edinger-Westphal nucleus. The motor pathways are via the parasympathetic fibers in the third cranial nerve. The reaction time for this reflex is about 0.2 sec. and involves both eyes. Even if the light is shined in only one eye, there is both a direct and a consensual reflex. In the tertiary stages of syphilis, the light reflex is lost even though the near reflex is retained. A person who has lost the light reflex but retained the near reflex is said to have an *Argyll Robertson pupil*.

Mechanisms for Accommodation

When a subject changes his attention from a distant object to a near object there is (1) a contraction of the ciliary muscle resulting in an increase in the convexity of the lens, (2) a convergence of the two eyes which results in the light from the near object focussing on corresponding points on the two retinas, and (3) constriction of the pupil. The stimulus for these changes is the sensation of a blurred image. The blurring is usually the result of (1) inadequate convexity of the lens and (2) the failure of the image formed on the right and left retinas to be on corresponding points. The response is the stimulation of parasympathetic neurons in the midbrain which go to the right and left ciliary muscles, and combined stimulation and inhibition of somatic efferent neurons in the midbrain which go to the appropriate muscles inserted on the sclera of the eyeball. In an erect person accommodating for an object directly ahead of him, the medial rectal muscles of the eye are stimulated and the lateral rectal muscles are inhibited. The contractions of both the ciliary muscles and the extrinsic muscles are coupled, occurring even when one eye is blind. The reaction time for this reflex is about one third of a second.

Strabismus. When the images from a single target form on the right and left retina in such a manner that they give rise to a single sensory impression, they are said to be fused. The inability to produce fusion may result from a weakness or some other abnormality of the extrinsic muscles of the eye. In response to this abnormality, the person looking at a near object converges the normal eye while moving the weak eye very little. The result is the formation of two images on the two eyes that do not fuse. The image formed on the sound eye is called the true image, and that on the affected eye, the false image. Initially this lack of parallelism of the visual axes of the two eyes (strabismus) results in the sensation of two images (diplopia), or blurring. If the strabismus is long-standing the diplopia disappears and the affected eye develops a peripheral fovea. When this happens, there is almost normal stereoscopic vision. In these cases surgical intervention correcting for the strabismus (squint) causes the subject to experience diplopia. In some cases the person with strabismus rejects the false image and sees only the true image. This results in the sensation of a clear image but decreases the visual field because of the rejection of impulses from the affected eye.

Role of the Vestibular Apparatus

The vestibular apparatus consists in part of (1) the semicircular canals, which are sensitive to body acceleration and deceleration, and (2) the utricles, which respond to the pull of both gravity and body acceleration and deceleration (Chap. 38). The importance of the vestibular apparatus in the control of the eye resides in its ability during acceleration and changes in body position to fix the pupil with respect to the external environment. For example, when the head is bent to the right, the eyes are rotated to the left in such a way that the vertical axis of the eye stays aligned with the direction of the gravitational pull. If, on the other hand, a subject sitting erect in a chair with his head directed forward is accelerated toward the left, the extrinsic muscles of the eye direct the pupils of the two eyes to the right. Under the same conditions the acceleration of the subject upward causes a rotation of the eyes downward.

Since the stimuli for the above eye reflexes are not visual ones, it makes little difference whether the eyes are open or closed. In either case the sensory messages for the reflex are carried from the vestibular apparatus in the vestibular branch of the eighth cranial nerve to the vestibular nuclei, where they are relayed to the somatic efferent neurons that innervate the extrinsic muscles of the eyes.

Nystagmus. The proper functioning of the vestibular apparatus and associated tracts is sometimes determined by producing an endolymph displacement in the semicircular canals and observing the resultant movement of the eye. This displacement may be produced by placing a solution in one of the ears which is either warmer or cooler than the body temperature (caloric stimulation) or by rotating the subject on a chair similar to a barber's chair or piano stool and then rapidly stopping the rotation and observing the movements of the eye. The direction of the eye movement depends upon the direction of rotation and the position of the head during rotation. If rotation is to the left and the head is erect, the endolymph in the lateral semicircular canal will be displaced to the left (displaced counterclockwise) when the rotation is stopped and the eyes will move slowly to the left. When the eyes have been rotated to the left as far as possible, they will quickly rotate back to the right and start their slow rotation to the left again. This type of eye movement is called vestibular nystagmus. The kind discussed above is characterized as being horizontal, but vertical and rotary nystagmus result if the head is placed in different positions during acceleration and deceleration. By changing the position of the head one changes the degree of endolymph displacement in each of the mutually perpendicular semicircular canals. Other types of nystagmus include (1) aural (due to a labyrinthine disturbance), and (2) congenital (due to a lesion sustained in utero or during birth or to an inherited factor).

Types of Eye Movements

Till now we have been discussing the two basic types of eye movement. In one type, *conjunctive eye movements*, the two eyes converge or diverge. We have noted that convergence is a mechanism for focussing a near object on corresponding points of the two retinas and that divergence is a mechanism for focussing a far object on corresponding points. In other words we use conjunctive eye movements to produce fusion, the sensation of a single image.

The second type of eye movement is a *conjugate eye movement*. Here both eyes move in the same direction. One type of conjugate movement is that of smooth pursuit. Here the eyes move in such a way as to follow a moving target as it passes by. The maximum speed at which this can occur is between 30° and 40° per second. In order to follow an object moving past faster than this, the individual must either move his head or jump from one visual position to the next. This type of eye movement is called saccadic and is the kind we use in reading.

LIGHT DEPRIVATION

Experiments with chimpanzees have clearly shown that long periods of light deprivation result in a reduction in visual acuity and difficulty in fixating objects, in visually following objects, and in distinguishing forms such as a square, a triangle, or a feeding bottle. These effects are reversible if the light deprivation lasts less than 7 months, but if it lasts from birth to over 16 months of age, many of the effects are irreversible. After such a period, the chimpanzee continues to show pupillary reactions to light, startle responses to turning on a light, and primitive pursuit movements of the eyes in response to a moving object, but for all other intents and purposes he is blind. In addition there is almost complete degeneration of the ganglion cell layer of the retina and extensive degeneration in the lateral geniculate nucleus. Similar changes have been noted in a normal chimpanzee deprived of light starting at 8 months of age and lasting for 16 months. Apparently light is necessary not only for developing a normal visual system but also for maintaining it. It is interesting to note that when light deprivation is experienced in only one eye, the chimpanzee can distinguish various forms when using the normal eye but not the light-deprived eye. Light deprivation also occurs in congenital cataract (loss of transparency of the lens or its capsule) patients. When the cataract is removed and corrective lenses used to correct for the optical abnormalities of the eye, the patient initially has difficulty distinguishing forms. He may, for example, have to count the number of sides of a figure to distinguish a square from a triangle. Generally, if the cataract is removed within a year of its appearance, good optical function develops in time. Longer periods of light deprivation, however, may result in permanent visual impairment. The discrimination of various patterns by man is a process that must begin at an early age.

Chapter 37

SOUND
THE EAR
NEURAL MECHANISMS

HEARING

The ear is the organ that gathers sound waves and converts them into displacements of perilymph and endolymph in the cochlea of the inner ear (Figs. 13-9 through 13-11). These displacements result in the stimulation of sensory neurons in the cochlear branch of the eighth cranial nerve, which in turn elicits both a conscious awareness of sound and the auditory reflexes. Also in the inner ear is the vestibular apparatus, about which we will have more to say in the next chapter.

SOUND

Vibrations set up on a violin string, the membrane of a drum, the vocal cords, a tuning fork, or any one of a number of other materials cause a series of compressions and rarefactions in the air, water, or solids surrounding them. These compressions and rarefactions travel in all directions from their source and can be characterized in terms of the velocity at which they move (c), the distance between adjacent rarefactions (λ), the frequency of rarefactions (f), and the difference in pressure (the amplitude) between a point of rarefaction and an adjacent point of compression. The relationship between c, λ, and f is as follows:

$$c(\text{cm./sec.}) = \lambda(\text{cm./wave}) \times f(\text{waves/sec.})$$

The capacity of a person to hear a particular vibration depends upon (1) the frequency of the vibration, (2) its amplitude, and (3) the character of the individual's auditory system (Fig. 37-1). Normally a young adult is capable of hearing frequencies ranging from 20 to 20,000 cycles per second (c.p.s.), and some children may hear frequencies up to 40,000 c.p.s., but the frequencies heard at the lowest amplitudes are those between 2,000 and 3,000 c.p.s. Under ideal conditions these frequencies can be heard at an amplitude of 2×10^{-4} dynes per square centimeter by a healthy young adult. A normal man of 40, on the other hand, requires an amplitude several times this value to hear the same frequencies.

The Decibel

We have noted that the amplitude or intensity of a sound wave can be measured in absolute terms, i.e., in dynes per square centimeter. This intensity varies with the amount of force that produces the vibration and can also be measured in terms of power (watts/cm.^{2} or dyne-cm./sec./cm.^{2}). It is more common, however, to measure the intensity or loudness of a sound in terms of the ratio between the absolute intensity of the sound being measured and a standard reference intensity which is approximately the threshold of audibility for a normal young adult. This reference intensity is 2×10^{-4} dynes per square centimeter (10^{-6} watts/cm.^{2}). The formula for computing the ratio (N) in decibels between the pressure of a particular sound (P_1 in dynes/cm.^{2}) and the standard pressure (P_k in dynes/cm.^{2}) is as follows:

$$N(\text{db.}) = 20 \log_{10} \frac{P_1 \text{ in dynes/cm.}^2}{P_k \text{ in dynes/cm.}^2}$$

The formula for computing the ratio from power data would be as follows:

$$N(\text{db.}) = 10 \log_{10} \frac{P_1 \text{ in watts/cm.}^2}{P_k \text{ in watts/cm.}^2}$$

In other words, if we increase the power of a sound to 10^6 times its reference value, we will have increased its intensity by 60 db. This is equivalent to a 10^3 increase in the pressure from its standard value:

$$60 \text{ db.} = 10 \log_{10} \frac{10^6}{1} = 20 \log_{10} \frac{10^3}{1}$$

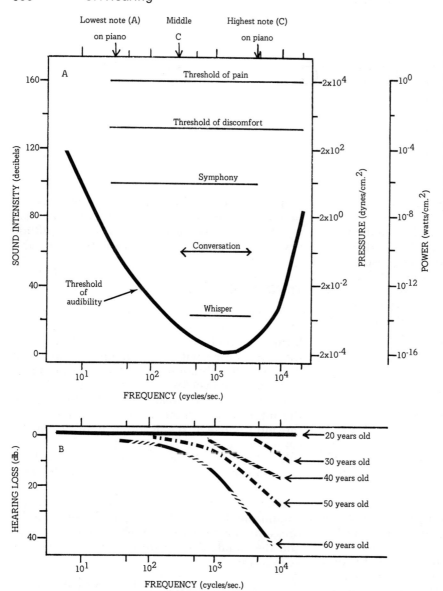

Fig. 37-1. The audibility of different frequencies. Note in A that the threshold for pain is relatively independent of frequency. Note in B that there is a progressive loss in men of the sensitivity to the higher frequencies as the men get older. (*In part from* Newby, H. A.: Audiology. 2nd ed., New York, Appleton-Century-Crofts, 1964; and Webster, J. C., et al.: San Diego County Fair Hearing Survey, J. Acoustical Soc. Am., 22:473–483, 1950)

An apparent peculiarity of this system is that a sound having an intensity of 2×10^{-4} dynes per square centimeter will be rated as being zero decibels:

$$N = 20 \log_{10} \frac{2 \times 10^{-4}}{2 \times 10^{-4}} = 20 \times 0 = 0$$

Table 37-1 compares the average threshold of audibility for various frequencies obtained by the U. S. Public Health Service in 1935 and 1936 with those obtained by the International Standards Organization (I.S.O.). The I.S.O. values are now generally recognized to be more representative and are being adopted both in the United States and throughout the world.

The Audiometer

The audiometer is an instrument used to measure a subject's hearing ability. It consists of an oscillator, an amplifier, an attenuator for controlling the intensity of the sound, and an earphone. The audiometer is generally capable of producing frequencies from 125 to 8,000 c.p.s. at intervals of one half the audiometer octave and at intensities up to 100 db. It should be re-

THE EAR

Sound waves are gathered by the external ear and transmitted to the tympanic membrane (Fig. 13-10). The tympanic membrane separates the outer ear from the middle ear and converts the vibrations in air to a displacement of the ossicles of the middle ear, which in turn sets up a displacement of the oval window. The oval window separates the middle ear from the inner ear and its displacement initiates waves in the perilymph and endolymp of the inner ear, which stimulate specialized receptors in the cochlea of the inner ear. These receptors initiate the transmission of action potentials to portions of the brain responsible for the conscious recognition of sounds and the integration of auditory signals into a number of reflexes. Each of these activities will be considered in more detail below.

The External Ear

The external ear consists of the pinna (auricle), the external auditory meatus, and the tympanic membrane. The pinna acts as a funnel to gather in the sound waves and provides a small amount of amplification, whereas the auditory meatus provides some protection to the 0.1-mm. thick tympanic membrane. It serves to protect the tympanic membrane from large foreign particles and to keep the air at the membrane's surface moist and near body temperature.

The Middle Ear

The middle ear is bounded at its lateral side by the conical tympanic membrane, and at its medial side by the oval and round windows of the cochlea. At its caudal surface is the pharyngo-

Frequency (c.p.s.)	1936 American Standards (db.)	1964 I.S.O. Standards (db.)
125	54.5	45.5
250	39.5	24.5
500	25.5	11
1,000	16.5	6.5
1,500		6.5
2,000	17	8.5
3,000		7.5
4,000	15	9
6,000		8
8,000	21	9.5

Table 37-1. Threshold of audibility of different frequencies. These values were obtained using 2×10^{-4} dynes per square centimeter as the standard and rounding all values to the nearest half decibel. They are used as a reference for audiometric zero in performing tests for hearing loss. Note that the American standard values are given at octave intervals (2 times the initial frequency) at frequencies between 125 and 8,000 c.p.s. The I.S.O. standards, on the other hand, are usually given at half octave intervals.

membered, however, that the zero-decibel level of the audiometer is not the 2×10^{-4} dynes per square centimeter discussed above, but rather the average threshold of hearing for young adults at each frequency studied. The audiometer, besides testing the hearing threshold for different pure tones (different frequencies), is sometimes used to test the threshold for certain words and to determine how well a patient can understand speech when it is presented at a comfortably loud level (speech audiometer). Generally a hearing threshold more than 25 db. above the zero-decibel level of the audiometer is regarded as a hearing handicap (Table 37-2).

Degree of Handicap	Hearing Loss	Ability to Understand Speech
Little or none	< 25 db.	Little or no difficulty with faint speech
Slight handicap	25-40 db.	Difficulty with faint speech
Mild handicap	40-55 db.	Difficulty with normal speech
Marked handicap	55-70 db.	Difficulty with loud speech
Severe handicap	70-90 db.	Can understand only shouted or amplified speech
Extreme handicap	> 90 db.	Usually cannot understand amplified speech

Table 37-2. Hearing loss. These data are based on the 1946 I.S.O. standards given in Table 37-1. (Modified from Davis, H.: Guide for the classification and evaluation of hearing handicaps in relation to the international audiometric zero. Trans. Am. Acad. Ophthal., 69:741, 1965)

tympanic tube (eustachian tube). Connecting the tympanic membrane to the oval window are three ossicles: the malleus (hammer), incus (anvil), and stapes (stirrup). When the tympanic membrane vibrates, it moves the attached malleus, which moves the attached incus, which moves the attached stapes. The footplate of the stapes is fastened to the oval window and when moved initiates a displacement of the perilymph of the scala vestibuli of the cochlea. The primary function of these ossicles is apparently to convert the movements of low-density air against the tympanic membrane into movements of the higher-density perilymph. This is accomplished by coupling the displacement of the relatively large surface of the tympanic membrane (0.7 cm.^{2} in man) with the smaller surface of the oval window (0.03 cm.^{2}). The result is that the force per unit area exerted on the oval window is from 15 to 20 times that exerted on the tympanic membrane.

Another role played by the middle ear is the protection of the inner ear and oval window from excessive movements of the ossicles. Two muscles, the *tensor tympani* and the *stapedius*, are important in this regard (Fig. 37-2). They are reflexly stimulated to contract in response to loud sounds. When the tensor tympani contracts, it pulls the malleus and tympanic membrane medially and into the middle ear. The stapedius muscle, on the other hand, pulls the stapes away from the oval window. Acting together, these two muscles tend to immobilize or stiffen the conduction system. This results in a decrease in the transmission of frequencies below 1,000 c.p.s. toward the oval window. Unfortunately, it provides little protection against frequencies greater than 2,000 c.p.s.

Some of the difficulties in depending upon this reflex for the protection of the ear against blows and loud, low-frequency sounds such as explosions are (1) that it has a reaction time of at least 10 to 20 msec. and (2) that it takes about 17 msec. of stimulation to obtain a maximum contraction. Fortunately, however, loud low-pitched sounds also change the axis of rotation of the stapes as soon as they reach the middle ear, thus providing an important and immediate protection for the ear. Because of mechanisms such as these, the ear can withstand pressures 10^7 times those of threshold level.

Conduction disturbances. In the absence of normal middle ear conduction due to the removal of the ossicles or their immobilization

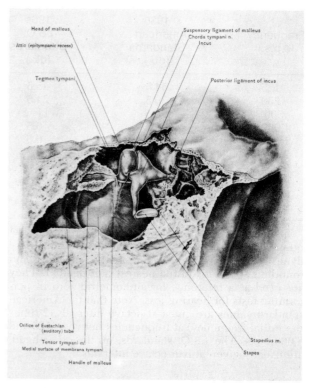

Fig. 37-2. View of the interior of the middle ear. (Deaver, J. B.: Surgical Anatomy of the Human Body. 2nd ed., p. 509; copyright 1926 by McGraw-Hill Book Company, New York; used with permission)

(stapes ankylosis, for example) a hearing loss as great as 65 db. can occur. Hearing losses of this magnitude or less can result from (1) obstruction of the pharyngotympanic tube, (2) infection in the middle ear (otitis media), or (3) ossification of the stapes and oval window (otosclerosis). The importance of the pharyngotympanic tube has to do with its role in equalizing the pressure in the middle ear with that in the atmosphere. Normally the situation in the middle ear is such that the surrounding tissues are slowly absorbing air. It is therefore important that the pharyngotympanic tube open periodically to allow additional air to enter the middle ear. This usually occurs during a yawn, during chewing, and during swallowing. If a partial vacuum is allowed to develop in the middle ear or if the pressure in the middle ear is allowed to exceed that in the atmosphere (as in ascending in an airplane), a stretching of the tympanic membrane may occur that dampens the conducting system and produces a hearing loss. Excessive differences in the

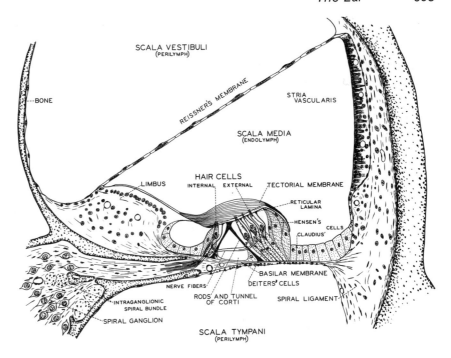

Fig. 37-3. Cross-section of the cochlear partition of the guinea pig in the lower part of the second turn. (Davis, H., et al.: Acoustic trauma in the guinea pig. J. Accoustical Soc. Am. 25:1182, 1953)

pressure across the membrane may also result in a rupture of the membrane. This in itself produces a hearing loss, as does the scar tissue that forms during the natural repair of the membrane.

Chronic *otitis media* and otosclerosis can produce permanent hearing losses. The former, now well controlled by the administration of antibiotics, does this by damaging the ossicles, sometimes to the extent that they must be replaced by a prosthesis. *Otosclerosis* tends more to immobilize the stapes and the oval window. At one time this condition was treated by the surgical technique of *fenestration*—the construction of an oval window in the horizontal semicircular canal. Recently, however, the techniques of choice have become (1) the dislodgment of the stapes or (2) the removal of the stapes and its replacement with a prosthesis.

The Inner Ear

The part of the inner ear responsible for the sensation of sound is the cochlea. It looks very much like a snail's shell, consisting of two half-turns which, if they could be uncoiled, would be about 32 mm. long (Fig. 37-5). At the base of the cochlea are two openings that connect with the middle ear. The superior opening is called the oval window and transmits the oscillations of the stapes to a chamber in the cochlea called the scala vestibuli (Fig. 37-3). The scala vestibuli is separated from an adjacent chamber, the scala media (cochlear duct, or cochlear partition), by a thin barrier, Reissner's membrane. The inferior opening between the inner ear and middle ear is called the round window and is in contact with the third chamber of the cochlea, the scala tympani. Both the scala vestibuli and the scala tympani contain perilymph and connect by means of a small opening at the apex of the cochlea, the helicotrema. The cochlear partition, on the other hand, contains endolymph.

The inner ear responds to sound waves in the following manner. First there is a displacement of the oval window produced by movement of the stapes. This causes a change in pressure in the scala vestibuli that is conducted through Reissner's membrane, into the cochlear partition, and into the scala tympani. This pressure may also be conducted from the scala vestibuli through the helicotrema and into the scala tympani. In either case increases in pressure in the scala tympani result in a movement of the round window outward; decreases in pressure result in a movement inward. These pressure changes are also associated with a displacement of the hair cells of the cochlear partition.

It is this displacement that is responsible for the stimulation of auditory neurons and therefore for the sensation of sound.

The cochlear partition. The cochlear partition contains hair cells that are closely associated with sensory neurons going into the eighth cranial nerve. The character of the displacement usually depends upon the frequency and amplitude of the sound waves that impinge upon the ear. Sound waves with frequencies greater than 4,500 c.p.s. produce their maximum displacement of the cochlear partition near its base (near the round window). Sound waves with frequencies less than 200 c.p.s., on the other hand, produce a maximum displacement toward the apex (near the helicotrema), where the compliance of the partition is approximately 100 times greater than at the base.

In Figure 37-4 we have an example of the response of the cochlear partition to a frequency of 200 c.p.s. Note that the maximum displacement occurs about 28 mm. from the stapes and involves almost the entire partition. In Figure 37-5, on the other hand, we see that the higher frequencies do not cause as much of the partition to be displaced. In other words, within limits, the higher the frequency of a tone (the smaller the wavelength), the closer the point of maximum displacement to the base of the cochlear partition, and the less extensive the involvement of the cochlear partition. It is apparently on this basis that each of us distinguishes the vast range of sounds to which the ear is exposed.

The hair cell. Apparently the first step in con-

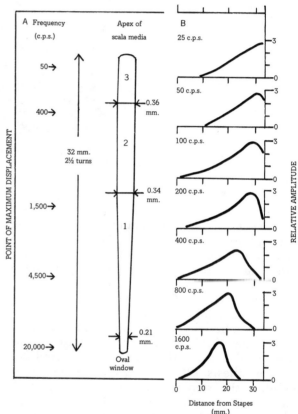

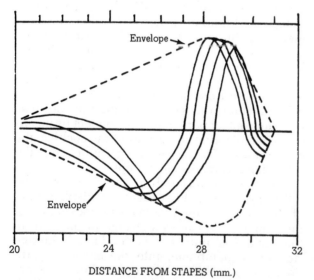

Fig. 37-4. Response of the basilar membrane to a 200-c.p.s. tone. The various waves shown above were obtained at a number of sequential instants during the cycle. The envelope represents the maximum displacement at each point along the basilar membrane. (*Modified from* Békésy, G. von: Experiments in hearing. P. 462; copyright 1960 by McGraw-Hill Book Company, New York; used with permission)

Fig. 37-5. Frequency discrimination along the cochlear duct (scala media). At A we have a diagram of the unraveled cochlear duct showing areas of maximal displacement for pure tones of different frequencies. At B we have a diagram of the displacement envelope produced on the basilar membrane by different frequencies. Note that the 20,000-c.p.s. tone produces its maximal displacement near the oval window (base) and is completely damped out at the apex. The 200-c.p.s. tone, on the other hand, displaces the whole basilar membrane, but produces its maximum deflection near the apex. (*Modified from* Stuhlman, O.: An Introduction to Biophysics. P. 286. New York, John Wiley & Sons, 1943; and Békésy, G. von: Experiments in Hearing. P. 448; copyright 1960 by McGraw-Hill Book Company, New York; used with permission)

verting a displacement of the cochlear partition into action potentials in the cochlear nerve is the deformation of hair cells. Bekesy has shown that at the point of maximum displacement of the cochlear partition the hair cells move up and down, whereas toward the stapes they move in a radial direction, and toward the helicotrema they move in a longitudinal direction. The mechanism whereby these deformations lead to the production of action potentials in sensory neurons, however, is not well understood.

Sensory disturbances. Disturbances of inner ear function may result from infections, certain drugs (streptomycin, for example), and trauma due either to a blow or to excessive intensities of sound. In the case of streptomycin toxicity, the damage is to the organ of Corti. In the case of infections and trauma the damage may be to either the organ of Corti or the auditory nerve. In the case of sound the degree of hazard varies directly, within limits, with its intensity, frequency, and duration. In industry, noise levels about 85 db. are considered a potential health hazard, but in order to produce a histological lesion along the basilar membrane with a sound of short duration, one must apply an intensity of at least 150 db. If the 150+ db. sound is a pure tone, the lesion it produces is restricted to a small area along the basilar membrane; otherwise it is more extensive.

Sound intensities of about 180 db. are attained close to a jet engine. This represents a power flux of 10 watts per square centimeter. Since the soft tissues of the body absorb high-frequency sounds and convert them into heat, this much energy represents a potential danger not only to the soft tissues of the ear but to the brain as well. For example, an energy flux of 1 watt per square centimeter at 20,000 c.p.s. will kill a mouse after 1 min. of exposure. This sort of information has been put to use in the field of ultrasonics. Using ultrasonic techniques a physician can destroy an abnormal growth in a small area in the brain by focussing waves of 10^6 c.p.s. and of the proper intensity on it.

NEURAL MECHANISMS

Pitch Differentiation

The ear is the transducer that converts sound waves into action potentials in the bipolar neurons of the eighth cranial nerve. This conversion occurs because different tones bend a different grouping of hair cells (Fig. 37-6) in the cochlear partition. Since different hair cells are innervated by different sensory neurons, it is believed that we at least in part differentiate one sound frequency from another on the basis of which family of neurons is stimulated. This perspective is called the *place theory* of Helmholtz. Also in its favor is the observation that the action potentials produced by separate tones remain segregated (tonotopic patterning) as they are transmitted through the auditory nerve, the cochlear nuclei of the medulla, the inferior colliculi of the midbrain, the medial geniculate nuclei of the thalamus, and the auditory cortex.

An alternate, but less satisfactory, concept of pitch differentiation is the *volley theory* of hearing. This is based on the observation that the frequency of action potentials in some sensory fibers from the cochlear partition is a faithful reproduction of the sound frequency that produces them. This relationship, however, is not found at frequencies above 3,000 c.p.s., or in the neurons activated by the sensory fibers in the eighth cranial nerve. In fact, the maximum frequency of action potentials in the auditory tracts of the brainstem is less than 150 per second, and the maximum frequency of discharge of the auditory neurons in the thalamus and cortex is even less.

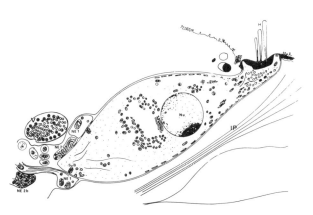

Fig. 37-6. A drawing of an inner hair cell in the cochlear partition of the guinea pig. H, hairs; Nu, nucleus; NE 1, afferent nerve endings and associated synaptic bar (SyB); NE 2, efferent nerve endings; V, synaptic vesicles consisting of small clear ones and larger "cored" ones; IP, inner pillar. The efferent neurons shown in the drawing come from the eighth cranial nerve. Their function is not understood. (Drawing supplied by Hans Engström)

Intensity Differentiation

As the loudness of a tone increases, there are within limits a number of changes in the auditory system: (1) the amplitude of deflection of the cochlear partition increases, (2) a greater number of neurons from the partition are activated (recruitment), i.e., there is a spread of activation to adjacent neurons, and (3) there is an increase in the frequency of firing of a number of auditory neurons in the brainstem. An example of the relationship between the intensity of a tone and the frequency of discharge for an auditory neuron in the superior olivary complex is seen in Figure 37-7.

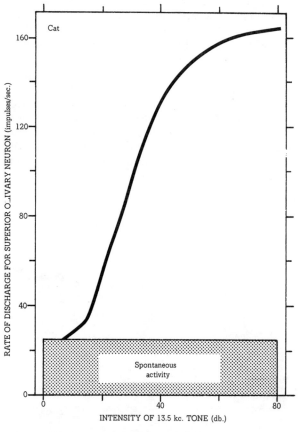

Fig. 37-7. Rate of discharge of a neuron in the superior olivary complex. The data were recorded between the seventh to tenth seconds after the onset of the tone. The tone used produced an optimal response on this neuron and was applied to the contralateral ear. (*Redrawn from* Goldberg, J. M., et al.: Response of neurons of the superior olivary complex of the cat to acoustic stimuli of long duration. J. Neurophysiol., *27:* 711, 1964)

The sensation of loudness, however, is not just a function of the intensity of the sound to which the ear is exposed. During the initial few seconds of exposure a constant intensity sound may seem much louder than during later periods of exposure. This decrease in the sensory experience associated with a constant stimulus has been observed for other sensory systems as well and is called *adaptation*. It is seen in two forms in the auditory system. In one type of neuron there is, in response to a constant intensity of sound, an initial high frequency of discharge followed by a lower frequency of discharge that stays at a fairly constant level during the remaining period of exposure to the sound. A smaller number of neurons, however, respond only at the initiation and termination of the sound. Other mechanisms that decrease the sensation of loudness include (1) a reflex contraction of the tensor tympani and stapedius muscles of the middle ear that dampens the deflection of the ossicles in response to low frequencies, (2) a reflex stimulation of neurons in the olivocochlear tract that depresses the sensory output in the afferent cochlear neurons, and thus decrease the intensity of the sensation by 18 to 25 db., and (3) a depression of the conscious recognition of sounds in the cerebral cortex.

A person's ability to concentrate on certain sounds while rejecting the conscious appreciation of others is similar to his ability in seeing to concentrate on light that is focussed clearly on the fovea centralis while rejecting the light that is not so focussed. In driving a car, for example, a person is generally conscious of only the objects directly in front of him. If under these circumstances a car should race toward the subject from his side (peripheral vision), he would normally become conscious of this. Apparently in the case of vision and audition there is a cortical mechanism for bringing danger signals as well as other signals of special interest into the area of consciousness. Another example of this is the following. A subject puts earphones on and is asked to concentrate on messages going into his right ear while ignoring those going into his left. Under these circumstances the subject generally responds to and remembers most of the information going into the right ear, while rejecting most of the information going into the contralateral ear. The information that is rejected, however, is apparently processed in some manner, since certain types of signals do reach the conscious level and most sensory information going into the left ear can be recalled if the subject is questioned about it within 10 sec. of its application. Some of the auditory signals that may reach the conscious level more readily than most include the individual's name, the cry of a child, or a call to dinner.

The Cochlear Nuclei

All the afferent fibers from the cochlear branch of the eighth cranial nerve synapse in cellular groupings found at the level of entry of the eighth nerve into the medulla. These cellular groupings have been divided into three nuclei: the dorsal, the anteroventral, and the posteroventral cochlear nuclei (Fig. 37-8). In each of these nuclei the cochlear partition seems to be completely represented, the high frequencies stimulating the more dorsal areas of each and the low frequencies stimulating the more ventral areas. The physiological significance of this triplication is not fully clear, but there does seem to be a general pattern in the auditory system of duplication and triplication. Additional examples of this will be given as our discussion progresses.

Another important aspect of the organization of the cochlear nuclei is that they exhibit a pattern of inhibition similar to that seen in some of the nuclei of both the visual and the somesthetic systems. In the case of the auditory system efferent neurons in the cochlear nerve apparently stimulate cell bodies in one neural field while *inhibiting* cell bodies in the adjacent fields. This apparently serves to sharpen and isolate an island of activity. It is not known whether this inhibition is of the pre- or postsynaptic type.

The action potentials that leave the vestibular nuclei go in many directions. Some descend to the anterior horns of the spinal cord and cause a subject to turn his head toward a sound. Others ascend the brainstem, sending fibers to the superior olive, the reticular formation, the cerebellum, the nuclei of the lateral lemniscus of the pons, and the nuclei of the inferior colliculus of the midbrain. In this ascent most of the fibers cross over to the contralateral side. In the inferior colliculus there is another apparent duplication, both the external and the central nuclei being stimulated by impulses originating from all parts of the cochlear partition. Some of the impulses received by the inferior colliculus are relayed to the medial geniculate nucleus. Some of the ascending impulses may not synapse here but may go directly to the medial geniculate nucleus. The medial geniculate nucleus, in turn, relays impulses to the cerebral cortex.

The Cerebral Cortex

The cerebral cortex contains numerous areas that can be activated by auditory stimuli (Fig. 37-9). The part of the human cortex that seems most important in distinguishing one pitch from another and, like many of the nuclei in the brainstem, segregates the impulses from the various areas of the cochlear partition, is an area in the right and left anterior transverse temporal gyrus of Heschl (Brodmann's areas 41 and 42 in Fig. 40-7, or area AI in Fig. 37-9). When this area is stimulated directly by an electric current, auditory illusions result. When it is destroyed on both sides of the cerebral cortex in man, almost complete bilateral deafness results. It can also be shown to respond to auditory stimuli applied to either the ipsilateral or contralateral ear, though its threshold for auditory stimuli in the contralateral ear is generally 5 to 20 db. lower than that for the ipsilateral ear.

One of the difficulties in studying the auditory system in man is that the function of the various auditory areas appears to be duplicated else-

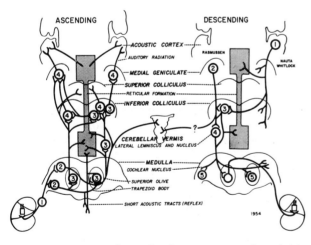

Fig. 37-8. An outline of auditory tracts and nuclei in the central nervous system. The numbers represent the approximate order of the neurons, starting with the origin of either the ascending or descending system. The shaded areas represent central portions of the brain stem, reticular formation, and diencephalon. (Galambos, R.: Neural mechanisms in audition. Laryngoscope, 68:393, 1958)

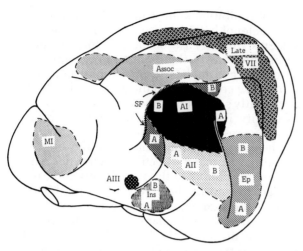

Fig. 37-9. Summary of areas of a cat's cerebral cortex that can be activated by an auditory input. These studies were performed by stimulating small groups of neurons from the cochlea and observing the resultant electrical activity on the cortex. Similar studies on man and other primates are more difficult because in these species the auditory areas are less accessible. The letters A and B are used to indicate loci that are stimulated by impulses from the apical (low frequencies) and basal (high frequencies) portions of the cochlear partition. Also shown are: auditory area I (A I), auditory area II (A II), the ectosylvian posterior area (Ep), the suprasylvian sulcus (SF), the insula (Ins), auditory area III (A III), motor area I, and visual area II (late VII). In this latter area responses to visual stimuli occur after a 100-msec. delay. Infratemporal fields lateral to A II and an insular field which also receive cochlear potentials are not shown. (Woolsey, C. N.: *In* Rasmussen, G. L., and Windle, W. F. (eds.): Neural Mechanisms of the Auditory and Vestibular Systems. P. 179, 1960. Courtesy of Charles C Thomas, Publisher, Springfield, Illinois)

where. Hence it requires a rare combination of lesions to produce a pronounced auditory disturbance. Studies of animals closely related to man, such as the primates, have also proven difficult since most of their auditory areas, like those of man, are located in fairly inaccessible spots. As a result, much of our information on the auditory system is based on the study of the cat and other carnivores. These animals, unlike man, do not experience deafness upon the bilateral removal of the primary auditory areas of the cerebral cortex. In them the subcortical areas are apparently much more important than they are in man. We find in both cats and monkeys, for example, that bilateral removal of the auditory areas of the cerebral cortex results in a loss of the ability to distinguish different frequencies, but that there is little change in either the threshold for a sound or the ability to distinguish different intensities.

In man the unilateral destruction of the primary auditory cortex (AI) results in only minor changes in hearing. In some cases, for example, there may be a decrease in the ability to localize a sound. There is also a unilateral retrograde degeneration of the ventral division of the pars principalis of the medial geniculate nucleus. A similar lesion in the cat or monkey, on the other hand, produces considerably less marked degeneration of the geniculate nucleus.

Surrounding the primary auditory cortex are a number of areas for which tonotopic localization is less precise and for which the auditory input is but one of a number of inputs. In the case of the association area there seems to be no tonotopic patterning and its input includes auditory, visual, and somesthetic neurons. The removal of any one of these cortical areas results in little or no degeneration in the medial geniculate nucleus, but the removal of all the cortical auditory areas produces profound degeneration of all the subdivisions of the medial geniculate nucleus.

Descending Pathways

The descending components of the auditory system are summarized in Figure 37-8. The neurons that constitute this system originate from the infratemporal and other cortical areas and pass through as well as synapse in the medial components of the medial geniculate complex and then the nucleus of the lateral lemniscus of the midbrain. They end in the cochlear nuclei. It is also speculated that they may send impulses to the neurons of the olivocochlear bundle, which apparently serves to inhibit the activation of the cochlear hair cells.

Chapter 38

THE VESTIBULAR APPARATUS

THE SEMICIRCULAR CANALS
THE OTOLITHIC ORGANS

The vestibular apparatus consists of a group of interconnected, fluid-filled ducts and sacs embedded in the temporal bone. It represents the nonauditory portion of the inner ear and includes the three semicircular canals, the utricle, and the saccule. Like the cochlea, these structures contain an inner, endolymph-containing membranous canal surrounded by perilymph (Fig. 38-1). In the membranous canal lie hair cells sensitive to endolymph displacement. In the case of each semicircular canal these hair cells are located in the crista ampullaris; in the case of the utricle and saccule they are in the macula of each structure. Also as in the cochlea, these hair cells are innervated by bipolar afferent neurons in the eighth cranial nerve which synapse in the vestibular nuclei. Their impulses have a pronounced effect on (1) the antigravity muscles, (2) the neck muscles, (3) the extrinsic muscles of the eye, and (4) the conscious sensation of one's position in space. Excessive stimulation of these receptors can lead to general activation of the autonomic nervous system causing nausea, vomiting, pallor, perspiration, bradycardia (decrease by 8-10 beats/min.), hypotension (decrease of 10 mm. Hg), and hyperventilation.

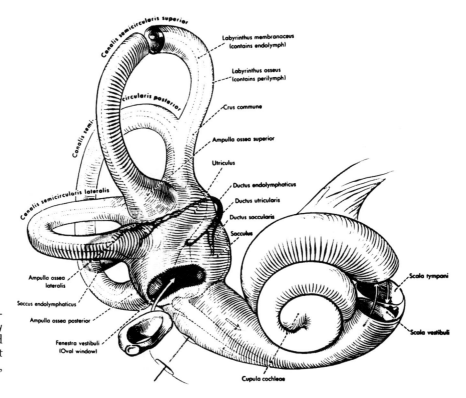

Fig. 38-1. The boney labyrinth. (*From a drawing by* Melloni, B. J. Reproduced by permission of Abbott Laboratories, North Chicago, Illinois)

THE SEMICIRCULAR CANALS

The semicircular canals include the right and left (1) horizontal (lateral, or external) canals, (2) superior (anterior vertical) canals, and (3) posterior (posterior vertical) canals. The ampullae of both horizontal and superior canals open into the utricle near its macula, while the ampulla of the posterior canal opens into the utricle at its opposite end. The position of these canals is such that when the head is vertical, the horizontal canal is oriented backward at an angle of 30° (Fig. 13-9) and the vertical canals make an angle of about 45° with the frontal and sagittal planes of the skull. In other words, to make the horizontal canal truly horizontal the head should be tilted forward 30°.

The Crista Ampullaris

The crista ampullaris (Fig. 38-2) is the receptor organ found at one end of each semicircular canal. It consists of neuroepithelium innervated by afferent and efferent neurons. Within the neuroepithelium lie the supporting cells and the hair cells, the cilia of which are embedded in a gelatinous structure (the cupula) that extends up to the roof of the ampulla. When a person is sitting or lying quietly, the cupula is erect and the afferent neurons from the hair cells of the crista conduct action potentials to the medulla. When there is an endolymph displacement from a horizontal semicircular canal through its ampulla (an ampullopetal movement), the cupula is bent toward the utricle, the hair cells in response to this, it is hypothesized, are depolarized, and the frequency of action potentials traveling over the sensory neurons from the crista increases. An ampullofugal displacement of endolymph (movement from utricle through ampulla), on the other hand, causes the crista to move in the opposite direction and results in hyperpolarization of the hair cells and a decrease in the firing of their sensory neurons. In other words, we have

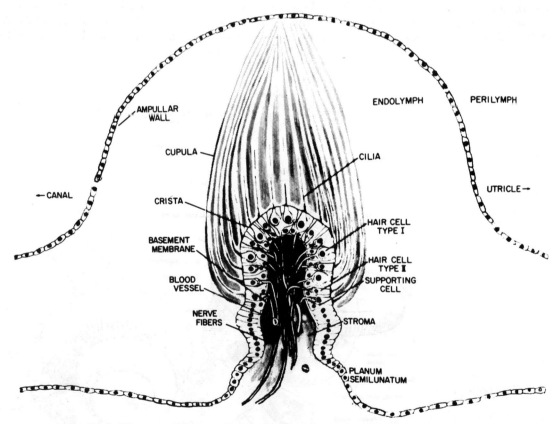

Fig. 38-2. Diagram of the crista ampullaris and cupula. (Hawkins, J. E., Jr.: *In* Best, C. H., and Taylor, N. B. (eds.): The Physiological Basis of Medical Practice. 8th ed., p. 102; copyright 1966, The Williams and Wilkins Co., Baltimore)

an organization in the horizontal semicircular canals that modifies the sensory input into the medulla in response to the displacement of its enclosed liquid.

In the case of the vertical semicircular canal the situation is similar but not identical. Here an ampullopetal displacement apparently results in hyperpolarization rather than depolarization of the hair cells, and produces depolarization and hence an increase in the discharge rate of the sensory neurons. The source of this difference between the horizontal and vertical canals can be guessed by comparing the hair cells in each. You will notice in Figure 38-3 a detailed drawing of two types of hair cells. One type (type I cell of Wersäll) is flask-shaped and innervated by an afferent neuron (or neurons), while the other (type II cell) is cylindrical and receives both afferent and efferent innervation. Both types of cell contain 40 to 60 hairs (stereocilia) plus a single, peripheral, somewhat longer hair (kinocilium). In the horizontal canals the kinocilium is on the side of each hair cell next to the utricle, whereas in the vertical canals the stereocilia lie nearest the utricle. On this basis Flock and Wersäll have postulated that a deflection of the hairs toward the kinocilium causes depolarization of the hair cells, and that a bending in the opposite direction causes hyperpolarization.

Response to Acceleration

On the basis of the material in the preceding paragraphs we see that the semicircular canals are well suited to relaying information concerning an acceleratory or deceleratory movement of the head in any plane. For example, if the head is suddenly rotated in the horizontal plane to the right (clockwise), the semicircular canals move with it, but the enclosed endolymph initially tends to lag in its movement. This lag is due to the inertial properties of the endolymph, and in the right semicircular canal has the same effect as an ampullopetal displacement of the endolymph. In other words, the cupula during the acceleratory movement of the head are bent toward the utricle, the hair cells are depolarized, and the frequency of discharge of the sensory neurons from the right horizontal semicircular canal is increased. In the case of the left horizontal semicircular canal, the endolymph displacement in reference to the canal is ampullofugal and the frequency of sensory impulses from here is decreased. An acceleration of the head to the left would have an opposite effect.

A quick bending of the head forward or backward or to one side, on the other hand, would produce maximal displacement of the endolymph in the vertical canals (Fig. 38-4). For example, an acceleratory movement of the head forward, as in nodding, produces ampullofugal endolymph displacement in the right and left superior semicircular canals (depolarization of hair cells) and

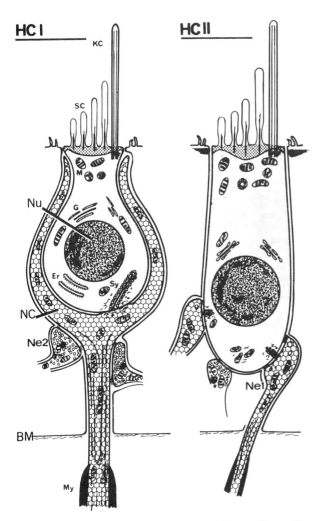

Fig. 38-3. A drawing of 2 types of hair cells in the crista. HC I and HC II, hair cells I and II; KC, kinocilium; SC, stereocilia; M, mitochondria; G, Golgi complex; Nu, nucleus; Er, endoplasmic reticulum; NC, nerve chalice of afferent neuron; Sy, synaptic bar; Ne 1, afferent nerve ending; Ne 2, efferent nerve ending; BM, basement membrane; My, myelinated afferent neuron. (Drawing supplied by Hans Engström)

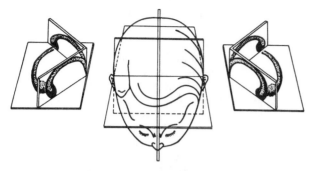

Fig. 38-4. Relationship of the semicircular canals to the horizontal, median sagittal, and transverse frontal planes of the head. (Hawkins, J. E., Jr.: In Best, C. H., and Taylor, N. B. (eds.): The Physiological Basis of Medical Practice. 8th ed., p. 100; copyright 1966, The Williams and Wilkins Co., Baltimore)

ampullopetal displacement in the posterior semicircular canals (hyperpolarization of hair cells). In other words, a deceleration or acceleration of the head in any direction normally produces a change in the sensory output from the various canals.

The cupulometer. The response of subjects to acceleratory and deceleratory head movements is studied by means of a cupulometer. This is a motor-driven rotating chair with a brake that permits abrupt stops. For studying the horizontal semicircular canals the subject is placed in the chair with his head 30° forward and is rotated with his eyes closed at a speed of 15, 30, or 60° per second. If the subject is rotated to his right there is an initial ampullofugal displacement of the endolymph during the initial period of acceleration that results in a deflection of the cupula to the right. As a constant angular velocity is reached, the cupula returns to the position it had prior to acceleration. In other words, modification of the sensory output from the canals is the result of acceleration or deceleration, not of a constant velocity. The next step in the study is to stop the cupulometer abruptly and have the subject open his eyes. This abrupt stop results in an ampullopetal displacement of the endolymph in the right canal and an ampullofugal movement in the left canal. As a result there is (1) a to-and-fro deflection of the eyes lasting about 20 sec., in which there is a rapid movement of the eyes to the left followed by a slow movement back to the right (a horizontal nystagmus to the left), (2) a sensation of vertigo in which the environment seems to be rotating to the left, (3) a tendency to fall to the right, and (4) a tendency to point a finger while the eyes are closed to the right of a target (past-pointing to the right).

These responses to deceleration are an indication of the role of the horizontal canals in the unconscious control of eye movements and muscle tone, as well as in conscious sensation. The error associated with past-pointing, for example, is an error in judgment during a voluntary act. You will note in Table 38-1 that it is always in a direction opposite to the sensation of vertigo. During acceleration to the right the horizontal canals also reflexly produce an extension of the limbs on the left and a flexion of the limbs on the right. During deceleration the reverse occurs. In other words we have here a response to deceleration that helps the subject maintain his balance. The endolymph displacement and the response of the individual to acceleration and deceleration vary, of course, with the position of the head during acceleration and deceleration. If the head is resting on one shoulder, vertical nystagmus results from rotational deceleration. If the head is bent 60° backward or 90° or more forward, the result is rotary nystagmus (Table 38-1).

Caloric stimulation. Displacement of the endolymph of the semicircular canals is also produced by cooling or warming the tympanic membrane of the ear. Cooling the ear sets up convection currents away from the ampulla (ampullofugal), while heating produces currents in an ampullopetal direction. In one technique the head is placed at the desired position and the ear canal is douched with about 240 ml. of water at 30° (cooling) or 44°C. (heating) for 40 sec. In a normal person the resulting nystagmus lasts for 90 to 130 sec., but in someone whose crista has been damaged due to streptomycin intoxication, for example, the duration is reduced. One advantage of this procedure over the one in which the cupulometer is used is that it tests each ear individually.

Neural Mechanisms

Sensory neurons from the semicircular canals, the utricle, and the saccule pass up the vestibular branch of the eighth cranial nerve and for the most part synapse in the vestibular nuclei. A few fibers, however, pass directly to the flocculonodular lobe and fastigial nuclei of the cerebellum. Fibers from the inferior vestibular nucleus, in turn, pass to the nuclei of the third, fourth, and sixth cranial nerves of both sides. These nuclei

Position of Head	Direction of Post-rotational Nystagmus	Sensation of Vertigo after Rotation	Past-Pointing after Rotation	Direction of Falling
Rotation to the right				
30° forward	Horizontal, left	Turning left	To right	To right
120° forward	Rotary, left	Falling to left	To right	To right
60° backward	Rotary, right	Falling to right	To left	To left
Right shoulder	Vertical, downward	Falling forward	Upward	Backward
Left shoulder	Vertical, upward	Falling backward	Downward	Forward
44°C. to right ear or 30°C. to left ear				
60° backward	Horizontal, right	Falling to right	To left	To left
120° forward	Horizontal, left	Falling to left	To right	To right

Table 38-1. Response of a person to displacement of the endolymph of the semicircular canals. The displacement was produced by (1) rotating the subject to his right (clockwise) while his eyes were closed and then abruptly stopping the rotation and having him open his eyes, (2) by warming the tympanic membrane to 44°C., or (3) by cooling the tympanic membrane to 30°C. Nystagmus by definition is the direction of the fast component. The sensation of vertigo, past-pointing (eyes closed), and the direction of falling were all tested after rotation with the subject's head erect.

have as part of their role the control of the extrinsic muscles of the eye, and apparently these connections in the brainstem are all that is necessary to produce nystagmus in response to acceleration or deceleration of the head in any plane. Evidence of this is that nystagmus is not eliminated by the ablation of the cerebellum or by the transection of the brainstem above the level of the nucleus of the third cranial nerve and below the level of the vestibular nuclei.

The lateral vestibular nuclei, on the other hand, send efferent neurons via the olivocochlear bundle back to the type II hair cells of the crista ampullaris, and internuncial fibers into the vestibulospinal tract. The latter fibers, along with some from the inferior vestibular nuclei and the reticular formation, impinge upon the anterior-horn cells of the cord and serve to facilitate and inhibit the contraction of some of the flexors and tensors of the limb and some of the muscles of the neck and trunk. The reticular formation also receives fibers from the vestibular nuclei, which synapse there with descending neurons in the reticulospinal tract. The cerebellum receives direct innervation not only from the vestibular apparatus but also from neurons originating in the vestibular nuclei. Together these fibers form the vestibulocerebellar tract. Apparently the vestibular nuclei send fibers activated by both the auditory and the vestibular apparatus of the inner ear to the temporal lobe of the cerebral cortex via the same tracts (see preceding chapter).

Loss of vestibular function. The vestibular apparatus is one of a number of systems in the body concerned with the control of skeletal-muscle tone, with postural reflexes, and with a person's conscious awareness of his relationship to the external environment. In man, bilateral destruction of the vestibular apparatus has been produced by the chronic use of streptomycin in treating tuberculosis. Under these conditions there generally results a transient vertigo and sometimes disabling ataxia (loss of coordination). Many persons, however, can compensate relatively well for the loss of the vestibular apparatus by a greater reliance on the joint and muscle receptors and the eyes. Such people have difficulty walking in the dark or on rough ground, however.

A disease in man which apparently affects the inner ear is Meniere's disease. Its symptoms are vertigo, nausea, vomiting, tinnitus (a ringing in the ears), and progressive deafness. It is sometimes treated by destruction of the vestibular nerve on the affected side, which results in loss of the symptoms of Meniere's disease and usually has only minor side effects. This is in marked contrast to the results obtained in animal experiments. Here unilateral removal or destruction of the vestibular apparatus causes (1) a movement of the eyes toward the side of the destruction, with the ipsilateral eye also moved downward and the contralateral eye moved upward, (2) a lateral flexion and rotation of the back of the head toward the side of the destruction, (3) flexion of the thorax on the pelvis, (4) increased extensor and abductor tone of the limb on the sound side of the body, (5) increased flexor and adductor tone of the limb on the side of the destruction and (6) a tendency to fall toward the side of the destruction.

The symptoms in animals that follow bilateral destruction of the vestibular apparatus through surgery or the use of an ototoxic agent such as streptomycin are similar to, though generally somewhat more pronounced than, those seen in man. These include (1) a diminution in muscle tone and (2) an ataxia that can cause loss of the ability to fly or to right oneself (Fig. 38-5).

Symptoms similar to some of those discussed above have also been noted in man after lesions have occurred in some of the vestibular centers and tracts in the brain. Past-pointing, for example, when not associated with acceleration or deceleration, is usually indicative of a cerebellar lesion. Spontaneous vertigo is commonly associated with labyrinthine disease but may also result from a medullary lesion, alcoholic intoxication, seasickness and other forms of motion sickness, abnormal eye movements, eyestrain, and watching motion pictures. Spontaneous vertical nystagmus frequently indicates a lesion in the brainstem. It may also occur in a person with a brainstem lesion when his horizontal semicircular canal is stimulated. In deaf-mutes, nystagmus and other reactions to acceleration and deceleration are usually absent.

 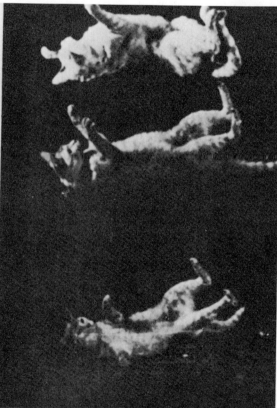

Fig. 38-5. Righting reactions of a normal cat (*left*) and a cat whose labyrinthine neuroepithelia have been destroyed by streptomycin (*right*). Both cats were dropped in the supine position. Only the normal cat immediately rights itself and lands on his paws. (Hawkins, J. E., Jr.: *In* Best, C. H., and Taylor, N. B. (eds.): The Physiological Basis of Medical Practice. 8th ed., p. 105; copyright 1966, The Williams and Wilkins Co., Baltimore)

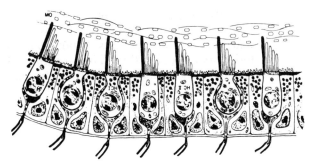

Fig. 38-6. Diagram of the neuroepithelium of the macula of the utricle. Note the similarity between this structure and the crista ampularis shown in Figure 38-2. Note also the presence here, but not in the crista, of a gelatinous otolithic membrane attached to the cilia. (Best, C. H., and Taylor, N. B.: The Physiological Basis of Medical Practice. 8th ed., p. 102; copyright 1966, The Williams and Wilkins Co., Baltimore)

THE OTOLITHIC ORGANS

The otolithic organs include the maculae in the utricle and saccule. The maculae, like the cristae of the semicircular canals, contain supporting cells, axons, and type I and II hair cells (Fig. 38-6). Unlike the cristae, however, they also have an otolith-containing membrane attached to the cilia. The otoliths are small particles of calcium carbonate that serve to increase the weight and hence the inertia of the maculae. When the head is vertical, the macula of the utricle is in the horizontal plane. In the saccule one macula lies on the medial wall and faces laterally while another lies on the roof and faces downward.

The Utricle

The hair cells of the utricle apparently function in much the same way as those of the crista. Apparently bending them in one direction produces depolarization and a resultant increase in the frequency of the action potentials conducted from it by its associated bipolar neuron. Bending in the other direction causes hyperpolarization and a decrease in the frequency of sensory impulses. In the cat, for example, when the head is level, the frequency of impulses from each utricle is about 6 per second. When the head is tilted 20° to the left, the frequency from the left utricle increases to 95 per second and that on the right decreases to zero per second. Since these receptors are of the slow-adapting type, these frequencies are maintained at a fairly constant level. Less marked lateral movements of the head produce milder changes in the frequency of impulses carried by the sensory neurons. One can also produce a reflex elevation of the head by pulling the snout down, or a reflex lowering of the head by pulling the snout up (Fig. 38-7), but the utricles are less sensitive to movements in this direction than to those in the lateral direction.

The above reflexes are sometimes characterized as static, since they tend to orient the head in terms of the gravitational pull of the earth. When the head is bent to the side, the combined weight of the otoliths and the associated membrane serve to bend the cilia of the hair cells. This is different from the situation in the semicircular canals, in which an endolymph displacement bends the cilia. On the other hand, the hair cells of the macula can also be bent in response to acceleration. For example, if the body is accelerated forward at a rate greater than 12 cm. per second per second (threshold), the frequency of discharge from the utricles changes, with resultant changes in muscle tone.

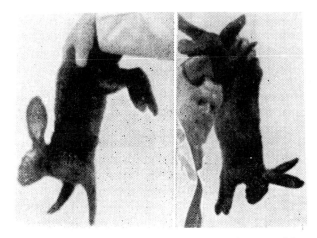

Fig. 38-7. Righting response of a thalamus rabbit (an animal in which the cerebral hemispheres have been removed but the optic thalami left intact). In such an animal the optical righting reflexes have been destroyed. The animal at the left has an intact labyrinth and the one on the right has had the labyrinth destroyed. When either the entire labyrinth or just the utricle is destroyed, the rabbit fails to orient his head with respect to the gravitational field. (*After* Magnus. *In* Best, C. H., and Taylor, N. B.: The Physiological Basis of Medical Practice. 8th ed., p. 92; copyright 1966, The Williams and Wilkins Co., Baltimore)

The Saccule

In certain lower vertebrates without a cochlea the saccule may serve an auditory function, and some have suggested that this may also be true in man. In the frog, for example, the denervation of all the structures of the inner ear except the saccule results in a complete loss of vestibular function. Adrian, on the other hand, has suggested that the saccule responds to both lateral tilting and linear acceleration. The question of whether or not this structure is vestigial in man remains unanswered.

Chapter 39

TASTE AND SMELL

TASTE
OLFACTION

Taste and smell are quite closely related physiologically. The flavors of a number of foods are the result of the stimulation of a combination of olfactory and gustatory receptors or sometimes of just gustatory receptors. It is common knowledge, for example, that many foods taste different when one has a head cold or when the nose is held closed. Under these circumstances one cannot tell the difference in taste between an apple and a potato. On the other hand, the pathways carrying impulses from the olfactory and gustatory receptors are quite different. Impulses from the former are carried over the first cranial nerve and are transmitted neither to the thalamus nor to the neocortex. Impulses from the latter stimulate fibers in the seventh, ninth, and tenth cranial nerves and are transmitted to the thalamus, which projects them to the postcentral gyrus of the cortex along with impulses from the mouth for touch and pressure.

TASTE

The major organ involved in taste is the tongue. Other structures, however, also have taste receptors. These include the palate, the uvula, the pharynx, the epiglottis, the tonsils, the mucosa of the lip and cheek, and the floor of the mouth.

It has been suggested that all the flavors we experience are due to the stimulation of a combination of four different gustatory receptors plus receptors for olfaction, touch, and pressure. The four types of gustatory receptors are those with their lowest threshold for (1) sweet (sucrose or saccharin), (2) salty (NaCl), (3) bitter (quinine, strychnine), and (4) acid or sour (hydrochloric acid) stimuli. The tip of the tongue is sensitive to all four modalities, but most sensitive to sweet and salt. The base of the tongue is most sensitive to bitter and the lateral margins to acid. The dorsum of the tongue in the midline is insensitive to any of the gustatory modalities.

Sensation	Substance	Threshold (%)
Sweet	Cane sugar	0.5
	Dulcin	1×10^{-3}
	Perillartine	1.7×10^{-4}
Salty	NaCl	2.5×10^{-1}
Acid	Hydrochloric acid	6.7×10^{-3}
Bitter	Quinine	0.5×10^{-4}
	Strychnine	4.0×10^{-5}

Table 39-1. Threshold concentrations for various sapid solutions. (Modified from Best, C.H. and Taylor, N.B.: The Physiological Basis of Medical Practice. 8th ed. Baltimore, Williams & Wilkins, 1966)

The Taste Buds

Taste buds are goblet-shaped clusters of cells found in the fungiform and circumvallate papillae of the tongue. In man there are about 12,000 of them, each having a diameter of approximately 50 μ, each containing both receptor and supporting cells, and each innervated by several different nerve fibers (Fig. 39-1). These cells can be studied by inserting a microelectrode slowly into a taste pore. As the electrode is advanced, potential differences appear, disappear, and reappear—apparently as a result of the electrode's penetrating a cell, passing through it, and then passing into another cell. In other words, the potentials recorded are thought to be the steady potentials of different receptor cells.

The steady potentials recorded by this technique vary from -50 to -95 mv., but can be made more positive (depolarization) by the application of a sapid (flavorful) solution to the taste bud. The amount of depolarization that occurs depends upon the concentration of the sapid molecule and its character (Fig. 39-2). Generally about 15 msec. after the depolarization of the receptor cells begins, the frequency of afferent impulses in the tympanic nerve also increases.

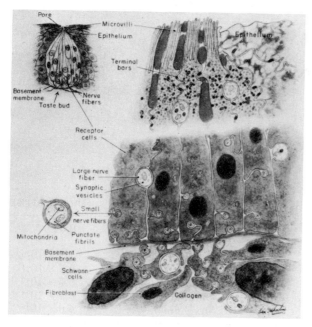

Fig. 39-1. A taste bud from the papilla foliata of the rabbit. (*Reprinted with the permission of* de Lorenzo, A. J. D.: Studies on the ultrastructure and histophysiology of cell membranes, nerve fibers and synaptic junctions in chemoreceptors. *In* Zotterman, Y. (ed.): Olfaction and Taste. P. 13. Oxford, Pergamon Press, 1963)

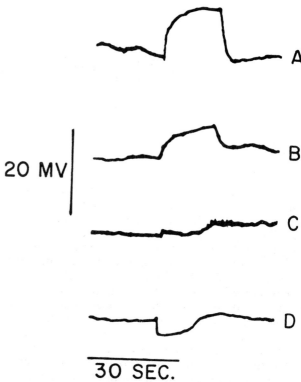

Fig. 39-2. Response of taste cells to a 1.0 molar NaCl solution. Record A was obtained when the microelectrode was at the surface of a taste cell. Records B, C and D were obtained from progressively deeper penetrations of the taste bud. The steady potentials recorded in record A, B, C, and D were 50, 45, 95, and 85 mv. respectively. (Tateda, H., and Beidler, L. M.: The receptor potential of the taste cell of the rat. J. Gen. Physiol., 47:481, 1964)

Neural Mechanisms

The nervous impulses from the taste receptors may be recorded by placing an electrode on a single neuron or by recording from an entire nerve. In the study of man the latter procedure is usually used, since during middle-ear surgery the physician exposes the chorda tympani. This nerve passes over the tympanic membrane and malleus and carries impulses from the anterior two thirds of the tongue to the seventh cranial nerve. Such surgery is usually done under local anesthesia, and therefore presents an opportunity to study taste in the conscious subject.

Using these procedures, investigators have been unable to demonstrate the specificity of a particular receptor cell or a particular neuron to a single type of gustatory stimulus; a single receptor cell may respond to both salty and bitter stimuli in a similar fashion, and a single sensory neuron probably innervates more than one receptor cell. This is of course in marked contrast to the rather well-defined segregation of sensory information passing into the central nervous system from the cochlea and its sensory nerve. At present the best guess as to how we distinguish one gustatory modality from another is that each modality initiates a characteristic pattern of activation in a neuron pool and that the various patterns are distinguished in the brain.

On the other hand, sensory messages in a single gustatory neuron or nerve can be distinguished with regard to intensity. As the concentration of a particular sapid molecule increases, so does the frequency of action potentials in a single neuron. We also find that adaptation occurs when a stimulus is maintained, and that with increasing concentrations an increased number of neurons are stimulated.

OLFACTION

At the roof of each nasal cavity, caudal to the cribriform plate of the ethmoid bone, is a 2.5-sq. cm. surface of yellow membrane, the olfactory membrane. This structure contains the major olfactory receptors of the body and is exposed to a small per cent of the odoriferous molecules we inspire during respiration. The number of inspired molecules coming in contact with the membrane can be increased by increasing the force of the inspiration, as in sniffing. This creates a turbulent flow in the nasal cavity which facilitates the movement of inspired air up into the relatively isolated part of the cavity containing the olfactory epithelium.

Molecules eliciting the sensation of smell have two characteristics—they are volatile and they are fat-soluble. Some of the most odoriferous ones are listed in Table 39-2. As should be apparent from this table, the olfactory epithelium is tremendously sensitive to these substances. Compare, for example, the sensitivity of the taste buds to strychnine, nearly the most potent sapid substance known (threshold stimulus, 4×10^{-5} %), with that of the olfactory receptors to chlorphenol (threshold stimulus, 3×10^{-7} %); the olfactory receptors are over 100 times more sensitive. On the other hand, compare our capacity to taste ethanol (threshold, 14%) and to smell it (threshold, 0.44%); the sensitivity of the olfactory receptors is 30 times greater than that of the taste buds. It is interesting to note that if we replace the OH group in ethanol with an SH group, we obtain ethyl mercaptan, a substance to which the olfactory membrane is 100 million times more sensitive than it is to ethanol. This substance can be detected by the olfactory apparatus when its concentration in the air is 1 molecule of ethyl mercaptan per 50 billion molecules of air. Other molecules of interest include methyl mercaptan (responsible for the odor of garlic) and skatole (produced in the intestine by the bacterial decomposition of tryptophan and responsible for much of the odor of feces). The former elicits olfactory awareness at a concentration as low as 0.04 μg. per milliliter of air. One milligram of the latter would produce an unpleasant odor in a hall 500 by 100 by 50 m.

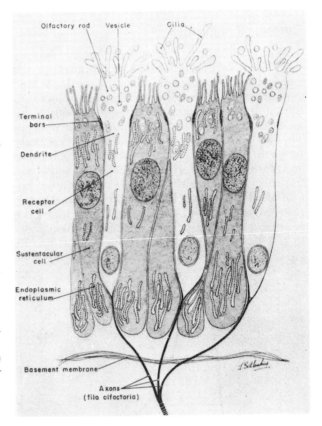

Fig. 39-3. The olfactory mucosa. This schematic drawing is based upon observations using the electron microscope. Note that the so-called receptor cells (receptor dendrites) shown here are merely an expansion of the axon terminals of a single bipolar neuron. (*Reprinted with the permission of* de Lorenzo, A. J. D.: Studies on the ultrastructure and histophysiology of cell membranes, nerve fibers and synaptic junctions in chemoreceptors. In Zotterman, Y. (ed.): Olfaction and Taste. P. 8. Oxford, Pergamon Press, 1963)

Odorivector	Threshold (mg./m.² of air)	(gm./gm. of air)
Phenol	1.2	1×10^{-6}
Diethyl ether	0.75-1	7×10^{-7}
Pyridine	4×10^{-2}	3×10^{-8}
Natural musk	7×10^{-3}	6×10^{-9}
Iodoform	6×10^{-3}	5×10^{-9}
Chlorophenol	4×10^{-3}	3×10^{-9}
Butyric acid	1×10^{-3}	8×10^{-10}
Mercaptan	4×10^{-5}	3×10^{-11}
Synthetic musk	5×10^{-6}	4×10^{-12}
Skatole	4×10^{-7}	3×10^{-13}
Vanillin	2×10^{-7}	2×10^{-13}

Table 39-2. Threshold concentrations for various odoriferous molecules. (From Moncrieff, R. W.: The Chemical Senses. London, Leonard Hill, 1967)

Attempts to classify the various odors have met with little success. Two reasons for this are (1) the large number of different and seemingly unrelated odors that exist, and (2) the failure of different investigators to agree on terminology. One classification divides the odors into six groups: (1) spicy (cloves, fennel, anise), (2) flowery (heliotrope, coumarin, geranium), (3) fruity (oil of orange, oil of bergamot, and citronellal), (4) resinous or balsamic (turpentine, eucalyptus oil, Canada balsam), (5) burnt (pyridine, tar), and (6) foul (sulfuretted hydrogen and carbon bisulfide). For other classifications the reader is referred to Chapter 9 of *The Chemical Senses*, by R. W. Moncrieff.

The Olfactory Mucosa

Figure 39-3 shows part of the olfactory mucosa. You will note that it consists of supporting cells (sustentacular cells) and receptor cells, which contain either microvilli or cilia that project into the nasal cavity. The receptor cell appears to be an expanded part of an axon in a bipolar neuron. This axon (receptor dendrite) and its cell body lie in the olfactory epithelium. When exposed to an odoriferous material to which it is sensitive, the receptor dendrite produces a generator potential which, within limits, is proportional to the concentration of the stimulating molecule. Like generator potentials produced in other receptors in the body, it (1) is not conducted, (2) adapts slowly in the presence of the stimulus, and (3) causes a series of action potentials to be conducted toward a second-order neuron (Figs. 39-4 and 39-5). Each of the receptor dendrites, in addition, is apparently odor selective. For example, in 1963 Gesteland et al., using frogs, reported that of the 25 odors they tested, each primary neuron showed strong responses to at least one of the odors and weaker responses to many of the others. They suggested that in the frog the olfactory receptors are grouped into eight categories, each with considerable overlap.

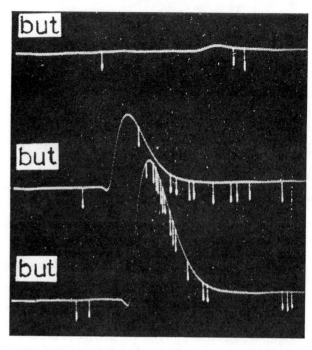

Fig. 39-5. Response of the frog to progressively higher concentrations of butyric acid vapor. The above patterns were obtained by superimposing upon the electroolfactogram of the olfactory mucosa the response of an olfactory neuron. Note that as the amplitude (in mv.) of the E. O. G. increases so does the frequency of action potentials in the olfactory neuron. (*Reprinted with the permission of* Gesteland, R. C., et al.: Odor specificities of frog's olfactory receptors. *In* Zotterman, Y. (ed.): Olfaction and Taste. P. 24. Oxford, Pergamon Press, 1963)

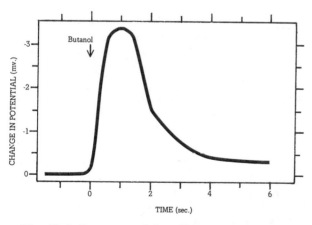

Fig. 39-4. Response of the olfactory epithelium to butanol. The above pattern is sometimes referred to as an electroolfactogram (E. O. G.). (*Redrawn from* Ottoson, D.: Analysis of the electrical activity of olfactory epithelium. Acta. Physiol. Scand. (Suppl. 122), 35: 72, 1956)

Neural Mechanisms

Apparently the generator potentials initiated in the portion of the bipolar neurons near the surface of the olfactory epithelium result in the production of a series of potentials in the more central portion of the neuron which are conducted through the cribriform plate and into an olfactory glomerulus in the olfactory bulb. It is in the glomerulus that the bipolar neurons (first-order neurons) synapse with mitral and tufted cells. Approximately 26,000 fibers from the olfactory membrane enter, and an average of 24 mitral and 68 tufted fibers leave, each glomerulus. These represent second-order neurons. The tufted fibers form an anterior commissure and synapse on the second-order neurons of the contralateral olfactory bulb, where they are for the most part inhibitory (Fig. 39-6). Other inhibitory fibers may also enter each bulb from the reticular formation and the hypothalamus. The mitral fibers, on the other hand, form the lateral olfactory tracts and carry their impulses to the archipallium.

In other words, the olfactory bulb acts as an area of synapse and integration. Apparently it also tends to segregate at least some of the messages it receives, the anterior bulb being electrically activated by fruity odors, and the posterior bulb by oily-smelling solvents such as benzene. On the other hand, benzene elicits no activity in the anterior lobe and fruity odors elicit no activity in the posterior lobe. It is of interest in this regard to note that apparently all sensory information concerning odors is not relayed to the central nervous system via the first cranial nerve. Although transection of the olfactory nerves abolishes olfactory sensations in response to cloves, lavender, benzol, and xylol, it does not abolish the recognition of the odor of camphor, pyridine, phenol, ether, and chloroform in the animals studied. These sensations are abolished only by section of the nasociliary and maxillary branches of the fifth cranial nerve.

The point of termination of the olfactory tract fibers has not been established, but there is evidence that some of the fibers go to the prepyriform area and parts of the amygdaloid complex, as well as to the olfactory tubercle. Unlike the other sensory modalities, however, olfaction does not appear to have thalamic representation.

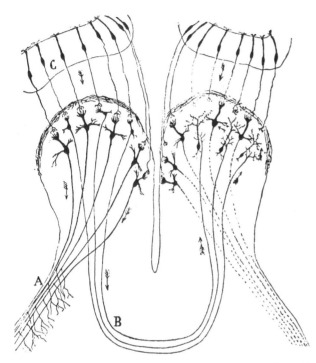

Fig. 39-6. Conduction in the olfactory bulb and tract. (C) Bipolar neurons in olfactory mucous membrane going to olfactory bulb, (A) lateral olfactory tract, (B) medial olfactory tract going into anterior commissure. (Ramón y Cajal, S.: Histologie du système nerveux de l'homme et des vertèbres. Vol. II, p. 665. Paris, Maloine, 1911)

Chapter 40

TOUCH, PRESSURE, AND
KINESTHETIC TRACTS
PAIN TRACTS
TEMPERATURE TRACTS
THE THALAMUS
THE CEREBRAL CORTEX

THE FOREBRAIN

The forebrain is essential for conscious sensation, cognition, remembering, calculating, planning, and judging. It also plays an important role in the initiation and coordination of skeletal muscle, smooth muscle, cardiac muscle, and glandular activity. It consists of the diencephalon (epithalamus, dorsal thalamus, ventral thalamus, and hypothalamus) and the cerebral hemispheres (cerebral cortex, basal ganglia, and rhinencephalon). We shall begin by discussing the fiber tracts that transmit impulses to the forebrain.

TOUCH, PRESSURE, AND KINESTHETIC TRACTS

Sensory impulses from the periphery enter the central nervous system via neurons in the dorsal roots, the cranial nerves, and in the case of visceral pain, the autonomic nerves. Some of the afferent fibers synapse with neurons at the level of their entry into the cord and some enter tracts in the white matter and pass to lower or higher levels of the central nervous system. Apparently all the fibers entering the *dorsal white column* (dorsal funiculus) are first-order group II (pp. 23-24 and Fig. 2-4) myelinated neurons from touch, pressure, joint, and ligament receptors, each fiber being stimulated primarily by one modality from a limited area of the body (Fig. 40-1). About 25 per cent of these fibers remain in the dorsal columns until they synapse in the dorsal column nuclei of the medulla. The rest leave the tract before entering the nuclei. Those remaining in the tract arrange themselves topographically, the fibers entering the cord most caudally being most medial (that is, in the fasciculus gracilis) and those entering most cephalad being most lateral (in the fasciculus cuneatus). In the medulla the dorsal column neurons synapse with second-order neurons (internuncial neurons) in the medial lemniscus, which cross over to form a topographically oriented tract whose fibers synapse in the *ventrobasal complex* of the contralateral thalamus. It is from here that third-order neurons pass to the cerebral cortex.

Other touch, pressure, and kinesthetic neurons carry their impulses to the thalamus via tracts in the lateral and ventral white columns. These tracts contain both unmyelinated and small myelinated fibers. Most but not all of these fibers, on entering the cord, synapse with neurons that pass to the contralateral side via the ventral white commissure (Figs. 40-1 and 40-2).

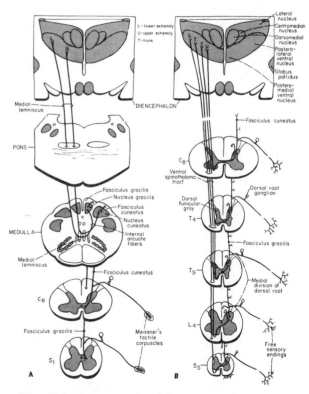

Fig. 40-1. Diagram of tactile tracts to the thalamus. (Crosby, E. C., et al.: Correlative Anatomy of the Nervous System. P. 86. New York, Macmillan, 1962)

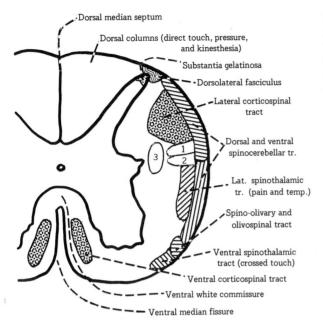

Fig. 40-2. Cross-section of a human spinal cord at the thoracic level to illustrate some of the spinal tracts. (1) Tegmentospinal tract, (2) tectotegmental spinal tract, (3) lateral reticulospinal tract. (*Modified from* Knighton, R. S., and Dumke, P. R. (eds.): Pain. P. 307. Boston, Little, Brown & Co., 1966)

Some form a spinobulbar tract (passing from spinal cord to medulla) that conducts impulses to the reticular formation in the medulla and midbrain. The reticular formation in turn relays these impulses via the reticulothalamic tract to the *intralaminar nuclei* of the thalamus. Some of the sensory neurons also form a paleospinothalamic tract which, like the spinobulbar tract, lacks topographic organization and impinges upon the intralaminar nuclei of the thalamus. A third tract formed by these fibers is the neospinothalamic tract. Unlike the other two tracts, it carries topographically patterned impulses.

In other words, the spinal cord in its dorsal columns carries touch, pressure, and kinesthetic impulses from the ipsilateral side of the body to the medulla, where these impulses are transmitted to a contralateral bulbothalamic tract. Touch, pressure, and kinesthetic impulses are also carried in the ventral and lateral columns—most of these, however, crossing to the contralateral cord before reaching the medulla. Similar sensory impulses from cranial nerves V, VII, and X, on the other hand, enter the brainstem and synapse with second-order neurons that cross over and ascend to the thalamus. Sensory impulses from taste receptors enter the brainstem via cranial nerves VII, IX, and X and synapse in the nuclei of the tractus solitarius with second-order neurons that apparently also cross over to the contralateral side and ascend to the thalamus.

PAIN TRACTS

Impulses eliciting the sensations of pain, heat, and cold reach the thalamus via somewhat different routes than do those for touch, pressure, and kinesthesia. One of the differences lies in the fact that there are apparently no pain or temperature fibers in the dorsal white columns. Other differences will be emphasized later in the discussion. In the case of pain there are two types of peripheral sensory fibers: delta fibers and C fibers. The delta fibers are small, myelinated, type A neurons which have a short period of latency and which are responsible for a well-localized type of pain characterized as a prickling sensation. The C fibers, on the other hand, are unmyelinated, conduct more slowly, and are responsible for a type of pain characterized as diffuse, long-lasting, and burning (see Table 2-2).

Many of the delta and C fibers conducting impulses associated with pain enter the spinal cord at the dorsolateral fasciculus (tract of Lissauer) via the dorsal root and synapse in the substantia gelatinosa. Most of the second-order neurons pass through the ventral white commissure to the contralateral side of the body, but a few ascend the cord in ipsilateral tracts. Apparently there are also connecting tracts between the right and left substantia gelatinosa that pass from one area to the next via the dorsal white commissures. The importance of the nociceptive (pain) fibers that cross over can be demonstrated by two surgical procedures. In one the ventral white commissure in man is cut and there results bilateral analgesia in the related dermatomes. In the second procedure there is transection of the ventral and lateral white columns ventral to the corticospinal tract, causing contralateral analgesia caudal to the transection (Fig. 40-3).

The more central distribution of impulses originating in delta and C fibers is not the same. Apparently most of the delta fibers transmit their impulses to contralateral neospinothalamic tracts, and most of the C fibers, to contralateral paleospinothalamic and spinoreticulodiencephalic tracts. That part of the neospinothalamic tract which projects to the nuclei of the ventrobasal

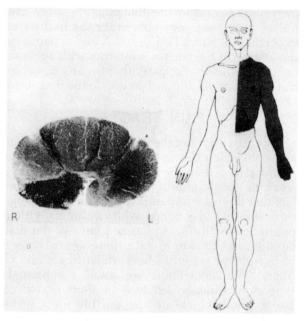

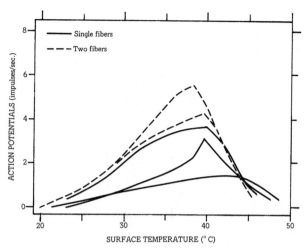

Fig. 40-4. Frequencies of maintained discharges in warmth fibers in the lingual nerve of the cat when the tongue is maintained at different temperatures. (*Modified from* Dodt, E.: Mode of action of warm receptors. Acta Physiol. Scand., 26:348, 1952)

Fig. 40-3. Response of man to an incision in the left ventral quadrant of the cervical cord (between C-1 and C-2). The degenerating fibers have been stained black in the cross-section of the cord seen at the left of the illustration. The response of the individual to this incision is shown at the right. Note that there was a complete loss of the sensation of pain and temperature (*dark area*) in the chest and arm on the side of the body contralateral to the lesion. There was also an incomplete loss of pain and thermal sensibility in the neck and abdomen. A more extensive loss of these modalities is produced by similar lesions which include both the ventral and lateral white columns. Apparently in the cervical portion of the cord, pain and thermal fibers from cervical and thoracic regions are concentrated in the ventral columns and similar fibers from the lumbar and sacral regions are concentrated in the more lateral portions of the white matter. (Knighton, R. S., and Dumke, P. R. (eds.): Pain. P. 48. Boston, Little, Brown & Co., 1966)

TEMPERATURE TRACTS

Temperature fibers, like the nociceptive fibers responsible for sharp pain, are small, myelinated neurons (type A, subgroup delta) that synapse at approximately the same loci and whose second-order neurons travel in the same fiber tracts as the second-order neurons innervated by the delta pain fibers. The warmth fibers conduct impulses when the areas they serve are between 25° and 45°C., firing most frequently at temperatures between 37° and 42°C. (Fig. 40-4). They cease all activity at temperatures above 45° and below 25°C. The cold fibers, on the other hand, conduct impulses at temperatures between 12° and 37°C., firing most frequently at about 20°C. (Fig. 40-5). They also fire at temperatures over 45°C., producing the sensation of cold. This is called paradoxical cold. Besides cold and warmth, there is a third sensation we experience. This is called hot and is due primarily to the stimulation of a combination of nociceptive and cold receptors.

THE THALAMUS

In the previous section I emphasized the importance of the thalamus in the sensation of pain and as an area of synapse for neurons transmitting impulses to the cerebral cortex. It (1)

thalamic complex retains a somatotopic pattern, while that part which projects to the dorsal nuclear complex of the thalamus retains only a vague somatotopic representation. Both the paleospinothalamic and the spinoreticulodiencephalic tracts lose their somatotopy, the former projecting to nuclei of the intralaminar group and the latter to the reticular formation in the medulla, pons, and midbrain. From the reticular formation the impulses are projected to the dorsal thalamus, the ventral thalamus, and the hypothalamus.

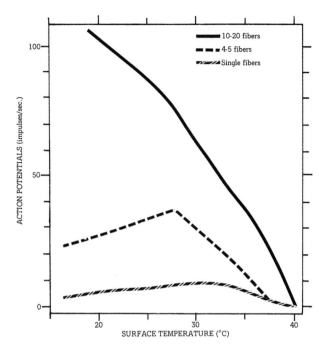

Fig. 40-5. Frequency of maintained discharges in cold fibers in the lingual nerve of the cat when the tongue is maintained at different temperatures. The paradoxical discharge that occurs in some cold fibers in response to heat is not shown. (*Modified from* Hensel, G., and Zotterman, Y.: Quantitative Beziehungen zwischen der Entladung einzelner Kältefasern und der Temperatur. Acta Physiol. Scand., 23:301, 1951)

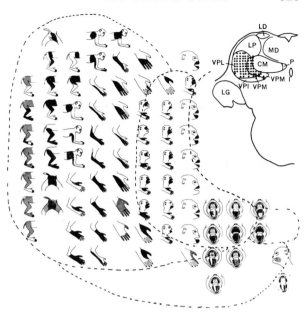

Fig. 40-6. Representation of cutaneous tactile sensibility in the ventrobasal complex (VPL, ventral posterolateral nucleus; VPM, ventral posteromedial nucleus; and VPI, ventral posteroinferior nucleus) of the thalamus of a monkey. The small drawing in the upper right-hand corner of the figure shows the location of these nuclei in the thalamus in relation to the dorsolateral (LD), posterolateral (LP), dorsomedial (MD), centromedian (CM), parafascicularis (P), and the lateral geniculate (LG) nucleus. The dots indicate points at which electrodes recorded a response to tactile stimulation at various points on the contralateral side of the body. The enlarged illustration at the left shows the area of the body which projected impulses to each of these dots. The black areas on the figurines projected the most intense stimulation and the cross-hatched areas the least intense stimulation. (Mountcastle, V. B. (ed.): Medical Physiology. 12th ed., Vol. II, p. 1386. St. Louis, C. V. Mosby, 1968)

receives auditory, visual, gustatory, touch, pressure, kinesthetic, temperature, and pain impulses which it relays to primary sensory areas in the cortex, (2) receives impulses which it relays to areas of the cortex in the temporal, parietal, and occipital lobes, but outside the primary sensory areas, (3) relays impulses from the basal ganglia and cerebellum to the motor cortex, (4) relays impulses from the hippocampus and amygdala to the limbic cortex (controlling visceral efferent and endocrine mechanisms and possibly also emotional behavior), (5) transmits to and receives from the hypothalamus, and (6) possibly receives impulses from the motor cortex and brainstem which it relays to the putamen and caudate nucleus. There are, in addition, many interconnections within the thalamus itself. In some areas of the thalamus somatotopic segregation is characteristic (Fig. 40-6), whereas in other areas it is not found.

THE CEREBRAL CORTEX

The cerebral cortex is a highly convoluted strip of gray matter which in the adult has a surface area of 2,000 cm.^{2}, a thickness of from 2.5 to 4 mm., and a total volume of 600 cm.^{3} It contains 12 to 14 billion neurons and a large number of glial cells and receives countless axons from the thalamus, from other parts of the ipsilateral cortex, and from homologous areas of the contralateral cortex. Some of its descending axons extend as far as the caudal portions of the spinal cord.

Lateral surface of the cerebral cortex:

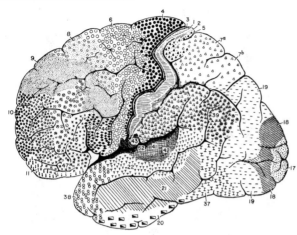

Medial surface of the cerebral cortex:

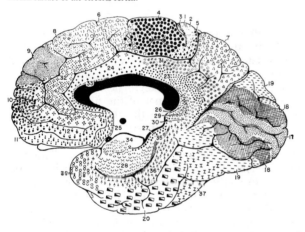

Fig. 40-7. Brodmann's map of the cytoarchitectural fields of the human cerebral cortex. Shown are both the lateral (*top*) and medial (*bottom*) surfaces of the cerebral hemisphere. (Brodmann, K.: Feinere Anatomie des Grosshirns. *In* Handbuch der Neurologie. Vol. I, p. 226. Berlin, Springer-Verlag, 1910)

Brodmann, on the basis of cytoarchitectural differences in the cerebral cortex, has divided it into a number of fields (Fig. 40-7). His classification is widely used, even though Vogt has suggested that it be replaced by a classification containing 200 fields, and Baily and von Bonin have suggested that the number of fields be decreased to 32. Between 90 and 92 per cent of the cerebral cortex contains six layers (isocortex, or neocortex) and the rest contains less than six layers. This latter part of the cortex lies medial to the rhinal sulcus and is called the allocortex.

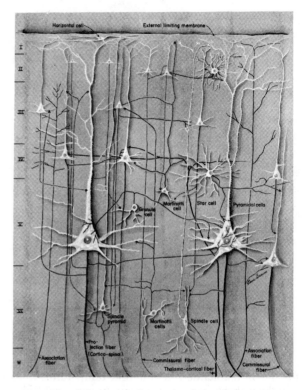

Fig. 40-8. Schematic diagram of the different types of neurons in the cerebral cortex. The Roman numerals indicate the various layers of the cortex and the W indicates the point where the white matter begins. (Elias, H., and Pauly, J. E.: Human Microanatomy. 3rd ed., p. 90. Philadelphia, F. A. Davis, 1966)

Cortical Layers

Figure 40-8 shows the cells making up the six layers of the neocortex. These cells are (1) the pyramidal cells, (2) the star or granule cells, and (3) the fusiform or spindle cells. The pyramidal cells have triangular or trapezoidal cell bodies varying in axial dimensions from 15 by 20 μ to 120 by 90 μ, the larger ones sometimes being called Betz cells. The star or granule cells, on the other hand, have small cell bodies and axons that usually terminate at an adjacent cell or in the superficial layers of the cortex. The fusiform cells differ from the other cells in that they have spindle-shaped cell bodies. The cortical cells are sometimes named according to their function: (1) association, (2) commissural, or (3) projection fibers. Examples of association fibers include granule cells and the Martinotti cells seen in Figure 40-8.

The most superficial layer of the cortex, layer I, is called the plexiform layer and contains pyramidal axons and dendrites and a few horizontal cells of Cajal. Layer II, the external granular layer, contains star cells (granule cells) and a number of small, tightly packed pyramidal cells. Layer III, the pyramidal cell layer, is similar to layer II but contains in addition some larger pyramidal cells which synapse near layer IV with thalamic afferent fibers. A few of the pyramidal neurons in this layer also send association and projection fibers out from the cortex. Layer IV, the internal granular layer, is small and poorly defined in the motor cortex (precentral gyrus), but quite pronounced in the sensory parts of the neocortex. It contains pyramidal cells and small granule cells. The granule cells synapse with afferent fibers from the thalamus. Layer V, the giant pyramidal layer, contains the giant pyramidal cells. Their axons may serve either a projection, commissural, or associative function. Layer VI, the fusiform layer, contains the spindle cells. These cells usually send projection axons from the cortex.

Cortical Columns

In the organization of the cortex illustrated in Figure 40-6 and discussed above there is a series of synaptic connections into which come the axons of the thalamus and from which leave the axons of the cortex. The synapses occur both between layers (vertical or columnar connections) and between vertical columns. The majority of the synaptic connections are limited to a single column, and the less numerous horizontal pathways are for the most part limited to layer I, the plexiform layer. In other words, the fundamental functional unit of the cortex is the vertical column. It apparently acts as an input-output link with a limited lateral spread of activation. And in fact, much of this lateral spread is inhibitory in nature, causing depression of adjacent columns.

Sensory Areas

In previous chapters we have discussed areas in the occipital cortex responsible for vision, and areas in the temporal cortex responsible for audition. In this section we shall concentrate on areas of the cortex which are stimulated by afferent fibers from the tongue, joint receptors, skeletal muscle, and skin. One such area, usually called the primary somatic sensory area of the *postcentral gyrus*, lies in the most anterior portion of the parietal lobe just posterior to Brodmann's area 4 and the superior sagittal sinus. Figure 40-9 compares its columnar segregation of afferent impulses with the segregation of motor impulses in the precentral gyrus (the primary motor cortex). In Figure 40-10 we see the location of the primary sensory area (S I) on the lateral surface of the cortex and its relationship with the secondary sensory area (S II), the audi-

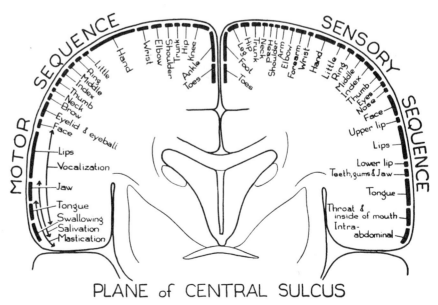

Fig. 40-9. Coronal section through the precentral gyrus (*left*) and the postcentral gyrus (*right*) of the cerebral hemispheres. Note the localization of motor control and sensory receptor areas on the cortex. Also compare the relatively small representation of some areas on the cortex (the trunk, for example) with the large representation of others (the hand). In general the degree of motor control and the fineness of 2-point sensory discrimination for an area are directly related to the amount of representation that area has on the cortex. (Rasmussen, T., and Penfield, W.: Further studies of the sensory and motor cerebral cortex of man. Fed. Proc., 6:455, 1947)

tory and visual sensory areas (A I, A II, V I, V II), the association areas (PCA, ALA, AMSA, PMSA), and the primary motor area of the cortex.

The postcentral gyrus. One means of studying the postcentral gyrus is by placing a microelectrode in its various columns. The investigator can then record the response of each column to different types of stimuli applied to different parts of the body. Studies such as these reveal that in the postcentral gyrus (1) all neurons in a single column are of the same modality type, (2) all neurons in a single column respond to nearly identical receptive fields, and (3) the column most sensitive to a particular stimulus from the skin will respond to this stimulus after a latency of only 2 to 4 msec., layers III and IV being activated earliest (Fig. 40-8).

The somatotopic representation on the postcentral gyrus is similar to that in the ventrobasal complex in the thalamus (compare Figs. 40-6 and 40-9). This is true in terms of both the order and the degree of representation. In both areas the modalities of touch, pressure, and kinesthesia seem to predominate, and impulses conducted over the dorsal columns represent almost 99 per cent of their input.

Somatic area II. A second sensory area lies in the parietal cortex superior to the sylvian fissure, in contact with the second auditory area (Fig. 40-10). Somatic area II (S II) is similar to somatosensory area I (S I) in its columnar organization and in the large number of impulses it receives from the dorsal columns. It differs, however, in that it receives substantially more impulses from the lateral and ventral columns of the cord S I and that it has little direct communication with the thalamus. Apparently many of the impulses transmitted from the thalamus to S II first synapse in the second auditory area.

Other differences between S I and S II are that the latter apparently contains no cells responsive to tendon or joint movements and receives impulses from surfaces on both sides of the body. Somatic area II also seems to be a more important receptive area for the sensations of temperature and the prickling type of pain. The more long-lasting and intense pain that originates from the stimulation of peripheral C fibers, however, may not have important cortical representation. There is some evidence that its conscious recognition may occur primarily at the level of the ventral thalamus and hypothalamus. We find, for example, that analgesics given at a dose known to depress the thalamus and hypo-

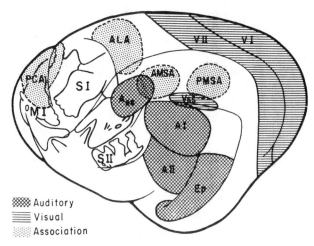

Fig. 40-10. Drawing of the lateral surface of the left hemisphere of a cat. Note the location of (1) the primary and secondary visual areas (V I, V II, and the visual area of the suprasylvian sulcus, V ss) in or near the occipital lobe, (2) the primary and secondary auditory areas (A I, A II, and the ectosylvian auditory area, Ep), (3) the primary and secondary somatic sensory areas (S I and S II), (4) the association areas (PMSA, the posterior middle suprasylvian association area; AMSA, the anterior middle suprasylvian association area; ALA, the anterior lateral association area; and PCA, the pericruciate association area), and (5) the motor cortex. A more complete presentation of the auditory areas will be found in Figure 37-9. (Thompson, R. F., et al.: Organization of auditory, somatic sensory, and visual projection to association fields of cerebral cortex in cat. J. Neurophysiol., 26: 345, 1963)

thalamus eliminate intense pain associated with the stimulation of the C fibers but do not affect the prickling type of pain associated with stimulation of the delta fibers.

Association Areas (The Homotypical Cortex)

Areas exist in the (1) frontal, (2) anterior temporal, and (3) parietotemporal lobes of the cerebral cortex of man which have been designated collectively the association or homotypical cortex. Structurally, they are transitional between the highly granular primary sensory areas and the almost agranular motor areas. They are most prominent in primates, and prior to 1930 were thought to be connecting links between the primary sensory and the motor areas (hence "association" area). It is now quite evident, however, that this concept is inadequate. Some of the evidence against it is that (1) the homotypic areas of the cortex continue to receive sensory information after the destruction of all the primary sensory areas, but not after the destruction of the thalamus, (2) some homotypic areas send fibers directly to such subcortical areas as the caudate nucleus and the hypothalamus, (3) both the motor and sensory areas of the cortex receive sensory impulses directly from the thalamus, and (4) with the possible exception of the temporal areas and some areas in the prefrontal lobe, homotypic areas receive fibers directly from the thalamus.

The frontal lobe. Occupying the anterior pole of the frontal lobe and merging posteriorly with olfactory structures is the frontal association area (prefrontal lobule, or orbitofrontal cortex). Its relationship to the personality has been recognized since the famous crowbar case of Phineas P. Gage in 1848. It was at this time that Mr. Gage, "an efficient and capable" foreman, had a tamping iron blown through the frontal region of his brain. After the accident his physician, Dr. Harlow, reported the following changes in personality (Boston Medical and Surgical Journal, 1848):

He is fitful, irreverent, indulging at times in the grossest profanity (which was not previously his custom), manifesting but little deference to his fellows, impatient of restraint or advice when it conflicts with his desires, at times pertinaciously obstinate yet capricious and vacillating, devising many plans for future operation which are no sooner arranged than they are abandoned in turn for others appearing more feasible. . . . His mind was radically changed, so that his friends and acquaintances said he was no longer Gage.

In 1935 Jacobsen performed an apparently similar operation, a *bilateral lobectomy*, on a highly emotional chimpanzee and found that after the operation the animal no longer had the anxieties and emotional responses to failure that he had prior to surgery. In Jacobsen's words, "It was as if the animal had . . . placed its burdens on the Lord." These observations encouraged neurosurgeons to perform prefrontal lobotomies on certain problem patients. The results were both gratifying and discouraging. The patient's intelligence was little affected and the anxieties and delusions that had incapacitated so many before the operation now, though they persisted, seemed to the patient remote and of little concern. Therefore in many the operation improved performance. On the other hand, the patient seemed to lose his sense of duty and any regard for the feelings of others. This operation became quite popular and undoubtedly saved a number of patients from suicide, drug addiction, and intense suffering. Unfortunately, the change in attitude that resulted was permanent and usually unpleasant. More recently, physicians have placed more reliance upon medication and prefrontal lobotomy has become less common.

The temporal lobes. Except for the primary auditory area the temporal cortex receives few thalamocortical projection fibers. On the other hand, it abounds with fibers linking other association areas. The stimulation of certain association areas in the temporal cortex elicits hallucinatory reenactments (flashbacks). These might include the sensation of hearing a specific rendition of a song or of seeing a particular orchestra performing. Temporal lobe disease may produce similar experiences. Patients with temporal lobe tumors, for example, often revert to a dreamworld of vivid visual and auditory hallucinations. They may also show a deteriorating sense of time and space.

Motor Areas

The precentral gyrus of the cerebral cortex receives impulses from (1) the visual, auditory, and somatosensory areas of the cortex, (2) the cerebellum and basal ganglia (Fig. 40-11), via the thalamus, and (3) the contralateral motor cortex, via the corpus callosum. Frequently a single cell in the motor cortex receives fibers from several of these areas. On the other hand, stimulation of a small zone (Fig. 40-9) on the motor cortex elicits a contraction limited to a group of synergistic muscles or a single muscle in a particular part of the body. This stimulation is also associated with the sensation of body movement even when the actual body movement is prevented by the use of a local anesthetic.

The motor neurons from the precental gyrus help form two parallel descending systems: (1) the pyramidal system and (2) the extrapyramidal system. The neurons in the pyramidal system originate in the precentral gyrus of the cerebral cortex, travel down the pyramidal tract, decussate in the medulla, help form the corticospinal tracts of the cord (Fig. 40-2), and impinge for the most part upon internuncial neurons in the brainstem and cord which in turn impinge upon somatic efferent neurons:

Pyramidal neuron → internuncial neurons →
    somatic efferent neurons

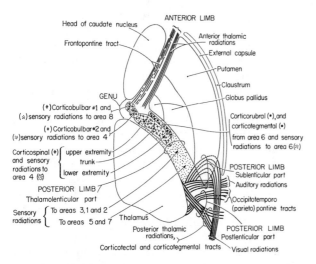

Fig. 40-11. A diagram showing the relationship between the thalamus, the anterior and posterior limbs of the internal capsule, and the basal ganglia. The basal ganglia are deeply placed areas of gray matter in the hemisphere and include the caudate nucleus, globus pallidus, putamen, the claustrum, and the amygdaloid nucleus (*not shown*). (Crosby, E. C., et al.: Correlative Anatomy of the Nervous System. P. 395. New York, Macmillan, 1962)

The cortical neurons that help form the extrapyramidal system, on the other hand, are shorter and synapse with neurons in the basal ganglia of the forebrain or the reticular formation of the brainstem; along with these synaptic areas they form a second set of descending tracts in the brainstem and cord. The pyramidal neuron, the basal gangila, the reticular formation, and the reticulospinal tracts collectively are part of the so-called extrapyramidal system.

In addition to the influences mentioned above we find other important descending tracts which affect the firing of the somatic efferent neuron and are also characterized as extrapyramidal. These include the rubrospinal tract, which is strongly influenced by impulses from the cerebellum, and the vestibulospinal tract, which is strongly affected by impulses from the vestibular apparatus of the inner ear. Both the vestibular apparatus and the cerebellum operate to integrate a single motor act into a total body reaction. The cerebellum does this by virtue of its afferent fibers from the cerebral cortex, the joint receptors, the skin, and the visual, auditory, and vestibular systems. The vestibular apparatus performs its functions by signaling changes in body position and in the velocity of movement.

Muscle tone. The maintenance of posture and attitude is to a large extent the result of the tension produced by the antigravity muscles of the body (extensors of the leg and thigh, elevators of the head, and the muscles that close the jaw). This tension results from the asynchronous firing of motor units (an alpha efferent neuron and the muscle fibers it innervates) at low frequencies and is, for the most part, initiated by a stretching of annulospiral endings in extensor muscles when the part the extensor muscle serves is flexed. This was discussed in Chapter 2; Fig. 2-12 provides a quick review of the role of the gamma efferent neurons and the annulospiral endings in the maintenance of muscle tone and muscle length. The importance of the annulospiral endings in maintaining muscle tone can be demonstrated by cutting the dorsal roots into the spinal cord. When this is done, sensory impulses from the annulospiral endings are prevented from entering the cord and muscle tone disappears. In other words, the myotatic reflex is destroyed.

Decerebration. The role of the forebrain in the maintenance of muscle tone and other reflex activity is not a simple one. Part of what it does is to send both inhibitory and facilitatory impulses to the brainstem and spinal cord. We find, for example, that a lesion in the pyramidal tract produces hypotonicity in skeletal muscles (due to the loss of facilitation). Lesions in the extrapyramidal system, on the other hand, usually produce hypertonicity (due to loss of inhibition) associated with spontaneous, aimless, unintentional movements. The overall effect of the loss of the forebrain can be studied by performing a transection in the midbrain between the superior and inferior colliculi (Sherrington procedure for decerebration). One of the symptoms resulting from such a transection (Fig. 40-12) is an increase in tone in the antigravity muscles, which can be eliminated by sectioning the dorsal roots. In other words, part of the overall function of the forebrain is apparently to depress certain extensor reflexes. Hence loss of the forebrain is followed by an exaggeration of these reflexes. It is important to realize, however, that after the loss of the forebrain (that is to say, in the decerebrate individual) the remaining brainstem and the spinal cord continue to function as a center for flexion reflexes and the integration of motor activity, as well as for the extension reflexes (Fig.

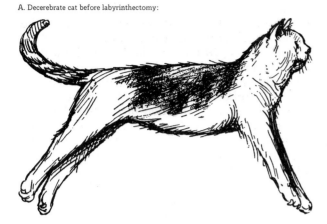

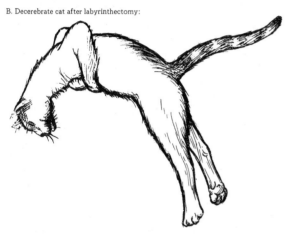

Fig. 40-12. A decerebrate cat (A) before and (B) after labyrinthectomy. In A the decerebrate animal shows extensor rigidity. In B after labyrinthectomy the head has dropped and this results in a reflex flexion of the forelimbs. (Pollock, L. J., and Davis, L. E.: The reflex activities of a decerebrate animal. J. Comp. Neurol., 50:384, 391, 1930)

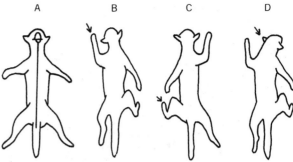

Fig. 40-13. Response of the decerebrate cat to stimulation. At A we see the cat prior to stimulation. At B we see his response to stimulation of his left forepaw (flexion of the forelimb and associated movements in the other 3 limbs), at C his response to stimulation of the left hindpaw (flexion of the hind limb), and at D his response to stimulation of the left ear (head turned to right). Note that none of these reflexes is restricted to a single part of the body. In other words, in the decerebrate cat the cord continues to integrate activity in one part with activity in other parts. (*Redrawn from* Sherrington, C. S.: Decerebrate rigidity and reflex coordination of movements. J. Physiol., 22:329, 330, 1898)

40-13). Witness, for example, the response of the decerebrate cat in Figure 40-13 to stimulation (B) of the forepaw, (C) of the hindpaw, and (D) of the ear. In each case there is withdrawal of the affected part with associated flexions and extensions throughout the body.

The symptoms of this type of decerebration are, for the most part, due to a withdrawal of inhibition to the gamma efferent (innervating intrafusal fibers) system, the evidence for this being that these symptoms (hypertonicity) can be eliminated by sectioning the dorsal roots. In 1923 and 1930, however, a form of decerebration was employed by Pollock and Davis which was followed by an extensor hypertonicity that could not be eliminated by dorsal root section. Apparently in this procedure there was a pronounced withdrawal of inhibition to the alpha efferent neuron (innervating extrafusal fibers). The procedure that produced these symptoms was ligation of the basilar artery, which resulted in infarction of the brainstem above the midcollicular level and infarction of a part of the cerebellum concerned with the control of the labyrinthine and neck reflexes. It has been suggested that this condition be called *labyrinthine extensor rigidity* and that the term *decerebrate rigidity* be reserved for symptoms resulting from a lesion similar to that produced by Sherrington.

Figure 40-14 summarizes some of the conclusions that have been reached on the basis of experiments such as those just discussed. Apparently there are areas in the cortex, the basal ganglia, the cerebellum, and the caudal portion of the reticular formation which depress muscle tone. There are also areas in the cephalad portion of the reticular formation and the vestibular nuclei which facilitate increases in muscle tone. Destruction of any of these areas may upset the

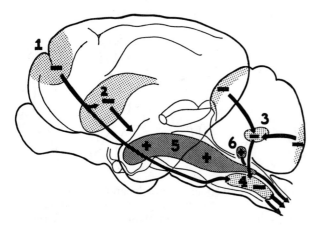

Fig. 40-14. Diagram of the facilitory and inhibitory systems in the brain of the cat which influence muscle tone. The suppressor pathways include (1) the corticobulboreticular system, (2) the caudatospinal system, (3) cerebelloreticular system, and (4) the reticulospinal system. The facilitory pathways include (5) a more cephalad reticulospinal system and (6) the vestibulospinal system. Apparently by producing a transection between areas 2 and 5 there is a sufficient loss of inhibition to result in a marked spasticity. (Lindsley, D. B., Schreiner, L. H., and Magoun, H. W.: An electromyographic study of spasticity. J. Neurophysiol., 12:198, 1949)

fine balance between facilitation and inhibition existing in the body.

Hemiplegia. Hemiplegia is paralysis of one side of the body, usually resulting from a cerebrovascular accident in the contralateral forebrain. Such an accident may result in the destruction of either an extrapyramidal motor projection at the level of the internal capsule or of a part of the frontal lobe. In both cases the symptoms are similar to those found bilaterally in decerebration. In man the symptoms are best characterized as unilateral *spasticity*—i.e., (1) muscle hypertonicity (muscle exerts an increased resistance to stretch), (2) hyperreflexia (exaggerated reflexes), and (3) clonus (oscillations in muscle tension in response to stretch).

Spinal transection. Kuhn, between 1947 and 1950, reported on 29 men who had experienced spinal transections between T-2 and T-12. His observations have proven helpful in our understanding of the care of such patients and of the capacity of the spinal cord to function in the absence of connections with the brain. The results of such a procedure are (1) complete loss of volitional control of movement below the level of the lesion, (2) almost complete loss of sensation below the level of the lesion, and (3) almost complete loss of reflex activity below the level of the lesion, which is temporary in most cases.

In 17 of the 29 men studied by Kuhn a dull, burning sensation in the buttocks, perineum, or lower abdomen was reported during pressure on the ischial tuberosities, stimulation of the anal or urethral canal, or distention of the bladder. Apparently these sensations were the result of afferent impulses traveling up splanchnic nerves and entering the cord over dorsal roots cephalad to the transection.

Immediately after the spinal transection there is a loss either of all reflex activity or of all except a few genital reflexes. The period of areflexia or hyporeflexia following the transection is sometimes referred to as *spinal shock*. The mechanism responsible for it is not well understood. It has been reported, however, that the greater the degree of cerebral dominance, the longer and more severe the spinal shock. In the frog, for example, the period of spinal shock lasts only a minute, in the dog several hours, in the monkey many days, and in man many months. Apparently it is not due to any form of irritation, since a second transection below the first results in only trivial reflex depression. Nor is it due to generalized hypotension, as was once thought, since the cord above the transection shows little or no reflex depression.

In time the spinal reflexes begin to reappear, the genital reflexes, if lost, being among the first to come back. Gentle stimulation of the frenulum of the penis may produce erection and in a few cases ejaculation. In addition, there is contraction of the bulbocavernous and anal sphincter muscles, and a delayed and slow contraction of the scrotal dartos muscle. Reflex emptying of the bladder and bowel associated with the relaxation of sphincter tone is usually seen 25 to 30 days after transection. These reflexes can be facilitated by dilation of the anal sphincter or stimulation of the skin served by the sacral portion of the spinal cord. The emptying, however, is not so complete as in the normal person, and in the case of the urinary bladder there is always a residue of urine. Such residue is an excellent medium for the growth of microorganisms and can result in a fatal urinary infection. This is usually prevented by frequent catheterizations which serve to empty the bladder completely.

Early in the recovery process a number of responses to stimulation of the plantar surface of the foot appear. One of the first is a dorsiflexion of the foot and a fanning of the toes (positive Babinski sign). This reflex is seen in normal infants, but in children and adults is symptomatic of pyramidal tract lesions. Apparently in normal persons the Babinski reflex is depressed as the central nervous system develops. If a positive Babinski reflex does not appear within 3 months after spinal transection, the period of areflexia will probably be permanent. Often this is due to the fact that the lesion has also resulted in destruction of the lower cord (possibly from an interference in blood flow).

After the appearance of a positive Babinski sign, other reflexes begin to appear. Stroking the plantar surface of the foot produces increasingly more extensive responses, including a flexor withdrawal of the foot and ankle initially and eventually flexion of the knee and hip. Two years after the transection the extensor reflexes in some patients may have developed to such a degree that the patient can stand for short periods without support, the major problem being loss of connection with the vestibular apparatus and hence difficulty in maintaining balance. At this time the tendon reflexes are also hyperactive, autogenic inhibition (see p. 34) returns, and, as a result, clonus is common. Trying to flex the arm of such a patient frequently causes in the stretched extensor muscle an alternate firing of afferent neurons from the intrafusal fiber (producing facilitation of alpha efferents) and from the Golgi tendon organ (producing inhibition of the alpha efferents). The result is an oscillating muscle movement called clonus.

The Electroencephalogram

Caton noted, in 1875, that he could record continuous electrical activity from the exposed cerebrum of animals. In 1929 Berger, using a string galvanometer, recorded electrical activity from the brains of men both awake and asleep by placing surface electrodes on the scalp. Since then the electroencephalogram (EEG, or electrocorticogram) has become a valuable tool for the study of physiology, epilepsy, and brain tumors and for the diagnosis of various neurological disorders. Unfortunately, however, our understanding of the underlying mechanisms has not progressed as far as it has for the electrocardiogram (ECG).

The EEG, like the ECG, may be recorded by either a bipolar or a unipolar technique. In the bipolar procedure two electrodes are placed on

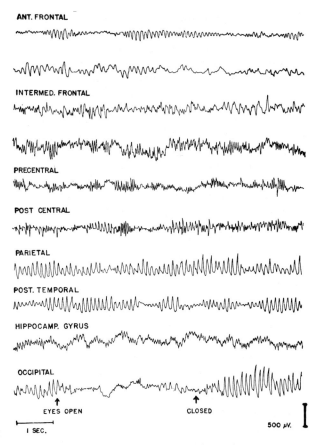

Fig. 40-15. Spontaneous electrical activity from different areas of the cortex of an alert man. Each of the above recordings was obtained from bipolar, silver chloride, cotton-wick electrodes placed directly on the exposed cortex. The time scale and sensitivity for the recording system are shown at the bottom of the illustration. Alpha rhythms (8–13 c.p.s.) are most prominent in (1) the parietal lobe with the exception of the postcentral gyrus, (2) the posterior temporal lobe, and (3) the occipital lobe. Note that when the subject opens his eyes the alpha rhythm disappears in the occipital lobe. (Penfield, W., and Jasper, H. H.: Epilepsy and the Functional Anatomy of the Human Brain. P. 187. Boston, Little, Brown & Co., 1954)

the scalp, each over an active electrogenic site, and the difference between the potentials at the two electrodes is recorded. In the unipolar technique, one electrode is placed over an active site and the second one serves as an indifferent electrode. In most cases the indifferent electrode is placed on the ear lobe, a position sufficiently distant from the heart and cerebral cortex to be little affected by the ECG or EEG potentials.

Figure 40-15 shows some of the patterns obtained from man when bipolar electrodes are placed at various points on the cortex. The waves recorded generally vary in frequency from 1 to over 50 per second and have an amplitude of between 50 and 200 μv. when recorded from the surface of the scalp. In a normal person with his eyes closed and in a quiet room the dominant rhythm from the occipital and most of the parietal lobe has a frequency of from 8 to 13 c.p.s. This is called the *alpha* or Berger *rhythm*. It varies in frequency, location, and amplitude from person to person, and in a few cases in which the brain is apparently otherwise normal, may not be seen at all. When found, it is a remarkably constant indication of an alert but relatively unoccupied brain and varies less than 1 c.p.s. During sensory stimulation or other mental activity, however, the alpha waves may disappear and be replaced by a higher frequency (13-25 c.p.s.), lower amplitude wave, the *beta wave*. Waves having a frequency lower than that of the alpha wave rarely occur in normal awake persons other than newborn infants. The *theta* (3-7 c.p.s.) and *delta* (0.5-3.5 c.p.s.) *rhythms* in an awake person are indicative of brain disease or injury. Delta waves are also seen when an individual sleeps.

Neural mechanism. The EEG waves are generally believed to result from electrical activity in the dentrites of the cortical pyramidal cells (Fig. 40-8). These dentrites are found in large concentrations in the upper molecular layer of the cortex. Apparently they undergo a series of polarizations and depolarizations that sets up the current producing the EEG. It is not, of course, just one set of dendrites that is reponsible for the EEG, but rather large numbers acting synchronously and going through a series of changes at regular intervals.

The mechanism for the synchrony apparently resides in the intrinsic characteristics of the cortex and its relationship with other parts of the brain. If, for example, we isolate a small part of the cortex from the rest of the brain but leave its blood supply intact, and then allow several days or weeks for recovery, we find that this isolated area of the cortex will be electrically silent. If we apply a single stimulus to it, on the other hand, it will develop a rhythmic pattern similar to that normally found—though this pattern disappears in several seconds without additional stimulation. It would seem, then, that the volley of electrical waves seen in the EEG is the result of a pacing mechanism located in some other part of the brain, probably the thalamus (Fig. 40-16).

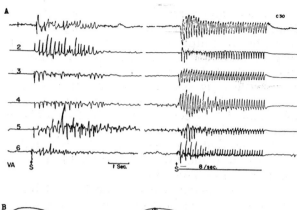

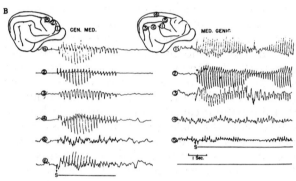

Fig. 40-16. Production of EEG waves by the stimulation of various nonspecific thalamic nuclei. In A, on the left, a volley of waves was initiated at points 1 through 6 on the cortex by a single stimulus applied to the ventral anterior nucleus of the thalamus. On the right we see the response of the cortex to a series of stimuli applied to the same thalamic nucleus. In B we note the response of the cortex to the stimulation of the central median (*left*) and the medial geniculate (*right*) nucleus. (Hanbery, J., and Jasper, H. H.: Independence of diffuse thalamocortical projection system shown by specific nuclear destructions. J. Neurophysiol., *16*:256, 1953)

Sleep and Wakefulness

During sleep there is usually (1) a mild reduction in alveolar ventilation, (2) a slight rise in venous and alveolar P_{CO_2}, (3) a decrease in heart rate, (4) a decrease in arterial pressure, (5) a decrease in the concentration of blood ketosteroids, (6) an increase in the concentration of eosinophils in the blood, (7) a decrease in rectal temperature which may be as much as 2°F., (8) an increase in the threshold for many reflexes, (9) a decrease in muscle tone, (10) constriction of the pupils owing to an increase in their parasympathetic tone, and (11) usually a divergence of the eyes upward. There are also numerous changes in the EEG. These can be divided into four stages:

Stage 1: Alpha waves are slowed slightly. Seen in relaxed wakefulness and during the initial phase of drowsiness.

Stage 2: A mixture of frequencies of 3 to 6 c.p.s. followed by volleys of waves with an amplitude of about 50 μv. and a frequency of from 14 to 15 c.p.s. (sleep spindles). A period of sleep from which one is easily aroused.

Stage 3: Primarily delta waves of 2 c.p.s. with occasional sleep spindles. A period of sleep of intermediate depth.

Stage 4: High-voltage, slow delta waves without associated spindles. Deep sleep associated with a high threshold for awakening.

In addition to these stages of sleep there occurs every 80 to 120 min. during sleep a period of rapid, low-voltage EEG activity similar to that seen in alert persons. Since the individual is in a considerably deeper stage of sleep during this period than during Stage 1, it is called the period of paradoxical sleep. Muscle tone is less than during the other stages, though there are occasional body movements. In addition there are rapid eye movements (REM's), an increase in heart rate, an increase in respiratory rate, and usually *dreaming*.

In 191 subjects awakened in the middle of their

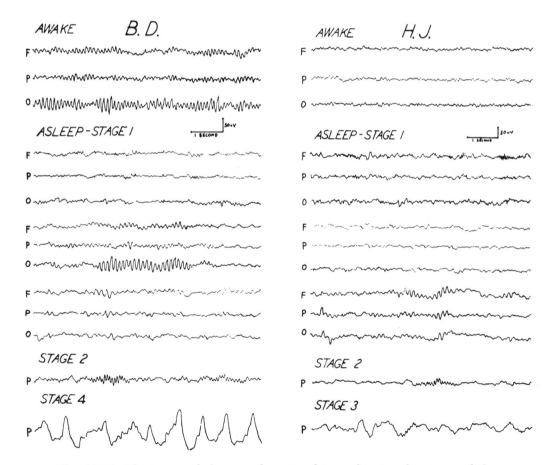

Fig. 40-17. Electroencephalograms from 2 subjects showing the stages of sleep. Records were obtained from the frontal (F), parietal (P), and occipital (O) lobes. (Dement, W. C., and Kleitman, N.: Cyclic variations of EEG during sleep and their relations to eye movements, body motility and dreaming. Electroenceph. Clin. Neurophysiol., 9:679, 1957)

period of paradoxical sleep (the period of REM), 80 per cent reported that they had been dreaming. On the other hand only 7 per cent of the subjects awakened during non-rapid eye movement sleep (NREM) reported experiencing dreams. Toward the end of the phasic period of REM there is marked bradycardia and a decrease in arterial pressure to 40 to 60 mm. Hg.

Figure 40-18 is a graph showing the quantities of REM sleep (paradoxical sleep) and non-REM sleep that people of different ages experience. On the basis of this chart it has been suggested that REM sleep might well be associated with the growth and maturation of the nervous system and a cause of the greater need of the infant for long periods of sleep throughout the day. Apparently REM sleep is also important for the adult. We find, for example, that when volunteers are prevented from getting paradoxical sleep (are awakened, that is, each time their EEG shows a paradoxical sleep pattern), they develop marked behavioral changes. Severe deprivation in animals leads to hallucination.

Neural mechanisms. The mechanisms that cause us to be tired, to go to sleep, to stay asleep, and to wake up are not understood. We do know, however, that there is an area in the lower brainstem which when stimulated produces sleep, and a part of the reticular formation which when stimulated arouses one from sleep. Some have suggested that these two areas act by exerting antagonistic influences on a cortical pacemaker cell in the thalamus. Others have hypothesized that the bulbar *sleep-activating system* exerts its action by a direct inhibition of the reticular formation.

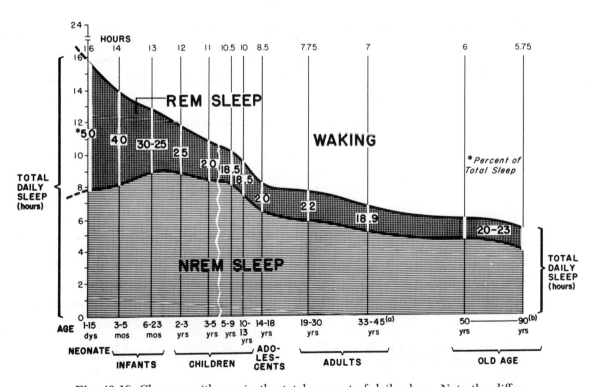

Fig. 40-18. Changes with age in the total amount of daily sleep. Note the difference in the amount of REM sleep in the infant (35% of sleep) and the adult (18 to 22% of his sleep). (Drawing supplied by Howard Roffwarg. *Based in part on data from* Roffwarg, H. P., Muzio, J. N., and Dement, W. C.: Ontogenic development of the human sleep-dream cycle. Science, *152*:608; copyright, April 1966 by the American Association for the Advancement of Science)

Lesions in the central nervous system have also helped us to understand some of the relationships existing there. For example, patients with tumors and vascular lesions in the brainstem frequently have as a part of their syndrome abnormal states of consciousness. In certain cases of encephalitis there may be an initial period of insomnia due possibly to an irritation of the brainstem followed by prolonged somnolence due to infarction. Transections of the cord at C-1 apparently do not disturb the normal sleep and wakefulness patterns, but if we also section the trigeminal nerves on both sides of the body, an animal will show an EEG characteristic of sleep. Apparently by sectioning the trigeminal nerve we have so lowered the sensory input to the reticular formation as to produce somnolence. This is despite the fact that sensory messages continue to enter the forebrain and brainstem via the olfactory, optic, and auditory nerves.

The importance of the reticular formation in maintaining consciousness has been demonstrated in man, in whom its destruction produces a comatose state. Apparently a transection of the midbrain just caudal to the third cranial nerve separates the forebrain sufficiently from the reticular formation to result in sleep. If, on the other hand, the section is performed in such a way as to destroy all the structures except the reticular formation, normal patterns of alertness are maintained. If the transection is performed in the pons (midpontine pretrigeminal), the result is insomnia. Prior to midpontine pretrigeminal transection cats were found to sleep 78 per cent of the time. After the operation the same cats slept only 37 per cent of the time. Apparently this lesion leaves enough of the reticular activating system intact to facilitate cerebral cortical activity, but separates the forebrain from the sleep-activating area.

REFERENCES

Best, C. H., and Taylor, N. B.: The Physiological Basis of Medical Practice. 8th ed. Baltimore, Williams & Wilkins, 1966.

Bishop, P. O.: Central nervous system: Afferent mechanisms and perception. Ann. Rev. Physiol., 29:427-484, 1967.

Brodmann, K.: Feinere Anatomie des Grosshirns. In Handbuch der Neurologie. Vol. I. Berlin, Springer-Verlag, 1910.

Crescitelli, F., and Dartnall, H. J. A.: Human visual purple. Nature (London), 172:195-197, 1953.

Crosby, E. C., Humphrey, T., and Lauer, E. W.: Correlative Anatomy of the Nervous System. New York, Macmillan, 1962.

Davis, H.: Excitation of auditory receptors. In Field, J. (ed.): Handbook of Physiology. Sec. 1 (Neurophysiology), vol. I, pp. 565-584. Washington, American Physiological Society, 1959.

———: Guide for the classification and evaluation of hearing handicap in relation to the international audiometric zero. Trans. Am. Acad. Ophthal. Otolaryng., 69:740-751, 1965.

Deaver, J. B.: Surgical Anatomy of the Human Body. 2nd ed. New York, Blakiston Division (McGraw-Hill), 1926.

de Lorenzo, A. J. D.: Studies on the ultrastructure and histophysiology of cell membranes, nerve fibers and synaptic junctions in chemoreceptors. In Zotterman, Y. (ed.): Olfaction and Taste. Pp. 5-17. Oxford, Pergamon Press, 1963.

Dement, W. C., and Kleitman, N.: Cyclic variations of EEG during sleep and their relations to eye movements, body motility and dreaming. Electroenceph. Clin. Neurophysiol., 9:673-690, 1957.

Denny-Brown, D.: The Cerebral Control of Movement. Liverpool, Liverpool University Press, 1966.

Dodt, E., and Zotterman, Y.: Mode of action of warm receptors. Acta Physiol. Scand., 26:345-357, 1952.

Duke-Elder, S.: A Textbook of Ophthalmology. Vol. I. St. Louis, C. V. Mosby, 1952.

Eldred, E., and Buchwald, J.: Central nervous system: Motor mechanisms. Ann. Rev. Physiol., 29:573-606, 1967.

Elias, H., and Pauly, J. E.: Human Microanatomy. Chicago, DaVinci, 1960.

Engström, H., Ades, H. W., and Hawkins, J. E., Jr.: Structure and function of the sensory hairs of the inner ear. J. Acoust. Soc. Am., 34:1356-1363, 1963.

———: The vestibular sensory cells and their innervation. In Szentagothai, J. (ed.): Modern Trends in Neuromorphology. Pp. 21-41. Budapest, Akadémai Kiadó, 1965.

Fry, G. A.: The image-forming mechanism of the eye. In Field, J. (ed.): Handbook of Physiology. Sec. 1 (Neurophysiology), vol. I, pp. 647-670. Washington, American Physiological Society, 1959.

Galambos, R.: Neural mechanisms in audition. Laryngoscope, 68:388-401, 1958.

Gesteland, R. C., Lettvin, J. Y., Pitts, W. H., and Rojas, A.: Odor specificities of frog's olfactory receptors. In Zotterman, Y. (ed.): Olfaction and Taste. Pp. 19-34. Oxford, Pergamon Press, 1963.

Gettes, B. C. (ed.): Refraction. Boston, Little, Brown & Co., 1965.

Goldberg, J. M., Adrian, H. O., and Smith, F. D.: Response of neurons of the superior olivary complex of the cat to acoustic stimuli of long duration. J. Neurophysiol., 27:706-749, 1964.

Hanbery, J., and Jasper, H. H.: Independence of

diffuse thalamocortical projection system shown by specific nuclear destructions. J. Neurophysiol., 16:252-271, 1953.

Heaton, J. M.: The Eye. Philadelphia, J. B. Lippincott, 1968.

Hensel, H., and Zotterman, Y.: Quantitative Beziehungen zwischen der Entladung einzlner Kältefasern und der Temperatur. Acta Physiol. Scand., 23:291-319, 1951.

Knighton, R. S., and Dumke, P. R. (eds.): Pain. Boston, Little, Brown & Co., 1966.

Kuhn, R. A.: Functional capacity of the isolated human spinal cord. Brain, 73:1-51, 1950.

Kuhn, R. A., and Macht, M. B.: Some manifestations of reflex activity in spinal man with particular reference to the occurrence of extensor spasm. Bull. Johns Hopkins Hosp., 84:43-75, 1949.

Kuhn, W. G., Jr.: The care and rehabilitation of patients with injuries of the spinal cord and cauda equina. J. Neurosurg., 4:40-68, 1947.

Laursen, A. M.: Higher functions of the central nervous system. Ann. Rev. Physiol., 29:543-572, 1967.

Lindsley, D. B., Schreiner, L. H., and Magoun, H. W.: An electromyographic study of spasticity. J. Neurophysiol., 12:197-205, 1949.

Magnus, R.: Animal posture. Proc. Roy. Soc. (Biol.), 98:339-353, 1925.

Marks, W. B., Dobelle, W. H., and MacNichol, E. F., Jr.: Visual pigments of single primate cones. Science, 143:1181-1182, 1964.

Moncrieff, R. W.: The Chemical Senses. London, Leonard Hill, 1967.

Mountcastle, V. B. (ed.): Medical Physiology. 12th ed. Vol. II, pp. 1057-1858. St. Louis, C. V. Mosby, 1968.

Mountcastle, V. B., and Henneman, E.: The representation of tactile sensibility in the thalamus of the monkey. J. Comp. Neurol., 97:409-439, 1952.

Newby, H. A.: Audiology. 2nd ed. New York, Appleton-Century-Crofts, 1964.

Ottoson, D.: Analysis of the electrical activity of olfactory epithelium. Acta Physiol. Scand. (Suppl. 122), 35:1-83, 1956.

Penfield, W., and Jasper, H. H.: Epilepsy and the Functional Anatomy of the Human Brain. Boston, Little, Brown & Co., 1954.

Pirenne, M. H. L.: Vision and the Eye. London, Chapman & Hall, 1967.

Pollock, L. J., and Davis, L. E.: The reflex activities of a decerebrate animal. J. Comp. Neurol., 50:377-411, 1930.

Polyak, S.: The Retina. Chicago, University of Chicago Press, 1941.

Ramón y Cajal, S.: Histologie du système nerveux de l'homme et des vertèbres. Vol. II, Chap. 23. Paris, Maloine, 1911.

Rasmussen, G. L., and Windle, W. F. (eds.): Neural Mechanisms of the Auditory and Vestibular Systems. Springfield, Charles C Thomas, 1960.

Roffwarg, H. P., Muzio, J. N., and Dement, W. C.: Ontogenic development of the human sleep-dream cycle. Science, 152:604-619, 1966.

Sherrington, C. S.: Decerebrate rigidity and reflex coordination of movements. J. Physiol., 22:319-332, 1898.

Stuhlman, O.: An Introduction to Biophysics. New York, John Wiley & Sons, 1943.

Tateda, H., and Beidler, L. M.: The receptor potential of the taste cell of the rat. J. Gen. Physiol., 47:479-486, 1964.

Thompson, R. F., Johnson, R. H., and Hoopes, J. J.: Organization of auditory, somatic sensory, and visual projection to association fields of cerebral cortex in cat. J. Neurophysiol., 26:343-364, 1963.

von Békésy, G.: Experiments in Hearing. New York, McGraw-Hill, 1960.

Woolsey, C. N.: Organization of cortico auditory systems. In Rosenblith, W. A. (ed.): Sensory Communication. Pp. 235-257. Cambridge, MIT Press, 1961.

Zotterman, Y. (ed.): Olfaction and Taste. Oxford, Pergamon Press, 1963.

INDEX

INDEX

A (angstrom), 3
A-band of muscle, 73
A fibers, 29
   see also Neuron
ABO blood groups, 262
Abortion, 562
Absorption, in allergy, 427
   of amino acids, 429
   of antibodies, 427
   of bile salts, 432
   of carbohydrates, 429
   from digestive tract, 427-440
   of electrolytes, 437
   of fat, 431
   of vitamins, 433
   of water, 436
   see also Reabsorption
Acceleration, 3, 22, 611
   of cerebrospinal fluid, 227
Accessory adrenal tissue, 521, 522
Acclimatization, to altitude, 348-354
   body temperature, 173, 498
Accommodation of the eye, 581, 596
   power of eye during, 582
ACD solution, 260
Acetate metabolism, 461, 522
Acetazolamine. *See* Carbonic anhydrase inhibitor
Acetoacetic acid, 459
Acetone, 459
Acetylcholine (ACh), 45
   release of, 44, 77
   response of smooth muscle to, 155
Acetylcholinesterase (AChE), 43, 45
   effect of denervation on concentration of, 68
Acetyl coenzyme A, 460, 515
Achalasia of esophagogastric junction, 415
Acidophilic, cells, 521
   tumors, 494
Acids, and blood flow, 200
   elimination of, 397
   organic, 439
   protein catabolism, 397
   volatile, 398
   weak, renal secretion of, 373
   see also pH
Acid-base balance, 397-403, 516
   bicarbonate concentration, 398
   kidney, 290

respiration, 290, 398
   tetany, 537
   urinary system, 398
   see also Buffer
Acid-base disturbances, 404
Acidity, chemoreceptors, 321
   hyperventilation, 348
   titratable, 401
Acidosis, 331, 397-403, 516
   ammonia production, 401-403
   compensated, 403
   diabetes, 516
   gluconeogenesis, 459
   metabolic, 403
   potassium excretion, 403
   respiratory, 403
   uremia, 405
   see also pH
Acne, 549
Acoustic tubercle, 319
Acromegaly, 494, 499
Actin, 74, 76, 149
   see also Actomyosin
Action potential, 38-41, 149
   all or none nature of, 40
   amplitude of, 50
   compound, 24
   duration of, 17
   of phrenic nerve, 296
   of respiratory nerve, 320
   role of ions, 39, 138
   role of sodium concentration, 39
   of smooth muscle, 149
   see also Potential
Activation, 67
Activation contraction coupling, 45, 71, 149, 536
Activation secretion coupling, 45, 536
Active reabsorption. *See* Transport
Active transport systems. *See* Transport
Actomyosin, 42, 45, 76
   cardiac, 134
Acuity, visual, 583
Adaptation, dark, 587
   receptor, 51
   see also Acclimatization
Addison's disease, 529, 530, 533
Adenine, 463
Adenohypophysis, 489
Adenoma, islet cell, 438

parathyroid, 539
Adenosine triphosphate (ATP), 86
   catabolism, 458
   clot retraction, 250
   effect of epinephrine on, 151
   muscle rigor, 80
   production by mitochondria, 14
Adipose tissue, 289, 460, 464, 511, 515
   quantity, 463
   see also Fat
Adrenal accessory tissue, 521, 522
Adrenal cortex, 205, 500, 521
   androgens, 529, 550
   development, 521
   fetus, 566
   hormones, 566
   hypofunction, 533
   secretion, 388
Adrenal dysfunction, 528-535
   heart, 116
Adrenalectomy, 528
Adrenal gland, 521-535
Adrenal medulla, 521-522
   development, 521, 522
   hormone production, 490
Adrenaline, 60-64
   see also Epinephrine
Adrenergic blocking agents, alpha, 68
   beta, 68
   dichloroisoproterenol, 68
   phenoxybenzamine, 68
Adrenergic receptors, 65
Adrenergic sympathetic neurons, 59
Adrenergic sympathetic tone, 478
Adrenocorticotrophic hormone (ACTH), 498, 525, 533
   see also Hormones
Adrenogenital syndrome, 529
Aesthenia, vasoregulatory, 109
Afibrinogenemia, 252
Afterbirth, 566
Afterdischarge, 33
Afterpotential, 25
   negative, 25, 29
   positive, 25, 29, 38, 41
Age, 8, 508
   accommodation of vision in, 502
   adipose tissue, 464
   alveoli, 317
   androgens, 542

642 Index

Age—(Cont.)
  antibody formation, 269
  arterial compliance, 203
  arterial cross-sectional area, 183
  atherosclerosis, 177
  axillary hair, 548
  blood volume, 201
  bone marrow, 271
  breasts, 544, 549-550
  breathing capacity, 317
  bronchioli, 317
  calcium in arteries, 175, 183
  capillary blood flow, 203
  collagen, 317
    in blood vessels, 175, 183
  compliance of arteries, 183
  compliance of lungs, 305
  coronary vessels, 115
  dead space, 317
  diffusing capacity, 317
  distention of aorta, 175, 203
  elastic fibers, 175
  elastin, 317
  erection, 554, 555
  estrogens, 542
  female, 544
  fetus, 237, 547
  follicle stimulating hormone, 500
  growth, 496
  hair, 549
  height, 542, 544
  hemopoiesis, 271
  homosexuality, 557
  hyperplasia, 8
  hypertrophy, 8
  impotence, 537
  intercourse, 557, 569
  larynx, 548
  lipid, 317
  lung, 305, 317
  lung capacity, 317
  luteinizing hormone, 500
  lymphatics, 548
  male, 545
  mammary duct, 549
  masturbation, 557, 569
  maximum breathing capacity, 317
  menarche, 550
  menopause, 568
  metabolic, 317
  metabolic rate, 468
  micturition, 407
  muscle, 549
  nearpoint, 582
  nervous system, 548
  nocturnal emission, 557
  orgasm, 557, 569
  ovulation, 549
  oxygen consumption, 317
  penis, 544
  permeability of digestive tract, 428
  petting, 557
  pregnancy, 549
  pressure, arterial, 184
  P-R interval, 139
  prostate, 549
  protein need, 429
  puberty, 542, 548-550
  pulse wave velocity, 183
  reproductive system, 548
  renal secretion of penicillin, 404
  residual volume, 302, 305, 317
  respiration, 317
  seminal vesicle, 549
  shoulders, 549
  skeletal, 549
  sleep, 636
  smooth muscle in arteries, 175
  stress, response to, 531
  sweat, 548
  teeth, 543
  testes, 496, 549
  thymus, 544
  veins, 178
  vital capacity, 317
  weight, 542
Agglutination, 263
Agglutinins, 262-270
  absorption, 428
Agglutinogens, 262-270
Agranulocyte, 273
  see also Lymphocyte, monocyte
Air swallowing, 415, 441
Airway, 294, 309
Alanine, 431
Albumin, 220
  analbuminemia, 255
  digestion, 430
  filtration of, 223, 360
  glomerular filtrate, 358
  osmotic pressure, 220
  plasma, 220
  plasma binding function, 258, 504, 525
  renal reabsorption, 366
  specific permeability to, 217
Albuminuria, 534
Alcohol, gastric secretion, 448
  intestine, 450
Aldocorticosterone. See Aldosterone
Aldosterone, 198, 205, 386, 500
  edema, 566
  fecal sodium and potassium, 441
  function, 388
  pregnancy, 566
  protein binding, 393
  secretion, 500, 528
  sodium deprivation, 446
  see also Hormone
Alimentary canal. See Digestion
Alkaline solution, 6
Alkalosis, 397, 403
Allergy, 427
All or none law, 82
  action potential, 40, 45
  end plate potential, 45
Alloxan, 515
Alpha cells of pancreas, 511
Alpha fibers, 27, 82
  see also Neuron
Alpha receptors, 62
  see also Receptor(s)
Altitude, atelectasis, 302
  atmospheric pressure, 349
  mountain sickness, 349
  Korotkow sounds, 348
Alveolar gasses, 279, 294
  atmospheric pressure, 350
Alveolus, 294, 309, 314
  concentration of gases, 348
  pH, 348
  ventilation, 317, 348
Amenorrhea, 549
Amines, renal excretion, 404
Amine hormones, 490, 521, 522
Amino acids, absorption, 428
  essential, 464
  hormones, 490
  metabolism, 462
  plasma concentration, 430, 462
  plasma nitrogen, 430
  renal transport, 366
  sequence, in glucagon, 513
    in growth hormone, 495
    in insulin, 511
  urinary nitrogen, 463
Amino acid nitrogen, lymph, 222
  plasma, 430
Aminopeptidase, 429
Ammonia, 331, 462
  renal secretion, 372, 401-403
  urine, 359, 401
Amnion, 566
Amniotic fluid, 237-241
Amphoteric, 398
Ampulla of semicircular canals, 609-613
Ampulla of Vater, 455
Amygdala, 624
  micturition, 408
Amylnitrite, 117
Anabolism, 463
  see also Metabolism
Anacrotic pressure limb, 119
Anal reflex, 339
Anal sphincter, 424
Anaphylactic shock, 219
Anastomosis, 173
  see also Arteriovenous anastomosis
Androgen, 501, 528
  adrenal, 522, 529
  calcification, 550
  female, 550
  hair, 550
  larynx, 550
  pregnancy, 566
  sebaceous glands, 550
  see also Hormone
Androstenedione, delta 4, 501, 524
  see also Hormone
Androsterone, 524
Anemia, macrocytic, 440
  pernicious, 440
  respiration, 328
  sickle cell, 212
Anesthesia, hypoxia, 195
  respiration, 294, 324
  vagal tone, 144

ventilation, 325
Aneurysm, 126
Angina pectoris, 116, 177
Angiocardiography, 113
Angiotensin, 198, 383, 388
   see also Hormone
Angiotensin I, 197
Angiotensin II, 534
Angiotensinase, 383
Angiotensinogen, 383, 388, 534
Angle, of incidence, 580
   of refraction, 580
Angstrom, 3
Anhidrosis, 67
Anion, 8
Anisotropic band in muscle, 74
Annulospiral endings, 28
Anode, 53
Anovulatory menstruation, 568
Anoxia. See Hypoxia
Antagonist, insulin, 516
Antiacetylcholine esterase, 43, 423
Antibiotics, 439, 441, 450
   action on fecal composition, 439
   hearing loss, 605
   intestinal flora, 441
Antibody, 267-269, 427
   growth hormone, 496
   insulin, 511, 518
   thyroid gland, 509
Anticholinergic agent, 43
Anticholinesterase, 43
Anticoagulant, 247
Antidiuretic hormone (ADH), 205, 369, 391, 493
   destruction, 529
   see also Hormone
Antidromic stimulus, 47
Antihistamines, 449
Antiperistalsis, 420
Antithrombin, 249
Antrum of stomach, 415
Antyllus, 174
Anxiety, 531
Aorta, 182, 203
   distention, 170
Aortic arch, 189
Aortic bodies, 196, 326
   cardiovascular control, 195
Aortic chemoreceptors, 327
Aortic insufficiency, 113, 123
Aortic stenosis, 113, 123
Aortic valves, 120
Aphagia, 469
Apneusis, 322
Apneustic center, 322
Apoferritin, 440
Appearance time for a dye, 132
Appetite, 470
Aqueous humor, 145, 233, 326
   pH, 397
   refractive index, 580
Arachidonic acid, 464
Arachnoid membrane, 226
Areas in brain association, 628
   auditory, 608

Brodman's, 626
   motor, 629
   sensory, 627
   somatic, 628
Argenine, 430
   renal reabsorption, 366
Argon, diffusion characteristics, 316
Argyll-Robertson pupil, 596
Aristotle, 491
Arousal, sexual, 553, 569, 571
Arrhythmia, effect of catecholamines, 62
Arterial pressure, 123, 207, 299
Arteriole, 122, 163, 186
   capillary pressure, 221
   efferent, 221
Arteriosclerosis, 136, 516
   heart, 175
   pulse pressure, 184
Arteriovenous anastomosis, 163, 169, 173-174, 186
   cutaneous, 474
   murmur, 127
Arteriovenous oxygen difference, 224
   chemoreceptors, 328
   heart, 114
Artery, 163
   aorta, 170
   arcuate, 355
   blood volume, 202
   common carotid, 175, 188, 325
   coronary, 115
   femoral, 123
   interlobular, 355
   occipital, 194
   pharyngeal, 194
   vertebral, 175, 325
Arthritis, urate crystals, 367
Artificial respiration, 300
   drowning, 345
Ascites, blood stasis, 202
   malnutrition, 221
Ascorbic acid. See Vitamin C
Aselli, Gasparo, 223
Asphyxia, 145
Association cortex, 628
Asthenia, water intoxication, 396
Asthma, expiration, 311
   respiration, 333
Astigmatism, 585
Astrocytes, 229
Ataxia, 613
   water intoxication, 396
Atelectasis, 308
   Herring-Breuer reflex, 331
   oxygen concentration, 353
Atherosclerosis, 177
   coronary, 116
Athlete, 89, 140
   electrocardiogram, 140
   heart, 139
Atmospheric pressure. See Pressure, barometric
Atresia, 561
Atrial diastole, 119
Atrial systole, 119, 120, 142

Atrioventricular dissociation, 142
Atrioventricular node, 95
   conduction rate, 94
Atrioventricular valves, 118
Atrium, 59
   artificial pacing, 101
   catecholamines, 64
   conduction rate, 95
   electrocardiogram, 99
   fibrillation, 100
   retrograde conduction, 100
Atrophy, adrenal, 521
   denervation, 68
   renal, 406
   skeletal muscle, 82
Atropine, 66, 68, 445
   gallbladder, 456
   hepatic secretion, 454
   pancreas, 452
   stomach, 144, 448, 449
   vagal tone, 456
   vomiting, 420
Audibility threshold, 601
Audiometry, 601
Audition, 22
Auditory areas in the cerebral cortex, 608
Auditory ossicles, 602
Auerbach's plexus, 410
Auricle. See Atrium
Autoantibodies, 509
Autonomic innervation, 57
Autonomic nervous system, 55-69
   see also Parasympathetic, Sympathetic
Antagonism, 57
Autoprothrombin I, 246
Autoprothrombin II, 246
Autoprothrombin III, 246
Autoprothrombin C, 243, 246
Autoregulation, 57, 134, 198-201
   heterometric, 134
   homeometric, 136
   renal, 380
Autosomes, 546
A-V. See Atrioventricular and Arteriovenous
aVf. See Electrocardiogram
aVl. See Electrocardiogram
A-V node. See Atrioventricular node
Avogadro's number, 4
aVr. See Electrocardiogram
a wave, 120
Axis of the heart, 104
Axis deviation of the heart, 105
Axon, 20, 44
Axoplasm, 44

b fibers, 29
   see also Neuron
Bacteria. See Microorganisms
Bainbridge, Francis A., 190
Barberio's test, 552
Barbiturate, 211
   excretion, 404
   vagal tone, 144

Barium meal, 420
Barodontalgia, 350
Barometeorism, 350
Barometric pressure, 348
Baroreceptor, 188, 327
  arterial, 335
  respiratory control, 324, 335
Barosinusitis, 350
Barotitis, 350
Barr body, 546
Barriers, alveolar-capillary, 315
  basement membrane, 217
  blood-aqueous, 233
  blood-brain, 230
  blood-cerebrospinal fluid, 329
  digestive tract, 427, 440
  endothelial cell, 217
  intercellular junction, 217
  placental, 566
  plasma membrane, 37
  skin, 579
  sperm, 568
  temperature, 472
Bartholin's glands, 572
Basal ganglia, 630
Basal metabolic rate (BMR), 467
Basement membrane, 213, 214
  capillaries, 435
Bases, weak, renal secretion, 373
Basilar membrane, 236
Basophil cell, 273, 521
Bayliss, William M., 199, 452, 491
Bed rest, 189
BEI. *See* Iodine, butanol extractable
Bernard, Claude, 43, 410, 487
Bernoulli, Daniel, 166
Bernstein, Julius, 39
Berthold, Arnold A., 491
Beta cells of pancreas, 511
Beta granule, 513
Beta receptors, 62, 63, 194
Betaine, 430
Bezold-Jarisch reflex, 190
Bicarbonate, acid-base balance, 398-400
  atmospheric pressure, 350
  carbonic anhydrase, 454
  chemoreceptors, 321
  concentration, 9, 437, 439
  exercise, 343
  fat digestion, 432
  pancreatic juice, 453
  renal reabsorption, 399
  treatment for drowning, 347
  urinary, 386
  urine, 359
Bifocal lenses, 583
Bigeminy, 106
Bile, 432
  composition, 433, 437
  fat digestion, 432
  function, 452
  secretion, 453
  volume, 443, 452
Bile duct, obstruction, 454
Bile pigments, 432

Bile salt, 428, 432, 433
  circulation and secretion, 454
Biliary stasis, 455
Bilirubin, atmospheric pressure, 350
Biliverdin, 433
Binding in amnionic fluid, 240-241
Binding in blood, 257, 527
  cholesterol, 460
  cortisol and corticosterone, 527
  free fatty acids, 460
  hemoglobin, 366
  inulin, 360
  phospholipids, 460
  protein in pregnancy, 566
  protein-bound iodine, 504-506
Binocular vision, 589
Biotin, 441
Bipolar neurons, 19
Black body, 474
Bladder. *See* Urinary bladder, and Gall bladder
Blastocoel, 562
Bleeding disorders, 253
Bleeding time, 252
Blindness, 516
  color, 589
  glaucoma, 235
  night blindness, 587
  oxygen toxicity, 353
  red green, 589
Blind spot, 587
Block, alveolar-capillary, 315
  atrioventricular, 102
  conduction, 43
  vagal, 338
Blockade, carbonic anhydrase, 376
  secretion of para-aminohippurate, 373
  secretion of phenol red, 373
Blocking agents, 68, 373
  phlorizin, 364
Blagden, Charles, 475
Blood, 178-180, 254-284
  acetone, 256
  albumin, 220, 366
  amino acids, 366, 462
  amino-nitrogen, 430
  atmospheric pressure, 349-350
  butyrate, beta-hydroxy-, 366
  calcium, 240, 538
  cholesterol, 178
  coronary sinus, 458
  corticosterone, 525
  cortisol, 525, 530
  density during drowning, 346
  estriol, 240
  estrogen, 560
  euglobulin, 220
  fatty acids, 461
  fibrinogen, 220
  follicle-stimulating hormone, 560
  formed elements, 270-274
  globulins, 220
  glucose, 470
    see also blood, sugar
  glycerol, 461

  gonadotrophin, chorionic, 564
  growth hormone, 542
  hemoglobin, 366
  hydroxycorticoids, 17-, 532
  ketones, 459
  lactate, 87, 366, 457
  lipids, 460
  malate, 366
  oxygen, 348
  pH, 397
  phosphate, 366
  potassium, 107, 240
  preservation, 260
  relaxin, 567
  reservoirs, 202
  stasis, 202
  sugar, 222, 366, 457, 470, 511, 518
  sulfate, 366
  sulfobromophthalein, 453
  testosterone, 566
  tonicity, 205
  uric acid, 366, 463
  vitamin C, 366
  vitamin D, 537
    see also Plasma
Blood aqueous barrier, 233
Blood brain barrier, 230
  cortisol, 266, 529
Blood cell aggregates, 212
Blood cell formation, 266
Blood cerebrospinal fluid barrier, 329
Blood clot, 242-249
  filtration, 202
  infarction, 202
Blood flow, 178-185
  abdominal viscera, 191
  adrenal, 521
  aneurysm, 126
  aorta, 114
  brain, 191, 194, 292
  bronchial artery, 175
  coronary, 62, 117, 133
  distribution, 191
  digestive organs, 292
  duration of ejection, 111
  effective renal, 375, 377
  effective renal plasma flow, 375
  exercise, 191
  heart, 191, 194, 292
  heart failure, 382
  hepatic artery, 194
  hepatoportal, 432
  kidney, 191, 194, 292, 378, 379-385
  liver, 292, 453
  mean ejection rate, 111
  mean ejection velocity, 111
  mesenteric artery, 194
  muscle, 194
  portal vein, 194
  pulmonary, 122, 315
  pulsatile, 182
  renal, 379
  retrograde, 175
  salivary gland, 444
  skeletal muscle, 191, 292
  skin, 191, 194, 292, 474

spleen, 194
stenosis, 126
stroke volume, 111
stroke volume index, 111
uterine, 240
velocity, 180
  peak rate of change of, 111
  in aorta, 120
venous, 177
see also Anastomosis and Ischemia
Blood groups, 261-266
Blood plasma, 255-261
see also Plasma
Blood platelets. *See* Platelet
Blood pressure, 122, 179
  arterial, 127
  filtration, 381
  heart, 123
  kidney, 362
  pulsatile, 182
  venous, 177
Blood pressure determination, 127
Blood reservoir, 186
Blood transfusion, 259-266
Blood vessels, 168-178
  cardiac, 115
  collateral, 175
  cross-sectional area, 180
  effect of catecholamines, 64
  hemostasis, 251
  intestinal, 169
  intramural injury, 178
  mesentery, 169
  skin, 169
  smooth muscle, 169
  stress, 169
  tone, 186
Blood volume, 201-206
  atmospheric pressure, 350
  distribution, 179, 202
Blurred image, 590
Body fluid, 9
  see also Fluid composition
Bonds, disulfide, 495, 511
  peptide, 511
Bone, 537, 543
  cartilaginous, 544
  membranous, 543
  see also Epiphyseal plate
Bone marrow, 271
  sinusoids, 213
Boney labyrinth, 609
Botulinum toxin, 45
Bouancy, 227
Bowditch, Henry Pickering, 82, 284
Bowel movements. *See* Defecation
Bradycardia, 142, 189
  see also Heart
Bradykinin, 219, 444, 476
Bradykinin-releasing enzyme, 444
Brain, blood flow, 191-194, 292, 353
  blood glucose, 458
  glycogen stores, 293
  inhibitory systems, 632
  oxygen, 291
  transection, 319

ventricles of, 266
Brain areas. *See* Areas in brain
Brain centers. *See* Centers
Brain damage, blood sugar, 518
  diabetes, 518
  ischemia, 292
  temperature, 477
Brain stem, circulatory control, 208
  temperature regulation, 478
  transection, 208, 209, 412
Braun Menendez, E., 383
Breast, 543, 546, 549, 550, 572
Breath holding, 343
Breathing, abdominal, 294
  Cheyne-Stokes, 326
  Kussmaul, 331
  mouth, 293
  work, 311
  see also Respiration
Bretyllium, 66, 68
Brodman's cortical fields, 626
Bromosulphalein, 453
Bronchi, 309
Bronchioles, 294, 309
Bronchiolitis, 311
Brownian movement, 9
Bruits, 126
Buccal gland, 443
Buffalo hump, 530
Buffer, 283, 397
  plasma, 257
Bulb, olfactory, 621
Bulk in the diet, 423
Bundle branch block, 99
  heart sounds, 125
Bundle branches of the heart, 95
  conduction rate, 94
Butanol, 620
Butyrate, beta hydroxy, 401
  renal reabsorption, 366
Butyric acid, 620

C fibers, 28
Caffein, 136
  gastric secretion, 448
Caisson's disease, 350
Calcification, 544
Calcitonin, 539
Calcium, 8, 137, 151, 389, 439, 536-540
  absorption, 428
  acetylcholine release, 44
  activation-contraction coupling, 45, 78
  as a procoagulant, 242, 248
  blood concentration, 390
  conservation, 537
  distribution, 536
  excretion, 537
  heart, 138
  hemostasis, 245
  hemostatic function, 243
  intracellular concentration, 151
  intracellular release, 76
  lymph, 222

plasma, 222
sequestration, 536
stabilization of cell membrane, 536
urinary concentration, 390
urine, 359
Calculi, urate, 367
Calorie, 3, 88, 465
  intake, 428, 469
  stimulation of ear, 612
  values, 465
Calorigenic hormones, 478
Calorimetry, 464-469
Canal, alimentary, 410
  central, 226
  membranous, 609
  semicircular, 610-614
Cancer, prostate, 552
Cannon, Walter, 68
Cannon wave, 141
Capacitance, 3, 170
  arterial, 183
  veins, 204
Capillary, 122, 212–224
  basement membrane, 435
  blood, 163
  blood volume, 202
  fragility, 252
  glomerular, 356
  heart, 116
  hydrostatic pressure, 219
  lymph, 164
  peritubular, 356
  permeability, 219
  sphincter, 168
Capsule, internal (brain), 630
Carbaminohemoglobin, 282
Carbohydrate (CHO), 428, 429, 457
  digestion, 428
  gastric secretion, 448
  kidney, reabsorption, 364
  metabolism, 506
  specific dynamic action, 469
Carbon dioxide, blood flow, 192, 200
  carotid bodies, 195
  concentration during sleep, 324
  diffusion characteristics, 316
  exercise, 343
  heart, 291
  oxygen toxicity, 354
  production, 291
  respiratory control, 321, 326
  tolerance to, 275, 344
  transport, 275, 282-283
  unconsciousness, 345
  vascular control, 199
  ventilation, 348
Carbon monoxide, 282
  diffusion characteristics, 316
Carbonic acid, 229
Carbonic anhydrase, renal, 399
  renal secretion, 376
Carbonic anhydrase inhibitor, 228, 399, 454
Carboxypeptidase, 429, 511
Carcinomatosis, expiration, 311, 429, 511

Cardiac cycle, 119
    see also Heart
Cardiac glycosides. See Digitalis
Cardiac index, 112, 129
Cardiac massage, 145
    drowning, 345
Cardiac output, 129-146, 186
    control of, 186-190
    determination, 130
    distribution of, 190-194
    effect of catecholamines, 62
    heart rate, 144
Cardiac sphincter, 413, 415
Cardiogreen, 130
Cardiopathy, 136
Caries, dental, 441
Carinamide, penicillin secretion by kidney, 374
Carotid artery, 325
Carotid bodies, 194, 326
Cardiovascular control, 195
    chemoreceptors, 327
Carotid sinus, 188
Carrier molecules, 10, 363
Cartilage, respiratory tract, 309
Cartilaginous plate, 497, 544
Conjunctiva, 581
Casein, 429, 430
Castration, 491, 501, 552
Catabolism, 410, 457-463
    nucleoproteins, 367
    pH, 397
    water formation, 390
Catacrotic pressure limb, 120
Catecholamine, 61, 521, 522
    exercise, 343
    venoconstriction, 204
    see also Epinephrine, Norepinephrine, Isoproterenol
Catechol-o-methyl transferase, 61
Cation, 8
Catheterization, cardiac, 123
Cathode, 53
Caudate nucleus, 629
Cecum, 6, 74
Cell, 6-14, 74
    acinar, 444
    amacrine, 586
    astrocyte, 229
    bipolar, 586
    cardiac muscle, 116
    endothelial, 214
    epithelial, 359
    follower, 17, 92
    ganglion, 586
    glial cell of Muller, 586
    goblet, 456
    hair, 605, 611
    heart, 116
    horizontal, 586
    Kupffer, 213
    liver, 7
    mast, 198, 449
    membrane, 8, 19, 278, 514
        action of insulin, 514
    mucous, 443

muscle, 73, 149
myoepithelial, 445
myofilaments, 149
nucleus, 513-515, 546
oxyntic, 447
pacemaker, 92
parenchymal, 213
parietal, 447
pericyte, 214
perivascular, 218
plasma, 218
potential. See Potential
proliferation, 440
Renshaw, 33
respiration, 291
Rouget, 213
satellite, 21
Schwann, 20
serous, 443
taste, 618
warm, 478
Cellulose, 441
Cement, 522
Center, 208
    apneustic, 322
    cardioinhibitory, 196
    defecation, 424
    descending reticular facilitatory area, 32
    descending reticular inhibitory area, 31
    erection, 554
    expiratory, 319, 322
    feeding, 469
    heat loss, 478
    heat maintenance, 479
    hypothalamus, 209
    inspiratory, 319, 322
    medullary, 319
    ossification, 554
    pneumotaxic, 319, 322
    pontine, 319
    rage, 211
    respiratory, 319
    reticular activating system, 323
    salivation, 446
    satiety, 469
    spasmodic respiratory, 322
    spinal sympathetic, 208
    swallowing, 412
    thermoregulatory, 477
    vasoconstrictor-cardioaccelerator, 208
    vasodepressor-cardioinhibitory, 208
    vasomotor, 196
    visual, 595
    vomiting, 420
    see also Areas in brain
Central nervous system. See Brain and Spinal cord
Cephalic phase of gastric secretion, 447, 448
Cephalin, 460
Cerebellar peduncles, 319
Cerebellum, 55, 607, 612, 632
    control of alpha efferent neuron, 31

Cerebral aqueduct, 228
Cerebral cortex, 625-637
    auditory areas, 608
    cardiovascular control, 210
    micturition, 407
    paracentral lobule, 407
    respiration, 324, 625
Cerebrospinal fluid (CSF), 226-231
    pH, 331, 397
    respiratory control, 329
Cerebrum. See Brain
Cervix of uterus, 553, 566
cgs system, 3
Chalone, 418
Chelating agent, 538
Chemoreceptor, 326-331
    central, 196, 321, 326
    peripheral, 194
    respiratory control, 321, 324
    stimulation, 328
    trigger zone, 420
    venous, 342
Chemotaxis, 267
Chenodeoxycholic salts, 433
Chewing, 411
Cheyne-Stokes breathing, 326
Chiasma, optic, 593
Chief cells of stomach, 429
Chlorea, 438
Chloride, 229, 385
    absorption, 428
    blood brain barrier, 230
    cell membrane permeability, 36
    drowning, 346
    equilibrium potential, 37
    lymph, 222
    plasma, 222
    renal reabsorption, 362, 428
    urine, 350
Chloroform, effect of Herring-Breuer reflex, 331
Chlorothiazide, renal secretion, 373
Cholecystectomy, 455
Cholecystokinin, 456
Choleretic, 455
Cholesterol, 432, 451, 459, 516, 522
    atherosclerosis, 178
    coronary heart disease, 178
    gallstones, 455
    metabolism, 507
    synthesis, 523
    transport, 223
Cholic salts, 433
Choline, 47, 552
    renal secretion, 373
Cholinergic, 68
Cholinergic sympathetic neuron, 478
    pancreas, 451
    salivary gland, 445
    see also Sympathetic neuron
Chondroitin sulfuric acid, 509
Chordae tendinae, 118
Chorion, 567
Chorionic gonadotropins (CG), 502, 562
Choroid membrane in eye, 232, 581

Index 647

Choroid plexus, 226
Choroidal epithelium, 228
Christmas factor, 246
Chromaffin cells, 521, 522
Chromatin, 546
Chromophore, 588
Chromosome, 12, 546, 551, 560
    abnormal, 547
Chronotropic agents, 140
    epinephrine, 197
Chvostek sign, 534
Chylomicrons, 434
Chyme, 435-436
Chymotrypsin, 429
Chymotrypsinogen, 429
Cilia, respiratory tract, 293
Ciliary body of eye, 233
Ciliary muscle, 581-582
C_{In}. See Clearance, inulin
Cineradiography, 132
Circle of Willis, 175
Circulation of blood, 163-225
    greater, 122
    lesser, 122
    systemic, 122
    see also Blood flow
Circulatory collapse, 532
Circus movement, heart, 109
Cisterna chyli, 222
Cisternae, cerebellomedularis
    (magna), 226
    chiamatis, 226
    magna, 226
    muscle, 73
    pontis, 226
    superior, 226
Citrate, 247
    kidney, 367
    renal reabsorption, 367
    sodium, activation of prothrombin, 243
Citric acid, 551
Claustrum, 630
Clearance (C), 362, 371
    alarm reaction, 381
    diodrast, 381
    exercise, 381
    glucose, 375
    heart failure, 382
    inulin, 360, 361, 370, 381
    para-aminohippurate, 375
    urea, 362
Climacteric, 568
    estrogens, 568
    follicle-stimulating hormone, 568
    hot flash, 568
    obesity, 568
    osteoporosis, 568
Clitoris, 554, 571
Clot. See Blood, clot
Clot formation, 242
Clot promoting factor, 246
Clot retraction, 250
Clotting time, 247, 251
Coagulation of blood, 242
Cobalt, 440

Cocaine, action on spinal cord, 208
    decentralization of stomach, 449
Coccygeal ligament, 226
Cochlea, 236
Cochlear aqueduct, 235
Cochlear duct, 603
Cochlear partition, 603
Coenzyme a, 460
Coffee, diuretic action, 393
Coitus, 574
Cold, 22
    exposure, 498
Colectomy, 425
Colic, renal, 406, 539
Collagen, 169
    heart, 134
Collateral circulation, 175
Colliculi of the midbrain, 323
Colloid osmotic pressure, 221
Colon, 410, 423, 437
    microorganisms, 439
    potassium absorption, 437
    sodium absorption, 437
Colonic juice, volume, 443
Color, 586-589
Colostrum, 427
Coma, 292, 516
Commisure, 621, 623
Compensation, acidosis, 403
    alkalosis, 403
    renal, 398
    respiratory, 398
Compensatory pause, 96
Competitive inhibition, 45, 530
Complex formation, 257
Compliance, 3, 170, 183, 303
    blood vessels, 203
    specific, 304
    urinary, 407
Composition, chylomicron, 435
    see also Fluid composition
Concave lens, 584
Concentration, 4, 23
    effect on diffusion, 10
    see also Fluid composition
Conditioning, respiration, 341
    salivation, 446
Conductance, 17, 201
    potassium, 41
    skin, 476
    sodium, 39
    vascular, 201
    see also Resistance
Conduction, 43
    atrioventricular delay, 140
    auditory, 602
    block in heart, 108
    heart, 95, 416
    intercellular, 43, 92, 108, 152
    intracellular, 12
    nervous, 25, 42
    neuromuscular junction, 43
    neuron, 23, 42
    olfactory bulb and tract, 621
    reentry, 109
    saltatory, 43

    stomach, 416
    synapse, 43
    velocity, 43
    velocity in heart, 95
    velocity of pressure wave, 184
    volume conductor, 103
    see also Transmission
Conduction velocity, role of ions in the heart, 137
Conductive heat loss, 473
Conductivity, skin, 473
    temperature, 472
Cones of retina, 586
    absorption of light, 588
Congenital defects, heart, 123
Congestion, 121
    pulmonary, 333
Conjunction, 581
Connective tissue, heart, 116
    muscle, 73
Consumption, oxygen, 313
Contact factor, 246
Contraception, 563
Contractility, 85
Contraction, 71, 79
    force, 138
    gastric, 415, 449
    heart, 140
    intestine, 416
    isometric, 71
    isotonic, 71
    isovolumic, of heart, 110, 380
    oviduct, 568
    premature, 96
    role of calcium, 45
    role of ions in the heart, 137
    stomach, 416
    time, 79, 81
Contracture of muscle, 85
Control, 17-69, 186-211, 318-339, 443-456, 487, 541, 594-637
    see also specific categories
    adrenal cortex, 198, 205, 388, 500, 525, 534
    aldosterone secretion, 533
    arterial pressure, 187, 189, 198, 205
    arterial tone, 204
    blood flow, 194
    blood pressure, 189
    blood sugar, 525
    blood vessels, 192
    blood volume, 205
    body temperature, 472, 477
    calcium in the blood, 540
    capillary filtration, 221
    carbohydrate metabolism, by pancreas, 519
    carbon dioxide in blood, 196
    cardiac output, 186, 189, 197
    cardiovascular centers, 196
    cardiovascular system, 207
    circulation, 186
    colon, 424
    defecation, 424
    endocrine system, 487
    extracellular fluid volume, 534

Control—(Cont.)
  filtration, 221
  gallbladder, 455
  gastric secretion, 446
  gastrointestinal tract, 419
  heart rate, 189, 208
  hormone release, 489, 518
  hypothalamus, 492, 500
  insulin secretion, 514, 517
  intestinal motility, 423
  iron absorption, 439
  kidney, 205, 380-393
  lactation, 502
  mammary glands, 502
  mastication, 412
  menstrual cycle, 563
  micturition, 408
  milk production, 502
  motor, 627
  osmotic pressure, 205
  ovary, 501, 562
  oxygen in blood, 195
  pancreatic secretion, 451
  peripheral resistance, 189, 201
  permeability, 218
  pH, 397-403
  phosphate in plasma, 539
  pituitary, 205, 489, 498
  plasma K levels, 388
  plasma Na levels, 388
  renin secretion, 388
  respiration, 318-345
  salivary glands, 443
  secretion, 443
  secretion of bile, 452
  secretion of insulin, 514
  sensory neurons, 188, 195
  sexual development, 502, 549
  skeletal muscle, 55
  sodium concentration, 198, 205
  stroke volume, 189
  swallowing, 412
  testis, 502
  thirst, 534
  thyroid, 498
  urinary bladder, 153, 407
  uterus, 153
  vasoconstriction, 194, 198
  venous tone, 204
  vomiting, 420
  water and electrolyte balance, 387-392
Control of blood vessels, axon reflex, 383
  metabolic theory, 383
  myogenic theory, 383
  tissue pressure theory, 383
Conus medullaris, 226
Convection, 474, 1404
Convergence, 27
Convergence of the eyes, 34, 591
Convex lens, 584
Convulsion, water intoxication, 396
Copp, 540
Copulation, 554, 569
Cord. See Spinal cord

Cord bladder, 408
Cori cycle. See Glycolysis
Cornea, 581, 585
Corrification, 568
Coronary blood flow, 117
Coronary sinus, 115, 458
Coronary spasm, 116
Coronary vessels, 115
  occlusion, 338
Corpora cavernosa, 554
Corpus albicans, 560
Corpus hemorrhagicum, 560
Corpus luteum, 560, 566
Corpus spongiosum, 554
Corpuscle, 590
  see also Blood cell
Corresponding points, 590
Cortex, Brodmann's fields, 626
Cortex, occipital, striate area, 594
Cortical columns, 627
Corticoid, 522
  see also Steroids
Corticosteroid-binding albumin, 525
Corticosteroid-binding globulin (CBG), 525
Corticosterone, 524, 528
  compounds B, E, and F, 528
  pituitary secretion, 388
Corticotropin, 625
  see also Hormone
Corticotropin-releasing factor (CRF), 500
Cortisol, 500, 524, 525
  see also Hormone
  pituitary secretion, 388
  secretion, 393, 528
Cortisone, 175, 524
  production, 530
  pregnancy, 566
  renal tubular maximum, 367
Cosmic rays, 579
Cothromboplastin, 246
Cough, 297
  cerebrospinal fluid pressure, 231
  control, 322
  respiration, 337
Coughing, 338
Coumarin, 245
Countercurrent heat mechanism, 473
Countercurrent multiplication, kidney, 369
Cournand, Andre F., 123
Cowper's gland, 552
C_{PAH}. See Clearance para-aminohippurate
C_{PK}. See Creatine phosphokinase
Cramps, 537
Cranio-sacral neurons. See Sympathetic neurons
Creatine phosphate (CrP), 78, 87, 462
Creatine phosphate concentration, effect of epinephrine, 151
Creatine phosphokinase, 459
Creatinine, 228, 462
  glomerular filtration, 360
  renal clearance, 371

renal secretion, 371
  urine, 359
Cremaster muscle, 556
Cretin, 508
Critical closing pressure, 172
Critical fusion frequency, 589
Crista ampullaris, 610
Crypt, of Lieberkuhn, 411
Cryptorchism, 551
Cumulus oophorus, 560
Cupula, 610
Cupulometer, 612
Curare, 43, 46, 68
  vomiting, 420
Current. See Conductance
Cushing, Harvey, 231
Cushing's disease, 499, 525, 530
C wave, 119
Cyanide, 327
Cyanide poisoning, carotid bodies, 195
Cyanosis, 329
Cybulski, 491
Cyclopentanoperhydrophenanthrene ring, 522
Cycloplegia, 67
Cylindrical lens, 585
Cystine, 430
Cystinuria, 408
Cystometrogram, 408
Cytochrome, 439
Cytochrome oxidase, 327
Cytofibrinokinase, 247
Cytoplasm, 6, 525
  concentration of ions, 37
Cytoskeleton, 14
Czermak, John, 188

D cells of pancreas, 511
Dale, Sir Henry, 43, 62
Dam, (Carl Peter) Henrik, 248
Dark adaptation, 587
d'Arsonval, Jacques A., 491
Dartos muscle, 551
Dead air space, 298
Death, biological, 145
  clinical, 145
Deceleration, 613
Decentralization, 68
  bladder, 408-409
  colon, 425
  effect on the development of a collateral circulation, 175
  heart, 145
  intestine, 69, 422, 456
  postganglionic, 69
  preganglionic, 69
  salivary gland, 443
  skeletal muscle, 54, 81
  stomach, 449
  ureters, 406
Decerebration, 32, 323, 630
  micturition, 409

Index

Decompression, 350
Defecation, 424, 442
  fiber intake, 442
  voluntary, 424
Defibrillation, 146
Deflation, 331, 424
Deglutition, 324, 412
  see also Swallowing
Dehalogenase, 504
Dehydration, 395
Dehydroepiandrosterone, 501, 524
Delayed heat, 90
Delirium, 537
Delta wave of ECG, 103
Delusions, 537
Demineralization of bone. See Osteoporosis
Denervation, bladder, 408-409
  law of, 68
  see also Decentralization
Density, 3
Deoxycorticosterone, 386
Deoxyribonuclease, 12, 429
Depolarization, 36, 48, 92, 155
  heart, 99
Deprivation, caloric, 461
  see also Starvation
  light, 597
  salt, 437
  sodium, 446
Descartes, Rene, 544
11 desoxycorticosterone (DOC), 522, 524, 528
Desoxyribonucleic acid (DNA), 12, 514, 515, 533
Desquamation. See Sloughing
Destruction, ACTH, 515
  antidiuretic hormone, 515
  epinephrine, 535
  glucagon, 515
  growth hormone, 515
  insulin, 515
  norepinephrine, 535
  oxytocin, 515
  prolactin, 515
Deuterium, 436
Developed tension. See Tension, active
Development, 508, 542
  see also Growth
Dextran, 395
Diabetes, heart, 116
  insipidus, 515
  mellitus, 515-518
  treatment, 518
Dialysis, feces, 439
Diameter. See Radius
Diamine oxidase, 449
Diapedesis, 218
Diaphragm, 92, 295, 318
Diarrhea, 424, 437, 438
Diastasis, 120
Diastole, ventricular, 120
Diastolic filling of the heart, 131
  see also Heart
Diastolic pressure, 184

Diastolic reserve volume, 132
  see also Heart
Dibenamine, 63
Dichloroisoproterenol (DCI), 63, 182
Dicrotic notch, 120
Dicrotic wave, 120
Dicumarol, 249
Dielectric strength, 9
Diet, bulk, 442
  casein, 451
  cholecystectomy, 455
  essential amino acids, 464
  fat free, 464
  fruit, 442
  high protein, 463
  iron, 439
  oxygen consumption, 466
  sodium restriction, 446
  specific dynamic action, 469
Diffusion, 9, 313-316
  block, 312
  rate, gases, 316
  trapping, kidney, 373
Diffusing capacity, 316
Digestion, 289, 410, 427
  enzymes, 429
  pH, optimum, 429
  substrate, 429
Digestive system, 410-470
  absorption, 428
  permeability to water, 435
  secretion, 428
Digestive tract, 410
Digitalis, intoxication, 138
  refractory period, 108
Diglutition, 414
  see also Swallowing
Diglyceride, catabolism, 433
Dihydroepiandrosterone, 523
Diiodotyrosine (DIT), 504
Dimethylglycine, 430
Diodrast, renal clearance, 371
Diopter, 580
Dipeptidase, 429
Dipertide, 429
Dipole, 102
  moment, 103
Disaccharide, 8, 410, 429
Disc, intercalated, 94
Discrimination, frequency of sound, 604
  intensity of sound, 606
Diseases, Addison's, 529
  Cushing's, 499, 525, 530
  ACTH, 499
  Graves', 509
  heart, 178
  Hirschsprung's 425
  see also Syndrome
Dissociation, 5
Dissociation constant, 157
Distensibility, 170, 203
  blood vessels, 170
  see also Compliance
Distention, accommodation, 52
  blood vessels, 170, 188

gastric, 449
heart, 132
intestine, 422
lung, 134
receptor potentials, 52
rectal, 425
skeletal muscle, 74, 83
smooth muscle, 157
stomach, 449, 447
velocity of, 53
Disulfide-disruption, 515
Diuresis, 529
Diuretics, 393
Diurnal rhythm, 477
  ACTH, 499
Divergence, 25
Diving, 335, 344
DOPA, 61
Dopamine, 61
Dorsal root, 155
  transection, 32
Dreams, 636
Dromotropic agents, 141
Dropsy, 387
Drowning, 345
  carbon dioxide, 345
  oxygen, 345
Drugs. See Adrenergic blocking agents, Cholinergic blocking agents, Glycosides, Muscle relaxants, and specific names of drugs
Ductus arteriosus, 123, 317
Ductus deferens, 552
Ductus epididymis, 149
Dumping syndrome, 435
Duodenal cap, 420
Duodenum. See Intestine, duodenum
Dura mater, 226-230
Dwarfism, 495, 508
Dye dilution techniques, 132
Dynamics, fluid, 165
  pulmonary, 302
Dyne, 3
Dysbarism, 602
Dyspnea, 108

Ectopic rhythm, 100
Eddies, 126
Edema, 221, 508
  pulmonary, 220, 316
Edinger-Westphal nucleus, 596
EDTA, 538
Effective hepatic blood flow (EHBF), 453
Effective hepatic plasma flow, 453
Effector neuron, 489
Effector organ, 29, 489
Efficiency, heart, 112
  respiratory system, 312
Einthoven, Willem, 104
Einthoven's triangle, 104
Ejaculation, 556
Ejection from heart, 119
Elastic fibers, 172
Elasticity, 171, 303

650   Index

Elasticity—(Cont.)
  blood vessels, 169-172, 175, 183-184, 203
  lens of the eye, 582
  pleura, 297
Elastic modulus, 171
Electrical axis of heart, 104
Electrocardiogram, 96-109
  abnormal, 100, 103, 106, 108
  augmented unipolar limb leads, 97
  bipolar limb leads, 104
  blood potassium, 107
  cardiac hypertrophy, 107
  cardiac injury, 106
  cardiac ischemia, 106, 459
  central terminal, 97
  delta wave, 103
  duration, 100
  duration of events, 101
  exercise ECG, 109
  heart rate, 139
  24-hour ECG, 109
  hypercalcemia, 537
  infarction, 106
  interpretation, 98
  notched P wave, 107
  peaked P wave, 107
  precordial lead positions, 98
  P-R interval, 100
  sequence, 100
  terminology, 99
  unipolar leads, 97
  U wave, 99
  voltage, 102-108
Electrodes, 92
Electroencephalogram (EEG), 348, 633, 634
  hyperventilation, 348
  hypoxia, 348
Electrolyte, 435
  balance, 384-390, 516
  concentration, 9
  kidney, 367
  renal reabsorption, 367
Electromotive force, 36
Electronmicrograph, alveolus, 314
  capillary, 11, 214, 215, 216
  endoplasmic reticulum, 13
  epithelial cell, 8
  glomerulus, 359
  intercalated disc, 94
  liver cell, 6
  muscle, 75, 116, 149, 152
  pinocytosis, 11
  plasma cell, 13
  sinusoid, 213
Electrophoresis, 257-259
  cardiac enzymes, 459
Elephantiasis, 223
Embden-Meyerhof pathway, cardiac infarction, 87, 458
Embryo, 238, 562
  hemopoiesis, 271
Emesis. See Vomiting
Emiocytosis, 511
Emmetropia, 306, 330, 584

Emphysema, 295, 306, 330, 347
End diastolic volume, 129, 135
  see also Heart
Encephalogram, air, 227
  electroencephalogram, 348, 633
Endocrine glands, 487-491
  see also Hormones
Endolymph, 235-237, 603, 609-612
Endolymphatic duct, 609
Endometrium, 560, 563
Endomyseum, 73
Endoplasmic reticulum, 12, 433
End plate, 44
  potential (EPP), 45, 87, 88
  miniature, 46
Energy, 3, 88
  gravitational, 118
  heart, 135
  kinetic, 118
Energy balance, 457
  see also Work
Enterocrinin, 456
Enterogastric reflex, 418
Enterogastrone, 418
Environment, 84, 137, 198
  altitude, 350
  temperature, 472-480
Enzymes, amylase, 451
  colon, 456
  digestive tract, 429
  intestine, 456
  lipase, 451
  mitochondrial, 459
  proteolytic, 451
  synthesis, 528
Eosinophil, 273
Epiandrosterone, dehydro, 501
Epidermal thinning, 528
Epididymis, 149
Epiglottis, 413
Epimyseum, 73
Epinephrine, 60-64, 534
  actions, 62, 64, 194, 520
  alpha receptor stimulation, 64
  artery, 151
  beta receptor stimulation, 64
  blood sugar, 457
  calorigenic action, 478
  cardiac irritability, 108
  carotid sinus, 188
  cold environment, 473
  effect on metabolites, 151
  effect on muscle tension, 151
  heart, 197
  intestinal motility, 423
  liver, 453
  renin secretion, 388
  respiration, 343
  salivary glands, 445
  sensitivity to, 507
Epinephrine reversal, 62
Epiphyseal plate, 8, 497, 545
Epiphysis. See Pineal gland
Epithelial cell, 8
Epithelial receptor, 335
Epithelium, alimentary, 411

Equilibrium, 611-615
Equilibrium length, 76
Equilibrium potential, bicarbonate, 37
  chloride, 37
  hydrogen, 37
  potassium (E_K), 37, 41
  sodium, 37
Equivalent, 4
Erectile tissue, 554
Erection, 554
ERG, 3, 88
Ergot, 62
Erythrocyte, 163, 275-284
  atmospheric pressure, 350
  fragility, 277
  oxygen consumption, 260
  storage, 260
Erythrophagocytosis, 267
Erythropoiesis, 266, 350
Erythropoietin, 349
E_s. See Potential, transmembrane
Esophagotomy, 390
Esophagus, 413-415
Essential amino acids, 464
Esterase, antiacetylcholine, 423
Estradiol, 523
Estrogen, 560, 565
  adrenal, 501, 502, 522
  atherosclerosis, 178
  water balance, 393
  see also Hormone
Ether, effect on Herring-Breuer reflex, 331
Ethylene, diffusion characteristics, 316
Etiocholanolone, 524
Euglobulin, 220
Eunuch, 543
Euthyroid, 503
Excitability, 17, 96, 155, 537
  heart, 138, 142
  negative afterpotential, 25
  smooth muscle, 149
  supernormal, for heart, 99
Excitatory postsynaptic potential (EPSP), 46, 47
Excretion. See Urine
Exercise, 87, 130, 145, 199, 395, 531
  arterial pressure, 130
  A-V oxygen difference, 224
  blood flow, 191
  blood lactate, 87
  body temperature, 474
  caloric intake, 469
  cardiac output, 130
  changes in the arterial blood, 343
  fatty acids, 461
  heart, 145
  heart failure, 382
  heart rate, 130
  heat loss, 474
  kidney, 381
  metabolism, 381, 461, 468
  muscle metabolism, 86
  oxygen consumption, 87, 130
  rectal temperature, 476

renal function, 381
respiration, 313, 341, 461
skeletal muscle, 292
stomach, 448
sweating, 475
urine production, 379
ventilation, 312
water loss, 390
Exocrine system, 429, 475
Exophthalmic goiter, 509
Exophthalmos, 509
Expiration, 297
Expiratory center, 319, 322
Expiratory reserve volume, 301
Extensor muscles. *See* Muscle
Exteroceptors, 22
Extraction, renal, 374
Extrafusal fibers, 27
Extrasystole, 96
Eye, 232, 580-594
effect of catecholamines, 64
movements, 597
refractive elements, 580
third, 544

Facilitation, 32, 34, 51, 67
duration, 51
hormone action, 529
muscle tone, 632
peripheral, 58
postsynaptic neuron, 46
respiration, 320
Factor, intrinsic, 440
see also Hormone
Fainting, 203
Fallot, tetralogy of, 123
Falx cerebri, 659
Faraday number, 37
Far sightedness, 584
Fasiculation, 622
P_{CO_2}, 348
Fasting, 529
Fat, 289, 431-434, 459-462
absorption, 431
catabolism, 429, 514
depots, 532
digestion, 431
gastric secretion, 448
metabolism, 529
specific dynamic action, 469
Fat-solubility, 8, 218
Fatty acid, 429, 433, 459
absorption, 428
free, 516
long chain, 432
unsaturated, 464
Fecaliths, 442
Feces, 441
bile salts, 455
clay-colored, 455
potassium, 437
salt deprivation, 437
tonicity, 439
Feeding center, 469
sham feeding, 448
Females, 560-575

urethra, 405
Feminization of the male, 529
Fenestra, 213
Fenestrated collars, muscle, 73
Fenestration, 214
Fenn effect, 89
Ferritin, 11, 439
Ferrous iron, 440
Fetus, 240, 547
hemopoiesis, 271
Fever, 472
respiration, 347
Fibrillation, 93, 109, 142
skeletal muscle, 81
ventricular, 100
Fibrin, 247-248
stabilizing factor (FSF), 246
Fibrinogen, 247-248, 267
Fibrinolysis, 247
Fibroplasia, retrolental, 353
Fibrosis, alveolar, 305
expiration, 311
lung compliance, 305
Fibrothorax, expiration, 311
Fick principle, 130, 375
liver, 453
Fields, Brodmann's cortical, 626
Filaments, 14
thick, *see* Myosin, 76, 149
thin, *see* actin, 76, 149
Filtration, 219
arterial pressure, 381
glomerular, 360
kidney, 356
Filtration fraction, 377
alarm reaction, 381
heart failure, 382
Five hydroxytryptamine, serotonin, 423
Flaccidity, muscular, 32, 81, 630
Flatus, 441
Flicker critical fusion frequency, 589
Florence reaction, 552
Flow, 3
air, 302, 310
aortic, 299
critical velocity, 126
effective renal, 375
laminar, 126, 178
turbulent, 126, 178
see also Blood flow
Flower-spray endings, 28
Fluid composition, amnionic, 240
aqueous humor, 229
bile, 432, 454
blood plasma, 240
blood plasma dialysate, 229
cerebrospinal, 229, 237, 240
digestive juices, 439
endolymph, 237
erythrocyte, 277
feces, 439
fecal dialysate, 439
gastric juice, 437
hepatic bile, 437
ileal secretion, 437

ileostomy fluid, 437
interstitial, 9, 37
intracellular, 9, 37
intraocular, 232
jejunal secretion, 437
nitrogen gas, 353
oxygen storage, 293
pancreatic juice, 437
perilymph, 237
saliva, 437
sweat, 475
urine, 359, 463
Fluid dynamics, 166
Fluid exchange, 204
Fluorocortisol, 9-alpha-, 528
Fluorodinitrobenzene, 86
Fluoroscopy, 123
Flutter, 96, 142
ventricular, 100
Flux, 10
focal point, 580
Focus, 579-586
Folic acid, 441
Follicle, Graafian, 561-562
thyroid, 505
Follicle-stimulating, hormone, (FSH), 92, 501
releasing factor (FRF), 501
see also Hormone
Follower cell, 92
Food, caloric value, 465
metabolic rate, 469
respiratory quotient, 465
Foot-pound, 88
Foramen ovale, 123, 317
Force, 3, 71
of contraction in heart, Ca/K ratio, 140
end diastolic volume, 140
hypertrophy, 140
parasympathetics, 140
sympathetics, 140
Forebrain, 622-637
Foreign substances, 404
see also Antibody
Formation. *See* Synthesis
Formed elements of the blood, 270
see also Erythrocyte, Leukocyte, Lymphocyte, Monocyte, Platelet
Forssman, Werner, 123
Fovea centralis, 586
Frank, Otto, 135
Free fatty acids, 460
Freezing point, 5
Frequency, audibility, 606
of contraction, cardiac force, 139
critical fusion, 81, 589
discrimination of sound, 604
sound, 125
see also Wavelength
Frontal lobe of cortex, 629
Fructose, 429, 457
Fruits, pH, 397
Functional residual capacity, 301
Fundus of stomach, 415

Funiculus, 622
  urinary bladder sensation, 408
Fusion, critical frequency, 589

Galactose, 429, 457
Gall bladder, 455
  bile, 455
  fat digestion, 432
Gallstone, 455
Galvani, Aloysio, 43
Gamma aminobutyric acid (GAEA), 47, 50
Gamma fibers, 27, 31
Gamma globulin, 427
Gamma loop, 334
Ganglion, basal, 630
  ciliary, 57
  collateral, 55
  intramural, 153
  parasympathetic, 411
  petrosal, 326
  stellate, 59, 69, 142
  submaxillary, 57
  superior cervical, 445
Ganglionic, block, 187
  blocking agents, 67
  intestine, 422
Gangrene, 69, 517
Gas constant, 12
Gases, diffusion rate, 315
  inert, 301
  partial pressure, 300
  viscosity, 310
  see also Carbon dioxide, Oxygen, Nitrogen
Gasping, 318
Gastrectomy, 430
Gastric. See Stomach
Gastrin, 449, 451
  intestinal, 450
  see also Hormone
Gastrocolic reflex, 424
Gastrointestinal tract, 410-456
Gastrojejunostomy, 441
Gastrone, 451
Gelatin, 464
Generator potential, 53
Geniculate nucleus, 593
Genital ridge, 521
Genitalia, 529, 551, 554, 570
Gerard, Ralph W., 92
Get set response, 136
Giant, pituitary, 498
Giantism, 494, 498
Gland, 411, 429, 490
Glaucoma, 235
Globulin, 258, 267
  absorption, 428
Globulin, alpha-2-. See Angiotensinogen
Globulin, beta, 515
Globus pallidus, 630
Glomerular capillary, 219
Glomerulonephritis, 376
Glomerulus, 355, 384
  filtrate, 359

Glossopharyngeal nerves, 326
Glottis, 413, 419
Glucagon, 518-520
  amino acid sequence, 513
  secretion, 519
Glucagonase, 513
Glucocorticoids, 522, 528-533
  see also Cortisol
Gluconeogenesis, 457, 516, 529
Glucosamine, 458
Glucose, 289, 429, 513
  filtration, 360
  kidney, 364, 518
  renal clearance, 371
  renal reabsorption, 366
Glucose concentration, 459
Glucose dependence, 459
Glucostat, 417
Glucosuria, 516
Glucuronides, 433
  renal secretion, 373
Glucuronide, 433, 527
Glutamic acid, 431
Glutamic oxaloacetic transaminase (SGOT), 459
Glutaminase, ammonia formation, 402
Glutathione-insulin-transhydrogenase, 511, 515
Glycerides, 429, 433
Glycerol, 433, 459
Glycerol-extracted muscle, 151
Glycine, 50, 433, 464
  renal reabsorption, 366
Glycogen, 289, 293, 457
  depletion during hypoxia, 293
  muscle, 73
  storage, 292, 458
Glycogenolysis, 63, 518
  effect of catecholamines, 65
  role of adrenergic sympathetic neurons, 60
Glycolysis, cardiac infarction, 87, 458
Goiter, 498, 509
  metabolic rate, 468
Goldblatt, Harry, 383
Golgi apparatus, 14
Golgi, Camillo, 14
Gonad. See Testis and Ovary
Gonadotropins, 500
Goose flesh, 473
Gout, 367, 463
  heart, 116
Graafian follicle, 561-562
Gradient-limited active secretion, 375
Gram-centimeter, 88
Gram equivalent, 5
Gram molecular weight, 4
Granulocyte, 266, 271-273
  fever, 477
Granulomatoses, expiration, 311
Graves disease, 509
Gravity, effect on blood pressure, 176
Grey matter, 21
Growth, 8, 464, 496, 508, 528, 542-551
  blood vessels, 174
  thyroxine, 507

  see also Age
Growth hormone (GH), 494-497
  amino acid sequence, 495
  see also Hormone
Growth hormone releasing factor (GRF), 490
Guanethidine, 66, 182
Guanine, 463

Half life, hormones, 527
  nitrogen, 353
Harvey, William, 122
Haustral segmentation, 423
Haversian canals, 543
Hearing, 599-608
Heart, 76, 92-149, 186-211
  arrest, 136
  arrhythmia, 100
  arterioluminal and arteriosinusoidal vessels, 115
  asystole, 121
  atrial systole, 142
  bradycardia, 142
  blood flow, 115
  calcium, 137
  cardiac output, 129
    see also Cardiac
  catabolism, 458
  catheterization, 123
  collateral circulation, 115
  conduction system, 95
  congestive failure, 387, 534
  cycle, 119
  decompensation, 136, 387
  defects, 123
  diastolic filling and reserve, 131
  dipole, 102
  disease, congestive failure, 387
  efficiency, 112
  ejection fraction, 136
  electrical axis, 104, 341
  electrical properties, 92-96
  electrocardiogram, 96-109
  electrolytes, 9, 107, 136-139, 458
    see also specific names
  end diastolic volume, 129, 135, 136
  failure, 112, 121, 136, 387
  force of contraction, 129
  frequency of stimulation, 137
  glucose extraction, 118
  glycogen stores, 293
  hypertrophy, 129
  infarction and the electrocardiogram, 106
  injury and the electrocardiogram, 106
  inspiratory tachycardia, 142
  intercellular conduction, 92
  intrinsic control, 92
  ischemia, 108, 459-460
  lactate, 118, 458
  lymph ducts, 116
  metabolism, 118, 458
  microcirculation, 116
  nutrition, 114, 458
  pacemaker, 102

position, 103
potassium, 137, 459
power, 113
pyruvate, 118
regurgitation, 118
residual capacity, 132
residual volume, 132
right heart vs. left heart, 121
scar tissue, 108
sinus arrhythmia, 142
size, 144
sodium, 137
stroke volume, 129, 136
systolic reserve volume, 132
systolic volume, 131, 136
veins, 115
venous return, 129
ventricular systole, 142
volume, 131
work, 110, 111
Heart attack, 177
Heart block, 99
   oxygen consumption, 115
Heartburn, 414
Heart failure, 382
Heart-lung preparation, 136
Heart rate, 111, 129, 142-145
   cardiac output, 143
   catecholamines, 62
   drowning, 346
   stroke volume, 143
Heart sounds, 123
Heat, calorimetry, 464-470
   of activation, maintenance and shortening, 88-90
   of vaporization, 475
   waves, 474
   isometric, 88
   loss, 472-476
   production, 90, 472
   receptors, 22
Heavy metals, 289
Heidenhain pouch, 449-450
Height, 463, 467, 542
Helicotrema, 603
Helium, 353
   diffusion characteristics, 316
Hematocrit, 254
   atmospheric pressure, 350
   viscosity of blood, 180
Hematopoiesis. See Hemopoiesis
Hematuria, 539
Hemianopsia, 594
Hemiballismus, 82
Hemicellulose, 441
Hemiplegia, 631
Hemochromatosis, 440
Hemodynamics, 165-185
Hemoglobin, 277-284, 280, 439
   abnormal, 280
   atmospheric pressure, 350
   buffers, 398
   drowning, 346
   fetal, 280
   filtration, 360
   oxygen storage, 293
   reduced, 329
   renal filtration and reabsorption, 366
Hemoglobinuria, 366
Hemolysis, 278
Hemophilia, 253
Hemopoiesis, 271-272
   altitude, 349
Hemorrhage, 385
Hemorrhoids, 178
Hemosiderin, 440
Hemostasis, 242
Henderson-Hasselbalch equation, 398
Henle, loop of, 358
Henry's law, 279
Heparin, 249
   production, 449
Hepatectomy, blood urea, 462
Hepatic flexure, 423
Hepatic secretion, vagus, 454
Hering-Breuer reflexes, 331
Hermaphrodite, 548
Hexamethonium, 67, 68
   renal secretion, 373
Hexosemonophosphate shunt (HMP), 458
Hill, A. V., 76, 86
Hindbrain. See Cerebellum, Medulla oblongata and Pons
Hippocampus, 624
Hirschsprung's disease, 425
Histamine, 449
   renal secretion, 373
   response of smooth muscle to, 154
Histidine, renal reabsorption, 366
Histidine decarboxylase, 449
Hodgkin, A., 39
Homeostasis, 163
Homosexuality, 569
Homotypical cortex, 628
Hooke, Robert, 171
Hooke's law, 171
Hormone, 490
   blood plasma concentration, 525
   calorigenic, 478
   chemistry, 490
   discovery, 490
   excretion, 432
   half life, 527
   secretion in 24 hours, 525
   secretion of by hypothalamus, 494
   target cells, 487
   transport in hypophyseal portal system, 494
   see also Antibody, Destruction, Synthesis, Sensitivity
Hormone list, 490
   adrenocorticotropic, 498, 525
   aldosterone, 198, 205, 388, 500
   androgens, 501, 529
   androstenedione, delta 4, 501, 524
   angiotensin, 198, 388, 534
   antidiuretic, 205, 369, 391, 493
   cholecystokinin, 456
   chorionic gonadotropins, 502, 562
   corticosterone, 528
   corticotropin, 625
   corticotropin-releasing factor, 500
   cortisol, 500, 525
   enterocrinin, 456
   enterogastrone, 418
   epiandrosterone, dehydro, 501
   epinephrine, 60-64, 534
   erythropoietin, 349
   estrogen, 501, 522, 560
   follicle stimulating, 92, 500
   follicle stimulating, releasing factor, 501
   gastrin, 445
   glucagon, 518-520
   glucocorticoids, 528
      see also Cortisol
   gonadotropins, 500
   growth hormone, 494-499
      releasing factor, 495
   inhibin, 502
   insulin, 496, 511-518
   intermediate lobe inhibiting factor, 490
   interstitial cell stimulating hormone (ICSH). See Luteinizing hormone
   lactogenic hormone. See Prolactin
   luteotropin. See Prolactin
   luteinizing, 501
      releasing factor, 501
   melanocyte stimulating, 544
      inhibiting factor, 544
      releasing factor, 544
   mineralocorticoid, 527, 587
      see also Aldosterone
   neurosecretory, 489
   oxytocin, 493, 502
   pancreozymin, 511
   parathormone, 389
   pitressin, 492
   progesterone, 501
   progestin, 522
   prolactin, 497-498
      inhibiting factor, 502
   relaxin, 566
   renin, 198, 385, 388
   secretin, 451-454
   testosterone, 501
   thyrocalcitonin, 539
   thyroid stimulating (TSH), 497-498, 504
   thyrotropin releasing factor (TRF), 498
   thyroxine, 497, 503
   triiodothyronine, 497, 505
   vasopressin, 492
Horner's syndrome, 69
Houssay animal, 517
Howell, William Henry, 493
Hr blood group, 265
Hue. See Color
Humidity, 475
   inspired air, 292
Hunger, 417, 470
Hunter, John, 174
Huxley, A. F., 78
Hyalinization, 562

Hyaluronic acid, 509
Hydration, potassium and sodium, 10
Hydrocephalus, 228
Hydrochloric acid, secretion, 449
Hydrocortisone, 566
Hydrogen, equilibrium potential, 37
Hydrogen ion, absorption and secretion, 428
Hydrogen sulfide, flatus, 441
Hydrostatic pressure, 219, 359
Hydroxyapatite, 536
Hydroxybutyrate dehydrogenase (SHBD), 459
Hydroxycholesterol, 20 alpha, 524
Hydroxycorticoid, 17-, 500, 522
  see also Cortisol, Steroids
Hydroxypregnenolone, 17-alpha-, 524
Hydroxyprogesterone, 17-beta-, 524
Hydroxyproline, 430
Hyperaldosteronism, 534
Hypercalcemia, 537
  treatment, 138
Hypercapnia, 200, 326, 329-331
  see also Carbon dioxide
Hyperemia, reactive, 200
Hyperglycemia, 417, 516, 534
Hyperinsulinism, 518
  heart, 116
Hyperkalemia, 517
  electrocardiogram, 138
  see also potassium
Hypermetropia, 584
Hypernatremia, drowning, 345
Hyperopia, 584
Hyperoxia, 302, 352
Hyperparathyroidism, 539
Hyperphagia, 469
Hyperplasia, 6, 16
Hyperpolarization, 36, 148
Hyperpotassemia, 107
Hyperreflexia, 507, 537
Hypersensitivity, 68
  catecholamine, 445
Hypertension, 145, 198
  coronary heart disease, 178
  essential, 383
  pulse pressure, 184
  renal, 383
  renoprival, 384
Hyperthermia, 479
Hyperthyroidism, 503, 539
Hypertonicity, 32
Hypertrophy, 6
  cardiac, 133, 139
  ECG, 108
  idiopathic myocardial, 113
  skeletal muscle, 82
  unilateral cardiac, 107
Hyperventilation, 347
  drowning, 345
Hypervolemia, 205
Hypocalcemia, 537
Hypocapnea, 347
  see also Carbon dioxide
Hypochlorhydria, 429
Hypoexcitability, 537

Hypoglycemia, 417, 511, 529
  stomach, 447
Hypokalemia, 534
  heart, 107, 138
  see also Potassium
Hypoparathyroidism, 539
  vitamin D, 538
Hypophysectomy, 496
  adrenal glands, 521
  glucocorticoids, 499
Hypophysial portal vessels, 492-494
Hypophysis. See Pituitary
Hypophysectomy, salivary glands, 446
Hypopolarization, 149
Hypopotassemia, 107, 138, 534
Hypoproteinemia, 255
Hypotension, orthostatic, 67
  respiration, 328
Hypothalamo-hypophysial tract, 492-494
Hypothalamus, 55, 492
  circulatory control, 208
  cold cell, 478
  endocrine production, 490
  food intake, 469
  gastric secretion, 447
  heat loss center, 478
  heat maintenance center, 479
  lesions, 469
  obesity, 469
  respiratory control, 324
  temperature regulation, 478
  warm cell, 478
Hypothermia, 479
Hypothyroid, 503
Hypothyroidism, 508
Hypovolemia, 391, 516
Hypoxia, 199, 292, 326-329
  cardiac, 458
  hypoxic, 292
  respiratory control, 326
  susceptibility of organs to, 292
  see also Partial pressure, oxygen
H zone, muscle, 74

I band, muscle, 73
Ichthyosis, 475
Ileal secretions, electrolyte concentration, 437
Ileocecal sphincter, 421
Ileocecal valve, 410
Ileostomy, 425, 436
Ileum, 421
Image, retinal, 553
Imino acids, 462
Impotence, 537
Impulses, 86
Incisura, 120
Indian wrestling, 35
Indicators, 394
Indigestible material. See Bulk in the diet
Indocyanine green, 130
Inertia, blood flow, 111
Infarction, 106, 116
  cardiac, 458

Infection, 266-269, 272
  fever, 477
Infundibulum, 489
Inhibin, 502
Inhibition, 32, 51, 67
  autogenic, 34
  duration, 51
  peripheral, 58
  postsynaptic neuron, 46
  presynaptic, 50
  reciprocal, 321
  respiratory control, 320
  skeletal muscle, 154
  smooth muscle, 154
Inhibitory postsynaptic potential (IPSP), 49
Inhibitory systems, 632
Initial heat, 90
Innervation, 58
  axodendritic, 27
  axodendrosomatic, 27
  axosomatic, 27
Inotropic agents, 140
Inspiration, 292-297
Inspiratory capacity, 301
Inspiratory center, 319, 322
Inspiratory reserve volume, 301
Inspiratory tachycardia, 142
  see also Heart
Insufficiency, aortic, 113, 123
  mitral, 113, 123
  pulmonary, 123
  pulse in aortic, 184
  tricuspid, 123
Insulin, 457, 496, 511
  amino acid sequence, 511
  antagonist, 516
  filtration, 360
  hypoglycemia, 449
  secretion, 511-514
  synthesis, 511
Intake, caloric, 428, 469
  fluid, 204-206
Integration of sensory messages, 577
Intensity, discrimination of sound, 606
Intercalated disc, 43, 93, 307
Intercellular fluid, 8
Intercostal muscles, 295
Intercostal nerves, 294
Intercourse, sexual, 554, 569
Intermediate lobe inhibiting factor, 490
Internal capsule, 630
Internuncial neuron, 33
Interoceptor, 22
Interstitial cell stimulating hormone. See Luteinizing hormone
Interstitial fluid, 9
  see also Fluid composition
Interventricular septum, 95, 124
Intestinal flora, 439, 441, 450
  see also Microorganisms
Intestinal glands, 429
Intestinal juice, 443
Intestinal phase of gastric secretion, 450

Intestinal secretions, 437
  electrolyte concentration, 437
Intestine, 154, 420-425, 429, 456
  duodenum, 410, 418, 420, 451
  effect of catecholamines, 64
  endocrine production, 490
  mucosal cells, 432
  secretory volume, 456
Intoxication, water, 396, 529
Intracellular fluid, 9
Intracranial pressure, 420
Intrafusal fibers, 28
Intraocular fluids, 232-234
Intrathoracic pressure, 294
Intrinsic factor, 440
Inulin, clearance, 360
  glomerular filtration, 360
Iodide, 229
  butanol-extractable, 504
  fate, 503
  protein-bound, 504
  secretion, 428
Iodide pump, 503
Iodinated compounds, 505
Iodoacetate, 86
Ion, 37
  see also Fluid composition
Iris, 581
Iron, 439-440
  ferric and ferrous, 282
  hemoglobin, 281
Irritability. See Excitability
Ischemia, cardiac, 458
  cerebral, 191, 292
  coronary, 116
  electrocardiogram, 106
  renal, 382
  see also Blood flow
Islets of Langerhans, 511
Isoelectric line, 99
Isoleucine, 464
Isometric contraction, 71
Isoproterenol, 63, 173, 178
Isotonic contraction, 71
Isotonic saline, 5, 12
Isovolumic ventricular relaxation, 120

Jarisch, 338
Jaundice, gallstones, 455
Jejunal secretions, electrolyte concentration, 437
Jejunum, 421
Jerk, knee, 33
Joule, 3
Junction, neuromuscular, 44
Juxtaglomerular apparatus, 383, 533

Kallikrein, 444
Karyotype, 546
Katz, B., 39
Kendall's compound A. (11-dehydrocorticosterone), 528
Kendall's compound B. (corticosterone), 528
Kendall's compound E. (cortisone), 528
Kendall's compound F. (cortisol), 459

Ketone bodies, 459, 460
Ketonemia, 460
Ketosis, 459, 516
Ketosteroid, 7-, 522, 529
  formation, 525
Kidney, 355-405
  autoregulation, 380
  blood supply, 355
  calculi, 539
  clinical evaluation, 376
  endocrine production, 490
  filtration, 356
  glucose production, 459
  glycogen stores, 293
  reabsorption, 361
Kilowatt-hour, 88
Kinesthetic tract, 622
Kinetic energy, 3
Kinin, 418
Knee jerk, 33
Korotkow sounds, 127
Krause end organs, 22
Krypton, diffusion characteristics, 316
Kupffer cells, 213
Kussmaul respiration, 331
Kyphoscoliosis, 311
Kyphosis, 303

Labial gland, 443
Labor, 493
Labyrinth, 235, 420
Labyrinth, bony, 609
Labyrinthectomy, 631
Lactase, 429
Lactate, cardiac production, 458
  exercise, 343
  renal reabsorption, 366
Lactate to pyruvate ratio, heart, 458
Lactation, 487, 493
  water loss, 390
Lactic acid, 86, 289, 457
  drowning, 346
Lactic dehydrogenase (LDH), 459
Lactogenic hormone. See Prolactin
Lactose, 429
Lambert, 587
Lamina propria, 410
Langerhans, islets of, 511
Laplace, Pierre Simon, 133
Laplace's law, 133, 415
Laryngectomy, 293
Laryngospasm, drowning, 345
Larynx, 413
Latency, 78, 96
  aldosterone, 533
  insulin, 515
  thyroid hormones, 473, 506
Law, Bohr effect, 281
  Dalton's, 279
  Henry's, 279
  Hooke's, 171
  Laplace, 133, 415
  Ohm's, 167
  Poiseuille's, 166-167
  Starling's, of the heart, 135

Lead I, 97
Lead II, 97
Lead III, 97
Lean body mass, 394
Lecithin, 432, 460
Lemmiscus, medial, 622
Lengthening. See Distention
Length-tension diagram, 135
  aorta, 170
  heart, 135
  skeletal muscle, 135
  smooth muscle, 157
Lens, nutrition, 233
  prescriptions, 585
Leucine, 506
Leukocytes, 273-274
Leusine, 464
Levarterenol. See Norepinephrine
Leydig cells of the testis, luteinizing hormone, 501
  see also Luteinizing hormone
Ligament, dentale, 227
  spiral, 236
  suspensory, of eye, 582
Light, 579, 587
Light deprivation, 597
Linoleic acid, 464
Lipase, 429, 433
Lipid, 8, 429-434, 459
  see also Fat
Lipid solubility, 233
Lipolysis, 516
Lipoprotein, hemostasis, 245
Lissauer, tract of, 623
Liver, 432, 455, 457, 518, 527
  cortisol, 527
  disease, 453
  endocrine production, 490
  glycogen stores, 293
  metabolism, 458
  secretion, 453
Liver cell, 7
Load-length diagram, skeletal muscle, 84
Load, skeletal muscle, 84
Local sign, 33
Loewi, Otto, 43
Long, Cyril Norman, 517
Long chain fatty acids, 432
Loop of Henle, 358
Loudness, 599
Lukens, Francis D. W., 517
Lundsquard, 86
Lung, 134, 295
  collapse, 303
  eosinophil sequestration, 529
  opening pressure, 303
  oxygen storage, 293
  surface tension, 304, 305
  tidal volume, 134
  volume, 134, 301, 311
    breath holding, 344
Luteinization, 501
Luteinizing hormone, 501
Luteinizing hormone releasing factor, 501

Luteotropin, 222
  see also Prolactin
Lymph, 222
  cholesterol transport, 451
Lymphatic system, 221-223, 460
Lymph duct, 164
Lymph node, 164, 529
  tracheo-bronchial, 115
Lymphoblast, 266
Lymphocyte, 273
Lysine, 464
  kidney, 367
  renal reabsorption, 366
Lysosome, 14

Macrophages, histiocytes, 11
Macula, 615
Macula densa, 533
Magnesium, 389, 439, 536
  blood concentration, 390
  urine, 359
  urinary concentration, 390
Magnus-Levy, Adolf, 491
Malate, renal reabsorption, 366
Male, 551-559
  urethra, 405
Malleus, 602
Malnutrition. See Starvation
Malpighian corpuscle. See Glomerulus
Maltose, 429
Mammary gland, 502
Mandelic (3-methoxy04-hydrosy) acid commonly called vanillymandelic acid or VMA), 61
Marrow, bone, 271
Masculinization of the female, 529
Masking. See Binding
Mass, 3, 71
Mass movements, 424
Mastication, 411
Mating, 554, 569
Maturation. See Growth and Puberty
Maxwell frame, 307
Mayer, Julius Robert, 173
Meals, osmolarity, 435
Measurement, 3
Mechanical efficiency, 112
Mechanoreceptors, 188
Medulla, adrenal, 521-522
Medulla oblongata, 393
  circulatory control, 208
  respiratory control, 318-322
  swallowing, 412
Medullary animal, 318
Megacolon, 425
Megaloblasts, 266
Meiosis, 551
Meissner's corpuscles, 23, 622
Meissner's plexus, 410
Melanin, 544
Melanocytes, 544
Melanocyte-stimulating hormone (MSH), 544
  inhibiting factor (MSH-IF), 544
  releasing factor, 544

Melanophore, 544
Melatonin, 544
Membrane of cell, 8
Menarche, 549
Meniere's syndrome, 613
Meningitis, respiration, 325, 330
Menopause, 568
  arterial pressure, 184
Menstrual cycle, 560
Mercury poisoning, 385
Mesencephalon, 323
  respiratory control, 323
Messenger ribonucleic acid (RNA), 533
Metabolic acidosis, 403
Metabolic alkalosis, 403
Metabolic rate, basal, 467, 479, 509
  body temperature, 468
  environmental temperature, 468
Metabolism, 289
  acetate, 461, 522
  anaerobic, 291
  body, 291
  brain, 291
  carbohydrate, 457, 506
  cholesterol, 507
  corticosterone, 527
  digestive organs, 291
  fat, 529
  heart, 291
  kidney, 291
  lactate, 117
  nucleoprotein, 367
  pH, 397
  protein, 506
  skeletal muscle, 291
  skin, 291
  water, 531
  water formation, 390
Metabolites, 198
Metals, heavy, 289
Metarteriole, 169
Methane, flatus, 441
Methemoglobin, 281
Methionine, 430, 464
Metric system, 3
Micelle, 433
Microelectrodes, 36
Microorganisms, digestive tract, 441
  histamine synthesis, 450
  stomach, 450
  see also Antibiotics
Microphages, neutrophils, 11
Micropuncture techniques, 360
Microsomes, 525
Micturition, 404-409
Midbrain, micturition, 409
  respiration, 323
  transection, 32
Milieu exterieur and interieur, 487
Milk production, 502
  see also Lactation
Mineralocorticoid, 387, 527, 533-534
  see also Aldosterone
Minerals. See Electrolyte, Calcium, Iron

Miniature end-plate potential (MEPP), 46
Minkowski, Oscar, 491
Minute work, 112
  respiratory, 312
Miosis, 69
Mitochondria, 14, 44, 506, 525
  heart, 116
  muscle, 73
Mitosis, 551
Mitral valve, 118
Mittelschmerz, 560
Modalities, 22
Mol, 4
Molar solution, 4
Molecular size, 360, 429
Mongolism, 547
Moncamine oxidase, 61
Monochromats, 589
Monocular vision, 589
  see also Vision
Monocytes, 274
Monoglycerides, 433
Monoiodotyrosine (MIT), 504
Monosaccharides, 457
Moonface, 529
Morphine, respiratory control, 325
Mosaicism, 547
Motility, bowel, 442
  eye, 597
  gastric, 418
  sperm, 553-554
Motion sickness, 420
Motor areas in brain, 629
Motor end-plate, 44
Motor neuron, 24
Motor point, 54
Motor unit, 27, 82
Mountain sickness, 349
Mouth, 440
Movement, passive in respiration, 334
  see also Motility, Mass
Mu, 3
Mucin, 440
  saliva, 443
Mucosa, intestine, 420
  muscularis, 422
Mucous, 419, 440
Muller, Johannes, 23
Murmur, 126
Muscle, 71-157
  abdominal, 298
    vomiting, 420
  acceleration, 82
  action potential, 81
  active state, 74, 79
  alae nasi, 295
  anal sphincter, 411
  antagonistic, 82
  antigravity, 32, 630
  antiparallel region, 76
  asynchronous firing, 82
  atrophy, 81
  back, 295
  brachialis, 71
  cardiac, 92

cheek, 295
chemistry, 86
contractile component, 76
contractility, 85
creeping phenomenon, 78
cross bridges, 76
dartos, 551
deceleration, 82
detrusor, 407
diaphragm, 318
digastric, 295
distention, 83
environment, 82
equilibium length, 76, 83
extensor, 32, 34
extensor digitorum longus, 80
external oblique, 297
extracts, 78
extrafusal fiber, 334
extrinsic eye, 82, 591
fasciculation, 81
fast, 81
fatigue, 82
fiber length, 83
fibrillation, 81
fixation, 82
flaccidity, 81
flexor, 33
force, 85
gamma loop, 32, 334
gastrocnemious, 79
glycerol-extracted, 78
glycogen stores, 293
heart, 92
hypopharyngeal, 413
inspiratory, 294
intercostal, 297, 322
intercostal and the gamma loop, 334
internal oblique, 297
internal rectus of the eye, 79, 271
intrafusal fiber, 334
laryngeal, 295
length, 83
levator ani, 408
levator palata, 295
movement, effect on respiration, 342
multifidus, 73
mylohyoid, 295
neck, 295
palatopharyngeal, 412
parallel fibers, 73
passive tension, 76
perineal, 408
pinnate fibers, 73
platysma, 295
rate of change of force, 85
rate of change of tension, 86
red, 76
resting length, 76
sartorius, 73, 255
scalene, 295
shortening, 76, 287
skeletal, 76
smooth, 148-157
soleus, 79, 273
spindle, 30, 53

stapedius, 602
sternomastoid, 295
stretch, 84
superior esophageal, 413
tensor tympani, 602
tetanus, 81
tibialis anticus, 77
tone, 171, 537, 630
tongue, 295
transversus abdominus, 297
trapezius, 295
twitch, 77-80
weakness, 529
white, 86
 see also Contraction, Heart, and Tetanus
Muscle end-plate, 29
Muscle spindle, 30
 diaphragm, 333
Muscularis mucosae, 410, 422
Mydriasis, 67
Myelin, 20, 51
Myelin sheath, 44
Myenteric plexus, 410, 449
Myoepithelial cells, 493
Myofibrils, 44, 73
 heart, 116
Myofilaments, 73, 149
Myogenic regulation, 199
Myoglobin, 73, 439
 filtration, 360
 oxygen storage, 293
Myoneural junction, 29
Myopia, 584
Myosin, 74, 76, 149
 crossbridges, 78
Myosis, 69
Myxedema, 508
 metabolic rate, 468

Nanomols, 3-4
Nausea, 419
Near sightedness, 584
Necrosis, 292, 459
 see also Infarction
Negative feed-back, 189, 472, 498
 thrombin production, 244
Negro, heat absorption, 474
Neostigmine, 43, 46
Nephrectomy, 405
Nephritis, 405
 ammonia production, 402
Nephron, cortical, 356
 juxtamedullary, 356, 369
Nephrosis, 405, 534
 renal filtration, 376
Nephrosclerosis, renal plasma flow, 376
Nephrotic syndrome, 358
Nernst equation, 37
Nerve, chorda tympani, 443
 colon, 424
 cranial, 319
 dorsal horn, 28
 fiber groups, 23

glossopharyngeal, 326, 329
hypogastric, 406, 556
impulse, 23
intercostal, 294
pelvic, 57, 406
phrenic, 26, 294, 296, 318
popliteal, 77
pudendal, 556
renal, 380
saphenous, 24
sensory, 23
skin, 57
vagus, bile secretion, 454
 block, 331, 449, 454
 cardiac sphincter of stomach, 414
 chemoreceptors, 329
 colon, 424
 exercise, 199
 heart, 108, 140, 142
 pancreas, 451
 respiratory centers, 322, 331
 sensory neurons, 448
 stomach and small intestine, 417, 418, 449
 see also Neuron, Reflex, Nervous system
Nerve net, 418, 456
Nervous reinnervation, 82
Nervous system, 19-35
 autonomic, 55-69, 206-211, 418-420, 422-424, 443-456
 central, 206-211, 318-326, 408-409, 424-425, 446, 469, 478-480, 531, 592, 594, 596, 606-608, 612, 621, 622-637
 peripheral, 17-35
Neural crest, 521, 522
Neurilemma, 20
Neurofibroma, 425
Neurohumor, 487
Neurohypophysis, 489
Neuromuscular junction, 29
Neuron, 17-35
 auditory, 606
 efferent, 92
 intrinsic, 449
 alpha, 27, 82
 parasympathetic, 64-67
  see also Parasympathetic neurons
 salivary gland, 64, 443
 postganglionic and preganglionic, 55, 58
 somatic efferent, and vomiting, 420
 stomach, 449
 sympathetic, 59-64
  see also Sympathetic neurons
 salivary gland, 444
 types in cortex, 626
 vasoconstrictor and vasodilator, 193
 vomiting, 420
 see also Nerve
Neutrophil, 273
Newborn, 317, 544
Newtonian fluids, 165
Nicotinic acid, 441
Night blindness, 587

Nitrate, 117
    muscle, 79
Nitrite, 117
Nitrogen, amino, in blood, 430
    bends, 350
    diffusion characteristics, 316
    washout, 351
Nitrogen-containing nutrient, 462
Nitroglycerine, 117
Nitrous oxide, diffusion characteristics, 316
    respiratory control, 325
Nocioceptive reflexes, 33
Nodal rhythm, 100
Node of Ranvier, 20
Nondisjunction, 547
Noradrenaline. *See* Norepinephrine
Norepinephrine, 60, 65, 145, 521, 522, 544
    actions, 62
    fate, 61
    respiration, 343
    salivary glands, 445
    sensitivity to, 507
Normality, 4
Normoblasts, 266
Nuclease, 429
Nucleic acid, 429, 462
Nucleosides, 463
Nucleus, brain, 533, 593, 607, 613, 622, 625, 630
    cellular, 513-515, 546
    hypothalamic, 492
    supraoptic, 500
Number, Avogadro's, 4
Nursing. *See* Suckling
Nutrients, 289, 465
    heart, 117
    see also Food, Vitamins
Nystagmus, 597

Obesity, 469
    respiratory compliance, 303
Obstruction, coronary heart disease, 178
    intestine, 442
    urinary, 405
    see also Stenosis
Occipital cortex, 594
Occlusion, 49
Oddi, sphincter of, 455
Odoriferous molecules, thresholds, 619
Ohm's law, 167
Olfaction, 619
Oliver, 491
Olivospinal tract, 623
Optic atrophy, 597
Optic chiasma, 593
Optics, 580
Oral temperatures, 476
Orgasm, 556, 573
Orgasmic platform, 572
Ornithine, 430
Orthodromic stimulus, 47
Orthostatic hypotension, 67

Orthosympathetic neurons, 55
    see also Sympathetic neurons
Osmol, 4
Osmolality, 4
Osmolarity, 4
Osmosis, 12
Osmoreceptors, 391
Osmotic pressure, 220, 278
    duodenum, 420
Ossicles, auditory, 602
Ossification, 543
Osteitis fibrosa, 389
Osteoclast, parathormone, 389
Osteoporosis, 529, 539
Otolith, 615
Ovary, 522, 561
    endocrine production, 491
Oviduct, 568
Ovulation, 501
Oxalate, 247, 248
Oxidation, role of mitochondria, 14
Oxygen, caloric value, 467
    chemoreceptors, 328
    diffusion characteristics, 316
    hyperventilation, 348
    hypoventilation, 348
    reserves, 289
    unconsciousness, 345
    see also Arteriovenous oxygen difference
Oxygen concentration, arteries, 123
    arteriovenous difference, 175
    chemoreceptors, 328
    heart, 123
    veins, 123, 292
    venous, 174
Oxygen consumption, 88, 289
    exercise, 343
    heart, 112, 291
    respiratory system, 291, 312, 380
    salivary gland, 445
    skeletal muscle, 291, 380
Oxygen debt, 87, 114
Oxygen therapy, surface tension of lung, 308
Oxygen toxicity, 353
Oxygen transport, 275, 279-281
Oxyhemoglobin, 279
    see also Plasma
Oxyntic cells, 440
Oxytocin, 157, 493, 502
    see also Hormone

Pacemaker, 71, 92
    dominant, 92
    ectopic, 92, 350, 351
    gastric, 416
    heart, 100, 137
    periodic nodal, 100
Pacinian corpuscles, 52
Pacing the heart, 101
    synchronous, 141
Page, Irving, 383
Pain, 22

heartburn, 339, 414
sympathectomy, 418
tracts, 623
vomiting, 420
Pain fibers, 623
P_{CO_2}, 347
Palatal gland, 443
Palate, 412, 419
Palpitations, 109
Pamaquine, methemoglobin, 281
Pancreas, 429, 511-520
    beta cell exhaustion, 515
    digestion, 432
    endocrine production, 490
    fat digestion, 432
    protein digestion, 430
    reserve, 432
    secretion, 451
Pancreatectomy, 432
Pancreatic amylase, 429
Pancreatic juice, electrolyte concentration, 437
    pH, 397
    volume, 443, 451
Pancreozymin, 451, 511
Panhypopituitarism, 496
Panting, 475, 478
Papanicolaou smear, 568
Papillary ducts of Bellini, 356
Papillary muscles, 118
Para-aminohippurate (PAH), renal clearance, 371
Paradoxical cold, 624
Paraffin, 242
Parallax, 590
Parallel circuit, 167
Parallel elastic and visco-elastic component, muscle, 76
Paralysis, diaphragm, 295
Paraplegia, micturition, 408
Parasites, filaria, 223
Parasympathetic neurons, 64-67, 192
    colon, 424
    gallbladder, 455
    heart, 140, 143
    intestine, 411, 423, 456
    pancreas, 451
    salivary gland, 443
    stomach, 416, 417
    uterers, 406
    vasodilation, 192
Parathormone (PTH), 389, 439
    renal filtration, 367
    renal tubular maximum, 367
    see also Hormone
Parathyroid glands, 536-540
    endocrine production, 490
Parathyroidectomy, 539
Parenchymal cell, 213
Parkinsonism, 82
Parotid gland, 443
Pars distalis, 489
Pars tuberalis, 498
Partial pressures, 279
    carbon dioxide, 201, 279, 347, 398
    effect on inspiration, 296

oxygen, 194, 200, 279, 281
oxyhemoglobin dissociation curve, 280
Parturition, 566
Passive reabsorption, kidney, 362
Passive secretion, kidney, 373
Passive transport. *See* Transport, passive
Past-pointing, 613
Patent ductus arteriosus and foramen ovali, 123
Pathway, optic, 593
  see also Tracts
Pavlov conditioning, 342
Pavlov, Ivan Petrovitch, 341
P_{CO_2}, 201, 279, 282, 347, 398
Pectin, 442
Pedicels, 214, 359
Pelvis, 544, 552, 570
Penicillin, excretion, 404
  renal secretion, 373
Penis, 552-559
Pentose, 429, 463
Pentose-phosphate pathway. *See* Hexosemonophosphate shunt
Pentothal, 211
Pepsin, 429
Pepsinogen, 447
Peptic ulcer, 440-441, 448
Peptide, 449
  fibrinogen, 247
Per cent, 4
Performance, heart, 110-114
  skeletal muscle, 85
Perfusion pressure, 201
Pericardium, 123, 135, 171
Pericyte, 214
Perilymph, 235-237
Perimeter, 589
Perimyseum, 73
Period of latency, 78, 96
  see also Latency
Peripheral nerves. *See* Nervous system, peripheral
Peripheral resistance unit (PRU), 3, 167
Peristalsis, 152, 414
Permeability, 8, 41, 216
  coronary capillary, 115
  plasma membrane, 37
  specific, 217
  see also Conductance
Permeability coefficient, 10
  chloride, 10
  potassium, 10
  sodium, 10
Permeability to water, intestine, 435
  stomach, 435
Peroxidase, 439
Perspiration, 475
  insensible, 475
  water loss, 390
  see also Sweat
Peru, atmospheric pressure, 350
Petechiae, 252
pH, 4, 11, 200, 397

aqueous humor, 229
  atmospheric pressure, 350
bile, 433, 454
blood plasma, 397
blood plasma dialysate, 229
body secretion, 397
digestive enzymes, 429
drowning, 346
duodenum, 418
exercise, 343
gastric antrum, 448
gastrin release, 449
heartburn, 414
intestinal juice, 456
mucus, 441
nephron, 376
oxyhemoglobin dissociation curve, 281
pancreatic juice, 432
plasma, 6
respiration, 329
urine, 6
see also Buffers and Acid-base balance
Phagocytosis, 11, 266-268
Pharyngoesophageal junction, 413
Phenol red, renal clearance, 371
  renal secretion, 373
Phenoxybenzyamine, 62
  atropine, 66
Phentolanine, 62
Phenylalanine, 61, 464
Phenyl diquanide, 338
Phenochromocytoma, 535
Phlorizin, 429
  transport systems, 364
Phosphate, 8, 389, 439, 536
  buffers, 398
  functions, 367
  kidney, 367
  parathormone, 389
  renal reabsorption, 366, 367
  tubular maximum, 389
  urine, 359
Phosphocreatine, 86
Phosphogluconate shunt. *See* Hexosemonophosphate shunt
Phospholipids, 8, 434, 460
Phosphoproteins, phosphoric acid, 463
Phosphorus, serum inorganic, 518
Phosphorylation, oxidative, 73
Photopic vision, 588
Phrenic nerves, 294, 318
Physical therapy, muscular atrophy, 82
Physiology, 1
Pia mater, 226, 227
Pickwickian syndrome, expiration, 311
Picrotoxin, 68
Pigmentation, 544
Pigments, visual spectra, 588
Piloerection, 473
Pineal gland, 544
Pinocytosis, 11, 218, 435
Pitch differentiation, 605
Pitressin, 492
  see also Hormone

Pituitary basophilism, 532
Pituitary gland, 489-502
  acidophilic tumors, 495
  anterior lobe, 493-502
  endocrine production, 490
  intermediate lobe, 544
  posterior lobe, 492
Pituitary stalk, 492
P_K, 401
Placenta, 502, 565
Plasma, 255-261
  buffers, 220, 222, 240, 257
  concentration of amino acids, 366
  concentration of carbohydrates, 365
  dialysate, 229
  loss, 216
  pH, 397
  procoagulants, 243, 246
  proteins, 220, 256, 346
  uric acid, 366
  vitamin C, 366
  see also Blood
Plasmablast, 270
Plasma expanders, 395
Plasma thromboplastin, 244, 246
Plasma thromboplastin antecedent (PTA), 246
Plasma thromboplastin component (PTC), 246
Plasma transglutaminase, 246
Plasmin, 247
Plasminogen, 247
Plasticity, 171
  skeletal muscle, 84
Platelet, 163, 274, 276
  aggregation, 212
  clot formation, 243
  plug formation, 249-250
Platelet cofactor I, 246
Platelet cofactor II, 246
Platelet concentration, thromboplastic generation, 244
Platelet depletion, 242
Platelet factor, 3, 243, 246
Pleura and pleural space, 295
Plexus, Auerbach's, 410
  carotid, 57
  intestinal, 152
  Meissner's, 410
  myenteric, 410, 449
  submucosal, 410
Pneumotaxic center, 319, 322
Pneumothorax, 296
P_{O_2}, 194, 206, 279, 281
Poikilothermia, 479
Poiseuille, Jean L. M., 166
Poiseuille's law, 166-167
Polarization, 92, 150
Poliomyelitis, respiration in, 325
Polycythemia, 254-255
Polydipsia, 516
Polynucleotides, 429
Polypeptides, 429
  hormones, 290
Polyphagia, 516
Polyuria, 370, 516, 534

660 Index

Pons, micturition, 409
  respiratory centers, 319, 322-324
Pores, 212
  renal capillaries, 357
Portal system, hepatic, 432
  hypophyseal, 494
Positive afterpotential, 38
Postcentral gyri, 627
Postganglionic neurons, 55, 58
Post-rotary reactions, 613
Posture, 91, 630
  blood pressure, 176
  blood volume, 203
Potassium, 8, 41, 137, 201, 385, 533
  cardiac infarction, 458
  cell membrane permeability to, 37-42
    death due to, 405
  charge carried by (Z_K), 37
  concentration, 236
  diabetes mellitus, 517
  effect on ECG, 107
  equilibrium potential, 37
  excretion, 403
  heart, 138
  loss, 438
  renal secretion, 371
  secretion, 428
  sodium restriction, 395
  starvation, 395
  transmembrane conduction gradient, 37
  urine, 359
  uremia, 405
Potassium conductance (g_K), 41
Potential, action, 17, 24, 38-41, 149
  atrial, 59
  cardiac, 99
  critical, 148
  duration of spike, 149
  effect of stretch on resting, 155
  end-plate, 45
  equilibrium, 37
  excitatory post-synaptic, 47
  follower, 92
  frequency of action, 155
  generator, 51, 53
  miniature end-plate, 46
  pacemaker, 92, 149
  plateau, 149
  postsynaptic surface, 47
  receptor, 51, 53
  responsiveness, 38
    see also Sensitivity
  resting, 17, 92
    see also Equilibrium potential
  role of ions in heart, 137
  sinusoidal, 149
  skeletal muscle, 77
  slow waves, 148
  spike type, 149
  transmembrane, 17, 36-54, 92, 153
  transmembrane, of smooth muscle, 149
  see also Action potential
Potentiation, 138, 445
  see also Synergism
Precapillary sphincter, 168
Precentral gyrus, 627
Prednisolone, 528
Preganglionic fibers. *See* Nervous system, autonomic
Preganglionic neurons, 55, 58
Pregnancy, 240, 550, 562
  mammary gland, 502
  regurgitation, 414
  respiratory compliance, 303
  uterine activity, 155
Pregnenolone-delta 5, 524
Preponderance, glomerulo-tubular, 378
Presbyopia, 582
Prescriptions, lens, 585
Pressoreceptors. *See* Baroreceptor
Pressure, 71
  arterial, 123, 207, 299
    angiotensin and renin, 383
    during drowning, 346
  atmospheric, and respiratory resistance, 310
  barometric, 348
  bile cannuliculi, 453
  capillary, 123
  cerebrospinal fluid, 232
  colloid osmotic, 359
  colon, 424
  critical closing, 172
  defecation, 425
  effect of gravity on blood, 176
  eye, 232
  filtration, 359
  gall bladder, 454
  glomerular capillary, 359
  heart, 133
  hydrostatic, 219
  intra-abdominal, 294
  intraocular fluid, 232-233
  intrapleural, intrapulmonary and intrathoracic, 294, 297, 311
  intraventricular, 111
  mean injection, 111
  opening, of lungs, 303
  osmotic, 220
  partial, 278-279
  peak systolic, 111
  perfusion, 201
  pulse, 181-184
  rate of change of, 111
  sensation of, 622
  stomach, 416
  swallowing, 412
  thoracic, 295
  tracts in central nervous system, 622
  transmural, of brain, 230
  ureter, 407
  urinary bladder, 407-408
  velocity of conduction, 184
  venous, 176
    during drowning, 346
  ventricular, 114
  see also Blood
Prethrombin, 243
Primaquine, methemoglobin, 281
Proaccelerin, 246
P-R interval, 99, 139
Probenecid, gout, 367
  penicillin secretion by kidney, 374
Procarboxypeptidase, 429
Procoagulants, 243
Proconvertin, 246
Profibrinolysin, 247
Progesterone, 501, 524
  effect on uterus, 152
  see also Hormone
Progestin, 526
Prolactin (LTH), 502
  see also Hormone
Prolactin inhibiting factor (PIF), 502
  see also Hormone
Proline, 430
Propagation, 45
Proprioceptors, 22
Proprioceptive receptors and respiration, 333
Prostate, 405, 551-552
Protamine sulfate, 249
Protanomaly, 589
Protanopia, 589
Protection, digestive tract, 440
Protein nitrogen, lymph, 222
  plasma, 222
Protein, 429, 457
  anabolism, 6, 506
    estrogen, 156
  buffers, 220, 283, 398
  catabolism, 514, 529
  digestion, 429
  gastric secretion, 448
  iodine, 462
  kidney, 366
  metabolism, 462
  plasma, 163
  specific dynamic action, 469
  synthesis, 12, 462
Protein binding, 258
  amnionic fluid, 240-241
  calcium, 536
  iodine, 504
  phosphate, 536
  see also Binding
Proteinuria, 358
Prothrombin, 242-247
Prothrombin accelerator, 246
Prothrombin concentration, 252
Prothrombin time, 251
  aberrant structures, 538
Protodiastole, 120
Prower factor, 246
Proximal tubule, active reabsorption, 368
P-R segment, 99
Pseudohermaphroditism, 528, 547
Psychomotor performance, electroencephalogram, 348
Ptyalin. *See* Salivary amylase
Puberty, 548, 549
  mammary gland, 502
  premature, 529

Pubic hair, 550
Puerperium, 567
Pulmonary blood flow and pressure, 182
Pulmonary circulation, 184
Pulmonary congestion, 123
Pulmonary valve, 117
Pulmonary ventilation, 298-300
Pulmonary insufficiency, 124
Pulse, arterial, 181-183
Pulse, pressure, 181-184
Pump, 363-376
  see also Transport
Pupils of eye, 595
  control, 58
  sign of brain ischemia, 146
Pupil size, 58, 146
Purine, 429
Purkinje fibers, conduction rate, 95
Purkinje shift, 588
Purpura fulminans, 253
P wave, 139
Pyelonephritis, renal necrosis, 376
Pyloric sphincter, 410, 417, 418
Pyramidal tract, 629
  control of alpha efferent neuron, 31
Pyrimidine, 429, 463
  absorption, 428
Pyrogens, 476, 479, 531
Pyruvic acid, 457

QRS complex, 99, 139
  see also Electrocardiogram
Q-T interval, 99
Quadrigeminy, 106
Quantification, 3
Queckenstedt-Stookey sign, 230
Quinidine, 43, 108, 117

Radiation, 474
  sickness, 420
Radio waves, 579
Radius, airways, 309
  blood vessels, 169
  erythrocyte, 142-145
  tension, 133
Raffenose, specific permeability, 217
Rami communicantes, 56
Ranvier, nodes of, 20
Rate, heart, 142-145
  see also Frequency
Raynaud's disease, 69
Reabsorption, active, 363, 368
  amnionic fluid, 237
  bicarbonate, 399
  cerebrospinal fluid, 229
  kidney, 361-370, 399
  passive, 362
Reaction time for eye, 596
Reactions. *See* Reflex
Reactive hyperemia, 200
Receptor(s), 22-23, 51-53, 62-68
  alpha and beta adrenergic, 62, 194
  baroreceptor. *See* Baroreceptor
  bronchial, 337
  bronchial wall, 336
  chemoreceptor, 194
  cholinergic, 66
  cold, 479
  coronary, 338
  delta, 66
  exteroceptors, 22
  gamma, 66
  Golgi tendon organ, 35
  inflation, 331
  interoceptors, 22
  kidney, 533
  Kraus end bulbs, 22
  lung, 331
  mechanoreceptor, 188
  Meissner's corpuscles, 22, 622
  muscle spindles, 22
  osmoreceptors, 387
  Pacinian corpuscles, 22
  P_{CO_2}, 347
  phasic, 52
  potential, 53
  pressoreceptor, 188
  proprioceptive, and respiration, 333
  proprioceptors, 22
  pulmonary, 338
  pulmonary stretch, 322
  Ruffini end organs, 22
  stretch, 22, 52, 331-335, 387
    lung, 331
    rectum, 424
    respiration, 331
    skeletal muscle, 22, 28, 51-53
    urinary bladder, 407
  threshold, 35
  tonic, 51
  tracheal, 337
  upper respiratory tract, 335
  urinary bladder, 407
  volume, 387
  warm, 479
Reciprocal inhibition, 321
Recirculation time, 132
Recoil, chest, 305
Recovery, 80
Recruitment, 53
  respiratory control, 320
Rectal temperature, 474
Rectum, 423
  distention, 424
Red blood cell, 163, 275-284
  see also Erythrocyte
Red muscle, 87
Reduced hemoglobin, 329
References, circulation, 285
  contraction, 157
  control (endocrine system), 575
  control (peripheral nervous system), 69
  introduction, 15
  nutrients and waste products, 480
  sensation and integration, 637
Reflex, 29-35
  anal, 339
  antagonism, 34
  autogenic inhibition, 34
  autonomic, 55
  baroreceptor, 188-189
  Bezold-Jarisch, 190
  calcium, 348
  chemoreceptor, 194-196
  closed chain circuit, 33
  cold water, 339
  consensual, 596
  contralateral, 33
  corneal, 331
  coronary chemoreflex, 338
  disynaptic, 27
  duration, 33
  enterogastric, 418
  extensor thrust, 33
  flexion, 34
  gastrocolic, 424
  gastroileal, 422
  Hering-Breuer, 331
  inflation, 332
  ipsilateral monosynaptic, 34
  knee-jerk, 33
  mechanoreceptor, 188, 196
  micturition, 407
  monosynaptic, 27, 31-33
  multiple chain circuit, 33
  multisynaptic, 33-35
  nociceptive, 33
  paradoxical inflation, 331
  paradoxical, of Head, 331
  P_{CO_2}, 348
  pulmonary chemoreflex, 338
  righting, 614
  somatic, 55
  stretch, 35
    see also Receptor
  synergism, 34
  thoracic, 338
  vasomotor, 208
  visual, 595
  von Bezold-Jarisch, 338
  withdrawal, 33, 34
Refraction of light, vitreous humor, 581
Refractive power, aqueous humor, 581
  cornea and lens, 581
Refractoriness to antidiuretic hormone, 534
Refractoriness to insulin, 516
Refractory period, 96
  atrioventricular node, 100
  heart, 92, 96, 99, 144
  orgasm, 556
  skeletal muscle, 78
Regulation. *See* Control
Regurgitation, 414, 420
  heart, 118
  heartburn, 414
  infancy, 414
  pregnancy, 414
  see also Control
Relative viscosity, 179
Relaxation, 79
Relaxation time, heart, 142
Relaxin, 566
Releasing factor, 490
  see also Hormone
Renal. *See* Kidney

Renal papilla, 356
Renin, 198, 383-385, 534
  see also Hormone
Renshaw cell, 33, 34
Repolarization, 36, 92
Reproduction, 542-575
Reserve, cortisol, 525
  epinephrine, 535
  glycogen, 292-293
  iodine, 504
  oxygen, 224, 293
  oxygen debt, 87, 114
  pancreatic, 432
  see also Binding
Reserve volumes, cardiac, 134
  respiratory, 300-302
Reservoir, cortisol, 525
  see also Binding
Residual capacity of the heart, 132
  see also Heart
Residual volume, 132, 301, 407
  see also Heart
Resistance, 3, 71, 303
  cell membrane, 54
  circulatory, 166, 201
  effect of catecholamines, 62
  electrical, of heart, 93
  renal blood flow, 380
  respiratory tract, 309
  turbulence, 168
Respiration, 289-354
  acid-base balance, 403
  artificial, 300, 345
  cell, 291
  control, 318
  depth, 325, 346
  heart sounds, 125
  Kussmaul, 331
  minute volume, 297
  proprioceptors, 333
  rate, 299, 907
  tidal volume, 299
  see also Breathing
Respiratory center, 319, 322
Respiratory gases, 279
Respiratory minute work, 312
Respiratory quotient (RQ), 117, 291, 465
  body, 759
  brain, 291
  digestive organs, 291
  heart, 291
  kidney, 291
  skeletal muscle, 291
  skin, 291
Respiratory rate, atmospheric pressure, 350
Respiratory tract, 292-294
  irritation, 335
Respiratory volumes, 301
Respiratory tree, 294
Resting potential. See Potential, resting
Resuscitation, cardiorespiratory, 301, 345
Retching, 324, 420

Rete Testis, 551
Reticular formation, 622
  descending, facilitatory and inhibitory areas, 32
  respiration, 324
  respiratory centers, 318
Reticulocytes, 266
  atmospheric pressure, 350
Reticuloendothelial cells, 247
Reticuloendothelial system, 266-267
Reticulospinal tract, urinary bladder control, 409
Reticulum, endoplasmic, 12, 513
  muscle, 73
  sarcoplasmic, 73
Retina, 586
Retinal image, 553
Retinene, 588
Retraction of a clot, 250-251
Reverberating circuit, 33
  respiratory control, 320
Reynold's number, 127
Reynolds, Sir Osborne, 127
Rh blood group system, 262
Rheumatoid spondylitis, respiration, 347
Rhodopsin, 588
Rhythm, ACTH diurnal, 525
  basal metabolism, 479
Riboflavin, 441
Ribonuclease, 429
Ribonucleic acid (RNA), 12, 506, 511, 514
  synthesis, 515
Ribose, 429
  production, 458
Ribosomes, 12, 506, 511, 533
Richards, A. N., 360
Rigidity, decerebrate, 630-631
Rigor, calcium, 138
Rigor mortis, 80
Rod, 586
Rotation, 611, 612
Ruminants, digestion, 414

Saccule, 615
Saliva, 444
  electrolyte concentration, 437
  volume, 443
Salivary amylase, 410, 429, 443
Salivary gland, 429, 443-446
  effect of catecholamines, 64
Salt deprivation, 437
Saltatory conduction, 43
Sandstrom, Ivar Victor, 538
Sapid solutions, thresholds, 617
Sarcolemma, 73
Sarcomere, 74
Sarcoplasm, 44, 73
Sarcosine, 430
Sarcoplasmic reticulum, 73, 149, 256
Sarcosomes, 73
Satiety center, 469
Scala media, tympani and vestibuli, 603

Stable prothrombin conversion factor, 246
Schaffer, 491
Schlemm, canal of, 234-235
Schwann cell, 20
Sclera, 232, 581
Scleroderma, respiratory compliance, 303
Scoliosis, respiratory compliance, 303
Scotopic vision, 586
Scrotum, 551
Scurvy, 251
Sebaceous glands, 549
Secretin, 451-454
  see also Hormone
Secretion, 443, 452
  buffers, 401
  digestive tract, 428
  gradient-limited active, 375
  hydrogen ions, 401
  iodine, 504
  kidney, 371-376
  role of calcium, 45
  Tm-limited active, 373
Secretogogues, 448
  i.v. injection, 449
Sedimentation rate, tabes dorsalis, 254
Segmentation, 421
Segments of body, 542
Sella tursica, 489, 497
Semen, emission of, 556
  pH, 551
Semicircular canal, 610-614
Semilunar valves, 117
Seminal fluid, 551
Seminal vesicles, 549, 552
Seminiferous tubule, 501
  dysgenesis, 547
Semipermeability, 212
Sensation, 577
  cutaneous, 22
  doctrine of specific nerve energy, 23
  modified pattern concept, 23
  muscle, 22
  special, 22
  visceral, 22
Sense organ, 22
Sensitivity to, epinephrine, 507
  hormones, 517
  insulin, 515
  thyroxine, 507
Sensitization, 262-270
Sensory areas of brain, 627
Sensory fibers, 22
Septal defects interatrial and interventricular, 123
Septum pellucidum, micturition, 408
Sequestered calcium, 536
Series circuit, 167
Series viscoelastic component, 76
Serosa, intestine, 420
Serotonin, 198, 420, 544
  intestine, 423
Serous cell, 443
Sertoli cells, 551
Serum, 250

Serum prothrombin conversion accelerator (SPCA), 246
7-ketosteroid, 522, 529
17-hydroxycorticoid, 500, 522
  see also Cortisol, Steroids
Sex flush, 557
Sex rash, 572
Sex skin, 572
Sexual characteristics of the male, 501
Sexual continence, 574
Sexual intercourse, 554, 559
Sexual practices, 557
Sexual response, 554, 557, 569-574
Sham feeding, 448, 451
Shearing force, 165
Shell temperature, 472
Sherrington, Sir Charles Scott, 207, 231
Shivering, 71, 473
  control, 478
Shock, 173
  anaphylactic, 198
  hypovolumic, 259, 385
  irreversible, 173
  psychic, kidney, 381
  spinal, 207-208
  surgical, 531
Shunt. See Arteriovenous anastomosis
Sickle cell, 212
Siderophilin. See Transferrin
Sigmoid colon, 423
Sign, Chvostek's, 537
  Queckenstedt-Stookey, 230
  Trousseau, 537
  see also Test
Silicone, 242
Sino-atrial (S-A) node, 93
  effect of catecholamine, 64
Sinus, 164
  dural, 227
  superior longitudinal, 226
Sinus arrythmia, 142
  see also Heart
Sinus, superior longitudinal, 227
Sinusoids, 164, 212
  adrenal, 521
  carotid body, 326
Sinus rhythm, 100
Sinus of Valsalva, 115
Skeletal muscle, 76
  annulospiral endings, 31
  control, 55
  extrafusal fibers, 31
  intrafusal fibers, 31
  see also Muscle
Skin conductivity, 473
  effect of catecholamine, 64
  electromagnetic waves, 579
Skin color, cyanosis, 329
  pallor, 329
Sleep, 634
  hypoxia, 195
  metabolic rate, 468
  respiration, 324
Sloughing, intestine, 432, 440, 456
  stomach, 440
  uterus, 563-564

Slow waves, 148
Sludge, 212
Small intestine, 420
  see also Intestine
Smell, 22, 619-621
Smoking, coronary heart disease, 178
Smooth muscle, 148-157
  activation-contraction coupling, 149
  multiunit, 150
  single unit, 151
  sliding filaments, 148
  visceral, 466
Sneeze, 297
  control, 322
Soda lime, 466
Sodium, 8, 39, 137, 229, 385, 533
  absorption, 428
  cell membrane, permeability to, 37
  conductance (g_{NA}), 39, 41
  diet, 446
  dietary restriction, 446
  equilibrium potential, 37
  excretion, 206, 386, 403
  height of action potential, 39
  intake, 437
  kidney reabsorption, 368
  loss, 438
  renal reabsorption, 368
  specific permeability, 217
  transcapillary movement, 218
  urine, 359
Sodium chloride, 618
  see also Chloride
Sodium potassium ratio, urine, 528
Sodium pump, 10, 363
Sol, 218, 242
Solubility, gases, 316
  lipids, 217
  nitrogen, 353
  water, 217
Solutions, dextrose, 12
  hypertonic, 12
  hypotonic, 12
Somatotrophic hormone, 494-499
Somatic area II, 628
Sound, 599
  audible frequencies, 125
  heart sounds, 125
  intensity discrimination, 606
  pressure, 125
  speech range, 125
  threshold of audibility, 125
Spasmodic respiratory center, 322
Spasm, carpopedal, 537
Spasmodic respiratory center, 322
Spasticity, 32, 537, 539, 631
Spatial recognition, 589, 594
Spatial summation, 48
Specific dynamic action, 469
Specific nerve energies, 23
Specific resistivity, blood, 103
  heart, 103
  lungs, 103
Sperm, 560
  motility, 554
Spermatic cords, 556

Spermatocyte, 551
Spermatogenesis, 501
Spermatogonia, 551
Spermatozoa, 553
Spermine, 552
Sphincter, anal, internal and external, 424
  of Boyden, 455
  cardiac, 415
  esophageal, 413-415
  hypopharyngeal, 413
  ileocecal, 421
  of Oddi, 455
  pharyngoesophageal, 420
  precapillary, 169, 186
  pyloric, 410, 417
  superior esophageal, 413
Sphingomyelin, 460
Spinal cord, 21, 408
  arterial pressure, 207
  autonomic neurons, 55
  cervical transection, 32, 34
  circulatory control, 207
  intermediolateral horn of, 211
  spinothalamic tract, 21
  temperature regulation, 477
  tracts, 622
  transection, 208, 319, 407, 624, 632
  urinary bladder control, 407, 409
  see also Center, Funiculus, Tracts
Spinal shock, 207
Spindle, muscle, 30, 53, 281, 334
Spiral ligament, 236
Spirometer, 311, 466
Splay, 366
  change in acidosis, 367
Spleen, blood reservoir, 202
  effect of catecholamines, 64
  eosinophil sequestration, 529
  sinusoids, 213
Splenic flexure, 423
Squint, 596
S-T interval, 99
S-T segment, 99
Stable factor, 246
Stable prothrombin conversion factor, 246
Staircase phenomenon, heart, 139
Standard error for electrolyte concentrations, 229
Standard temperature, pressure, and dryness (STPD), 466
Stapedius muscle, 602
Stapes, 602
Staphylokinase, 247
Starch, 428
Starling, E. H., 135, 452, 491
  capillary filtration, 219
Starling's law of the heart, 135
Starvation, 395, 461
  cardiac metabolism, 117
  edema, 221
  fatty acids, 461
  gluconeogenesis, 459
Stasis, intestine, 442

Steady potential, 17, 36
  see also Potential
Stearic acid, transport, 223
Steatorrhea, 432
Stefani, 175
Stellate ganglion, 142
Stenosis, aortic, 113, 123
  esophageal, 447
  mitral, 113, 123
  pulmonary, 123
  pulse pressure in aortic, 184
  tricuspid, 113
Stereopticon, 590
Sterility, 553
Steroid, 241, 457, 521, 522, 565
  adrenocortical, 522-534
  C-19 steroids, 522
  C-21 steroids, 522
  hormones, 491
  potencies, 528
  secretion, 526
  synthesis, 522
Stethoscope, chest positions for heart sounds, 126
Stewart-Hamilton technique, 131
Stimulus, 38, 77
  afterpotential, 26
  antidromic, 47
  artifact, 47
  chronaxie, 54
  conditioned, 446
  discharge zones, 49
  electrical, 53
  frequency, 26, 81, 138
  intracellular events, 78
  muscle, smooth vs. skeletal, 60
  occlusion, 49
  orthodromic, 47
  posterior hypothalamus, 209
  preoptic region of the hypothalamus, 210
  rheobase, 54
  sensory neurons, 33
  somatic afferent, 60
  square wave, 54
  subliminal, 48
  subthreshold, 38, 78
  superthreshold, 38
  strength, of glands, 60
    of hypothalamus, 209
    of muscle, 60, 81
    of nerve, 38, 128
    of receptors, 53
    of uterus, 153
  sympathetic, 60
  orthodromic, 47
  threshold, 38
  unconditioned, 447
  utilization time, 54
  vibratory, 53
Stomach, 415-420
  effect of catecholamines, 64
  empty, 416
  endocrine production, 490
  fat digestion, 430, 431
  motility, 415-420

  satiety, 390
  secretion, 443, 451
    electrolytes, 437, 441
  secretory control, 447
  vomiting, 419
Stones, calcium phosphate, 367
  gallstones, 455
  kidney, 367
Stool, 438, 455
  see also Feces
Stop flow analysis, 371
Storage. See Reserves
Strabismus, 596
Straining, 297, 425
  cerebrospinal fluid pressure, 230
Streptomycin, 613
Stress, 447, 531
  ACTH, 499
  medical, 531
  surgical, 169, 531
Stretch. See Distention
Stretch receptors. See Receptors, stretch
Stretch reflex. See Reflex, stretch
Stroke, 178
Stroke volume, 111, 131-142
  catecholamines, 62
  peripheral resistance, 136
Stroke volume index, 111
Stroke work, 112
Stroke work index, 112
Strontium secretion, 428
Strychnine, 68
Stuart factor, 246
Stuart-Power factor, 246
Subliminal fringe, 49
Sublingual glands, 443
Submaxillary gland, 443
Submucosa, 410
Submucosal plexus, 410
Substantia gelatinosa, 623
Suckling, 493, 502, 567
  sexual drive, 575
Sucrase, 429
Sucrose, 429
  filtration, 360
Sugar absorption, 428
Sugar in blood, effect of catecholamines, 62
Sulfate, absorption, 428
  renal reabsorption, 366
  urine, 359
Sulfobromophthalein (BSP), 453
Summation, spatial, 48
  temporal, 48
Supraoptic nucleus, 500
Suprarenal gland, 521
Surface area, 467
Surface factor, 246
Surface tension, 432
  lung, 307
Surfactant, 432
  lung, 307
Surgery, effect of, 500
  open heart, 292
  pacemaker implant, 102

  shock, 531
Survival, drowning, 345
Suspensory ligaments, 582
Swallowing, 324, 412-414
  air, 441
Sweat glands, 475
  control, 478
  water loss, 390
  see also Perspiration
Sweat production, 395
Sympathectomy, 65, 69
  pain, 418
  urinary bladder, 407, 409
Sympathetic neuron, 56, 59-64, 141, 144, 192, 204
  atrial depolarization, 59
  cholinergic, 211, 445, 475
  colon, 424
  duration of action, 60
  heart, 141
  intestine, 411, 423
  liver, 453
  pancreas, 451
  salivary gland, 444
  stomach, 416, 448
  ureters, 406
  vasomotor pathways, 56
  venous constriction, 204
  see also Nervous system
Sympathomimetic agents, 25, 61
  transmission, 46
Synalbumin, 516
Synapse, 21
  delay, 25
  gastric, 422
  unidirectional transmission, 25
Synaptic transmission, 46
Synaptic troughs, 11
Syndromes, adrenogenital, 529
  Cushing's, 499, 532
  dumping, 435
  Horner's, 69
  Meniere's, 613
  uremic, 405
  Wolff-Parkinson-White, 103
  see also Disease
Synergism, 529
Synthesis, 513
  adrenocortical, 522
  ammonia, 401
  cholesterol, 523
  desoxyribonucleic acid, 515
  enzymes, 528
  liver glycogen, 528
  protein, 516
  ribonucleic acid, 515
Syphilis, 596
System, metric, 3
Systemic circulation, 122
Systole, ventricular, 119
Systolic arterial pressure, 127
Systolic reserve volume, 132
  see also Heart
Systolic volume of heart, 132
  see also Heart
Syzmonowicz, 491

Szent Gyorgyi, Albert, 85, 86

T system, 73
T-3. See Hormone list, triiodothyronine
T-4. See Hormone list, thryoxine
Tabes dorsalis, 596
Tachycardia, 100, 109, 144
    paroxysmal ventricular, 100
Tactile. See Touch
Taenia coli, 154
Tamponade, 123
Taste, 22, 617-618
Taurine, 433
Taurocholate, 432
T_a wave, 99
Techniques. See Test
Teeth, 411, 543
    decalcification, 441
Tela choroidea, 226
Teleology, 1
Teloglia, 44
Temperature, 80
    ambient, 480
    body, 472
    core, 476
    environmental, 473
    fever, 477
    muscle twitch, 79
    oral, 476
    rectal, 477
    tracts, 623
Temporal lobe, 629
Tendon stretch, 22
Tenotomy, 76
Tension, 71, 79, 133
    active, 76
    aorta, 170
    elastic vs. viscous structures, 171
    heart, 133
    passive, 76, 279
    stomach, 415
    surface, 432
    surface of lung, 307
    total, 83
Tension muscles, 172
Tension time indices, 86
Tensor tympani muscle, 602
Tentorium cerebelli, 659
Test, 522
    aortic insufficiency, 183
    aortic stenosis, 184
    arteriosclerosis, 184
    Barberio's, 552
    barium meal, 420
    bleeding time, 252
    blood pressure, 123
    blood type, 263-266
    butanol extractable iodine, 504
    capillary fragility, 252
    cardiac output, 130
    Chvostek sign, 534
    clot retraction time, 250
    clotting time, 251
    glomerular filtration, 360
    glucose tolerance, 518
    hearing, 601
    hematocrit, 254
    hemostasis, 251
    hypertension, 184
    inulin clearance, 360
    micropuncture, 360
    pregnancy, 564
    protein bound iodine, 505
    renal clearance, 371
    sedimentation rate, 254
    stop flow, 371
    thyroid function, 505
    turbulence, 126
    visual acuity, 583
    vital capacity, 311
    see also Sigh
Testis, 549, 552
    development, 551
    descent, 551
    endocrine production, 491
    temperature, 551
Testosterone, 501
    see also Hormone
Tetanus, 81
    complete, 81
    incomplete, 81
Tetanus toxin, 68
Tetanus:twitch ratio, 81
Tetany, 81, 537
    uremia, 405
Tetraethylammonium chloride, 67, 68
    atropine, 66
Tetrahydrocortisol glucuronide, 527
Tetrahydrocortisone, 527
Tetrahydroglucuronides, 527
Tetralogy of Fallot, 123
Tetrodotoxin, 68
Thalamus, 621, 624
    stimulation, 634
Thalamus animal, 615
Thermoreceptors, 479
Thin filament, 76
Thirst, 390
Thoracic duct, 222, 259
Thoracic pressure, 295
Threshold, 17, 38, 48
    audibility, 601
    sapid solutions, 617
Thrombin, 244
Thrombocyte. See Platelet
Thrombolysis, 268
Thromboplastin, 243-246
Thromboplastin generation test, 244
Thrombosthenin, 251
Thymus, 529, 544
    involution, 528
Thyrocalcitonin, 539
    see also Hormone
Thyroglobulin, 504
Thyroid, 503-510
    colloid, 505
    endocrine production, 490
Thyroid disease, heart, 116
Thyroidectomy, 507, 539
Thyroid-stimulating hormone, 497, 504
    see also Hormone

Thyroparathyroidectomy, 539
Thyrotoxicosis, 539
    see also Hyperthyroidism
Thyrotropin releasing factor, 498
    see also Hormone
Thyroxine, 497, 503
    see also Hormone
Thyroxine binding globulin (TBG), 504
Thyroxine binding prealbumin (TBPA), 504
Tidal volume, respiratory, 299
Timed expiratory volume, 311
Tinnitus, 613
Tissue, adipose. See Adipose tissue
Tissue extracts in clot formations, 244
Tissue juice, 244
Tone, autonomic, 67
    muscle, 537, 630
    smooth muscle, 171
Tongue, 411
Tonic activity, parasympathetic neurons, 140
Touch, 22
Touch tract, 622
Tourniquets, blood stasis, 202
Toxicity, oxygen, 353
    see also Osmotic pressure
Tracer substance, 394
Track records, 89
Tract, auditory, 607
    Lissauer, 623
    neospinothalamic, 623
    olfactory bulb conduction, 621
    pain, 623
    paleospinothalamic, 623
    reticulothalamic, 622
    spinal cord, 622
    temperature, 623
    touch, 622
    vestibulo-cerebellar, 613
T_P wave, 99
Transamination, 431
Transcortin, 525
Transducers, 29
    receptors, 51
    Walton-Brodie, 197
Transection, spinal, 632
Transferase, 61
Transferrin, 439, 440
Transfusion, 259-266
Transmembrane potential, 9, 17, 36
    see also Potential
Transmission, neuromuscular, 43, 71
    synaptic, 25, 46
Transmitter, neuro, 29
Transmural pressure, brain, 230
Transport, active, in kidney, 363-376
    amino acid, 430
    chylomicron, 435
    competitive, 373
    digestive tract, 428
    fatty acid, 435
    insulin action, 514
    intestinal, 428
    iodide, 504

Transport—(Cont.)
  ions in kidney, 363
  lipids, 434
  oxygen, 73
  passive, 362
  pinocytosis, 11
  sodium-potassium, 10, 36
  see also Binding, Portal systems
Transport maximum sulfobromophthalein, 453
Transverse tubule, muscle, 73
Trauma, 499, 531
Tremor, 82
Treppe, 85, 139
  heart, 138
Triad muscle, 73
Trichromatic theory, 588
Trichromats, anomalous, 589
Tricuspid valve, 118
Triglyceride, 516
  absorption, 428
  acetyl-Co A catabolism, 515
Triiodothyronine, 497, 503, 505
  see also Hormone
Triple response, 197
Tritanopia, 589
Tropomyosin, 429, 511
Trousseau sign, 537
Trypsin, 429, 511
Trypsinogen, 429
Tryptamine, 5-hydroxy, 155
Tubocurarine, 45
Tubular maximum (Tm), 365-368, 373
  bromsulphalein, 453
  glucose (Tm_G), 365
  limited active secretion, 373
  para-aminohippurate (Tm_{PAH}), 375-378
Tubule, longitudinal, of muscle, 73
  renal, 358
    see also Nephron
  transverse, 78
  transverse, of heart, 116
Tumors, acidophilic, 494
  adrenal, 529
  adrenal medulla, 535
  adrenocortical, 544
  pineal, 544
Tunica dartos, 551
Turbulence, 126
  peripheral resistance, 168
T wave, 99
  see also Electrocardiogram
Tympanic membrane, 601
Typhoid fever, 120
Tyrosine, 61, 504
Tyrosyl radical, 504

Ulcer, gastrointestinal, 440-441, 448
Ultraviolet waves, 579
Umbilical vessels, 238
Unconsciousness, 230
Unit membrane, 8
Urea, 362, 462
  plasma, 222
  specific permeability, 217

Uremia, 405
Uremic syndrome, 405
Ureter, 405
Urethra, 405-408
Uric acid, 367
Urinary bladder, 405-409
  automatic, 406, 409
  catecholamines, 64
  control, 153, 406-409
Urinary concentration, 462
  androgens, 542
  chorionic gonadotropin, 564, 565
  estriol, 565
  estrogens, 542
  gonadotropins, 565
  pregnanedione, 565
Urination, 404-409
Urine, 355-409
  acidity, 404, 533
  epinephrine, 535
  hemoglobin, 277
  norepinephrine, 535
  potassium, 533
  sodium, 368, 533
  volume, 204-206, 363, 530
Urodenital diaphragm, 556
Urokinase, 247
Uterus, 153, 240, 563-567
  catecholamines, 64
  pregnancy, 156
Utilization time, 54
Utricle, 615
U wave, 99
  see also Electrocardiogram

V1-V6 leads, 97, 98
  see also Electrocardiogram
Vagal escape, 93, 144
Vagina, 553, 568, 570
Vagotomy, 65, 142, 199, 418
Vagus nerve, 57, 65
  bile secretion, 454
  block, 331, 449, 454
  cardiac sphincter of stomach, 414
  chemoreceptors, 328
  colon, 424
  exercise, 199
  heart, 108, 140, 142
  pancreas, 451
  respiratory centers, 322, 331
  sensory fibers, 448
  stomach and small intestine, 417, 418, 449
Vagus stuff, 43
  see also Acetylcholine
Valence, 5
Valsalva maneuver, 299
Valves. See Atrioventricular and Semilunar
Valvular incompetence, veins, 178
Vanilmandelic acid, 535
Vaporization, 475
Varicose veins, 178
Vas deferens, 552
Vasoactive agents, angiotensin, 198
  epinephrine, 197

  histamine, 154, 449
  norepinephrine, 60
  peptides, 198
  renin, 198
  serotonin, 198
  vasopressin, 198
Vasoconstriction, glucocorticoids, 532
Vasoconstrictor fiber, 56, 193
Vasodilator fiber, 193
  capillary, 220
Vasomotor center, 196
Vasopressin, 492
  see also Hormone
Vasum vasorum, 115
Vater, ampulla of, 455
Vectorcardiogram (VCG), 105
  isoelectric point, 105
  ventricular hypertrophy, 108
Vein, 164, 169, 176-182
  antecubital, 123
  anterior cardiac, 115
  congestion, 221
  jugular, 227, 230
  ophthalmic, 227
  saphenous, 123
  varicose, 177
  vertebral, 227
Velocity, 3
Vena cava, 121
  pressure, 182
Venous tone, baroreceptor reflex, 203
Ventilation, 298-312
  exercise, 312
Ventricle, 110
  conduction rate, 95
  wall thickness, 96
  work, 110
Ventricular diastole, 119
Ventricular endocardium, conduction rate, 95
Ventricular force, 210
Ventricular pressure, P-R interval, 141
Ventricular septal defect, 124
Ventricular systole, 119, 142
  see also Heart
Venule, 169
Veratridene, 338
Veratrine, 190
Vesicles, capillary endothelium, 218
  pinocytotic, 73
  synaptic, 151
Vessel. See Blood vessel and Lymph duct
Vestibular apparatus, 596, 609-615
Vestibular nuclei, 612, 632
VF, 97
  see also Electrocardiogram
Villi, 410, 422
Virilization, 533
Visceral, 148-157
Viscoelasticity, 171
Viscometer, Ostwald, 179
Viscosity, 3, 171, 179
  tension, 171
Vision, 22, 579-597
  accommodation of, in aging, 582

acuity, 583
  binocular, 589
  blind spot, 590
  color, 587-589
  field of, 592
  near vision, 582
  pigments, 588
  photopic, 588
  scotopic, 586
  stereoscopic, 590
Visual purple, 587
Vital capacity, 299-302
Vitamin, absorption, 428
  fat soluble, 433
Vitamin A, 588
Vitamin B_{12}, absorption, 428
Vitamin C, 251, 441
Vitamin D, 439, 538
  cortisol, 532
Vitamin K, 249, 441
Vitreous humor, 232
Volt, 3
Volume receptors, 205
Vomiting, 324, 419
von Mering, Baron Joseph, 491
von Willebrand's syndrome, 253
VR, 97
  see also Electrocardiogram
V wave, 120

Wakefulness, 631
Waste products, 289
  elimination, 397
Water, 435
  absorption, 428
  balance, 390-396
  digestive tract, 435-439
  load, 529
  metabolism, 531
  partial pressure, 279, 467
  reabsorption, 362
  specific permeability, 217
  transcapillary movement, 218
Water-solubility, 8
Watt, 3
Wave, electromagnetic, 579
  Mayer, 195
  peristaltic, 413-417, 421-422
  Traube-Herring, 195
Wavelength, 579, 587
  see also Frequency, Light
Wave, peristaltic, 413-417, 421-422
Waves, electromagnetic, 579
Weight, 71, 463, 467, 542
  altitude, 350
  pregnancy, 567
Wheal, 197
White blood cell, 163
White matter, 21

White muscle, 87
Wolff-Parkinson-White syndrome, 103
Work, breathing, 6, 311
  cardiac, 112-113
Wound healing, 532

X chromosome, 546
Xerostomia, 67, 441
XO genotype, 547
X-ray, 579
X wave, 119
XXY genotype, 547
Xylocaine, 145
Xylol, 335
Xylose, 429

Y chromosome, 546
Yolk sac, 271
Young-Helmholtz trichromatic theory, 588
Young's modulus of elasticity, 171

Z bands, 74
Zona, fasciculata of adrenal gland, 521
  glomerulosa of adrenal gland, 521
  reticularis of adrenal gland, 521
Zone, chemoreceptor trigger, 420
Zwitterion, 283
Zymogen granules, 14